Lecture Notes in Electrical Engineering

Volume 388

About this Series

"Lecture Notes in Electrical Engineering (LNEE)" is a book series which reports the latest research and developments in Electrical Engineering, namely:

- Communication, Networks, and Information Theory
- Computer Engineering
- Signal, Image, Speech and Information Processing
- Circuits and Systems
- Bioengineering

LNEE publishes authored monographs and contributed volumes which present cutting edge research information as well as new perspectives on classical fields, while maintaining Springer's high standards of academic excellence. Also considered for publication are lecture materials, proceedings, and other related materials of exceptionally high quality and interest. The subject matter should be original and timely, reporting the latest research and developments in all areas of electrical engineering.

The audience for the books in LNEE consists of advanced level students, researchers, and industry professionals working at the forefront of their fields. Much like Springer's other Lecture Notes series, LNEE will be distributed through Springer's print and electronic publishing channels.

More information about this series at http://www.springer.com/series/7818

Jiadong Sun · Jingnan Liu
Shiwei Fan · Feixue Wang
Editors

China Satellite Navigation Conference (CSNC) 2016 Proceedings: Volume I

CSNC

Springer

Editors
Jiadong Sun
Chinese Academy of Sciences
China Aerospace Science and Technology Corporation
Beijing
China

Jingnan Liu
Wuhan University
Wuhan
China

Shiwei Fan
China Satellite Navigation Office
Beijing
China

Feixue Wang
National University of Defense Technology
Changsha
China

ISSN 1876-1100 ISSN 1876-1119 (electronic)
Lecture Notes in Electrical Engineering
ISBN 978-981-10-0933-4 ISBN 978-981-10-0934-1 (eBook)
DOI 10.1007/978-981-10-0934-1

Library of Congress Control Number: 2016937343

Printed on acid-free paper

This Springer imprint is published by Springer Nature
The registered company is Springer Science+Business Media Singapore Pte Ltd.

Preface

BeiDou Navigation Satellite System (BDS) is China's global navigation satellite system which has been developed independently. BDS is similar in principle to global positioning system (GPS) and compatible with other global satellite navigation systems (GNSS) worldwide. The BDS will provide highly reliable and precise positioning, navigation and timing (PNT) services as well as short-message communication for all users under all-weather, all-time, and worldwide conditions.

China Satellite Navigation Conference (CSNC) is an open platform for academic exchanges in the field of satellite navigation. It aims to encourage technological innovation, accelerate GNSS engineering, and boost the development of the satellite navigation industry in China and in the world.

The 7th China Satellite Navigation Conference (CSNC2016) is held during May 18–20, 2016, Changsha, China. The theme of CSNC2016 is Smart Sensing, Smart Perception, including technical seminars, academic exchanges, forums, exhibitions, and lectures. The main topics are as follows:

S1 BDS/GNSS Application Technology
S2 Navigation and Location-Based Services
S3 Satellite Navigation Signals
S4 Satellite Orbit and Clock Offset Determination
S5 BDS/GNSS Precise Positioning Technology
S6 Atomic Clock and Time-frequency Technology
S7 BDS/GNSS Augmentation Systems and Technology
S8 BDS/GNSS Test and Assessment Technology
S9 BDS/GNSS User Terminal Technology
S10 Multi-sensor Fusion Navigation
S11 PNT System and Emerging Navigation Technology
S12 Standardization, Intellectual Properties, Policies, and Regulations

The proceedings (Lecture Notes in Electrical Engineering) have 176 papers in ten topics of the conference, which were selected through a strict peer-review process from 440 papers presented at CSNC2016. In addition, another 193 papers were selected as the electronic proceedings of CSNC2016, which are also indexed by "China Proceedings of Conferences Full-text Database (CPCD)" of CNKI and Wan Fang Data.

We thank the contribution of each author and extend our gratitude to 237 referees and 48 session chairmen who are listed as members of editorial board. The assistance of CNSC2016's organizing committees and the Springer editorial office is highly appreciated.

The 7th China Satellite Navigation Conference (CSNC 2016) Committees

Scientific Committee

Chairman
Jiadong Sun, China Aerospace Science and Technology Corporation

Vice-Chairman
Rongjun Shen, China
Jisheng Li, China
Qisheng Sui, China
Changfei Yang, China
Zuhong Li, China Academy of Space Technology
Shusen Tan, Beijing Satellite Navigation Center, China

Executive Chairman
Jingnan Liu, Wuhan University
Yuanxi Yang, China National Administration of GNSS and Applications
Shiwei Fan, China

Committee Members (By Surnames Stroke Order)
Xiancheng Ding, China Electronics Technology Group Corporation
Qingjun Bu, China
Liheng Wang, China Aerospace Science and Technology Corporation
Yuzhu Wang, Shanghai Institute of Optics and Fine Mechanics, Chinese Academy of Sciences
Guoxiang Ai, National Astronomical Observatories, Chinese Academy of Sciences
Shuhua Ye, Shanghai Astronomical Observatories, Chinese Academy of Sciences
Zhaowen Zhuang, National University of Defense Technology
Qifeng Xu, PLA Information Engineering University
Houze Xu, Institute of Geodesy and Geophysics, Chinese Academy of Sciences
Guirong Min, China Academy of Space Technology

Xixiang Zhang, China Electronics Technology Group Corporation
Lvqian Zhang, China Aerospace Science and Technology Corporation
Junyong Chen, National Administration of Surveying, Mapping and Geoinformation
Benyao Fan, China Academy of Space Technology
Dongjin Luo, China
Guohong Xia, China Aerospace Science and Industry Corporation
Chong Cao, China Research Institute of Radio Wave Propagation (CETC 22)
Faren Qi, China Academy of Space Technology
Sili Liang, China Aerospace Science and Technology Corporation
Shancheng Tu, China Academy of Space Technology
Rongsheng Su, China
Zhipeng Tong, China Electronics Technology Group Corporation
Ziqing Wei, Xi'an Institute of Surveying and Mapping

Organizing Committee

Secretary General
Haitao Wu, Navigation Headquarters, Chinese Academy of Sciences

Vice Secretary General
Wenhai Jiao, China Satellite Navigation Office Engineering Center
Jianjun Wu, National University of Defense Technology
Weina Hao, Navigation Headquarters, Chinese Academy of Sciences

Committee Members (By Surnames Stroke Order)
Qun Ding, The 20th Research Institute of China Electronics Technology Group Corporation
Miao Yu, China Academy of Space Technology
Li Wang, International Cooperation Research Center China Satellite Navigation Engineering Office
Ying Liu, China Satellite Navigation Office Engineering Center
Shuhua Zhang, National University of Defense Technology, Changsha
Xiuwan Chen, Peking University
Xiangan Zhao, China Defense Science and Technology Information Center
Ouyang Guangzhou, Academy of Opto-electronics, Chinese Academy of Sciences
Gang Hu, Beijing Unicore Communications, Inc.
Min Shui, National Remote Sensing Centre of China
Zhong Dou, National Time Service Center, Chinese Academy of Sciences

Editorial Board

Contents

Part I
BDS/GNSS Application Technology

The Research on Four-Dimensional Water Vapor Tomography Based on Real-Time PPP Technique

Qingzhi Zhao, Yibin Yao and Chaoqian Xu

Abstract With the development of International GNSS Service (IGS) real-time pilot project (RTPP) acquiring precipitable water vapor (PWV) with high accuracy has become a reality based on the real-time precise point pointing (RT-PPP) technique. The accuracy of zenith total delay (ZTD) and PWV derived from RT-PPP have been validated using observed global positioning system (GPS) data and meteorology data from Satellite Positioning Reference Station Network (SatRef) in 2014. The ZTD comparison with that from afterwards PPP and GAMIT software shows that the relative coefficients are 0.9786 and 0.9687, respectively. The PWV comparison with that from radiosonde shows that the relative coefficient and RMS are 0.9512 and 2.13 mm, respectively. It is a clear evidence that the RT-PPP technique has a similar accuracy with the result calculated using afterwards IGS products. However, PWV is mean of water vapor information of many GNSS signal rays during a period of time over the station, which cannot reflect the three-dimensional water vapor distribution. Slant water vapor (SWV) can be obtained by mapping PWV at different elevation and azimuth angles. The tomographic experiment has been performed using SWVs of twelve stations from SatRef as tomographic observation and compared with result from radiosonde. The comparison shows a good agreement and the RMS, SD, Bias, and MAE of integrated water vapor (IWV) are 3.60, 2.78, 2.29, and 2.92 mm, respectively, the root mean square (RMS), standard deviation (SD), Bias, and mean absolute error (MAE) of calculated water vapor density are 1.08, 1.03, −0.21, and 0.77 g/m^3, respectively. The above result makes it possible that acquiring the real-time three-dimensional water vapor distribution using tomography approach with SWVs derived from RT-PPP technique, which has an important influence on short-term disastrous weather and now-casting precipitation forecasting.

Keywords IGS · Real-time pilot project · Real-time PPP technique · Slant water vapor · Tomography approach

Q. Zhao (✉) · Y. Yao · C. Xu
School of Geodesy and Geomatics, Wuhan University, Wuhan 430072, China
e-mail: zhaoqingzhia@163.com

J. Sun et al. (eds.), *China Satellite Navigation Conference (CSNC) 2016 Proceedings: Volume I*, Lecture Notes in Electrical Engineering 388,
DOI 10.1007/978-981-10-0934-1_1

1 Introduction

In recent years, acquiring three-dimensional/four-dimensional water vapor distribution based on ground-based GNSS tomographic technique has developed significantly [1–5]. The wet refractivity field or water vapor field with high spatial-temporal resolution can be reconstructed by tomography technique using a large number of SWVs which penetrate the whole tomographic area derived from the network of ground-based GNSS. There are two methods for obtaining PWV/SWV using GNSS observed data: PPP technique based on undifferenced observations [6] and baseline/network solving pattern based on double-differenced observations. With the proposed of IGS RTPP [7], users can get real-time products of precise satellite orbit and clock online [8], which promotes the rapid development of RT-PPP technique [9, 10] and makes RT-PPP technique becomes one of the most important tools to get real-time PWV/SWV [11, 12].

The ZTD derived from double-differenced observations is compared with that from numerical forecasting model and radiosonde data, and the accuracy is 6–13 mm [13, 14]. The accuracy of ZTD derived from RT-PPP technique using data provided by GPS system is 11–16 mm, about 2–3 mm PWV error, the accuracy of BDS is slightly lower than that of GPS, but combining data of two systems can get 1.3–1.8 mm PWV error [15]. So, the accuracy of PWV obtained by RT-PPP technique is similar with that of double-differenced technique. However, PWV is an average value of water vapor information of many GNSS signal rays during a period of time over the station, which cannot reflect the three-dimensional water vapor distribution. The emerging of GNSS water vapor tomography technique solves this tricky problem which constrains the application for GNSS meteorology [16]. The real-time three-dimensional wet refractivity or water vapor distribution with high spatial-temporal resolution can be obtained by tomography approach using SWVs derived from RT-PPP technique. This has very important practical significant for improving the accuracy of numerical weather forecasting model, quantified precipitation forecasting and short-term disastrous weather forecasting [17–20].

In this paper, a four-dimensional water vapor monitoring system was designed and developed based on RT-PPP and tomography technique. Simulation experiment has performed based on the observed data of 12 stations from SatRef and obtained a good result.

2 The Method for Acquiring SWV and Theory of Tomography

2.1 The Method for Acquiring SWV

ZTD can be obtained by undifferenced or double-differenced observations; and the popular software package of data processing includes GAMIT/GOBK, Bernese,

GIPSY, EPOS, TriP, and so on. ZTD is the average value which is mapped by many STDs from the slant path to zenith path based on mapping function and consists of two parts: zenith hydrostatic delay (ZHD) and zenith wet delay (ZWD). ZWD is about 10 % of ZTD and ZHD is about 90 % of ZTD [21].

ZWD is relevant with the change of water vapor in troposphere and cannot be modeled accurately. On the contrary, ZHD can be obtained accurately with surface pressure by model, such as Saastamoinen model. The error of ZHD is 0.2 mm when the variation of surface is 1 hPa using Saastamoinen model [22], which meets the requirement of meteorological application. So the ZWD can be extracted from ZTD by minus ZHD. The formula for calculating ZHD using Saastamoinen is as follows: [23]:

$$\mathrm{ZHD}=\frac{0.002277\cdot P_s}{1-0.00266\cdot\cos(2\varphi)-0.00028\cdot H} \tag{2.1}$$

where P_s is surface pressure (unit: hPa), φ is the latitude of station, H is the height of station (unit: km).

Slant wet delay (SWD) can be calculated using mapping function like VMF1, GMF, NMF, the formula as follows:

$$\mathrm{SWD}_{\mathrm{A,e}} = \mathrm{M}_{\mathrm{wet}}(\mathrm{e})\cdot\mathrm{ZWD} + \mathrm{M}_{\Delta}(\mathrm{e})\cdot(\mathrm{G}^{\mathrm{w}}_{\mathrm{NS}}\cdot\cos\mathrm{A} + \mathrm{G}^{\mathrm{w}}_{\mathrm{WE}}\cdot\sin\mathrm{A}) + \mathrm{R}_{\mathrm{e}} \tag{2.2}$$

where A is azimuth, e is elevation angle, $\mathrm{M}_{\mathrm{wet}}$ is wet mapping function, M_{Δ} is the mapping function of horizontal gradient, $\mathrm{G}^{\mathrm{w}}_{\mathrm{NS}}$ and $\mathrm{G}^{\mathrm{w}}_{\mathrm{WE}}$ are the horizontal gradient components in East–West and South–North directions, R_{e} is unmodeled postfit residuals.

SWD can be converted to SWV by conversion factor as follows: [24]:

$$\mathrm{SWV} = \Pi\cdot\mathrm{SWD} \tag{2.3}$$

where $\Pi = 10^6/((k_2' + k_3/T_m)\cdot R_v\cdot\rho), k_3 = (3.776\pm0.014)\times10^5\ \mathrm{K}^2\ \mathrm{hPa}^{-1}$, $k_2' = 16.48\ \mathrm{K\ hPa}^{-1}$, are constants, $R_\omega = 461\ (\mathrm{J\ kg}^{-1}\ \mathrm{K}^{-1})$ represents water gas ratio constant, ρ is water vapor, T_m represents weighted mean temperature.

2.2 The Theory of Tomography

SWV is relevant with the integrated wet profile of slant path, and represents the total water vapor content of unit area along slant path column (unit: kg/m^2), also equals to the total content of liquid water vapor for the same height (unit: mm) [25]. The definition of SWV is as follows:

$$\mathrm{SWV} = \int_{S} \rho_v \mathrm{d}s \tag{2.4}$$

where s represents the path of the satellite signal ray, ρ_v represents water vapor density (unit: g/m^3). A linear equation between SWV and water vapor density can be established:

$$\mathrm{SWV}^p = \sum_{ijk} (\mathrm{A}^p_{ijk} \cdot \mathrm{X}_{ijk}) \tag{2.5}$$

where SWV^p is the slant water vapor amount of the pth signal ray (unit: mm); A^p_{ijk} is the distance of the pth signals ray inside the (i, j, k) voxel; X_{ijk} is the water vapor density in the voxel (i, j, k) (unit: g/m^3). The above equation can then be written in matrix form as follows:

$$\boldsymbol{y} = \boldsymbol{A} \cdot \boldsymbol{x} \tag{2.6}$$

where $\boldsymbol{y}$ is a column vector with a set of SWV measurements, $\boldsymbol{A}$ is a coefficient matrix with the element of distance, $\boldsymbol{x}$ is a column vector of unknown water vapor density. In our study, an elevation cut off angle of 10° for signals of every receiver is selected in order to ignore the influence of ray bending [26, 27].

It can be seen from above equation that the nature of solving the unknown water vapor density is an issue of inversion algorithm. The design matrix $\boldsymbol{A}$ in Eq. (2.6), however, is a large sparse matrix and many voxels are not crossed by signal rays. The normal equation of $\boldsymbol{A}$ is singular which leads to severe numerical problems when applying a direct inversion method [28]. Some constraint conditions are often introduced to the tomography equation [16, 29]. In our study, the Gauss-weighted functional method is used in the horizontal direction [30], while in the vertical direction the function relationship of exponential distribution is established [31]. Finally, the tomography model is then acquired as

$$\begin{pmatrix} \boldsymbol{A}^{\mathrm{slant}}_{l\times n} \\ \boldsymbol{A}^{\mathrm{zenith}}_{m\times n} \\ \boldsymbol{H}_{l\times n} \\ \boldsymbol{V}_{l\times n} \end{pmatrix} \cdot \boldsymbol{x}_{n\times 1} = \begin{pmatrix} \boldsymbol{SWV}_{l\times 1} \\ \boldsymbol{PWV}_{m\times 1} \\ \boldsymbol{0}_{l\times 1} \\ \boldsymbol{0}_{l\times 1} \end{pmatrix} \tag{2.7}$$

where $\boldsymbol{A}^{\mathrm{slant}}_{l\times n}$ is mapping function in the slant direction, $\boldsymbol{A}^{\mathrm{zenith}}_{m\times n}$ is mapping function in the vertical direction, $\boldsymbol{H}$ and $\boldsymbol{V}$ are the coefficient matrixes for the horizontal and vertical constraints, respectively.

3 Data Introduction and Accuracy Analysis of RT-PPP

3.1 Data Introduction

Data of twelve stations derived from SatRef in Hong Kong has been selected for tomographic experiment, and the distance between two neighboring stations is about 10–15 km, the geographic location for every station is shown by ▲ in Fig. 1. All stations are equipped with meteorological sensors and the real-time meteorological parameters can also be obtained. It is reported that accuracies of surface pressure and temperature are better than ±0.08 hPa and ±0.2 °C, respectively, and surface relative humidity is better than ±02 % at 25 °C (More information in http://www.paroscientific.com/pdf/met4DataSheet.pdf). The real-time observed GPS data and meteorological data can be downloaded from the website (http://www.geodetic.gov.hk/smo/gsi/programs/en/GSS/satref/satref.htm) with the interval of 5 s and 1 min, respectively. There is a radiosonde station in King's Park which is shown by ● in Fig. 1, and the water vapor profile in the vertical direction with high accuracy can be obtained using recorded radiosonde data [21, 32, 33]. These data can be freely downloaded afterwards from website (ftp://ncdc.noaa.gov/pub/data/igra/), which recorded the data of UTC 00:00 and 12:00 every day.

3.2 Accuracy Analysis of RT-PPP ZTD

The accuracy of ZTD processed by GAMIT software using GPS data is better than ±1 cm [34], in this paper, it is regarded as a reference to validate the ZTD calculated by RT-PPP technique. In addition, the result calculated by PPP technique using afterwards precise products of orbit and clock provided by IGS is also used to test the result of RT-PPP technique. Data of twelve stations are processed using

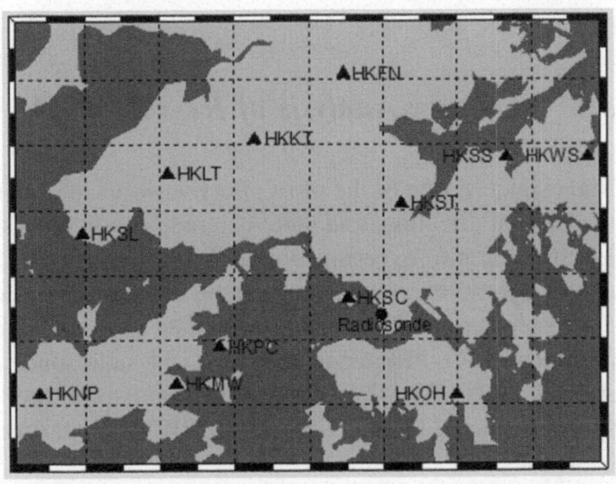

Fig. 1 Location of stations, *filled triangle* represents GPS station and *filled circle* represents radiosonde station

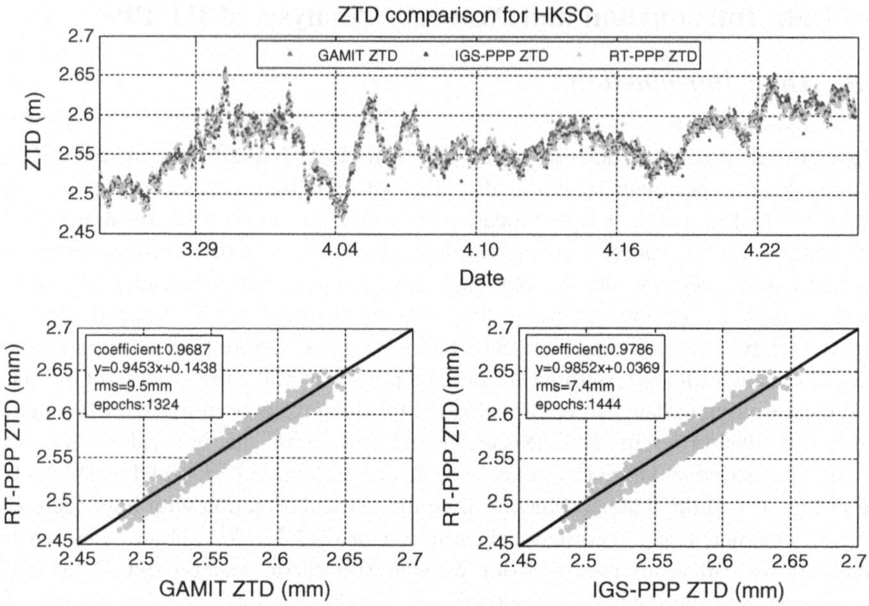

Fig. 2 Comparison of ZTD derived from real-time PPP, GAMIT software and IGS-PPP for HKSC (Color figure online)

GAMIT10.5 software, Fig. 2 gives the ZTD comparison once every half an hour derived from different methods for HKSC station.

It can be seen from Fig. 2 that the changed trend of ZTD is consistent between RT-PPP, IGS-PPP, and GAMIT. For the statistical result of 32 days, the coefficient of ZTD calculated by RT-PPP technique with IGS-PPP technique and GAMIT software are 0.9687 and 0.9786, respectively, and the RMS are 9.5 and 7.4 mm, respectively. This shows that the ZTD accuracy of RT-PPP technique is the same with afterwards processed result.

3.3 Accuracy Analysis of RT-PPP PWV

Radiosonde is one of the most effective ways to obtain PWV with high accuracy in the vertical direction, and there is a radiosonde 45,004 about 2.4 km far away from HKSC station in experiment area. It has been proved that the horizontal change of water vapor in the troposphere can be neglected between two stations [4]. Therefore, the result derived from radiosonde data is regarded as a reference to validate the accuracy of RT-PPP technique. First, the radiosonde data from March 25, 2014 to April 25, 2014 for UTC 00:00 and 12:00 is processed and the PWV is obtained for specific epochs. Figure 3 shows the PWV comparison between different methods.

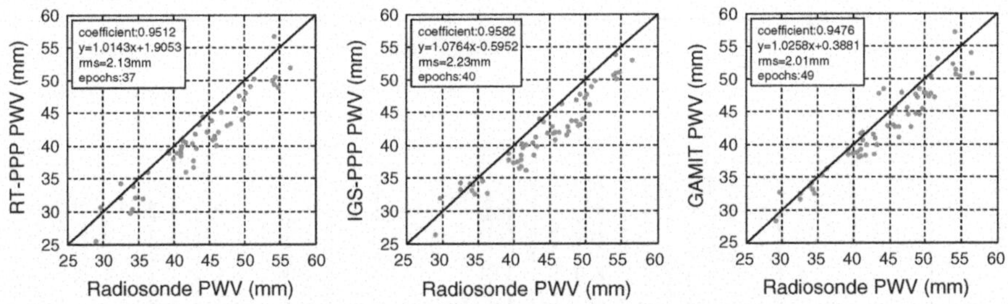

Fig. 3 Comparing PWV retrieved from GPS data with that calculated from radiosonde data for HKSC

Figure 3 is an evidence that the PWV calculated by RT-PPP technique has a consistency with that from radiosonde. The statistical result shows that the coefficients of radiosonde with RT-PPP, IGS-PPP and GAMIT are 0.9512, 0.9582, and 0.9476, respectively, and the RMS are 2.13, 2.23, and 2.01 mm, respectively. This result reaches the accuracy level that can be achieved internationally [25, 35, 36]. With this we can conclude that the PWV accuracy derived from RT-PPP technique is the same with the result that is processed afterwards.

4 Comparison of Tomographic Result

4.1 Introduction of Tomographic Experiment

The range of the tomographic area is as follows: latitude from 22.19°N to 22.54°N and longitude from 113.87°E to 114.35°E. Voxel division: horizontal resolution in longitude and latitude are 0.06° and 0.05°, respectively, and vertical resolution is 0.8 km. The total number of voxels is $7 \times 8 \times 13$. The observed data of twelve stations (as shown by ▲ in Fig. 1) are selected for tomographic experiment.

4.2 IWV Comparison

Radiosonde data can provide accurate water vapor density profile in vertical direction [21], and IWV is the integral of water vapor along the vertical direction above the radiosonde station. The IWV derived from radiosonde and tomographic result is compared for the location of radiosonde (see Fig. 4). It can be seen from Fig. 4 that the changed trend of IWV for two methods is similar. By calculation, the RMS, SD, Bias and MAE of IWV are 3.6, 2.78, 2.29, and 2.92 mm, respectively.

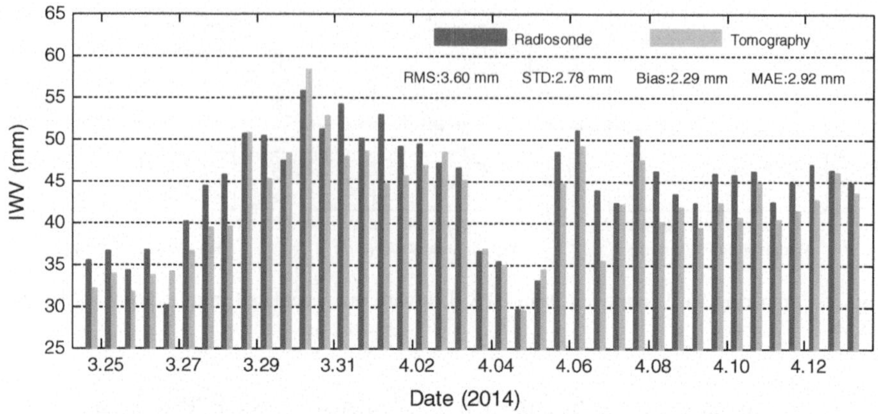

Fig. 4 Comparison of IWV time series derived from radiosonde and tomographic result (Color figure online)

4.3 Comparison of Layered Water Vapor

Although IWV result derived from tomographic result agrees well with those from the radiosonde, while also showing lower RMS, SD, MAE, and Bias, one may not conclude that the vertical water vapor distribution profile has been obtained correctly. For instance, if two vertical layers are exchanged arbitrarily, the calculated IWV value remains unchanged, but the vertical distribution of water vapor changes. Therefore, in order to further test the accuracy of vertical water vapor density, 40 epochs of tomographic period are analyzed. Figure 5 gives the tomographic result with that of radiosonde for different layers. We can see that the distributed water vapor density agrees well with that from radiosonde data. By calculation, the RMS of water vapor density is 1.08 g/m^3, which means tomographic water vapor density has high accuracy in the vertical direction.

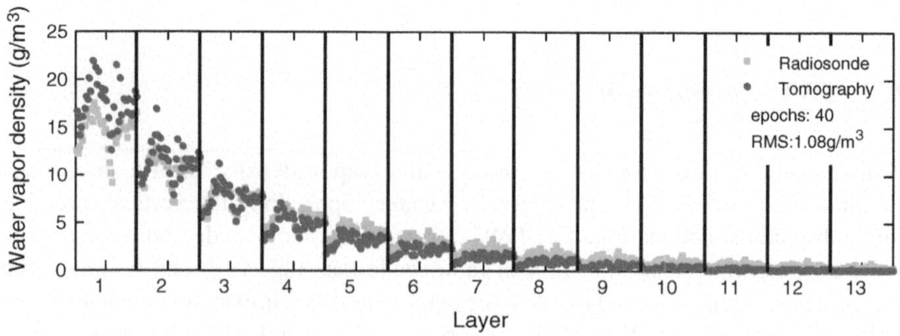

Fig. 5 Tomographic result comparison with water vapor density from radiosonde (*green*) for various layers, against water vapor density from tomography (*blue*) for 40 epochs spanning from March 25, 2014 to April 13, 2014 (color figure online)

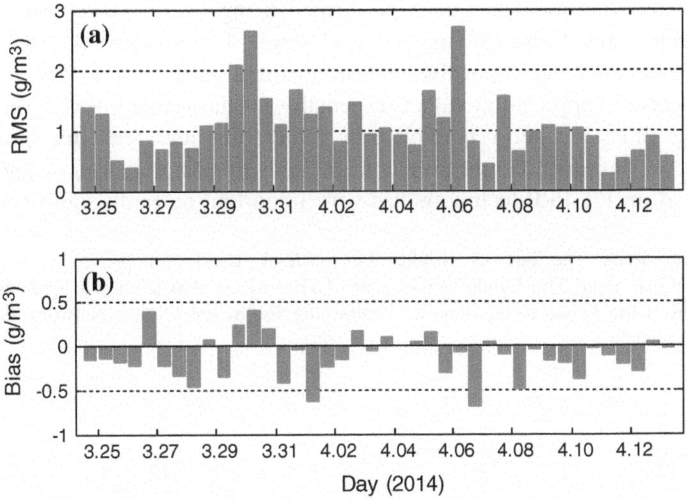

Fig. 6 Distribution of RMS and Bias for tomographic result

Table 1 Statistical result between radiosonde and tomographic result (unit: g/m^3)

Data comparison	RMS	SD	Bias	MAE
Tomography versus radiosonde	1.08 [0.28 2.70]	1.03 [0.25 2.70]	−0.12 [−0.68 0.41]	0.77 [0.23 1.84]

In addition, the tomographic result for every day is also analyzed, and the RMS and Bias (see Fig. 6) and statistical result for 20 days (see Table 1) are obtained. The average RMS, SD, Bias, and MAE of tomographic water vapor density are 1.08, 1.03, −0.21, and 0.77 g/m^3, respectively, which means that the accuracy of tomographic result is good.

5 Conclusion

In this paper, a four-dimensional water vapor tomography method has been proposed based on the RT-PPP technique. The accuracy of ZTD and PWV derived from RT-PPP technique is first validated using data from March 25, 2014 to April 25, 2014. On comparing the results of PPP technique, GAMIT software, and radiosonde, it shows that RT-PPP technique is capable of acquiring ZTD and PWV with the same accuracy of results that are processed afterwards. SWVs are calculated based on RT-PPP technique, and regarded as tomographic observation.

Tomographic experiment has been performed using twenty days' data of twelve stations from SatRef and obtained a good result. It is evident that the real-time three-dimensional water vapor distribution obtained based on RT-PPP technique is of many potential application values, especially for those timeliness strong weather forecasting like short-term precipitation forecasting and real-time precipitation forecasting. Consequently, the research on four-dimensional water vapor tomography based on RT-PPP technique has very important research significance.

Acknowledgements The authors would like to thank IGAR for providing access to the web-based IGAR data. The Lands Department of HKSAR is also acknowledged for providing GPS data from the Hong Kong Satellite Positioning Reference Station Network (SatRef) and meteorological data.

References

1. Bastin S, Champollion C, Bock O, Drobinski P, Masson F (2005) On the use of GPS tomography to investigate water vapor variability during a Mistral/sea breeze event in southeastern France. Geophys Res Lett 32(5)
2. Lutz SL (2008) High-resolution GPS tomography in view of hydrological hazard assessment. Doctoral dissertation, Diss., Eidgenössische Technische Hochschule ETH Zürich, Nr. 17675
3. Champollion C, Flamant C, Bock O, Masson F, Turner DD, Weckwerth T (2009) Mesoscale GPS tomography applied to the 12 June 2002 convective initiation event of IHOP 2002. QJR Meteorol Soc 135:645–662
4. Jiang P, Ye SR, Liu YY, Zhang JJ, Xia PF (2014) Near real-time water vapor tomography using ground-based GPS and meteorological data: long-term experiment in Hong Kong. In: Annales Geophysicae, vol 32(8). Copernicus GmbH, pp 911–923
5. Adavi Z, Mashhadi-Hossainali M (2014) 4D tomographic reconstruction of the tropospheric wet refractivity using the concept of virtual reference station, case study: northwest of Iran. Meteorol Atmos Phys 126(3–4):193–205
6. Zumberge JF, Heflin MB, Jefferson DC, Watkins MM, Webb FH (1997) Precise point positioning for the efficient and robust analysis of GPS data from large networks. J Geophys Res Solid Earth (1978–2012) 102(B3):5005–5017
7. Dow JM, Neilan RE, Rizos C (2009) The international GNSS service in a changing landscape of global navigation satellite systems. J Geod 83(3–4):191–198
8. Caissy M, Agrotis L, Weber G, Hernandez-Pajares M, Hugentobler U (2012) Coming soon: the International GNSS real-time service. GPS World 23(6):52–58
9. Li X, Zhang X, Ge M (2011) Regional reference network augmented precise point positioning for instantaneous ambiguity resolution. J Geod 85(3):151–158
10. Li X, Ge M, Zhang H, Nischan T, Wickert J (2013) The GFZ real-time GNSS precise positioning service system and its adaption for COMPASS. Adv Space Res 51(6):1008–1018
11. Li X, Dick G, Ge M, Heise S, Wickert J, Bender M (2014) Real-time GPS sensing of atmospheric water vapor: precise point positioning with orbit, clock, and phase delay corrections. Geophys Res Lett 41(10):3615–3621
12. Dousa J, Vaclavovic P (2014) Real-time zenith tropospheric delays in support of numerical weather prediction applications. Adv Space Res 53(9):1347–1358
13. Haase J, Ge M, Vedel H, Calais E (2003) Accuracy and variability of GPS tropospheric delay measurements of water vapor in the western Mediterranean. J Appl Meteorol 42(11): 1547–1568

14. Gendt G, Dick G, Reigber C, Tomassini M, Liu Y, Ramatschi M (2004) Near real time GPS water vapor monitoring for numerical weather prediction in Germany. J Meteorol Soc Jpn 82 (1B):361–370
15. Lu C, Li X, Nilsson T, Ning T, Heinkelmann R, Ge M, Schuh H (2015). Real-time retrieval of precipitable water vapor from GPS and BeiDou observations. J Geod 1–14
16. Flores A, Ruffini G, Rius A (2000) 4D tropospheric tomography using GPS slant wet delays. In: Proceedings of Annales Geophysicae, vol 18(2). Springer, Heidelberg, pp 223–234
17. Emanuel K, Raymond D, Betts A, Bosart L, Bretherton C, Droegemeier K, Thorpe A (1995) Report of the first prospectus development team of the US weather research-program to NOAA and the NSF
18. Bauer HS, Wulfmeyer V, Schwitalla T, Zus F, Grzeschik M (2011) Operational assimilation of GPS slant path delay measurements into the MM5 4DVAR system. Tellus A 63(2):263–282
19. Chen B, Liu Z (2014) Voxel-optimized regional water vapor tomography and comparison with radiosonde and numerical weather model. J Geod 88(7):691–703
20. Yuan Y, Zhang K, Rohm W, Choy S, Norman R, Wang CS (2014) Real-time retrieval of precipitable water vapor from GPS precise point positioning. J Geophys Res Atmos 119 (16):10044–10057
21. Niell AE (2001) Preliminary evaluation of atmospheric mapping functions based on numerical weather models. Phys Chem Earth Part A 26(6):475–480
22. Tregoning P, Herring TA (2006) Impact of a priori zenith hydrostatic delay errors on GPS estimates of station heights and zenith total delays. Geophys Res Lett 33(23)
23. Saastamoinen J (1972). Atmospheric correction for the troposphere and stratosphere in radio ranging satellites. Use Artif Satell Geod 247–251
24. Bevis M, Businger S, Chiswell S, Herring TA, Rocken C, Ware RH (1994) GPS meteorology: mapping zenith wet delays onto precipitable water. J Appl Meteorol 33:379–386
25. Benevides P, Catalao J, Miranda PMA (2015) On the inclusion of GPS precipitable water vapour in the nowcasting of rainfall
26. Bevis M, Businger S, Herring T, Rocken C, Anthes R, Ware R (1992) GPS meteorology: remotesensing of atmospheric water vapor using the global positioningsystem. J Geophys Res 97(D14):15787–15801
27. Mendes VB (1999) Modeling the neutral-atmosphere propagation delay in radiometric space technique. PhD dissertation, University of New Brunswick, Frederiction, New Brunswick
28. Bender M, Stosius R, Zus F, Dick G, Wickert J, Rabe A (2010) GNSS water vapor tomography-Expected improvements by combing GPS, GLONASS and Galileo observations. Adv Sp Res 47:886–897
29. Bender M, Dick G, Ge M, Deng Z, Wickert J, Kahle HG, Tetzlaff G (2011) Development of a GNSS watervapour tomographysystem using algebraic reconstruction techniques. Adv Sp Res 47:1704–1720. doi:10.1016/j.asr.2010.05.034
30. Song SL, Zhu YW, Ding JC et al (2005) Shanghai GPS network tomography vapor three-dimensional distribution to improve numerical forecast humidity field. J Chin Sci Bull 50(20):2271–2277
31. Elósegui P, Ruis A, Davis JL, Ruffini G, Keihm SJ, Bürki B, Kruse LP (1998) An experiment for estimation of the spatial and temporal variations of water vapor using GPS data. Phys Chem Earth 23(1):125–130
32. Adeyemi B, Joerg S (2012) Analysis of water vapor over nigeria using radiosonde and satellite data. J Appl Meteorol Climatol 51:1855–1866. doi:10.1175/JAMC-D-11-0119.1
33. Liu ZZ, Wong MS, Nichol J, Chan PW (2013) A multi-sensor study of water vapour from radiosonde, MODIS and AERONET: a case study of Hong Kong. Int J Climatol 33:109–120. doi:10.1002/joc.3412
34. Gusfarienza H, Yuwono BD, Awaluddin M et al (2015) Penentuan zenith tropospheric delay dan precipitable water vapor Menggunakan Perangkat Lunak Gamit. Jurnal Geodesi Undip 4 (2):78–86

35. Seco A, Ramírez F, Serna E, Prieto E, García R, Moreno A, Priego JE (2012) Rain pattern analysis and forecast model based on GPS estimated atmospheric water vapor content. Atmos Environ 49:85–93
36. de Ortiz Galisteo JP, Bennouna Y, Toledano C, Cachorro V, Romero P, Andrés MI, Torres B (2014) Analysis of the annual cycle of the precipitable water vapour over Spain from 10-year homogenized series of GPS data. Q J R Meteorol Soc 140(679):397–406

A Spread Spectrum Communication Method Based on GNSS Positioning and Timing System

Jie Zhang, Zhaorui Wang and Qingtao Wan

Abstract Application of spread spectrum technology can effectively solve the problem of power constraints in satellite communication. A difficulty of spread spectrum communication is fast acquisition of spread spectrum signal. For the lack of priori information, code phase and frequency offset are both random quantities, it usually takes a long time in signal acquisition, which would reduce the efficiency of the communication especially for the burst communication. Traditional methods realized spread spectrum signal fast acquisition at the expense of acquisition algorithm complexity and hardware resources cost. This article presents a low cost new code phase and carrier frequency of spread spectrum signal correction method that is easy to implement based on GNSS positioning and timing system. In order to correct the frequency offset and code phase difference, position and time information given by GNSS system are used to build the network-wide time synchronization system and indirect loopback correction system in ground station. At the same time, with the aid of satellite timing, the influence of clock error is eliminated in the way of DDS. With the above measures, the spread spectrum signal fast acquisition could be realized in the network-wide time synchronization system.

Keywords Spread spectrum communication · Timing · Time synchronization

1 Introduction

Due to the high orbit satellite, propagation loss of satellite communication system, especially in the GEO satellite communication system is great. Meanwhile, in order to avoid interference between satellite and ground communication systems, low satellite antenna EIRP and weak signal power reaching the ground are obstacles to terminal miniaturization and communication capacity expansion. The spread spectrum technology can effectively solve the problem of limited power in satellite

J. Zhang (✉) · Z. Wang · Q. Wan
National Astronomical Observatories, Chinese Academy of Sciences, Beijing, China
e-mail: zhangjie05@mails.ucas.ac.cn

J. Sun et al. (eds.), *China Satellite Navigation Conference (CSNC) 2016 Proceedings: Volume I*, Lecture Notes in Electrical Engineering 388,
DOI 10.1007/978-981-10-0934-1_2

communication channel with sufficient bandwidth. To comply with International Telecommunication Union (ITU) regulations on signal strength (see [1]), spread spectrum communication technology is used to reduce the wave flux density in the unit band by broadening signal bandwidth for the power. At the same time, the spread spectrum communication has the advantage of confidential, and strong antijamming capability.

Synchronization is a key problem for spread spectrum signal receiving processing. PN code and carrier, at both transmitting and receiving ends should be kept under synchronization, including the synchronization of phase and frequency. At the receiving end, the rough code phase and carrier frequency offset estimation are given by capturing and coarse synchronization. Then the results of coarse synchronization are used to initialize tracking loop and precise synchronization, finally restore the information. As a result, the capture of the spread spectrum signal with high rate and long PN code period is very difficult in high dynamic, low SNR environments.

There are several capture normal methods [2, 3]: sliding correlator, capture time is proportional to the square of PN code period; matching filter, with a short capturing time at great hardware expense. These two methods are 2-D search for code phase and carrier frequency, and affected by the frequency offset. Besides the above two methods, FFT based on carrier frequency is able to overcome the frequency offset on the influence of PN code, but adaptive threshold adjustment for large signal strength varies and increases the complexity of the system.

From the above, for the lack of prior information, the capture of PN code phase and frequency offset value which is random costs a large amount of computation and hardware resources, repeated threshold judgement and adjustment are necessary for the probability of signal capture. Therefore, capture process with a long time reduces the communication efficiency especially for burst communication.

In this study, a new PN code phase and carrier frequency offset correction method based on the link of "Ground station—GEO satellite—Remote station" is proposed. In order to spread spectrum signal fast acquisition, the code phase difference, carrier Doppler frequency offset and clock error are corrected with the aid of GNSS satellite navigation system positioning and timing functions.

2 Construction of Network-Wide Time Synchronization System

In order to improve communication efficiency and user management, network-wide time synchronization technology has been widely used in ground-based 3G and 4G mobile communication system. Time synchronization between the base station (BS) is achieved by GPS or IEEE 1588v2 time transfer technology [4]. The mobile station (MS) achieves time synchronization with BS by time correction parameter in synchronous channel. The 3G and 4G mobile communication networks' time

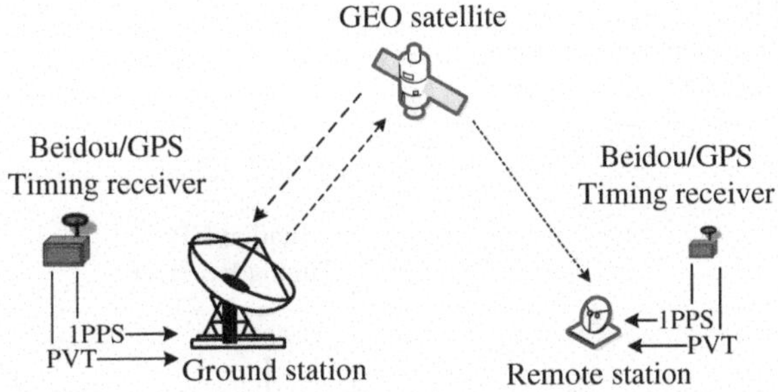

Fig. 1 Network-wide time synchronization system

synchronization precision index is only for μs [4], which is inadequate for the military and security spread spectrum communication with high spreading gain and PN code rate (Mcps). Timing function of Beidou/GPS satellite navigation system with the precision of 20–100 ns could be used as a unified time reference. The signal is transmitted and received under the unified time reference (1PPS). The capture process is accelerated by PVT (position, velocity, time) parameters. The original ground station and remote station are equipped with Beidou/GPS timing receivers to construct the network-wide time synchronization system (Fig. 1).

3 Synchronization of Spread Spectrum Signal

Prerequisites for spread spectrum communication are carrier and PN code synchronization, which means synchronization of both frequency and phase. Code phase, carrier frequency offset, and clock error are corrected, respectively, based on Beidou/GPS positioning and timing system.

3.1 Correction of PN Code Phase Offset

Code phase consists mainly of link propagation delay and equipment delay, which is corrected in different ways, respectively, at ground station and remote station.

3.1.1 Code Clock Round-Trip Link Indirect Correction in Ground Station

In Fig. 1, an example of a communication link composed of "Ground station—GEO satellite—Remote station", round-trip correction link consists of transceiver baseband, RF equipments at ground station, (up/down converter, PA, LNA, and antenna) and GEO satellite, as shown in Fig. 2.

Fig. 2 Principle of code clock indirect loop back correction

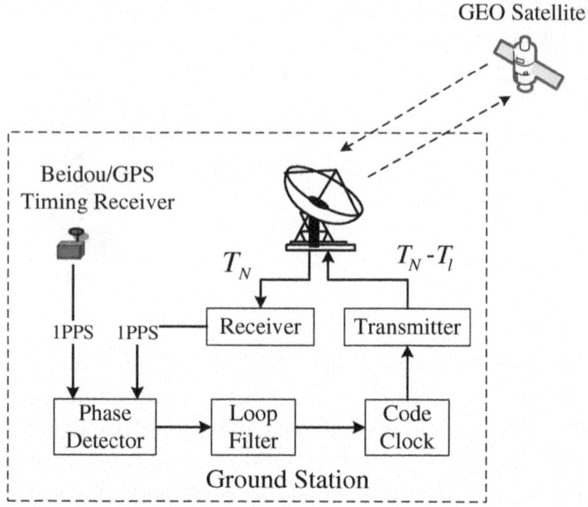

The ground station transmits the spread spectrum signal and then receives the signal through the round-trip link. The receiver baseband output 1PPS on the start rising at the edge of each 1 s frame. The phase difference between 1PPS of Beidou/GPS receiver and 1PPS the receiver output is given by phase detector, which is used to feedback the correction of transmitting code clock after loop filter. Through the above indirect adjustment method, data transmitted from the ground station arrives at the receiving end on the rising edge of 1PPS, which means that the data transmitted at the time of $T_N - T_l$ arrives at T_N through round-trip link (T_N is integer second, T_l is the link propagation and equipment delay).

The experiments based on satellite ground link composed of equipments in Fig. 1 verify the above program. The Beidou/GPS timing receiver HM-1103C from Hwa Create Corporation [5]; the ground station located in navigation and communication central station of National Astronomical Observatory in Wuqing district, Tianjin; GEO satellite is a retired commercial satellite. The measured results of 1PPS synchronization precision in actual satellite ground round-trip link are shown in Fig. 3.

The 1PPS synchronization error that is corrected by the ground-satellite round-trip loop gradually converges with 1σ error of 34 ns.

3.1.2 Code Phase Correction in Remote Station

For the communication system in Fig. 1, the 1PPS in the ground-satellite round-trip loop is synchronized with Beidou/GPS 1PPS, but for signal transmitted from the ground station, there is time difference (ΔT) between the time arriving at the ground station and the remote station as the two different down links ("satellite—ground station" and "satellite—remote station").

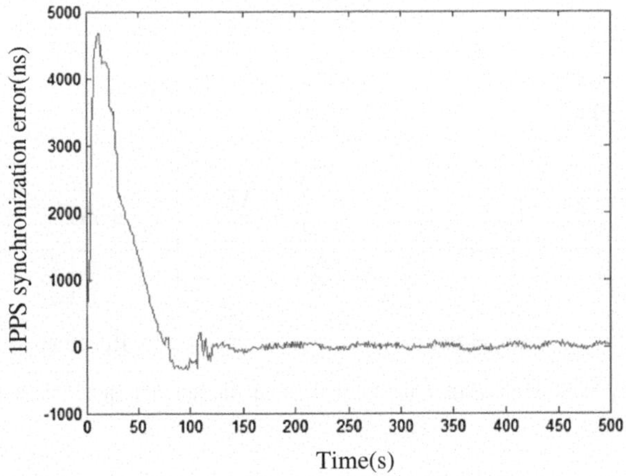

Fig. 3 Receiver-Beiou/GPS output 1PPS synchronization error of the round-trip link

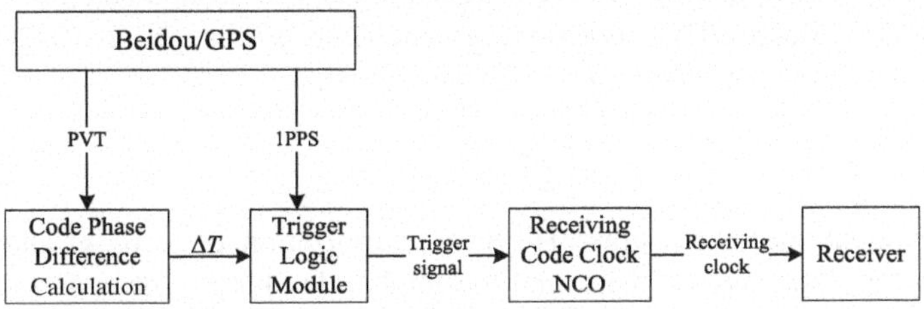

Fig. 4 Correction of propagation time delay difference correction

According to the output of the Beidou/GPS timing receiver position coordinates, precise satellite ephemeris parameters and ground station coordinates, the remote station can calculate the down link propagation delay difference ΔT.

Modules for propagation delay difference correction are shown in Fig. 4.

The propagation time delay difference ΔT can be calculated by real-time computing with Beidou/GPS 1PPS output. The work timing sequence is shown in Fig. 5. The ground station transmits the signal at the time of $T_N - T_l$ (T_N is integer second, T_l is the ground-satellite round-trip propagation and equipment delay), the remote station starts receiving the clock to correlate at the time of $T_N + \Delta T$.

For the "Ground station—GEO satellite—Remote station" link, the code phase difference correction of whole link consists of the above two processes. The remote station code clock starts at a fixed time to receive, so as to, improve the efficiency without signal acquisition process.

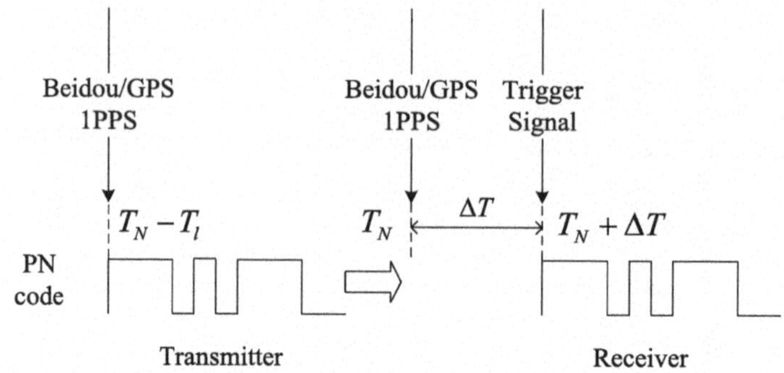

Fig. 5 Propagation time delay difference correction of remote station

3.1.3 Total Code Phase Error

Total code phase error ΔT_s is composed of timing error caused by Beidou/GPS positioning error ΔT_p and 1PPS synchronization error of ground-satellite round-trip link loop ΔT_l. Beidou/GPS positioning precision ΔP is about 10–20 m, c is propagation velocity of electromagnetic wave, then

$$\Delta T_p = \Delta P / c \tag{1}$$

According to the test results, 1PPS synchronization error of ground-satellite round-trip link loop ΔT_l is 34 ns, and the total code phase error is as follows:

$$\Delta T_s = \sqrt{(\Delta T_p)^2 + (\Delta T_l)^2} \tag{2}$$

According to Eq. (2), total code phase correction error ΔT_s is 50–70 ns. The code phase synchronization error should be within $\frac{1}{2}$ code chip for the capture of the PN code. Thus, this program meets the highest code rate 10 Mcps spread spectrum communication capture requirements.

3.2 *Carrier Frequency Offset Correction*

Carrier frequency offset is composed of Doppler, transformer, and clock error.

3.2.1 Frequency Offset Correction in Ground-Satellite Round-Trip Link

The same way as the code phase correction in Fig. 2, the frequency offset of the forward link is corrected in the ground-satellite round-trip link. In order to keep the

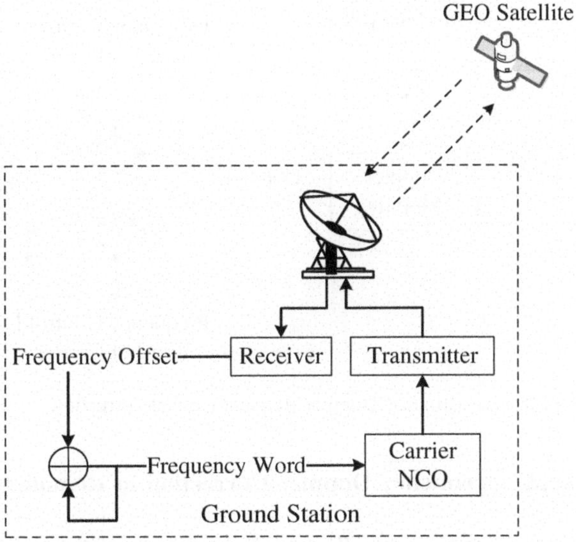

Fig. 6 Principle of frequency offset correction

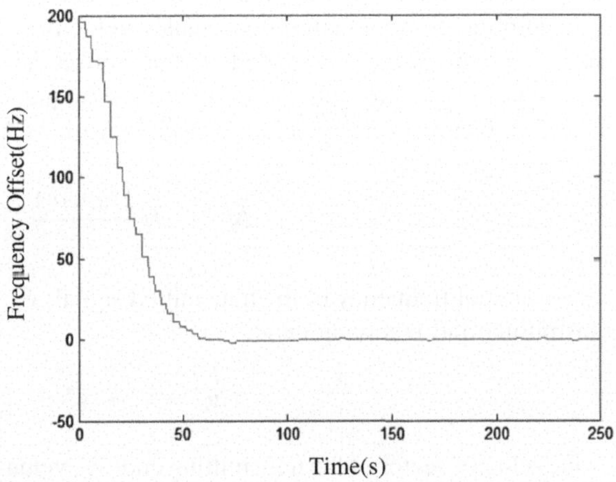

Fig. 7 Results of frequency offset correction

down link frequency at nominal value, as shown in Fig. 6, the receiver outputs the frequency offset to the transmitter for transmitting frequency pre-bias by the closed loop correction.

The results of frequency offset round-trip loop correction finally converge to 0 Hz, as shown in Fig. 7:

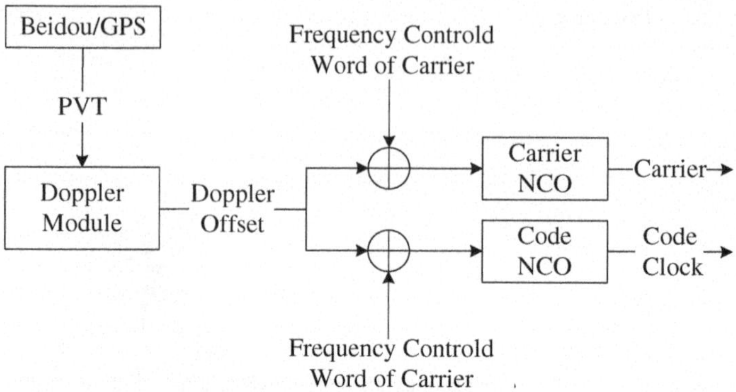

Fig. 8 Consisting of Doppler frequency offset correction

3.2.2 Frequency Doppler Correction in Remote Station

Doppler frequency offset is derived from the relative motion of the carrier. For "Ground station—GEO Satellite—Remote Station" the relative motion of the carrier including the remote station motion and satellite drift can result in frequency offset of the carrier and the code clock.

According to the location coordinates velocity of Beidou/GPS timing receiver output, Doppler carrier and code frequency offset Δf could be calculated, and then, Δf and precise satellite ephemeris are used to frequency pre-bias of carrier and code NCO. The Doppler carrier and code frequency offset Δf is calculated as follows: [6]:

$$\Delta f = -f_T \frac{(\overrightarrow{v_d} \cdot \overrightarrow{a})}{c} \tag{3}$$

f_T, nominal frequency of the transmitted signal; $\overrightarrow{v_d}$, relative velocity vector of the transmitting and receiving ends.

$$\overrightarrow{v_d} = \overrightarrow{v_t} - \overrightarrow{v_r} \tag{4}$$

$\overrightarrow{v_t}$, velocity vector of the transmitting end; $\overrightarrow{v_r}$, velocity vector of the receiving end; $\overrightarrow{a}$, unit vector along the straight line from the receiving end to the transmitting end.

$$\overrightarrow{a} = (\overrightarrow{u_t} - \overrightarrow{u_r}) / \|\overrightarrow{u_t} - \overrightarrow{u_r}\| \tag{5}$$

$\overrightarrow{u_t}$, location vector of the transmitting end; $\overrightarrow{u_r}$, location vector of the receiving end; $\|\cdot\|$, distance between transmitting and receiving ends; $\overrightarrow{v_d} \cdot \overrightarrow{a}$, radial component of relative velocity vector along the straight line from the receiving end to the transmitting end; c, propagation velocity of electromagnetic wave.

The structure of the Doppler frequency offset correction module as shown in Fig. 8.

For the "Ground station—GEO Satellite—Remote Station" link, frequency of uplink and downlink are as follows:

$$f'_u = f_u + \Delta f_u \tag{6}$$

$$f'_d = f_d + \Delta f_d \tag{7}$$

f_u, nominal frequency of uplink; f_d, nominal frequency of downlink; Δf_u, frequency offset of uplink; Δf_d, frequency offset of downlink. Frequency of uplink is corrected to nominal value f_u frequency pre-bias at the transmitting end; Frequency of downlink is corrected to actual value f'_d at the receiving end.

The above methods apply to "Ground station—GEO Satellite—Remote Station" link, for the reverse link, the code phase difference and frequency offset can be calculated according to PVT parameters at remote station.

3.2.3 Clock Error Correction

Clock error of oscillator at transceiver both ends is also an important cause of frequency offset besides Doppler motion. OCXO is of high stability, but higher cost is not suitable for the terminal equipment. TCXO is of short-term stability but frequency drift for long time, while GNSS timing short-term stability is poor, but of high long-term stability. Therefore, combining the advantages of both TCXO and GNSS timing can be used to the frequency offset correction, which is "discipline". The principle of "discipline" is to adjust the local oscillator according to the frequency error of clock signal generated by local oscillator and determined frequency signal, for both short-term and long-term stability [7, 8].

In this study, a correction method of low cost based on direct digital synthesizer (DDS) with aid of satellite timing is proposed. Specific as follows: GNSS output 1PPS as gating signals, the local clock error Δx of the actual and nominal value could be estimated according to the clock cycles between the two adjacent 1PPS. Then code clock and carrier frequency offset caused by clock error Δx could be corrected by pre-bias. Principle of DDS is as follows [9]:

$$\text{FTW} = \frac{f_o}{2^M}(f_{\text{sys}} + \Delta x) \tag{8}$$

FTW, the frequency tuning word; f_o, the output frequency; M, the length of the phase accumulator; f_{sys}, nominal frequency of the oscillator.

Structure of clock error correction module is shown in Fig. 9: the local clock error Δx could be estimated according to the clock cycles between gating Beidou/GPS 1PPS signals. In order to avoid the short-term unstability of Beidou/GPS 1PPS, filter the counting results using weighted filtering algorithm. The principle of weighted filtering algorithm is as follows [10]:

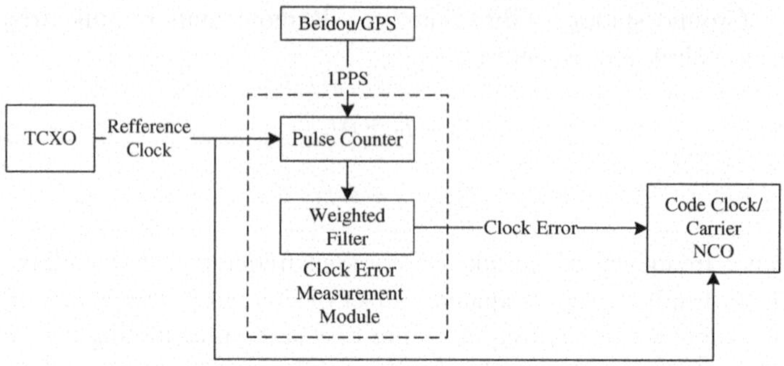

Fig. 9 Consisting of clock error correction

$$\hat{x} = \sum_{i=0}^{n-1} W_i(n)x_{n-i} \tag{9}$$

x_{n-i}, 1PPS count number; $W_i(n)$, weight coefficient; n, length of measurement window, the longer the length, the higher the precision.

4 Conclusions

In this study, a new spread spectrum code phase and carrier frequency correction method based on Beidou/GPS timing is proposed. For the key of spread spectrum signal acquisition: code phase and carrier frequency offset, network-wide time synchronization system is constructed and to correct from three aspects based on Beidou/GPS timing: code phase error, carrier frequency offset, and clock error. The acquisition of spread spectrum signal is greatly accelerated due to the correction of code phase and carrier frequency. The method of this paper is simple with low cost, not only for satellite spread spectrum communication system, also can be applied to other ground mobile or fixed station communications.

References

1. Cui J, Shi H et al (2008) The transmission link of CAPS navigation and communication system. Sci China Ser G Phys Mech Astron 38(12):402–411. doi: 10.1007/s11433-009-0062-9
2. Tsui JBY (2008) Fundamentals of global positioning system receivers a software approach, 2nd edn. (trans: Chen J et al). Publishing House of Electronics Industry, Beijing
3. Ricai T (2007) Spread spectrum communication. Tsinghua University Press, Beijing
4. Chu Y (2012) Application of time synchronization equipment in mobile communication network. Mod Transm 4:58–62. doi:10.3969/j.issn.1673-5137.2012.08.004

5. Hwa Create Corporation Ltd. (2013) HM-1103C navigation/timing module data sheet V1.1. Beijing
6. Kaplan ED, Hegarty CJ (2007) Understanding GPS principles and applications, 2nd edn. (trans: Kou Y). Publishing House of Electronics Industry, Beijing
7. Ma X, Kong W, Sun H (2011) Design of a frequency calibration system based on GPS disciplined rubidium clock. Telecommun Eng 51(10):109–112. doi:10.3969/j.issn.1001-893x.2011.10.022
8. Shan Q, Yang J (2009) Review of satellite disciplined clock system. J Test Meas Technol 23(5):396–401. doi:10.3969/j.issn.1671-7449.2009.05.005
9. O'Leary P, Maloberti F (1991) A direct-digital synthesizer with improved spectral performance. Commun IEEE Trans 39(7):1046–1048. doi:10.1109/26.87209
10. Zhao Y (2001) Arithmetic mean method and weighting mean method of digital filter. Instrum Technol 4:41–44. doi:10.3969/j.issn.1006-2394.2001.04.017

The Effect of GNSS Sites Distribution on TEC Derivation

Xiaolan Wang, Guanyi Ma, Qingtao Wan, Jinghua Li, Jiangtao Fan and Jie Zhang

Abstract With worldwide increased Global Navigation Satellite System (GNSS) receivers, it is possible to obtain the ionospheric total electron content (TEC) and hence monitor the ionosphere with GNSS. Using a thin layer assumption of the ionosphere and dual-frequency Global Positioning System (GPS) observations from 16 geomagnetically quiet days in four seasons of 2006, this paper adopts the spherical harmonic model to fit TEC and investigates the effects of two network constitutions on global TEC derivation, one with 275 GPS receivers and the other with 125 GPS receivers. The results show that the data can be well fitted for both network constitutions. The derived TECs are consistent with each other for four seasons. This is especially true for TECs at low- and mid-latitude. The derived satellite and receiver biases are stable during the year. The standard deviation of the satellite and the receiver biases are less than 0.5 and 3 ns, respectively.

Keywords GPS · TEC · Sites distribution · Spherical harmonic model

1 Introduction

As a dispersive medium, the ionosphere can cause a time delay to transionospheric radio waves, which is proportional to the total electron content (TEC) along the ray path of radio wave propagation. TEC is one of the important parameters of studying the ionospheric characteristics, and the measurement of TEC is an important task for ionospheric corrections in transionospheric radio system [1]. With the extensive development and application of GNSS, the ionospheric TEC has been measured with GNSS receivers economically and effectively. The derived TEC has been

X. Wang (✉) · G. Ma · Q. Wan · J. Li · J. Fan · J. Zhang
National Astronomical Observatories, Chinese Academy of Sciences, Beijing, China
e-mail: xlwang@nao.cas.cn

X. Wang
University of Chinese Academy of Sciences, Beijing, China

J. Sun et al. (eds.), *China Satellite Navigation Conference (CSNC) 2016 Proceedings: Volume I*, Lecture Notes in Electrical Engineering 388,
DOI 10.1007/978-981-10-0934-1_3

used to construct models for improving the positioning precision of GNSS user and monitor the ionospheric activity for space weather.

Since the civilian use of Global Positioning System (GPS) and GPS receivers being set up worldwide, TEC derivation methods have been studied continuously [1–12]. Center for Orbit Determination in Europe (CODE), the Jet Propulsion Laboratory (JPL), the European Space Agency (ESA), Canadian Energy, Mines and Resources Center (EMR), Polytechnic University of Catalonia in Spain (UPC), National Institute of Information and Communication Technology in Japan (NICT), and institute of geology and geophysics, Chinese academy of sciences (IGGCAS) have completed the relevant study on the algorithm of the ionosphere TEC products and released the analysis results. In the early times, the ionosphere was observed with a single GPS receiver, the instrumental biases and the TEC were assessed by representing the TEC as a polynomial of latitude and longitude [2]. Later using observation data from several GPS receivers, the TEC and instrumental biases were derived by random walk stochastic process and solved by Kalman filter method [3]. With the global network of over 100 GPS receivers, JPL models the vertical TEC to bi-cubic splines on a spherical of 2.5° by 5.0° latitude and longitude, then by Kalman filter method to solve the TEC and instrumental biases [4]. Using around 300 GPS receivers selected homogeneously from GEONET, and assuming that the TEC is identical at any point within 2° by 2° grid block, the TEC over Japan and instrumental biases were determined least-squares fitting method [7]. However, instrumental biases exist in the hardware of the GPS satellite and receiver which affects the accuracy of the TEC estimation greatly. For accurate TEC derivation, satellite and receiver biases should be removed properly.

Based on a thin layer assumption of the ionosphere and the spherical harmonic model, this paper adopts the algorithm spherical harmonic model to derive TEC for two network constitutions, one with 275 GPS receivers and the other with 125 GPS receivers. With the observation data selected from 16 geomagnetically quiet days in 2006, 5–8 March, 23–26 June, 10–13 September, and 26–29 December, covering a period of 10 months, we analyze the effect of the sites distribution on TEC derivation.

2 Transionospheric Radio Propagation

A GPS receiver measures two pseudoranges (P_1 and P_2) and two carrier phases (L_1 and L_2) corresponding to two signals at 1575.42 MHz (referred to as f_1) and 1227.6 MHz (referred to as f_2). The slant TEC along the ray path can be calculated as

$$\mathrm{STEC}_p = \frac{2(f_1 f_2)^2}{k(f_1^2 - f_2^2)}(P_2 - P_1) \tag{1}$$

$$\mathrm{STEC}_l = \frac{2(f_1 f_2)^2}{k(f_1^2 - f_2^2)}(L_1\lambda_1 - L_2\lambda_2) \tag{2}$$

where $k = 80.62\ \mathrm{m}^3/\mathrm{s}^2$, λ_1 and λ_2 are the wavelengths corresponding to f_1 and f_2, respectively. To reduce the effects of the pseudorange error on TEC, a baseline or offset between STEC_p and STEC_l can be used to obtain an absolute slant TEC with higher precision [7]

$$B_{rs} = \frac{\sum_{i=1}^{N}(\mathrm{STEC}_{p_i} - \mathrm{STEC}_{l_i})\sin^2\alpha_i}{\sum_{i=1}^{N}\sin^2\alpha_i} \tag{3}$$

$$\mathrm{STEC} = \mathrm{STEC}_l + B_{rs} \tag{4}$$

where α_i is the elevation of the satellite. The square sine of the elevation $\sin^2\alpha_i$ is included as a weighting factor, as the pseudorange with low elevation apt to be affected by the multipath effect and the reliability decreases.

When studying the ionosphere, the vertical TEC (refers to TEC hereafter) is generally used, which refers to the TEC along the path that passes through the earth center and cross the ionosphere perpendicularly. To convert the STEC to TEC, we assume the ionosphere as a single thin layer at height h above the earth as shown in Fig. 1.

The intersection of the slant path from the satellite S to the receiver R through the ionosphere is referred to as piercing point P. The zenith angle χ is expressed as the following:

$$\chi = \arcsin\left(\frac{R_e \cdot \cos\alpha}{R_e + h}\right) \tag{5}$$

Fig. 1 Geometry of single layer model

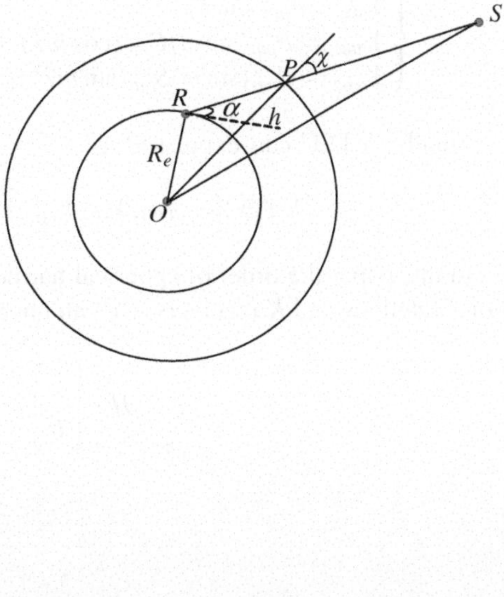

where R_e is the mean radius of the Earth. The vertical TEC at the piercing point can be obtained

$$\text{VTEC} = (\text{STEC} - b_s - b_r) \cdot \cos\chi \tag{6}$$

with b_s and b_r corresponding to the satellite and receiver biases, respectively. The $\cos\chi$ is the oblique-to-zenithal factor or the mapping function. With the spherical harmonic model, the vertical TEC can be expressed as follows [13]

$$\text{VTEC}(\varphi, \lambda) = \sum_{n=0}^{n_{\max}} \sum_{m=0}^{n} \tilde{P}_{nm}(\sin\varphi)(\tilde{C}_{nm}(\cos m\lambda) + \tilde{S}_{nm}(\sin m\lambda)) \tag{7}$$

where φ and λ are the geocentric latitude and the sun-fixed longitude of the ionosphere pierce point, $n_{\max}$ is the max order of the spherical harmonic, $\tilde{P}_{nm}(\sin\varphi) = \text{MC}(n, m)P_{nm}(\sin\varphi)$ is the normalized Legendre polynomials, $\text{MC}(n, m)$ is the normalization function, $\tilde{C}_{nm}$ and $\tilde{S}_{nm}$ are the coefficient of the spherical harmonic.

3 TEC Derivation

For the line of sight from satellite j to receiver k piercing through the ionosphere at time t, referring to Eq. 7 and putting the unknowns in the left side, we can write the following equation:

$$\sec\chi_{jk}\,\text{VTEC}_{jk} + b_{sj} + b_{rk} = \text{STEC}_{jk} \tag{8}$$

Using Eq. 7 we can obtained the following:

$$\begin{cases} Y_{n0} = \tilde{P}_{n0}(\sin\varphi) \\ Y_{nmc} = \tilde{P}_{nm}(\sin\varphi)C_{nm}(\cos m\lambda) = \text{MC}_{nm}C_{nm}P_{nm}(\sin\varphi)(\cos m\lambda) \\ Y_{nms} = \tilde{P}_{nm}(\sin\varphi)S_{nm}(\sin m\lambda) = \text{MC}_{nm}S_{nm}P_{nm}(\sin\varphi)(\sin m\lambda) \end{cases} \tag{9}$$

Finally, VTEC can expressed as,

$$\text{VTEC} = [Y_{00}, Y_{10}, Y_{11c}, Y_{11s}, Y_{20}, Y_{21c}, Y_{21s}, \ldots] \tag{10}$$

In this paper the order of spherical harmonics expansion is taken to be $n_{\max} = 8$. For J satellites and K receivers, we can have a matrix form of equations from Eq. 8,

$$H \cdot \begin{bmatrix} E \\ B \end{bmatrix} = \text{STEC} \tag{11}$$

where $E = [e_{00}, e_{10}, e_{11c}, e_{11s}, \ldots, e_{87c}, e_{87s}, e_{88c}, e_{88s}]$ is the coefficient of the spherical harmonic, $B = [b_{s1}, \ldots, b_{sJ}, b_{r1}, \ldots, b_{rK}]$ is the biases of satellites and receivers $H = \begin{bmatrix} \cdot & \cdot & \cdot & \cdot & \cdot & \cdot & \cdot \\ \sec\chi_{jk}P_{nm}(\sin\varphi)(\sin m\lambda) & \sec\chi_{jk}P_{nm}(\sin\varphi)(\sin m\lambda) & 0 & 1 & 0 & 1 & 0 \\ \cdot & \cdot & \cdot & \cdot & \cdot & \cdot & \cdot \end{bmatrix}$ is the matrix of coefficients with a dimension of $M \times N(N = 81 + J + K,\ M > N)$.

With a least-squares fitting technique, the solution to the set of equations can be obtained by the singular value decomposition (SVD), and the matrix of coefficients can be decomposed as the following [14]:

$$H = [U] \cdot \begin{bmatrix} \omega_1 & & & \\ & \omega_2 & & \\ & & \cdots & \\ & & & \omega_N \end{bmatrix} \cdot [V^T] \tag{12}$$

where U and V are orthogonal matrix, the dimension is $M \times M$ and $N \times N$, respectively, and $\omega_n (n = 1, \ldots, N)$ is the singular values of a diagonal matrix. Then the least-squares solution of Eq. 11 can be computed,

$$\begin{bmatrix} E \\ B \end{bmatrix} = [V] \cdot [\mathrm{diag}(1/\omega_n)] \cdot [U^T] \cdot \begin{bmatrix} \cdot \\ \cdot \\ \mathrm{STEC}_{jk} \\ \cdot \\ \cdot \end{bmatrix} \tag{13}$$

Because it is not possible to determine unambiguously all of the satellites and receiver biases absolutely, one of them should be set to be 0, as a reference. Here, the reference is taken to be the satellite with pseudorange number of 1. Subsequently, the spherical harmonic coefficients, satellite biases, and receiver biases can be obtained by solving Eq. 13. Then, VTEC is computed with Eq. 7.

4 Network Constitution and Results

We use two network constitutions to investigate the effect of site distribution TEC derivation. One network constitution consists of 275 GPS receivers of International GNSS Service (IGS), the other one selects 125 receivers from IGS. The site distribution of the two network constitutions are shown in Fig. 2.

Latitudinal distribution of the sites is shown in Fig. 3, where the horizontal axis refers to the sites number, the vertical axis shows latitude ranges, HN: 60°N–90°N, MN: 30°N–60°N, LN: 0°N–30°N, HS: 60°S–90°S, MS: 30°S–60°S, LS: 0°S–30°S. As seen in Fig. 3, a main difference of 115 receivers for the two network constitutions happens in MN area. This indicates that the GPS receivers are mostly dense in northern middle latitudes.

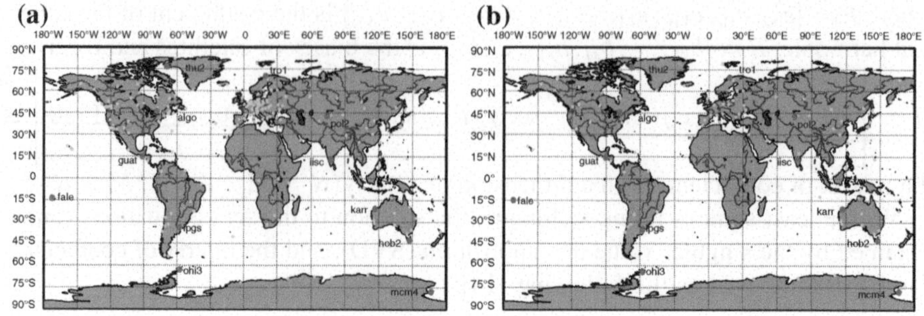

Fig. 2 GPS sites map (*red dots* are the location of selected 12 sites distribution, *pink dots* are the location of the rest of the sites distribution). **a** 275 GPS sites map, **b** 125 GPS sites map (Color figure online)

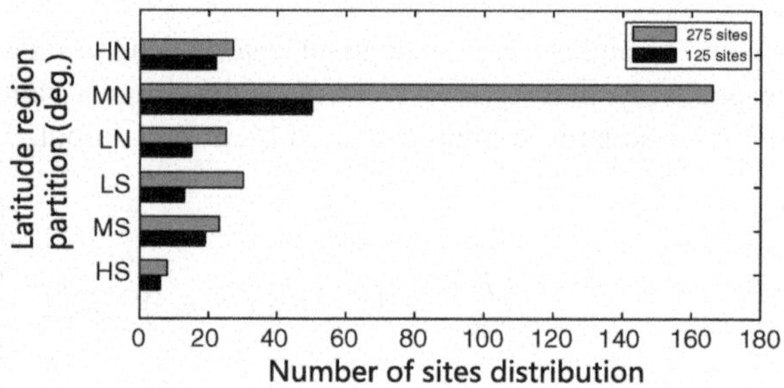

Fig. 3 Distribution of the number of sites with latitude changes (Color figure online)

Table 1 Geographic coordinates of the 12 GPS sites

Site	Latitude(°)	Longitude(°)	Site	Latitude(°)	Longitude(°)
THU2	76.54°N	68.82°W	FALE	13.83°S	171.99°W
ALGO	45.96°N	78.07°W	LPGS	34.91°S	57.93°W
GUAT	14.59°N	90.52°W	OHI3	63.32°S	57.90°W
IISC	13.02°N	77.57°E	KARR	20.98°S	117.09°E
POL2	42.68°N	74.69°E	HOB2	42.80°S	147.44°E
TRO1	69.66°N	18.94°E	MCM4	77.84°S	166.67°E

In order to examine the results and make comparison, we also selected 12 sites whose locations are listed in Table 1. With the observation data selected from 16 geomagnetically quiet days in 2006, 5–8 March, 23–26 June, 10–13 September, and 26–29 December, covering a period of 10 months, we analyze the effect of the sites distribution on TEC derivation. Here, the ionospheric height is taken to be 450 km and the time resolution of GPS observation data is 15 min.

It consists of two processes for TEC derivation. One process is to read and filter the observation data, and then constitute equations. The other process is to solve the equations. The calculating time of derived TEC from 275 GPS receivers is about 1 h 45 min 56 s and 7 min 44 s, and it is 41 min 55 s and 2 min 58 s from 125 GPS receivers. There is 1 h 8 min 47 s time difference with the two network constitutions.

4.1 Derived TEC

Figure 4 shows TEC maps with the two network constitutions at 6:00 UT on March 8, June 23, September 13, and December 29, the left column corresponds to 275 receivers, the right column is from 125 receivers. Not much difference can be seen from the two network constitutions for four seasons TEC peaks at low mid-latitude. Meanwhile, the TEC value is larger in the spring and autumn, and it is smaller in the summer and winter. The differences between maximum TECs are 2.24 TECU, 0.95 TECU, 0.15 TECU, and 0.84 TECU for the four seasons, respectively. The differences between locations where the maximum TECs occur are very similar to one another and it is less than 2° in latitude and not more than 10° in longitude. Meanwhile, the derived TEC are compared to those obtained from CODE, the results indicate that the derived TEC in 2006 are in agreement with those obtained from CODE.

Figure 5 shows derived TEC in 16 days for the 12 observation stations in Table 1, where the red curves represent the derived TEC from 275 sites, and black curves represent the derived TEC from 125 sites. As seen in the figure, the derived TEC tends to be consistent with each other and the TEC peaks at the same time every day. This is especially true for TECs over middle and low latitude sites. The TECs at high latitude are much noisy. Though why this is so is not the research topic of this paper, we should point out that the ionosphere is much more variable at high latitudes than that at middle and low latitudes.

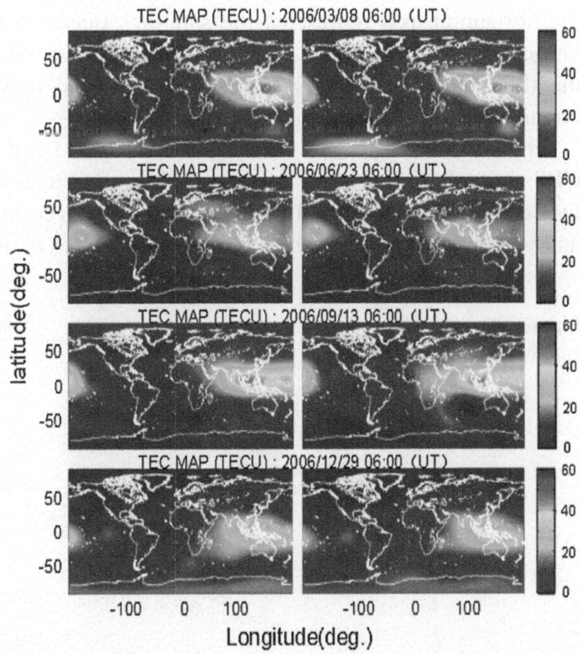

Fig. 4 The TEC maps at 6:00 UT (Color figure online)

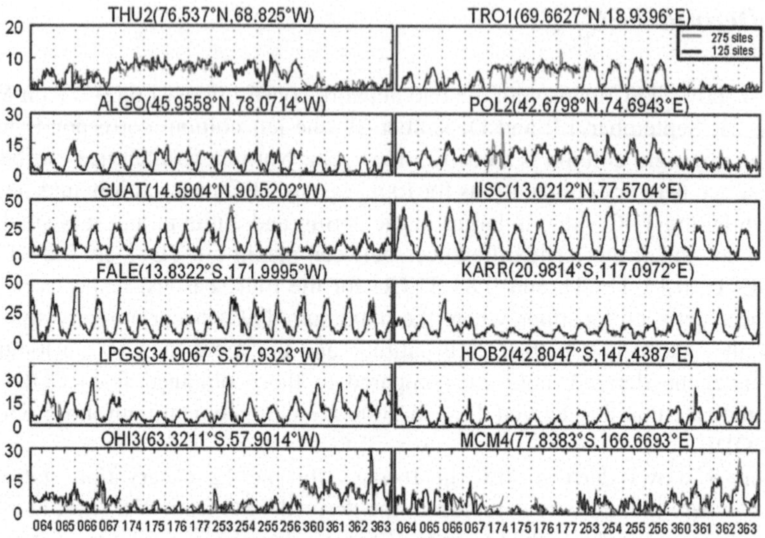

Fig. 5 TEC contrast figure of 12 sites (Color figure online)

4.2 The Satellites Biases

Figure 6 shows the comparison of estimated satellites biases for the 16 days from the two network constitutions which one is 275 sites and another is 125 sites, where the horizontal axis represents the satellites biases derived from 275 sites and the vertical axis represents the satellites biases derived from 125 sites. As can be seen, the difference of satellites biases from the two network constitutions is almost unanimously.

Figure 7 shows the 31 GPS satellites biases for the 16 days derived from 125 GPS sites. Vertical dashed lines divide non-consecutive days. As seen in the figure, two satellites (PRN 12 and PRN 31) were observed in 4 days in winter (360–363), and were not observed in the remaining 12 days. On the days 176 and 177, the bias

Fig. 6 Satellites biases contrast figure

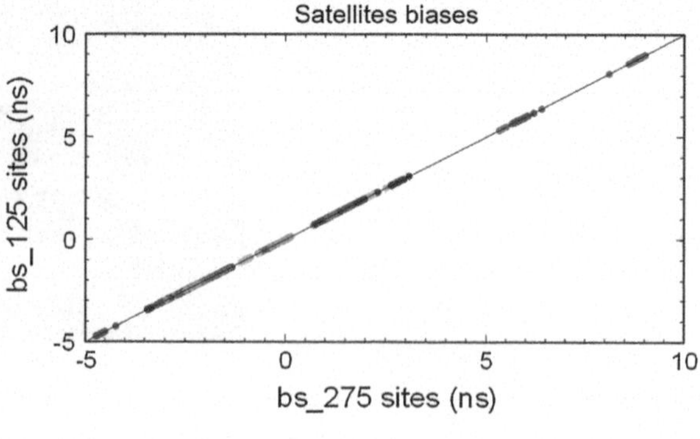

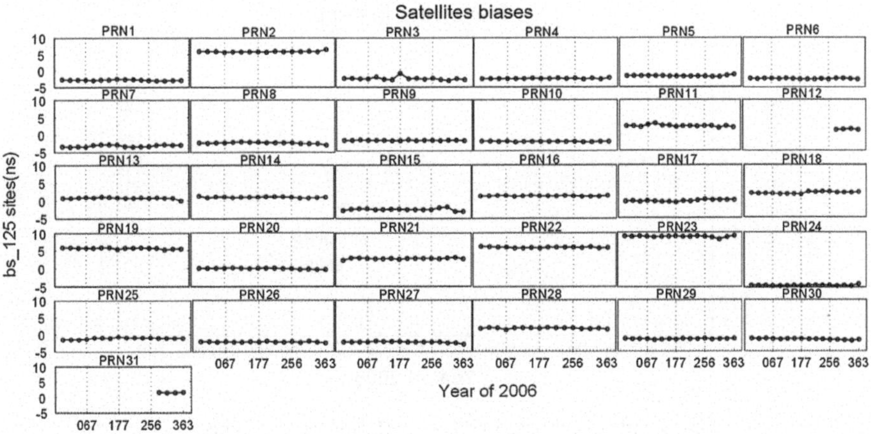

Fig. 7 Satellites biases derived from 125 GPS sites

of PRN3 shows a large variation. The difference is 1.9 ns between the two days. The remaining satellites biases show small changes in 16 days. For the 16 days, the standard deviation of 31 GPS satellites biases ranges 0.0796–0.4881 ns.

4.3 The Receivers Biases

Figure 8 shows the comparison of receivers biases derived from 275 sites and 125 sites, where the horizontal axis represents the receivers biases derived from 275 sites and the vertical axis represents the receivers biases derived from 125 sites. As can be seen, the difference of receivers biases with different network constitution is extremely close and not big changed.

Figure 9 shows the biases of 12 receivers from Table 1 in 16 days derived from 125 GPS sites. Vertical dashed lines divide non-consecutive days. Shown in the

Fig. 8 Receivers biases contrast figure

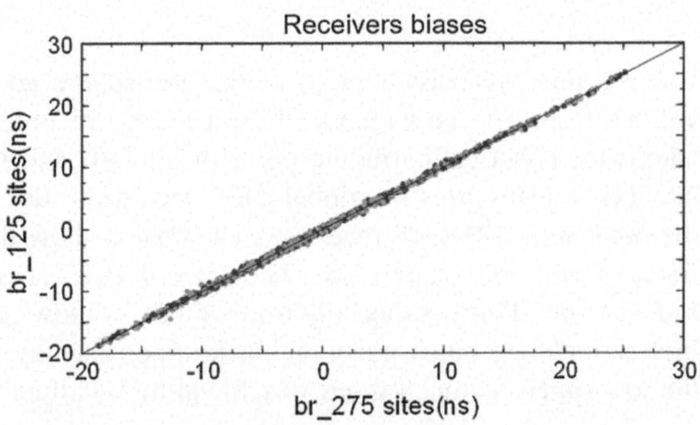

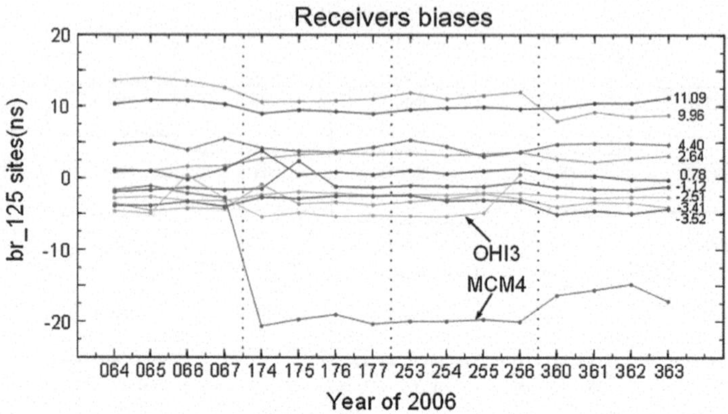

Fig. 9 Receivers biases derived from 125 GPS sites (Color figure online)

right are the mean of the receiver biases. As seen in the figure, the types of the 12 receivers are ASHTECH Z18, TRIMBLE 4000SSI, ASHTECH Z-XII3, AOA BENCHMARK ACT. The different hardware is likely to cause the different biases. For example, the difference can be achieved 10.3 ns between the types of AOA BENCHMARK ACT and ASHTECH Z-XII3. Except three receivers, MCM4, OHI3, FALE, the remaining nine receivers are relatively stable for different seasons. For the 16 days covering 10 months, the standard deviation of the receiver biases ranges from 0.3673 to 0.9759 ns. As for the large variation of MCM4, we found that its receiver was in type AOA SNR-12 ACT for spring, whose standard deviation is 0.6953 ns, and it was changed to ASHTECH Z-XII3 in summer, autumn, and winter, whose standard deviation is 2.053 ns. Therefore, the receiver biases curve of MCM4 has a greater change. The receiver biases of OHI3 station do not get on the days of 360–363 of 2006, the reason is, the data of IHO3 station are removed when the observation data is preprocessing. For the same reason, the receiver biases curve of FALE station is not plotted in Fig. 9.

5 Conclusion

Using a thin layer assumption of the ionosphere and dual-frequency GPS observations from 16 geomagnetically quiet days in four seasons of 2006, this paper adopts the spherical harmonic model to fit TEC and investigates the effects of two network constitutions on global TEC derivation, one with 275 GPS receivers and the other with 125 GPS receivers. The results show that the data can be well fitted for both network constitutions. The derived TECs are consistent with each other for four seasons. This is especially true for TECs at low- and mid-latitude. The TECs at high latitude are much noisy and lack of consistency. This can be attributed to that the ionosphere is much more variable at high latitudes than that at middle and low latitudes. The derived satellite and receiver biases are stable during the year. The standard deviation of the satellite and the receiver biases are less than 0.5 and 3 ns, respectively. Concerning the processing time of TEC derivation, the difference is 1 h 8 min 47 s between the two network constitutions. We can conclude that it can not only derive TEC effectively but also reduce the calculating time by selecting the appropriate network constitution. Further work should be done on finding a practical method to constitute GPS receivers network.

Acknowledgments This work is supported by the National Natural Science Foundation of China (Grant Nos. 11473045, 11403045).

References

1. Lanyi GE, Roth T (1988) A comparison of mapped and measured total ionospheric electron content using global positioning system and beacon satellite observations. Radio Sci 23(4):483–492
2. Coco DS, Coker C, Dahlke SR, Clynch JR (1991) Variability of GPS satellite differential group delay biases. IEEE Trans Aerosp Electron Sys 27(6):931–938
3. Sardon E, Rius A, Zarraoa N (1994) Estimation of the transmitter and receiver differential biases and the ionospheric total electron content from global positioning system observations. Radio Sci 29(3):577–586
4. Mannucci A, Wilson B, Yuan D, Ho C, Lindqwister U, Runge T (1998) A global mapping technique for GPS-derived ionospheric total electron content measurements. Radio Sci 33(3):565–582
5. Iijima B, Harris I, Ho C, Lindqwister U, Mannucci A, Pi X, Reyes M, Sparks L, Wilson B (1999) Automated daily process for global ionospheric total electron content maps and satellite ocean altimeter ionospheric calibration based on global positioning system data. J Atmos Sol Terr Phys 61(16):1205–1218
6. Rideout W, Coster A (2006) Automated GPS processing for global total electron content data. GPS Solut 10(3):219–228
7. Ma G, Maruyama T (2003) Derivation of TEC and estimation of instrumental biases from GEONET in Japan. Ann Geophys 21:2083–2093
8. Li Z, Yuan Y, Li H, Ou J, Huo X (2012) Two-step method for the determination of the differential code biases of COMPASS satellites. J Geod. doi:10.1007/s00190-012-0565-4
9. Gao W, Ma G, Chen Y, Shen H, Li J, Huang W, Li Z (2008) Derivation of GPS-TEC and instrumentao biases in the equatorial anomaly region. Chin J Space Sci 28(6):541–546
10. Ma G, Gao W, Li J, Chen Y, Shen H (2014) Estimation of GPS instrumental biases from small scale network. Adv Space Res 54:871–882
11. Kaplan ED, Hegarty CJ (2006) Understanding GPS: principles and applications. Artech House Publishers, London
12. Liu J, Chen J, Zhang Y, Li S, Ge M (1999) WADGPS principle and method. Surveying and mapping Press, Wuhan
13. http://gpspp.sakura.ne.jp/, Cited 20 Mar 2015
14. Zhang X (2004) Matrix analysis and applications. Tsinghua University Press, Beijing

GNSS-R Aircraft Flight Test Based on Digital Beam Forming

Kangning Zhang, Fenghui Li, Xingguo Li, Linfei He, Qijia Dong, Wenliang Zhao and Fuzhan Yue

Abstract The technology can be used to measure the wind and the height of the sea based on the GNSS-R, It can be complementary to the other advantages of sea measurements, has more advantages than other methods when it can be used to constitute constellation. When to deal with the environments of the sea surface, only need is to install the receiver, so compared to traditional means of remote sensing, GNSS-R technology reduce costs, weight, and the power, at the same time, the technology can improve the resolution and accuracy, It can be used to the new way at remote sensing. This paper designs a new GNSS-R receiver based on the Digital beam forming (DBF), the receiver through the dbf to synthesis the receive signal. The receiver has 31 dbf channels, so it can complete the synthesis a plurality of different direction's signal at the same time. In order to effectively improve the detection area, the receiver can achieve the angle scanning between the fixed angle and the dynamic angle. In order to test the receiver, we did an experiment by using the plane on November 3, 2015 at Bohai Bay, flying height about 1200 meters, in the experiment, we successfully captured the reflected signals at the high altitude (1.2 km) dynamic scenes, after that we got the delay-doppler mapping through to deal with the reflected signals. By calculating the phase difference between the reflection channel's DDM and direct access, we got the high information of the receiver, after that we compared the high information that can be got through the receiver, the answer showed that the deviation is less than 4 m.

Keywords Delay-doppler mapping · Digital beam forming · Sea surface altimetry

K. Zhang (✉) · F. Li · X. Li · L. He · Q. Dong · W. Zhao · F. Yue
Space Star Technology Co., Ltd., Beijing 100086, China
e-mail: ning_zhang_139@163.com

J. Sun et al. (eds.), *China Satellite Navigation Conference (CSNC) 2016 Proceedings: Volume I*, Lecture Notes in Electrical Engineering 388,
DOI 10.1007/978-981-10-0934-1_4

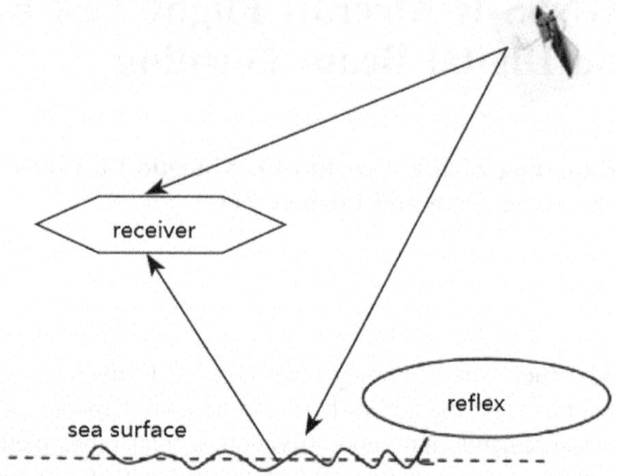

Fig. 1 Detective of the GNSS-R

1 Introduction

It is the important physical parameter of the Marine Science Subject based on the high level of the sea, wave height, and wind field of the sea surface [1]. As a new type of passive detection technology, The GNSS-R technology has developed rapidly. GNSS-R technology does not require additional signal source transmitter, Because GNSS-R only used the navigation satellite signal. As shown in Fig. 1, GNSS-R receiver at the same time receives the navigation satellite direct signal, but also received from the sea scattered signal. The reflection signals from the sea surface carry the characteristics of the sea surface, and the sea surface features are mainly reflected by the reflection wave, polarization characteristics, amplitude, phase, frequency, and other parameters. So, the physical characteristics of the reflecting surface can be effectively reflected by the receiving and processing of the reflected signal. This paper designs a new GNSS-R receiver based on the digital beam forming, and using the receiver did an experiment by using the plane on November 3, 2015 at Bohai Bay. In the experiment, the GNSS-R receiver captures the reflected signal from the sea, and generated the DDM. The experiments get the receiver's height information by calculating the phase difference between the reflection channel and the direct channel.

2 The Principle of GNSS-R

It is an important direction in the current GNSS-R research to measure the height of the receiver by using the GPS sea surface reflection signal. The basic principle is to use the time difference between the sea surface reflection signal and the direct signal to calculate the different distance. So, the value can help the receiver to get the

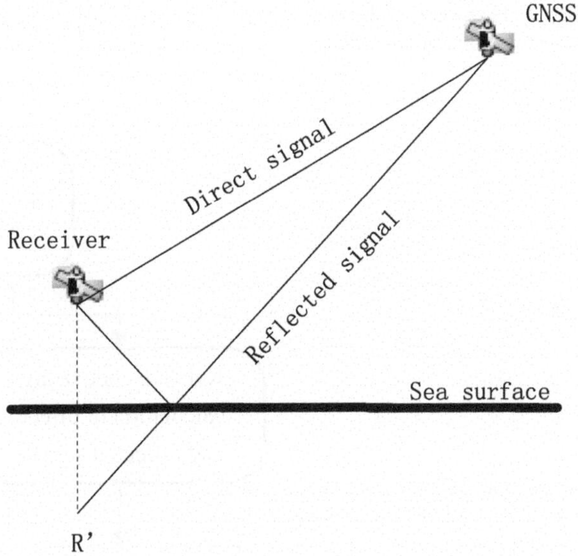

Fig. 2 Altimetry principle of GNSS-R

height from the sea surface to the receiver [2]. The principle is as shown in Fig. 2. In this diagram, R stands for the position of the receiver, R′ stands for the position of the mirror image of receiver. As shown in Fig. 2, we can get the geometric relationships as $\Delta\rho = 2h\sin(\varepsilon)$.

The flow chart in Fig. 3 demonstrates how to measure the height from the sea surface to the receiver.

The specific process is as follows:

1. First got the measured data, including L1 carrier, CA code data, receiver and satellite position, and speed information.
2. According to the observation time, or the position and velocity information of the receiver and the satellite, calculated the position of the specular reflection point and reflection angle.
3. Using carrier and code data modified the delay due to the ionosphere.
4. Acquisition mode or real-time atmospheric refractive index profile, calculated neutral atmospheric additional delay.
5. By analyzing the change of the wave front gradient, the position of the specular reflection point is determined, according to the answer, the delay between the reflection wave and the direct wave is calculated, at last the sea surface height can be retrieved after deducting the additional delay.

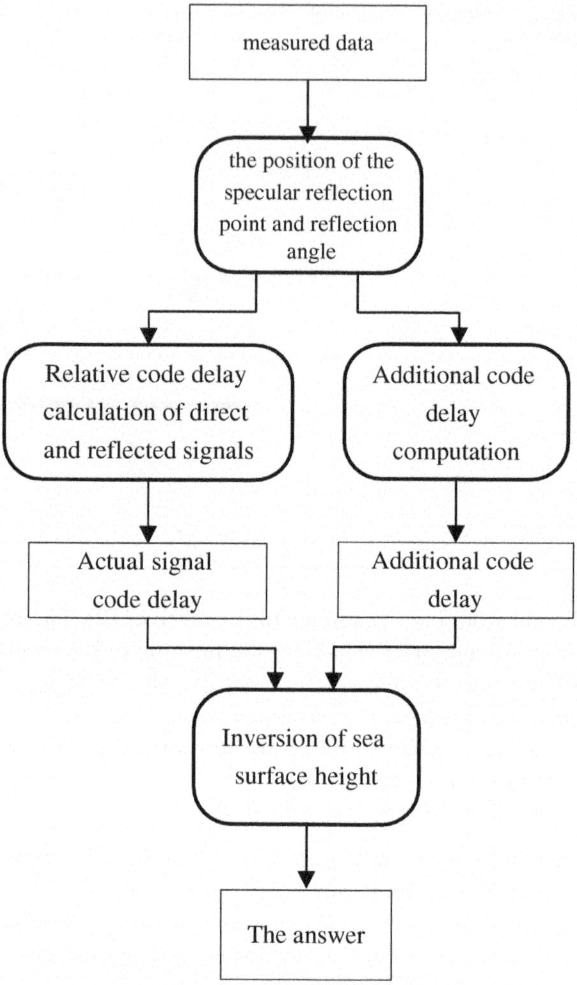

Fig. 3 Flow chart of the way to get the high of sea surface

3 The Composition of GNSS-R Detection System Based on dbf

3.1 System Components of the GNSS-R Receiver

The GNSS-R receiver system is made of the navigation and positioning antenna, the reflector antenna, GNSS-R receiver and the data logger instrument. As shown in Fig. 4.

The navigation and positioning antenna mounted on the top of the airplane, in order to receive the multiple frequency point direct signal from GPS and BD2 navigation constellation. Output the RF signal to GNSS-R receiver. The reflector

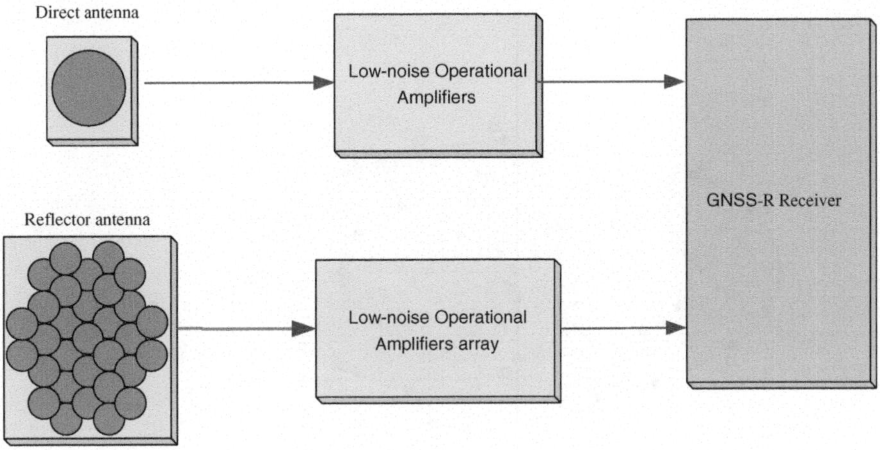

Fig. 4 System components of the GNSS-R receiver

antenna mounted on the bottom of the airplane, in order to receive the signal from the ocean reflection signal, Output the RF signal to the Low-noise Operational Amplifiers array. After that, the amplified signal output can be sent to the radio frequency unit of the GNSS-R receiver. Sent to the GNSS-R receiver's navigation signal, In the radio frequency processing channel, the signal is amplified, filtered, and downconvertedn, after that the analog signal is formed. In the baseband processing module, the receiver using the digital correlates to complete the digital intermediate frequency signal, gets the original measurement and navigation message. After that, the DSP deals with the information to obtain the navigation positioning information and ephemeris and almanac information and reflection signal.

3.2 *Digital Beam Forming*

The principles of digital beam synthesis [3], as shown in Fig. 5,

$d_m = -(x_m \cdot \cos\varphi_m + y_m \cdot \sin\varphi_m) \cdot \sin\theta_m$, x_m is the x-axis coordinate value of the m antennas, y_m is the y-axis coordinate value of the m antennas, φ_m is the azimuth angle of the m antennas, and θ_m is the pitch angle of the m antennas. So, the array response vector is

$$a(\theta,\varphi) = [1, \mathrm{e}^{-j\xi_2}, \mathrm{e}^{-j\xi_3}, \ldots, \mathrm{e}^{-j\xi_{31}}]^{\mathrm{T}}$$

Hypothesis that the first antennas receive the signal that can be credited to $x_1(n) = s(n)$, So, all the antennas received signal can be written as:

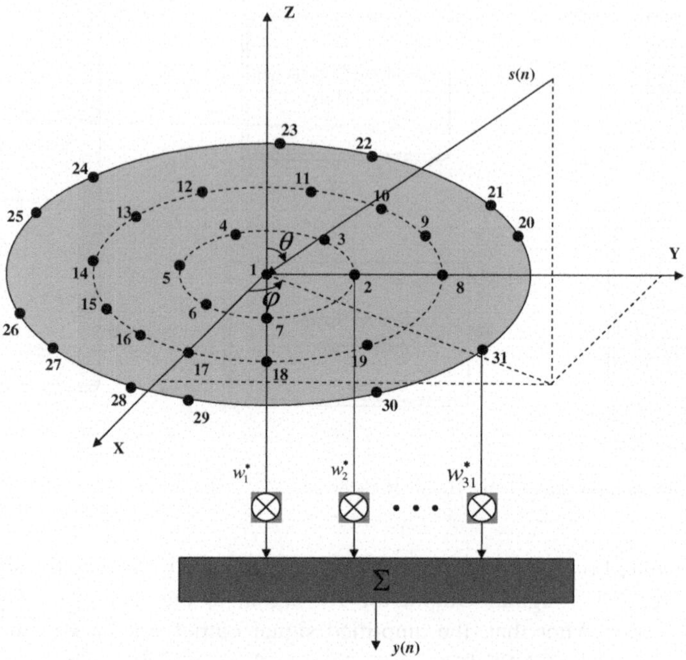

Fig. 5 The theory of the dbf

$$x(n) = [x_1(n), x_2(n), \ldots, x_{31}(n)]^{\mathrm{T}} = a(\theta, \varphi)s(n)$$

After that, the dbf output is $y(n) = w^H x(n) = w^H a(\theta, \varphi)s(n)$.

3.3 Anechoic Chamber Test

In order to test the results of the dbf, we put all the equipment into the anechoic chamber, tests the dbf effect in the real environment. The schematic diagram can be shown in Fig. 6. In the anechoic chamber, we were suspending the reflector antenna. In order to reduce the lost, the transmitting antenna selects the linear polarization antenna, and the 1575.42 MHz carrier signal is transmitted from the transmitting antenna. After treatment, the actual effect is analyzed.

Through the experiment can be seen the phase difference between the signals before and after calibration comparison which is obvious. Status contrast between before and after calibration is as shown in Fig. 7, and the phase difference between the signals before and after calibration comparison is as shown in Fig. 8.

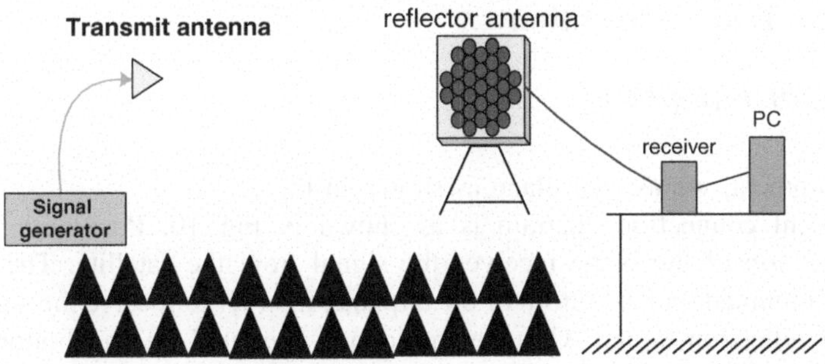

Fig. 6 Schematic diagram of anechoic chamber test

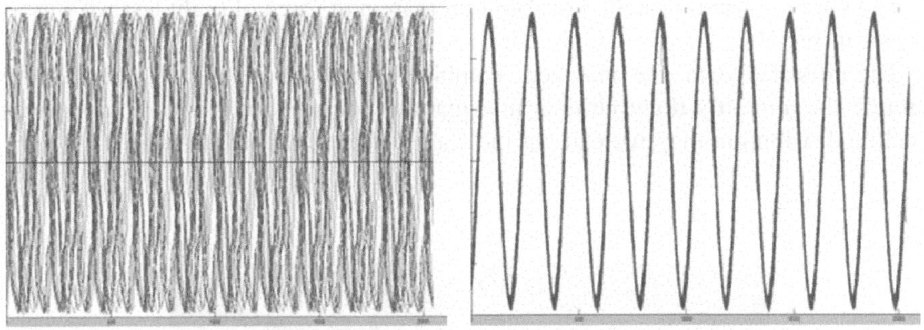

Fig. 7 Status contrast between before and after calibration

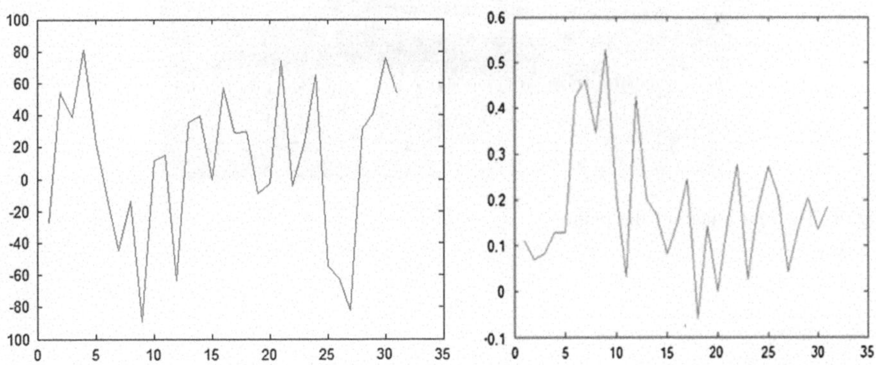

Fig. 8 The phase difference between the signals before and after calibration comparison

4 Flight Test Verification

4.1 Flight Equipment

In the equipment, we use the plane as shown in Fig. 9.

Equipment connection diagram is as shown in Fig. 10. Positioning antenna installed on top of the plane receives the signal from the satellite. The reflector antenna is mounted on the bottom of the airplane, in order to receive the signal from the ocean reflection signal. Optoelectronic pod mounted on the bottom of the airplane too needs to close the reflector antenna. The receiver completes the signal processing between the direct and reflected signals. The direct signal is used to the position calculation, and the reflector channel mainly used to complete the dbf.

The GPS navigation satellite distribution map in the area of flight test work is as shown in Fig. 11.

The collected data are analyzed, combined with the flight path map of the aircraft, the aircraft's flight attitude information (course), and the GPS navigation satellite distribution map we can get the status of the captured satellite.

Fig. 9 The real image of the plane

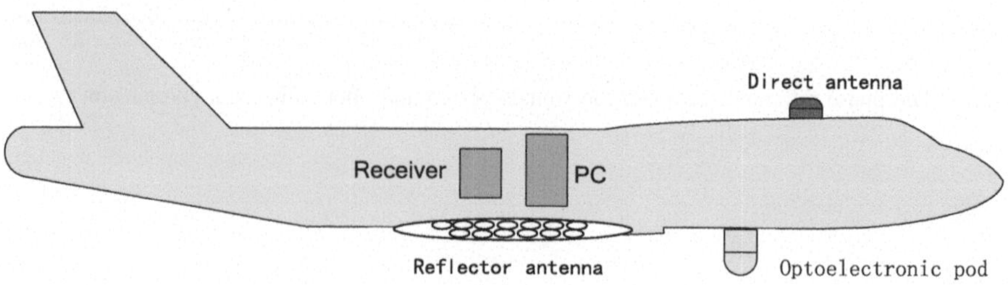

Fig. 10 Equipment connection diagram

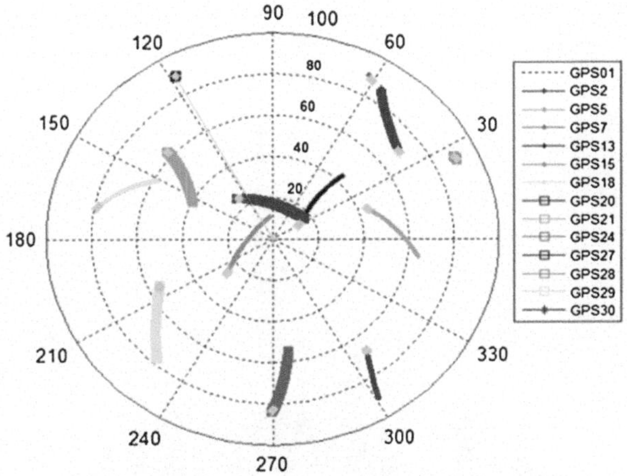

Fig. 11 Prediction of GPS satellites position (Color figure online)

4.2 Experimental Study

Through the analysis of the test data can be judged that the receiver gets the ocean reflection signal from the 13 satellite. After processing, the DDM shown is shown in Figs. 12, 13, 14.

According to the measure, the system gets the high information, compared with that of the receiver as shown in Fig. 15. The measurement error is less than 4 m.

This paper detailed introduced the receiver's work principle based on digital beam forming, calibration mode, and the height measurement theory. Experiment

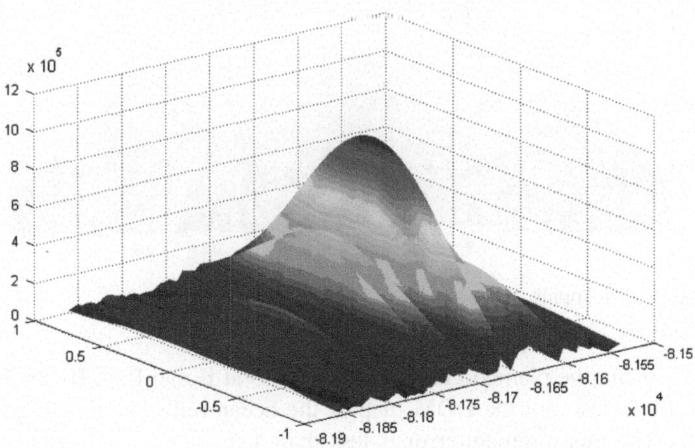

Fig. 12 The DDM of the satellite 13

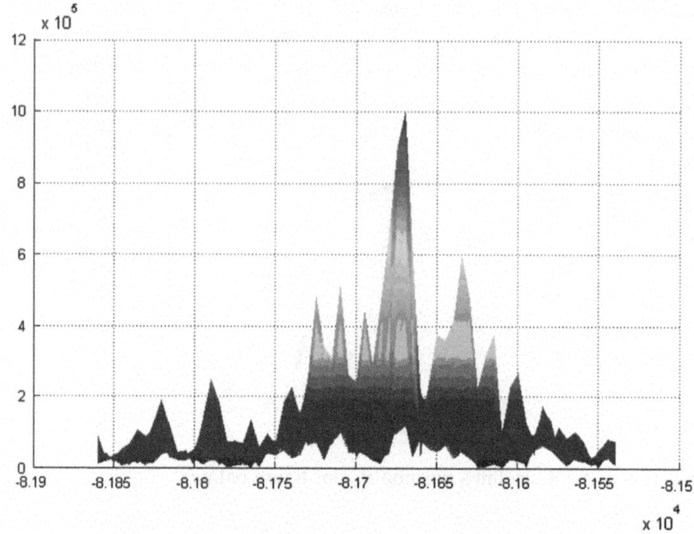

Fig. 13 The doppler mapping of the satellite 13

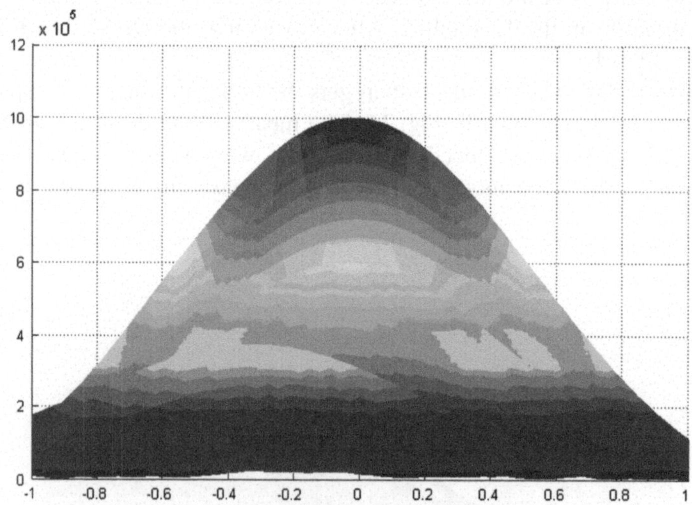

Fig.. 14 The delay mapping of the satellite 13

was done by using the plane on November 3, 2015 at Bohai Bay. By analyzing the data of the flight test, got the DDM map of the ocean reflected signal; go the high information. The measurement error is less than 4 m.

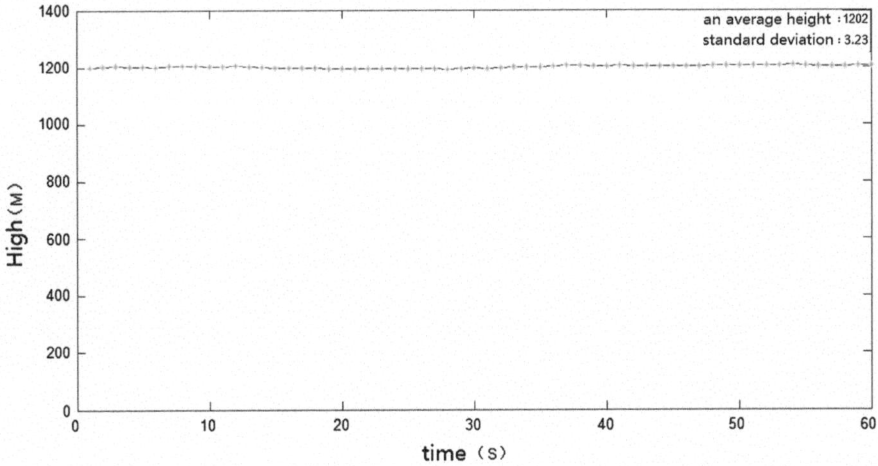

Fig. 15 Answer

References

1. Wang X, Cui H, Fu Y et al (2002) GNSS-R marine remote sensing technology. In: Twenty-fifth national conference on space exploration
2. Martin-Neira M, Caparrini M, Font-Rossello J et al (2001) The PARIS concept: an experimental demonstration of sea surface altimetry using GPS reflected signals. IEEE Trans Geosci Remote Sens 39(1):142–150
3. Steyskal H, Rose JF, Steyskal H et al (1988) Digital beamforming for radar systems. Microw J 32(1):121–123

GPS Radio Occultation Inversion Analysis of Neutral Atmospheric Detection Technology Based on Dual-Frequency Dual-Mode GNSS Occultation Receiver

Lu Zhang, Xingguo Li, Fenghui Li, Junbo Han, Qijia Dong, Yingna Liu and Meng Wang

Abstract Compared with the traditional radio atmospheric detection and radar detection, GNSS occultation technique provides a powerful tool for atmospheric detection, the GNSS occultation technique which has the characteristics of without calibration, all-weather, high precision, high vertical resolution, and global uniform coverage. The GNSS occultation technique based on the atmosphere GPS/LEO radio occultation, GPS satellite radio signal is obscured by Earth's atmosphere, GPS satellite radio signal cutting the Earth's atmosphere and ionosphere sectional, then the radio signal refracted before reach LEO satellites. The received GPS signal contains information of the Earth's atmosphere and ionosphere. We can get atmospheric and ionosphere electron density profile, the total electron content of the ionosphere and atmospheric bending angle profiles, index of refraction, temperature and barometric pressure through the relevant atmospheric inversion model. This paper introduces GPS radio occultation atmospheric detection technology based on mountain-based experimental verification, test is located at Wuling Mountain Peak of Yanshan Mountains (located in Chengde in Hebei Province, 117.48°E, 40.60°N, ~2118 m), and made a detailed presentation for signal capture, signal track, and data quality. Different from prediction algorithm of spaceborne occultation, the test used the mountain-based occultation prediction based on the fixed position, getting the forecast results of a week about occultation event. Because of mountain conditions and other factors, occultation signal capture is different because of different azimuths, and made a comparison between different azimuths making a analysis against signal to noise ratio (SNR), the consistency between the Carrier phase and the pseudorange, the reason of signal loss of lock. In order to verify the accuracy of GPS radio atmospheric occultation, this article compares the refractive index profile obtained by inversion with meteorological data, test results were better than 5 %.

Keywords GPS radio atmospheric occultation · Refractive index profile · Inversion algorithm · Meteorological data

---

L. Zhang (✉) · X. Li · F. Li · J. Han · Q. Dong · Y. Liu · M. Wang
Germany Space Star Technology Co., Ltd., Beijing, China
e-mail: zhanglumengcun@163.com

J. Sun et al. (eds.), *China Satellite Navigation Conference (CSNC) 2016 Proceedings: Volume I*, Lecture Notes in Electrical Engineering 388,
DOI 10.1007/978-981-10-0934-1_5

1 Introduction

GNSS radio occultation technique is a remote sensing technology detecting the Earth's ionosphere and the atmosphere through the occultation phenomenon between GNSS Navigation Constellation—Earth—LEO satellite [1], the GNSS occultation detection schematic diagram is as shown in Fig. 1.

GNSS occultation receiver can be used to carry out radio occultation observations for GPS/BD2 navigation constellation, to acquire the ionosphere environment information of the global region, including ionospheric electron density profile, ionospheric total electron content contour, atmospheric bending angle, refractive index, temperature, pressure, humidity, etc.

The inversion product of GNSS occultation detection system is the ionosphere and neutral atmosphere parameters, ionosphere inversion product including ionospheric electron density profile and ionospheric scintillation index; neutral atmospheric inversion products including atmospheric refraction ratio, density, pressure, temperature, and humidity profiles [2]. Mountain-based test, as a ground validation method of GNSS occultation detection, can provide technology and experience for spaceborne GNSS occultation exploration, this paper introduces a kind of experimental verification of mountain-based GPS atmospheric occultation technique, to support space-borne GNSS occultation exploration.

2 Section Heading GPS Occultation Detection Technology of Atmospheric Mountain-Based

2.1 GPS Occultation Forecast Mountain-Based

Occultation event occurs depends on the geometric positional relationship of GPS and LEO satellites. The occultation if exists, It is necessary to forecast occultation events and input the forecast result to receiver. Then the receiver can be acquisition and tracking occultation signal. As shown in Fig. 2 is the occultation forecast schematic.

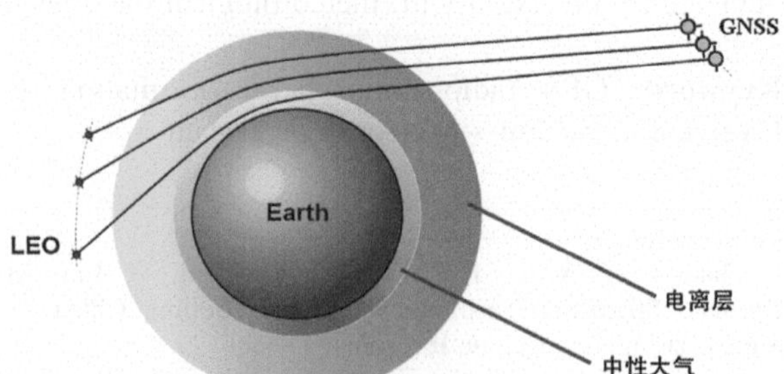

Fig. 1 Radio occultation schematic

Fig. 2 GNSS-R probe schematic

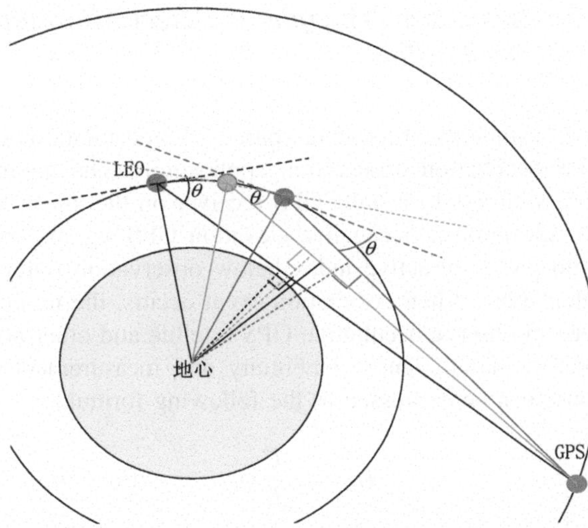

θ is the angle of elevation. ϕ is the zenith angle complementary angle between LEO satellite and a geometry line which is the connection of LEO satellites and GNSS navigation satellite. If $\phi \leq 90°$, then $\theta \geq 0°$ is positive elevation. If $\phi > 90°$, then $\theta < 0°$ is negative elevation. The algorithm to forecast mountain-based occultation is based on the height of tangent latitude point. Let the height of tangent latitude point be set to 0–2 km. In the experiment the mountain-based occultation forecast is based on fixed point to get a week result of occultation event forecast. The mountain-based occultation forecast result is as shown in Table 1. The forecast result is between September 16, 2015 and September 23, 2015. The forecast result showed the satellites number when occured occultation event. Also the begin time, end time, angle of elevation, and azimuth can be seen in the forecast result. Based on the forecast result the receiver can be acquisition and tracking the occultation signal [3].

Table 1 Occultation forecast results of September 16, 2015

预报发生掩星事件GPS卫星号	掩星开始时间			掩星结束时间			开始仰角	结束仰角	开始方位角)	结束方位角
	天次	时	分	天次	时	分				
22	1	2	46	1	3	16	6.72	−3.83	181.96	176.91
27	1	3	29	1	3	57	−3.83	6.84	187.86	185.02
21	1	3	59	1	4	44	−3.79	6.86	110.12	94.93
14	1	4	23	1	4	49	6.7	−3.71	149.09	151.01
8	1	4	39	1	5	6	−3.99	6.65	200.07	199.92
32	1	4	47	1	5	24	6.96	−3.78	215.5	205.77
19	1	5	23	1	5	50	−3.9	6.85	203.05	203.79
4	1	6	22	1	6	53	−3.79	6.65	177.84	172.19

2.2 Inversion Principle of Detect Mountain-Based Occultation

The technology of mountain-based GPS occultation observation is one of the radio GPS occultation observation applications. The mountain-based GPS occultation observation is to installa GPS receiver on the top of hill. The receiver can get the low elevation and negative elevation GPS signal. Then the information of lower atmosphere refractive index below observation point can be got through inverse calculation. When occultation event occurs, the mountain-based occultation receiver can observe occultation GPS satellite and other non-occultation reference GPS satellite signal, ignore ambiguity and measurement noise [4]. GPS observation phase can be expressed as the following formula:

$$L_K = -\frac{C}{f_k}\varphi_k = \rho + \mathrm{d}A_k + \mathrm{d}I_k + c(t - T) \tag{1}$$

Using a single differential technology, the difference between occultation GPS observation L_k^0 and reference GPS observation L_k^0 can eliminate the receiver clock error.

$$L_k^0 - L_k^r = \rho_k^0 - \rho_k^r + c(t^0 - t^r) + \mathrm{d}I_k^0 - \mathrm{d}I_k^r + \mathrm{d}A_k^0 - \mathrm{d}A_k^r A_k^r \tag{2}$$

The additional phase delay term for occultation observation can be written as below:

$$\mathrm{d}A_k^0 + \mathrm{d}I_k^0 = (L_k^0 - L_k^r) - \rho_k^0 + \rho_k^r - c(t^0 - t^r)\mathrm{d}I_k^r + \mathrm{d}A_k^r \tag{3}$$

The subscript k = 1, 2. φ_k is the signal observation phase. c is the speed of light in vacuum. L is the distance between transmitter and receiver. $\mathrm{d}A_k$ is the delay of neutral layer phase. t, T is transmitter and receiver clock error, respectively.

Suppose the atmosphere ball is symmetry, according Snell's law to determine the impact parameter [5]

$$a = n_2 r_2 \sin\varphi_2 = n_1 r_1 \sin\varphi_1 \tag{4}$$

n_1, n_2 is the refractivity of receiver and transmitter. r_1, r_2 is the distance between refraction point and geocentric. α is bending angle. By Abel integral transform, part of the bending angle profiles can be converted into atmospheric refraction index profile [6].

$$n(a) = n_1 \exp\left[\frac{1}{\pi}\int_a^{n_1 r_1} \frac{\tilde{a}(x)}{\sqrt{x^2 - a^2}}\mathrm{d}x\right] \tag{5}$$

Note that positive elevation bending angle subtracts the negative elevation bending angle, simultaneously which eliminates the influence of the ionosphere (ionosphere ball symmetry suppose) [7].

2.3 *Refractivity Inversion Process*

Use occultation receiver that receives the GPS occultation data to inversion atmospheric refractive index profile. The data processing as following: get ready the navigation files and observation files of Rinex format, calculation of all satellite orbital position at the time of observation using the method of extrapolation track. Extract the receiver position coordinates in the head file of observation file, then calculate all satellite elevation at the receiver's coordinate. Choose the highest elevation satellite as reference, and the satellite which has the positive and negative elevation as occultation. Extract the reference satellite data in the observation files, those data include observation epoch time, L1CA code pseudorange, L1P code pseudorange, L2P code pseudorange, L1 carrier, and L2 carrier. Extract the L1 carrier and L2 carrier of the reference satellite and the occultation then use single difference method to eliminate local clock error. Based on navigation files to extrapolation the satellite position coordinates the distance between occultation and reference satellite. Based on the clock error factor in the navigation files calculate the satellite clock errors between occultation and reference satellite. And then calculate the clock error and tropospheric delay of the reference satellite and occultation. Finally, according to measuring phase pseudorange observation equation calculate the L1, L2 carrier additional phase delay. Translation atmosphere additional phase delays filter to calculate the additional Doppler shift. The L1 impact parameter is obtained in accordance with Doppler shift optical geometry principle. By founding the maximum number from impact parameter sequence, and calculating the corresponding refractive index,calculating bending angle sequences and splice bending angle, according to Abel inverse calculation index of refraction. The data processing step is as shown in Fig. 3.

3 Components and Results of Atmospheric Occultation Inversion System

3.1 *Mountain-Based Test System Components*

Test site on Wuling Mountain Peak of Yanshan mountain main peak(located in Chengde, Hebei Province, east longitude 117.48°, north latitude 40.60°, altitude 2118 m), the antenna beam center pointing south west 15°, beam range is ±35°,

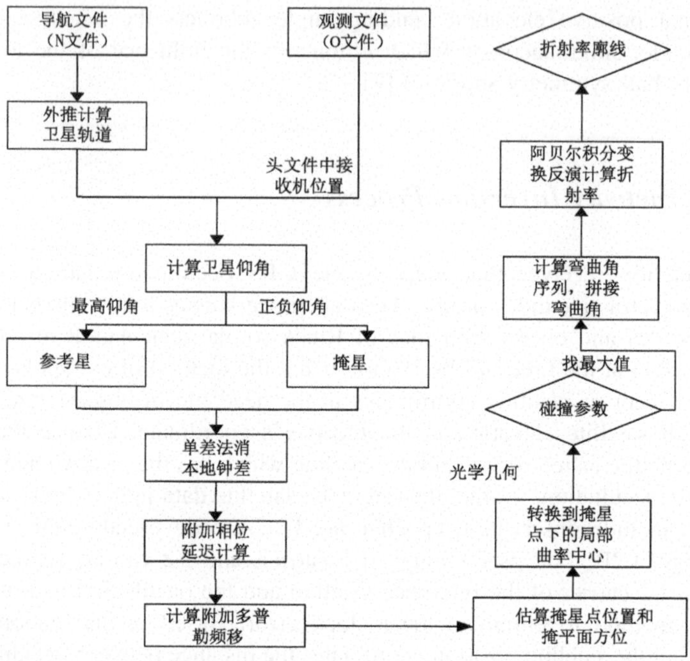

Fig. 3 Occultation data processing flow chart

Fig. 4 Mountain-based test equipment connection diagram

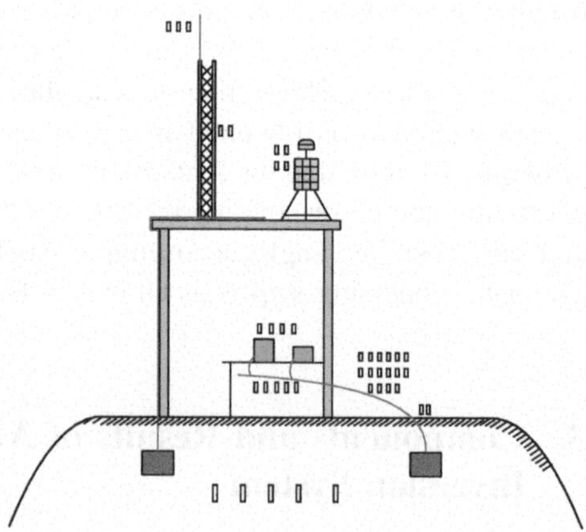

pitch angle is ±3°, receiving atmospheric occultation signal from the South. Test equipment connection is diagram shown in Fig. 4. Positioning antenna is used to receive GPS/BD2 navigation signal that comes from zenith direction, direct observation satellite data is to provide a reference observation data for positioning and speed measurement.

3.2 Occultation Receiving Data Analysis and Inversion Results

Not all the occultation receiving data analysis and inversion result of all the occultation data that mountain-based occultation receiver received can be applied to inversion, analyse SNR and pseudorange continuity of occultation data and select the GPS occultation whose tracked status is well. Before the experimental data applicated, analysis GPS occultation pseudorange and SNR trend with elevation, then determine whether they can be applied to inverse the refractive index. Figure 5a–c, respectively, show pseudorange and SNR trends with the elevation, Fig. 5a–c, respectively, are September 21 received No. 20, 2 GPS occultation and September 21 No. 9 GPS occultation, the figure shows that non-satellite tracking status, the pseudorange observations are lost during the observation process. Figure 5b shows pitch angle must be between $-1°$ and $3°$ continuous, the data can be used to get refractivity. The pitch angle is between $0°$ and $3°$ of No. 9 GPS occultation, it cannot get valid data of occultation observation side, and therefore it cannot be applied to the inversion.

Figure 5d–f, respectively, show pseudorange and SNR trends with the elevation, Fig. 5d–f, respectively, are September 28 received No. 9,15 GPS occultation and September 21 No. 2 GPS occultation, the scheme shows that tracking status of pseudorange and SNR is well, the pitch angle is between $-1°$ and $3°$, and SNR is relatively high, so these data can be used to inverse refractive index. When the occultation data are processing, if the observed negative elevation cannot less than $-1°$, that cannot get enough refraction signal of the occultation side, then cannot get refractive index value of low altitude position. On the other hand, if the occultation observation data obtained pitch angle between $-1°$ and $3°$exists the phenomenon of loss of lock, the entire occultation data will not be used for occultation inversion.

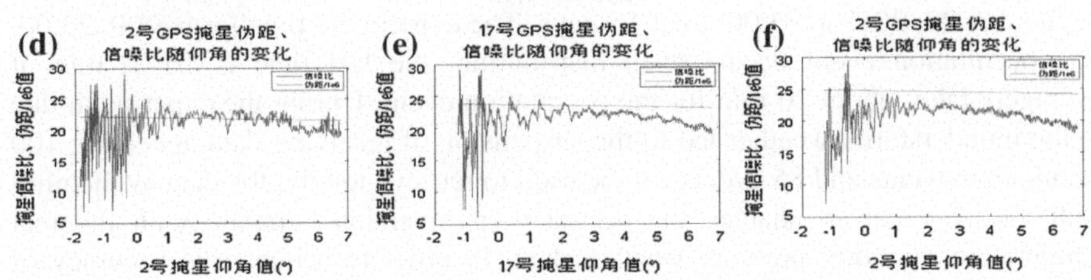

Fig. 5 The change trends of GPS occultation pseudorange and SNR with the pitch angle of the occultation

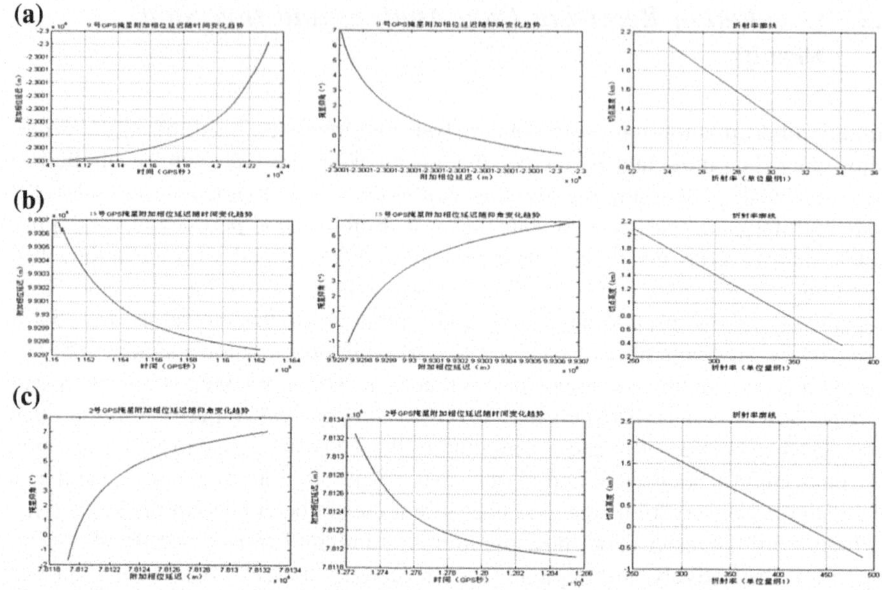

Fig. 6 GPS occultation inversion result. **a** September 28, 2015 No. 9 GPS occultation inversion results. **b** September 28, 2015 No. 15 GPS occultation inversion results. **c** September 21, 2015 No. 2 GPS occultation inversion results

Through the above data analysis, exclude noncontiguous occultation data whose acquisition and tracking status is not well. Use occultation data to inverse the refractive index profile of September 21, 2015 and September 28, 2015,analysis occultation additional phase delays varying trends with time and occultation additional phase delays varying trends with elevation, the results are shown in Fig. 6a–c.

4 Verification and Summary

The observation experiment starts at eight o′ clock from September 15, 2015 to September 29, 2015 at 23:00, total 15 days. The experiment time form 8:00–23:00. The occultation event occurs even distribution, the last time is 16–21 min, it averagely takes about 20 min for the occultation event. During the experiment due to the thunderstorm, weather led to the suspension of receiving data, receiving 100 occultation events and 85 events can be used to retrieve totally, the data availability is 85 %. In order to analyze the accuracy of inversion compare with the real temperature humidity pressure weather data. In order to achieve the accuracy of judgment, two assessments are required. Comparing the two contour lines which require that the occultation event occur time needs to coincide with the

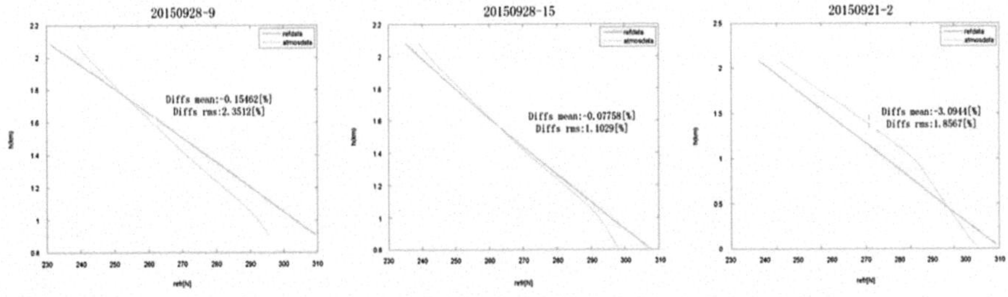

Fig. 7 The refractive index profile results contrast with meteorological inversion results

meteorological data time, also the distance needs to close between occultation signal cut position and the analysis field geographic coordinates. Besides, it is better to consider the retrieval accuracy influence of space matching. Second, to assess the algorithm accuracy of calculating refractivity based on meteorological data, when the accuracy of atmospheric inversionis judged, it is needed to deduct the error of calculating the refractive index forward from the total error (Fig. 7).

Use GPS occultation data of September 21 and September 28 to inverse the refractive index profile and compare with meteorological data, as shown below figure. The average of difference was less than 1.0 %, the RMS difference is 2.3512, 1.1025, 1.8567 %, respectively, less than 5 %, which verifies the accuracy of GPS occultation atmospheric probe fully. Mountain-based test as ground validation method of GNSS radio occultation provided technical support and experience for spaceborne GNSS radio occultation.

References

1. Yan H, Guo P, Zhang Z et al (2003) GPS occultation technique situation in shanghai observatory. Shanghai Astronomical Observatory, Xuhui, pp 39–47
2. Hu X, Zeng Z, Zhang XX et al (2005) Atmospheric inversion methods of GPS radio occultation. Chin J Geophys 48(4):768–774
3. Wickert J, Reigber C, Beyerle G et al (2001) Atmosphere sounding by GPS radio occultation: first results from CHAMP. Geophys Res Lett 28(17):3263–3266
4. Zus F, Grunwaldt L, Heise S et al (2013) Atmosphere sounding by GPS radio occultation: first results from TanDEM-X and comparison with TerraSAR-X. Adv Space Res 53(2):272–279
5. Wickert J, Beyerle G, Hajj GA et al (2002) GPS radio occultation with CHAMP: atmospheric profiling utilizing the space-based single difference technique. Geophys Res Lett 29(8):28-1–28-4
6. Shubin VN, Karpachev AT, Tsybulya KG (2014) Global model of the F2 layer peak height for low solar activity based on GPS radio-occultation data. J Atmos Solar Terr Phys 104(2):106–115
7. Stolle C, Jakowski N, Schlegel K et al (2004) Comparison of high latitude electron density profiles obtained with the GPS radio occultation technique and EISCAT measurements. Ann Geophys 22(6):2015–2028

Research on BD Nanosecond One-Way Timing Method

Longxia Xu, Dandan Li, Xiaohui Li and Feng Zhu

Abstract Different from other GNSS navigation systems, the constellation of BD navigation system consists of GEO, MEO, and IGSO satellites which is the characteristic of BD. The altitude of GEO orbit is higher than that of MEO. Therefore, the ephemeris and ionosphere delay errors are more dependent in the paths from GEO satellites to two users involved in time comparison. Therefore, the results of common view time transfer based on GEO satellites are better than that of MEO. A method of changing the generation of UTC time parameters is proposed which makes full use of the advantages of GEO satellites in time transfer. The time difference between system time of navigation system and UTC(K) is monitored by user terminal which referenced with a ultrastable frequency signal. The UTC time parameters are generated based on these time differences. Users with the new UTC time parameters not only can get the time difference between their local time and UTC(K), but also can farther eliminate the errors in one-way timing method. Without aggravating the burden of users, the precision of this method is equivalent to that of common view time transfer which is superior to the existed GPS and BD one-way timing method.

Keywords BDS · Common view time transfer · One-way · UTC time parameters

L. Xu (✉) · D. Li · X. Li · F. Zhu
National Time Service Center, CAS, Xi'an, China
e-mail: xulongxia@ntsc.ac.cn

L. Xu · D. Li · X. Li · F. Zhu
Key Laboratory of Precision Navigation Positioning and Timing, CAS, Xi'an, China

L. Xu
State Key Laboratory of Geo-Information Engineering, CAS, Xi'an, China

J. Sun et al. (eds.), *China Satellite Navigation Conference (CSNC) 2016 Proceedings: Volume I*, Lecture Notes in Electrical Engineering 388,
DOI 10.1007/978-981-10-0934-1_6

1 Introduction

Timing together with positioning and navigation is the important function of a satellite navigation system. The common used timing method is one-way timing which is implemented in three steps. First, the distances from satellites to receivers are measured by receiving signals transmitted by navigation system. Then, estimating all the delays in the receiving path, such as geometrical delay, by using parameters in the navigation message transmitted from navigation satellites. Finally, the receivers get the time difference between their local time and the system time of navigation system. If users further want to know how long its local time delays with respect to UTC, the UTC time parameters which provide the correlation between UTC time and GNSS system time can help them. At present, the precision of existed one-way timing method is at the level of tens of nanoseconds. This cannot meet the requirements of nanosecond users.

Different from other navigation satellite systems, BD constellation contains five GEO satellites which is the specialty of BD navigation system [1]. Based on this specialty, this paper proposes a method for BD to generate the UTC time parameters that broadcast in navigation message. With the parameters provided by this new method, users with the same receiver configuration as that of one-way timing can acquire timing results equivalent to that of common view time transfer. Therefore, users can realize timing with the precision of better than 3 ns via GEO satellites in BD navigation system.

2 Method of Generating UTC Time Parameters

2.1 Existed Method of Generating UTC Time Parameters

There are two kinds of time parameters broadcasted by navigation systems. The first one is the SV clock correction parameters which are used to correct the satellite time to navigation system time. The other is the UTC time parameters which are used to correct the navigation system time to the local realization of UTC time.

The time references of system times of different navigation systems are different. For example, the system time of GPS navigation system is steered to UTC(USNO), GLONASS is steered to UTC(SU), and BD is steered to UTC(NTSC). The generation of the UTC time parameters of these navigation systems can be expressed by Fig. 1. As shown in this figure, the common view time transfer and TWSTFT are two primary ways for generating UTC time parameters presently. The time differences between UTC(K) and the time keeping center of navigation system are monitored and then computed to generate UTC time parameters. Then these parameters are uploaded to satellites and then broadcasted to users together with other parameters in navigation message.

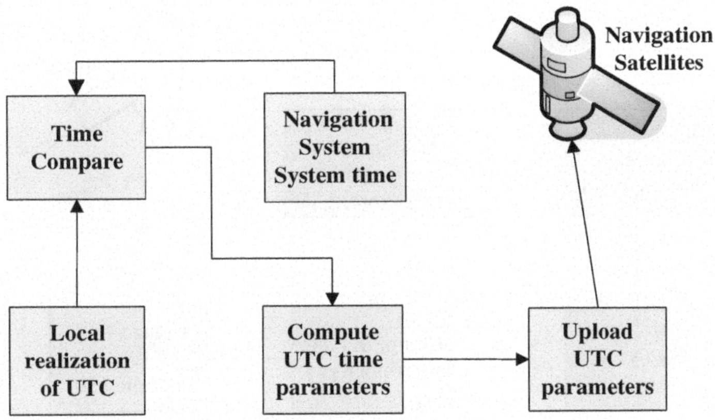

Fig. 1 Generation of GNSS UTC time parameters

When mentioned to BD, the compare link is established between BDT and national time service center and the time differences between BDT and UTC (NTSC) are measured. According to the time differences, the UTC time parameters are computed and then arranged in the BD navigation message.

2.2 *New Method of Generating UTC Time Parameters*

Applying the UTC time parameters generation method mentioned in the above section, users get the time differences relative to UTC(NTSC) with at best ten nanoseconds accuracy [2]. To take full advantages of BD GEO satellites, a new method for monitoring and generating the UTC parameters is proposed. With this new method, users can improve the accuracy of timing with the precision of about three nanoseconds.

The principle of the new method for generating UTC time parameters is shown in Fig. 2. A timing receiver is placed at National Time Service Center and used to monitor the timing error of BD GEO satellites. The differences between UTC (NTSC) and BDT that obtained through BD satellites are determined and analyzed. Then these data are sent to the master control station of BD navigation system where these data are processed into UTC time parameters. Finally, the upload station transmits the parameters up to BD GEO satellites.

The time differences between UTC(NTSC) and BDT are got through the following steps. By measuring the time delay from each satellite (signal transmitter) to the receiver the pseudoranges are measured. After deducting the geometrical distance, SV clock correction, ionosphere delay, and troposphere delay from the pseudoranges, the time biases between the user's local time and BDT are obtained which contain the ephemeris error, ionosphere error, troposphere error, and SV

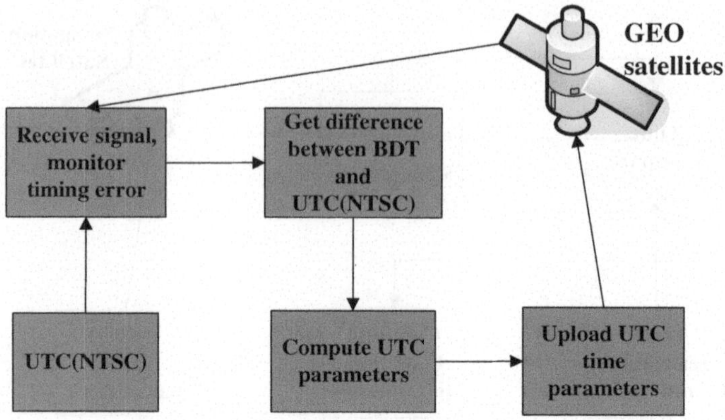

Fig. 2 New method for generating BD UTC time parameters

clock error. Since the timing received is placed at NTSC, the user's local time is UTC(NTSC).

Based on these time differences, the UTC time parameters of BD will be incorporated in the BD's navigation message and then transmitted to users. Users applying these parameters equivalent to achieving one time comparison between users and UTC(NTSC). According to the precision of common view time transfer, this method can provide a timing service with the precision of about 3 ns. Considering the advantages of BD GEO satellites in time transfer, the precision of the results will be better.

Besides the superior precision of this method, it also brings other benefits. The obvious point is that positioning users can get more accurate positioning result after using the UTC time parameters generated by this method.

3 Error Analysis of Common View Time Transfer

Supposing that there are two stations, respectively, receiving the navigation signal, each station monitors the time difference between GNSST and the local time of the station. The time differences are influenced by ephemeris error τ_{eph}, SV clock error τ_{clk}, ionosphere error τ_{ion}, and all the other errors τ_{oth} which can be expressed by Eq. (1).

$$\tau_A = T_A - T_S + \tau_{\text{eph}}(A) + \tau_{\text{clk}}(A) + \tau_{\text{ion}}(A) + \tau_{\text{oth}}(A) \tag{1}$$

The process to get the result in the left side of equation one is one-way timing [3, 4]. The local time of station A is expressed with T_A and the system time of navigation system is denoted T_S (Fig. 3).

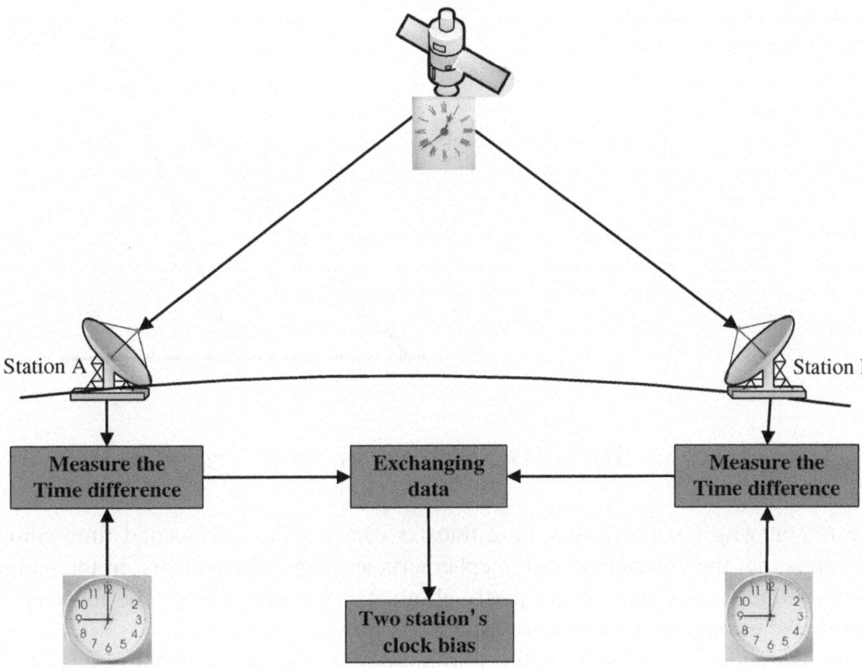

Fig. 3 Principle of common view time transfer

According to Eq. (1), it can be seen that the precision of one-way timing method that based on navigation system is at most at the level of ten nanoseconds. This cannot meet the requirements of users with the time synchronization precision of several nanoseconds. As a result, they have to adopt others means, such as common view time transfer or TWSTFT at the cost of non real-time and high cost [5–7].

Based on the same principle as station *A*, station *B* can also get the time difference expressed in Eq. (2).

$$\tau_B = T_B - T_S + \tau_{\text{eph}}(B) + \tau_{\text{clk}}(B) + \tau_{\text{ion}}(B) + \tau_{\text{oth}}(B) \tag{2}$$

We have Eq. (3) after data exchanging and subtracting Eq. (2) from Eq. (1). This is exactly the process of common view time transfer.

$$\begin{aligned}\tau_A - \tau_B = T_A - T_B \\ + \left(\tau_{\text{eph}}(A) - \tau_{\text{eph}}(B)\right) \\ + \left(\tau_{\text{ion}}(A) - \tau_{\text{ion}}(B)\right) + \left(\tau_{\text{oth}}(A) - \tau_{\text{oth}}(B)\right)\end{aligned} \tag{3}$$

The SV clock error is totally eliminated and the errors of ephemeris and ionosphere are partly removed. Taking the GPS satellites as an example, the accuracy of the time difference between station *A* and *B* is about 3 ns when the baseline between the two stations are 2000 km.

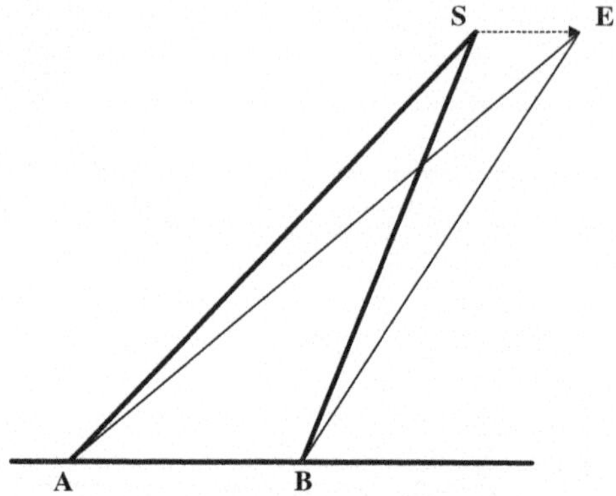

Fig. 4 Relation of ephemeris error and two stations *A* and *B*

4 Advantages of BD GEO Satellites in Time Transfer

The reason why common view time transfer can provide nanosecond time comparison is that the correlative errors, ephemeris and ionosphere errors, in the paths from satellite to two stations are partly eliminated via subtracting. The more correlative the errors, the better the results are.

The altitudes of BD GEO satellites orbit are higher than that of MEO satellites. They are almost stationary and almost always can be seen for stations involved in time comparison. Besides these advantages, there is no jump in timing results when using GEO satellites for time transfer. There is no need for users to frequently switch satellites as the GEO satellites can be seen almost a whole day. All these advantages will be discussed in the following sections.

4.1 More Correlative Ephemeris Error

The impact of ephemeris error on the result of time comparisons is determined by its projection on the path of satellite-to-station [8, 9]. Figure 4 shows the relation of ephemeris error and ground stations. Provided point *S* is the true position of satellite, and *E* is the broadcast position. Vector *SE* represents the ephemeris error and *AB* is the baseline of station *A* and *B*.

According to the law of cosines, it has [2–4].

$$EB^2 = EA^2 + AB^2 - 2 \cdot AB \cdot EA \cdot \cos(\angle SAB - \angle SAE) \quad (4)$$

$$SB^2 = SA^2 + AB^2 - 2 \cdot AB \cdot SA \cdot \cos(\angle SAB) \quad (5)$$

Combing Eqs. (4) and (5), we can get Eq. (6).

Table 1 Effect of 10-m ephemeris error on the result of time comparison under different baselines

Baseline		100 km	500 km	1000 km	2000 km	3000 km
Error in time comparison result (ns)	GEO	0.1	0.47	0.93	1.87	2.77
	MEO	0.13	0.8	1.6	3.17	4.77

$$\begin{aligned}|EB - EA - (SB - SA)| &\leq \angle SAE \cdot AB \cdot \sin(SAB) \\ &\approx \left(\frac{SE\sin(\angle SAB)}{SA}\right) \cdot AB \cdot \sin(\angle SAB) \\ &= \left(\frac{SE}{SA}\right) \cdot AB \cdot \sin^2(\angle SAB) \leq = \left(\frac{SE}{SA}\right) \cdot AB\end{aligned} \tag{6}$$

It can be seen that the impact of ephemeris error on user's timing results becomes obvious with the increase of baseline between the two stations A and B. Provided the ephemeris error is ten meters ($SE = 10$ m), and the baseline between A and B is 1000 km ($AB = 1000$ km). For GEO satellites, the orbit altitude is 36,000 km ($SA = 36{,}000$ km). Then the impact of ephemeris error on time difference of station A and B is:

$$\left(\frac{10\,\text{m}}{36{,}000\,\text{km}}\right) \cdot 1000\,\text{km} = 0.28\,\text{m}$$

That equals to about 0.9 ns error in time comparison. To compare the influences of same ephemeris error on time comparison results based on GEO and MEO satellites, a table lists the value of effect under different baselines (Table 1).

4.2 Variation of Ionosphere Error

For measurements obtained by pseudorandom code, the time delay introduced by ionosphere $\varepsilon^{\text{Iono}}$ is correlated with signal carrier frequency, elevation of ionospheric pierce point and the total electron content [9–11].

$$\varepsilon^{\text{Iono}} = \frac{1}{\sin\phi'} \cdot \frac{40.3}{f^2} \cdot \text{TEC} \tag{7}$$

The effect of ionospheric delay on two stations A and B apart p can be expressed as following:

$$\begin{aligned}\varepsilon^{\text{Iono}} &= \frac{1}{\sin\phi_A} \cdot \frac{40.3}{f^2} \cdot \text{TEC}_A - \frac{1}{\sin\phi_B} \cdot \frac{40.3}{f^2} \cdot \text{TEC}_B \\ &\approx \frac{p}{d_A} \cdot \left|\frac{p}{d_A} - \cos\phi_A\right| \cdot \frac{40.3}{f^2} \cdot \text{TEC}_A\end{aligned} \tag{8}$$

Table 2 Results of ionospheric delay correction of the new method

Station *B*	Baseline (km)	Based on MEO		Based on GEO	
		ID of Station *B*	ITD (ns)	ID of Station *B*	ITD (ns)
Jinan	749	8	1.3	7.7	0.5
Changsha	775	13	3.6	8.5	1.3
Guiyang	896	14.7	5.3	8	0.8
Lhasa	1776	12.8	3.5	6	1.2
Yushu	1179	10.5	1.1	6.2	1
Yinchuang	526	7.6	1.8	6.6	0.6
Chongqing	593	12.5	3.1	7.5	0.3
Nanchang	891	12.4	3	8.7	1.5
Sanya	1807	17.8	9.9	10.5	3.3
Shanghai	1177	10.4	1.2	8.8	1.6
Changchun	1740	5.3	4.1	7.7	0.5
Urumqi	2142	6.1	4.8	4.1	3.1

Wherein d_A is the distance from satellite to station A, p is the baseline between station A and B. ϕ_A, ϕ_B are, respectively, the angle between station-to-satellite and station A-to-station B. According to Eq. (8), we can see that the closer the two stations are the smaller the effect of ionospheric delay on the time comparison results.

Applying the ionospheric product released by IGS, the seven-day variations of ionosphere delay are calculated for thirteen places in China. The ionospheric delay of Xi'an is taken as station *A* and others are regard as station B. The column "ID of S B" means the ionospheric delay of station *B* when it is commonly observing the same satellite with station A. The column "ITD" is the ionospheric time difference after station *B* using the ionospheric delay of station *A* to correct its own. These two kinds of results are, respectively, calculated based on one GPS MEO satellite and one BD GEO satellite.

Based on the results in Table 2, we can get the conclusions that the residual ionospheric delays of station *B* at most test places are less than 3 ns when the baseline is smaller than 2000 km.

5 Time Transfer Experiment Based on BD GEO Satellites

To validate all the advantages of GEO satellites in the above section, an experiment of time transfer is carried out between Beijing and Xi'an based on the BD satellites. Figure 5 shows the standard deviations of time comparison results of three different types of BD satellites. It is obvious that the standard deviations of GEO satellites (1–5) are superior relative to the other two kinds of satellites. This proves the unique superiority of BD GEO satellites in time transfer.

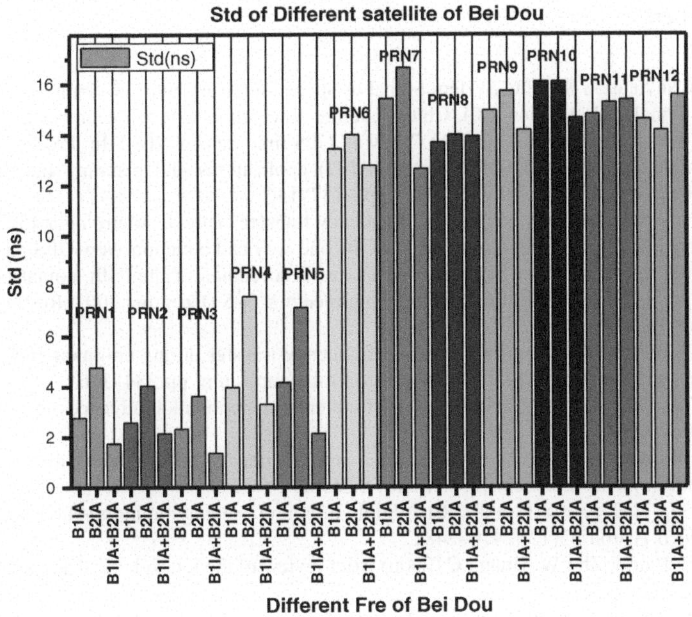

Fig. 5 Stds of the common view time transfer results between Xi'an and Beijing based on different types of BD satellites

6 Conclusions

As far as known, the navigation system of BD has GEO satellites in its constellation. Due to the high orbit altitude and the long time being visible, the GEO satellites have distinct advantages in time comparison when compared with MEO satellites.

This paper takes full advantages of BD GEO satellites and proposes a new method for generating the UTC time parameters. Essentially speaking, it is a new method for users to achieve one-way timing. The only difference is that its precision is about three times accurate than that of the traditional method especially when the satellites used for time transfer are GEO ones.

Acknowledgment Funded by the West Ph.D Program of West Light, CAS (2013BS24), National Natural Foundation (11503030) and the State Key Laboratory of Geo-information Engineering, No. SKLGIE2014-M-2-5.

References

1. BeiDou Navigation Satellite System Signal in Space Interface Control Document (2012) Open service signal B1I (Version 1.0). China Satellite Navigation Office, December 2012
2. Gifford A, McNeff J, Pace S (2001) One-way GPS time transfer 2000. In: Proceedings of the 32nd annual precise and time interval (PTTI) applications and planning meeting, 28–30 November 2000, Reston, Virginia, USA, pp 137–146
3. Levine J (2008) A review of time and frequency transfer methods. Metrologia 45:S162–S174
4. Lombardi MA, Novick AN Comparison of the one-way and common-view GPS measurement techniques using a known frequency offset. In: Proceedings of the 34th annual precise and time interval (PTTI) application and planning meeting, 3–5 December 2002, Reston, Virginia, USA, pp 39–51
5. Allan D, Weiss MM, Accurate time and frequency transfer during common-view of a GPS satellite. In: Proceedings 1980 frequency control symposium, pp 334–346
6. Parker TE, Matsakis D (2004) Time and frequency dissemination advances in GPS transfer techniques. GPS World 11:32–38
7. Lombardi MA, Nelson LM (2001) Time and frequency measurements using the global positioning system. Cal Lab: Int J Metrol 8:26–33
8. Xu L X, Li X H, Xue Y R et al (2012) Study of a new one-way timing method. Sci China G: Phys Mech Astron 55(12):2476–2481
9. Zhang H, Ping J, Zhu W, Huang C (2006) Brief review of the ionospheric delay models, Progr Astron 24(1):16–26
10. Feng W, Chen X, Wu Xing, Ye X (2011) Influence of space latitude on differential ionospheric grid system. J Geod Geodyn 31(4):135–138
11. Yuan Y, Ou J (2003) Results and analysis on determining global TEC with the GPS data provided by IGS. Prog Nat Sci 13(8):885–888

Preliminary Research on Snow Depth Monitoring with GPS SNR

Kaiyang Dai, Qin Zhang, Shuangcheng Zhang, Ning Zhang, Kai Liu and Xiaowei Hou

Abstract The GNSS observations quality can be affected by the environment around the station. Using the signal-to-noise ratios to correct the observation influenced by multipath effect has attracted attention of many scholars. Meanwhile, using GNSS-MR technology based on SNR information to study environmental changes around the station has become a new research subject. And that makes the GNSS active in the field of remote sensing. First, the relationship between GPS multipath and SNR is analyzed in this paper. Then, based on the SNR and the signal amplitude variation, the basic principle of GPS-MR technology is given. GPS observation data in P351 and SNOTEL station snow data are used to evaluate the effectiveness of the algorithm. Also, satellite reflection point trajectory and satellite selection of snow depth detection are further analyzed. The experimental results show that: GPS-MR can be used to measure the snow depth, and its detection precision is 0.08 m. GPS-MR technology not only can make full use of multipath effect, but also provides a development space for GPS technology using in the surface environment monitoring.

Keywords GNSS-MR (GNSS multipath reflectometry) · SNR (signal-to-noise ratio) · Snow depth · Lomb-Scargle spectrum analysis

1 Introduction

GPS receiver antenna receives the direct satellite signal, in addition, it received reflected signal reflected by the surroundings. The superposition of two kinds of signal will cause the change of antenna phase center position, and bring error for

K. Dai · Q. Zhang (✉) · S. Zhang · N. Zhang · K. Liu · X. Hou
College of Geology Engineering and Geomantic,
Chang'an University, Xi'an 710054, Shanxi, China
e-mail: zhangqinle@263.net.cn

K. Dai
e-mail: daikaiyang@126.com

J. Sun et al. (eds.), *China Satellite Navigation Conference (CSNC) 2016 Proceedings: Volume I*, Lecture Notes in Electrical Engineering 388,
DOI 10.1007/978-981-10-0934-1_7

navigation and positioning. It is called the multipath error [1]. With the continuously deeper research and application on GNSS, the mechanism of multipath error is studied more thoroughly. Zhang et al. [2] used the signal-to-noise ratio to evaluate the quality of observation, and reduced the multipath error by lowering the weight of observation influenced by multipath effect. Based on the principle of multipath and its correlation with SNR, Wu et al. [3] proposed a method that used SNR to weaken the multipath effect, and got the "clean" observed value. Bilich et al. [4] researched on the direct and reflected GPS signal and the relationship between environment and reflectivity, and further established the mapping relation between signal-to-noise ratio and multipath environment.

Therefore, using the SNR information to research surrounding environment around GPS station became a new research field. A method called GPS-R that used two antennas to receive direct and reflected signals respectively, and then obtained the surface parameters already has a wide research and application. But it must use special antenna device to get the reflection information, which limits its development [5]. Then, Professor Larson innovatively proposed GPS-MR technology. That is, using conventional GPS receiver to detect surface parameters information, and further expanded the development and application of GNSS in the field of remote sensing. Since then, the research group used SNR that multipath influenced components to do research respectively on snow depth, soil humidity, vegetation coverage et al., and made a series of research and application achievements of great value [6–9].

Scholars in domestic started relatively late in this area. Ao et al. [10, 11] used GPS multipath reflected components of SNR observation to detect soil moisture. Simulated the changing trends and analyzed how to choose the experiment satellite and how to evaluate the quality of inversion results. Wu et al. [12] proposed a method that used a single GPS receiver reflected signal to measure the height of water surface. And the result was in accord with the gauge results. Yang et al. [13] analyzed 10 years GPS data in one station, and found that there were annual cycle between GPS multipath effect and vegetation coverage around the station. Wan et al. [14] used SNR observation researched on the water content of desert grassland, wheat and alfalfa, and verified the effectiveness of the GPS-MR for vegetation water content detection by comparing with measured data.

Snow is an important freshwater resources, and it is also one of the important factors that affect global climate. The information of snow cover range, snow depth, and the snow water equivalent have the significant meanings for the monitoring of global water resources protection and utilization, snow disaster monitoring and prevention and that of climate science. Traditional snow depth monitoring methods mainly used discrete meteorological site or laser detector. Its data quality is affected by the number and distribution of the sites. Satellite remote sensing technology is used in the different spatial and temporal resolution of snow information extraction, while its accuracy is affected by factors such as clouds and atmospheric refraction, and it is powerless to the thickness of snow. In the recent years, based on the SNR, a new means of snow depth-detecting technology was found. It is GPS-MR. It has advantages that converges large area, high time resolution, all-weather, real-time, automation, and making full use of existing GNSS monitoring stations provides a new and powerful sensor for the meteorological department.

At present, there are less studies on GPS-MR for snow detection in domestic. There are more and deeper research to do, to promote the development of GPS-MR technology. In this paper, based on the achievements of above scholars, the relationship between SNR and multipath and the basic principle of GPS-MR for snow depth detecting technology is discussed in details. Moreover, combined with the measured data, the validity of this algorithm is analyzed.

2 Principle of GPS-MR Technology

2.1 Relationship Between Multipath and SNR

The signal that GPS receiver antenna receives is the synthesis signal. According to electromagnetic theory, synthetic signal is track by receiver PLL loop, and the signal can be expressed as plural form, j is the plural unit. There are:

$$\vec{r_d} = A_d e^{j\varphi_d} = A_0 A_\alpha e^{j\varphi_d} \tag{1}$$

$$\vec{r_m} = \sum_{i=1}^{m} \alpha_i A_0 A_\alpha e^{j(\varphi_d + \Psi_i)} \tag{2}$$

$$\vec{r_c} = A_c e^{j\varphi_c} = A_c e^{j(\varphi_d + \delta\varphi)} \tag{3}$$

A_d, φ_d are for the direct signal amplitude and phase, respectively. And $A_d = A_0 A_\alpha$. A_0 is for the fixed part of direct signal amplitude. And A_α is called the standard antenna gain, which is the function of antenna gain, satellite elevation angle and azimuth angle and so on. α_i is for the ith satellite reflected signal amplitude coefficient relative to the A_d. Ψ_i is for the ith satellite reflected signal phase delay relative to the φ_d. $\alpha_i A_0 A_\alpha$ is for the reflected signal amplitude. A_c, φ_c is for the amplitude and phase of synthesis signal respectively. $\delta\varphi$ is the phase shift caused by multipath. Due to the satellite movement, $\delta\varphi$ changes over time, there are [3]:

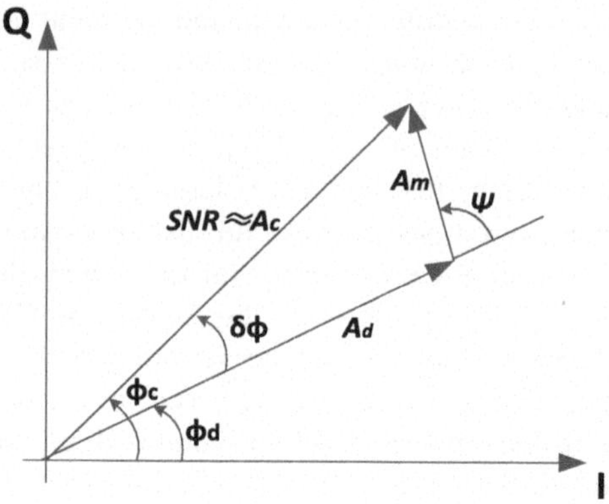

Fig. 1 Influence of multipath effects

$$\overrightarrow{r_c} = \overrightarrow{r_d} + \overrightarrow{r_m} \quad (4)$$

The phase and amplitude of synthesized signal is affected by the phase shift and amplitude of multipath signal. From the multipath effect influence diagram (Fig. 1). When the phase delay of multipath signal is $\Psi = 0°$, direct wave and reflected wave are in phase, the combination of them lead to the signal power increased. Therefore, the multipath is beneficial to receiver in this case; When the phase delay of multipath signal is $\Psi = 180°$, the power of synthesized signal is weaker. The reflected phase delay is vary in different times and places, so the receiving signal amplitude and phase will appear unstable phenomenon [15].

SNR is the ratio of carrier signal intensity and noise intensity. Most receiver mark SNR as C/N0, and its unit is dB Hz. It is mainly influenced by the antenna gain parameters, the state of receiver correlator, and the multipath effect. When the multipath occurs, the satellite signal quality will be lower, and SNR values will change accordingly. When only considered the amplitude of SNR, SNR can be used as a substitute for SNR A that represent the amplitude of signal rate. SNR A_c, SNR A_d, SNR A_m are for synthetic, direct and reflected signals, respectively. And there are [3]:

$$\mathrm{SNR}\,A = A_c + \mathrm{noise} \approx A_c = \mathrm{SNR}\,A_c \quad (5)$$

$$\mathrm{SNR}\,A_d = A_0 A_\alpha \quad (6)$$

$$\mathrm{SNR}\,A_m = A_0 A_\alpha \sum_{i=1}^{m} \alpha_i \quad (7)$$

The principle of using GPS signal to get surface environment parameters around the station is based on the SNR that included in the influence of carrier phase multipath, by separating the multipath components of synthetic signal, and then using it to achieve the extraction of information parameter.

2.2 Algorithm and Data Processing

GPS receiver antenna received the reflected signal and direct signal. The synthesis signal have amplitude or phase change relative to the direct signal. Figure 2 shows diagram of GPS-MR for snow depth.

The height of antenna is H, It is the distance from receiver antenna phase center to soil surface. h is for the distance from receiver antenna phase center to snow surface, and it is called vertical reflection distance in the following paper. h_{snow} is for the thickness of snow, E is for the angle between signals and snow surface, A_d, A_m are for the amplitude of the direct and reflected signal respectively.

For geodesy GPS receiver antenna, in order to effectively restrain the reflected multipath error, the amplitude of the direct and reflected signal have the relation as follows:

$$A_d \gg A_m \tag{8}$$

According to the Fig. 1, we can get that:

$$\text{SNR}^2 = A_c^2 = A_d^2 + A_m^2 + 2A_d A_m \cos\psi \tag{9}$$

SNR information caused by GPS multipath should be separated from the received SNR observation value. Because of the big gap in the numerical between A_d and A_m. Generally, the low-order polynomials can be used to detrend the A_d.

Detrend SNR variation from doy 110–140 in 2015 at P351 are showed in Fig. 3. The data sample rate is 15s. SNR under the condition of low elevation angle have big fluctuations, which is mainly because of the influence of multipath in low elevation angle. With the rising of satellite, multipath effect decreases. Accordingly, the quality of SNR improves. Therefore, SNR is usually used to evaluate the

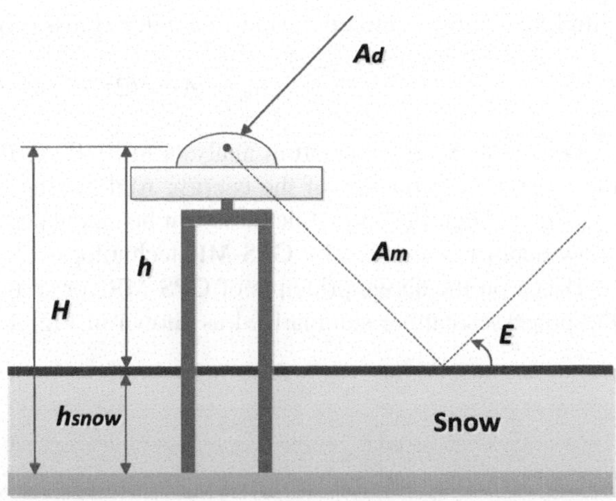

Fig. 2 Diagram of GPS-MR for snow depth

Fig. 3 SNR variations of PRN 04 in P351

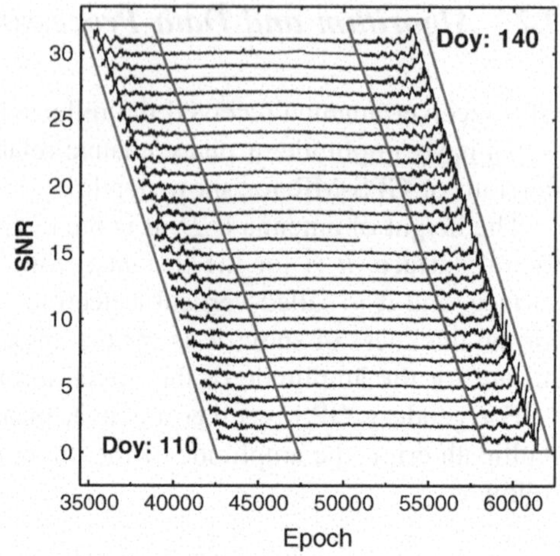

multipath and model multipath error. In turn, the GPS multipath reflected signal will directly affect the GPS SNR, and using SNR data after the spectrum analysis can get the environment parameters, namely the GPS-MR technology. GPS satellites cycle is 11:58 min. From Fig. 3, we can also see the SNR data have an about 4 mins deviation every day.

Multipath in low elevation angle have more serious influence on SNR. So this article use the SNR data of 5°–20°. The residual sequence of reflected signal amplitude can be expressed as [8]:

$$A_m = A\cos\left(\frac{4\pi h}{\lambda}\sin E + \phi\right) \tag{10}$$

Combined with Fig. 2, λ is for carrier wavelength. E is for satellite elevation angle. h is the vertical reflection distance. If set $t = \sin E, f = \frac{2h}{\lambda}$, Eq. (10) can be simplified as the standard cosine function expressions:

$$A_m = A\cos(2\pi ft + \phi) \tag{11}$$

Do Lomb-Scargle spectrum analysis to SNR residual sequence [16]; Calculate the maximum frequency of the energy; And get the vertical reflection distance by $f = 2h/\lambda$; Then the snow thickness can be calculated by $h_{\text{snow}} = H - h$. Thus, the snow depth is measured by GPS-MR technology.

Based on the above principle of GPS-MR for snow depth detecting technology, the processes can be summarized as shown in Fig. 4:

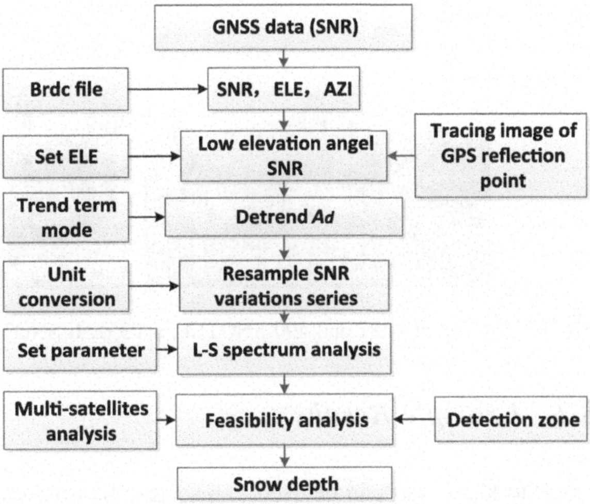

Fig. 4 Flow graph of GPS-MR for snow depth

3 Analysis of Examples

3.1 Data Source

Plate boundary observation (PBO) is one component of the scientific project "Earthscope". PBO spans the North American continent, and it is constructed and maintained regularly. Therefore, PBO can provide long-term, stable, and reliable data. In addition, PBO station mainly established in remote rural areas, and multipath error can be assumed that mainly comes from the surface environment (surface soil, vegetation, seasonal snow, or sea level, etc.) around GPS station.

The experiment site is in P351 station. The site is located in Montana Ketchum in United States. Longitude is 1144308.98, and latitude is 435227.88. Its altitude is 2692.60 m. This area is relatively broad, and covered with snow in winter. The P351 station established in 2008. Receiver has been used TRIMBLE NETRS, and antenna is used TRIMBLE 29659.00. The height of antenna is 2 m. Figure 5a is the environment around P351 station.

SNOTEL is run by natural resource conservation service organizations (NRCS) of United States department of agriculture (USDA). It is an automatic climate monitoring system in operation and management in remote areas in the western United States. It is composed of more than 600 sites, the network provides a very valuable data of snow depth, snow water equivalent, temperature for high altitude area. These data were used to predict the annual water supply, flood and as well as some research related to climate. SNOTEL data published on the Internet in different format. Figure 5b is the condition of 490 SNOTEL which is near by P351.

Fig. 5 Site map at P351 and 490 SNOTEL. **a** P351. **b** SNOTEL

3.2 *Example Results*

Satellite Sky Chart can be used to research the influence of satellite space geometric distribution to positioning or the baseline accuracy. While for the GPS-MR technology, studying satellite tracing image of GPS reflection point is more meaningful. The satellite signal reflection point trajectory is the function of satellite elevation angle, azimuth angle, and the height of antenna, and it reflects the satellite location relative to the receiver. Through signal reflection point trajectory, we can clearly see the direction that of GPS-MR snow depth represented. So as to get information about the spatial distribution of snow. Figure 6 shows the 0–90° and 0–20° tracing image of GPS signal reflection point of doy 120 in P351.

(0, 0) in Fig. 6 is the projection point of the receiver antenna phase center on the ground. The horizontal and vertical axis represent the distribution of the signal reflection point around the station. The dotted circle in the graph from inside to outside is elevation angle of 20°, 15°, 10°, 5° of the reflection point trajectory respectively. Since we aim at the low elevation angle condition, SNR residual

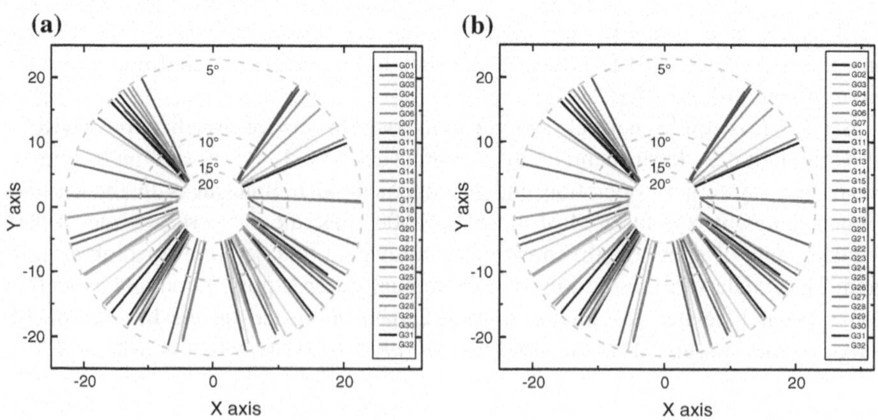

Fig. 6 Tracing image of GPS reflection point. **a** 5°–90°. **b** 5°–20°

sequence affected by multipath, Fig. 6b shows the 5–20° satellite signal reflection point trajectory, namely eliminating more than 20° segmental arc and some low elevation angle segmental arc has less data in Fig. 6a.

According to the reflection point projection trajectory, the data form doy 110 to doy 140 are processed. Limited to space, the residual error sequence and its spectrum of PRN24 for four days in P351 station are given in Figs. 7 and 8 respectively.

The horizontal axis in Fig. 7 is the sine value of elevation angle, the vertical axis is detrend SNR residual error sequence linearized by formula $(snr' = 10.\hat{}(snr/20))$. The unit is volt, rather than dB Hz in the original observation file. In Fig. 8, the horizontal axis shows frequency after Lomb-Scargle spectrum analyzed, and the vertical axis is L-S spectrum amplitude of each frequency. The frequency of L-S spectrum peak amplitude is regarded as the effective frequency f, and vertical reflection distance h can be calculated by $h = f\lambda/2$. Finally, the thickness of snow is got by $h_{snow} = H - h$. The "snow" in Fig. 7 or Fig. 8 is SNOTEL measured snow depth.

With the decrease of the thickness of snow depth, oscillation frequency of SNR residual sequence increases gradually (Fig. 7). The frequency of peak amplitude spectrum also increases gradually (Fig. 8). Snow thickness change is closely related to the energy spectrum change of SNR in low elevating angle. It is further proved

Fig. 7 SNR variations

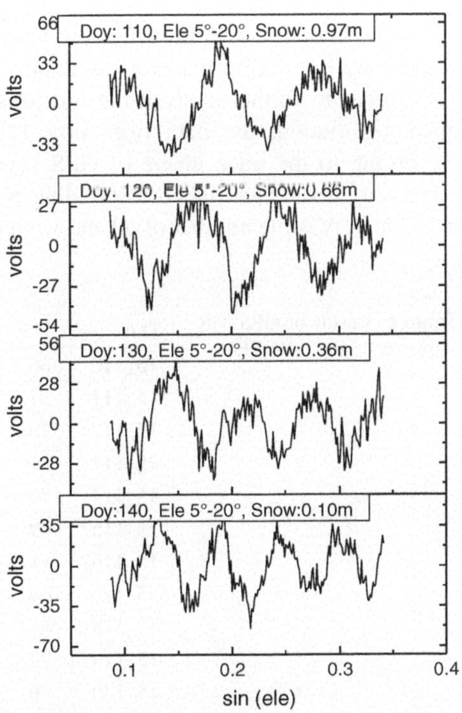

Fig. 8 L-S results for SNR variations

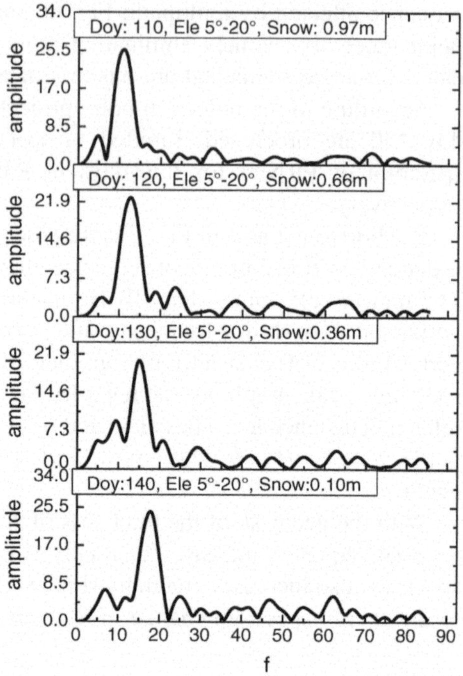

that snow thickness will influence the SNR frequency spectrum, and GPS reflected signal may be used to detect snow depth.

In order to further analyze the relationship between snow depth and SNR frequency variation, the data from doy 110 to doy 140 in 2015 are calculated. According to the trace image of GPS signal reflection point (Fig. 6), partial data processing results and its corresponding SNOTEL measured snow depth are listed in Table 1. *N* is the number of all snow depth from trace image, and *T* is the number

Table 1 Result of GPS-MR snow depth and SNOTEL records

Doy	N	T	%	GPS-MR/m	SNOTEL/m
15_110	66	59	0.89	0.946	0.97
15_111	71	63	0.89	0.894	0.91
15_112	70	65	0.93	0.8543	0.89
15_113	71	67	0.94	0.8368	0.84
15_114	70	65	0.93	0.813	0.81
15_115	69	59	0.86	0.8505	0.84
15_116	71	68	0.96	0.9046	0.84
15_117	69	62	0.9	0.8523	0.84
15_118	66	60	0.91	0.8206	0.81
15_119	72	66	0.92	0.7785	0.71
15_120	70	60	0.86	0.7207	0.66

of effective snow depth which ignores snow depth whose deviation is greater than 1.5 times of mean square error. At last, the average of effective snow depth on behalf of the day snow depth.

The difference between GPS-MR snow depth and the measured snow depth is showed clearly in Fig. 9. The general trend is good. But with the reducing of snow depth, the error increases gradually. In other words, its accuracy is reduced. The reason may be that the snow is thinner, and the GPS signal through the falling snow. The difference of material causes big error to the algorithm.

Furthermore, the correlation coefficient between GPS-MR snow depth and SNOTEL snow depth is greater than 0.95 (in Fig. 9). It is significant.

Combining with Table 1, Figs. 9 and 10, the following conclusion can be obtained

1. SNR data can reflect the quality of GPS signal, and can be used for environmental surface monitoring research.

Fig. 9 Comparison of GPS-MR and SNOTEL at P351

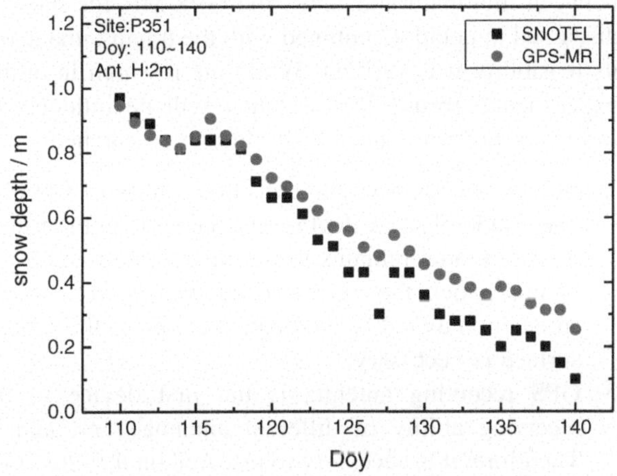

Fig. 10 Correlation analysis of GPS-MR and SNOTEL

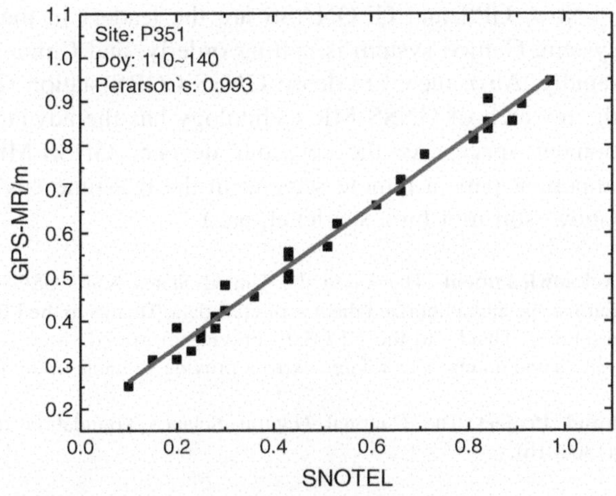

2. Combined with the tracing image of GPS signal reflection point, the snow depth of different direction around station can be obtained.
3. Through the correlation analysis, GPS-MR technique can be used for snow depth monitoring.
4. Data processing results show that: the mean residual is 0.08 m, and the standard deviation is 0.536. Therefore, the precision of GPS-MR for snow detecting technology can reach to 0.08 m.

4 Conclusion

SNR data can reflect the signal quality. Using SNR information can not only modify the observation influenced by multipath error, but also play an important role in environment surface monitoring. In this paper, based on the relationship between multipath and SNR, the GPS-MR for snow depth detecting technique is discussed in detail. Combined with the tracing image of GPS signal reflection point, its feasibility was verified. While the research in domestic is not yet matured, the experimental results in this paper will be influenced by many factors. And the following problems need to be further researched:

1. In order to improve the detection ability of GPS-MR technology in a different magnitude of snow depth, especially the light or moderate snow depth detection, its data model remains to be further improved.
2. In this paper, the experimental data is from relatively flat area. In site measurement may not be an ideal situation, so the research on GPS-MR model of tilt surface is necessary.
3. GPS receiving antenna is the first device to process satellite signal. The receiving ability of different antenna vary, and its SNR quality differ, too. Therefore, it is necessary to research on the effects of different type GPS receiver antenna to GPS-MR.

Now, GPS and GLONASS are the leaders in the field of satellite navigation system. Galileo system is netting orderly, and China's BDS system is developing rapidly. Also, there are dense GNSS CORS station (IGS, PBO, CMONOC, etc.), the research of GNSS-MR technology has the advantageous resources and development space. As the research deepen, GNSS-MR technology will play an important part in remote sensing in the real-time environment surface monitoring (snow, soil moisture, sea level, etc.).

Acknowledgment Thanks to the United States NSF, USGS, and NASA who launched the Earthscope and authorized the use of GPS data. Thanks to the PBO H2O team to provide reference resources. Thanks to the NRCS to provide measured snow depth data. Heartfelt thanks to the editors and anonymous referee experts provide valuable opinions and suggestions for this article.

Fund Project The National Natural Science Foundation of China (41104019, 41202189, 41304016).

References

1. Zhang Q, Li J et al (2005) GPS measurement principle & application. China Science Publishing & Media Ltd, Beijing
2. Zhang B, Huang J, Su L (2003) A research on using SNR values to mitigate carrier phase multipath in GPS surveying. Sci Surv Mapp 28(3):32–35
3. Wu Y, Chen X, Wu C (2008) Mitigation of muti-path effect using SNR values. Geomat Inf Sci Wuhan Univ 33(8):842–845
4. Bilich A, Larson KM, Axelrad P (2004) Observations of signal-to-noise ratios (SNR) at geodetic GPS site CASA: implications for phase multipath. Proc Cent Eur Geodyn Seismol 23:77–83
5. Liu J, Shao L, Zhang X (2007) Advances in GNSS-R studies and key technologies. Geomat Inf Sci Wuhan Univ 32(11):955–960
6. Larson, KM, Gutmann, ED, Zavorotny VU et al (2009) Can we measure snow depth with GPS receivers? Geophys Res Lett 36(17). doi:10.1029/2009GL039430
7. Gutmann ED, Larson KM, Williams MW et al (2012) Snow measurement by GPS interferometric reflectometry: an evaluation at Niwot Ridge, Colorado. Hydrol Process 26(19):2951–2961
8. Larson KM, Small EE, Gutmann ED et al (2008) Use of GPS receivers as a soil moisture network for water cycle studies. Geophys Res Lett 35(24). doi:10.1029/2008GL036013
9. Small EE, Larson KM, Braun JJ (2010) Sensing vegetation growth with reflected GPS signals. Geophys Res Lett 37(12). doi:10.1029/2010GL042951
10. Ao Minsi H, Hu Y, Liu Y et al (2012) Inversion of soil moisture fluctuation based on signal-to-noise ratio of global positioning system. J Geomat Sci Technol 29(2):140–143
11. Ao M, Zhu J, Hu Y, Zeng Y, Liu Y (2015) Comparative experiments on soil moisture monitoring with GPS SNR observations. Geomat Inf Sci Wuhan Univ 40(01):117–121
12. Wu J, Yang R (2012) Measuring water surface height by using reflected signal of geodetic-quality GPS receiver. J Geod Geodyn 32(6):135–138
13. Yang Y, Hao X, Peng B et al (2014) Study on vegetation coverage with GPS mutipath effects. J Geomat 39(2):52–54
14. Wan W, Larson KM, Small EE et al (2015) Using geodetic GPS receivers to measure vegetation water content. GPS Solut 19(2):237–248
15. Xie G (2009) Principles of GPS and receiver design. Publishing House of Electronics Industry, Beijing
16. Xu B, Yang T, Tan B et al (2011) The simulate study of detection based on Lomb-Scargle algorithm. Nucl Electron Detect Technol 31(6):702–705

The Research of High-Frequency Single-Epoch GNSS Kinematic Differential Positioning Technology

Dashuai Chai, Shengli Wang, Xiushan Lu and Bo Shi

Abstract The paper proposed the mathematical model, the processing method, and the basic process for the kinematic differential positioning. First, the single epoch is used for initializing, and the wide-lane ambiguity is searched and fixed to calculate the basic ambiguity during this process. Then, the baseline solution is calculated by the least squares method, and gross error is processed by using robust estimation. Moreover, the residual error is considered to correct the basic ambiguity. Finally, epoch is updated. If there is a situation that new satellite appears or cycle slip happens, the corresponding basic ambiguity is needed to be calculated again, otherwise the basic ambiguity is only transmitted. An example, in which 20 Hz GNSS data is sampled by V-Surs I vehicle-borne three-dimensional mobile surveying system, is presented in this paper. The method mentioned above is used to calculate this data and compare it with the results calculated by the GNSS/INS tightly coupled of Inertial Explorer 8.60 software. The experimental results demonstrate that the method has high stability and accuracy in the medium-short baseline, and RMS in the north direction is less than 2 cm, RMS in east direction is less than 3 cm, and RMS in the up direction is less than 5 cm.

Keywords High frequency · Single epoch · Kinematic differential positioning · On-board data

D. Chai · B. Shi
College of Geomatics, Shandong University of Science and Technology, Qingdao 266000, China

S. Wang (✉) · X. Lu
Institute of Ocean Engineering, Shandong University of Science and Technology, Qingdao 266000, China
e-mail: victory_wsl@vip.126.com

J. Sun et al. (eds.), *China Satellite Navigation Conference (CSNC) 2016 Proceedings: Volume I*, Lecture Notes in Electrical Engineering 388,
DOI 10.1007/978-981-10-0934-1_8

1 Introduction

With the development of BeiDou Navigation Satellite System (BDS) and the Global Navigation Satellite System (GLONASS), together with mature Global Positioning System (GPS), there will be enough satellites under operation. Compared with a single system, combination of the three systems will improve the geometric structure of satellites, avoid generating ill-conditioned normal equation coefficients, and ensure the stability and reliability of the satellite navigation system [1]. In recent years, the static positioning of GNSS is almost mature, however, kinematic

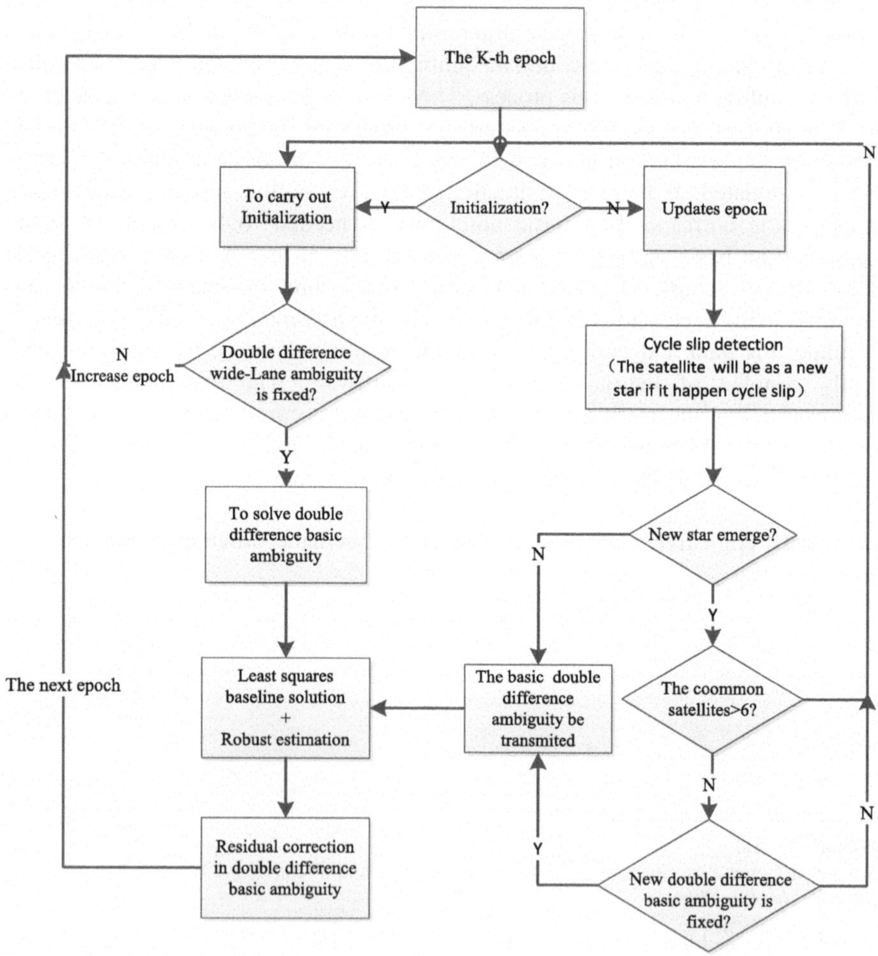

Fig. 1 Kinematic differential process

positioning, especially the high-accuracy real-time positioning is still under development [2]. As is well known that the positioning accuracy using observation of carrier phase is 2–3 orders of magnitude higher than using observation of pseudorange [3], so it is necessary for us to use the observation of carrier phase. However, it is difficult to calculate the ambiguity in kinematic positioning because of the satellite unlock and cycle slip caused by the complex environment and the movement of receivers. To solve this problem, many scholars have already made an intensive study [4–7]. In this paper, our study focuses on the kinematic differential positioning technology of GNSS based on single epoch in the least squares method with broadcast ephemeris and dual-frequency data, and then analyzes the accuracy and stability. The process can be shown in Fig. 1.

2 Mathematical Model

The measurement data composed of GPS, BDS, and GLONASS was used in this paper, and the across-satellite difference was just used in respective system. If the satellites number is n, there will be $n - 3$ double difference observation equations available including 3 position parameters and $n - 3$ double difference ambiguities. As a result, the number of unknown parameters is more than the number of equations. When using the least squares method for kinematic positioning, we need to search the ambiguity in rapid first, according to the wide-lane ambiguity to fix basic ambiguity that was used in this paper. Therefore, using the linear combination between dual-frequency P-code and carrier phase can obtain the floating value of double difference wide-lane ambiguity. The equation is shown as follows [8]:

$$\nabla\Delta N_{WL} = \frac{f_1 - f_2}{f_1 + f_2}\left(\frac{\nabla\Delta P_1}{\lambda_1} + \frac{\nabla\Delta P_2}{\lambda_2}\right) - (\nabla\Delta\varphi_1 - \nabla\Delta\varphi_2) \tag{1}$$

where $\nabla\Delta$ represents double difference operator, N_{WL} represents the floating value of wide-lane ambiguity, f represents frequency, P represents the observation of code pseudorange, λ represents wavelength of frequency, and φ represents the observation of carrier phase. When using this method to solve wide-lane ambiguity instantaneously, the key is that the code pseudorange needs enough accuracy. So, some satellites with poor quality of code pseudorange can be eliminated in the process of preprocessing. After the floating value of wide-lane ambiguity is obtained, the integer solution of wide-lane ambiguity will also be obtained by searching it. In this paper, the ratio value and unit weight of standard error were used to determine whether the wide-lane ambiguity was successfully fixed. After the wide-lane ambiguity is fixed, the floating value of basic ambiguity can be obtained by wide-lane observation equation and L1 observation equation.

The double difference observation equation of carrier phase for GPS and BDS is as follows [9]:

$$\begin{aligned}\lambda_i \nabla\Delta\varphi_i = \nabla\Delta\rho - \lambda_i \nabla\Delta N_i - \nabla\Delta I + \nabla\Delta T + \nabla\Delta(H + h_i)\sin E \\ + \nabla\Delta O + \nabla\Delta m_{i,\varphi} + \nabla\Delta d_{\text{other}} + \nabla\Delta\varepsilon_{i,\varphi}\end{aligned} \tag{2}$$

where $i = 1, 2$ represents the band of signal, ρ represents the distance of satellite receiver, N represents ambiguity, and I and T represent ionosphere delay and troposphere delay on propagation path, respectively, H represents receiver's instrument height, h_i represents the correction of receiver's antenna phase center, E represents elevation angle of satellite, O represents the errors of satellite orbit, $m_{i,\varphi}$ represents multipath effects in carrier phase surveying, d_{others} represents other errors, including the earth rotation, relativistic effects, etc., $\varepsilon_{i,\varphi}$ and represents noise of carrier phase. Other symbols represent the same as Eq. (1).

Because GLONASS uses a signal system of frequency division multiple access (FDMA), the ambiguity needs to be decomposed into a double difference ambiguity and an across-receiver difference ambiguity of reference satellite.

Here, assuming that the satellite b and reference satellite r carry out across-satellite difference after across-receiver difference, the process of decomposition is as follows [9]:

$$\begin{aligned}\lambda_i^b \Delta N_i^b - \lambda_i^r \Delta N_i^r &= \lambda_i^b \Delta N_i^b - \lambda_i^b \Delta N_i^r + \lambda_i^b \Delta N_i^r - \lambda_i^r \Delta N_i^r \\ &= \lambda_i^b (\Delta N_i^b - \Delta N_i^r) + (\lambda_i^b - \lambda_i^r)\Delta N_i^r \\ &= \lambda_i^b \nabla\Delta N_i + (\lambda_i^b - \lambda_i^r)\Delta N_i^r\end{aligned} \tag{3}$$

The double difference observation equation of carrier phase for GLONASS is as follows [9]:

$$\begin{aligned}\nabla\Delta\lambda_i\varphi_i = \nabla\Delta\rho - \lambda_i^b \nabla\Delta N - (\lambda_i^b - \lambda_i^r)\Delta N^r - \nabla\Delta I + \nabla\Delta T + \nabla\Delta(H + h_i)\sin E \\ + \nabla\Delta O + \nabla\Delta d_{i,\varphi} + \nabla\Delta m_{i,\varphi} + \nabla\Delta d_{\text{other}} + \nabla\Delta\varepsilon_{i,\varphi}\end{aligned} \tag{4}$$

where $d_{i,\varphi}$ represents receiver's hardware delay of carrier phase, ΔN represents the ambiguity of across-receiver difference. Other symbols represent the same as Eq. (2). Because medium-short baseline was resolved in this paper, only the troposphere delay was considered.

Similarly, the two kinds of wide-lane observation equation of carrier phase are as follows:

$$\nabla\Delta\lambda_W\varphi_W = \nabla\Delta\rho - \lambda_W^b \nabla\Delta N_{WL} + \nabla\Delta T \tag{5}$$

$$\nabla\Delta\lambda_W\varphi_W = \nabla\Delta\rho - \lambda_W^b \nabla\Delta N_{WL} - (\lambda_W^b - \lambda_W^r)\Delta N_{WL}^r + \nabla\Delta T \tag{6}$$

where the other errors are not listed compared with Eqs. (2) and (4), λ_W represents wavelength of wide lane, φ_W represents carrier phase of wide lane. Other symbols represent the same as Eq. (2). The single difference ambiguity of Eqs. (4) and (6) can be resolved by the observation equations of pseudorange and carrier phase.

Here, we just use Eqs. (2) and (5) to show the solution of basic ambiguity. The floating value of basic ambiguity can be obtained via the following equation that uses simplified Eq. (2) to subtract simplified Eq. (5).

$$\nabla\Delta N_i = \frac{1}{\lambda_i}(\lambda_W \nabla\Delta N_W - \lambda_W \nabla\Delta\varphi_W + \lambda_i \nabla\Delta\varphi_i) \tag{7}$$

The integer basic ambiguity can be obtained by rounding Eq. (7) directly. The integer basic ambiguity is substituted into the observation equation, and then error equation can be obtained by the linearization of observation equation. Thus Eq. (8) is formed.

$$V = Bx - L \tag{8}$$

where V represents error correction of observation, B represents coefficient matrix, x represents coordinate increments, and L represents the difference between observation value and approximate theoretical value.

According to the principle of least squares, the normal equation can be written as the following equation:

$$B^T PBx = B^T PL \tag{9}$$

where P represents weight matrix, and the ratio of weight among systems are 1:1:1.

$$P = \begin{pmatrix} P_{\mathrm{GPS}} & & \\ & P_{\mathrm{GLONASS}} & \\ & & P_{\mathrm{BDS}} \end{pmatrix} \tag{10}$$

where P_{GPS} represents weight matrix of GPS, P_{GLONASS} represents weight matrix of GLONASS, and P_{BDS} represents weight matrix of BDS.

The following equations can be obtained by resolving Eq. (9):

$$x = \left(B^T PB\right)^{-1} B^T PL \tag{11}$$

$$\sigma_0 = \sqrt{(V^T PV)/(n - t)} \tag{12}$$

where σ_0 represents unit weight of standard error, n represents the number of observations, and t represents the number of essential observations.

The above ambiguity could have incorrect ambiguity, which will make the observation equation to contain gross error. Furthermore, the quadratic form of residual error, $V^T PV$, is particularly sensitive to gross errors, and the presence of single gross error can lead to collapse of the estimated values [10]. In order to resist gross error, robust estimation was used. The IGG-III scheme of iteration and variable weights was used in this paper, and the equation is as follows [11]:

$$\overline{P_i} = \begin{cases} P_i, |\bar{v}_i| \leq k_0 \\ P_i \frac{k_0}{|\bar{v}_i|} \left(\frac{k_1 - |\bar{v}_i|}{k_1 - k_0} \right)^2, k_0 < |\bar{v}_i| \leq k_1 \\ 0, |\bar{v}_i| > k_1 \end{cases} \tag{13}$$

where $\overline{P_i}$ represents equivalent weight, $\bar{v}_i$ represents standardized residual error, k_0 represents quantile parameter, and k_1 represents elimination point. The robust estimation is through constant iteration and weight making to decrease the effect of gross error on parameter estimation. Considering the efficiency and preventing filter divergence, the number of iterations is generally 3–5 times [11]. In addition, the use of robust estimation requires sufficient original data.

3 Case Analysis

In order to verify this method, this paper applied the GNSS 20 Hz data which had been collected from the V-Surs I vehicle-borne three-dimensional mobile surveying system for about 3 h. This system has been researched and developed by Shandong University of Science and Technology and the company of Supersurs mobile surveying service (shown in Fig. 2). It is equipped with a three-system receiver of NovAtel Propak6 and inertial measurement unit (IMU) of span LCI type. The type

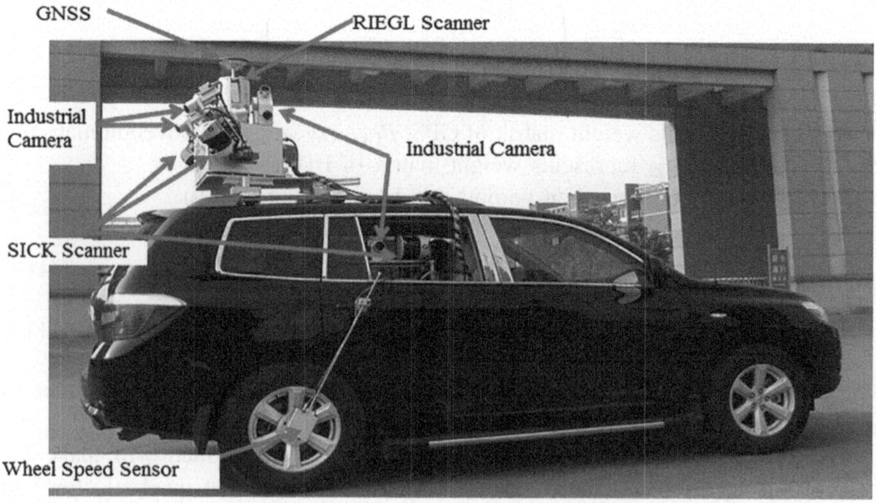

Fig. 2 V-Surs I vehicle-borne three-dimensional mobile surveying system

of antenna and the sampling rate of the receiver are NOV703GGG and 0.05 s, respectively. The base station of this experiment was set up on the roof of S2 laboratory in Shandong University of Science and Technology.

The ratio >2 and $\sigma_0 < 0.01$ were used as the conditions that the wide-lane ambiguity was fixed successfully in this paper. When the robust estimation was used to process gross error, k_0 equaled 1.2 and k_1 equaled 2.8. In addition, when the degree of freedom was less than 3, the robust estimation was not used.

The trajectory of this experiment is shown in Fig. 3. The real-time baseline length between base station and user station is shown in Fig. 4. As shown in Fig. 4, the baseline length is more than 10 km most of the time.

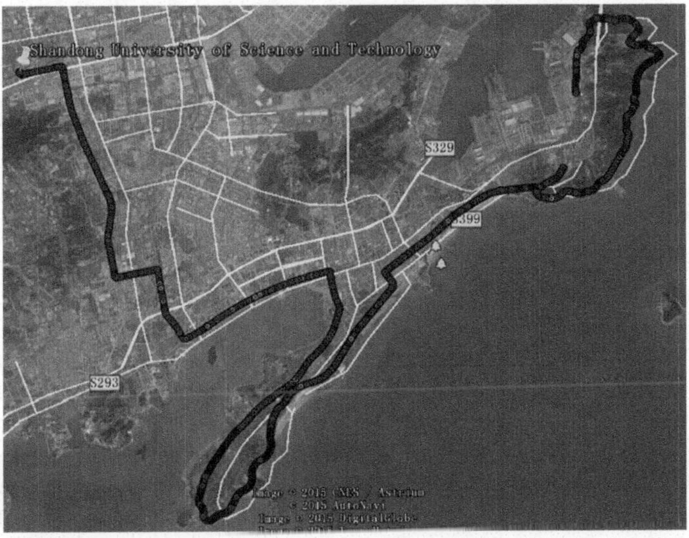

Fig. 3 The trajectory of this experiment

Fig. 4 Baseline length between base station and user station

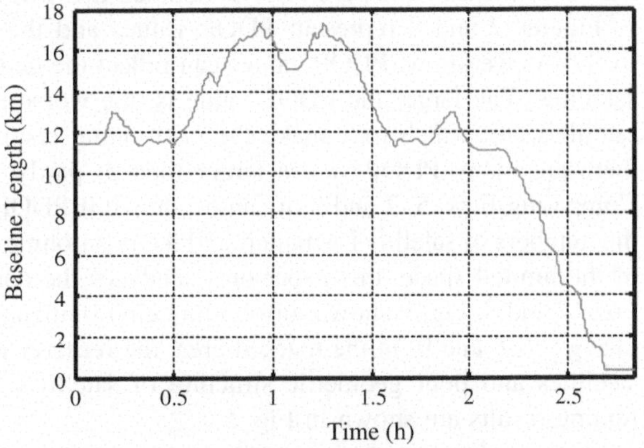

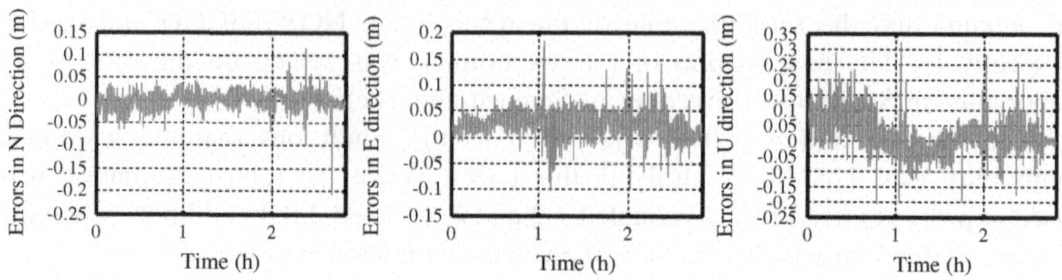

Fig. 5 Position error between this method and benchmark in different directions

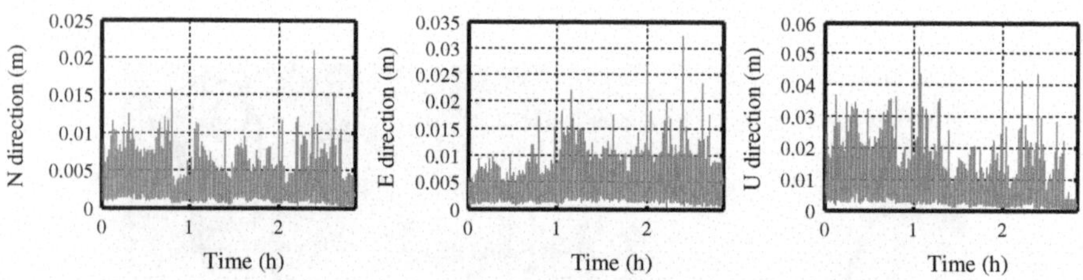

Fig. 6 Standard error of this method in different directions

We used GNSS/INS tightly coupled resolution of Inertial Explorer 8.60 software (IE 8.60) to resolve this data, and the results were used as true values. The differences in *N*, *E*, and *U* directions between this method and IE 8.60 are shown in Fig. 5, as well as the standard errors in *N*, *E*, and *U* directions are shown in Fig. 6. As shown in Fig. 5, the differences are mostly within 5 cm in *N* direction, and *E* direction is slightly worse than *N* direction, however, the differences largely change in *U* direction. From Fig. 6, it can be noticed that standard errors are basically within 1 cm in *N* direction, within 2 cm in *E* direction, mostly within 3 cm in *U* direction. As thus, it can be concluded that this method has higher precision. Comparing Figs. 5 and 6, it is noticed that if this time has a larger difference in Fig. 5, at the same time there has been a larger standard error in Fig. 6. Based on these results, it can be concluded that this method has higher reliability.

Figures 7 and 8 represent PDOP values and the number of satellites, respectively. As we know, PDOP values can reflect the geometric structure of the visible satellites. The larger the PDOP value is, the worse the geometric structure of the satellites are, which may make the positioning result worse. As shown in Fig. 7, there are many PDOP values larger than 3, partly larger than the acceptable 6. Comparing Figs. 5, 7 and 8, it can be shown if PDOP value is larger at some time, the numbers of satellite is smaller, and the positioning results are also poor. Because of the limited space, this paper only analyzes the condition of the marked area in Figs. 7 and 8. Fig. 9 shows surrounding environment of the marked area. As shown in Figs. 7, 8 and 9, in the marked area, the receiver is under the corridor with less satellites and poor geometric structure of satellites, at the same time poor positioning results are shown in Fig. 5.

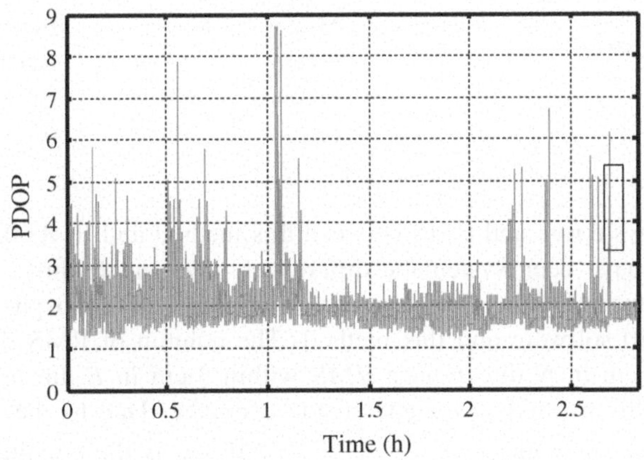

Fig. 7 PDOP values of observation epoch

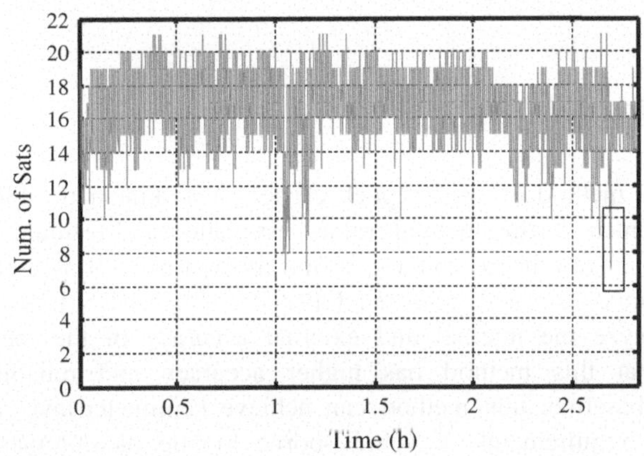

Fig. 8 The number of satellites

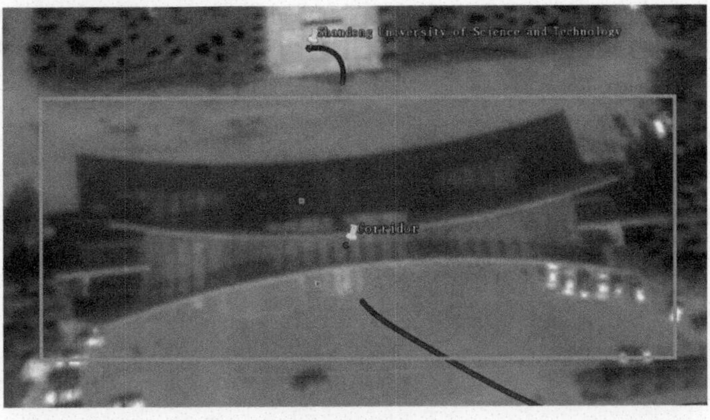

Fig. 9 The surrounding environment of the marked area

Table 1 The accuracy statistics of single-epoch positioning results

Statistics	Max/m	Min/m	Mean/m	RMS
N	0.1155	−0.2066	−0.0003	0.0123
E	0.1858	−0.1016	0.0238	0.0276
U	0.3228	−0.2054	0.0183	0.0461

Table 1 shows errors and RMS between this method and IE 8.60. As shown in Table 1, there have been systematic differences in E and U directions from this column of mean, and this is because the error model of troposphere is different between IE 8.60 software and this method. The column of RMS demonstrates a RMS within 2 cm in N direction, a RMS within 3 cm in E direction, and these illustrate that this method has high outside accuracy. Due to the medium-short baseline used in this paper, $\sqrt{N^2 + E^2 + U^2} < 10$ cm is the condition that ambiguity was fixed successfully. Experimental results show that the successful rate of ambiguity fixed is 94.87 % in this data, and single epoch is convergence at the beginning of this experiment.

4 Conclusion

In this paper, a method of single-epoch carrier-phase kinematic differential positioning is presented. In this method, robust estimation and residual correction are used to further ensure the correctness of the method as well as the accuracy. To verify this method, the real collected GPS/GLONASS/BDS kinematic data are applied to analyze the internal and external accuracy of the method. Results demonstrate that this method has higher accuracy and reliability. For the medium-short baseline, this method can achieve centimeter-level precision and fully meet the requirements of vehicle-borne mobile measurement. Thus, this method can be used for high-accuracy positioning, and it has great practical significance. However, this paper shows that under some severe environment (i.e., PDOP values become larger), the accuracy decreases. Moreover, there are certain epochs when the number of satellites is not sufficient to use the least squares estimation before the receiver lock enough satellites again, this often needs several epochs to fully fix the ambiguities, which needs further study.

Acknowledgments This work is supported by State Key Laboratory of Geodesy and Earth's Dynamics (Institute of Geodesy and Geophysics, CAS) (No. SKLGED2015-3-1-E), together with a Project of Shandong Province Higher Educational Science and Technology Program (No. J13LH04).

References

1. Fan L, Zhong SM, Ou JK et al (2013) Precision analysis of standard point positioning by COMPASS and GPS observations. The 4th China satellite navigation conference (CSNC), Wuhan
2. Li YH, Shen YZ, Li BF (2011) GPS kinematic relative positioning based on LAMBDA. Eng. Surveying Mapp 20(1):6–10
3. Li ZH, Huang JS (2010) GPS surveying and data processing, 2nd edn. Wuhan University Press, Wuhan
4. Tang WM, Sun HX, Chen J (2008) Method of ratio accumulation of GNSS and fast double-differenced ambiguity resolution. J Geodesy Geodyn 28(5):68–72
5. Liu LL, Liu JY, Li GH (2005) Rapid ambiguity resolution on-the-fly for single frequency receiver. Geomatics Inf Sci Wuhan Univ 30(10):885–887
6. Liu XD (2011) Research on data processing method of dynamic GPS precise positioning. Master thesis, The PLA Information Engineering University, Henan
7. Wang RQ (2004) The theoretical research of kinematics positioning of GPS. PhD thesis, Central South University, Hunan
8. Li ZH, Zhang XH (2009) New techniques and precise data processing methods of satellite navigation and positioning. Wuhan University Press, Wuhan
9. Wang SL (2013) Research on key technology of multi-GNSS ground based augmentation system. PhD thesis, Southeast University, Jiangsu
10. Wang SL, Wang Q, Pan SG et al (2011) New cycle-slip detection and repair method for CORS network. J Chin Inertial Technol 19(5):590–593
11. Zhang XH, Guo F, Pan Li et al (2012) Real-time quality control procedure for GNSS precise point positioning. Geomatics Inf Sci Wuhan Univ 37(8):940–944

Performance Analysis of Real-Time BDS Clock Estimation with Different Orbit Accuracy

Wenju Fu, Qin Zhang, Guanwen Huang, Pingli Li, Bobin Cui and Rui Tu

Abstract Based on un-differenced model of clock estimation, the mathematic model of real-time clock estimation influenced by orbit error is derived, and then the estimation accuracy of BDS real-time clock and real-time positioning accuracy are analyzed. The results show that real-time BDS clock accuracy achieves 0.30 ns using the rapid orbit product and the corresponding 2-D and 3-D real-time positioning accuracy are 0.36 and 0.48 m, respectively. With 0.1 and 0.2 m orbit error influenced, the decreased accuracy of real-time clock estimation is 0.05 and 0.14 ns, and the decreased positioning accuracy is 0.057 and 0.154 m, respectively.

Keywords BDS · Real-time clock estimation · Orbit accuracy · Un-differenced

1 Introduction

Real-time satellite clock error is a necessary product for real-time applications. Its availability and accuracy directly affect real-time applications, such as real-time precise point positioning [1–3]. In recent years, many scholars had done a lot of research about

W. Fu (✉) · Q. Zhang · G. Huang · P. Li · B. Cui
School of Geology Engineering and Geomatics, Chang'an University,
No.126, Yanta Road, Xi'an, China
e-mail: fuwenju2009@163.com

Q. Zhang
e-mail: zhangqinle@263.net.cn

G. Huang
e-mail: huang830928@163.com

P. Li
e-mail: 707696277@qq.com

B. Cui
e-mail: 626665594@qq.com

R. Tu
Chinese Academy of National Timing Service Center, Shu Yuan Road, Xi'an, China
e-mail: turui-2004@126.com

J. Sun et al. (eds.), *China Satellite Navigation Conference (CSNC) 2016 Proceedings: Volume I*, Lecture Notes in Electrical Engineering 388,
DOI 10.1007/978-981-10-0934-1_9

the real-time clock error estimation algorithm, and achieved many results which show that the real-time clock error estimation accuracy can reach about 0.1 ns [4–11]. However, only GPS clock was evaluated in the above results. Therefore, performance assessment of real-time BDS clock estimation should be interesting.

In order to achieve an average of 8–9 visible satellites in the Asia-Pacific region, BDS is designed as a constellation that includes three kinds of satellite orbit, such as GEO, IGSO, and MEO. As a result, the difficulty of estimating precise orbit is increased, especially GEO satellites. Therefore, an influence model of orbit error for real-time clock estimation is derived, and preliminary results of BDS satellite real-time clock estimation are evaluated, and the estimation accuracy of BDS real-time clock error with different orbit accuracy is analyzed, and then the real-time precise point positioning is carried out using corresponding clock results.

2 The Error Equation of Real-Time Clock Estimation

The un-differenced model is simple and easy to implement, and it has a high estimation accuracy. Usually, the model is used to estimate real-time clock with ionosphere-free combined observations [4, 6, 8, 9]. For this model, station coordinates and satellite orbits are fixed in the observation equation while receiver clock, satellite clock, tropospheric wet delay, and ambiguity parameters are estimated. During a given time, tropospheric wet delay parameters change slowly, and ambiguity parameters are constants when there are not cycle slips, so they are treated as piecewise time-invariant parameters. Based on satellite elevation angle, the ionosphere-free linear combined observations of pseudorange and carrier phase are given with different prior weights. Therefore, the error equation of real-time clock estimation can be expressed

$$V = AX + BY - L \quad P \tag{1}$$

where X is the receiver and satellite clock parameter vector, Y is the troposphere and ambiguity parameter vector, A and B are the corresponding coefficient matrixes of X and Y, L and P are the observation vector and weight matrixes, and V is the residual vector. The observation vector can be represented as follows:

$$L = L_{IF} - P_0 + (T - \mathrm{REL} + \mathrm{SPCV} + \mathrm{RPCV} + \mathrm{PWP}) \tag{2}$$

where L_{IF} denotes ionosphere-free combined observation, P_0 represents the geometric distance between satellite phase center and station coordinates after the corrections of Earth solid tide, ocean tide, pole tide, receiver antenna phase center offsets, the Earth rotation and satellite antenna phase center offsets. T denotes troposphere delay corrections, REL denotes the relativistic corrections, SPCV denotes satellite antenna phase center variations. RPCV denotes receiver antenna phase center variations, PWP represents phase windup corrections.

The observation vector not only contains troposphere error, phase windup error, receiver and satellite phase center offsets and variations, tidal error, relativistic effects, and so on, but also includes station coordinates error and satellite orbit error. However, as an essential product for the real-time clock estimation, the real-time satellite orbit accuracy directly determines the real-time clock accuracy. Therefore, it is necessary to analyze the impact of satellite orbit error on real-time clock estimation. For the limited space, the influence of orbit radial error to clock estimation is analyzed.

3 Components of Orbit Radial Error in Space Rectangular Coordinate System

Orbit radial error belongs to the orbit coordinate system while BDS precise satellite orbit products are space rectangular coordinate system. Therefore, in order to obtain three-dimensional components of orbit radial error in the space rectangular coordinate system, conversion formula between two coordinate systems should be derived. Assuming that $R = (X \quad Y \quad Z)^T$ and $V = (V_x \quad V_y \quad V_z)^T$ stand for satellite position and speed, the unit vectors of orbit along, cross, and radial direction are

$$I_a = \frac{R}{|R|}, \quad I_c = \frac{R \times V}{|R \times V|}, \quad I_r = I_a \times I_c \tag{3}$$

where "×" is cross product symbol. Assuming that $\delta_{\mathrm{xyz}} = (\delta_x \quad \delta_y \quad \delta_z)^T$, $\delta_{\mathrm{acr}} = (\delta_a \quad \delta_c \quad \delta_r)^T$ stand for error vector and $M = (I_a \quad I_c \quad I_r)^T$ represents rotation matrix. For M is the unit orthogonal matrix, there is $M^{-1} = M^T$, the relationship can be expressed

$$\begin{aligned} \delta_{\mathrm{acr}} &= M \cdot \delta_{\mathrm{xyz}} \\ \delta_{\mathrm{xyz}} &= M^T \cdot \delta_{\mathrm{acr}} \end{aligned} \tag{4}$$

where "·" is dot product symbol. When the radial error is introduced, there is $\delta_{\mathrm{acr}} = (0 \quad 0 \quad \delta_r)^T$. Assuming that $I_r = (r_1 \quad r_2 \quad r_3)^T$, the relationship can be expressed

$$\delta_{\mathrm{xyz}} = M^T \cdot \begin{pmatrix} 0 \\ 0 \\ \delta_r \end{pmatrix} = \begin{pmatrix} r_1 \cdot \delta_r \\ r_2 \cdot \delta_r \\ r_3 \cdot \delta_r \end{pmatrix} \tag{5}$$

In order to evaluate the influence of orbit radial error on BDS real-time clock estimation, satellite position and velocity are computed based on the satellite precise ephemeris. Then, rotation matrix is calculated according to formula (3). Finally, three-dimensional components of radial error in space rectangular coordinate system are calculated according to formula (5), and are added to the corresponding precise orbit. The influence of the radial error on real clock error estimation can be evaluated using the orbit data with the introduced error.

4 The Influence Model of Orbit Radial Error on Clock Estimation

In order to investigate the impact value of the orbit radial error on the real-time clock estimation, the influence model of orbit error is derived based on error equation of the real-time clock estimation. First, error equation of the real-time clock error estimation is introduced into orbit error, and the error equation with orbit radial error can be express as

$$V = AX + BY - (L + \Delta L) \quad \mathrm{P} \tag{6}$$

The normal equation of real-time clock estimation can be obtained using the least squares principle.

$$\begin{bmatrix} N_{XX} & N_{XY} \\ N_{YX} & N_{YY} \end{bmatrix} \begin{bmatrix} \hat{X} \\ \hat{Y} \end{bmatrix} = \begin{bmatrix} U_X + U_{\Delta X} \\ U_Y + U_{\Delta Y} \end{bmatrix} \tag{7}$$

where ΔL denotes the constant matrix caused by orbit radial error, and $N_{XX} = A^T PA$, $N_{XY} = A^T PB$, $N_{YX} = B^T PA$, $N_{YY} = B^T PB$, $U_X = A^T PL$, $U_Y = B^T PL$, $U_{\Delta X} = A^T P\Delta L$, $U_{\Delta Y} = B^T P\Delta L$. Assuming that $\overline{N}_{YY} = N_{YY} - N_{YX}N_{XX}^{-1}N_{YX}$, $\overline{U}_Y = U_Y - N_{YX}N_{XX}^{-1}U_X$, $\overline{U}_{\Delta Y} = U_{\Delta Y} - N_{YX}N_{XX}^{-1}U_{\Delta X}$, receiver clock and satellite clock parameters $\hat{X}$, and troposphere and ambiguity parameters $\hat{Y}$ can be calculated by the elimination parameter method.

$$\begin{aligned} \hat{Y} &= \overline{N}_{YY}^{-1}\overline{U}_Y + \overline{N}_{YY}^{-1}\overline{U}_{\Delta Y} \\ \hat{X} &== N_{XX}^{-1}\left(U_X - N_{XY}\overline{N}_{YY}^{-1}\overline{U}_Y\right) + N_{XX}^{-1}\left(U_{\Delta X} - N_{XY}\overline{N}_{YY}^{-1}\overline{U}_{\Delta Y}\right) \end{aligned} \tag{8}$$

Obviously, it can be seen from formula (8) that the impact value of orbit radial error on the real-time clock error estimation is $N_{XX}^{-1}\left(U_{\Delta X} - N_{XY}\overline{N}_{YY}^{-1}\overline{U}_{\Delta Y}\right)$.

5 Numerical Examples

In order to analyze the BDS real-time satellite clock error estimation accuracy with different orbit error, 20 stations with 1 s sampling interval are adopted. The data were provided by MGEX (the IGS multi-GNSS experiment network) for September

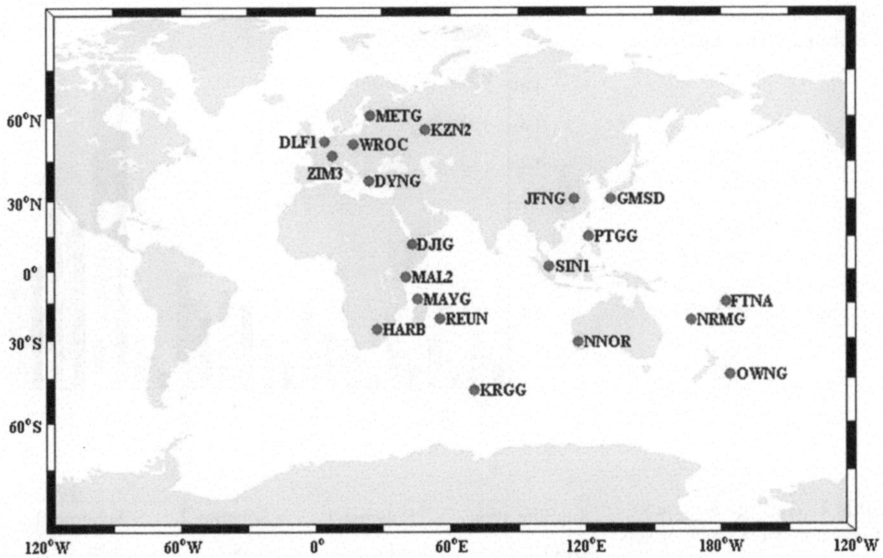

Fig. 1 Distribution of station used for real-time clock estimation (Color figure online)

20, 2015 (DOY 263). Then the rapid orbit products provided by the German Research Centre for Geosciences (GFZ) and the ones introduced different orbit radial error are used to estimate real-time satellite clock error, respectively. Taking GFZ rapid clock products for reference, standard deviation (STD) of real-time clock error estimation is investigated. Real-time positioning accuracy of three stations involved in clock estimation and three stations not involved in clock estimation are carried out using the rapid clock and the calculated real-time clock. The stations used for real-time clock estimation are shown in Fig. 1.

For real-time clock estimation, sequential least squares algorithm is used in real-time [14], and the station GMSD receiver clock is taken for the reference. The observation data are excluded by 7° elevation mask, and the TurboEdit algorithm is used to detect cycle slips in real-time [12, 13], and the median method is used to detect and eliminate gross data and larger noise data according to post-fit residuals of the observations. In order to analyze the estimation accuracy of real-time clock under different orbit accuracy, three schemes are designed. Scheme 1 uses GFZ rapid orbit products, and scheme 2 uses the one which is introduced 0.1 m radial error, and scheme 3 uses the one which is introduced 0.2 m radial error. The accuracy results of real-time satellite clock estimation for three schemes are shown in Fig. 2 and Table 1.

It can be seen from Fig. 2 that estimation accuracy of GEO and IGSO satellite clock is better than the MEO satellites, especially the C11 satellite, for station number in view of the GEO satellites and IGSO satellites is more and stable, and station number in view of MEO satellites is less, even no stations sometimes. Table 1 shows that the average accuracy of all BDS satellites is 0.42 ns, and the

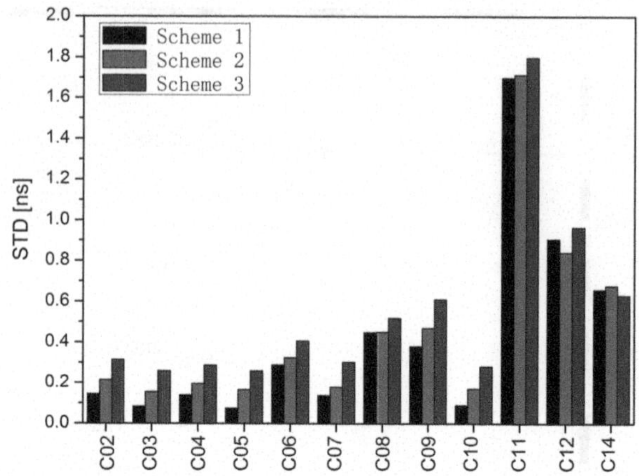

Fig. 2 Estimation accuracy of real-time BDS clock offsets (Color figure online)

Table 1 Average accuracy of real-time BDS clock offsets

Satellite type	Scheme 1	Scheme 2	Scheme 3
GEO	0.11	0.18	0.28
IGSO	0.27	0.32	0.42
All satellites	0.42	0.46	0.55
Without C11	0.30	0.35	0.44

average accuracy of the GEO satellites is 0.11 ns. If the C11 satellite is removed, the average accuracy of BDS satellites is 0.30 ns. The real-time satellite clock estimation accuracy of three schemes is 0.30, 0.35, and 0.44 ns, respectively. It can be seen that the clock estimation accuracy decreases 0.05 and 0.14 ns if the orbit radial error is 0.1 and 0.2 m, respectively.

In order to verify the real-time clock estimation performance of three schemes furtherly, GFZ rapid clock products and the calculated BDS real-time clock products of three schemes are used to evaluate real-time precise point positioning of three stations (GMSD, JFNG, PTGG) involved in the clock estimation and three stations (SHA1, KUN1, LHA1) not involved in the clock estimation. Figure 3 is real-time kinematic positioning error at station GMSD, and the other stations are the same.

It can be seen from Fig. 3 that the convergence time of real-time positioning with GFZ rapid clock is the fastest. The convergence time in vertical direction with real-time clock under three schemes is significantly worse than that of GFZ rapid clock, but the positioning accuracy is nearly the same after the convergence time, especially in the horizontal direction. The positioning accuracy of each clock product can achieve 0.5 m after two-hour convergence time. Real-time kinematic positioning accuracy after convergence time at six stations is investigated in Fig. 4 and the positioning accuracy of horizontal (2-D) and three-dimensional (3-D) directions is shown in Table 2.

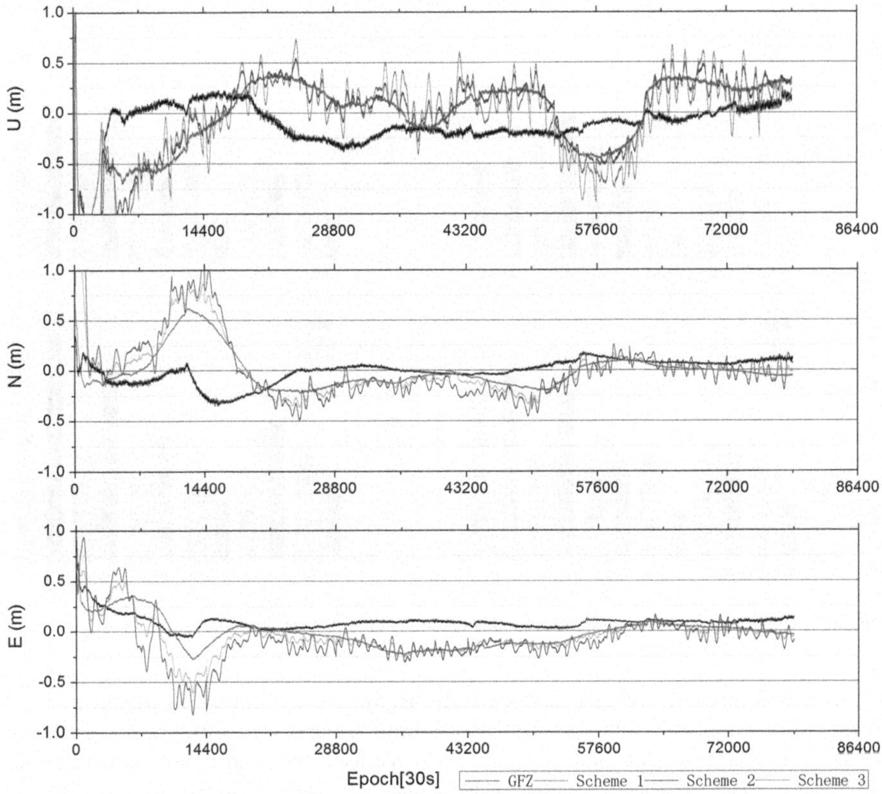

Fig. 3 Real-time kinematic positioning error at station GMSD

Table 2 Real-time kinematic positioning accuracy of 2-D and 3-D

Station	GFZ		Scheme 1		Scheme 2		Scheme 3	
	2-D	3-D	2-D	3-D	2-D	3-D	2-D	3-D
GMSD	0.137	0.216	0.221	0.352	0.292	0.414	0.368	0.498
JFNG	0.240	0.320	0.368	0.448	0.457	0.520	0.541	0.602
PTGG	0.087	0.183	0.173	0.318	0.213	0.360	0.280	0.447
SHA1	0.263	0.416	0.381	0.463	0.455	0.522	0.524	0.608
KUN1	0.466	0.563	0.528	0.611	0.572	0.638	0.628	0.702
LHA1	0.312	0.413	0.466	0.659	0.570	0.738	0.700	0.916
Average	0.251	0.352	0.356	0.475	0.427	0.532	0.507	0.629

It can be seen from Fig. 4 that real-time kinematic positioning accuracy in East, North, and up directions of GFZ rapid clock products is about 0.20, 0.14, and 0.24 m. The one of real-time clock products is about 0.26, 0.23, and 0.31 m, of which the stations involved in clock estimation are slightly better than the stations

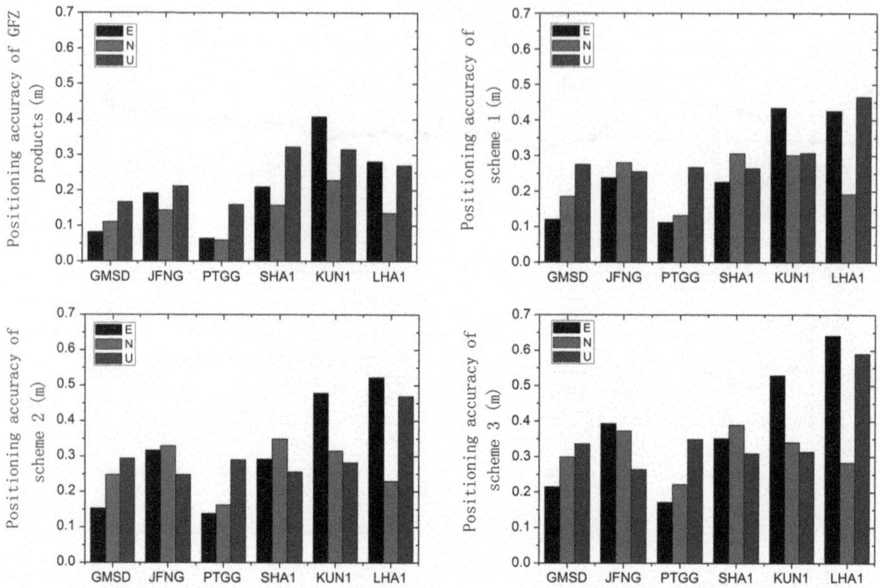

Fig. 4 Real-time kinematic positioning accuracy at six stations (Color figure online)

not involved in clock estimation, especially in the vertical direction. It can be seen from Table 2 that the 2-D and 3-D positioning accuracy of GFZ rapid clock are 0.251 and 0.352 m while the accuracy of real-time clock are 0.356 and 0.475 m, respectively. With the introduction of different radial orbit error, the performance of real-time clock is gradually reduced in the positioning performance. When the orbit radial error is 0.1 and 0.2 m, the 2-D positioning accuracy decreases 0.071 and 0.151 m, and 3-D positioning accuracy decreases 0.057 and 0.154 m, respectively.

6 Conclusions

Based on the observation data of MGEX stations, a real-time BDS satellite clock estimation algorithm is achieved, and preliminary results of real-time BDS satellite clock are obtained, and the accuracy of real-time satellite estimation clock under different orbital radial error is evaluated, and the corresponding real-time kinematic precise point positioning estimate is carried out. Some conclusions can be obtained: (1) The real-time BDS clock estimation accuracy can be achieved 0.30 ns with 20 stations and the real-time positioning accuracy can reach 0.36 and 0.48 m in 2-D and 3-D, respectively. (2) The influence of 0.1 and 0.2 m orbit radial error is about 0.05 and 0.14 ns on the real-time clock estimation, and is about 0.057 and 0.154 m on real-time kinematic positioning.

Only the influence of orbit radial error on the real-time clock estimation is analyzed, however, it is necessary to analyze the influence of orbit error in along and cross directions in order to obtain a comprehensive result. The improvement of real-time BDS clock estimation accuracy is limited due to the number of stations which is less. In addition, for the orbit error is introduced by random, the corresponding positioning error curve is a random fluctuation. Similarly, when the rapid clock products and real-time estimated clock products are used for positioning, the difference values of the corresponding positioning results are likely to be associated with the clock products. Therefore, it should be interesting to reveal the numerical relationship further.

Acknowledgements Thanks for MGEX provided BDS observation data and GFZ provided rapid orbit and clock product, this work was supported partly by the Program of the National Natural Science Foundation of China (41104019, 41104022, 41304033 and 41504006), and the Grand Projects of the Beidou-2 System (GFZX0301040308), the Special Fund for Basic Scientific Research of Central Colleges (Grant No. 2014G1261051 and 310826165014, Chang'an University), Chinese Academy of Sciences (CAS) programs of "Pioneer Hundred Talents", "Light of West China" and "Youth Innovation Promotion Association".

References

1. Zumberge JF, Heflin MB, Jefferson DC et al (1997) Precise point positioning for the efficient and robust analysis of GPS data from large networks. J Geophys Res 102(B3):5005–5017
2. Mohamed E, Salim A (2015) Performance of real-time precise point positioning using IGS real-time service. GPS Solut. First online: 09 June 2015. doi:10.1007/s10291-015-0467-z
3. Hadas T, Bosy J. IGS RTS precise orbits and clocks verification and quality degradation over time. GPS Solut 19(1):93–105. doi:10.1007/s10291-014-0369-5
4. Ge MR, Chen JP, Douša J, Gendt G, Wickert J (2012) A computationally efficient approach for estimating high-rate satellite clock corrections in realtime. GPS Solut 16(1):9–17. doi:10.1007/s10291-011-0206-z
5. Bock H, Dach R, Jäggi A, Beutler G (2009) High-rate GPS clock corrections from CODE: support of 1 Hz applications. J Geodesy 83(11):1083–1094. doi:10.1007/s00190-009-0326-1
6. Zhang XH, Li XX, Guo F (2011) Satellite clock estimation at 1 HZ for realtime kinematic PPP applications. GPS Solut 15(4):315–324. doi:10.1007/s10291-010-191-7
7. Hauschild A, Montenbruck O (2009) Kalman-filter-based GPS clock estimation for near realtime positioning. GPS Solut 13(3):173–182. doi:10.1007/s10291-008-0110-3
8. Lou Y, Shi C, Shou X, Ye S (2009) Realization and analysis of GPS precise clock products. Geomatics Inf Sci Wuhan Univ 34(1):88–91
9. Li X, Xu Y, Wang L (2010) Undifferenced precise satellite clock error estimation and precision analysis. Geomatics Inf Sci Wuhan Univ 35(6):661–664
10. Li L, Zhu J, Chen Y, Kuang C et al (2011) Convergent analysis of real-time estimation of satellite clock bias. J Geodesy Geodyn 31(4):80–84
11. Li H, Wang J, Wang H, Li B (2010) Precise clock error estimation of GPS satellite and analysis based on GNSS network. Geomatics Inf Sci Wuhan Univ 35(8):P1001–P1003
12. Blewitt G (1990) An automatic editing algorithm for GPS data. Geophys Res Lett 17(3): 199–202

13. Liu N, Xiong Y, Xu S (2011) Detection and repair of cycle slips using improved turboedit algorithm and Chebyshev polynomial method. Geomatics Inf Sci Wuhan Univ 36(12):1500–1503
14. Yang YX, He HB, Xu GC (2001) Adaptively robust filtering for kinematic geodetic positioning. J Geodesy 75(2–3):109–116. doi:10.1007/s001900000157

Correlation Analysis Between Ionosphere Anomaly and Earthquake

Yulong Kong, Hongzhou Chai and Huarun Wang

Abstract This document took the long sequence of IGS ionospheric VTEC (vertical total electron content unit) products as original data, based on the theory research of various abnormally detection methods, we could select the optimal method to process VTEC products, combined with earthquake cases of which the magnitude greater than Ms5.0 in global range from 2003 to the present. According to the principle of mathematical statistics, we found the relationship between the abnormal of ionosphere and the factors of earthquakes. Through calculating the correlation coefficient and the single factor variance analysis between earthquake factors and the ionosphere anomalies, we not only confirmed that the earthquake has an effect on ionosphere anomalies, but also proved the conclusion that the magnitude has the positive correlation with the positive ionosphere anomalies, and under the condition of single factor analysis, we confirmed that the magnitude made a significant impact on the ionosphere anomalies before the earthquake happened.

Keywords Ionosphere anomaly · Earthquake · Correlation

1 Introduction

As an important part of the human leaving environment, ionosphere has an important impact and significance on the lives of modern society. Earthquake, because of its unexpected and enormous destructive power, has always been one of the most important natural disasters that we human have to face. A wide range of organizations and scholars is working on the earthquake monitoring and forecasting. At present, the forecasting methods for earthquake generally rely on the research of the earthquake precursor, such as hydro-geochemical anomaly, gravity anomaly, magnetic anomaly and other natural phenomena. It is the first time that the scientist Barnes found the disturbances in the ionosphere when the earthquake

Y. Kong (✉) · H. Chai · H. Wang
Institute of Surveying and Mapping, Zhengzhou 450001, China
e-mail: 492327387@qq.com

J. Sun et al. (eds.), *China Satellite Navigation Conference (CSNC) 2016 Proceedings: Volume I*, Lecture Notes in Electrical Engineering 388,
DOI 10.1007/978-981-10-0934-1_10

occurred in Alaska in 1964. It is the regional study of the ionospheric precursors of earthquake [1]. The theme of this article is to analyse the relationship between the magnitude factors of 18 worldwide earthquake factors and the ionosphere anomalies through the anomalous detection with the ionosphere VTEC products of IGS. In this article, we make an improvement of the traditional anomalous detection method, and for the first time we try to quantitate the ionosphere anomalies, combined with the mathematical statistics method, to find out the exact statistical relationship between ionosphere anomaly and the earthquake.

2 Introduction and Improvement of Anomalous Detection Method

2.1 Traditional Anomalous Detection Method

Recently, there are four frequently used detection methods to detect the anomaly of VTEC. They are the average method, the median method, interquartile range method and slippage windows method [3]. In fact, the slippage windows method is the most widely used of them.

When the slippage windows method is used for processing the ionosphere VTEC data, the time as well as the width of the window which is used to calculate is limited. To select a set of data from the time windows as the sample data to calculate the slippage average and the slippage standard deviation we have the functions as follow:

$$\bar{X}_k = \frac{1}{N}\sum_{i=k+1}^{k+N} X_i \tag{1}$$

$$\sigma_k = \sqrt{\frac{\sum_{j=1}^{N}\left(X_{i,j} - \bar{X}_k\right)^2}{n-1}} \tag{2}$$

where N is the width of the time windows, k is recorded as the position of the detection point among all of the original VTEC data, X_i means that it is the 'i'th data in certain time window. Similarly, we can understand the meaning of $X_{i,j}$, where 'i' means the moment and 'j' means the day of current time window. Then the slippage average $\bar{X}_k$ is used to calculate σ_k the standard deviation of the VTEC in certain time window. The upper and lower threshold values can be computed with the $\bar{X}_k$ and σ_k, as follow:

$$\mathrm{VTEC}^i_{\mathrm{up}} = \bar{X}_k + 2\sigma_k \tag{3}$$

$$\mathrm{VTEC}^i_{\mathrm{down}} = \bar{X}_k - 2\sigma_k \tag{4}$$

Then the final anomaly detection functions are

$$F_1 = \mathrm{VTEC}^i_{\mathrm{down}} - X_{N+1+k} \quad (5)$$

$$F_2 = X_{N+1+k} - \mathrm{VTEC}^i_{\mathrm{up}} \quad (6)$$

where F_1 represents the negative anomalies, oppositely, F_2 represents the positive anomalies. Namely, the detection result in the current time window is the negative anomaly when $F_1 > 0$, while it is the positive anomaly when $F_2 > 0$.

2.2 *Improvement of Anomalous Detection Method*

Due to the fact that the sample data which are calculated by traditional anomalous detection are the continuous and factual observation data. That is to say, the data which are used to calculate $\bar{X}_k$ and σ_k, include anomalous sequence itself. Then it will make the upper and lower threshold values ($\mathrm{VTEC}^i_{\mathrm{up}}$ and $\mathrm{VTEC}^i_{\mathrm{down}}$) deviate from the actual range of VTEC.

To eliminate the influence of anomaly values to calculate $\bar{X}_k$ and σ_k, two improvements about slippage windows method are made to acquire more stationary sequence.

1. Detecting the data of initial window to assure whether the anomalies are existent or not. For the detected anomalies, the neighbouring normal VTEC value replaces the anomalous one as the interpolation [4], thereby, a stabilized adjustment is accomplished from the start.
2. With the time window slipping, the another day's detected VTEC data will be added into the new window to compose a new window sequence of data to detect the anomalies of the next day. According to this fact, we can replace the current anomalous value with the normal neighbouring value so as to keep the inner data stationary in the window while it is slipping.

In this document, the improved method is used for ionosphere anomaly detection of all the earthquakes.

3 Principle of Correlation Analysis

3.1 *Calculation of Correlation Factor*

For two-dimensional random variable (*X*, *Y*), ρ_{XY} and $\mathrm{Cov}(X, Y)$ are defined as correlation coefficient and covariance of random variable *X* and *Y*, determined by [2]

$$\mathrm{Cov}(X,Y) = E\{[X - E(X)][Y - E(Y)]\} \tag{7}$$

$$\rho_{XY} = \frac{\mathrm{Cov}(X,Y)}{\sqrt{D(X)}\sqrt{D(Y)}} \tag{8}$$

To assume that a linear function of X (a + bX) is the approximate expression of Y, in this way, the mean squared error 'e' could be used to measure the degree of approximation between a + bX and Y, the differential equations are as follow:

$$\begin{cases} \frac{\partial e}{\partial a} = 2a + 2bE(X) - 2E(Y) = 0 \\ \frac{\partial e}{\partial b} = 2bE(X^2) - 2E(XY) + 2aE(X) = 0 \end{cases} \tag{9}$$

Then we have

$$\begin{cases} a_0 = E(Y) - E(X)\frac{\mathrm{Cov}(X,Y)}{D(X)} \\ b_0 = \frac{\mathrm{Cov}(X,Y)}{D(X)} \\ e_{\min} = (1 - \rho_{XY}^2)D(Y) \end{cases} \tag{10}$$

According to the Eq. (3.4), the MSE is the monotonically decreasing function of $|\rho_{XY}|$, that is to say, when the $|\rho_{XY}|$ is bigger, 'e' is smaller, and the linear relationship is closer between X and Y. Namely, the Eq. (3.2) could be used to compute the correlation between the ionosphere anomaly and the earthquake magnitude.

3.2 *One-Way Analysis of Variance*

However, the factors influencing the ionosphere anomaly is complex and nonlinear in fact. It means that just figure out the correlation coefficient is insufficient to prove certain factor has the influence on ionosphere anomaly. Using one-way ANOVA statistical analysis could obtain the significant effect of certain factor on experimental result.

For one-way ANOVA statistical analysis, we suppose that the factor A has s levels in an experiment, denoted as $A_1, A_2, \ldots A_s$. Performing independent experiments for n times (n > 2) would draw the results as follow: (Table 1).

Table 1 Single factor analysis of variance

Level result	A_1	…	A_s
	X_{11}	…	X_{1s}
	$\vdots$	$\vdots$	$\vdots$
	X_{n1}	…	X_{ns}
Sample mean	$\bar{X}_1$	…	$\bar{X}_s$

Firstly, to judge the mean values of s ensembles $(N(\mu_1, \sigma^2)\ldots N(\mu_s, \sigma^2))$ are equal or not, namely

$$H_0 : \mu_1 = \mu_2 = \ldots = \mu_s$$

$$H_1 : \mu_1, \mu_2 \ldots \mu_s \text{ are not all equal.}$$

Then, we should estimate the unknown parameters $\mu_1, \mu_2, \ldots \mu_s, \sigma^2$. Three statistic values (S_E, S_A, S_T) are introduced to confirm the refused domain of the assumption above. They are denoted as follow:

$$\begin{cases} S_E = \sum_{j=1}^{s}\sum_{i}^{n} (X_{ij} - \bar{X}_j)^2 \\ S_A = \sum_{j=1}^{s} n\bar{X}_j^2 - n\bar{X}^2 \\ S_T = S_A + S_E \end{cases} \tag{11}$$

where S_E called sum of squared errors, S_A is efficacy square sum, and according to principle of statistic, we have

$$\begin{cases} \frac{S_E}{\sigma^2} \sim \chi^2(n-s) \\ E\left(\frac{S_E}{n-s}\right) = \sigma^2 \end{cases} \tag{12}$$

When H_0 is true.

$$\begin{cases} \frac{S_A}{\sigma^2} \sim \chi^2(n-s) \\ E\left(\frac{S_A}{s-1}\right) = \sigma^2 \end{cases} \tag{13}$$

Above all, we can get a fraction $F = \frac{S_A/(s-1)}{S_E/(n-s)}$, of which the denominator and the numerator are mutual independence. For the denominator, whenever the assumption is true or not, its mathematical expectation is σ^2. We know that if H_0 is true, the mathematical expectation of numerator is σ^2 as well, and if H_0 is false, the value of numerator would be lager. Meanwhile, when H_0 is true

$$F = \frac{S_A/\sigma^2}{s-1} / \frac{S_E/\sigma^2}{n-s} \sim F(s-1, n-s) \tag{14}$$

Hereby, we can get the rejection region of hypothesis testing problem

$$F = \frac{S_A/(s-1)}{S_E/(n-s)} \geq F_\alpha(s-1, n-s) \tag{15}$$

where α is the significance level. Namely, if $F > F_{crit}$ that we regard the factor as an important one which has significant influence on the result under the confidence level of $1 - \alpha$.

4 Experiment and Analysis Combined with Earthquakes

4.1 Index Quantification of Anomaly and the Criterion of Selection of Earthquakes

To figure out the statistical relationship between ionosphere anomaly and earthquake, we should make an index quantification of the ionosphere anomaly at first.

In this article, we select the maximum of the positive anomaly $\mathrm{Max}^{(+)}$ before an earthquake, the maximum of the negative anomaly $\mathrm{Max}^{(-)}$, and the sum of the both (called fluctuation denoted as dist) as the index of the anomaly level before an earthquake, and select the magnitude as the level of an earthquake.

As the solar activities and geomagnetic disturbances are the main factors which have an effect on the ionosphere anomaly so that such data should be eliminated while the solar and geomagnetic activities are violent during an earthquake. For example, if the condition $\mathrm{Kp} \geq 5$ or $\mathrm{Dst} \leq -50$ nt exist. We would not take such earthquake in consideration [5].

4.2 Detection and Statistic Analysis of Ionosphere Anomaly Before Earth Quakes

First, we select 18 earthquakes of which magnitude over Ms.5.0 as requested since 2003, as shown in following table (Table 2).

Second, compute the correlation coefficient among magnitude and three other indexes, as shown below.

As shown in Table 3, it is obvious that the magnitude shows positive correlation to anomaly indexes. Comparing with values, we can get the point that the correlation between magnitude and positive anomaly before earthquake is the strongest, and the correlation with the negative anomaly is the weakest.

Then, we set the Ms.6.5 and Ms.7.5 as the boundaries to separate magnitude into three levels of which each level contains six earthquakes, they are the sample data that we selected for each level of the factor. After that, we choose $\alpha = 0.1$ as the significance level so that single factor variance analysis among magnitude factor and ionosphere positive anomaly before earthquakes could be figured out as follow.

From Table 4, $F = 3.259 > F_{crit} = 2.695$ so that the assumption H_0 is refused, that is to say, different levels of the magnitude factor have a significant effect on ionosphere positive anomaly before earthquakes.

Table 2 Magnitude and indicators of ionosphere anomalies

Epicentre	*Ms*	$Max^{(+)}$/TECU	$Max^{(-)}$/TECU	dist/TECU
Qinghai Yushu	5	3.7	0	3.7
Szechwan Shifang	5.5	0	0	0
Qinghai Zhiduo	5.5	3.4	0	3.4
Taiwan Yilan	6.3	2.7	3	5.7
Taiwan Taidong	6.5	5.4	0	5.4
Tibet Zhongba	6.5	5	1.3	6.3
Taiwan Gaoxiong	6.7	8.4	0	8.4
Taiwan nantou	6.7	4.9	10.2	15.1
Burma	6.8	6	9	15
Taiwan Hualian	6.8	2.9	1	3.9
Tibet Gaize	6.9	1.2	0	1.2
Szechwan lushan	7	7.7	0	7.7
Harris	7.6	7.5	0	7.5
Solomon Islands	8.1	6.1	1.2	7.3
Chile	8.2	5.3	5.3	10.6
Kurile	8.3	12.3	1.5	13.8
Sumatra	8.6	3.1	2.4	5.5
East Japan	9	10.2	0	10.2

Table 3 Correlation coefficient among magnitude and indicators

	Ms	$Max^{(+)}$	$Max^{(-)}$	dist
Ms	1			
$Max^{(+)}$	0.571235	1		
$Max^{(-)}$	0.095644	−0.01799	1	
dist	0.47401	0.69677	0.704641	1

Table 4 The single factor analysis of variance between magnitude and positive anomalies

One-way ANOVA					
Group	Obs num	Sum	Average	Variance	
Lower than 6.5	6	20.2	3.367	3.739	
6.5–7.5	6	31.1	5.183	7.702	
Upper than 7.5	6	44.5	7.417	11.290	
Var analysis					
Difference	SS	df	MS	F	F_{crit}
Interblock	49.381	2	24.691	3.259	2.695
Intra-group	113.65	15	7.577		
Total	163.03	17			

Similarly, the result of single factor variance analysis among magnitude factor with the other two indexes could be calculated using the same method. For the negative anomaly, we have $F = 1.108 < F_{crit} = 2.695$, while for the fluctuation anomaly, the result is $F = 2.98 > F_{crit} = 2.695$. The fluctuation anomaly has been defined as the sum of the positive and negative anomaly in the above. We can regard that the influence caused by fluctuation anomaly is contributed to the positive anomaly in fact.

Above all, we proved that conspicuous positive correlation is existent between magnitude and ionosphere positive anomaly, in contrast, magnitude does not have obvious effect on ionosphere negative anomaly.

5 Conclusions

In this paper, the main work could be divided into two part. In the first part, we improved the traditional anomalous detection method by making the data which are used for detecting more stationary. In the second part, we used the improved method to detect the ionosphere anomalies before 18 earthquakes of which magnitudes were over Ms.5.0 all over the world since 2003. Then, three indexes were set as the quantification of the detected result computing with earthquake magnitude factor to figure out the correlation coefficient and the one-way analysis of variance among them. Finally, we reached the conclusion that magnitude has significant effect on ionosphere positive anomaly before an earthquake, while it does not have obvious influence on negative anomaly.

Acknowledgments This work is supported by China National Science Foundation of China (No: 41274045, 41574010).

References

1. Liu J (2011) Study of ionosphere prediction models and discussion of pre-seism ionosphere anomalous. Zhengzhou Information Engineering University, Zhengzhou
2. Sheng Z, Xie S, Pan C (2008) Probability theory and mathematical statistics. Beijing Higher Education Press, Beijing
3. Zhao Y, Zhang X, Liu J (2010) Perturbation analysis of the ionospheric TEC before and after the Wenchuan earthquake. Prog Geophys 25(2):447–453
4. Lu GY, Wong DW (2008) An adaptive inverse-distance weighting spatial interpolation technique. Comput Geosci 34:1044–1055
5. Yiyan Z, Yun W, Xuejun Q et al (2010) Anomalous variations of ionospheric VTEC before Ms 8.0 Wenchuan earthquake. Chin J Geophys 53(3):556–566 (in Chinese)

The Satellite Selection Algorithm of GNSS Based on Neural Network

Jinben Wei, Anmin Ding, Kezhao Li, Leijie Zhao, Yunkai Wang and Zhiwei Li

Abstract With the development of Global Navigation Satellite System (GNSS), there is an increase in the number of visible satellite in navigation and positioning. Calculating all visible satellite not only makes the hardware design of the receiver harder, but also can't guarantee to improve the accuracy of the positioning results, so it is significant to take a certain satellite selection algorithm to the visible satellites optimization. The traditional optimal satellite selection algorithm is complicated, and it also influences the real-time performance of the navigation and positioning. The subsequent improved satellite selection algorithm has effectively enhanced in robustness, accuracy and real-time. The satellite selection algorithm by using Generalized Regression Neural Network (GRNN) to predict GDOP is proposed. The simulation results showed that the algorithm has improved in robustness, accuracy and real-time, and it has certain exploring value for the GNSS satellite selection algorithm with machine learning.

Keywords GNSS · Satellite selection · GDOP · GRNN

1 Introduction

With the rapid development of the global navigation satellite system, the number of visible satellite is rapidly increasing at the same time, which greatly improves the availability and reliability of the system. But it would enhance the difficulty of the receiver hardware design, occupying massive computing resources in the solution process and thus reducing the real-time performance of navigation and positioning, if visible satellites are all involved in the solution. After a certain number of satellites are involved in the positioning solution, excessive satellites participating in the solution cannot improve the positioning accuracy significantly. Some

J. Wei (✉) · A. Ding · K. Li · L. Zhao · Y. Wang · Z. Li
Schools of Surveying and Land Information Engineering,
Henan Polytechnic University, Jiaozuo 454000, Henan, China
e-mail: weijinben2009@163.com

J. Sun et al. (eds.), *China Satellite Navigation Conference (CSNC) 2016 Proceedings: Volume I*, Lecture Notes in Electrical Engineering 388,
DOI 10.1007/978-981-10-0934-1_11

observation data with large errors participating in solving step may even reduce the accuracy of the positioning. Therefore, on the premise of high positioning accuracy, using a certain satellite selection algorithm for the rational selection of visible satellites is very meaningful.

GNSS positioning accuracy depends on two factors: the accuracy of the pseudorange observation; and the geometric dilution of precision (GDOP) [2]. The precision of pseudorange observations is determined by the measurement accuracy of relevant code, the satellite ephemeris precision, the atmospheric refraction, and many other factors, while GDOP depends on the geometry of visible navigation satellite distribution. The traditional optimal algorithm of satellite selection grouped all visible satellites according to the number of receiver channels, and calculated the GDOP value of each subset, then selected one of the subsets which has the minimum GDOP value as the final minimum satellite subset.

The satellite selection algorithm can be assessed according to the following three principles: precision, timeliness, and robustness. Precision is the difference of chosen satellite subset GDOP value between the modified satellite selection algorithm and traditional optimal satellite selection algorithm. Real-time refers to the proper satellite subset which is participated in solving step that is speedy gained by the satellite selection algorithm. Robustness refers to the stability of satellite selection algorithm, whether the algorithm can be used for receiver with any channel to select proper satellite subset in any place at any time for navigation and positioning.

Take the traditional optimal algorithm as an example, it performs the best in precision and robustness, but it is time-consuming, and cannot conform with the real-time performance, especially in the modern civil service of GNSS. When the accuracy of navigation and position reaches a certain level, it is more urgent to obtain the positioning results faster than higher accurate positioning. The calculation of GDOP value, especially the calculation of matrix inversion, occupies a lot of computing space and time during the traditional optimal algorithm. The following sub-optimal satellite selection algorithms mostly try to reduce the number of GDOP calculation or to avoid the calculation of GDOP value to achieve the timeliness of the algorithm. Yarlagadda et al. [7] proved the tetrahedron volume that is composed of four satellites is directly proportional to the GDOP value approximate in four-satellite-positioning, thus forming the maximum polyhedron volume method, which most optimized satellite selection algorithm is based on. But the weak robustness of the proposed method is proved by theoretical calculation and experiments [7]. The closed-form formula for GDOP computation was proposed [5]. But this method has too many floating-point operations, and the accuracy of the final GDOP value calculation is difficult to be guaranteed. The use of neural network in GPS satellite selection algorithm was proposed by Dan Simon, and the robustness and accuracy of the proposed method was analyzed [3]. Neural network for GDOP regression analysis was proposed [2, 4], but the simulation method is computing all the visible satellites, and the robustness of the proposed method needs to be improved. So, how to apply the neural network to GNSS satellite selection algorithm needs further study.

After the summary of satellite selection algorithm principles, two parts of work are done in the following step. Analyzing the accuracy evaluation index of GDOP, and providing the GNSS satellite selection algorithm idea based on neural network; analyzing the simulation of the satellite selection algorithm.

2 GNSS Satellite Selection Algorithm Based on Neural Network

2.1 *Analyze of GDOP Value*

GDOP is an important index to measure the accuracy of satellite navigation, which can be expressed by the following equation [1]

$$\mathrm{GDOP} = \sqrt{tr\left(\boldsymbol{H}^{\mathrm{T}}\boldsymbol{H}\right)^{-1}} \tag{1}$$

where $\boldsymbol{H}^{\mathrm{T}}\boldsymbol{H}$ can be expressed as

$$\boldsymbol{H}^{\mathrm{T}}\boldsymbol{H} = \begin{bmatrix} \sum_{i=1}^{n} a_{xi}^2 & \sum_{i=1}^{n} a_{xi}a_{yi} & \sum_{i=1}^{n} a_{xi}a_{zi} & \sum_{i=1}^{n} a_{xi} \\ \sum_{i=1}^{n} a_{xi}a_{yi} & \sum_{i=1}^{n} a_{yi}^2 & \sum_{i=1}^{n} a_{yi}a_{zi} & \sum_{i=1}^{n} a_{yi} \\ \sum_{i=1}^{n} a_{xi}a_{zi} & \sum_{i=1}^{n} a_{yi}a_{zi} & \sum_{i=1}^{n} a_{zi}^2 & \sum_{i=1}^{n} a_{zi} \\ \sum_{i=1}^{n} a_{xi} & \sum_{i=1}^{n} a_{yi} & \sum_{i=1}^{n} a_{zi} & n \end{bmatrix} \tag{2}$$

In the above formula, $\left[a_{xn} \quad a_{yn} \quad a_{zn}\right]^{\mathrm{T}}$ is the direction cosine of n th satellite in the station-origin coordinate system. $\boldsymbol{H}$ is the order observation matrix while tr represents the trace of matrix. Since each global navigation satellite system has the system time deviation, the following solution was taken if multisystem is involved: using the system time deviation value broadcasted by GNSS navigation message and regarded as a known value [8]. Thus, $\boldsymbol{H}$ is $n \times 4$ order observation matrix even in multisystem.

2.2 *Grnn*

Generalized Regression Neural Network (GRNN) is a kind of radial basis function neural network, which has strong nonlinear mapping capability, flexible network structure, high degree of fault tolerance, and robustness. The network converges to the optimal regression plane in which more samples were gathered. The BP neural

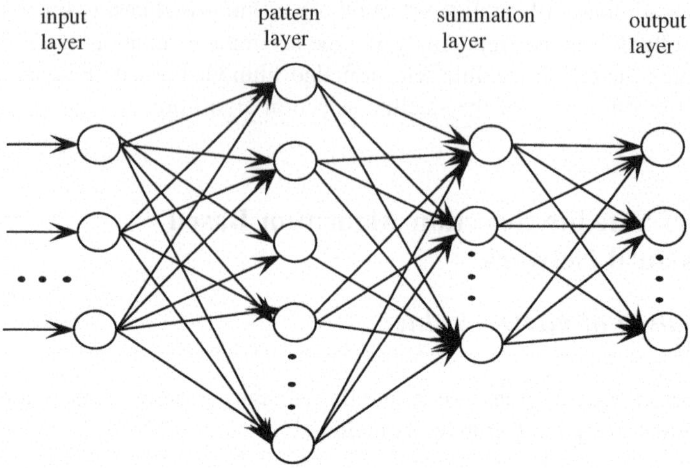

Fig. 1 The structure of generalized regression neural network

network, mainly used in the current study, has shortage in slow convergence, long training time, easy to fall into local minimum and big error in dealing with some complex nonlinear function mapping model.

As shown in Fig. 1, the GRNN consists of an input layer, a pattern layer, a summation layer and output layer [2].

1. input layer
   The number of neurons in input layer is same as the dimensions of input characteristics of training samples. The input characteristics are directly delivered to the pattern layer.
2. pattern layer
   The number of neurons in pattern layer is equal to the number of training samples. Different neurons correspond to different training samples. The transfer function of neuron in pattern layer can be expressed as $p_i = \exp\left[-\frac{(X-X_i)^T(X-X_i)}{2\sigma^2}\right]$, where X is the input variable, X_i is the training sample which i neuron corresponding to, σ is smoothing factor. The output of ith neuron is the form of squared index of squared Euclidean distance between the input variable X and corresponding sample.
3. summation layer
   Two kinds of neurons are used to calculate sums in summation layer. The first formula is the sum of output of the neuron of pattern layer, that is $\sum_{i=1}^{n} \exp\left[-\frac{(X-X_i)^T(X-X_i)}{2\sigma^2}\right]$. The linked weights between pattern layer and summation layer are 1. The transfer function is $S_D = \sum_{i=1}^{n} P_i$. The second formula is making weighted summation to the output neuron of pattern layer, that is

$\sum_{i=1}^{n} Y_i \exp\left[-\frac{(X-X_i)^T(X-X_i)}{2\sigma^2}\right]$, denotes the summation of *i*th neuron in pattern layer and *j*th neuron in summation layer. The connection weight of neurons is the *j*th element of *i*th output sample Y_i. The transfer function is $S_{Nj} = \sum_{i=1}^{n} y_{ij} P_i$ $(j = 1, 2, \ldots k)$.

4. output layer
   The number of neurons in output layer is the same as the dimensions of output in training sample. Since each neuron divides the output of summation layer, the output to neuron j is the j th element of $\widehat{Y}(X)$, that is $y_j = \frac{S_{Nj}}{S_D} (j = 1, 2, \ldots k)$.

2.3 The Usage of Neural Network in Satellite Selection

Equation (2) can be expressed as follows

$$\boldsymbol{H}^{\mathrm{T}}\boldsymbol{H} = \begin{bmatrix} h_1 & h_2 & h_3 & h_4 \\ h_2 & h_5 & h_6 & h_7 \\ h_3 & h_6 & h_8 & h_9 \\ h_4 & h_7 & h_9 & h_{10} \end{bmatrix} \tag{3}$$

Supposing $\boldsymbol{H}^{\mathrm{T}}\boldsymbol{H}$ have four eigenvalues $\lambda_i(1, 2, 3, 4)$, the four eigenvalue of $(\boldsymbol{H}^{\mathrm{T}}\boldsymbol{H})^{-1}$ are $\lambda_i^{-1}(1, 2, 3, 4)$, thus GDOP value can be expressed as follows

$$\mathrm{GDOP} = \sqrt{(\lambda_1^{-1} + \lambda_2^{-1} + \lambda_3^{-1} + \lambda_4^{-1})} \tag{4}$$

By the feature of matrix, we can get the following formula [3]

$$f_1(\overrightarrow{\lambda}) = \lambda_1 + \lambda_2 + \lambda_3 + \lambda_4 = \mathrm{trace}(\boldsymbol{H}^{\mathrm{T}}\boldsymbol{H}) \tag{5}$$

$$f_2(\overrightarrow{\lambda}) = \lambda_1^2 + \lambda_2^2 + \lambda_3^2 + \lambda_4^2 = \mathrm{trace}\left[(\boldsymbol{H}^{\mathrm{T}}\boldsymbol{H})^2\right] \tag{6}$$

$$f_3(\overrightarrow{\lambda}) = \lambda_1^3 + \lambda_2^3 + \lambda_3^3 + \lambda_4^3 = \mathrm{trace}\left[(\boldsymbol{H}^{\mathrm{T}}\boldsymbol{H})^3\right] \tag{7}$$

$$f_4(\overrightarrow{\lambda}) = \lambda_1\lambda_2\lambda_3\lambda_4 = \det(\boldsymbol{H}^{\mathrm{T}}\boldsymbol{H}) \tag{8}$$

Thus get the following two mapping relations

1. $\Re^4 \rightarrow \Re^4$
   input: (f_1, f_2, f_3, f_4); output: $(\lambda_1^{-1}, \lambda_2^{-1}, \lambda_3^{-1}, \lambda_4^{-1})$
2. $\Re^4 \rightarrow \Re^1$

input: (f_1, f_2, f_3, f_4); output: GDOP

If regarded $H^T H$ as an input variable, another two mapping relations are as follows

3. $\Re^{10} \rightarrow \Re^4$

   input: $(h_1, h_2, h_3, h_4, h_5, h_6, h_7, h_8, h_9, h_{10})$; output: $(\lambda_1^{-1}, \lambda_2^{-1}, \lambda_3^{-1}, \lambda_4^{-1})$
4. $\Re^{10} \rightarrow \Re^1$

   input: $(h_1, h_2, h_3, h_4, h_5, h_6, h_7, h_8, h_9, h_{10})$; output: GDOP

The main advantage of GRNN is a highly nonlinear mapping capability. The method of applying it in the process of GDOP solution is as follows take $(h_1, h_2, h_3, h_4, h_5, h_6, h_7, h_8, h_9, h_{10})$ or (f_1, f_2, f_3, f_4) as input layer which is solved easily, take $(\lambda_1^{-1}, \lambda_2^{-1}, \lambda_3^{-1}, \lambda_4^{-1})$ or GDOP as output layer. Training a GRNN which can solve the output layer speedily, applying the neural net to the process of satellite selection can greatly enhance the efficiency of satellite selection.

3 Simulation Test

The proposed satellite selection algorithm is verified in programming based on STK and MATLAB. Simulation parameters are as follows: the satellite orbit files of BDS and GPS broadcast by NORAD are used as the satellite orbit data in simulation scenario; the virtual observation stations are set up in Beijing and Jiaozuo; the elevation angle of receiver in observation station is settled as 20 degrees; the simulation time lasts for 24 h; the data are collected every 2 min, thus 720 observation epochs are sampled. In addition, the fourth mapping relation mentioned above is used to verify the proposed algorithm in simulation calculation in terms of its convenience and practicability.

Taking Jiaozuo station as an example, a GRNN model is constructed using the data of first five observation epochs. In the training course, abandon the satellite subset which the determinant of $\boldsymbol{H}^T\boldsymbol{H}$ is less than 0.12 and normalized the sample data [4]. The trained neural network is used for satellite selection, then giving comparative analysis about satellite selection algorithm with optimal GDOP value.

The blue dotted line in Fig. 2 is the GDOP value change of selecting four satellites at Jiaozuo observation station using the trained GRNN, while the red solid line represents GDOP value variation using optimal algorithm. As seen in Fig. 2, the GDOP value of the selected satellite subset is approximately equal in both algorithm, the difference maintain on a smaller scale. Figure 3 shows the time–cost comparison of two algorithms. As seen in Fig. 3, the algorithm proposed in this article spent shorter time than the optimal algorithm did, while the timeliness increased. The number of visible satellite was up to 24 from epoch 628 to epoch 633, at this moment the difference of time cost is larger than other epoch. With the gradual improvement of GNSS systems, number of visible stars is increasing, and the timeliness of this algorithm would be further improved.

The robustness of the machine learning model is weak in Ref [6]. The precise GDOP value cannot be predicted if the test place is far from the place where the machine learning model is trained. The comparison of GDOP value changes in Beijing observation station selecting four satellites using the GRNN above and optimal algorithm is given in Fig. 4. The maximum difference value is 0.38, which is in the acceptable range. Figure 5 shows the comparison of time-consuming in satellite selection at Beijing observation station. The proposed algorithm used less time than optimal algorithm in almost 80 % epochs while the time–cost is approximately same in the rest epochs.

In Figs. 2 and 4, there is an inaccuracy in predicted GDOP value of GRNN model when the selected GDOP value is small in optimal algorithm. The GDOP value shows a linear state in GRNN model, while the GDOP value changes in optimal algorithm. This is because the smaller GDOP value is located in the ending part of a mapping function in GRNN training. It is worthy of attention in the follow-up study.

The reason why the existing machine learning mode still has weak robustness in predicting GDOP value is as follows: the machine learning model is used as a

Fig. 2 GDOP value comparison of selecting four satellites in Jiaozuo observation station

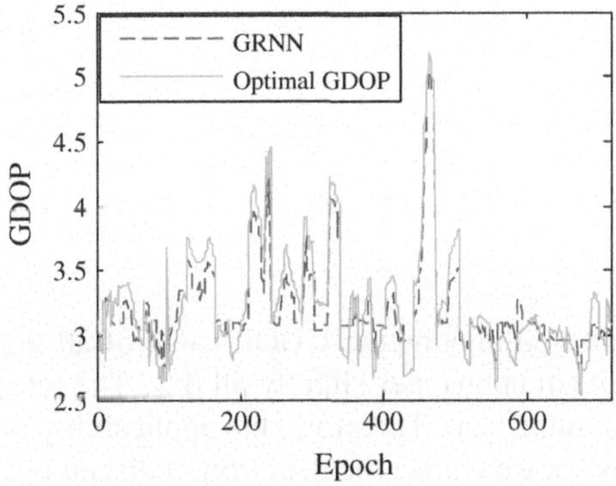

Fig. 3 Time comparison of satellite selection in Jiaozuo observation station

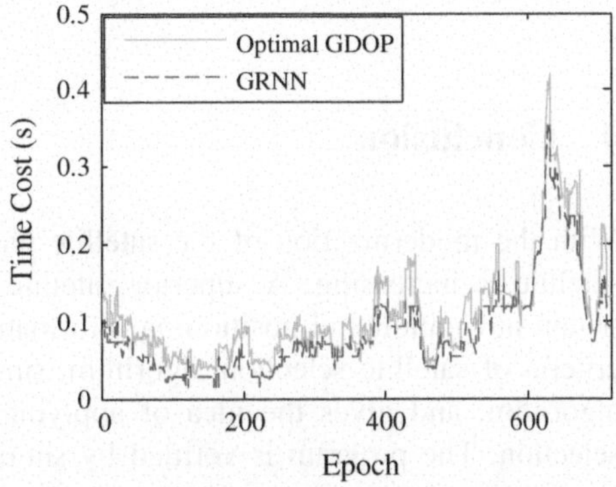

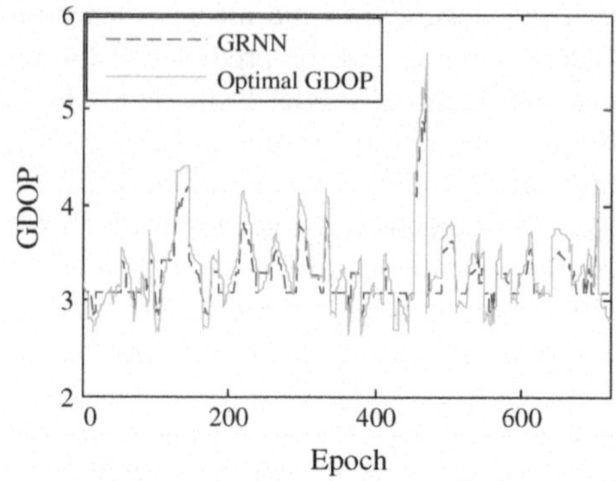

Fig. 4 GDOP value comparison of selecting four satellites in Beijing observation station

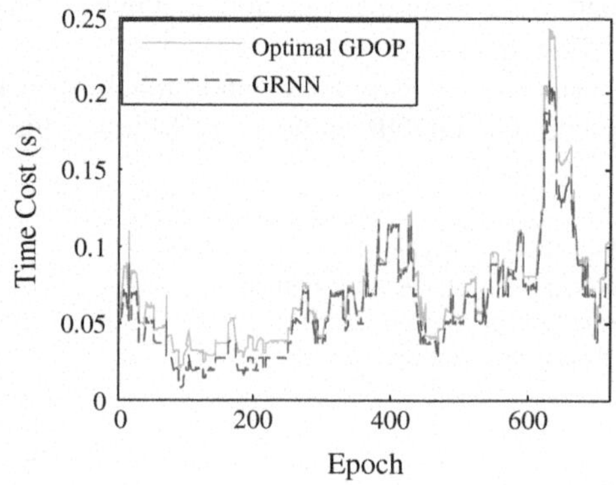

Fig. 5 Time comparison of satellite selection in Beijing observation station

method to compute GDOP value of all the visible satellites, but the GDOP in this situation changes slightly all day. The interval of input and output is narrow in the training step. Therefore, the applicability of the machine learning model gets worse when we collect the data from a distant place. The training data in GRNN model of this paper is selected from different combination of satellite subsets which have wide GDOP change, therefore the robustness of the GRNN model enhanced obviously.

4 Conclusion

With the modernization of the satellite navigation system, the number of visible satellite is increasing. A superior satellite selection algorithm is the key step to obtain navigation and positioning information. This article provides the evaluation criteria of satellite selection algorithm, summarizes the development status of the algorithm, and gives the idea of applying the neural network to GNSS satellite selection. The program is verified by simulation experiment and the results show higher real-time performance than optimal algorithm, with better robustness and precision than existing neural network.

Acknowledgements This work was supported by National Science Foundation of China (Grant Nos. 41202245, 41272373). The authors would like to thank Liu Yao of Xiangtan University, Shi Yifang of Twente University for their valuable suggestions.

References

1. Blanco-Delgado N, Nunes FD et al (2011) Relation between gdop and the geometry of the satellite constellation. In: International conference on IEEE localization and GNSS (ICL-GNSS), pp 175–180
2. Jwo DJ, Lai CC (2007) Neural network-based GPS GDOP approximation and classification. GPS Solutions 11(1):51–60
3. Simon D, El-Sherief H (1995) Navigation satellite selection using neural networks. Neuro Comput 7(3):247–258
4. Tafazoli S, Mosavi MR (2011) Performance improvement of GPS GDOP approximation using recurrent wavelet neural network. J Geogr Inf Syst 3:318–322
5. Teng Y, Wang J (2015) A closed-form formula to calculate geometric dilution of precision (GDOP) for multi-GNSS constellations. GPS Solutions 2015:1–9
6. Wu CH, Su WH et al (2011) A study on GPS GDOP approximation using support-vector machines. Instrum Measur IEEE Trans 60(1):137–145
7. Yarlagadda R, Ali I et al (2000) Gps gdop metric. IEEE Proc Radar Sonar Navig 147 (5):259–264
8. Zhang X, Zhang H, et al (2014) GNSS system time offset determination and its effect on positioning result. J Time Freq (1):57–64

Stability Analysis of the CORS Reference Station Network and Terrestrial Network

Cai-ya Yue, Ya-min Dang, Qiang Yang and Xue-li Han

Abstract The changes of the Earth rotation, the plates' movement and the environment of observation can cause the point shift in the GPS network, even if the observation period is a month, the displacement of the two times cannot be ignored. In this paper, take the terrestrial network station and CORS station network in Sichuan province as an example, using the method of limit error ellipse, the horizontal direction of the station is compared and analyzed. And using the stability of the station to indicate the degree of displacement, so as to identify the stability of the fixed point and unstable fixed point; at the same time, the stability of the elevation direction of the station is analyzed by the method of mathematical statistics. It provides a theoretical basis for the analysis of the stability of the network and a reference basis for the selection of the stability of the station to do a permanent reference station.

Keywords Limit error · Displacement · Stability · GPS net

1 Introduction

With the development and improvement of GPS technology, the application of GPS in dynamic positioning and static positioning has been mature, and the application is more . The basic application of static positioning is to establish various types and accuracy of measurement control network by using GPS measurement technology [1].

C. Yue (✉) · X. Han
Shandong University of Science and Technology,
Qingdao 266590, Shangdong, China
e-mail: 1462453505@qq.com

C. Yue · Y. Dang · Q. Yang · X. Han
Chinese Academy of Surveying and Mapping, Beijing 100830, China

C. Yue
Chinese Academy of Surveying and Mapping,
No. 28, Lotus Pond West Road, Haidian District, Beijing 100036, China

J. Sun et al. (eds.), *China Satellite Navigation Conference (CSNC) 2016 Proceedings: Volume I*, Lecture Notes in Electrical Engineering 388,
DOI 10.1007/978-981-10-0934-1_12

At present, the State Bureau of Surveying and mapping geographic information, the State Seismological Bureau and the Provinces have established the GPS network, and the high accuracy of the measurement method has been provided the measurement results to each unit. Due to the relative motion between the plates of the Earth's crust movement and the interior of a plate and atmospheric loading and ocean loading effects can lead to GPS sites shifts. Therefore, monitoring the stability of the GPS network, and finding whether the displacement of the station and network is significant, provide a high accuracy of the station coordinates and prove the stability of the network. At present, the main analysis method of network stability included the whole network test of the average gap method, the test method and point by point of the single point of displacement method, the limit of the error method [2]. In this thesis, the stability of the terrestrial network and the CORS reference station network is analyzed based on the limit error ellipse method.

2 Principle of Limit Error Ellipse and Error of Point Position

2.1 The Principle of Limit Error Ellipse

The limit error ellipse is a practical method for measuring the point by point. The principle is to measure the amount of the site through the observation of a significant test site; the interval can be 10 days, 1 month, or a year. The standard is to calculate the difference between two times of the error ellipse and the survey site displacement value in two periods, if the difference is positive, then the displacement is not significant, the point is stable; conversely, if the difference is negative, the displacement is significant, the point is not stable that is, GPS network is not stable.

Any direction ψ difference:

$$Q_{\varphi\varphi} = Q_{xx}\cos^2\varphi + Q_{yy}\sin^2\varphi + Q_{xy}\sin 2\varphi \tag{1}$$

$$\sigma_{\varphi}^2 = \sigma_0^2 Q_{\varphi\varphi} \tag{2}$$

In formula (2), Q_{yy} and Q_{xx}, respectively, express the coordinate difference direction cofactor of X and Y axis, Q_{xy} is to express coordinate mutual difference cofactors, σ_0^2 is unit weight mean error, $Q_{\psi\psi}$ and σ_{ψ}^2, respectively, direction for the cofactor and parallax.

The difference between maximum and minimum value of [3]:

$$\begin{cases} E^2 = \sigma_0^2 Q_{EE} = \frac{\sigma_0^2}{2}(Q_{xx} + Q_{yy} + K) \\ F^2 = \sigma_0^2 Q_{FF} = \frac{\sigma_0^2}{2}(Q_{xx} + Q_{yy} - K) \\ \tan \varphi_E = \frac{y}{x} = \frac{Q_{EE} - Q_{xx}}{Q_{xy}} \\ \tan \varphi_F = \frac{y}{x} = \frac{Q_{FF} - Q_{xx}}{Q_{xy}} \end{cases} \quad (3)$$

In formula (2), E and F is the parallax maxima value and the parallax minimal value, the direction ψ_E and ψ_F express, respectively, the parallax maxima E and the parallax minima F, other parameters are same as before.

Error ellipse equation:

$$\frac{x^2}{E^2} + \frac{y^2}{F^2} = \frac{(E\cos\tau)^2}{E^2} + \frac{(E\cos\tau)^2}{F^2} = 1 \quad (4)$$

Among them, τ is the angle form vertical axis E along clockwise when the coordinate axis is made by maximum value E, original point O, and minimum value F. We can get arbitrary direction parallax by point error ellipse, but this method can only to a point analysis, if you want to analyze every point in the GPS network you should make the point error ellipse for each site [4].

2.2 *Point Error*

When estimating the displacement of the sites, the first is to make sure the accuracy of the point of the solution whether or not meets the requirements of the judgment, only in the case of the accuracy of the license can be judged. In determining the position of a point in a plane, however, a series of observations is usually carried out, and the average value of these observations is used as the true position of the point. Such processing can eliminate partial random error, but there are still some differences between the true position and the measured value of points, and the point of the error can indicate the difference between the two.

The calculating formula of the error in point position:

$$\begin{cases} \sigma_x^2 = \sigma_0^2 \frac{1}{p_x} = \sigma_0^2 Q_{xx} \\ \sigma_y^2 = \sigma_0^2 \frac{1}{p_y} = \sigma_0^2 Q_{yy} \end{cases} \quad (5)$$

$$\sigma_p^2 = \sigma_0^2 (Q_{xx} + Q_{yy}) \quad (6)$$

σ_x^2 is the longitudinal coordinate variance of the unknown point, σ_y^2 is the transverse coordinate variance of the unknown point. σ_p^2 is the mean square error of the point to be solved. In Table 1, the errors of the two-dimensional and three-dimensional points of the measurement station of the terrestrial network and CORS reference

Table 1 Test site in the error

Terrestrial network station	Error in two dimensional point (m)	Error in three-dimensional point (m)	Terrestrial network station	Error in two dimensional point (m)	Error in three-dimensional point (m)
SHAO	0.0009	0.0009	SHAO	0.0011	0.0011
PIMO	0.0031	0.0032	PIMO	0.0014	0.0015
BJFS	0.0013	0.0015	BJFS	0.0015	0.0018
SCSN	0.0022	0.0023	NANJ	0.0030	0.0032
SCJU	0.0019	0.0021	XUYG	0.0025	0.0027
SCSP	0.0015	0.0017	YANT	0.0021	0.0022
SCTQ	0.0024	0.0026	ZHGJ	0.0014	0.0015
SCNN	0.0059	0.0064	MUCH	0.0020	0.0022
SCML	0.0043	0.0049	LESH	0.0014	0.0015
SCGZ	0.0018	0.0021	MINS	0.0017	0.0019
SCXC	0.0029	0.0033	HUIL	0.0027	0.0030
YNZD	0.0032	0.0035	BAIY	0.0011	0.0011
LHAZ	0.0034	0.0037	LHAZ	0.0030	0.0033
URUM	0.0016	0.0022	URUM	0.0014	0.0017
POL2	0.0015	0.0021	POL2	0.0006	0.0007

station network are listed. The errors in the two-dimensional point position of the land network measurement station are less than 6 mm, and the three-dimensional point position errors are less than 7 mm; the error of the two-dimensional points in this two station networks is less than 4 mm, and the three-dimensional point error is less than 3 mm. The errors in the measurement sites of these two kinds of networks are less than 1 cm, which can meet the requirement of judging the displacement of the site. In addition, the error of the two-dimensional point and the three-dimensional point is better than the error in the measurement station of the province CORS. The error of the province is better than that of the terrestrial network. The value is in 1–2 mm.

3 Network Stability Analysis

There are many influence factors to the network stability, which include observational factors, such as GPS satellite signal by ionospheric refraction, troposphere delay, multipath effect, and GPS receiver noise; the crustal movement, such as rigid body motion, as well as non rigid motion [5], Their principles are different; The solving benchmark, such as broadcast ephemeris and precise ephemeris, global reference frame selection, and network shape.

3.1 Stability Analysis of Horizontal Direction of Station

In order to have the same influence factor, we used the observation data of terrestrial network in Sichuan province and the data of CORS station in Sichuan province in the network stability analysis. As shown in Fig. 1, a is the land network of the Sichuan and Yunnan region, and b is a part of the CORS reference station network in Sichuan province. The data used are divided into two time periods, one is from January 1st, 2014 to January 10, 2014; the other is from February 1, 2014 to February 10th, 2014, and the interval is one month. By using GAMIT/GLOBK software to solve the data of the two periods of the baseline solution and adjustment, the control station is selected from the Chinese territory and IGS station surrounding with China, after processing get the final measurement station coordinates and accuracy.

Through above error ellipse theory for offset of the maxima and minima and deformation error, ellipse is composed of spindle long radius and short radius and azimuth angle of the spindle. The practical application of the deformation error ellipse method is to make the deformation error ellipse at each point, and take K times the error as the limit deformation error ellipse, according to whether the point displacement vector is dropped in the oval, so as to determine whether the displacement is significant [6–8]. In this paper, two times the error is taken as the limit value. According to Δx and Δy either of the two coordinates of the difference calculated si = sqrt($\Delta x^2 + \Delta y^2$) and expressed in extreme the direction dimension, when si is less than the σ_i, indicating that the displacement of points is not significant. Otherwise, the point displacement. Table 2 lists the difference of the limit error of σ_i and the test site error of si, point stability and stability.

In Table 2, the stability of the CORS reference station network and the terrestrial network is listed. Δs is the difference of the site potential difference and limit error in two parallax, If the difference is positive, indicating that the site of the disparity is

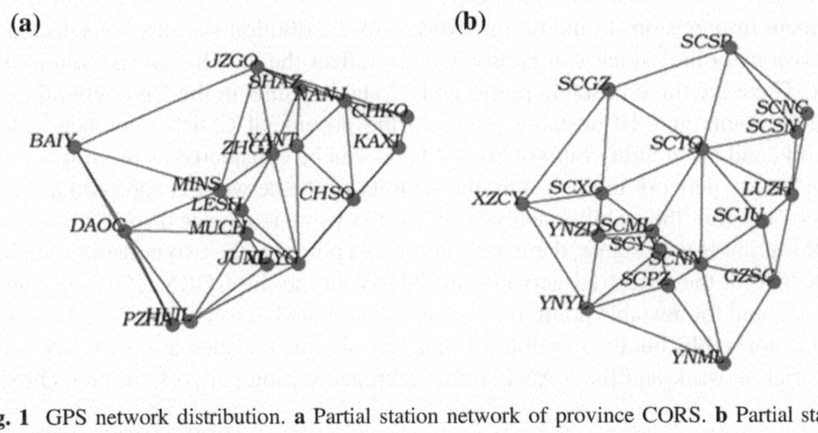

Fig. 1 GPS network distribution. **a** Partial station network of province CORS. **b** Partial station network of terrestrial network

Table 2 Comparative analysis of stability of the measuring station

Stability analysis of partial measurement station of road state network				Stability analysis of some stations in the CORS reference station network			
Station	Δs (mm)	Stability	Stability ratio	Station	Δs (mm)	Stability	Stability ratio
XZCY	−0.0030	Stable	8.81	JZGO	−0.0237	Stable	4.60
YNML	−0.0720	Stable	1.03	SHAZ	−0.0700	Stable	8.90
LUZH	0.0026	Stable	2.36	KAXI	−0.0073	Stable	0.77
SCSN	−0.0440	Stable	1.47	NANJ	0.0140	Instable	−0.39
SCJU	−0.0450	Stable	3.34	XUYG	−0.0013	Stable	0.05
SCSP	−0.0089	Stable	0.35	YANT	0.0032	Instable	−0.19
SCTQ	−0.0127	Stable	0.29	ZHGJ	−0.0131	Stable	1.58
SCNN	0.0330	Instable	−0.85	MUCH	0.0040	Stable	0.25
SCML	0.0250	Instable	−0.75	LESH	−0.0186	Stable	2.35
SCGZ	0.0001	Stable	0.05	MINS	−0.0178	Stable	1.54
SCXC	−0.0501	Stable	0.31	CHSO	0.0074	Instable	−0.25
YNZD	−0.0742	Stable	1.01	CHKO	−0.0212	Stable	4.29
GZSC	0.0078	Instable	−0.72	HUIL	0.0246	Instable	−0.67
YNYL	0.0474	Stable	0.19	DAOC	0.0004	Stable	0.04
YNZD	−0.0117	Stable	1.03	BAIY	−0.0159	Stable	12.24

greater than the limit of error, and the site is not stable, and vice versa, the stability of the site and the difference between them are bigger and more able to reflect the site displacement significance. In this paper, the use of site stability is to indicate the extent of the displacement of the site. Calculated the ratio of Δs and potential difference, the stable rate is positive, indicating that the stability of the site, and the larger value, indicating that the stability is better; value is negative, indicating that the site is not stable, the smaller of the value, the stability is worse.

For two kinds of network as a whole, the calculated point displacement below 8 mm, Δs is less than submillimeter, meets the site displacement significance judgment in precision. In the two networks, 15 distribution stations are selected as the obvious point, which can approximately reflect the stability of the whole network. There are three unstable points and 12 stable points in the land network, five unstable points and 10 unstable points in the provincial CORS reference station network, and the stability ratio of 80 and 60 % can be considered as the single point of the station network in the 8 mm, the stability of the network is approximately 80 and 60 %. From the stability analysis, when the value is greater than 0, it indicates that this point is very stable, there are some stable points in the two networks, such as the XZCY in the terrestrial network and SHAZ in the the CORS reference station network, and for unstable point, if its value is less than 0, it can be explained that this point is not stable, but the instability is not very significant, such as the SCNN in the terrestrial network and the YANT in the reference station network in the CORS.

3.2 Stability Analysis of Station Height

GGPS measurement and analysis work in station elevation direction can be calculated by the elevation direction of time series of station single day solution in the time sequence diagrams, and we can clearly see which day the solution quality is better, eliminate the bad solutions, use the principal component analysis to eliminate network stations of common mode error, after which can be used to judge the stability of the grid [9, 10]. In this paper, we use the method of mathematical statistics to determine the stability of the measuring station.

In Fig. 2, the vertical axis for the two periods before and after the observation and difference for the elevation direction coordinate difference and the horizontal axis is the analysis of 15 stations. Due to the atmospheric loading, ocean tide loading and the influence of the Earth's environment, measuring station have annual oscillation, semiannual oscillation, seasonal cycle oscillation in elevation direction [11]. Taking these factors into account, the choice of sampling interval is 1 month, so the effect of the seasonal cycle of the height of the station will be negligible. According to Fig. 2, it can be seen that the terrestrial network and the part of the station network in the CORS reference station network are stable, while some of the fluctuations in the station is relatively large, but its value is not more than 1 cm. CORS reference station network in the elevation direction is slightly stable than the terrestrial network in the elevation direction, whether or not it is due to the province CORS reference station network is basically in the same plate and influenced by the rigid body motion and non rigid motion, whether the direction of elevation more stable remains to be further proof.

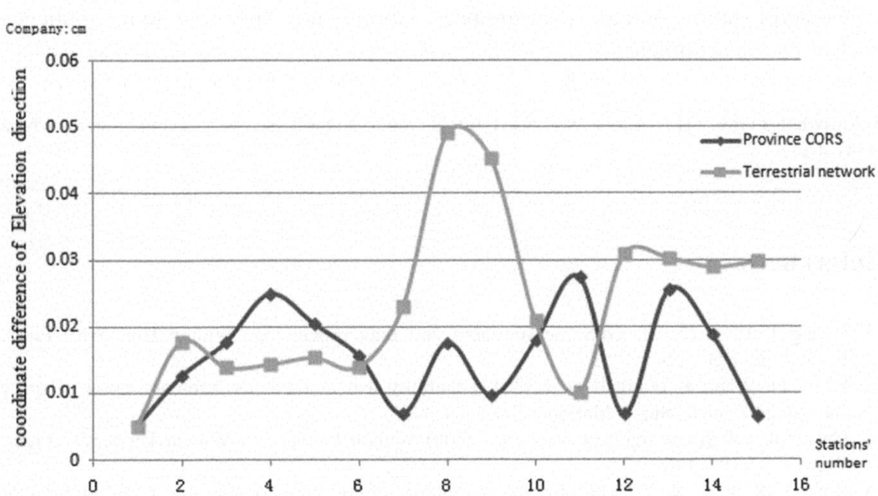

Fig. 2 Elevation direction coordinate difference

4 Conclusion and Discussion

In this paper, the stability analysis of the terrestrial network and the horizontal and vertical stability of the CORS station in Sichuan Province based on the limit error ellipse method, and the selected stations distribution uniformity is in the whole network, which can reflect the stability of the point shift in the 1 cm. The conclusion is summarized as follows.

(1) Through calculation point position potentiometer and limit error point by point, it is found that points position potentiometer and limit error were below 8 mm in the two observations, so overall,the movement of the two phases is very small, and the stability of the network is basically satisfied with their applications.
(2) When the point displacement is below 1 cm, the horizontal displacement of the station is significantly greater than the height. Some survey site though unstable, but unstable degree does not exceed 40 % of a potentiometer, indicating that the degree of instability is low.
(3) Using statistical methods, the stability of the station height is analyzed. Most of the stations is stable in elevation direction, and the stability of the CORS reference station network is slightly higher than the terrestrial network. However, the stability of the network and the network in the survey site whether in the same plate, what impact will be cause, is to be further explored.
(4) This paper concisely expressed the stability rate of the survey station. Such as the XZCY in the terrestrial network station and the SHAZ of the provincial CORS reference station network, their stability rate reached 8.81 and 8.90, indicating that these two stations are very stable, suitable for permanent control station, and this measurement station is not vulnerable to the influence of the environment.

Acknowledgments This study is funded by National Natural Science Foundation of China (41474011).

References

1. Huang J, Li Z (2010) GPS measurement and data processing. Wuhan University Press, Wuhan, pp 193–195
2. Wang D, Wang J, Wang B et al (2013) Stability test of GPS net reference point in space coordinate system. Surve Mapping Sci 2:87–89
3. University of Surveying and mapping (2008) Wuhan University. Wuhan University Press, Wuhan, pp 183–190
4. Yang Y, Song L, Xu T (2002) Robust estimation for correlated observation based on bifactor equivalent weights. J Gedesy 76:353–358
5. Chen Y (2009) Theory and method of modern measurement data processing. Beijing Surveying and Mapping Press, Beijing, pp 65–67

6. Cheng X, Chen T (2013) Analysis of mapping (12):12–14
7. Summer K (2005) Pretreatment method and stability of the GPS plane control network data in Nanjing. Hohai University, Nanjing, pp 49–61
8. Meryer Y (1993) Wavelets algorithms and applications. SIAM, New York
9. Cai J (2009) A new kind of point position error metric. J Surve mapp 38(3):276–279
10. Yang R, Yin H, Chen P (2010) Stability test method and application of the monitoring network reference point. Wuhan J Wuhan Univ Technol 32(6):861–863 (Information and Engineering Edition)
11. Huang L (2006) GPS reference station coordinate component time series noise characteristics analysis. Wuhan Geodes Geodyn 2:31–33

Author Biography

Yue Caiya (1990–), People in Taiqian County, Henan Province, Master, mainly engaged in GNSS data processing and analysis.

A New Modified Method for Cycle-Slips Detection Based on Polynomial Fitting Method

Y.K. Wang, Kezhao Li, Leijie Zhao, Zhiwei Li and Jinben Wei

Abstract The polynomial fitting method is used widely to detect cycle slips of carrier phases of GNSS for its easily computer programming and simple algorithm. But, this method is influenced by the observation errors and sampling rates. In this contribution, the polynomial fitting method and the Newton interpolation method are combined to solve those influences. In this paper, we introduce the new method theory and its mathematical model. Finally a test is taken to prove the reliable of the new method. The result shows that the new method does well in cycle-slips detection, and it reduces the calculation amount.

Keywords The polynomial fitting method · Cycle-slips · Newton interpolation method · Standardization

1 Introduction

GNSS is used to get the information of user position widely. We using carrier phase positioning to get high accuracy result of positioning [1, 2]. But due to receiver clock-offsets, ionosphere refraction and so on, there are cycle slips on carrier phases, therefore the cycle slips detection becomes a key step of precise navigation and positioning [3, 4]. So far, there are many methods that have been developed to detect cycle slips, such as the high-order difference method, the polynomial fitting method, wavelet analysis, and so on. But every methods has its limits, such as the high-order difference method is too difficult to make computer program, and many of them cannot be used to detect cycle slips real-time [5, 6].

To overcome these shortcomings, the polynomial fitting method and the Newton interpolation method are combined to solve this solution. The Newton interpolation is used to transform low-frequency data to high frequency data, and the polynomial

Y.K. Wang (✉) · K. Li · L. Zhao · Z. Li · J. Wei
School of Surveying and Landing Information Engineering
of Henan Polytechnic University, Jiaozuo, China
e-mail: 1325010015@qq.com

J. Sun et al. (eds.), *China Satellite Navigation Conference (CSNC) 2016 Proceedings: Volume I*, Lecture Notes in Electrical Engineering 388,
DOI 10.1007/978-981-10-0934-1_13

fitting method is used to detect cycle slips. The literature [7] shows that the Newton interpolation method does the arc approximation well, so it is a way to solve the problem caused by big errors of observations when we using the polynomial fitting for cycle slips detections, and make full use of the advantages of polynomial fitting. Finally, a test is taken to prove the validity of this method.

2 The Polynomial Fitting for Detecting Cycle-Slips

2.1 The Polynomial Fitting Model

The model of polynomial fitting as follows [8]:

$$\begin{aligned}\tilde{\varphi}_i &= a_0 + a_1(t_i - t_0) + a_2(t_i - t_0)^2 \\ &+ \cdots + a_n(t_i - t_0)^n\end{aligned} \tag{1}$$

In Eq. (1), $i = 1, 2, \ldots, m$; $m \geq n + 1$, m is the number of epochs which participate the polynomial fitting; t_i is the epoch of observation time; t_0 is the initial time; n is the order of fitting.

The step to detect and repair cycle slips is as follows:

(1) The first m successive observations are substituted into Eq. (1) to calculate coefficients.
(2) The equation of the mean errors is as follows:

$$\sigma_j = \sqrt{\frac{V_i V_i}{m - (n+1)}} \tag{2}$$

In Eq. (2), σ_j is the mean error; V_i is the residual error.
(3) Using the coefficients that we get in the step (1) to extrapolate the next epoch phase observation, and got differences between actual values and extrapolated values.
(4) The result of step (3) will follow the judgment as follows:
If the result follows the equation is as follows:

$$\left|\varphi_j - \varphi_j^z\right| < 3 \cdot \sigma_j \tag{3}$$

It means that there is no cycle-slips on carrier phases, and then the first observed value will be discarded and the next observe value is joining the next polynomial fitting.
If the result follows the equation is as follows:

$$\left|\varphi_j - \varphi_j^z\right| \geq 3 \cdot \sigma_j \tag{4}$$

It means that there is a cycle-slip on carrier phases, then the integer part of the real values will be instead by the integer part of the extrapolated value, and the fractional part of the real value is retained.

(5) Do the steps of (1)–(4) until to the end of observations.

It should be noted that as for the fourth derivative or the fifth derivative of the distance between satellites and Earth is close to zero [9]. So, the order of the polynomial of this paper is 4. $m = 14$, on the basis of literature [10].

The literature [11] shows that the polynomial fitting curve which will be divergent because of the accumulation errors and time. So, the method of the cycle-slips detection based on the polynomial fitting must be modified.

2.2 *Method Improvement*

The original observations that we get directly from receivers contain a lot of errors. Mainly includes, clock errors, ionosphere refraction, troposphere delay, multipath, etc. These errors can be eliminated by using difference method between adjacent epoch observations. Eq. (1) could be written as:

$$\begin{aligned}\Delta\varphi_i = b_0 + b_1(t_i - t_0) + b_2(t_i - t_0)^2 \\ + b_3(t_i - t_0)^3 + b_4(t_i - t_0)^4\end{aligned} \tag{5}$$

In Eq. (5) $\Delta\varphi_i = \varphi_{i+1} - \varphi_i$.

The equation shows that the difference between $(t_i - t_0)$ and $(t_i - t_0)^n$ is getting bigger and bigger, and accumulation errors are getting bigger too. So, differences between the real values and the extrapolations are also getting bigger. In this paper, we use one of the time series analysis to solve this situation. The equation is as follows:

$$u_i = \frac{t_i' - \min_{1 \leq j \leq k}\left\{t_j'\right\}}{\max_{1 \leq j \leq k}\left\{t_j'\right\} - \min_{1 \leq j \leq k}\left\{t_j'\right\}} \tag{6}$$

In Eq. (6) $t_i' = (t_i - t_0)^r (r = 1, 2, 3, 4)$; $t_j' = (t_i - t_0)^r$.

Eq. (5) could be written as:

$$\tilde{\varphi}_i = b_0 + b_1 u_i + b_2 u_i^2 + b_3 u_i^3 + b_4 u_i^4 \tag{7}$$

3 Newton Polynomial Interpolation Method

The Newton polynomial interpolation could be expressed by the following equation [12]:

$$\begin{aligned} P_n(t_i) &= c_0 + c_1(t_i - t_0) + c_2(t_i - t_0)(t_i - t_1) \\ &\quad + \cdots + c_n(t_i - t_0)(t_i - t_1)\cdots(t_i - t_n) \end{aligned} \tag{8}$$

In Eq. (8) $c_i(i = 1, 2, \ldots, n)$ is the undetermined coefficients and $P_n(t_i) = \Delta\varphi_i(i = 1, 2, \cdots, n)$. So, the difference quotient of Newton polynomial is as follow equation:

$$\begin{cases} c_0 = \varphi(t_0) \\ c_1 = \varphi[t_0, t_1] = \frac{\varphi(t_0) - \varphi(t_1)}{t_0 - t_1} \\ c_2 = \varphi[t_0, t_1, t_2] = \frac{\varphi[t_0, t_1] - \varphi[t_1, t_2]}{t_0 - t_2} \\ \vdots \\ c_2 = \varphi[t_0, t_1, \ldots, t_n] = \frac{\varphi[t_0, t_1, \ldots, t_{n-1}] - \varphi[t_1, t_1, \ldots, t_n]}{t_0 - t_2} \end{cases} \tag{9}$$

Carrier phases are collected equally spaced. So, the special case of the Newton interpolation is taken for this solution. In this case the equation is more concise form and keeping high accuracy. The Newton forward difference equation is as follows [13]:

$$\begin{aligned} P_n(t) &= \varphi(t_0) + k \cdot \Delta\varphi(t_0) + \frac{k(k-1)}{2!}\Delta^2\varphi(t_0) \\ &\quad + \cdots + \frac{k(k-1)(k-2)\ldots(k-1+n)}{n!}\Delta^n\varphi(t_0) \end{aligned} \tag{10}$$

In Eq. (10) $t_k = t_0 + kh$; $k = 0, 1, 2, \ldots, n$; $h = \frac{t_n - t_0}{n}$.

$$\begin{aligned} \Delta^m\varphi_{(t_k)} &= \Delta^{m-1}\varphi_{(t_{k+1})} - \Delta^{m-1}\varphi_{(t_k)} \\ &= \sum_{i=0}^{m}(-1)^i C_m^i f_{(t_{k+n-i})} \end{aligned} \tag{11}$$

The normal interpolations can be obtained by the Eqs. (9) and (10). Then we can determine the value of the corresponding time interval of interpolation.

4 The New Method to Detect Cycle-Slips

In order to eliminate or abate influences by receiver clock errors, ionosphere refraction, troposphere delay, multipath, etc., we get differences of observation values. The ionosphere refraction and troposphere delay of the process are substantially eliminated, and other kinds of errors are so small that can be ignored. So, this is more conducive to improve the polynomial fitting accuracy.

The literature [9] shows that the way to detect cycle slips by polynomial fitting is influenced by carrier phase measurement errors and sampling frequencies. The difference method is used to abate observation errors, and the Newton polynomial interpolation is used to solve the problem caused by the sampling frequency. We use the Newton interpolation to interpolate low sampling frequency to higher sampling frequency of carrier phases, which helps to improve the accuracy of polynomial fitting.

The new method in this paper using the piecewise interpolation method is to use the Newton interpolation to transform low-frequency data to high frequency data, and to avoid the problem that caused by using the whole date. However, when the interpolated sampling frequency is too high, it will increase the amount of calculations. So, combining with the literature [11], the low-frequency observations are interpolated to 5 s, and when the observing environment is more severe they will be interpolated to 1 or 0.1 s.

The flow chart of the new method is shown in Fig. 1.

5 Experiment

In order to verify the correctness of the method, the data selected from IGS which offered in February 11, 2015 are used as an example. The test by TEQC shows that there is no cycle slips on carrier phases. We use the polynomial fitting to deal with the observations which sampling interval is 1, 5 s and get deviations between extrapolated value and actual values. Results are shown in Fig. 2a, b:

Figure 2 shows that most of the deviations between extrapolated values and actual values are under 0.15 cycles. This means that when the data sampling frequency is 1 or 5 s, the method that detecting cycle slips by the polynomial fitting can judge the cycle slips that over 1 cycle. So, we transform the data in which the data sampling frequency is 10, 15 or 30 s to the data whose data sampling frequency is 5 s by Newton interpolation method, and it is the way to detect cycle slips on the carrier phases. The deviation of extrapolated values and actual values is shown in Fig. 3:

Figure 3 shows that with the increase of sampling rate, deviations between extrapolated values and actual values are become bigger, but most of the deviations under 0.3 cycles. Among them the result of the data whose sampling rate is 10 or 15 s is approaching the result of the data whose sampling rate is 5 s, then we can

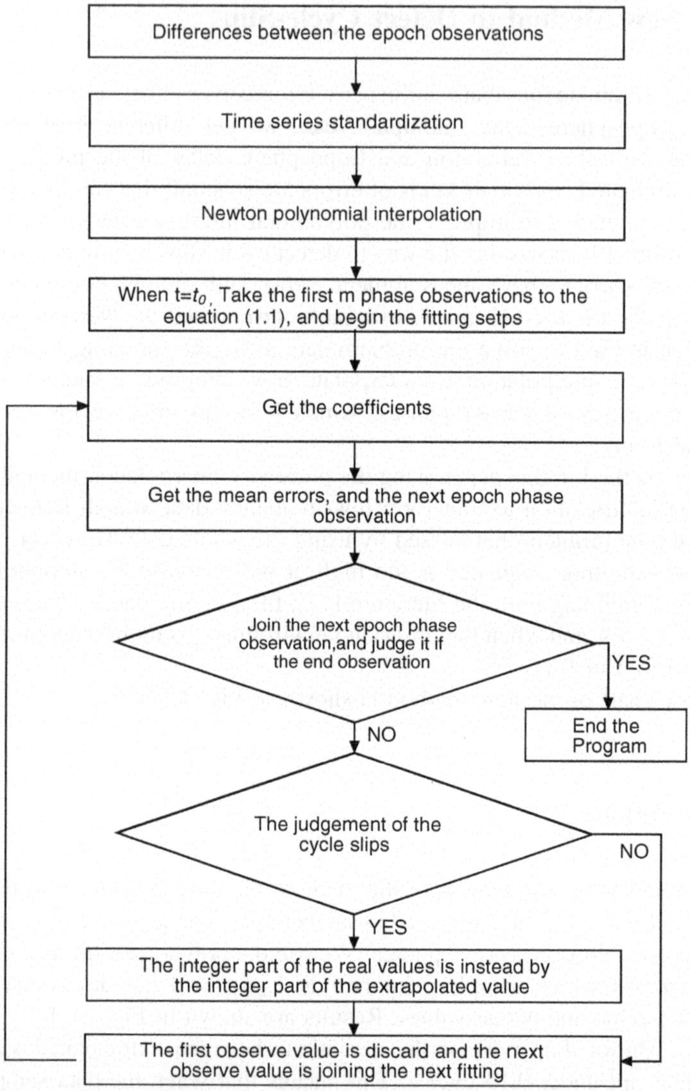

Fig. 1 The algorithm flow chart of the cycle-slip detection and restoration with the polynomial fitting method and Newton interpolation method

detect cycle slips that over 1 cycle. For the result of the data whose sampling rate is 5 s, thought there are some big errors in the result of the polynomial fitting, but they are controlled in less than 0.3 cycles, so it also does well in detecting cycle slips.

To test the performance of the new method on the cycle-slips detection, we join some simulated cycle slips on the data which sampling rate is 10, 15 and 30 s. The simulated cycle slips are shown in Table 1:

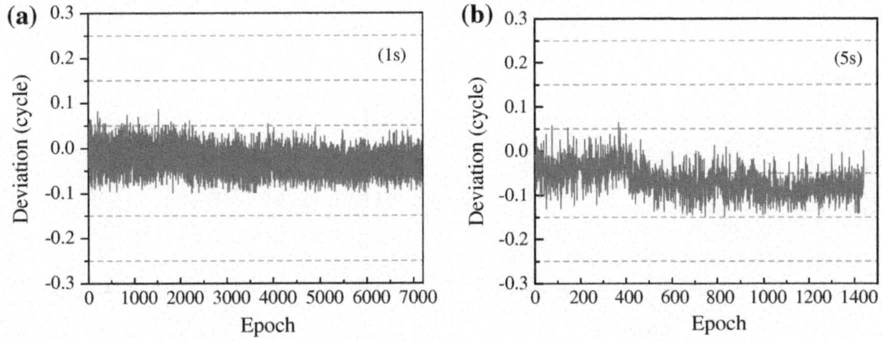

Fig. 2 The deviation between extrapolated values and the actual value

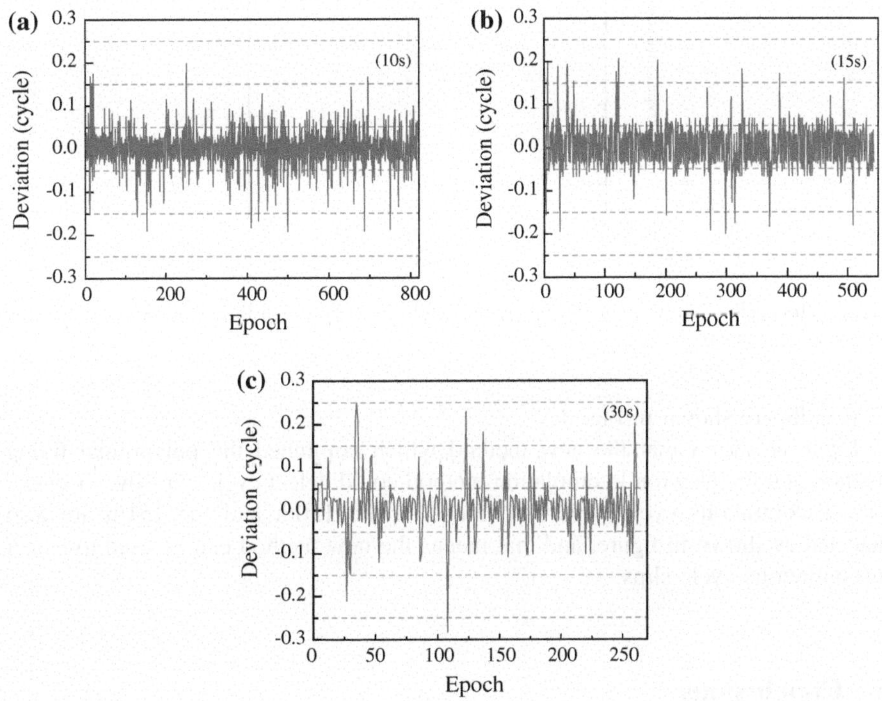

Fig. 3 The deviation between extrapolated values and the actual value

Table 1 Cycle slips added by manual simulation

Epoch	50	150	151	200
Cycle slips	1	10	−10	−10

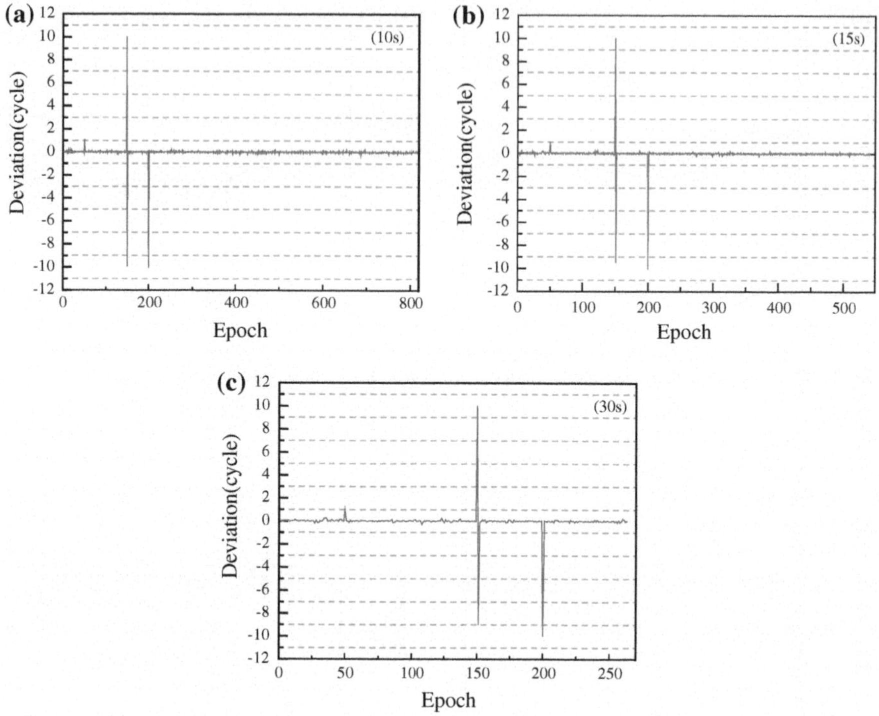

Fig. 4 The cycle-slip detection and restoration with the polynomial fitting method and Newton interpolation method

Results are shown in Fig. 4:

Figure 4 shows that the new method which combined the polynomial fitting method and the Newton interpolation method could detect cycle slips over 1 cycle. The discontinuous cycle slips that joined in the 150th and the 151th are also detected as shown in figure, and that means the new method can be used to detect discontinuous cycle slips.

6 Conclusions

In this paper we introduced a new method which combined the polynomial fitting method and the Newton interpolation method. The result of the data test and analysis shows that it could detect cycle slips well. As for the polynomial fitting could detect cycle slips which over 1 cycle when the data's sampling rate is 5 s. So, when the sampling rate is over 5 s, we use the Newton interpolation method to transform the data to the sampling rate 5 s. It could detect cycle slips over 1 cycle, this is the way to improve the accuracy of fitting and to avoid more computations. In actual, we should use the Newton interpolation method to transform the data to suitable rates by different environments, and get the suitable accuracy on the detection of cycle slips.

Acknowledgements This research was funded by the National Natural Science Foundation of China (Nos. 41202245, 41272373).

References

1. Banville S (2014) Improved convergence for GNSS precise point positioning. University of New Brunswick
2. Zhang XH, He XY (2015) BDS triple frequency carrier-phase linear combination models and their characteristics. Sci China Earth Sci 45(5):601–610
3. Kezhao L, Li Y, Lin C (2014) Positioning principle of GNSS. The Press of Coal Industry, Beijing
4. Stienne S, Reboul B (2014) Choquel B Cycle slip detection and repair with a circular on-line change-point detector. Signal Process 100:51–63
5. Yang J, Li JW, Wu H (2013) Processing cycle-slip of discontinuous epoch in single-frequency GPS data. Sci Surv Mapp 38(4):59–61
6. Huo G, Miao L (2012) Cycle detection of GPS carrier phase with methodology of SA4 multi-wavelet transform. Chin J Aeronaut 25:227–235
7. Liu YH (2014) Orbit determination of spacecraft–interpolation method and fitting method of satellite coordinates. XIDIAN University, pp 29–51
8. Li ZH, Huang JS, Zhang Q (2013) GPS measurement. Wuhan University Press, Wuhan, pp 28–29
9. Yuan B, Bai Z, Fu C (2011) A new method for cycle-slip determination in single-frequency phase observation based on polynomial fitting method. Geotech Invest Surv 12:63–66
10. Luo F, Yao YB, Song WW (2007) Detection for cycle slips using polynomial fitting and carrier phase rate method. GNSS World China 5(3):9–13
11. Zhang Y The report of the experimentations of cycle-slips detection of GPS
12. Lin X, Jiang Y, Wen B (2015) Determination of optimal wind power acceptance level based approaching and Newton interpolation. Power Syst Protect Control 43(18):13–17
13. LI Z, Gao J, Wang H (2014) Inertial dynamic model of GPS/INS integrated navigation based on newton interpolation. Geomat Inf Sci Wuhan Univ 39(5):591–595

The Short-Term Forecast of BeiDou Satellite Clock Bias Based on Wavelet Neural Network

Qingsong Ai, Tianhe Xu, Jiajing Li and Hongwei Xiong

Abstract According to nonlinear and nonstationary characteristics of BeiDou satellite clock bias time series, this paper proposed a method using the wavelet neural network (WNN) based on the first-order difference of adjacent epoch to predict the satellite clock bias. Experimental data with sampling interval of 15 min rapid and ultra-rapid satellite clock bias provided by Wuhan University is used to test the validation of the method. The results show that the forecast precision of 6 h for BeiDou satellite can reach 1–2 ns, and the 24 h can reach 2–4.6 ns using the proposed method. The test results also show that the accuracy and stability of the model prediction can be improved greatly using the proposed method compared to the traditional gray model and quadratic polynomial model.

Keywords Satellite clock bias · First-order difference of adjacent epoch · Wavelet neural network

Q. Ai · J. Li · H. Xiong
School of Geology Engineering and Surveying, Chang'an University, No.126, Yanta Road, Xi'an, Shanxi, China
e-mail: cadx_aqs@yeah.net

T. Xu (✉)
State Key Laboratory of Geo-Information Engineering, Xi'an, Shanxi, China
e-mail: thxugfz@163.com

T. Xu
Xi'an Research Institute of Surveying and Mapping, Xi'an, Shanxi, China

H. Xiong
School of information engineering, China University of Geoscience, Beijing, China

J. Sun et al. (eds.), *China Satellite Navigation Conference (CSNC) 2016 Proceedings: Volume I*, Lecture Notes in Electrical Engineering 388,
DOI 10.1007/978-981-10-0934-1_14

1 Introduction

It is well known that the high precision satellite clock is the basis of the navigation system, and the forecast accuracy of satellite clock parameters directly influence the performance of the navigation system services. Research shows that spaceborne atomic clock is quite sensitive, easily disturbed by the outside environment or their own physical properties. This characteristics makes it difficult to understand the detailed changes, and hard to predict the satellite clock bias with high accuracy [1]. Based on the properties of satellite atomic clocks, many domestic and foreign scholars have put forward a lot of prediction models in order to improve the forecast accuracy of satellite clock bias. The most common forecast model is quadratic polynomial model, and gray model GM (1, 1); In addition, spectrum analysis (SA) model, autoregressive integrated moving average (ARIMA) model, Kalman filtering (KF) model, and relevant improvement models are also widely used. The quadratic polynomial model is first proposed by Allan to forecast the rapid satellite clock bias [2]. Although quadratic polynomial model can generally reflect the physical properties of the atomic clock, it is inevitable to be influenced by the periodic term and random term. On the other hand, the prediction accuracy will be greatly decreased with the increase of the step size using quadratic polynomial model. Gray model can take advantage of less information to forecast in short term, but it is hard to use longer historical data to improve the prediction accuracy further. Its essence is to use index function as predict model, and the prediction value is easily affected by the coefficients of index function. It is difficult to determine a set of reasonable coefficients to predict satellite clock bias using gray model [3].

In terms of the BeiDou Navigation Satellite System (BDS), the satellite is equipped with rubidium (Rb) atomic clock. In order to provide the user with high precision navigation, positioning and timing services, the prediction of BeiDou satellite clock bias becomes more and more important. Because of the shortage of the traditional forecast models above-mentioned, a new method of combining the wavelet neural network (WNN) and the first-order difference of adjacent epoch for satellite clock bias prediction is developed in this paper. The proposed method absorbs not only the advantages of the wavelet commendable time-frequency localization characteristics and strong nonlinear mapping ability [4], but also the advantages of the first-order difference with good numerical stability and weakening influence of system error.

2 First-Order Difference Algorithm for Satellite Clock Bias

Assume the satellite clock time series is expressed as $X = \{x^{(0)}(1), x^{(0)}(2), \ldots, x^{(0)}(m)\}$, then a new sequence $\Delta X^{(0)} = \{\Delta x^{(0)}(k), k = 1, 2, \ldots, m-1\}$ can be obtained after making a first-order difference between the adjacent epoch.

The novel algorithm just uses the new clock series that forecasts the difference in the next period and restores the undifference clock time series which used the predicted value. The formula can be expressed as

$$x^{(0)}(\tau) = x^{(0)}(m) + \sum_{m+1}^{n} \Delta x^{(0)}(\tau - m), \quad m+1 \leq \tau \leq m+n$$

It improves the numerical stability of satellite clock bias, eliminates part of influence of the system error, and enhances the fitting and prediction accuracy.

3 The Basic Model of Neural Networks

Numerous studies and mathematical theory have proved that artificial neural networks (ANNs) can successfully approximate any complicated nonlinear time series, which provides a great support for the study of nonlinear problems [5, 6]. Currently, the widely used method is the feed-forward neural network in which the most famous is BP (Back Propagation) networks. The ideal result of the BP network is to be used in forecasting the satellite clock bias series of nonlinear, nonstationary. In mathematics, BP algorithm is a local search optimization method, but its aim at seeking general extremum of complex function that may fall into local minimum and lead to network training failure. In view of the above shortcomings, many scholars have improved the primary arithmetic, which mostly focus on increasing the rate momentum of learning; Although these proposed methods can improve the predict accuracy to a certain extent, but still unable to resolve the problem of local extremum. In order to overcome the above shortcomings, wavelet theory has been gradually introduced [7].

3.1 The Principle of Wavelet Neural Network

Wavelet analysis uses the multi-resolution processing signal or function by translation and dilation. It has great advantage in the local information extraction and analysis about signal and function. The method combines the self-learning approximation ability, adaptive and self-organizing, fault tolerance, and other characteristics of ANN. The WNN not only inherits the good time–frequency localized feature of wavelet transform, but also retains the self-learning and many other features of ANNs. The main difference between the WNN and the BP network is the difference in the neuron excitation function of the hidden layer. In WNN the wavelet function replaces the role of sigmoid function in the hidden unit, and the wavelet parameters and wavelet shape are adaptively computed to minimize an energy function for finding the optimal representation of the signal. Simultaneously,

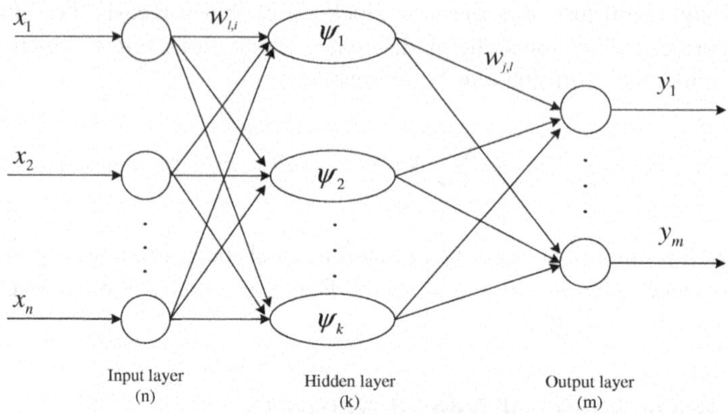

Fig. 1 Network structure of wavelet neural network

the WNN's excitation function is introduced in the dilation and translation coefficients. The network is trained with BP algorithm in batch way [8, 9]. In the following, a three-layer WNN model with both "sigmoid neuron nonlinearity" and "wavlon nonlinearity" is proposed and described for the purpose of satellite clock bias series prediction [6] (Fig. 1).

n , k , m is the sum of input, hidden and output layer nodes
$x_i(i = 1, 2, \ldots, n)$ is input vector,
$y_j(j = 1, 2, \ldots, m)$ is output vector
$\psi_l\,(l = 1, 2, \ldots, k)$ is wavelet basic function, η is learning rate
a_l, b_l is dilation and translation coefficient of wavelet hidden layer, respectively
$w_{l,i}$ is the connection weight between the hidden unit, l and input unit i
$w_{j,l}$ is the connection weight between the output unit, j and the hidden unit l
The output vector could be expressed as $y_j = (\sum_{j=1}^{m} w_{j,l}\psi_{(a,b)}(\text{net}_l)$
$\text{net}_l = \sum_{i=1}^{n} w_{l,i}x_i$ the error function can be written as $E = \frac{1}{2}\sum\limits_{j=1}^{m}(y_j - d_j)^2$
d_j is the desired target output

Under the idea of gradient descent, the derivation procedure of the corresponding connection weights, dilation, and translation coefficient could be referred to the literature [10, 11]. The network is trained by the gradient algorithm based on the forward and backward propagation, and is adjusted until the error function value fall into a limited range (Fig. 2).

3.2 Determine the Network Structure of Wavelet Neural Network

A very important step of WNN application is to determine the suitable network architecture. The hidden layer nodes of the network have a great impact on the forecast accuracy of WNN. If the nodes are few, the network learning cannot be very well which needs to increase the number of training, and the accuracy is also affected. If the nodes is too many, the training time increases a lot and the network is likely to over-fitting. The optimal selection of hidden layer nodes can refer to [12]. Based on the range of the nodes, the iteration of learning process can be stopped if it meets the error conditions [refer to the formula(E)]. If it still does not satisfy the error condition after reaching the maximum number of iterations, the number of nodes should be increased by one layer, then the implicit iterative process is repeated until the error meet the requirement. In practice, the number of hidden layer nodes can be adaptively determined using the above-mentioned method. In this paper, the WNN architecture was finally chosen as net 5-8-1(one input unit, eight hidden units and one output unit) through trial-and-error processes of practical examples of data test. The connection weights, the parameters of wavelet dilation, and translation were initialized as [10].

4 Experimental Result

Now BDS has provided continuous navigation positioning and timing services to China and its surrounding areas. At present there are three types of BeiDou satellite in orbit, GEO satellite represented by C1–C5, IGSO satellite represented by C6–C10 and MEO satellite represented by C11–C14, respectively [13, 14]. The satellite clock bias series presents the overall approximate one-way increasing or decreasing trend, which is shown in Figs. 3, 4, and 5. With the purpose of analyzing the applicability of WNN in the satellite clock bias prediction, the C03, C09, and C11 were selected for the experiment. Test data is the rapid and actual measured ultra-rapid satellite clock bias provided by Wuhan University in the second and third day of BeiDou weeks of 490. In order to evaluate the accuracy of the prediction, we adopt final clock bias of corresponding time provided by Wuhan University. The following two schemes is performed for the experiment.

The first scheme is: Using the clock bias sequence of the second day modeling and forecasting the next 6 h.

The second scheme: Using the clock bias sequence of the second day modeling and forecasting the next 24 h.

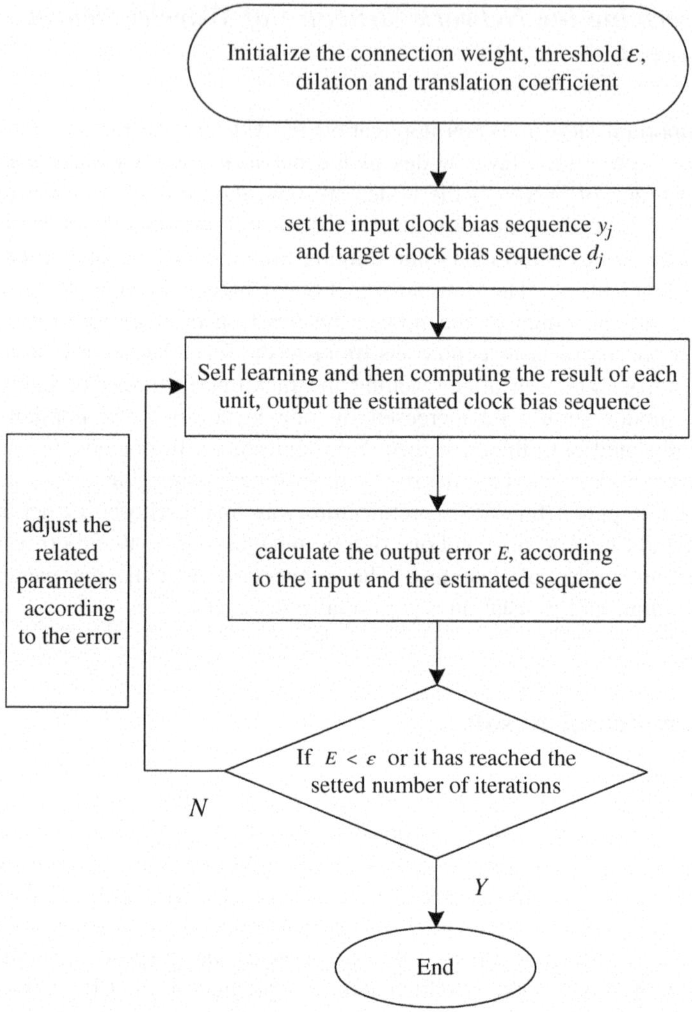

Fig. 2 The flow chart of wavelet neural network forecasting satellite clock bias

Fig. 3 C03 satellite clock bias

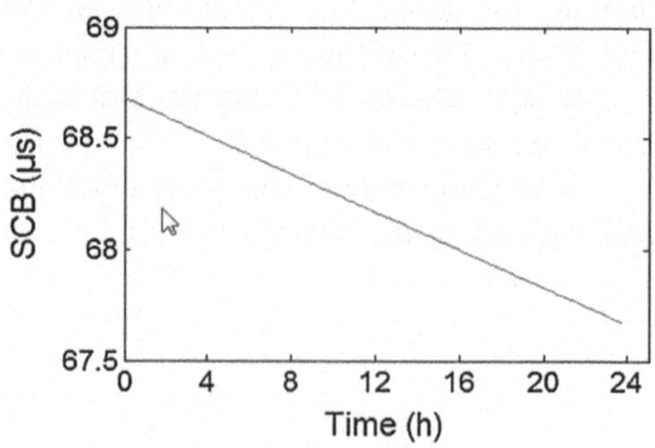

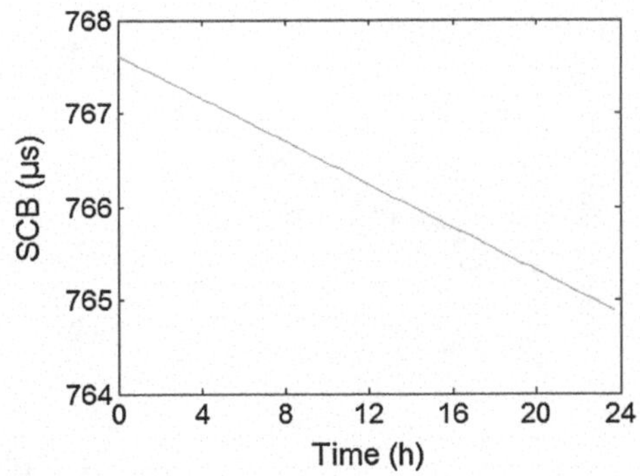

Fig. 4 C09 satellite clock bias

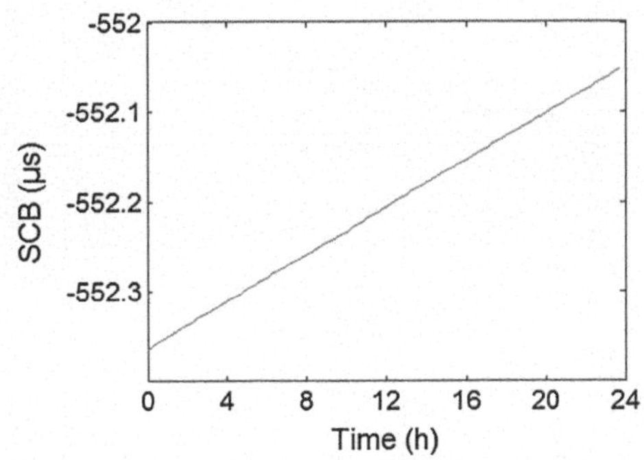

Fig. 5 C11 satellite clock bias

4.1 6 h Prediction Scheme

This scheme uses the traditional quadratic polynomial, the gray model (GM), and WNN based on the first-order difference to forecast the clock bias in the next 6 h. Due to reference clock of the ultra-rapid, rapid, and the final product is different, so there exits a systematic deviation among the three types of products. In order to eliminate the impact of different reference clock, C01 is taken as the reference satellite of clock bias to carry out the precision evaluation [15]. The following Fig. 6a–c show the prediction errors and Table 1 shows the prediction precision of 6 h.

From the statistical results in Table 1 and the prediction errors from Fig. 6a–c, it can be seen that both the prediction accuracy and stability of the proposed method are obviously superior to those of GM and QP for the satellite C03. As for the C09 and C11 satellite, there is still a certain improvement, although the effect is not so significant as C03. It is because that GM and quadratic polynomial have a relatively considerable fitting ability in a short time, which lead the error accumulation is not obvious. The prediction accuracy and stability of rapid products are significantly better than those of the ultra-rapid product. It is because that the actual measurement of rapid products is only 3 h delay, while the rapid products' delay time is 17 h. In this situation, the discrepancy of the amount of data and the number of stations will cause the difference of prediction accuracy.

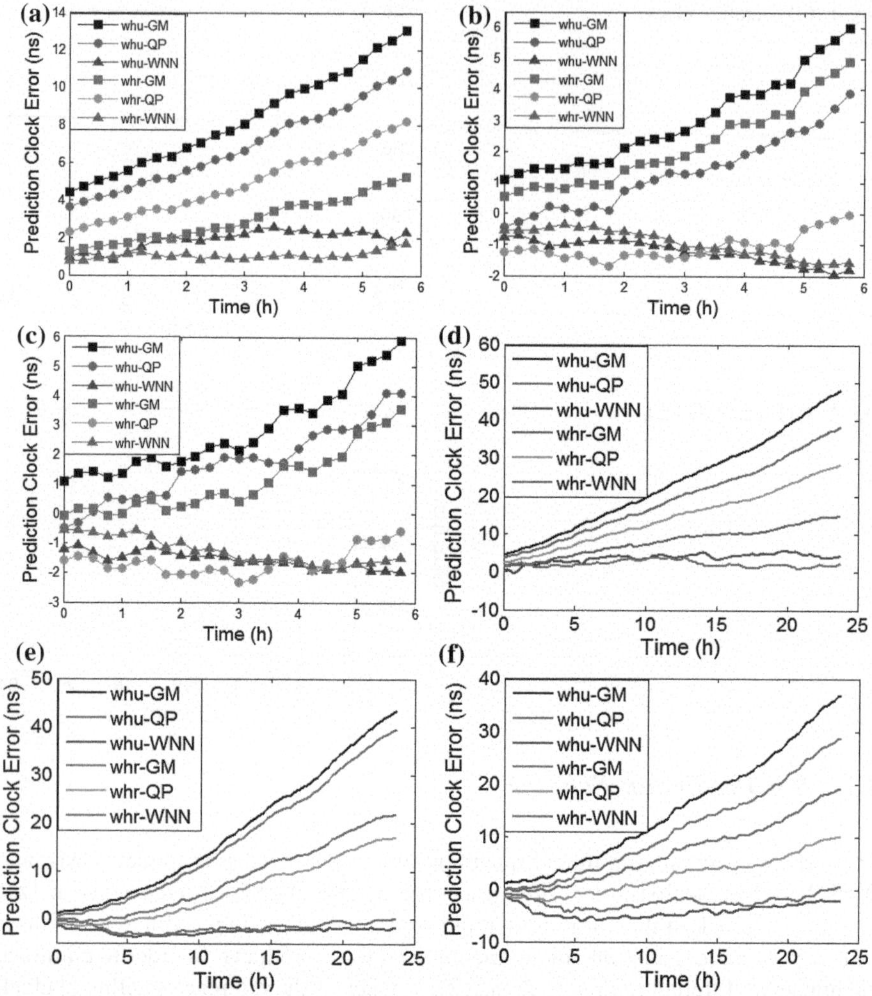

Fig. 6 **a** 6 h prediction error of C03. **b** 6 h prediction error of C09. **c** 6 h prediction error of C11. **d** 24 h prediction error of C03. **e** 24 h prediction error of C09. **f** 24 h prediction error of C11

Table 1 The prediction precision in 6 h

Type/method		GM		QP		DWNN	
		RMS	Max	RMS	Max	RMS	Max
whu	C03	8.685	13.067	7.189	10.897	1.972	2.543
	C09	3.325	6.001	1.779	3.891	1.233	1.951
	C11	3.154	5.880	2.113	4.101	1.594	1.970
whr	C03	3.185	5.270	5.219	8.236	1.078	1.711
	C09	2.541	4.919	1.162	1.679	1.064	1.584
	C11	1.537	3.537	1.696	2.321	1.349	1.845

Table 2 The prediction precision in 24 h

Type/method		GM		QP		DWNN	
		RMS	Max	RMS	Max	RMS	Max
whu	C03	27.068	48.234	21.769	38.390	3.881	5.931
	C09	22.623	43.416	11.179	21.910	2.567	3.872
	C11	19.005	37.058	9.310	18.939	4.558	6.105
whr	C03	8.937	14.891	16.240	28.566	2.272	3.863
	C09	20.496	39.622	8.552	17.018	1.988	3.961
	C11	14.335	28.723	4.475	10.010	2.634	4.448

4.2 24 h Prediction Scheme

In order to analyze the forecasting accuracy of WNN in 24 h, the paper chose all the 96 epoch to forecast next 24 h' clock basis. Figure 6d–f show the prediction errors and Table 2 shows the prediction precision of 24 h.

From Table 1, it can be seen that the cumulative error will increase with the extension of the forecast time and the stability will decrease obviously for quadratic polynomial and gray model. Nevertheless, the forecast precision and the stability of WNN are changed relatively small with the extension of the forecast time. This is related to the stability of the WNN and the good approximation ability for the nonlinear sequence. Making a difference between the adjacent epoch could improve the numerical stability of the clock bias sequence and weaken the influence of system error. The prediction accuracy and stability of rapid products is significantly better than the ultra-rapid product.

5 Conclusions

The characteristics of associative memory and wavelet analysis of the neural makes the WNN to have great approximation capability for nonlinear clock bias sequence. In this view, this paper proposed a WNN algorithm based on the first-order difference to modeling and prediction of BDS satellite clock bias. The ultra-rapid and rapid products are used in the prediction test for the proposed method and its accuracy is verified by the final products. The results show that the forecast precision of 6 h can reach 1–2 ns, and 24 h can reach 2–4.6 ns. The prediction accuracy of rapid products is better than the ultra-fast product, which conforms to the actual quality and precision level of the BeiDou satellite clock products. It achieves a higher accuracy than that of the gray model and quadratic polynomial model.

It should be noted that, the connection weight and the correlation parameter can be randomly initialized in the course of network learning. Through several training adjustments and corrections, the different initial values will have an influence on the prediction results, but they are all in the acceptable accuracy. How to optimally determine the initial value is the point of the further study.

Acknowledgments Thanks to BDS satellite clock bias products provided by the analysis center of Wuhan University. This study is supported by the foundation of natural science of china (Grant No. 41174008 and 41574013) and open foundation of state key laboratory of aerospace dynamics (Grant No. 2014ADL-DW0101).

References

1. Cui XQ, Jiao WH (2005) The application of grey model in satellite clock bias prediction. J Wuhan Univ 30(5):447–450 (Information Science Edition)
2. Allan DW (1987) Time and frequency (time-domain) characterization, estimation, and prediction of precision clocks and oscillators. IEEE Trans Ultrason Ferroelectr Freq Control 34 (6):647–654
3. Zhu LF, Wu XP, Li C (2007) Defect analysis of grey model in the prediction of satellite clock bais. J Astronaut Metrol Meas 27(04):42–44
4. Mosavi MR (2011) Wavelet neural network for corrections prediction in single-frequency GPS users. Neural Process Lett 33(2):137–150
5. Hornik K (1993) Some new results on neural network approximation. Neural Netw 6 (09):1069–1072
6. Pati YC, Krishnaprasad PS (1990) Analysis and synthesis of feedforward neural networks using discrete affine wavelet transformations. IEEE Trans Neural Netw 4(1):73–85
7. Wang YP, Lu ZP (2013) Wavelet neural network algorithm for the prediction of satellite clock bais. J Surveying Mapp 42(3):323–330
8. Feng ZY (2007) The application and comparative study of wavelet neural network and BP network. Chengdu University of Technology
9. Jin XF (2012) The application of wavelet neural network in time series. Shanxi Medical University
10. Chen Z, Feng, TJ, Meng QC (1999) The application of wavelet neural network in time series prediction and system modeling based on multiresolution learning. In: IEEE conference on systems, man, and cybermetics, vol 1, pp 425–430
11. Wan L, Yang J (2012) The application of wavelet neural network in prediction of short time traffic flow. J Comput Simul 29(9):352–355
12. Guo FH, Zhang YX. Ying H (2014) Design the prediction of ozone demand based on BP neural network. J Green Sci Technol 7:213–215
13. Yang Y, Li J, Xu J et al (2011) Contribution of the compass satellite navigation system to global PNT users. Chin Sci Bull 56(26):2813–2819
14. Zhang S, Wang L, Huang G (2010) New challenges and opportunities in GNSS. Geotech Invest Surveying 38(8):49–53
15. Yu HL, Hao JM, Liu WP (2014) A method for evaluating the accuracy of satellite clock error. Hydrogr Surveying Charting 02(2):11–13

Comparative Analysis of Different Filtering Methods for the Tight Integration of GPS and Strong-Motion Records

Junqiang Han, Rui Tu, Qin Zhang and Guanwen Huang

Abstract For the tight integration of GPS and the strong-motion records, in order to obtain high-precision co-seismic waves, the precise separation of the moving state and the baseline shift is much important and also difficult for the data solution. In this paper, the experienced CA/CV models and improved CA/CV models are analyzed, and some meaningful conclusions are got from the results comparison of different validations.

Keywords GPS · Strong-motion sensor · Tight integration · Baseline shift

1 Introduction

Currently, combination of GPS and seismic earthquake to monitor and warn seismic has become a hotspot in GPS seismology. The combination not only can complement both high frequency information, but also can further improve the accuracy of deformation monitoring. There are mainly about loose integration and tight integration in these methods. Loose integration is the use of high-precision GPS displacement to effectively correct baseline drift strong-motion to get the co-seismic wave by restoring (such as [1, 6, 8]). Tight integration refers to put the acceleration earthquake observation directly into the GPS positioning model, at the same time to solve the baseline drift and with the shock wave through parameter

J. Han · Q. Zhang · G. Huang
College of Geology Engineering and Geomantic, Chang'an University,
Xi'an 710054, Shaanxi, China
e-mail: chd_hjq@126.com

R. Tu (✉)
National Time Service Center, Chinese Academy of Sciences,
Xi'an 710600, Shaanxi, China
e-mail: turui-2004@126.com

G. Huang
National Key Laboratory of Geographic Information Engineering, Xi'an 710054, China

J. Sun et al. (eds.), *China Satellite Navigation Conference (CSNC) 2016 Proceedings: Volume I*, Lecture Notes in Electrical Engineering 388,
DOI 10.1007/978-981-10-0934-1_15

estimation, this approach can not only effectively identify and correct earthquake baseline drift, but accuracy of result can be further improved by high-resolution acceleration observations (e.g., [3, 5, 7]). So it attracted widespread attention.

In tight integration process of GPS and strong-motion, besides the conventional positioning parameters, a parameter named baseline drift strong-motion is introduced. This baseline drift is caused by the tilt rotary of instruments and other factors; the order can be attained to thousandth or ten thousandth of itself, but there is a big deviation magnitude between speed and displacement after integration. However, in co-seismic session, baseline drift is a complicated process of rapid change, while it is difficult to express the problem about the baseline drift characteristics and state of motion.

In current study, there are two approaches. As Li et al. [5], process PPP tightly integrated positioning corrected equation empirical method with the strong-motion observation data, like there is no baseline drift in strong-motion, only with a high-resolution motion acceleration constraint stations, so it can further accelerate the convergence speed and accuracy PPP. But at the co-seismic phase, baseline drift of Strong-Motion must do completely exist; there is a different bias solved by empirical methods, and not real-time operation, seriously affecting the real-time earthquake monitoring. Another method is to estimate the baseline drift and movement state, provided baseline drift due to the acceleration of movement and strong correlation, and their separation is the key to obtain accurate precision with shock wave. As Geng et al. [3] Using Strong-Motion baseline drift as a random walk process to access high-precision with shock wave together with other positioning parameters in real-time. Tu et al. [7] likewise strong-motion. The baseline shift of strong-motion is estimated as unknown parameters, but the conventional filter state equation is revised, instead of solving the filtering velocity or acceleration information after the original baseline strong-motion velocity or acceleration deduction solving drift, high-precision, and real time to estimate the baseline drift and shock waveform information.

Based on this background, this paper compares and analyzes CA model and CV model experience in tight integration; and obtained that using CV model for solving the state, noise can be efficiently reduced and the results can be more accurate.

2 CA Model

2.1 Observation Equation

GPS and tightly integrated process often use single PPP positioning mode, after considering the strong-motion acceleration baseline drift, its tightly integrated observation equation model are as follows [7]:

$$\begin{bmatrix} L_p^g \\ L_l^g \end{bmatrix}_k = \begin{bmatrix} A^g & B^g & 1 & 0 \\ A^g & B^g & 1 & 1 \end{bmatrix} \begin{bmatrix} s \\ z \\ dt^g \\ N^g \end{bmatrix}_k + \begin{bmatrix} \varepsilon_p^g \\ \varepsilon_l^g \end{bmatrix}_k, \quad \begin{matrix} \varepsilon_p^g \sim N(0, \delta_{p,g}^2) \\ \varepsilon_l^g \sim N\left(0, \delta_{l,g}^2\right) \end{matrix} \tag{1}$$

$$[L_a^{sm}]_k = [a]_k + [u^a]_k + [\varepsilon_a^{sm}]_k, \quad \varepsilon_a^{sm} \sim N\left(0, \delta_{a,sm}^2\right) \tag{2}$$

Among them, L_p^g, L_l^g, respectively, from observation and minimum values for the pseudophase observations between satellite and receiver, the antenna phase deviation, relativistic effects, tides, turn around, etc., phase error correction uses the existing model (e.g., [2, 4]). L_a^{sm} Acceleration earthquake observation instrument, A^g is a unit vector between the satellite and the receiver, B^g is the troposphere projection coefficients, s, a, u^a, respectively, displacement stations, as well as baseline drift acceleration vector, z, dt^g, N^g, respectively, Zenith direction of the tropospheric delay, the receiver clock error, and phase ambiguity, ε for the observation noise variance δ^2 for the angle marked g represents the GPS system, *sm* represents accelerograph system; k for the epoch number.

2.2 *Equation of State*

2.2.1 Empirical Equation of State of CA Model

In the three-dimensional Cartesian coordinate system, the equation of state of the movement of objects is available for empirical CA model representation [9], such as formula (3):

$$\begin{bmatrix} s \\ v \\ a \end{bmatrix}_k = \begin{bmatrix} 1 & \tau & 0.5\tau^2 \\ 0 & 1 & \tau \\ 0 & 0 & 1 \end{bmatrix}_k \begin{bmatrix} s_{k-1} \\ v_{k-1} \\ a_{k-1} \end{bmatrix} + \omega_k, \omega_k = \tau \begin{bmatrix} \frac{\tau^4}{20} q_a^2 & \frac{\tau^3}{8} q_a^2 & \frac{\tau^2}{6} q_a^2 \\ \frac{\tau^3}{8} q_a^2 & \frac{\tau^2}{3} q_a^2 & \frac{\tau^4}{2} q_a^2 \\ \frac{\tau^2}{6} q_a^2 & \frac{\tau^4}{2} q_a^2 & q_a^2 \end{bmatrix} \tag{3}$$

wherein, s, v, a stations are displacement, velocity, acceleration of three-dimensional vectors, for the sampling interval, dynamic noise power spectral density for the acceleration, respectively

Its baseline drifts equation of state is as follows:

$$[u^a]_k = [u^a]_{k-1} + \omega_{u_k}, \quad \omega_{u_k} \sim N\left(0, q_u^2\right) \tag{4}$$

wherein, q_u for the baseline drift of the power spectral density. The state of the troposphere, clock error, ambiguity, and other location parameters passed with a consistent routine PPP positioning method.

2.2.2 Improved CA Model Equation of State

In GPS and Strong-Motion tight integration, acceleration and baseline drift have a strong correlation, and at co-seismic, its movement and baseline drift are changing rapidly, so it is difficult to accurately determine the dynamic noise. Tu et al. [7], an improved CA model, using the baseline drift acceleration instead solving the corrected acceleration, can effectively reduce the complex dynamic changes rapidly between epochs, and contribute to an accurate separation of its acceleration and baseline drift, such as formula (5).

$$\begin{bmatrix} s \\ v \\ a \end{bmatrix}_k = \begin{bmatrix} 1 & \tau & 0.5\tau^2 \\ 0 & 1 & \tau \\ 0 & 0 & 1 \end{bmatrix}_k \begin{bmatrix} s_{k-1} \\ v_{k-1} \\ [L_a^{sm}]_k - \hat{u}_{k-1} \end{bmatrix} + \omega_k \tag{5}$$

which $\hat{u}_{k-1}$ is on a baseline drift epoch solving valuation.

Based on the above observation and state equations CA model, using the Kalman filter parameters to solve, you can get combination system displacement, velocity, acceleration, and baseline drift.

3 CV Model

CV model is one of the easiest, and of the moving model. In this case, the original acceleration sequence of Strong-Motion needs to speed the integration sequence, and then solved parameters with combination GPS observations tight.

3.1 Observation Equation

Its strong-motion observation equation is as follows:

$$[L_v^{sm}]_k = [v]_k + [u]_k + [\varepsilon_v^{sm}]_k, \quad \varepsilon_v^{sm} \sim N\left(0, \delta_{v,sm}^2\right) \tag{6}$$

wherein L_v^{sm} is a speed sequence analyzer earthquake observation points after the original acceleration, when the speed of the drift baseline drift.

3.2 Equation of State

3.2.1 Empirical Equation of State of CV Model

Its equation of state experienced CV model is as follows:

$$\begin{bmatrix} s \\ v \end{bmatrix}_k = \begin{bmatrix} 1 & \tau \\ 0 & 1 \end{bmatrix}_k \begin{bmatrix} s \\ v \end{bmatrix}_{k-1} + \omega'_k, \omega'_k = \tau \begin{bmatrix} \frac{\tau^2}{3} q_v^2 & \frac{\tau^1}{2} q_v^2 \\ \frac{\tau^1}{2} q_v^2 & q_v^2 \end{bmatrix} \tag{7}$$

wherein, q_v is the power spectral density of the velocity.

3.2.2 Improved CA Model Equation of State

Its state equation improved CV model is as follows:

$$\begin{bmatrix} s \\ v \end{bmatrix}_k = \begin{bmatrix} 1 & \tau \\ 0 & 1 \end{bmatrix}_k \begin{bmatrix} s_{k-1} \\ [L_v^{sm}]_k - \hat{u}_{k-1} \end{bmatrix} + \omega'_k, \omega'_k = \tau \begin{bmatrix} \frac{\tau^2}{3} q_v^2 & \frac{\tau^1}{2} q_v^2 \\ \frac{\tau^1}{2} q_v^2 & q_v^2 \end{bmatrix} \tag{8}$$

wherein, ω' is the dynamic noise. Based on the above observation and state equations CV model parameters using the Kalman filter to solve, you can get combination of system displacement, velocity and baseline drift.

4 Exemplification

For comparison, both simulated data and the experimental data were used and analyzed. In the simulation data, GPS data from the static observation, sampling interval is 1 Hz, simulation of strong-motion data is 100 Hz, 42,300 s in a day seconds to 42,600 s, simulate a constant acceleration of baseline drift, and the remaining time is zero mean the white noise sequence, a total data length of two hours. The experimental data is derived from [6]. The first experimental data, GPS sampling interval is 1 Hz, strong-motion data is 100 Hz, within 43,280–43,340 s, there is a simulation of the movement. It should be noted that the simulation data and experimental data were analyzed only in one dimension.

4.1 Analysis of CA Model

4.1.1 Simulation Data Analysis

Figure 1 reflects the comparison of the experience of CA model and improved CA model resulted by simulation data, where a1 and b1 are the original acceleration (black) and solving the baseline drift (red); it can be seen that solving basic baseline drift acceleration is consistent with the original analog. a2, b2, and a3, b3 specifically is speed and acceleration outputted by filtering system. It shows that the baseline drift acceleration are well corrected, but in the presence of baseline drift occurs jumping two endpoints solving some of hysteresis. The reason is that the simulated baseline drift jump changes and its state equation cannot be accurately stated; and therefore need to go through a shorter time to get corrected. It can be found from comparison in this static state, baseline drift can get a better correction, both two effects are fine and there is no significant difference.

4.1.2 Real Data Analysis

Figure 2 reflects comparative result of empirical CA model and improved CA model solution results of e, where a1 and b1 are the original acceleration (black) and solving the baseline drift (red), which can be seen in a stationary state. The use of empirical models and improved CA model for solving basic baseline drift acceleration simulation is consistent with the original, but in the motion the empirical model implicated acceleration state that exists to solve the baseline drift

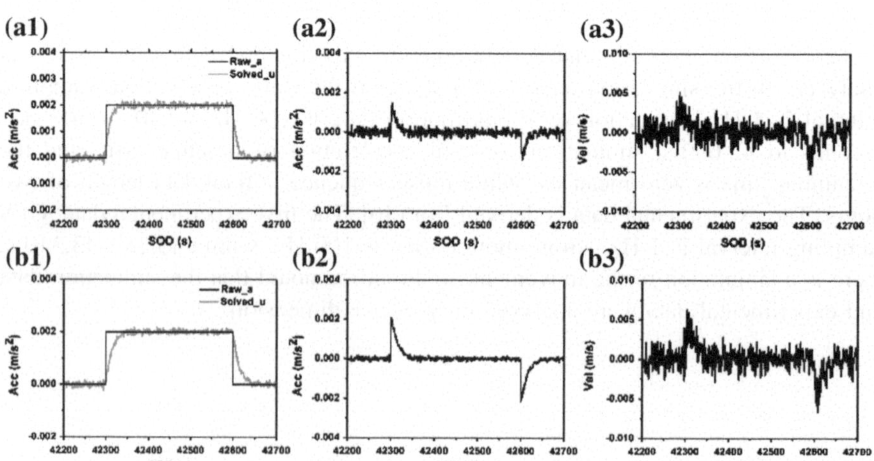

Fig. 1 Comparison of the results of the empirical CA model and the improved CA model by using simulation data (Color figure online)

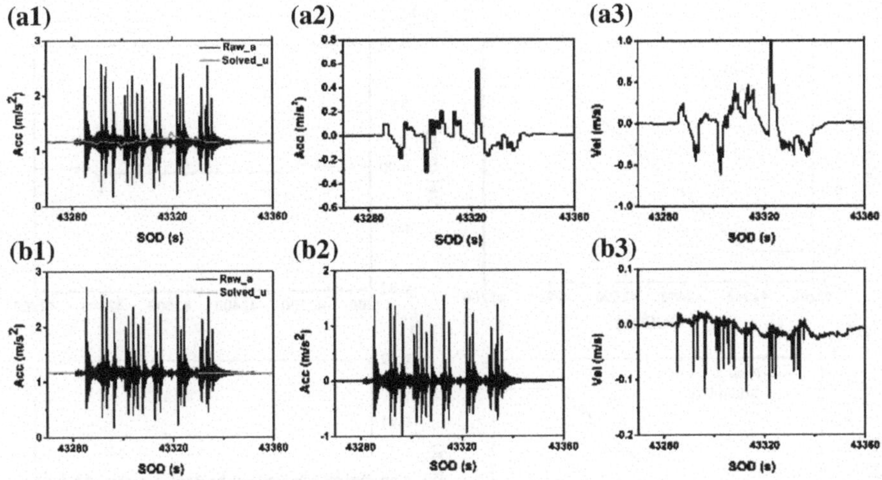

Fig. 2 Comparison of the results of the empirical CA model and the improved CA model by using measured data (Color figure online)

disturbance. Therefore, a2, b2, and a3, b3 can be seen, the acceleration and speed of CA empirical model solution are not reflecting the actual real movement. Because, the GPS sampling frequency is 1 Hz, strong-motion instrument is 100 Hz, and the simulation of the movement of a few Hz to several Hz, during this one second time, the shift of conventional model cannot accurately describe the state of motion within each Hz, so that a2 and a3 result of acceleration and velocity are seriously abnormalities.

4.2 Analysis of CA Model

4.2.1 Simulation Data Analysis

Figure 3 reflects the comparative result of experience CV model and improved CV model solution results of simulation data, where a1 and b1 to the original speed (black) and baseline drift velocity solved (red) can be seen; solving basic baseline drift velocity is consistent with the original analog a2, b2 speed sequence deducted after baseline drift can be seen from the baseline drift velocity are well corrected. However, experience in the CV model, the baseline drift exists in jumping hysteresis solving some of the two endpoints. The reason is that the simulated changes in baseline drift jump occurs, which is not an accurate representation of the equation of state, requiring short time to get corrected. Meanwhile, in the improved CV model, strong-motion instrument integration with velocity after deducting baseline drift instead of the real speed, can effectively reduce the dynamic noise speed,

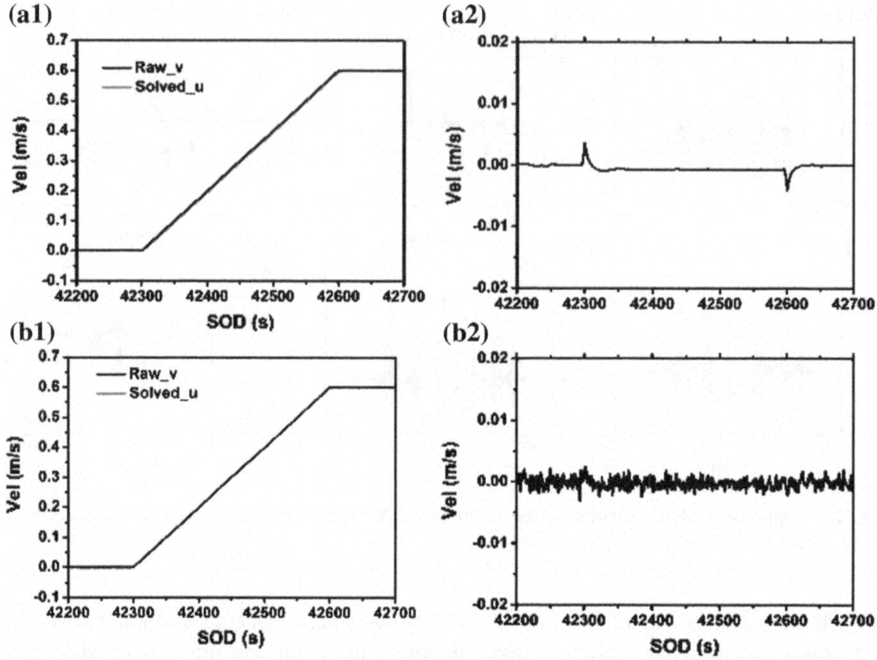

Fig. 3 Comparison of the results of the empirical CV model and the improved CV model by using simulation data (Color figure online)

convenience and reasonable given accurate dynamic noise to solve the parameters. But using formula (6) relative to baseline drift velocity dynamic noise increases, as a result there is a big noise, so that noise recovery speed is slightly larger than the CV model experience, but does not affect the results accurate property.

4.2.2 Test Data Analysis

Figure 4 is a comparison of results solved by experience CV model and improved CV model solution with the actual response data, where a1 and b1 are the original speed (black) and baseline drift velocity solved (red). It can be seen, in the absence of movement under baseline drift, two options of baseline for solving the drift velocity is consistent with the actual basic. In the presence of motion there is baseline drift conditions, baseline drift due to the strong correlation between speed, CV model experienced only a part of baseline drift correction, so that with the speed of recovery there is a deviation, as shown in a2. Improved CV model well separate speed and baseline drift, the speed of recovery is more accurate. Meanwhile, in the experience of CV model, the abnormal states of filter lead to the existence of divergence in result.

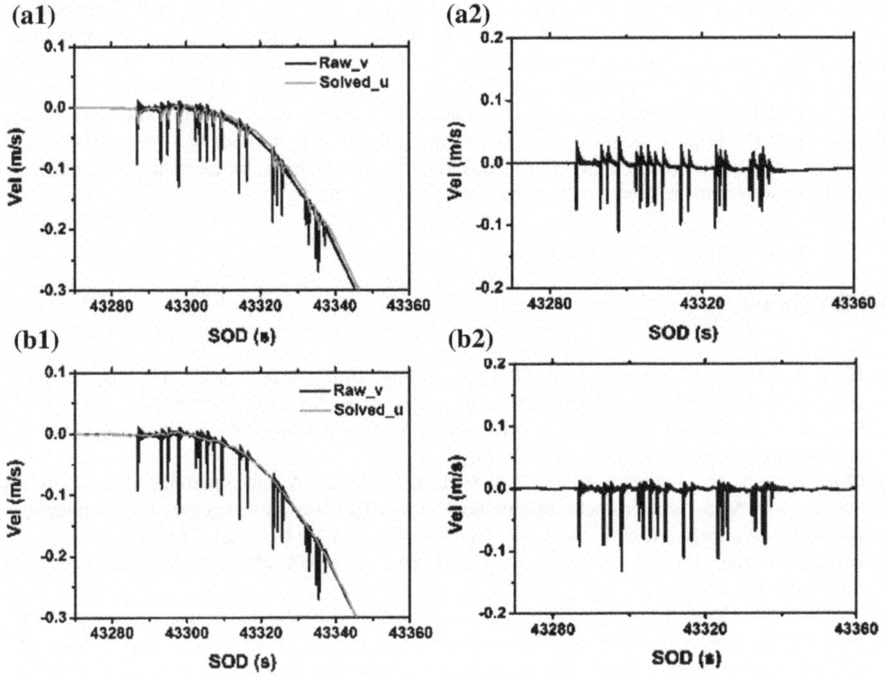

Fig. 4 Comparison of the results of the empirical CV model and the improved CV model by using measured data (Color figure online)

5 Conclusion

How to effectively separate the baseline drift error and motion state dealing with the GPS and strong-motion tight integration is the problem discussed in this article. It analyzes the different between CA model and CV model and improved them, and the numerical results were verified in cases. The conclusion is that because the seismic wave movement is with a high frequency, conventional CA model and CV model, it is difficult to accurately describe the complex state of motion as well as its strong correlation of baseline drift error, which results in a big disturbance exception between baseline drift and restore solving co-seismic information. However, with improved CA model and CV model, considering the continuity of the baseline drift error and effective deduction, the corrected acceleration or velocity can be very accurate to descript the complex dynamic changes, and effectively reduce the dynamic noise. When given accurate and reasonable noise to solve dynamic noise, makes the results less influenced by complex dynamic state. Meanwhile, it can be found that in comparative CA model and CV model, which is sensitive to the high frequency acceleration, small errors can lead to solving the divergence of their results. Because it is easier to control the status of noise of solving, the results are more accurate.

Acknowledgments This research was supported by the project of National Natural Science Fund (41104019, 41104022, 41304033 and 41504006), Project of Chinese Hydropower Engineering Consulting Group of science and technology (GW-KJ-2012-21). The authors would like to thank The open fundation of National Key Laboratory of Geographic Information Engineering (2013ADL-DW0103). The open fund of Sate Key Laboratory of Geographic Information Engineering (SKLGIE2013-Z-2-1) and Central University Research Funds (2014G1261051), The National Natural Science Foundation of China, Light of West China, for their support, and to the editor and anonymous referees for their valuable suggestions.

References

1. Bock Y, Melgar D, Crowell BW (2011) Real-time strong-motion broadband displacements from collocated GPS and accelerometers. Bull Seism Soc Am 101:2904–2925
2. Dach R, Hugentobler U, Fridez P, Michael M (2007) Bernese GPS software version 5.0. Astronomical Institute, University of Bern, Bern
3. Geng J, Bock Y, Melgar D, Crowell BW, Haase S (2013) A new seismogeodetic approach applied to GPS and accelerometer observations of the 2011 Brawly seismic swarm: Implications for earthquake early warning. Geochem Geophys Geosyst 14(14):2124–2142
4. Kouba J, Héroux P (2001) Precise point positioning using IGS orbit and clock products. GPS Solutions 5(2):12–28
5. Li X, Ge M, Zhang Y, Wang R, Klotz J, Wicket J (2013) High-rate co-seismic displacements from tightly-integrated processing of raw GPS and accelerometer data. Geophy J Int 195 (1):612–624
6. Tu R, Wang R, Ge M, Walter TR, Ramatschi M, Milkereit C, Bindi D, Dahm T (2013a) Cost effective monitoring of ground motion related to earthquakes, landslides or volcanic activities by joint use of a single-frequency GPS and a MEMS accelerometer. Geophys Res Lett 40 (15):3825–3829
7. Tu R, Ge M, Wang R, Walter TR (2013b) A new algorithm for tight integration of real-time GPS and strong-motion records, demonstrated on simulated, experimental, and real seismic data. J Seismol 18(1):151–161
8. Wang R, Parolai S, Ge M, Jin MP, Walter TR, Zschau J (2013) The 2011 Mw 9.0 Tohoku earthquake: comparison of GPS and strong-motion data. Bull Seismol Soc Am 103(2B): 1336–1347
9. Yang YX (2006) Adaptive navigation and kinematic positioning. Surveying and mapping press, Beijing

Application of Low Cost GPS/BDS Receiver in UAV Precise Navigation and Positioning

Lei Wang, Xiaohua Wen and Changbao Huang

Abstract Recently, unmanned aerial vehicle (UAV) technology has been growing rapidly and widely used in civil applications, such as aerial mapping, precise agriculture, power line patrolling. However, most of high precision position and orientation systems (POS) are still heavy and expensive, so they are still unpractical to be equipped on UAV. In this study, we investigated the feasibility of providing precise navigation and positioning service for UAV with single frequency GPS/BDS receiver. The challenges of precise positioning for UAV with low cost receivers are threefold: (1) signal captured with low cost receivers and antennas has poorer quality and interference immunity. (2) UAV moves fast and flexible. (3) UAV navigation requires real time and high reliable positioning solution. GPS-based single frequency can provide centimeter accuracy positioning in friendly environments, while its availability and reliability is difficult to meet the requirement. In this study, we introduced GPS/BDS combined RTK positioning, which increases the redundancy number and improved the precision and reliability of float solution. Meanwhile, we analyzed the signal and antenna gain characteristics of low cost receiver and antenna, and established a refined stochastic model for GPS and BDS observations. We also made use of the Doppler observations to improve the kinematic model and quality control, which significantly improves the success rate of carrier phase ambiguity resolution in single frequency RTK case. In order to validate the performance of our algorithm, we carried out in-flight tests with Tersus single frequency GPS/BDS receiver and analyzed the test results. The results indicate that our RTK algorithm can improve the fix rate of ambiguity resolution from 36 % to around 60 % in challenging environments. We also validated that GPS/BDS RTK has better availability and reliability than GPS stand-alone RTK.

Keywords Precise navigation · UAV · BDS · Low cost receiver · RTK

L. Wang (✉) · X. Wen · C. Huang
Shanghai Skywell Network Technology Co. Ltd., Shanghai, China
e-mail: lei.wang@wellcores.com

J. Sun et al. (eds.), *China Satellite Navigation Conference (CSNC) 2016 Proceedings: Volume I*, Lecture Notes in Electrical Engineering 388,
DOI 10.1007/978-981-10-0934-1_16

1 Introduction

The unmanned aerial vehicle (UAV) has been developed rapidly in recent years. Meanwhile, its industrial and agricultural applications also become increasingly attractive, such as the package delivering, the airborne photogrammetry, the spraying, and the power line patrolling. The key technology of UAV applications is its flight control system, which heavily depends on its positioning and altitude sensors. On the other hand, GNSS positioning technology can provide global, real-time, continuous positioning service and consequently be appealing to flight control system. As GPS chips become affordable, the GNSS-based positioning sensor becomes a de facto standard technology of the UAV flight control system. Currently, most flight control systems employ a stand-alone GPS chip which only provides several meter level positioning solutions. This precision can only meet the requirement of amateur applications. Many industrial applications require much more precise positioning solutions, such as spraying and power line patrolling. In this study, we attempt to improve the accuracy of UAV positioning with an affordable GPS/Beidou receiver by employing real time kinematic (RTK) technology. Corresponding challenges come from both antennas and receivers, as RTK requires high-quality carrier phase observations. In this study, we attempt to use non-geodetic level antennas, which lead to more severe multipath effects. Due to vehicles dynamics and signal obstructions, the carrier phase observations are frequently contaminated by cycle slips or suffer from intermittent disruption. Another challenge is that it is difficult to positioning with stand-alone GPS in less friendly environment. Fortunately, we tracked Beidou signals as well, which increases visible satellite number and the positioning availability.

In this study, a number of key issues in low cost RTK data processing are addressed, including the impact of non-geodetic antennas, the multipath effect mitigation, the cycle slip detection, etc. Then, an experiment is carried out with low cost GNSS receiver and non-geodetic antenna. The experiment results indicate that the low cost GNSS can also achieve fairly good RTK results, but it still can be improved.

2 Low Cost Miniature GNSS Antennas

With regarding to the requirement of UAV, only miniaturized, lightweight and affordable GNSS antenna can be equipped on UAV. Therefore, most geodetic level GNSS antennas cannot meet the requirement, although they are relatively mature and high performance. Literally, smaller size means lower antenna gain, which has adverse impact on GNSS signal acquisition and tracking. Moreover, miniaturized GNSS antenna is more appealing to multipath effect, since it has more chance to receive the reflected signals.

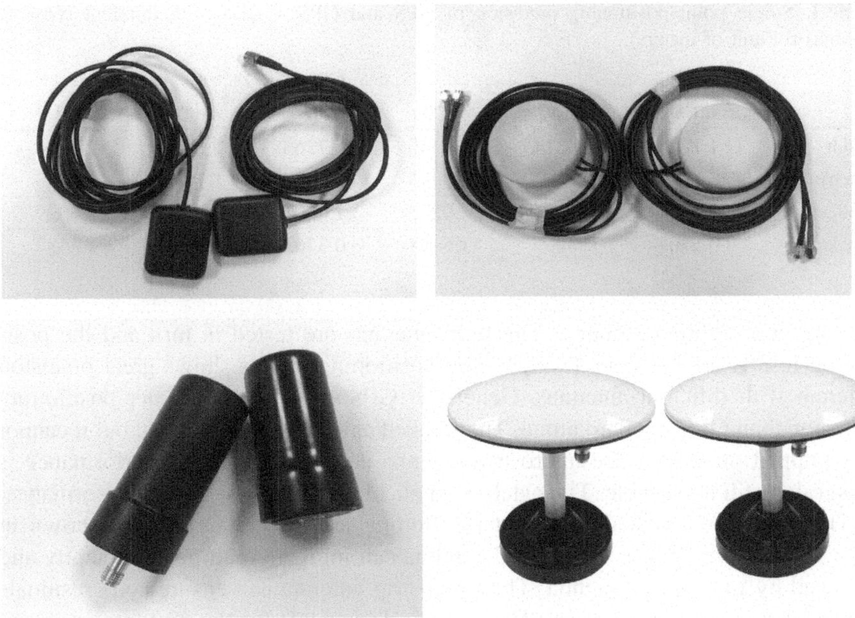

Fig. 1 Antennas used in low cost antenna test, which are L1/B1 patch antenna, L1/B1/G1 microstrip antenna, L1/B1/B3 helical antenna, and full-band stacked antenna

Considering above disadvantages of miniaturized antennas, we investigate the impact of antennas on GPS/BDS signals quality. In our study, four types of mainstream GNSS antenna types are selected, which are known as patch antenna, microstrip antenna, helical antenna and stacked antenna, which are shown in Fig. 1. The first three types are miniaturized antennas which potentially can be used in UAV precise positioning and the stacked antenna is the geodetic level antenna which is used as the reference.

There are many performance measures of GNSS antennas from communication perspective, but most of them are not directly associated with positioning performance. From positioning point of view, the antenna performance can be evaluated from two aspects: the capability of multipath effect mitigation and the stability of carrier phase center.

The impact of multipath on code measurement can be up to tens of meters [8] and its mitigation mainly depends on antenna design, e.g., adding choke ring or metal plate on antenna to suppress the reflective signals. For dual-frequency case, multipath can be approximately extracted by linear combination [12], but it is not applicable for single frequency case. In this study, the multipath effect is assessed by single point positioning and corresponding residuals. We set up the same receivers at the same location and change antennas in different tests. The observation environment is relatively open, but there are some trees and low buildings around. The collected data are processed with the standard least squares without any

Table 1 Single point positioning precision of GPS and GPS + BDS with different types of antennas (in unit of meters)

	GPS			GPS + BDS		
	std(*E*)	std(*N*)	std(*U*)	std(*E*)	std(*N*)	std(*U*)
Patch	1.1729	2.0214	3.4440	0.7268	1.1222	2.5454
Microstrip	0.8162	0.9542	2.0945	0.6570	0.8042	2.1788
Helical	1.6637	1.5791	4.0213	1.0501	1.1303	2.6029
Stacked	0.5704	0.8209	1.5468	0.3765	0.4391	1.2337

filtering or weighting strategy. The four antennas are tested in turn and the positioning results are listed in Table 1. The positioning results shows great precision different with different antennas. Generally, GPS + BDS has a better positioning precision than GPS standard alone. The stacked antenna performs best, but it cannot be equipped on UAV. The microstrip antenna comes next and its performance is poorer, but still acceptable. The patch and helical antennas have lower performance.

The posteriori residuals of point positioning are also analyzed and shown in Fig. 2. The figure shows the stacked antenna can mitigate multipath efficiently and followed by microstrip antenna. The microstrip antenna can ensure code residuals smaller than 5 m in most cases. The posteriori residuals for the patch antenna and the helical antenna reach about 20 m in the worst case. The helical antenna has a comparable performance with the stacked antenna for elevation angle higher than

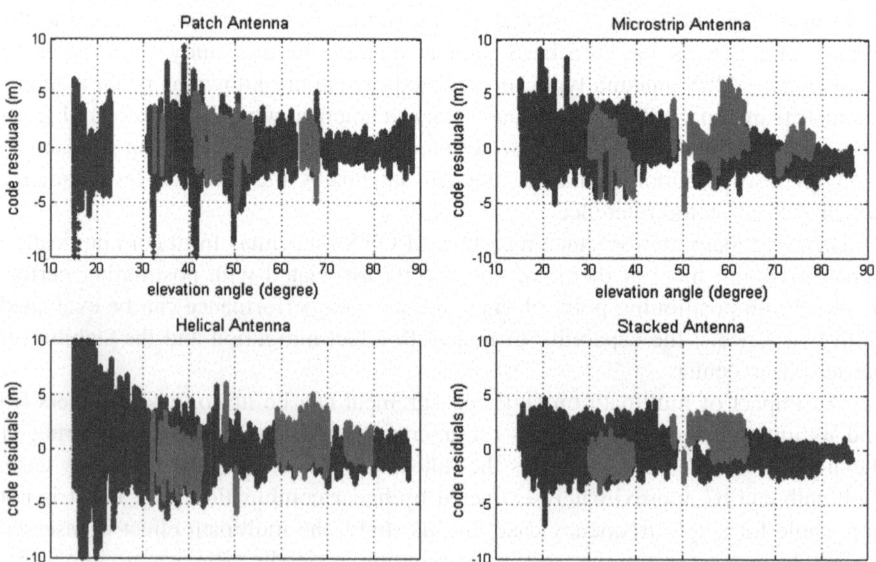

Fig. 2 GPS + BDS standard point positioning residuals with different antennas (*blue* GPS, *red* BDS) (Color figure online)

30° case, but it suffers from severe multipath effect in low elevation angle. This characteristic is determined by its shape and spherical radiation pattern. On the other hand, the microstrip antenna benefits from its lower gain design in low elevation angle, which mitigates multipath efficiently.

Another issue of the miniaturized antenna is its stability of carrier phase center. The antenna phase variation can be calibrated in absolute or relative manner. The absolute antenna phase variation can be calibrated in the anechoic chamber or on the robotics. The relative antenna phase variation can be calibrated by field test without any professional equipment [6, 10]. Especially, a spherical harmonic function can be used to fit phase center variation (PCV) with the posteriori residual of relative positioning [11]. In this study, the antenna PCV has not been considered yet, but we will calibrate these non-geodetic antennas in near future.

3 Handling Multipath Error

The multipath effect weakens the GNSS signals and degrades the code tracking precision, and consequently leads a poor positioning precision. Therefore, the multipath effect has to be considered when low-end receiver is employed. In static GNSS data processing, the multipath effect can be mitigated with the sidereal filter [9] or the adaptive filter [5]. For kinematic positioning case, the multipath effect characteristics can be much more complex. Therefore, a real-time multipath detection and adaption has to be employed.

With regarding to the multipath detection, a real-time multipath extraction can be performed with dual frequency observations [15], but it is not applicable for single frequency users. Fortunately, it is still possible to use C/N0 variation as a multipath effect indicator, which is also referred as SIGMA-Δ model by Brunner et al. [3]. C/N0 is the ratio of the signal carrier to noise power in 1 Hz bandwidth. C/N0 decreases when the direct signals are inferred by reflective signals, so it can be used to measure the multipath effect. Normally, C/N0 also depends on the antenna design, low-noise amplifier (LNA) and satellite elevation angle. An example of C/N0 in static and kinematic case is demonstrated in Fig. 3. The figure shows C/N0 changes smoothly except signals from low elevation satellite in static case, but its variation becomes more complex in kinematic case. The sharp variation is contributed by obstructions, such as canopy, high-rise buildings along the trajectories.

With regarding to multipath handling, there are a number of ways. At first, the multipath contaminated code measurement can be identified as outlier. A DIA procedure can be employed to handle single outlier case [1], but UAV often faces more complex environment. In this study, an adaptive Kalman filter is used [14]. A fixed length moving window average is used as the C/N0 template. If the coming C/N0 is significantly lower than the template, then the weight of corresponding observation can be reduced to incorporate the potential multipath effect.

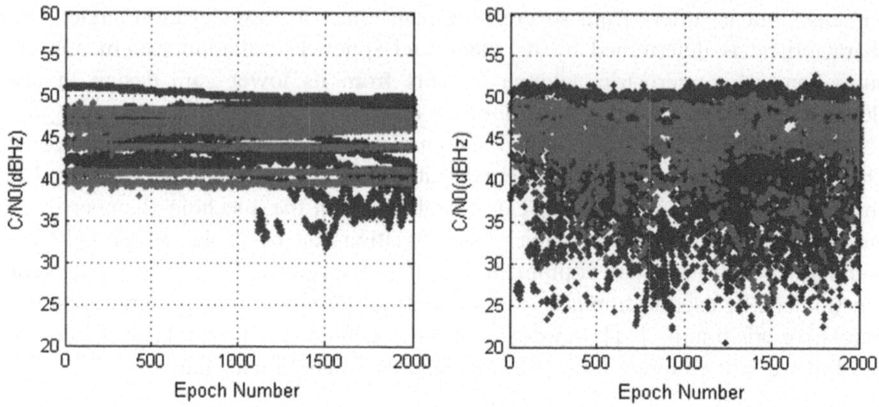

Fig. 3 Comparison of C/N0 variation for static case (*left*) and kinematic case (*right*)

4 Single Frequency Cycle Slip Detection

Another key issue in single frequency kinematic data processing is the cycle slip event. The multiple frequency (including dual frequency) cycle slip detection algorithm has been well documented [2, 7, 13], but single frequency cycle slip is still challenging [4]. A time difference method and Doppler measurement can be used to detect single frequency cycle slip. In this study, a Doppler-aided cycle slip detection method is used. The relationship between the carrier phase and Doppler measurement can be given as:

$$\phi_t - \phi_{t-1} \approx D_{t-1}\Delta t \tag{4.1}$$

where ϕ_t and ϕ_{t-1} are carrier phase observation (in cycle) on t and $t-1$ epoch. D_{t-1} is Doppler observation (in Hz) on $t-1$ epoch. Δt is the time difference between epoch t and $t-1$. The carrier phase observation can be viewed an integrated form of Doppler observation. The cycle slip occurs between the epoch t and $t-1$ would cause the carrier phase difference abnormal, and therefore can be detected by

$$|\phi_t - \phi_{t-1} - D_{t-1}\Delta t| > \varepsilon \tag{4.2}$$

where ε is the threshold of cycle slip detection. An example of Doppler-aided cycle slip is demonstrated in Fig. 4. Doppler observation is immune from cycle slip, but it still is affected by loss of lock. Therefore, Doppler observations also have outliers, but the baseband algorithm normally can catch these events. Therefore, Doppler-aided cycle slip should be combined with receiver event indicator to handle cycle slip in real-time data processing.

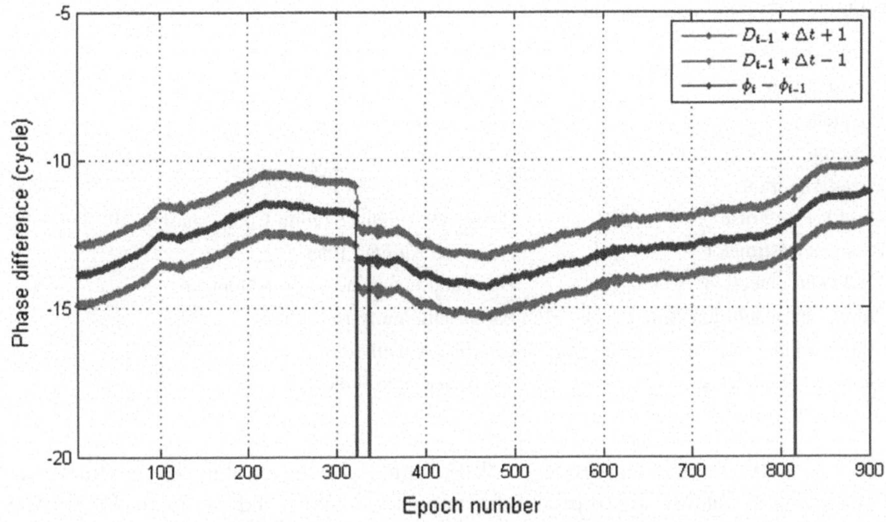

Fig. 4 A demonstration of Doppler-aided single frequency cycle slip detection

5 Experiment and Discussion

In order to verify the feasibility of low cost GPS/BDS RTK for UAV navigation, an experiment is carried out. A DJI phantom 3 drone is used as the test platform. A microstrip antenna and a Tersus GNSS board is tethered on the drone, which is shown in Fig. 5. According to the previous discussion, microstrip antenna is equipped on the drone. A compact, affordable GNSS board named Tersus is used to track GPS L1 and Beidou B1 signals. This GNSS board supports SD card and WiFi connection. It allows to process GNSS data in both real-time and post-process mode. A Trimble BD982 board and a geodetic level antenna are set up around 150 m away and serve as reference station. A RTCM 3 data stream is transmitted

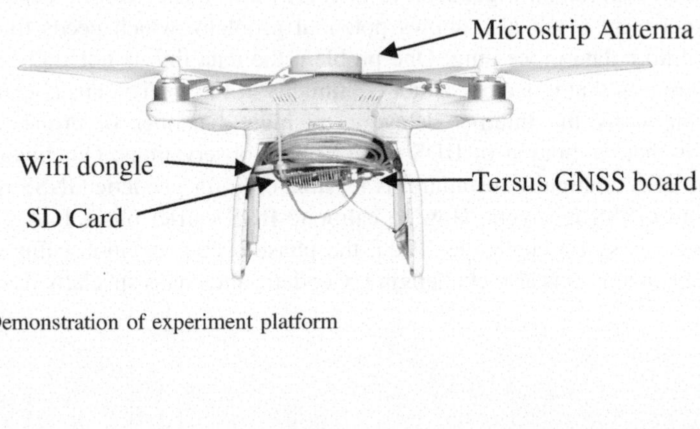

Fig. 5 Demonstration of experiment platform

Table 2 RTK data processing strategy

Configuration item	Value
Constellation	GPS/GPS + BDS
Frequency	L1/L1 + B1
Data rate	5 Hz
Elevation cutoff	15°
Observation noise	Phase: 10 + 3/sin(*E*) mm, Code: phase = 100:1
Parameter estimator	Adaptive Kalman filter
Cycle slip detection	Baseband event indicator + Doppler
Ambiguity resolution strategy	Partial ambiguity resolution
Integer estimation	LAMBDA method
Ambiguity validation	Ratio test (Threshold:3)

over 4G network and converted to WiFi signals by a portable 4G module. The Tersus board obtains differential information via WiFi and performs RTK. The drone flies around a small football field and there are tall trees on the south side. In this experiment, the data is collected for post-process analysis.

The RTK data processing strategy can be summarized in Table 2. Considering poor observation quality, a partial ambiguity resolution strategy is adopted. More specifically, the BDS ambiguities are left as float and only observation from high elevation satellite is attempted to fix. A LAMBDA method is used to search the optimal integer ambiguities. The RTK results with GPS stand-alone and GPS + BDS are presented in Fig. 5. The figure indicates that BDS is helpful for single frequency RTK data processing, although it is not used in ambiguity resolution. With the help of BDS observations, the RTK ambiguity fix rate increases to 55.7 % from 36.7 %. The participation of BDS significantly increases redundant observation number. Most BDS satellites have high elevation angle, so they do not suffer from severe multipath effect. As a result, GPS + BDS can improve the precision of float solution and thus helpful for RTK ambiguity resolution. Considering the low cost receiver and non-geodetic antennas, the single frequency RTK results in high dynamic scenarios is acceptable, but it still can be further improved. With proper modeling and calibration, a geodetic level RTK performance can be expected.

The experiment result also shows potential problem, which needs to be considered in future data processing. One problem the data link is not stable enough, that is, the reason that trajectory is not continuous in Fig. 6. The unstable data link may be caused by the Internet delay or by high dynamics of drone. Another challenge is that calibration of BDS carrier phase observations. Our investigation shows that BDS ambiguity resolution is feasible only for geodetic GNSS receivers or calibrated GNSS receivers. How to calibrate BDS carrier phase biases for low cost receivers is still a challenge. Then, the phase center variation calibration for non-geodetic antennas is also challenging. Geodetic antenna is specially designed to

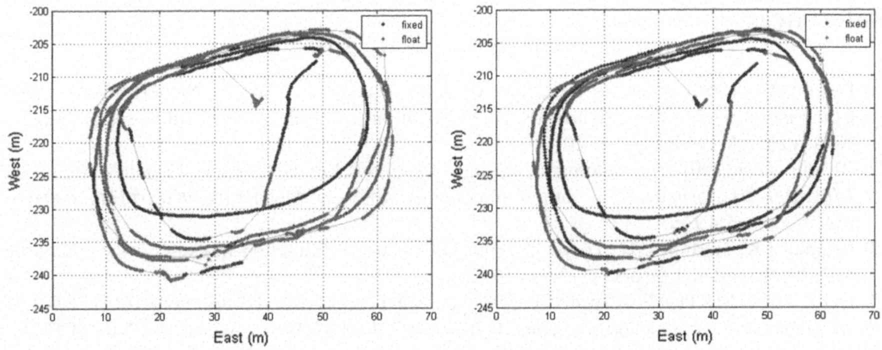

Fig. 6 RTK solution of drone (*left* GPS RTK, *right* GPS + BDS RTK)

acquire stable phase center. For example, Trimble Zephyr antenna has eight symmetric feed points, but most non-geodetic antennas only have one or two feed points. Therefore, the phase center of non-geodetic antennas may not be so stable. In kinematic data processing, the altitude, especially yaw, varies dramatically, which amplifies the impact of phase center variation (PCV) and prevent the integer ambiguity recovery.

6 Concluding Remarks

In this study, we investigated the possibility of applying low cost RTK to UAV. A number of challenges are addressed. The miniature, non-geodetic grade antenna exposes severe multipath problem and dramatic phase center variation. In order to incorporate strong multipath case, a C/N0-based Sigma-Δ model is used to weight the observations. Single frequency cycle slip detection is another challenging issue and we attempt to use baseband event flag and Doppler measurement to detect it. As a result, the four types of observations: code, carrier phase, Doppler and C/N0 are all used in single frequency RTK data processing to enhance its performance. In order to validate the feasibility of precise UAV navigation with low cost RTK system, an on-board experiment is carried out. The experiment results indicate that the fix rate of GPS + BDS single frequency RTK reaches about 56 %, which is acceptable. The performance of the single frequency RTK still can be further improved by calibrating antenna phase variation and calibrating BDS carrier phase biases.

References

1. Baarda, W (1968) A testing procedure for use in geodetic networks. Netherlands Geodetic Commission, Delft, The Netherlands. Available at: http://adsabs.harvard.edu/abs/1968QB296.N4A3v2n5. Accessed 4 Dec 2013
2. Blewitt G (1990) An automatic editing algorithm for GPS data. Geophys Res Lett 17(3):199–202. Available at: http://onlinelibrary.wiley.com/doi/10.1029/GL017i003p00199/full. Accessed 4 Dec 2013
3. Brunner FK, Hartinger H, Troyer L (1999) GPS signal diffraction modelling: the stochastic SIGMA-Δ model. J Geodesy 73(5):259–267
4. Dai Z (2012) MATLAB software for GPS cycle-slip processing. GPS Solutions 16(2):267–272. Available at: http://link.springer.com/10.1007/s10291-011-0249-1. Accessed 8 Nov 2013
5. Ge L, Han S, Rizos C (2000) Multipath mitigation of continuous GPS measurements using an adaptive filter. GPS Solutions 4(2):19–30
6. Görres B et al (2006) Absolute calibration of GPS antennas: laboratory results and comparison with field and robot techniques. GPS Solutions 10(2):136–145. Available at: http://dx.doi.org/10.1007/s10291-005-0015-3
7. Kim D, Langley RB (2001) Instantaneous real-time cycle-slip correction of dual frequency GPS data. In: Proceedings of the international symposium on kinematic systems in geodesy, geomatics and navigation, pp 255–264. Available at: http://gauss.gge.unb.ca/papers.pdf/kis01.kim.pdf
8. Langley RB (1997) The GPS error budget. GPS World 8(3):51–56
9. Park K-D et al (2004) Site-specific multipath characteristics of global IGS and CORS GPS sites. J Geodesy 77(12):799–803. Available at: http://dx.doi.org/10.1007/s00190-003-0359-9
10. Schmid R et al (2005) Absolute phase center corrections of satellite and receiver antennas. GPS Solutions 9(4):283–293. Available at: http://dx.doi.org/10.1007/s10291-005-0134-x
11. Takasu T, Yasuda A (2008) Evaluation of RTK-GPS performance with low-cost single-frequency GPS receivers. In: Proceedings of international symposium on GPS/GNSS, pp 852–861. Available at: http://www.gnss-pnt.org/symposium2008/abstract/oral/B12a/7-727-a.pdf
12. Uti S, Estey LH, Meertens CM (1999) TEQC: the multi-purpose toolkit for GPS/GLONASS data. GPS Solutions 3(1):42–49
13. Wu Y et al (2010) Cycle slip detection using multi-frequency GPS carrier phase observations: a simulation study. Adv Space Res 46(2):144–149
14. Yang Y, Gao W (2006) An optimal adaptive Kalman filter. J Geodesy 80(4):177–183. Available at: http://link.springer.com/10.1007/s00190-006-0041-0. Accessed 4 Dec 2013
15. Wang L, Feng Y, Wang C (2013) Real-time assessment of GNSS observation noise with single receivers. J Global Positioning Syst 12(1):73–82

Using the BDS-R Signal for Soil Moisture Estimation

Xueqian Luo, Songhua Yan, Juan Shan, Hui Yan and Hao Wang

Abstract BeiDou Navigation Satellite System Reflectometry (BDS-R) is an emerging area of BeiDou (BD) applications that use reflected signals in microwave remote sensing. Soil moisture (SM) estimation by using the BD GEO signals are more favorable than the GPS signals, since the fixed elevation angle, the fixed height of orbit, the fixed satellite coverage. In this study, the principle of SM measuring by BDS-R is described. First the signal-to-noise ratio (SNR) data of BD directed and reflected signals are collected though the right-hand circular polarization (RHCP) and left-hand circular polarization (LHCP) antenna, then the SNR data is extracted and the reflection coefficient is computed, at last, the variance of the reflection coefficient is calculated and the empirical model between in situ SM and the variance is established. One month of experimental data are collected at BaoXie, WuHan, and analyzed for further inversion. Experimental results show that the variance of reflection coefficient increases when SM increases and decreases when SM decreases. We can conclude that using BDS-R to retrieve SM is feasible, which will expand the application field of the BD system.

Keywords BDS-R · SM · Reflection coefficient · Variance · SNR

1 Introduction

Soil moisture (SM) is fundamental to land surface hydrology, flooding, groundwater recharge, and evapotranspiration. It plays a critical role in the interaction between the land surface and the atmosphere, and it can also be considered as a key state variable that influences the distribution of short-wave solar energy, long-wave radiant energy, the runoff generation, and the percolation of water into the soil.

X. Luo (✉) · S. Yan · J. Shan · H. Yan · H. Wang
School of Electronic Information, Wuhan University, Wuhan, China
e-mail: luoxueqian@whu.edu.cn

J. Sun et al. (eds.), *China Satellite Navigation Conference (CSNC) 2016 Proceedings: Volume I*, Lecture Notes in Electrical Engineering 388,
DOI 10.1007/978-981-10-0934-1_17

In recent years, remote sensing using Global Navigation Satellite System (GNSS) signals has the advantage of high stability, being global, and high penetration [1–3]. GNSS-R can been used to sense SM [4], ocean altimetry, and ice properties [5, 6]. Some experiments have been conducted by employing GPS-R signals [7].

The BDS system has been developed by China for over a decade. It currently consists of five Geostationary Earth Orbit (GEO) satellites, five Inclined Geosynchronous Orbit (IGSO) satellites, and four Medium Earth Orbit (MEO) satellites. Several remote sensing methods using BDS-R have also been developed such as wet refractivity field assessment [8], Typhoon Wind Speed Observation, etc. However, more experiments are needed to expand its application field [9].

Recently, Shanghai Ocean University has conducted an experiment to detect coastal sea ice using the BD GEO reflected signals [10]. They emphasized that GEO satellites could provide stable geometry for GNSS-R technique with lower angular velocities. They did the Bohai Bay experiment on January 24, 2015. This experiment succeeded in detecting the change of sea ice concentration using the reflected power ratio for 11 h. This was the first coastal experiment using GEO reflected signals to detect sea ice in China.

In this study, a new estimation model using the variance of reflection coefficient is presented. More than 2 months of BDS-R signals have been analyzed. The variance of reflection coefficient is strongly correlated with in situ SM and the correlation coefficient between them is 0.72. The experiment result shows that the variance can be used to SM estimation. In the rest of the paper, we illustrate the method that derails and validate its feasibility for SM estimation. The experimental principle is presented in Sect. 2; the data processing and results are shown in Sect. 3; and the future directions are summarized in the last section.

2 Experimental Equipment

2.1 BeiDou Signal Receiving Device

BD GEO satellites have become the selecting object of communications, meteorology, reconnaissance, navigation and position, due to its unique high orbit and static characteristics. The BDS GEO satellite communication has experienced years of development, whose ability in the mobile communication system is gradually improving in many ways, such as the coverage area, communication capacity, and communication ability. The spatial distribution of BDS satellites and GPS satellites and the space constellations of BDS are shown in Fig. 1a and Table 1.

In Fig. 1b, two patch antennas are employed. They mainly receive the direct signals and reflected signals, namely RHCP and LHCP antenna. The antennas are 2 m high above the ground and the beam widths of them are 120°. The two antennas are controlled by the central control unit, the main part of which is DTU

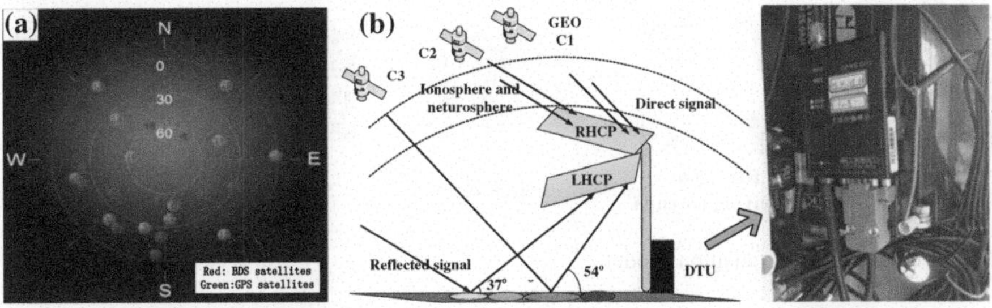

Fig. 1 **a** constellation view of BDS and GPS in graphical interface of UM220 receiver; **b** illustration of GEO experiment, the *gray ellipses* on the ground stand for the reflection area (footprint) of the GEO satellites, and the picture on the right side is the central control unit (*DTU*)

Table 1 The space constellation of BD navigation satellite system

Satellite	Number	Altitude (km)	Position
GEO	5	35,786	58°E, 80°E, 110°E, 140°E, 160°E
IGSO	7	35,786	\
MEO	6	21,528	\

(Data Transfer Unit). It can receive signals from the antennas, and transmit the SNR data to the server. In this study, a long-term ground-based BDS-R SM monitoring station has been established, which is located at the latitude and longitude coordinates of 30.46995°N, 114.52709°E.

The function of the DTU can be divided into two types: (1) convert the data from the terminals into IP flow data; (2) convert the center control instruction into serial data. DTU is the terminal equipment transmitting data through the wireless communication network. With the development of wireless communication network and the expansion of service scope, DTU has been applied in various fields.

The gray ellipses on the ground in Fig. 1b represent the Fresnel zones of the GEO satellites. The Fresnel zone is the area where the reflected antenna (LHCP) can receive the reflected signal, which is restrained by the antenna beam widths and the Fresnel radius. The semimajor axis l and semiminor axis s of the ellipses have been displayed as follows:

$$l = \sqrt{2hc\tau \sin\gamma}/\sin^2\gamma, \quad s = \sqrt{2hc\tau \sin\gamma}/\sin\gamma \qquad (2.1)$$

Among them, h is the depth of BD satellite receiver, c represents the speed of light, γ is the satellite elevation angle and τ is the length of the chips. The formula of reflection area is $S = \pi l s$. The reflection area of BDS C05 is the largest, and the BDS C03 is the smallest. Comparing with the elevation angle, the BDS C05 satellite has the lowest elevation angle around 20°, and the C03 has the highest elevation angle around 55°. The elevation angles of the other three satellites are 43°, 37°, and 28°.

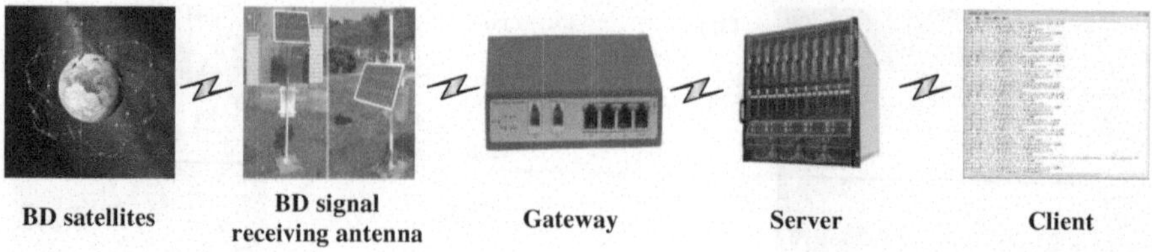

Fig. 2 BDS-R data acquisition model

2.2 The BDS-R Data Acquisition

In this study, the experimental data are received in real time, and transmitted in real time, as is shown in Fig. 2. The original BDS-R observation acquisition can be divided into four steps: (1) the Baoxie BDS-R SM monitoring station receives observations from the BDS GEO satellites. One can see that, the experimental site is a flat, bare field without much tall buildings, trees, or forests in the surroundings; (2) the station adopts the data terminal unit module to transmit the data to the gateway through the general packet radio service technology; (3) the gateway sends the observations to the data server on the Internet; (4) clients acquire the observation through the software application. The data are received in NMEA 0183 format, which defines an electrical interface and data protocol for communications between marine instrumentation.

2.3 The Power Supply System

The observation devices are installed in the wild with no electronic power facilities. Therefore, the power supply is composed of a storage battery and a solar panel to ensure the integrity of data. In this experiment, we need a hardware module to manage the storage battery and solar panels, which can keep coordination between them.

In Fig. 3, the black box on the left side is the power control module (the Vectra solar charge controller). The model can utilize the pulse width modulation to manage the storage battery and the solar panel. It mainly administrates analog circuit by using a digital output, and adjusts the duty cycle and period of the pulse. It can also check the storage battery by the state of LED. When the battery reaches saturation, the controller can prolong the interval of charging by increasing the pulse cycle appropriately to shortening the life cycle of storage battery.

With the power supply system, we also need to consider the stability of voltage at the same time. The output voltage of power control module is 12 V, but the rated voltage of the acquisition module and transmission module is 4.5–5.5 V. DC/DC transfer module is applied to ensure system working under rated voltage.

Fig. 3 Power supply system, power control module, and solar panel

3 Experiment Results

3.1 The Variance Value

During this experiment period, the ground temperature was recorded to be 10°–20° and a ground SM meter was used to collect in situ SM measurements for comparison of our BDS-R retrieved results. There are two humidity sensors buried under the ground. We use the average values of the two hygrometers as the in situ SM value in this experiment. The BDS-R signals are received at every second, which means a large data volume. After much experiment, a one-hour averaging of the SM values is chosen to be performed since SM does not change much in only one second.

The SNR value of the BD satellites is shown in Fig. 4, the blue columnar stands for the SNR values, the numbers in the rectangle box represent the satellite serial number been captured.

Fig. 4 SNR data of the BD signals, the blue columnar stands for the SNR values

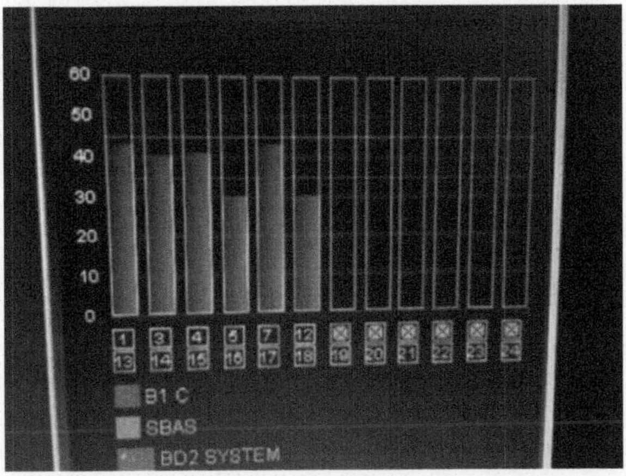

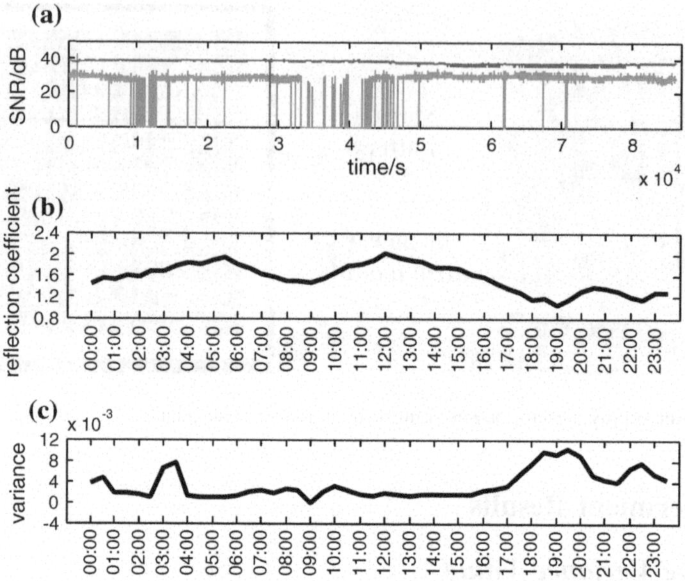

Fig. 5 **a** SNR data of direct and reflected signal; **b** the average reflection coefficient values from 00:00:00 a.m. to 23:59:59 p.m. on November 15, 2014; **c** the variance values of every one hour

Figure 5 shows the SNR data, the reflection coefficient values of every one hour and the variance values from 00:00:00 a.m. to 23:59:59 p.m. on November 15 (Nov. 15), 2014. In Fig. 5a, the blue line represents the SNR data of the reflected signal (SNR_r) and the SNR_rvalues mainly vary between 25 and 30 dB. In addition, the red line represents the SNR data of direct signal (SNR_d). It is clear that the fluctuation of SNR_d is more smooth than the SNR_r, the values of SNR_d vary between 38 and 40 dB. The red vertical lines or blue vertical lines (e.g., there are no SNR values of reflected signal from 30,000 to 50,000 s) illustrate that the short-term disruptions occur in this second, which is probably caused by power system instability.

Figure 5b shows the values of reflection coefficient, it mainly varies from 1.8 to 2.4. The formula of reflection coefficient is $R = 10^{(\mathrm{SNR}_r - \mathrm{SNR}_d)/20} = \sqrt{10^{\mathrm{SNR}_r/10}/10^{\mathrm{SNR}_d/10}}$.

Figure 5c depicts the values of variance which varies from 0 to 0.008. Three peaks appeared at 03:00, 19:00 and 22:00, which mean that the fluctuation of reflection coefficient is noticeable at these moments.

3.2 *Data Processing*

The data processing can be divided into 6 steps: (1) BD original data acquisition, including the satellite serial number, UTC, SNR data, elevation angle and azimuth

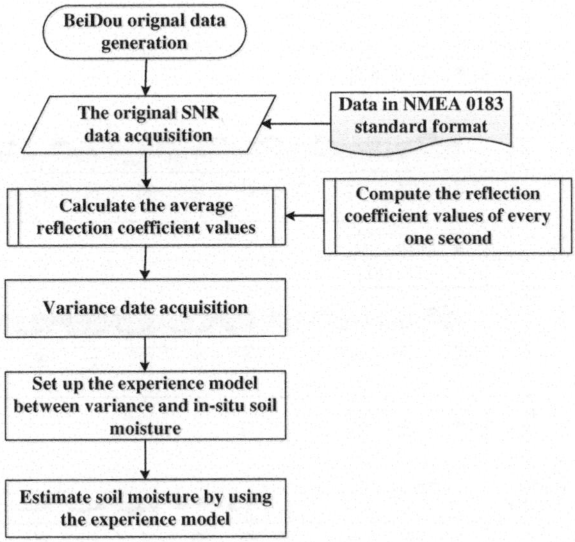

Fig. 6 The flowchart of data processing

angle, which has been introduced in Sect. 2.2. In the following data processing, we need the UTC information and the SNR data of every GEO satellite; (2) SNR data and UTC information acquisition; (3) calculation of the reflection coefficient values *R* in every second; (4) variance values generation, the resulting values of variance is a two-dimensional array, which can be compared with the in situ SM; (5) the experience model establishment, we fit the relationship between the variance and the in situ SM, and we get a logarithmic relation curve; (6) SM fluctuations estimation according to the experience model. For example, if we know the SNR data of a certain period in one day, then we can estimate the in situ SM values according to the experience model between SM and variance (Fig. 6).

3.3 *Results*

The BDS signals collected from November 14, 2014 00:00:00 to November 27, 2014 23:59:59 at the long-term Baoxie BDS-R SM monitoring station are used in this experiment. During this period, there were significant rainfall events on November 23 and 24.

In Fig. 7, the solid black circles represent the values of in situ SM (%), the hollow red, green, and pink stars stand for the variance of BDS C01, BDS C02, and BDS C03 respectively. The labels marked in *x*-axis are to signify the date in 2014; for example, 14 represents November 14, 2014. The left *y*-axis represents the value of in situ SM, while the right *y*-axis signifies the variance.

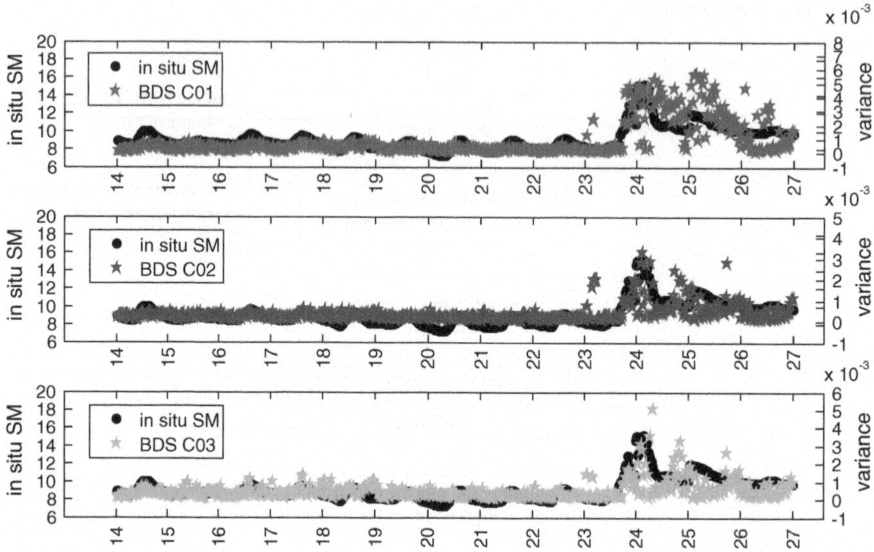

Fig. 7 Variation of in situ SM and variance of reflection coefficient, the labels marked in *x*-axis are to signify the date in 2014

For BDS C01, the in situ SM values increase to 15 % when a rainstorm occurs on November 23, and then decease to 8 %. The fluctuation of variance is consistent with the SM, the variance values increase to 0.006 when the rainfall event occurs, and decrease to 0.001 after this rainstorm. It shows that the tiny changes in variance can reflect the changes in SM. The correlation coefficient between them is 0.75.

For BDS C02 and C03, the inversion results are basically the same. Before the date November 23, the variance values basically keep stable until the rainfall event occurs on November 23. In a word, the overall trend of variance and SM is consistent, and the correlation coefficient can reach to 0.65.

Figure 8 shows the in situ SM values and variance of reflection coefficient for BDS C01. It conforms that the variance is consistent with the in situ SM. The solid red stars and solid black circles represent the variance and in situ SM, respectively. The variance indicates a same fluctuation on each day during the period from November 14 to 23. For example, on November 14, the in situ SM first increases from 8 to 10 % and then decreases to about 8 %. The variance of coefficient shows the same fluctuation, increasing from 0.0002 to 0.0008 and decreasing to around 0.0001. However, after a heavy rain on November 23, during the period from November 24 to November 26, the in situ SM values show a downward trend from 16 to 8 %, and the variances mainly decrease from 0.006 to 0.0001. But the variance fluctuates dramatically at higher SM values. It seems that the error is bigger than lower SM values. Further research will focus on the SM inversion using variance.

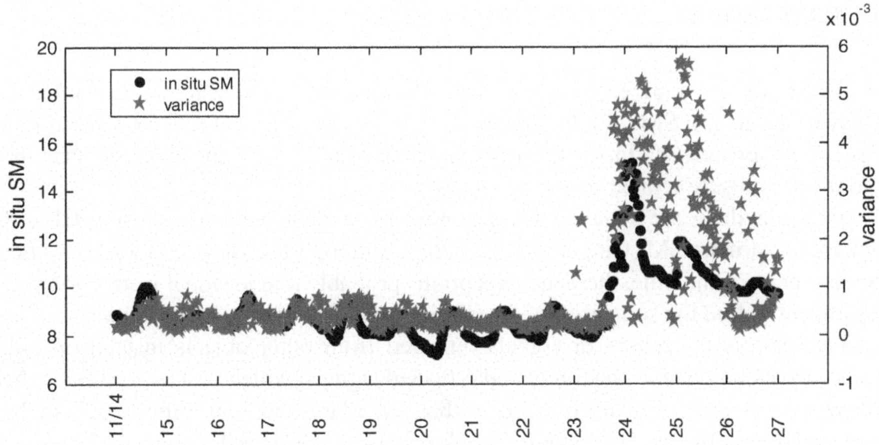

Fig. 8 Variation of in situ SM and variance of reflection coefficient for BDS C01

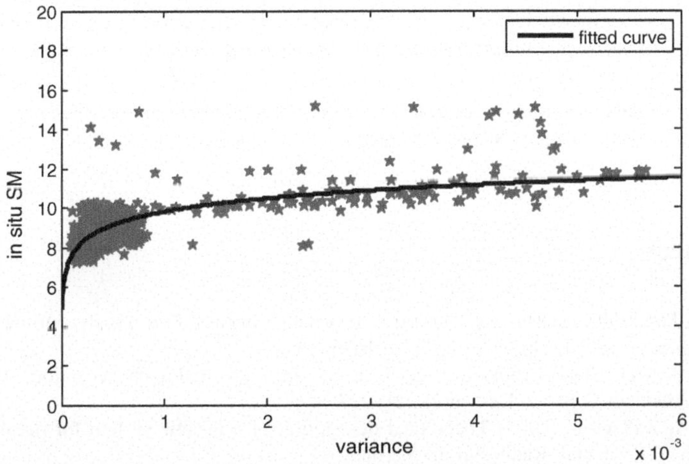

Fig. 9 Logarithmic relationship between the variance and in situ SM

Figure 9 shows the logarithmic relation between the variance and in situ SM of BDS C01. The variance and in situ SM are plotted out in red stars for comparison. The labels marked in *x*-axis are to signify the variance, while the right *y*-axis signifies the in situ SM values. As we can see, the red solid stars mainly focus on the smaller values of the SM.

4 Conclusion

In this study, several experiments were conducted to investigate the relation between the in situ SM and the variance, about 13 days of BD reflected and direct signals are processed. From the experimental results, we can conclude that the variance is closely related to the in situ SM.

However, there are some issues that need to be dealt with when using BDS-R signals to estimate SM. One is the instability; although the variance increases after the rain most of the time, there are exceptions probably due to signal instability. It is significant to find the source of the instability to realize more robust estimation. The second issue is the noises or signals reflected from other objects instead of soil. These noises will distort the direct and reflected signals, which will be studied in the following research. The third issue is that abundant and long-time data will be analyzed to fit the more accurate relation between SM and variance. Complex terrain, such as vegetation-covered ground, will also be considered in BDS-R-based SM estimation.

For future research, we will investigate the characteristics of BD GEO reflected signals on instability, analyze the reasons behind it, and find the ways to maintain the stability to achieve reliable SM estimation. Furthermore, we will also collect more data to verify the feasibility of the inversion method.

Acknowledgements The authors would like to acknowledge the support of the National Natural Science Foundation of China (NSFC 41571420).

References

1. Jin SG, Feng GP, Gleason S (2011) Remote sensing using GNSS signals: current status and future directions. Adv Space Res 47(10):1645–1653
2. Jacobson MD (2010) Snow-covered lake ice in GPS multipath reception—theory and measurement. Adv Space Res 46(2):221–227
3. Hajj GA, Zuffada C (2003) Theoretical description of a bistatic system for ocean altimetry using the GPS signal. Radio Sci 38(5):1089
4. Yan S, Li Z, Yu K, Zhang K (2014) GPS-R L1 interference signal processing for soil moisture estimation: an experimental study. EURASIP J Adv Signal Process 2014(1):1–13
5. Larson KM, Braun JJ, Small EE, Zavorotny VU, Gutmann ED, Bilich AL (2010) GPS multipath and its relation to near-surface soil moisture content. IEEE J-STARS 3(1):91–99
6. Masters D, Zavorotny V, Katzberg S, Emery W (2000) GPS signal scattering from land for moisture content determination. In: Editor proceedings of the international geoscience and remote sensing symposium. Pub Place, Honolulu, USA, pp 3090–3092
7. Larson KM, Löfgren JS, Haas R (2013) Coastal sea level measurements using a single geodetic GPS receiver. Adv Space Res 51(8):1301–1310
8. Lu C, Li X, Nilsson T, Ning T, Heinkelmann R, Ge M et al (2015) Real-time retrieval of precipitable water vapor from GPS and BeiDou observations. J Geodesy 1–14

9. Yang Y, Li J, Wang A, Xu J, He H, Guo H et al (2014) Preliminary assessment of the navigation and positioning performance of BeiDou regional navigation satellite system. Science China (Earth Sciences) 01:144–152 (01 Jan 2014)
10. Fernandez-Moran R, Wigneron JP, Lopez-Baeza E, Al-Yaari A, Coll-Pajaron A, Mialon A et al (2015) Roughness and vegetation parameterizations at L-band for soil moisture retrievals over a vineyard field. Remote Sens Environ 170:269–279

The Potential to Estimate Soil Moisture Based on sn_rnx Data

Juan Shan, Songhua Yan, Xueqian Luo, Xingxing Li and Hancheng Yuan

Abstract Using Global Navigation Satellite System (GNSS) signals of BeiDou navigation satellite system (BDS) and global position system (GPS) for soil moisture estimation is an emerging technology in microwave remote sensing since soil moisture is an important index in regional water cycle research. Based on the developed forward multipath model, this study first investigates the feasibility of soil moisture estimation based on GNSS interference signals; then compares the difference between the Signal-to-Ratio (SNR) data and the signal strength (sn_rnx) data from GPS. After that, this paper analyzes the experimental data from BDS receivers and CORS stations. The analyzed results show that both fitted phase of BDS SNR data and GPS sn_rnx data will increase when the soil moisture increases, and the phase will decrease when soil moisture decreases, and the method of using sn_rnx data can obtain the change trend of soil moisture and can be used as an alternative to solve the problem that the SNR data is unavailable in many GNSS data sets.

Keywords GNSS · Soil moisture · Interference signal · SNR · Sn_rnx

1 Introduction

Global Navigation Satellite System (GNSS) signals have been an emerging area of satellite navigation applications as these signals have the advantages of high stability and being global. For example, the geodetic GPS receivers recorded multipath signals have potentials for geophysical parameters detections, e.g., soil moisture, ocean altimetry, and ice properties [1, 2]. Both BDS and GPS satellites have made achievements in remote sensing field at present.

BeiDou satellite navigation system was built on October 25, 2012 with the last one Geostationary Earth Orbit (GEO) satellite launched, and officially began on

J. Shan (✉) · S. Yan · X. Luo · X. Li · H. Yuan
School of Electronic Information, Wuhan University, Wuhan, China
e-mail: shanjuan@whu.edu.cn

J. Sun et al. (eds.), *China Satellite Navigation Conference (CSNC) 2016 Proceedings: Volume I*, Lecture Notes in Electrical Engineering 388,
DOI 10.1007/978-981-10-0934-1_18

December 27, 2012 as a new generation of 1. Like GPS, BDS is composed of three parts, including the space constellation part, ground control part and user receive part. The space system part is current made up of 14 satellites, including five GEO satellites, five Inclined Geosynchronous Orbit (IGSO) satellites and four Medium Earth Orbit (MEO) satellites [3].

Developed for over a decade, BDS has made significant improvements in current GNSS remote sensing research recently. For example, an important typical application is relative positioning and troposphere delay estimation [4]. Besides, L-land signal of BDS have been used to sense soil moisture because the signal is sensitive to soil moisture [5]. Thus, regional soil moisture monitoring capabilities can be improved in support of multi-sensor collaboration [6]. In this paper, the interference BDS signal comprising direct and reflected BDS signal from the soil surface is investigated and SNR data is mainly analyzed for soil moisture estimation.

GPS multipath signal has been employed for remote sensing. For example, GPS instruments can be used to infer soil moisture using information contained in the ground-reflected or multipath signal [7]; signals routinely could be used to provide a global network of soil moisture sensors [8], and electrodynamics single-scattering forward model used to test three GPS interferogram metrics are affected by soil moisture [9]; The relationship between the SNR multipath delay phase and the soil moisture can well described by using exponential function in Fresnel reflection area [10].

Though SNR is usually used as measurement quality indicators for GNSS measurements, the SNR data are not always available because it is no actual effect and just is used as a supplementary data for signal quality assessment in GNSS position. Based on this condition, signal strength indicators, i.e., sn_rnx, another type of data will be analyzed and investigated in this paper to estimate the variation of soil moisture since the sn_rnx has properties similar to those of SNR data.

The remainder of the paper is organized as follows. Section 2 introduces the distribution of constellation of the BDS regional system and the effect of using interference BDS signal on soil moisture estimation. Section 3 shows the experiment field of using sn_rnx data from GPS to determine soil moisture and the experiment results, and the final section gives discussions and conclusions.

2 Analyze SNR Data of Interference BDS Signal

2.1 Description of the Constellation Distribution of BDS Satellites

As one of only four navigation and positioning system in the world, BDS is a regional navigation system and only providing operational positioning services over Asia-Pacific region at present. As shown in Fig. 1, the service area of BDS was marked with the red circle, the details of which can be found in the literature (China office, 2013).

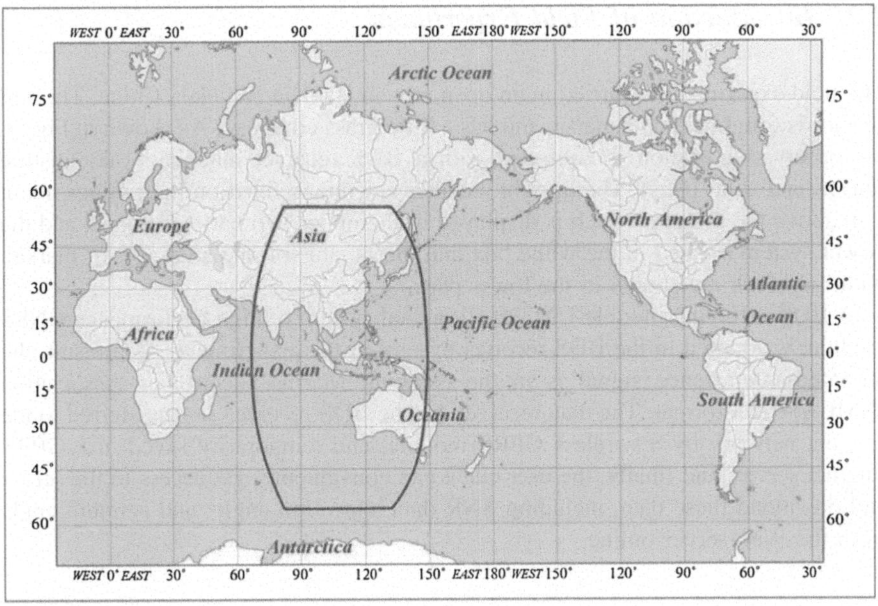

Fig. 1 The BDS service area

There is a significant difference in satellite visibility of BDS regional system from that of GPS system because of the special constellation distribution of BDS, as is shown in Fig. 2. The ground tracks of the BeiDou satellites were described. In the figure of 2.2, large ellipsoid line is the trajectory of GEO satellite such as C01 and C02, while like the line of the eight is the trajectory of IGSO satellites such as C07 and C08. Since these satellites located in the southern sky of China, the interference phenomenon of IGSO satellites occurs when the satellite elevation angle is changed between 0° and 30°, so the interference signal of IGSO satellites will be used for deeper research in this paper.

Fig. 2 Ground tracks of the BeiDou satellites

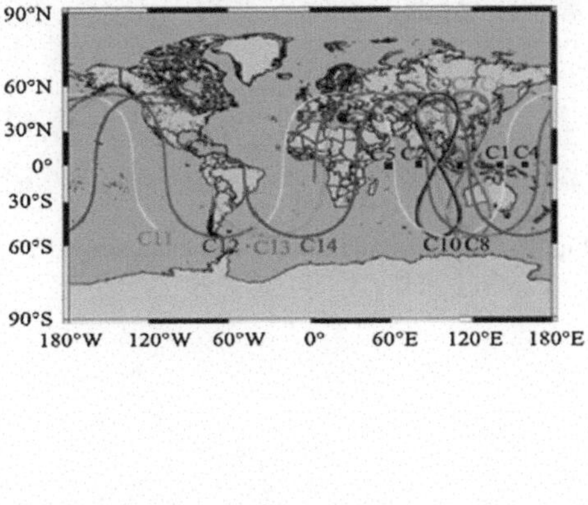

2.2 Introduction of Field Experiment

The field experiment is carried in an open area in BaoXie, WuHan, China. The soil surface is roughness and sparse, and has a little grass coverage. As shown in Fig. 3, the observation station is equipped with a BDS receiver antenna, soil moisture sensor, and solar panel. The upright receiver antenna is fixed on the railings about 2 m above the ground, which is displayed in the upper two parts of Fig. 3; and the power system located in the white box and the in situ soil moisture sensor marked with red block are shown in the lower part.

A BDS receiving module-UM220-III, a dual system of high-performance GNSS module, is adopted in the BDS receiver system in this experiment. As a result, the recorded interference signal is in the National Marine Electronics Association (NMEA) 0183 format. The data received by the BDS receiver are transferred to the Internet network by a wireless GPRS module, and temporarily saved in a GPRS Internet server; and finally, the user client can conveniently get access to the server and download these data, including SNR data, elevation angle, and azimuth angle from the Web server online.

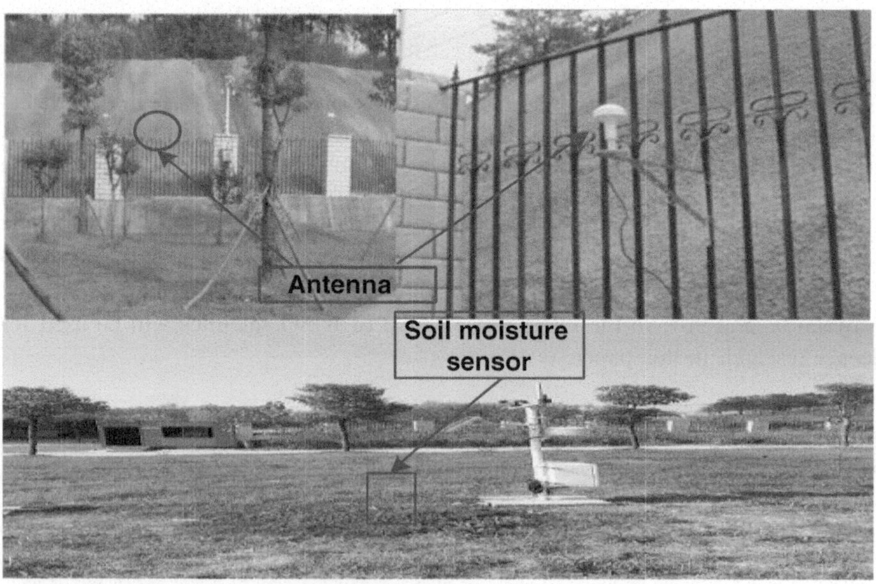

Fig. 3 Antennas and the experimental site

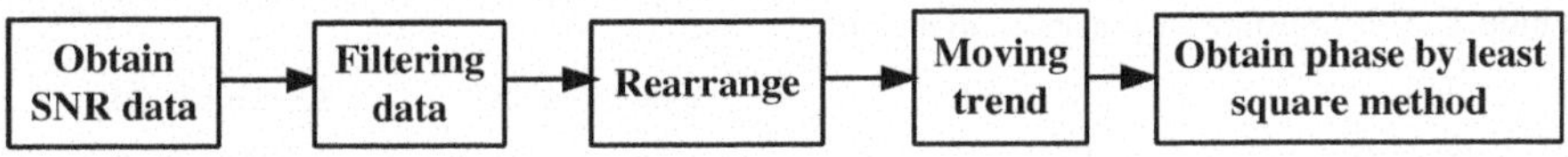

Fig. 4 Block diagram of the data processing

2.3 *Experiment Results*

In the paper, more than 2 months of observation data are collected and processed. The introduction of principle and methodology can be referred to the literature where GPS signal were described [8]. A block diagram of the data processing is shown in Fig. 4, which involves the following steps:

After obtaining the phase of interference signal, the relationship between the phase of BDS and the soil moisture will be determined. Figure 5 shows the analysis results from September 6 to October 14, 2014 (about 30 days). The upper and the lower part displays the relationship between the phase of C07/C08, and the soil moisture, respectively. The *x*-axis signifies the date in 2014; for example, 6/10 represents June 10, 2014. The blue triangles represent the value of soil moisture measured by in situ sensor and the full cyan circles stand for the fitted phase. In the lower part of Fig. 5, the phase data are discontinuous because of power shortage of solar battery.

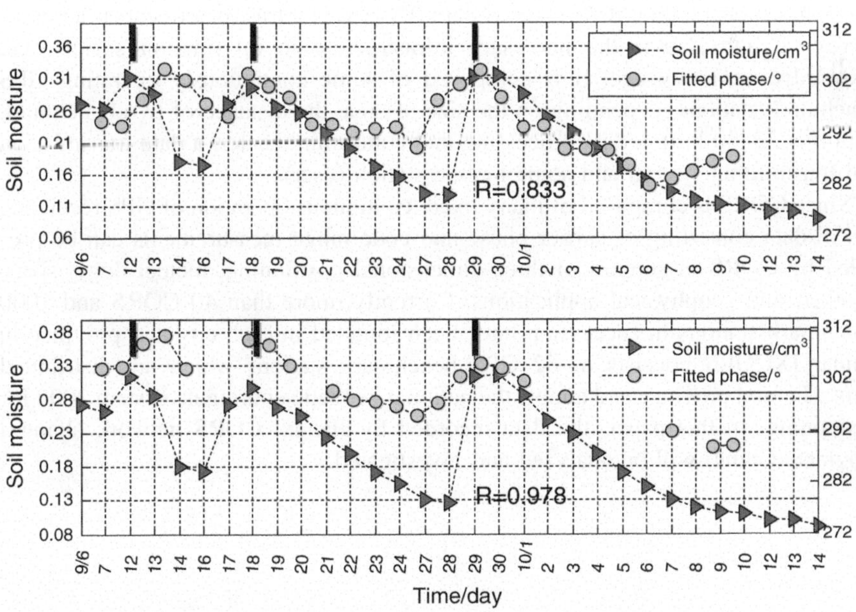

Fig. 5 Variation of soil moisture and phase of SNR

Table 1 Soil moisture information after rainfall

Date(/day)	September 12	September 18	September 29
Soil moisture (cm^3)	0.315	0.296	0.314

During the 2-month observation interval, there are three distinct precipitation events (storm totals >10 mm) which occurred on September 12, 18, and 29, respectively, and marked in black bars in the figure. Therefore, the maximum points of soil moisture are shown in the following Table 1.

In Fig. 5, each measured rainfall event is followed by a rise in soil moisture; and by a decrease over a period of days or weeks. Both BDS C07 and C08 satellites phase closely track the value of in situ soil moisture observed by soil moisture sensor. For example, the third rainstorm took place on September 29 leads to a high soil moisture of 0.314, and the phases of the above two satellites peak, i.e., 304.6° of BDS C07 and 305.2° of C08 on September 30. The BDS phases then decrease all the way. The correlation coefficient of both satellite of C07 and C08 reach 0.833 and 0.978, respectively.

3 Analyze sn_rnx Data of Interference GPS Signal

3.1 Composition of the CORS

Continuous Operational Reference System (CORS) is infrastructure construction of surveying, and also is the necessary infrastructure in information society and knowledge economy age. It is composed of some continuously operating GPS benchmark stations (depending on the size of area, the distance of two benchmark station is 20–80 km), a data processing and controlling center, a data transmission and transmitting system and some mobile station (user).

Since CORS is composed of many fixed, continuous operation of GPS receivers, GNSS data consisting of carrier phase and code range measurements can be provided by CORS in support of three-dimensional positioning, meteorology, space weather, and geophysical applications. Currently, more than 40 CORS and 1000 benchmark stations in succession have been founded in each city and province in China. JXCORS consists of 62 GPS benchmark stations which are distributed across Jiangxi province and adopts the national geodetic coordinate system 2000 for positioning. In this paper, the data recorded by Jiangxi CORS network (Jiangxi province, China) will be analyzed and investigated.

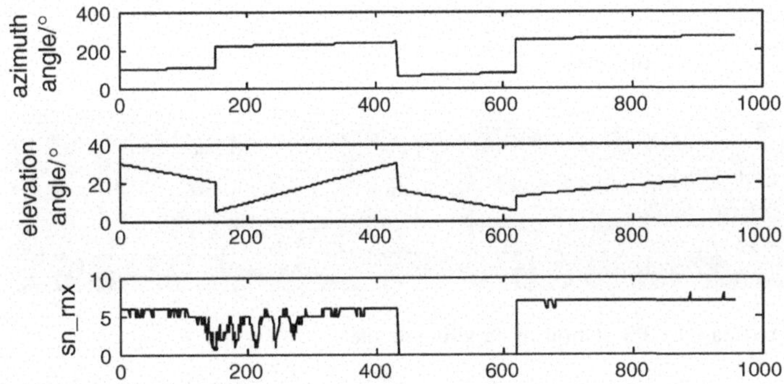

Fig. 6 The sn_rnx data

3.2 *The Characteristics of sn_rnx Data*

The generation of the RINEX (The Receiver Independent Exchange Format Version) signal strength indicators sn_rnx in the data records is a standardization of SNR data (RINEX vision 3.00). RINEX data uses discrete or discontinuous values to represent information. Here, sn_rnx is projected into interval 1–9, and every integer represents different signal strength. When the value of sn_rnx is greater, the signal strength is bigger. Figure 6 shows sn_rnx data which manifests a quasi-periodic oscillating pattern.

In Fig. 6, sn_rnx changes in the lowest figure between 3 and 7 and manifests an obvious quasi-periodic oscillating pattern when elevation angle are from 7° to 40°, during which the azimuth angle remains nearly the same (around 226°).

Since it is a quantization of SNR, the sn_rnx has the same period and the nearly same phase as the sine wave of SNR. As a result, the same model of SNR can also be used to obtain the phase of sn_rnx data.

3.3 *Description of Field Experiment*

Since JXCORS has 62 stations, including some roof-based GPS stations and ground-based stations whose experiment scene are shown in Fig. 7. The ground-based station was founded in the meteorological observatory. In the figure, the geodetic antenna is placed on the top of tripod at a height of 2 m, which is a microstrip GPS antenna. The area is a flat filed with short sparse grass and no building within 300-m radius around.

Fig. 7 Ground-based GPS station of experiment site

3.4 Experiment Results

Due to the absence of the in situ soil moisture in JXCORS, 2 months rainfall data in Yugan station and Jingdezhen station (both stations are in Jingdezhen city, Jiangxi province, China) are collected for experiment comparison.

Figure 8 shows the result from May 7 to June 10, 2013 from Yugan station. The blue and cyan triangles stand for the fitted phase of GPS01 and GPS10, respectively. During the 2-month observation interval, the three distinct black bars represent the precipitation events(storm totals >20 mm) occurred on May 7, 25 and June 6, respectively.

In Fig. 8, each rainfall event is followed by a rise in both the phase of GPS01 and GPS10. For example, the first rain took place on May 25 leads to an increase in phase from 167.7° to 215.1° of GPS01 and from 31.5° to 51.6° of GPS10.

Another experiment is conducted in Jingdezhen station. The same two satellites, GPS01 and GPS10 are selected for data analysis, and the result is shown in Fig. 9. In Fig. 9, the three precipitation events lead to increases in the phase of both GPS01 and GPS10, especially the first rain and the third rain event.

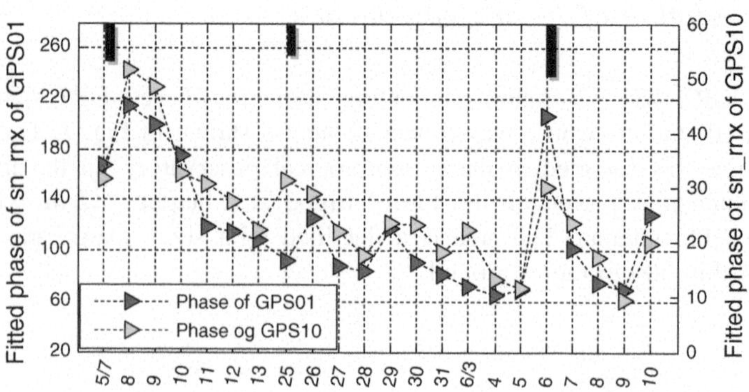

Fig. 8 The relationship of the phase of GPS01 and GPS10 (from Yugan station)

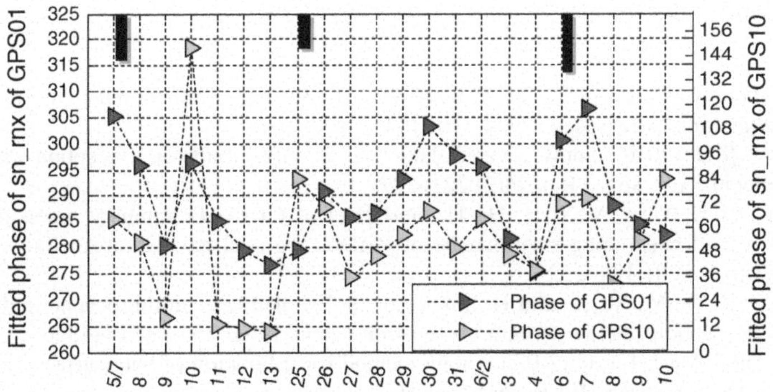

Fig. 9 The relationship of the phase of GPS01 and GPS10 (from Jingdezhen station)

Although the phase values are agreement with the rainfall, there are still some abnormal data when there is no rain. For example, the change between the two fitted phase of GPS 01 and GPS10 is inconsistent from June 8 to 10. This may be caused by the noise or other reason and need to be investigated further.

4 Discussions and Conclusions

Since BDS is expected to make a contribution to GNSS remote sensing for it has a highly signal structure similar to GPS in theory. SNR characteristics of BDS satellites are first introduced and then more than 30 days of BD interference signals are investigated and processed in this paper. Experimental results show that the phase of SNR and in situ soil moisture are strongly correlated. As a result, interference signals of BDS may be used to retrieve soil moisture as well as GPS signals.

Owning to the lack of SNR data in most CORS stations, sn_rnx data, which is a quantization of SNR data, is expected as an alternative for soil moisture estimation. The similar data processing method used on the SNR data is also used to identify the phase change of sn_rnx data from CORS stations. From these experimental results, we can also conclude that the change in interference phase of GPS is related to the precipitation events.

Based on the existing technology and method, future work will concentrate on conducting more experiments to develop absolute retrieval algorithms for soil moisture estimation by sn_rnx phase when considering the antenna gain pattern, soil types and soil moisture profiles. Investigating the instability of interference signals and analyzing the reasons behind it will also be explored.

Acknowledgments The authors would like to acknowledge the support of the national natural science foundation of China (NSFC 41571420).

References

1. Jin SG, Feng GP, Gleason S (2011) Remote sensing using GNSS signals: current status and future directions. Adv Space Res 47(10):1645–1653
2. Jacobson MD (2010) Snow-covered lake ice in GPS multipath reception—theory and measurement. Adv Space Res 46(2):221–227
3. Yang Y, Li J, Wang A, Xu J, He H, Guo H et al (2014) Preliminary assessment of the navigation and positioning performance of BeiDou regional navigation satellite system. Sci Chin (Earth Sci) 01:144–152
4. Xu A, Xu Z, Ge M, Xu X, Zhu H, Sui X (2013) Estimating zenith tropospheric delays from BeiDou navigation satellite system observations. Sens Basel 13(4):4514–4526
5. Guo P, Shi J, Du J, Liu Q (2012) A new method for estimation of bare surface soil moisture with L-band radiometer. IEEE, pp 658–661
6. Xiang Z, Chen N, Chen Z (2014) Spatial pattern and temporal variation law-based multi-sensor collaboration method for improving regional soil moisture monitoring capabilities. Remote Sens Basel. 6(12):12309–12333
7. Rodriguez-Alvarez N, Bosch-Lluis X, Camps A, Vall-Llossera M, Valencia E, Marchan-Hernandez JF et al (2009) Soil moisture retrieval using GNSS-R techniques: experimental results over a bare soil field. IEEE Trans Geosci Remote Sens 47(11):3616–3624
8. Zavorotny VU, Larson KM, Braun JJ, Small EE, Gutmann ED, Bilich AL (2010) A physical model for GPS multipath caused by land reflections: toward bare soil moisture retrievals. IEEE J Sel Topics Appl Earth Obs Remote Sens 3(1):100–110
9. Chew CC, Small EE, Larson KM, Zavorotny VU (2014) Effects of near-surface soil moisture on GPS SNR data: development of a retrieval algorithm for soil moisture. IEEE Trans Geosci Remote 52:537–543
10. Wu ASS, Yang L (2012) Soil moisture retrieval by active/passive microwave remote sensing data. SPIE Remote Sensing

The Application of Kinematic Relative Positioning Technology with BDS in Aerospace TT&C Equipment Precision Verification

Wei Zhou, Zhulong Sun, Yaqi Ren and Feijuan Yao

Abstract The aerospace TT&C system plays an important role in tracking spacecraft, and it is necessary for maintaining its good performance and precision. This paper focuses on the application of BeiDou satellite navigation system (BDS) in the precision verification of aerospace TT&C equipment, an algorithm model of the BDS kinematic relative positioning is proposed, and the feasibility of this algorithm is verified by an actual flight experiment. Based on above, the BDS precision verification system is constructed, and the composition, function, information interaction, and working principle of the system are introduced, several methods are studied such as kinematic relative positioning data processing method with BDS, data processing method for TT&C equipment and comparison analysis method, then a complete data processing flow is formed. Based on this system, the precision verification experiment of a TT& C equipment is completed, and various index of precision is obtained, which plays an important role in analyzing the performance of the aerospace TT&C equipment.

Keywords Aerospace TT&C equipment · Precision verification · BDS · Carrier phase · Kinematic relative positioning

1 Introduction

Aerospace TT&C equipment plays an important role in determination, control and maintain of spacecraft orbit, and it is necessary for maintaining good performance and precision of TT&C equipment [1]. However, the accuracy and performance of TT&C equipment will be reduced because of device aging, installation error, algorithm defect, etc. Therefore, before the use of a developed or transformed TT&C equipment, the performance test and precision verification should be carried out to determine whether it can meet the requirements [2].

W. Zhou (✉) · Z. Sun · Y. Ren · F. Yao
LuoYang Electronic Equipment Testing Center, Luoyang, China
e-mail: zhouwei_0611@163.com

J. Sun et al. (eds.), *China Satellite Navigation Conference (CSNC) 2016 Proceedings: Volume I*, Lecture Notes in Electrical Engineering 388,
DOI 10.1007/978-981-10-0934-1_19

In the late 1960s, for the first time in China the cinetheodolite is used for the precision verification of large radio system, until 1990s, using the cinetheodolite measurements to obtain the precision of the TT&C equipment has become a typical application model of the precision verification [3]. But this model is difficult to meet the needs of precision verification gradually because of the shortcomings such as limited tracking ability, long data processing cycle, higher meteorological conditions requirement. With the rapid development of satellite positioning technology, experts and scholars in China has carried on the research on the application of GPS kinematic relative positioning technology in aerospace TT&C equipment precision verification.

At present, the development of BeiDou satellite navigation system (BDS) has been transferred to a new stage of construction and application simultaneously. This paper focuses on the application of BDS in the Aerospace TT&C equipment precision verification. The algorithm of kinematic relative positioning is studied, through data analysis we found that the precision of BDS carrier phase kinematic relative positioning is centimeter level, which can meet the needs of current TT&C equipment precision verification, based on above, the composition, working mode and test method of BDS precision verification system are discussed, a complete process of precision verification is achieved by an actual flight experiment, the feasibility and effect of this method are analyzed, which provides a new solution for aerospace TT&C equipment precision verification.

2 Composition and Working Principle of BDS Precision Verification System

The working principle of BDS precision verification system is shown in Fig. 1: the plane as the tracking target equipped with BDS receiver and cooperation target (responder, beacon, light source) for TT&C equipment will fly according to the planning route, and flight trajectory will be obtained by the difference resolving of BDS observation from onboard receiver and reference receiver, which will be converted to the TT&C equipment topocentric coordinate as a comparative standard, through the experimental process, TT&C equipment will tracking the plane at the same time. After the experiment is completed, the observation of TT&C equipment will be corrected and compared with standard data, ultimately the error characteristic of TT&C equipment will be obtained.

BDS precision verification system contains the onboard measuring subsystem, ground measuring subsystem, track display subsystem and data processing subsystem [4], the composition and function of subsystem are shown as follows (Fig. 2):

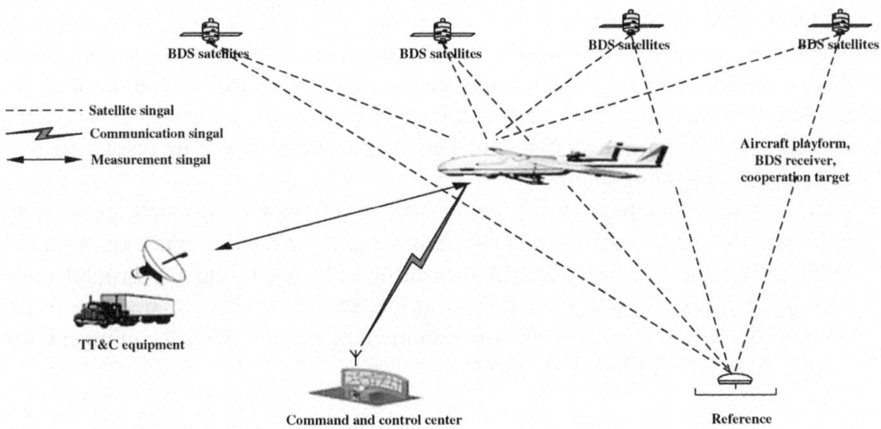

Fig. 1 Working principle of BDS precision verification system

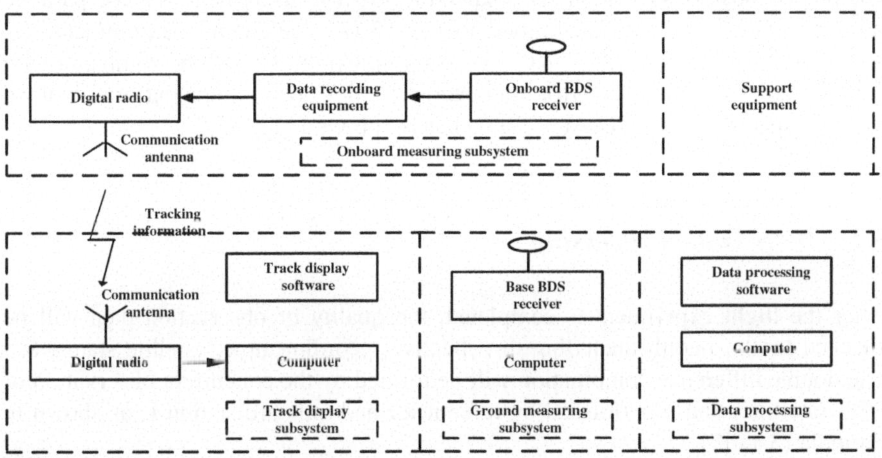

Fig. 2 BDS precision verification system

1. Onboard measuring subsystem
   Onboard subsystem mainly includes the BDS receiver and antenna, it is mainly to complete the raw BDS data reception, and the single point positioning results is sent to the track display subsystem.
2. Ground measuring subsystem
   Ground measuring system as the reference is set on the point with known coordination and close to the TT&C equipment, which includes the BDS receiver and antenna.

3. Track display subsystem
   Track display subsystem includes digital radio, computer, power supply, track display software, etc. Onboard measuring subsystem will transmit the positioning information to the track display subsystem, after data processing, the position, velocity, course angle and yaw angle of target will be displayed.
4. Data processing subsystem
   Data processing subsystem is composed of computer and data processing software, the main task of this subsystem is realizing the difference resolving of BDS carrier phase from onboard measuring subsystem and the ground measuring subsystem, and getting the position and velocity of the tracking target (plane). After compared with the measurement data of the TT&C equipment, the precision index will be obtained [5].

3 Data Processing Method

The precise positions of target are obtained by BDS measurement as a comparative standard, which will be transformed to the TT&C equipment measurement elements. After preprocessing, TT&C equipment tracking data will be compared with the standard data, then the error characteristics of TT&C equipment are obtained [6]. The specific data processing flow char is shown as Fig. 3.

3.1 BDS Data Processing

After the flight experiment is completed, the quality of observation data will be checked firstly, mainly including the effective recording time, satellite status, etc. The double difference combination will be formed by the concurrent observation of BDS satellite j and n performed by reference i and onboard station k, as shown in Formula (3.1) [7].

$$\begin{aligned}\nabla\Delta\varphi_{ik}^{jn} - \nabla\Delta N_{ik}^{jn} = & \left(\frac{\partial\rho_{ik}^{jn}}{\partial X_k}\right)_0 \delta X_k \\ & + \left(\frac{\partial\rho_{ik}^{jn}}{\partial Y_k}\right)_0 \delta Y_k + \left(\frac{\partial\rho_{ik}^{jn}}{\partial Z_k}\right)_0 \delta Z_k + V_{ik}^{jn}\end{aligned} \tag{3.1}$$

Expressed as a matrix form:

$$L_k = H_k \delta R_k + V_k \tag{3.2}$$

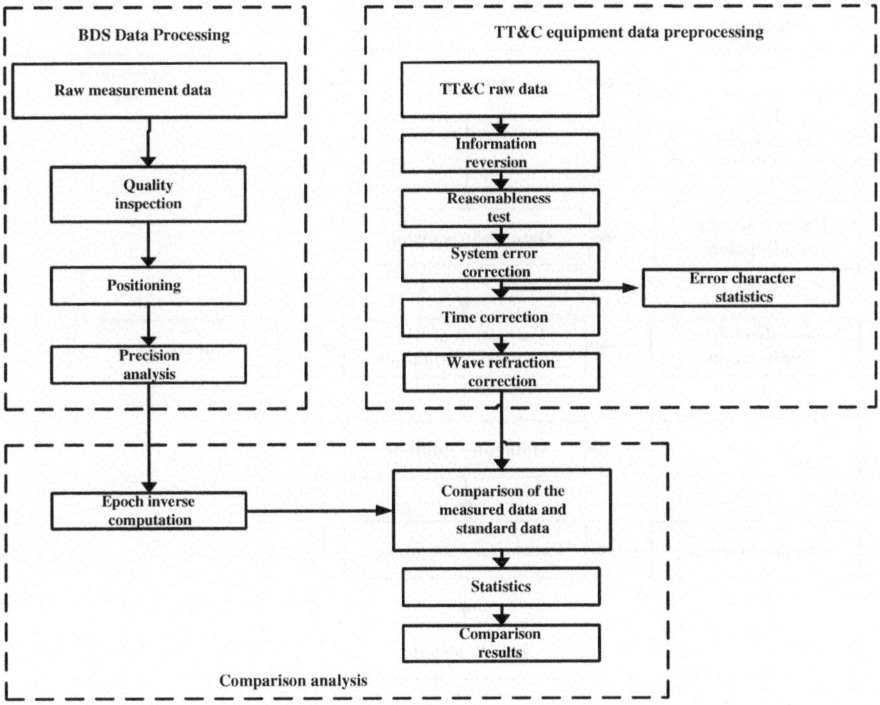

Fig. 3 Data processing flow char

The status equation of Kalman filtering is:

$$X_k = \Phi_{k,k-1}X_{k-1} + \Gamma_{k-1}W_{k-1} \tag{3.3}$$

For the corresponding the observation equation is:

$$L_k = H_kX_k + V_k \tag{3.4}$$

In Formula (3.3), (3.4)

$$X_k = \delta R_k \tag{3.5}$$

The recursive equation of Kalman filtering is:

$$\hat{X}_{k,k-1} = \Phi_{k,k-1}\hat{X}_{k-1,k-1} \tag{3.6}$$

$$P_{k,k-1} = \Phi_{k,k-1}P_{k,k-1}\Phi^T_{k,k-1} + \Gamma_{k-1}Q_{k-1}\Gamma^T_{k-1} \tag{3.7}$$

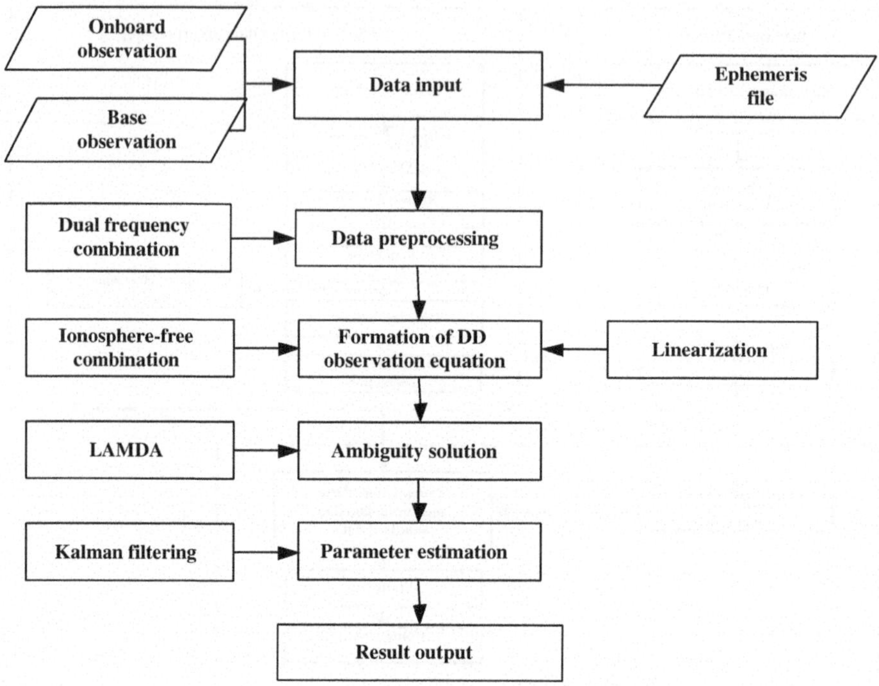

Fig. 4 BDS data processing strategy and flow

$$K_k = P_{k,k-1}H_k^T(H_kP_{k,k-1}H_k^T + R_k)^{-1} \tag{3.8}$$

$$P_{k,k} = (I - K_kH_k)P_{k,k-1} \tag{3.9}$$

The resolving strategies such as ambiguity fix method, parameter estimation method are shown in Fig. 4.

3.2 TT&C Equipment Data Processing

TT&C equipment data processing mainly include the following contents: data format convert, outliers detecting and error correction (system error, time error and atmospheric refraction error) [8], the TT&C equipment data processing flow is shown in Fig. 5.

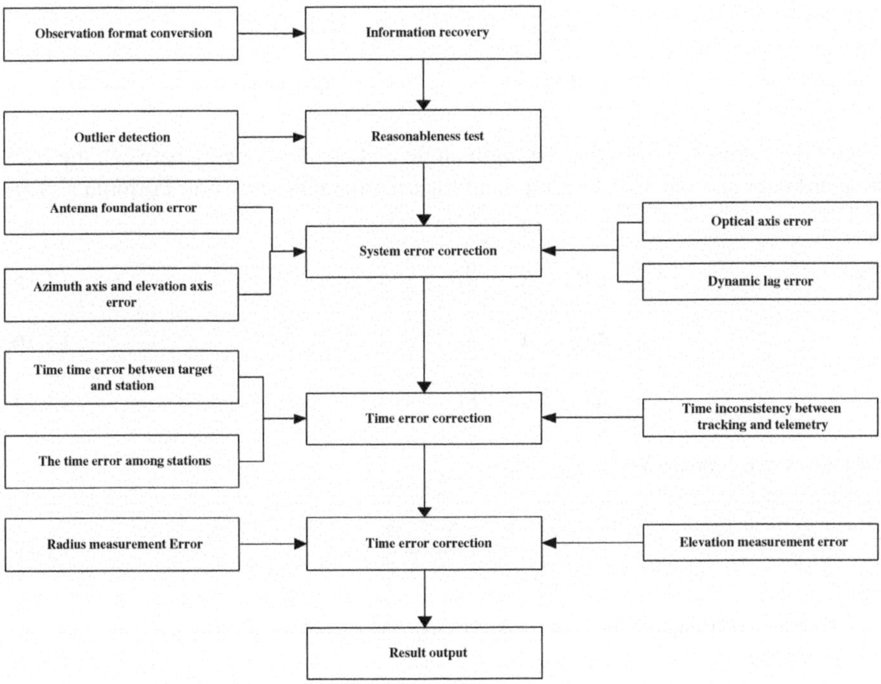

Fig. 5 TT&C equipment data processing flow

3.3 *Analysis Method*

The position of TT&C equipment is expressed as: x_{0k}, y_{0k}, z_{0k}, $k = 1, 2, \ldots, m$; in the launch coordinate system, and the TT&C equipment tracking information is expressed as: $\bar{R}_j$, $\bar{E}_j$, $\bar{A}_j$, while the target position obtained by BDS is expressed in launching coordinate system as follows: x_{bj}, y_{bj}, z_{bj}, the transformation principle is shown as follows:

$$\begin{bmatrix} \bar{x}_j \\ \bar{y}_j \\ \bar{z}_j \end{bmatrix} = \begin{bmatrix} x_{bj} - x_{0k} \\ y_{bj} - y_{0k} \\ z_{bj} - z_{0k} \end{bmatrix} \tag{3.10}$$

$$\bar{R}_j = \sqrt{\bar{x}_j^2 + \bar{y}_j^2 + \bar{z}_j^2} \tag{3.11}$$

$$\bar{E}_j = \sin^{-1}\left(\bar{y}_j / \bar{R}_j\right) \tag{3.12}$$

$$\bar{A}_j = tg^{-1}\left(\bar{z}_j / \bar{x}_j\right) + \begin{cases} 0 & \text{when} \quad \bar{x}_j > 0, \bar{z}_j \geq 0 \\ \pi & \qquad \bar{x}_j < 0 \\ 2\pi & \qquad \bar{x}_j > 0, \bar{z}_j < 0 \end{cases} \tag{3.13}$$

$$\overline{A}_j = \begin{cases} 0 & \\ \pi/2 & \text{when} \\ 3\pi/2 & \end{cases} \begin{matrix} \bar{x}_j = 0, \bar{z}_j \geq 0 \\ \bar{x}_j = 0, \bar{z}_j > 0 \\ \bar{x}_j = 0, \bar{z}_j < 0 \end{matrix} \tag{3.14}$$

Get the difference of radius, azimuth angle and elevation angle between the BDS standard data and the TT&C equipment tracking data, as shown in Formula (3.15)–(3.17):

$$\Delta R_j = R_j - \bar{R}_j \quad j = 1, 2, \ldots, N \tag{3.15}$$

$$\Delta A_j = A_j - \bar{A}_j \quad j = 1, 2, \ldots, N \tag{3.16}$$

$$\Delta E_j = E_j - \bar{E}_j \quad j = 1, 2, \ldots, N \tag{3.17}$$

The average error $\overline{X}$ is:

$$\overline{X} = \frac{\sum_{j=1}^{N} X_j}{N+1} \tag{3.18}$$

Among Formula (3.15)–(3.17), X_j is ΔR_j, ΔE_j , ΔA_j.
Total error σ_Z:

$$\sigma_Z = \sqrt{\frac{1}{N-1} \sum_{N=1}^{N} X_j^2} \tag{3.19}$$

Random error σ:

$$\sigma = \sqrt{\frac{1}{N-1} \sum_{j=1}^{N} (X_j - \bar{X}_j)^2} \tag{3.20}$$

System error δ:

$$\delta = \sqrt{\sigma_Z^2 - \sigma^2} \tag{3.21}$$

4 Experimental Analysis

A precision verification experiment for one TT&C equipment is performed in this paper. In this experiment, the plane equipped with BDS receiver, cooperation target, and communication device flied according to the planning route. The standard data was obtained by BDS measurement, which was compared with the TT&C

equipment measurement, then the error characteristics of radius, azimuth and elevation angle are obtained.

4.1 BDS Data Processing

In certain flight, for example, NovAtel 628 were used to obtain the raw data of GPS and BDS. The observed time is 1 h 50 min, and the sampling rate is 1s. After the carrier phase kinematic relative positioning is carried out, the difference of target position between BDS and GPS are shown in Fig. 6 and Table 1. The results show that the precision verification system based on the technology of BDS carrier phase kinematic relative positioning can access to the precision of centimeter level, several times higher than TT&C equipment, which can completely replace the GPS to provide the standard for TT&C equipment precision verification.

4.2 Precision Verification Results

The purpose of the precision verification is to check the problem and analyze whether the precision can meet the design requirements [9]. Four flights are designed in this experiment according to the working mode of the TT&C

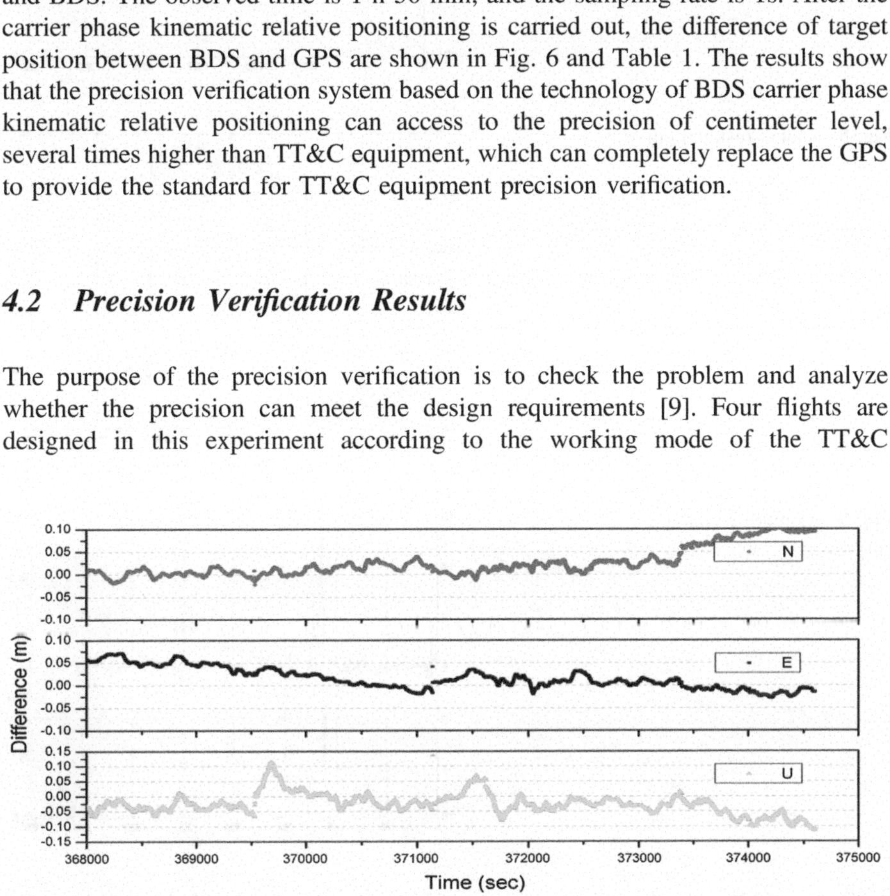

Fig. 6 Position difference between BDS and GPS

Table 1 Positioning precision statistics (m)

Direction	Mean	RMS
N	0.018	0.029
E	0.023	0.026
U	−0.025	0.033

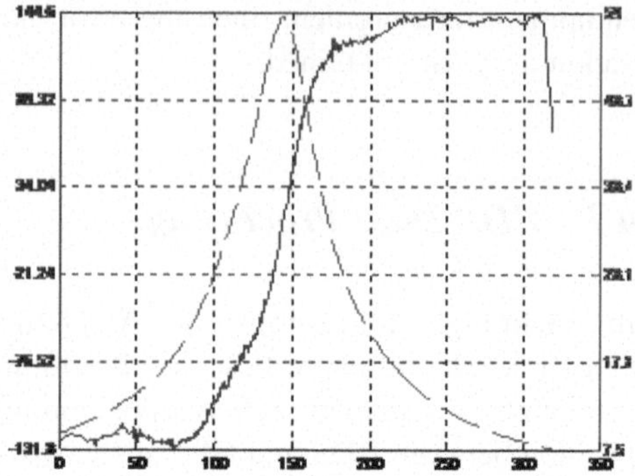

Fig. 7 Radius residual curve

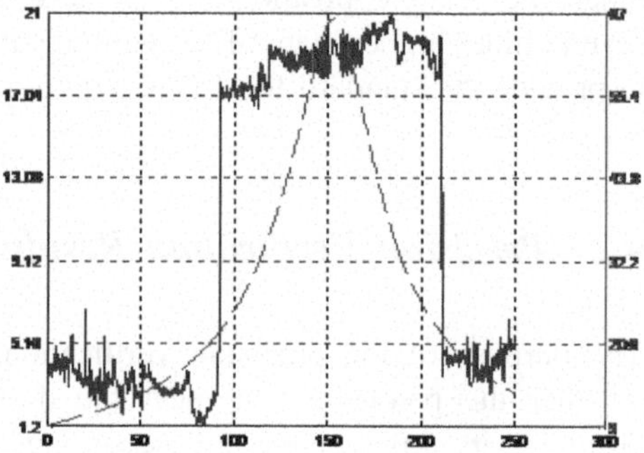

Fig. 8 Radius residual curve

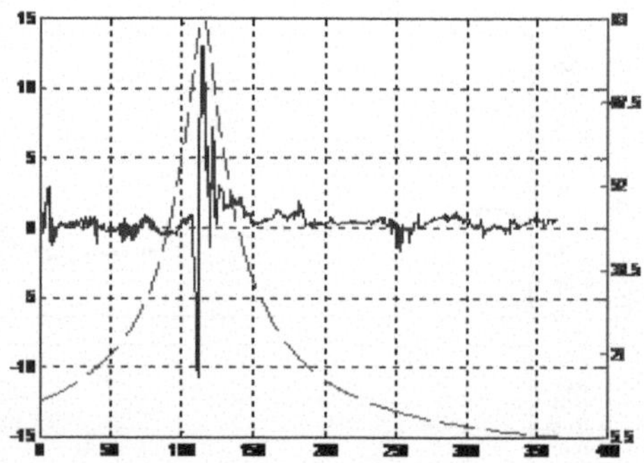

Fig. 9 Azimuth residual curve

equipment. After the first flight, some problems are found by the comparison of position information of BDS observation and TT&C equipment measurement, such as: "drifting" of radius residual (Fig. 7), "jump" of radius residual near the approach point (Fig. 8), significant variation of angle residual near the approach point (Figs. 9 and 10). The problems above have been solved by modifying the data processing software and control software, and the radius residuals with disappearance of the above phenomenon are shown as Fig. 11.

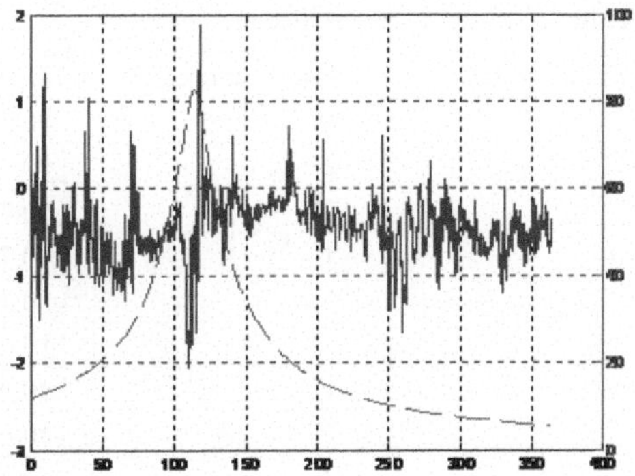

Fig. 10 Elevation angle residual curve

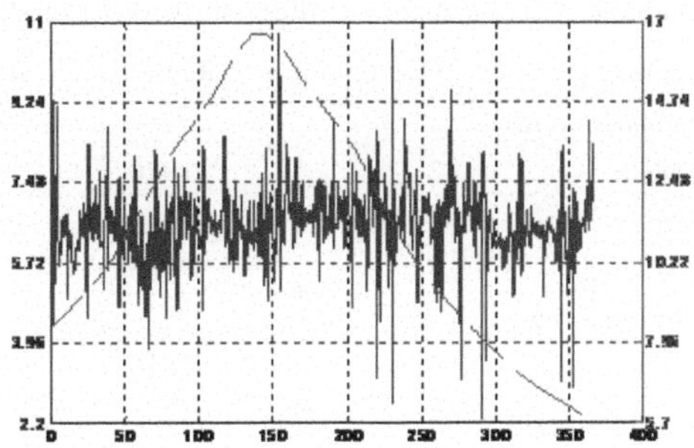

Fig. 11 Elevation angle residual curve

Table 2 Random error statistics

Flight number	Entry number	σ_R (m)	σ_A (mrad)	σ_E (mrad)
2	1	*0.863	*0.387	*0.216
	2	*0.059	*0.432	*0.288
	3	*0.049	*0.336	*0.238
3	1	*0.971	*0.449	*0.280
	2	*0.052	*0.389	*0.272
	3	*0.943	*0.296	*0.266
4	1	*0.825	*0.268	*0.178
	2	*0.778	*0.275	*0.136
	3	*0.771	*0.259	*0.111

On the basis of the subsequent flight experiments, the error characteristics of *R*, *E*, *A* are investigated in three work modes, respectively. The random error is expressed as σ_R, σ_A, σ_E, and δ_R, δ_A, δ_E are used to describe the system error. The random error, system error and total error are shown as Tables 2, 3, 4.

Table 3 System error statistics

Flight number	Entry number	δ_R (m)	δ_A (mrad)	δ_E (mrad)
2	1	*0.639	*0.313	*0.353
	2	*0.208	*0.221	*0.375
	3	*0.737	*0.392	*0.363
3	1	*0.705	*0.228	*0.271
	2	*0.346	*0.277	*0.245
	3	*0.655	*0.218	*0.237
4	1	*0.903	*0.550	*0.562
	2	*0.811	*0.487	*0.591
	3	*0.830	*0.542	*0.580

Table 4 Error statistics in different working mode

Working mode	Flight number	Radius (m)	Azimuth angle (mrad)	Elevation angle (mrad)
Random error				
Mode 1	2	*0.976	*0.403	*0.262
Mode 2	3	*0.968	*0.249	*0.267
Mode 3	4	*0.786	*0.266	*0.139
System error				
Mode 1	2	*0.516	*0.288	*0.385
Mode 2	3	*0.553	*0.284	*0.251
Mode 3	4	*0.849	*0.532	*0.566
Total error				
Mode 1	2	*0.572	*0.495	*0.466
Mode 2	3	*0.608	*0.378	*0.366
Mode 3	4	*0.912	*0.595	*0.583

5 Conclusion

A new method of TT&C equipment precision verification based on BDS is proposed in this paper, the principle and mathematical model of precision verification are introduced in detail, and the method of engineering application is discussed. The actual flight experiment is carried onto achieve a complete verification process and analyze the verification effect, the results show that:

1. The precision verification system based on the technology of BDS carrier phase kinematic relative positioning can access to the precision of centimeter level, which is capable to provide excellent standard data for TT&C equipment precision verification.
2. Through the analysis of experimental data the main problems of equipment have been found and solved. After the equipment status are adjusted, various precision indicators are obtained, which played an important role in analyzing the tracking performance of the aerospace TT&C equipment.

References

1. Xie H, Zhang SX, Dong X (1999) Research on the method of GPS precision assessment aerospace measurement and control equipment. J Inst Command Technol 10(3):P30–34
2. Zhang Y (2007) A satellite self-validation method under perturbations. J Spacecraft TT&C Technol 26(6):P63–69
3. Lisheng L (2001) External trajectory measurement data processing. National Defence Industry Press, Beijing
4. Zhou W, Hao JM (2013) The precision assessment system of TT&C equipment based on unmanned aerial vehicle. In: China satellite navigation conference 2013 proceedings
5. Jin Z, Luo W, Zhu J, Li D (2001) Radar precision assessment using GPS on an unmanned aerial vehicle. J Tsinghua Univ 41(9):P40–43
6. Luo X (2008) Radar field measurements for the dynamic DGPS calibration system design and implementation. Master thesis, National Defense Science and Technology University
7. Wei Z, Ge M (1998) Radar field the mathematical model of GPS relative position. Surveying and Mapping Press, Beijing
8. You H (2006) Radar data processing and application. Electronics Industry Press, Beijing
9. Zhou W (2013) Research and realization on theories and methods of precise position base on BeiDou navigation satellite system. Dissertation, Degree of Doctor, PLA Information Engineering University

Part II
Navigation and Location Based Services

An Efficient Strategy for Large-Scale CORS Data Processing

Bolin Xiong and Dingfa Huang

Abstract With the development of large-scale continuous operation reference stations (CORS) network technology, the rapid solution of massive data via network has become a hot research topic. When taking the monitoring of the crustal movements and the solution of the earth model parameters as examples, the existing data processing and analysis modes exhibit the low processing efficiency. In this study, a data processing framework for large-scale CORS network is proposed based on cloud computing. An elastic compute service with distributed storage and parallel computing ability is built to implement the high-efficiency solution of the large-scale CORS network data, according to the temporal–spatial characteristic of the CORS data. The results show that the accuracy obtained by the proposed method is equivalent to that of the Gamit/Globk solution. However, the efficiency of the proposed method is obviously superior to the current processing method. For the test of CORS data from 10 stations over 577 days, the efficiency of this method is increased by 25.74 % compared with the Gamit/Globk. With the increase of the CORS data and the number of nodes, the efficiency of proposed method will be further increased. Finally, this study demonstrates that the data processing framework greatly improves the efficiency in processing the large-scale CORS network data. Also, it provides a high efficiency data solution for the applications in view of the large-scale CORS network data, such as the crustal movements, crustal motion and deformation, instantaneous global plate motion, the earthquake cycle, coseismic slip distribution, and the interpolate fault slip.

Keywords Large-scale CORS network data · Cloud computing · Elastic compute service · Data processing framework · Distributed storage · Parallel computing

B. Xiong · D. Huang (✉)
Faculty of Geosciences and Environmental Engineering, Southwest Jiaotong University, Chengdu 611756, Sichuan, China
e-mail: dfhuang@swjtu.edu.cn

B. Xiong
e-mail: xbl123fm@126.com

J. Sun et al. (eds.), *China Satellite Navigation Conference (CSNC) 2016 Proceedings: Volume I*, Lecture Notes in Electrical Engineering 388,
DOI 10.1007/978-981-10-0934-1_20

1 Introduction

The global navigation satellite system (GNSS) has been used increasingly in geodetic surveying and geodynamics research. Many continuous operation reference stations (CORS) have been built and operated all over the world. On the one hand, a huge number of data have been obtained from the large-scale CORS stations. It has been processed by the International GNSS Service (IGS) to obtain high-accuracy and high-reliability International Terrestrial Reference Frame (ITRF) parameters [1]. On the other hand, the large-scale CORS network data can also be used for the scientific researches, such as the crustal movements along the Dead Sea Fault [2], crustal motion and deformation in Greece [3], instantaneous global plate motion [4], the model of the earthquake cycle along the San Andreas Fault System [5], coseismic slip distribution of the 2011 off the Pacific Coast of Tohoku Earthquake [6], and the interpolate fault slip along the Japan Trench [7]. However, the approaches significantly depend on the high-accuracy and large-scale CORS network data solutions.

Although the large-scale CORS network data can provide high-accuracy results, the procedure is time-consuming. For example, the Research Facility for Space Geodesy (FESG) in Munich Technique University, cooperated with the Institute for Planetary Geodesy (IPG) of Technical University Dresden, takes about 3 months of computation on the LINUX cluster to reprocess 11 years of GPS data [8]. Obviously, major challenge comes from the processing efficiency of the large-scale CORS network data.

Generally speaking, large-scale CORS network data can be processed efficiently throughout the simplified algorithms and optimized processing mode. Ge et al. [9] removed inactive parameters from the normal equation (NEQ) to enhance the computational efficiency. However, the parameter elimination at each epoch is still time-consuming. Blewitt et al. [10] proposed a new approach for processing large-scale CORS network data by converting carrier-phase observations to so-called "carrier range" using the double-differenced integer ambiguities resolved based on the fixed point theorems when using the precise point positioning (PPP) technique. Karimi et al. [11] proposed a grid-based geoprocessing for integrated GNSS simulation. All these algorithms mentioned above are very helpful for the high-efficiency data processing. However, optimized mode becomes an important aspect to improve the efficiency of large-scale CORS network data processing. The traditional strategies for analyzing large-scale CORS network data is to divide the whole network into several subnetworks and then combine them on (NEQ) level where some of common stations are needed in order to connect the subnetwork together [12]. Blewitt et al. [13] established continuously operating regional GPS clusters to study crustal deformation at tectonic plate boundaries and to serve as active geodetic networks. Working group 1 of International Association of Geodesy (IAG) develops software that runs in the form of a distributed process for processing large-scale CORS network data [14]. However, when long-period observations are processed, the optimized modes become ineffective.

With the development of cloud computing, it is possible to resolve the solution of the large-scale CORS network data efficiently, which has been increasingly exploited in the geospatial services [15, 16]. Cloud computing is a supercomputing using Internet, which distributes the tasks into a resource pool composed by large number of computer nodes. The users are able to select different services according to their demands. It contains three service modes: SaaS (software as a service), PaaS (platform as a service), and IaaS (infrastructure as a service). The core technology of the PaaS is the distributed processing, which is responsible for resolving the collaborative problems among the cluster, and also provides the mass storage and large-scale processing capabilities in the cluster.

In this paper, a new PaaS strategy is proposed for the large-scale CORS network data processing. To realize the high-efficiency solution of the large-scale CORS network data, the elastic compute service (ECS) for the single data analysis center is developed by PaaS in parallel. The new strategy is developed, namely the CGOC (Common GPS Office on the Cloud), to demonstrate the accuracy and efficiency. Finally, some useful conclusions are presented.

2 CORS Data Processing Mode

2.1 Data Processing Mode in China

The CORS networks in China are the individual network GNSS positioning systems, which were built on the basis of the organization demands. They aim to provide high-quality national-level GNSS CORS network [17]. This processing procedure is described as follows: the observations of the CORS stations are delivered to the data center through Internet. Then, the large servers are used to store the CORS data in a centralized manner. In addition, the number of servers will be increased with the increase of the CORS data. When Gamit/Globk is used in the CORS center [17], several computers independently process the task during data analyzing. Although the mode is applied to the CORS network in China, it cannot resolve the disperse situation in the CORS center. Furthermore, with the development of the multi-constellation GNSS, the data storage and computational burden will be increased gradually.

2.2 Data Processing Mode in IGS

The IGS establishes the CORS stations over the world, which aims to obtain high-accuracy and high-reliability ITRF parameters and GNSS data products through processing the large-scale CORS network data. Its data processing mode was established in several data analysis centers to process the large-scale CORS network data in the form of distribution, which resolves the status of the CORS center [13]. However, the data analysis centers have a high computational burden with the increase of the data. Therefore, Work Group 1 of IAG initiated three projects (i.e., Digger, Dancer, and Dart) to change the insufficiency of the IGS mode [14, 1]. Digger aims at fast reprocessing geodetic data; Dancer builds a peer-to-peer network of the GPS receivers to allow least squares solution for network of unlimited size; Dart offers an interface to the Dancer reference frame for real-time kinematic users [1]. However, Digger and Dart are still under construction, and Dancer has been built, but only processes a short-period observations.

The content described above has introduced data processing mode in China and IGS. The computational burden of the processing tasks will be increased when the CORS data becomes larger.

3 Data Processing Framework of Large-Scale CORS Network

3.1 ECS Framework

The ECS is a type of computing server, which has an elastically expandable processing capability. It is a computer cluster which plays the role of a server. Owing to the individual CORS center in China, the slave node of the ECS is used for managing the regional CORS data in the CORS center. Then, the slave node is connected to the Internet to realize data sharing. The IP addresses of all the slave nodes are mapped to the Master of the cluster, and the Master checks and manages the slave nodes in real time. The administrator monitors the health condition of the cluster through the Master, and checks the data storage and status of the parallel tasks. The ECS framework is shown in Fig. 1.

3.2 Distributed Storage of the CORS Data

Assuming that the CORS data set is $M = \sum_{i=1}^{n} M_i$, where M_i represents the daily CORS data. The cluster capacity is V and the data storage address is $D = \sum_{i=1}^{n} D_i$, where D_i indicates the route of M_i in the ECS. When V is larger than M, then M_i is distributed to the ECS. In addition, each M_i has replication to prevent loss of data due to broken nodes in the ECS. The relationship between M_i and D_i is a peer-to-peer relationship, and D_i will be returned to the Master for storage purpose. When V is less than M, M_i completely occupies storage space at first, and then the administrator only needs to add the nodes to the ECS. Nevertheless, the data stored previously remained unaffected.

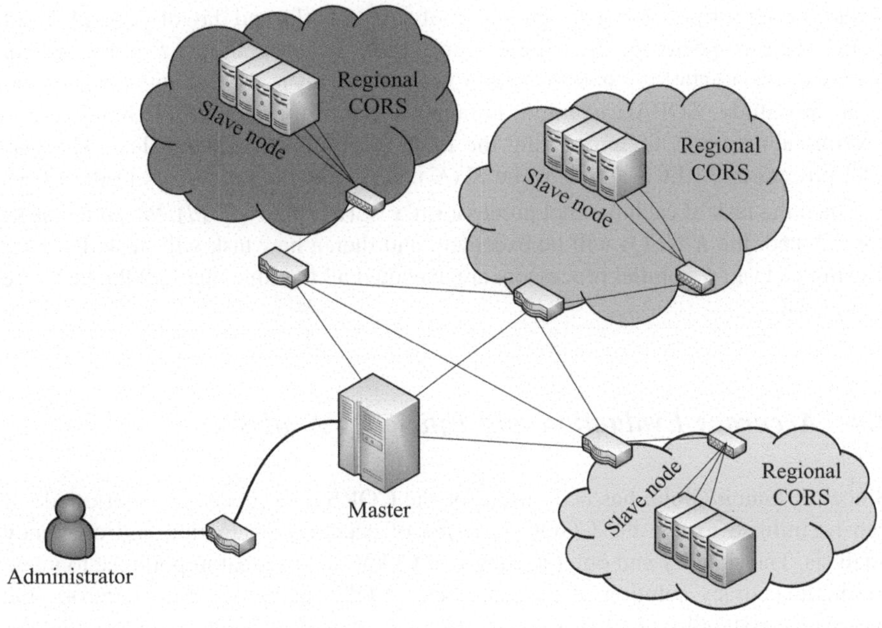

Fig. 1 Overview of ECS framework

Due to the CORS data confidentiality, the security of the large-scale CORS network data is particularly important. However, the CORS data are usually stored in a centralized manner without taking into account of the data security when processing data routinely. The distributed storage proposed in this study is not only able to realize data sharing in the ECS but also has the ability to improve the data security.

3.3 Implementation of Parallel Processing Tasks

Since the large-scale CORS network data has a temporal–spatial characteristic, the data processing has a good parallel procedure. Regarding to the large-scale CORS network data, the SOPAC (Scripps Orbit and Permanent Array Center) divides the global IGS stations into different regional subnetworks. Therefore, the NEQ building is the essence of the large-scale CORS network data processing tasks. When N days of the CORS data are divided into M subnetworks, NEQ can be simply expressed as follows:

$$x_{i,j} = (B_{i,j}^T P_{i,j} B_{i,j})^{-1} (B_{i,j}^T P_{i,j} l_{i,j}) \tag{1}$$

where the subscripts i and j are the index of subnetworks and day of year (DoY) of CORS data, respectively; B is the design matrix of observations; P is the weight matrix of observations; l is observation vector; and $x_{i,j}$ is the unknown parameter.

In this study, SOPAC example is taken to adopt the method of dividing into several subnetworks for processing the large-scale CORS network data. The parallel processors of ECS are set to be K (K is less than the processors in the ECS), and then the task of each parallel processor is $x = (B^T PB)^{-1}(B^T Pl)$. When the tasks are initiated, the K NEQs will be executed, and then a new task will be added after finishing a task. K parallel processors are executed all the time, until all the tasks are completed.

3.4 Accuracy Evaluation and Efficiency Analysis

Since the Gamit/Globk has been used for the CORS data processing worldwide, it can be utilized to test the CGOC in terms of accuracy evaluation and efficiency analysis. The stability and outer accuracy of CGOC are of great importance to judge the high-accuracy solution of the large-scale CORS network data. Therefore, the baseline repeatability and the baseline difference between CGOC and Gamit/Globk are also presented for evaluating the CGOC accuracy. The baseline repeatability represents the dispersion degree of the observations in several sessions, which is important to show the CGOC stability. Its definition is described in Eq. (2) [18]:

$$R = \sqrt{\frac{\frac{n}{n-1} * \sum_{i=1}^{n} \frac{(C_i - C_m)^2}{\delta_{C_i}^2}}{\sum_{i=1}^{n} \frac{1}{\delta_{C_i}^2}}} \tag{2}$$

where n is the total of sessions of the same baseline; C_i is the baseline vector in a session; $\delta_{C_i}^2$ is the variance of C_i; and C_m is the mean of all baseline vectors in a session.

Since the large-scale CORS network data processing framework is closely related to the ECS and CORS data, the ECS node and CORS data volume are necessary to analyze the efficiency of CGOC. Speedup ratio is the most common standard for judging the efficiency of parallel algorithms [19], as a result, it will be selected to present the CGOC efficiency. Its definition is described as follows:

$$S = T_1 / T_p \tag{3}$$

where S represents the speedup ratio; T_1 represents the consumption time of single node; T represents the consumption time of nodes; and p represents the number of nodes.

The computation time of large-scale CORS network data is the basic performance measurement, and then the experiment compares the execution time of

CGOC and Gamit/Globk. The computation time of CGOC can be obtained from the baseline reports, while the time of Gamit/Globk can be obtained by the following:

$$t = \max(\{t_n\}) \tag{4}$$

where t_n represents computation time of the nth computer node; and $\{t_n\}$ represents the set from t_1 to t_n.

4 Experiment

4.1 Experiment Environment Configuration and Data Preparation

To implement the large-scale CORS network data processing framework, we use the open-source Apache Hadoop to build the ECS with four computer nodes. The four nodes are referred to as A, B, C, and D, respectively, where A is a laptop, B, C, and D are desktop computers. A is chosen as the host computer of the experiment with its CPU of an Intel(R) Core(TM) i5-3200M, a dual-core processor, a main frequency of 2.5 GHz, and a memory of 4 GB. The CPU of the B computer is an Intel(R) Pentium(R) G2030, with a dual-core processor, a main frequency of 3.0 GHz, and a memory of 2 GB. Both C and D computers have the same hardware configuration, with CPUs of Pentium(R) E5300, dual-core processors, main frequencies of 2.59 GHz, and memory of 2 GB. The operation system environment of the CGOC and Gamit/Globk is Ubuntu11.10. The experimental data are obtained from the CORS center of the Seismological Bureau of Sichuan Province in China. A total of 577 days of data (from 2009 to 2010) with a sampling rate of 30s are selected. There are 10 stations with the data size of about 13.28 GB. These data are divided into eight groups (72, 144, 216, 288, 360, 432, 504, and 577 days) during the experiment. The daily data is regarded as a session. When the CGOC is initiated, all groups of data are processed using the four combinations of ECS: A, AB, ABC, and ABCD. Meanwhile, the Gamit/Globk is also processed in accordance with the combinations mentioned above.

4.2 Accuracy Analysis

4.2.1 Baseline Repeatability

To compare the stability between CGOC and Gamit/Globk, baseline repeatability of the CGOC and Gamit/Globk solutions from the aspects of the baseline component (*X*, *Y*, *Z*) and length are shown. Figure 2a, b, c, and d represents the baseline repeatability of the CGOC in the *X*, *Y*, and *Z* directions, as well as length. In

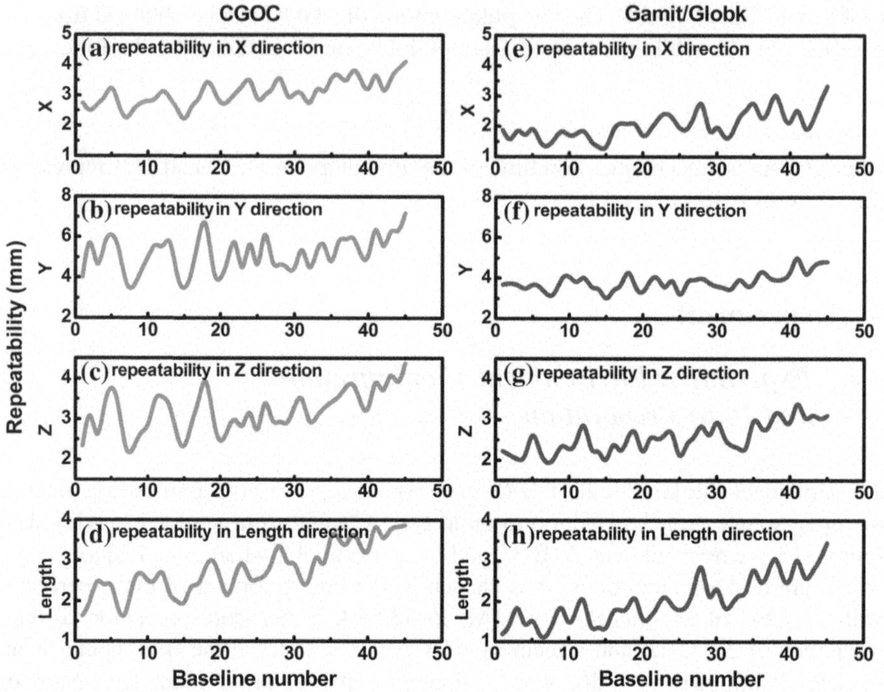

Fig. 2 Baseline repeatability

addition, Fig. 2e–h indicates the baseline repeatability of Gamit/Globk solutions. The repeatability of all the baselines (see Table 1 for the baseline information) is shown in Fig. 2. It is seen that CGOC has a similar trend compared with Gamit/Globk, and the repeatability of the CGOC is slightly larger than that of the Gamit/Globk. However, the baseline repeatability of both solutions has a smaller difference within a range of millimeter level. Therefore, the results indicate that the CGOC has a good stability.

4.2.2 Baseline Difference

To test outer accuracy, Fig. 3a–d shows the distribution of the differences in the *X*, *Y*, and *Z* directions, as well as length. It is found that most of the differences are concentrated within the range of ±1, ±1.5, and ±1 cm for the *X*, *Y*, and *Z* directions, respectively. Also, most of the differences for length are distributed within the range of ±1 cm. The difference is increased with the increase of baseline length. As a result, it is a good outer accuracy that CGOC can meet the needs of high-accuracy processing.

Table 1 Baseline information

Baseline number	From-to	Length (km)	Baseline number	From-to	Length (km)
1	CHDU-RENS	48.81	24	CHDU-ROXI	135.74
2	QLAI-YAAN	50.23	25	QLAI-ZHJI	139.07
3	CHDU-ZHJI	61.45	26	QLAI-ROXI	147.45
4	LESH-ROXI	66.85	27	ROXI-YAAN	149.42
5	NEIJ-ROXI	68.77	28	CHDU-NEIJ	151.95
6	ROXI-YBIN	74.81	29	RENS-YBIN	162.13
7	LESH-RENS	78.05	30	NEIJ-ZHJI	163.27
8	QLAI-RENS	78.56	31	ROXI-ZHJI	172.00
9	CHDU-QLAI	79.39	32	LESH-ZHJI	176.98
10	LUZH-YBIN	80.16	33	LESH-LUZH	178.58
11	LESH-YAAN	85.53	34	YAAN-ZHJI	186.13
12	LUZH-NEIJ	87.69	35	NEIJ-QLAI	192.94
13	RENS-ROXI	88.31	36	LUZH-RENS	194.45
14	LESH-QLAI	97.68	37	YAAN-YBIN	202.21
15	RENS-ZHJI	98.94	38	NEIJ-YAAN	207.69
16	NEIJ-YBIN	104.28	39	CHDU-YBIN	210.48
17	RENS-YAAN	108.07	40	QLAI-YBIN	213.05
18	LUZH-ROXI	115.31	41	CHDU-LUZH	235.35
19	NEIJ-RENS	117.32	42	YBIN-ZHJI	244.82
20	LESH-YBIN	117.97	43	LUZH-ZHJI	250.95
21	CHDU-LESH	122.80	44	LUZH-QLAI	262.02
22	CHDU-YAAN	124.92	45	LUZH-YAAN	263.56
23	LESH-NEIJ	132.22			

The experiment tests the baseline repeatability and the baseline difference between CGOC and Gamit/Globk, it is concluded that the solutions of the CGOC and Gamit/Globk have equivalent accuracy.

4.3 *Efficiency Analysis*

4.3.1 Relationship of ECS Node and Speedup Ratio

In this study, the ECS node is gradually increased to judge the CGOC efficiency. The results are shown in Fig. 4.

It can be seen from Fig. 4 that when the CORS data volume is unchanged, the speedup ratio of the CGOC gradually grows with the increase in the ECS node. This indicates that the efficiency of the CGOC will be improved with the increase of the number of ECS nodes.

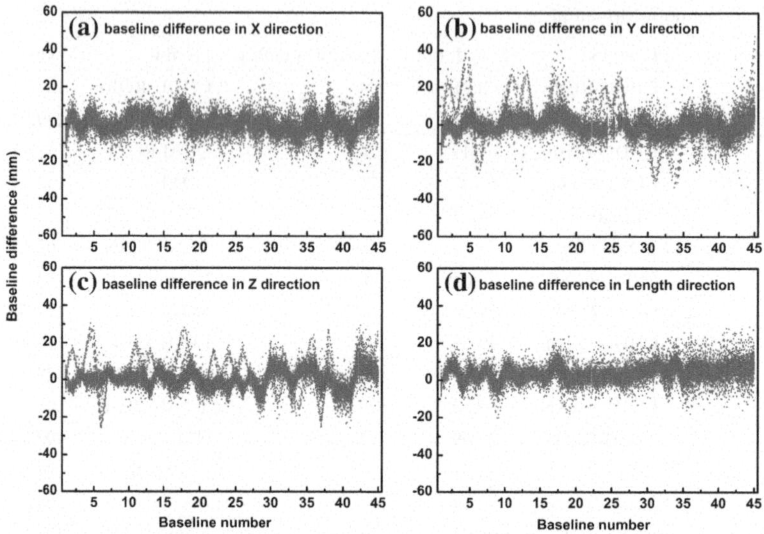

Fig. 3 Baseline difference between CGOC and Gamit/Globk

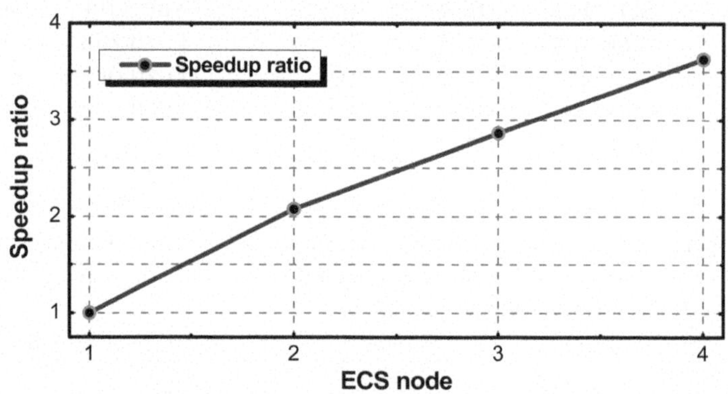

Fig. 4 The change of speedup ratio with the increase of ECS node

4.3.2 Relationship of Data Volume and Speedup Ratio

To illustrate how the CORS data volume affects the speedup ratio of the CGOC, the above-mentioned eight groups of data and the four combinations are selected to record the speedup ratio under the framework. Figure 5 shows that the speedup ratio has a minor change with the increase of CORS data. It is because that the computer is running slightly slower due to the long-period processing. This indicates that when the number of ECS node is unchanged, the speedup ratio of the CGOC basically is stable with the increase of the CORS data.

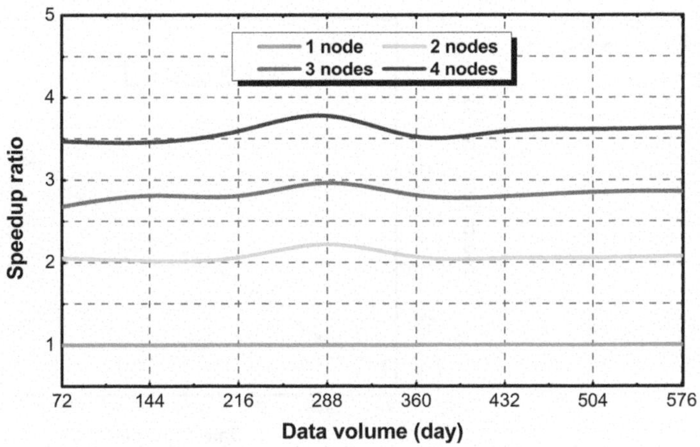

Fig. 5 The change of speedup ratio with the increase of CORS data

4.3.3 Comparison of Computation Time of the CGOC and Gamit/Globk

The computation time between the CGOC and Gamit/Globk is shown in Table 2. The computation time of the CGOC is less than that of the Gamit/Globk, when any combinations are used for processing the CORS data. The relationship between the time difference compared with the Gamit/Globk and CORS data volume is shown in Fig. 6. It can be seen that time difference becomes larger with the increase of the CORS data. The efficiency of the CGOC is increased by 25.74 %, when processing the 577 days of CORS data from 10 stations.

Table 2 Computation time of CGOC and Gamit/Globk

Data volume (day)	CGOC computation time (s)				Gamit/Globk computation time (s)			
	1 node	2 nodes	3 nodes	4 nodes	1 node	2 nodes	3 nodes	4 nodes
72	9360	4563	3493	2697	10,483	5327	4271	3738
144	18,285	9060	6504	5289	21,751	10,483	8338	6955
216	26,697	12,964	9506	7433	31,326	16,432	12,463	9981
288	38,899	17,549	13,126	10,302	41,257	21,751	16,940	13,631
360	45,605	22,091	16,202	12,940	51,076	26,386	21,411	16,831
432	54,710	26,688	19,541	15,253	60,981	31,326	25,475	20,059
504	63,763	31,007	22,349	17,643	71,370	36,306	29,669	23,401
577	73,284	35,301	25,608	20,193	82,668	41,387	33,737	27,191

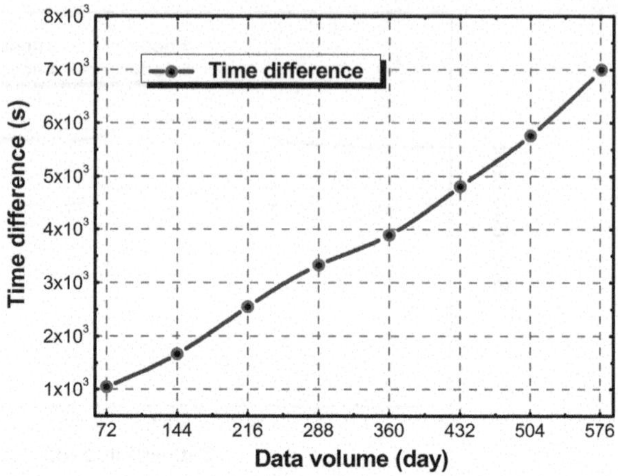

Fig. 6 The change of time difference with the increase of CORS data

5 Conclusions

This study proposed a new large-scale CORS network data processing framework. In this framework, the ECS with distributed storage and parallel processing are developed to establish a CGOC system. For large-scale CORS network, only one data analysis center in the ECS is required, where computers are added as nodes. The accuracy and efficiency of the CGOC are tested using the large-scale CORS network measurements.

The experimental results show that the accuracy of the CGOC is equivalent to the Gamit/Globk. However, the efficiency of the CGOC becomes higher, when ECS nodes are increased. When processing the CORS data from 10 stations over 577 days using CGOC, it is evident that the efficiency of the CGOC is increased by 25.74 % compared with the Gamit/Globk. This study demonstrates that the new data processing strategy has the ability to greatly improve the efficiency in large-scale CORS network data processing. This new strategy can provide a high-efficiency data processing method for the applications based on large-scale CORS network data.

Acknowledgments This work is supported by the National Natural Science Foundation of China (Nos. 41374032 and 41104020). The authors hereby acknowledge with thanks to the Seismological Bureau of Sichuan Province for the CORS data supports. Thanks are extended to Apache Hadoop and Gamit/Globk, which are used in this contribution.

References

1. Boomkamp H (2013) Distributed processing for large geodetic solutions. In: Altamimi Z, Collilieux X (eds) Reference frames for applications in geosciences. Springer, Berlin, pp 13–18
2. Wdowinski S, Bock Y, Baer G et al (2004) GPS measurements of current crustal movements along the Dead Sea Fault. J Geophys Res 109:B05403
3. Hollenstein C, Müller MD, Geiger A et al (2008) Crustal motion and deformation in Greece from a decade of GPS measurements, 1993–2003. Tectonophysics 449:17–40
4. Prawirodirdjo L, Bock Y (2004) Instantaneous global plate motion model from 12 years of continuous GPS observations. J Geophys Res 109:B08405
5. Smith BR, Sandwell DT (2006) A model of the earthquake cycle along the San Andreas Fault System for the past 1000 years. J Geophys Res 111:B01405
6. Iinuma T, Hino R, Kido M et al (2012) Coseismic slip distribution of the 2011 off the Pacific Coast of Tohoku earthquake (M9.0) refined by means of seafloor geodetic data. J Geophys Res 117:B07409
7. Suito H, Nishimura T, Tobita M et al (2011) Interplate fault slip along the Japan Trench before the occurrence of the 2011 off the Pacific coast of Tohoku earthquake as inferred from GPS data. Earth Planet Space 63:615–619
8. Steigenberger P, Rothacher M, Dietrich R et al (2006) Reprocessing of a global GPS network. J Geophys Res 111:B05402
9. Ge M, Gend G, Dick G et al (2006) A new data processing strategy for huge GNSS global networks. J Geodesy 80:199–203
10. Blewitt G, Bertiger W, Weiss JP (2010) Ambizap3 and GPS carrier-range: a new data type with IGS applications. In: Proceedings of IGS workshop and vertical rates, Newcastle, UK. http://research.ncl.ac.uk/IGS2010/abstract.htm
11. Karimi HA, Zimmerman B, Roongpiboonsopit D et al (2011) Grid-based geoprocessing for integrated Global Navigation Satellite System simulation. J Comput Civ Eng 26:19–27
12. Chen H, Jiang W, Ge M et al (2014) An enhanced strategy for GNSS data processing of massive networks. J Geodesy 88:857–867
13. Blewitt G, Bock Y, Gendt G (1993) Regional clusters and distributed processing. In: Proceedings of the IGS analysis center workshop, Ottawa, Canada, pp 62–91
14. Boomkamp H (2010) Global GPS reference frame solutions of unlimited size. Adv Space Res 46:136–143
15. Addair TG, Dodge DA, Walter WR et al (2014) Large-scale seismic signal analysis with Hadoop. Comput Geosci 66:145–154
16. Evangelidis K, Ntouros K, Makridis S et al (2014) Geospatial services in the cloud. Comput Geosci 63:116–122
17. Dang Y, Zhang P, Jiang Z et al (2011) The data processing and analysis of national GNSS CORS network in China. In: The XXV general assembly of IUGG Melbourne, Australia, 2007–2010 China National Report on Geodesy, Report No. 4
18. Blewitt G (1989) Carrier phase ambiguity resolution for the global positioning system applied to geodetic baselines up to 2000 km. J Geophys Res 94:10187–10203
19. Xie J (2012) Implementation and performance optimization of a parallel contour line generation algorithm. Comput Geosci 49:21–28

Research and Implementation of BDS/GPS Coarse Time Navigation Algorithm

Shaolei Peng, Dongqing Zhao, Zhiyong Huang, Yijun Tian, Hao Wu and Caijie Zhu

Abstract When the receiver cannot demodulate satellite signal transmission time, positioning can be completed with a combination of code phase measurement and auxiliary information, that is, coarse time navigation (CTN). Assistant information includes ephemeris, priori position, and coarse signal transmission time. For the GPS systems, requirement of the position and time precision is 100 km and 1 min. This paper attempts to determine the allowed priori position and coarse time error range of BDS and has realized the GPS/BDS coarse time navigation algorithm and carried out experimental verification using the measured data of IGS CUT0 station. The results show that BDS, GPS, and BDS/GPS coarse time navigation positioning precision is equivalent to their pseudorange positioning precision. Dual system CTN can effectively improve positioning precision, which is better than the precision of single system pseudorange positioning or coarse time navigation. Conclusion is that BDS/GPS coarse time navigation is an effective method without pseudorange and that dual system navigation is easier to receive enough visible satellites.

Keywords BDS · GPS · Coarse time navigation · Positioning · Precision

1 Introduction

Traditionally, receivers obtain the precise time by decoding the handover word (HOW), which gives the time of week (TOW), but it takes a few seconds. In weak signal conditions, even if the signal is captured by the receiver, navigation data bit demodulation appears to be higher in error rate. Then the receiver may not be able to complete bit synchronization, frame synchronization, and also cannot demodulate the TOW for satellite position computation and complete ephemeris parameters required satellite position, which results in positioning cannot be completed.

S. Peng (✉) · D. Zhao · Z. Huang · Y. Tian · H. Wu · C. Zhu
Institute of Geospatial Information, Information Engineering University, Zhengzhou, China
e-mail: cug_peng@163.com

J. Sun et al. (eds.), *China Satellite Navigation Conference (CSNC) 2016 Proceedings: Volume I*, Lecture Notes in Electrical Engineering 388,
DOI 10.1007/978-981-10-0934-1_21

For this situation, Frank Van Diggelen put forward a kind of nontraditional auxiliary positioning algorithm containing five state variables [1–5], which makes it possible that the receiver completes positioning precisely before bit synchronization and frame synchronization and greatly improves the receiver time to first fix (TTFF) performance. To a certain extent, the algorithm which solved the problem of that receiver cannot realize positioning under the condition of weak signal.

With conventional four state pseudorange positioning, this method increases the number of satellites required. However, under the condition of weak signals, such as urban canyon environment, for a single system it may be difficult to reach five visible satellites. But with the development of satellite navigation and positioning technology, many countries are building their own satellite navigation and positioning systems. GEO satellites altitude angle is higher. So the visible probability is bigger. This paper introduced the theory of coarse time navigation algorithm into BDS and realized GPS/BDS navigation system coarse time navigation algorithm. Thus the scope of application of coarse time navigation is expanded greatly. What is more, when visible satellites are more than five, dual system can effectively improve the positioning accuracy.

2 Coarse Time Navigation

2.1 Transmit Time of Satellite Signals

For conventional pseudorange positioning, the receiver decodes the navigation message to obtain TOW and uses TOW to assemble transmit time of satellite signals [3, 4] just like Eq. 2.1.

$$t_t = \text{TOW} + (30w + b) \times 0.020 + (c + \text{CP}/C_0) \times 0.001 \tag{2.1}$$

- W is the word number of navigation message received. Each word contains 30 bit;
- b is the bit number of the word which is not integrated one in the navigation message received. Each word lasts 20 ms corresponding to 20 cycles of C/A code;
- c is the C/A code number of the bit which is not integrated one in the navigation message received;
- CP is the phase of C/A code in the sampling time. An integrated C/A code contains C_0 chips. C_0 is 1023 for GPS and 2046 for BDS.

In weak signal conditions, because TOW cannot be decoded for high error rate of navigation data, the satellite signal transmit time and the satellite position cannot be obtained. However, the receiver still can measure the phase of C/A code and the satellite signal submillisecond transmit time. And coarse time navigation is a method which uses the phase of C/A code with assistant information including position, time, and ephemeris to solve positioning problem.

2.2 Coarse Time Navigation Equation

The relative satellite velocity causes the coarse time navigation problem. We can compute the relative velocity of each satellite and use navigation equations to obtain position, receiver common bias, and coarse time error. That is, we solve for the time, instead of waiting to decode it, and TTFF is reduced.

Conventional pseudorange positioning equation is as follows:

$$\rho^{(i)} = c(t_r - t_t^{(i)}) = ||\overrightarrow{s\left(t_t^{(i)}\right)} - \vec{u}|| + b - c\lambda^{(i)} + \varepsilon^{(i)} \tag{2.2}$$

i	stands for satellite number;
t_r	stands for the reception time of satellite signal;
t_t	stands for the transmit time of satellite signal;
$\overrightarrow{s(t_t)}$	stands for the satellite position (S_x, S_y, S_z) at t_t;
$\vec{u}$	stands for the receiver position (x, y, z) at t_r;
b	stands for the common bias (like receiver clock error and instrumental bias);
λ	stands for error correction including the ionospheric delay, satellite clock bias, and so on;
ε	stands for the measure error.

Supposing that we have a set of less accurate satellite signal transmit time $t_{t0}^{(i)} (i = 1, 2, 3\ldots)$ and they have a common bias compared with signal transmit time from Eq. 2.1.

$$t_t^{(i)} = t_{t0}^{(i)} + \delta_{t_t} \quad (i = 1, 2, 3\ldots) \tag{2.3}$$

$\vec{u}$ stands for the receiver real position and there is an equation as follows:

$$\vec{u} = \vec{u_0} + \overrightarrow{\delta u} \tag{2.4}$$

$\vec{u_0}$ stands for the receiver prior position (x_0, y_0, z_0), $\overrightarrow{\delta u}$ stands for the position correction $(\delta_x, \delta_y, \delta_z)$. Combine Eqs. 2.3 and 2.4 with Eq. 2.1 and the result is as follows:

$$\begin{aligned}\rho^{(i)} &= c(t_r - (t_{t0}^{(i)} + \delta t_t)) \\ &= ||\overrightarrow{s\left(t_{t0}^{(i)} + \delta t_t\right)} - (\vec{u_0} + \overrightarrow{\delta u})|| + b - c\lambda^{(i)} + \varepsilon^{(i)}\end{aligned} \tag{2.5}$$

Equation 2.6 is linearization of Eq. 2.5

$$\rho^{(i)} = c(t_r - (t_{t0}^{(i)} + \delta t_t)) = ||\overrightarrow{s\left(t_{t0}^{(i)}\right)} - \overrightarrow{u_0}|| \\ - \overrightarrow{e^{(i)}} \cdot \overrightarrow{\delta u} + v^{(i)} \cdot \delta_{t_t} + b - c\lambda^{(i)} + \varepsilon^{(i)} \tag{2.6}$$

Equation 2.6 can be converted to Eq. 2.7 with equation ($\hat{\rho}^{(i)} = ||\overrightarrow{s\left(t_{t0}^{(i)}\right)} - \overrightarrow{u_0}|| - c\lambda^{(i)}$), and Eq. 2.7 is just coarse time navigation equation.

$$\delta\rho^{(i)} = \rho^{(i)} - \hat{\rho}^{(i)} = -\overrightarrow{e^{(i)}} \cdot \overrightarrow{\delta u} + v^{(i)} \cdot \delta_{t_t} + b - c\lambda^{(i)} + \varepsilon^{(i)} \tag{2.7}$$

$\delta\rho$ stands for residual error of pseudorange;
$\vec{e}$ stands for unit vector from the receiver prior position to the satellite position;
$\overrightarrow{u_{xyz}}$ stands for the receiver position correction;
b stands for the common bias (like the receiver clock error);
v stands for change rate of distance from the receiver prior position to the satellite position at a coarse satellite signal transmit time t_{t0};
δ_{t_t} stands for correction of the coarse time;
ε stands for the measure error.

If there are k satellites, equations can be set up as follows:

$$\delta\rho = H\delta x + \varepsilon \tag{2.8}$$

In equations, $\delta_x = (\overrightarrow{\delta_u}, b, \delta_{t_t}) = (\delta_x, \delta_y, \delta_z, b, \delta_{t_t})$ and

$$H = \begin{bmatrix} -e^{(1)} & 1 & v^{(1)} \\ \vdots & \vdots & \vdots \\ -e^{(k)} & 1 & v^{(k)} \end{bmatrix}$$

if H has five independent rows, δ_x can be calculated. That is, at least five satellites are observed, all unknown parameters including position $\vec{u}$, common bias b, and coarse time correction t_t can be calculated.

However, precondition to establish coarse time navigation equations is that coarse signal transmit time of all satellites has one common bias and ephemeris at the observation time has been obtained.

Ephemeris can be provided by A-GNSS servers. How to get coarse signal transmit time of all satellites which have one common bias can be solved through the technology of reconstructing pseudorange.

2.3 Reconstructing Pseudo-range

In coarse time navigation, because navigation message has not been decoded, the receiver can only measure the submillisecond pseudorange. Frank Van Diggele put forward reconstructing pseudorange algorithm to solve how to establish coarse time navigation equations with submillisecond pseudorange information. Figure 1 is the algorithm flowchart.

The algorithm process is as follows:

1. Select the satellite as a reference satellite (the best choice of reference satellite is the highest satellite), specify an integer millisecond ambiguity $N^{(1)}$ for the satellite, the measured code phase value $z^{(1)}$, thus obtaining the first satellite pseudorange measurements:

$$\rho^{(1)} = N^{(1)} + z^{(1)} \tag{2.9}$$

2. After the measured pseudoranges were linked to geometric distance, we get

$$\begin{aligned} N^{(1)} + z^{(1)} &= r^{(1)} - \delta_t^{(1)} + b + \varepsilon^{(1)} \\ &= \hat{r}^{(1)} - d^{(1)} - \delta_t^{(1)} + b + \varepsilon^{(1)} \end{aligned} \tag{2.10}$$

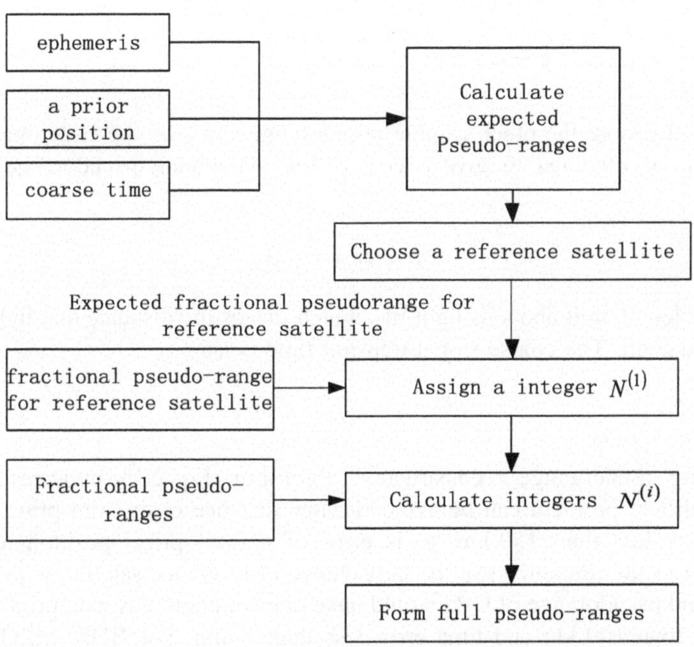

Fig. 1 Flowchart of reconstructing pseudorange

Among them, $r^{(1)}$ is the true geometric distance, $\hat{r}^{(1)}$ is rough geometric distance between the satellite position under coarse satellite signal launch time to get and a priori receiver position, $d^{(1)}$ is the error present in $\hat{r}^{(1)}$, caused by a priori position and time error, $\delta_{\mathrm{t}}^{(1)}$ is known satellite clock offset, b is the common bias. (receiver clock error), $\varepsilon^{(1)}$ is a measurement error. For other satellites, it is assumed for its integer millisecond ambiguity $N^{(k)}$, empathy can get the following equation:

$$N^{(i)} + z^{(i)} = \hat{r}^{(i)} - r_d^{(i)} - \delta_t^{(i)} + b + \varepsilon^{(i)} \tag{2.11}$$

3. Subtract (2.6) from (2.5), elimination of the same common bias, we get

$$\begin{aligned} N^{(i)} = N^{(1)} + z^{(1)} - z^{(k)} + (\hat{r}^{(i)} - r_d^{(i)} - \delta_{\mathrm{t}}^{(i)} + b + \varepsilon^{(i)}) \\ - (\hat{r}^{(1)} - r_d^{(1)} - \delta_{\mathrm{t}}^{(1)} + b + \varepsilon^{(1)}) \end{aligned} \tag{2.12}$$

To get the correct value, you must make a priori position and time error satisfies the following formula:

$$\left| -r_d^{(i)} + \varepsilon^{(i)} + r_d^{(1)} - \varepsilon^{(1)} \right| < 150\,\mathrm{km} \tag{2.13}$$

so that we can obtain the following expression of $N^{(k)}$:

$$N^{(i)} = \mathrm{round}\left(N^{(1)} + z^{(1)} - z^{(i)} + (\hat{r}^{(i)} - \delta_{\mathrm{t}}^{(i)}) - (\hat{r}^{(1)} - \delta_{\mathrm{t}}^{(1)})\right) \tag{2.14}$$

This will restore the other satellite pseudorange value by the code phase value, which can be obtained to give a coarse time navigation pseudorange residual equation:

$$\delta\rho = N + z - \hat{\rho} \tag{2.15}$$

The all length unit above is light-ms, which means the distance that light travels in a millisecond. The coarse signal transmit time is t_{t0}.

$$t_{\mathrm{t0}}^{(i)} = t_{\mathrm{r}} - (N^{(i)} + z^{(i)}) \tag{2.16}$$

In deriving pseudorange reconstruction algorithm, Eq. 2.13 indicates that the integer rollover problem can be avoided when distance error from priori position and time is less than 150 km. r_d is error of r from priori position and time. According to the reference [6], the best choice of reference satellite is the highest satellite and pseudorange of GPS would have one common bias with priori position error less than 100 km and time error less than 1 min. For BDS, MEO satellite pseudorange rate is less than 700 m/s, IGSO satellite pseudorange rate is less than

400 m/s, and GEO satellite pseudorange rate is very small [3, 4, 6–8]. r_d from time error is proportional to the maximum of satellite pseudorange rate. So we set position and time error limit as 100 km and 1 min and the integer rollover problem can be avoided for BDS. So far, the solution of GPS and BDS coarse time navigation has been elucidated fully.

2.4 BDS/GPS Coarse Time Navigation

Different satellite navigation and positioning systems use different space-time bases. A unified space-time base is the precondition of combined positioning. BDS uses CGCS2000 national geodetic coordinates and BDST (Beidou time), while the GPS uses WGS-84 coordinate system and the corresponding GPST (GPS time). According to the related discussion of references [9], difference of two coordinate systems has a negligible effect on pseudorange positioning. So pseudorange positioning only needs to unify the time base.

BDST and GPST both use atomic time system. The two time systems have a common second and all have not leap second. However, the beginning time of BDST and GPST are different. The GPST began at 0:0:0 on January 1, 1986, while the BDST began at 0:0:0 on January 1, 2006. So, there is a 1356 weeks' time bias between the two systems. Moreover, there is a 14 s' time bias between GPST and UTC (universal time coordinated) at 0:0:0 on January 1, 2006. Therefore, there is a 14 s' time bias between GPST and BDST too. The conversion relation of them is as follows:

$$\mathrm{BDST} = \mathrm{GPST} - \Delta T_1 - \Delta T_2 \tag{2.17}$$

In Eq. 2.17, $\Delta T_1 = 1356\,\mathrm{weeks}$, $\Delta T_2 = 14\,\mathrm{s}$

Because carrier frequencies of BDS and GPS are different, the receiver hardware delays of them are different. We put hardware delay and the receiver clock error together as common bias. $b1$ and $b2$ are BDS and GPS common biases. $h1$ and $h2$ are BDS and GPS hardware delays. And there are equations as follows:

$$b1 = h1 + \delta t_r \tag{2.18}$$

$$b2 = h2 + \delta t_r \tag{2.19}$$

Therefore, combined pseudorange positioning needs at least five satellites and combined coarse time navigation needs at least six satellites. If reference satellite and reconstructing satellite belong to different systems when reconstructing pseudorange in BDS/GPS combined coarse time navigation, that common biases of different systems are different which would lead Eqs. 2.12 and 2.13 to be as follows:

$$N^{(i)} = N^{(1)} + z^{(1)} - z^{(i)} + (\hat{r}^{(i)} - r_d^{(i)} - \delta_t^{(i)} + b1 + \varepsilon^{(i)}) \\ - (\hat{r}^{(1)} - r_d^{(1)} - \delta_t^{(1)} + b2 + \varepsilon^{(1)}) \tag{2.20}$$

$$\left| -r_d^{(i)} + \varepsilon^{(i)} + b2 + r_d^{(1)} - \varepsilon^{(1)} - b1 \right| < 150\,\text{km} \tag{2.21}$$

$h1$ and $h2$ are very small, so that they nearly have no effect on Eq. 2.21.

3 Experiments

The data this paper used are 100-epoch (epoch interval 30 s) GPS L1 and BDS B1 pseudorange observation of IGS Australia CUT0 station at 2:00 to 4:00 on September 1, 2015. We conducted GPS CTN experiment, BDS CTN experiment, and BDS/GPS CTN experiment with these data and compared the result of CTN with the result of pseudorange positioning. There are 13 GPS and 10 BDS satellites during observation time. If we denote the modulus by square brackets, the sub-millisecond pseudorange is pseudorange [1 light-ms]. In addition, we computed the position on this premise that all observations' weights were equal.

The experimental results are shown in Figs. 2, 3, 4, 5, 6, 7, and 8. Among them, Figs. 2, 4, and 6 are BDS, GPS, BDS/GPS coarse time navigation error sequence

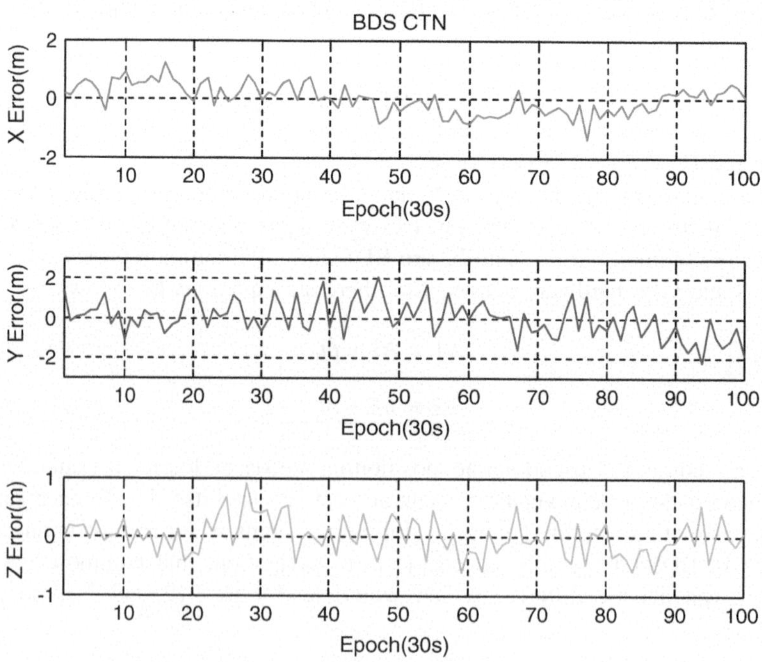

Fig. 2 BDS coarse time navigation error sequence diagram

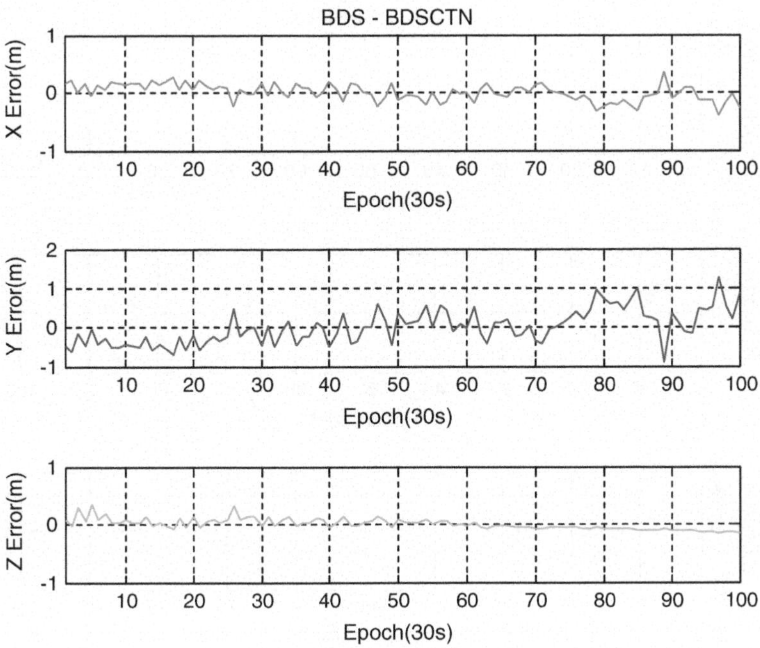

Fig. 3 BDS coarse time navigation and pseudorange positioning mutual difference sequence diagram

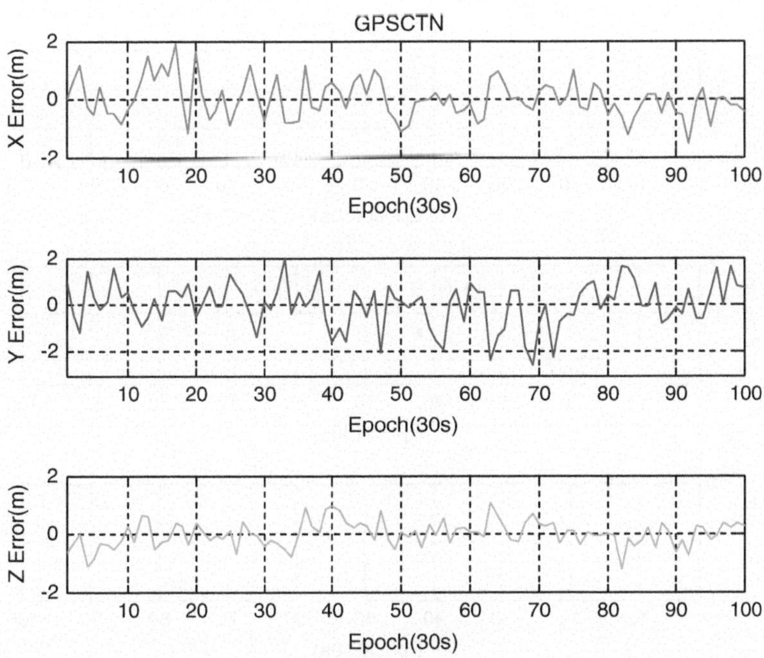

Fig. 4 GPS coarse time navigation error sequence diagram

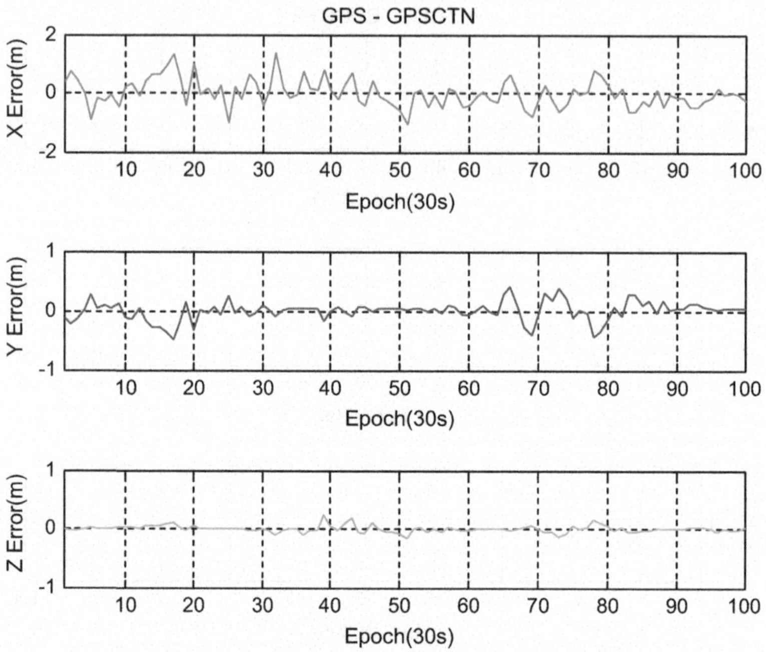

Fig. 5 GPS coarse time navigation and pseudorange positioning mutual difference sequence diagram

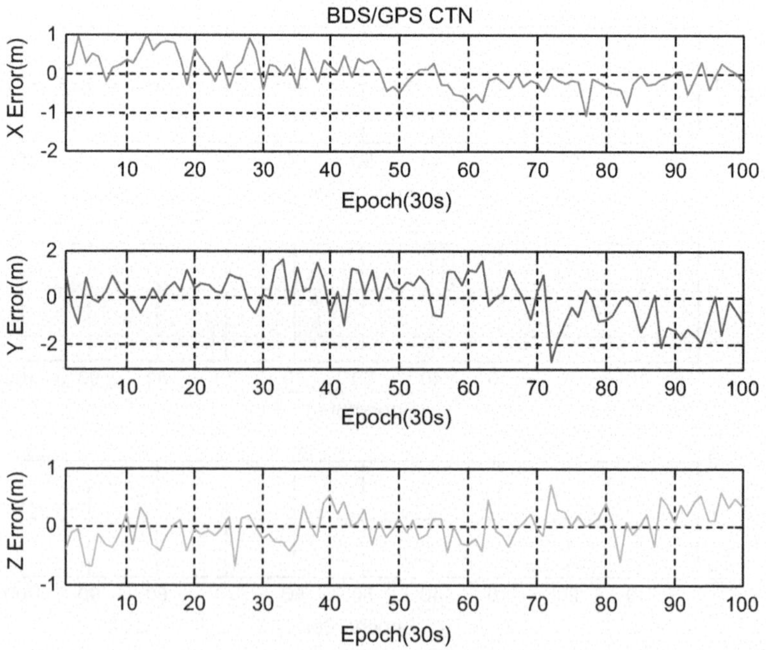

Fig. 6 BDS/GPS coarse time navigation error sequence diagram

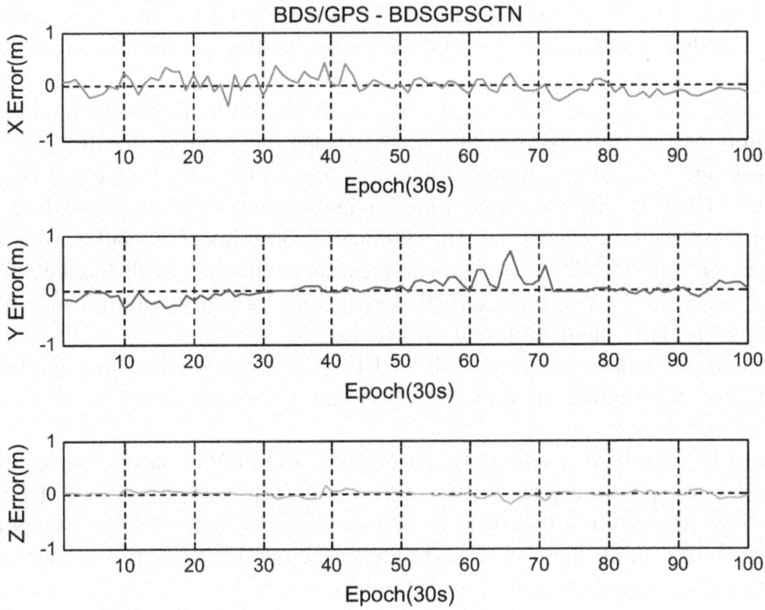

Fig. 7 BDS/GPS coarse time navigation and pseudorange positioning mutual difference sequence diagram

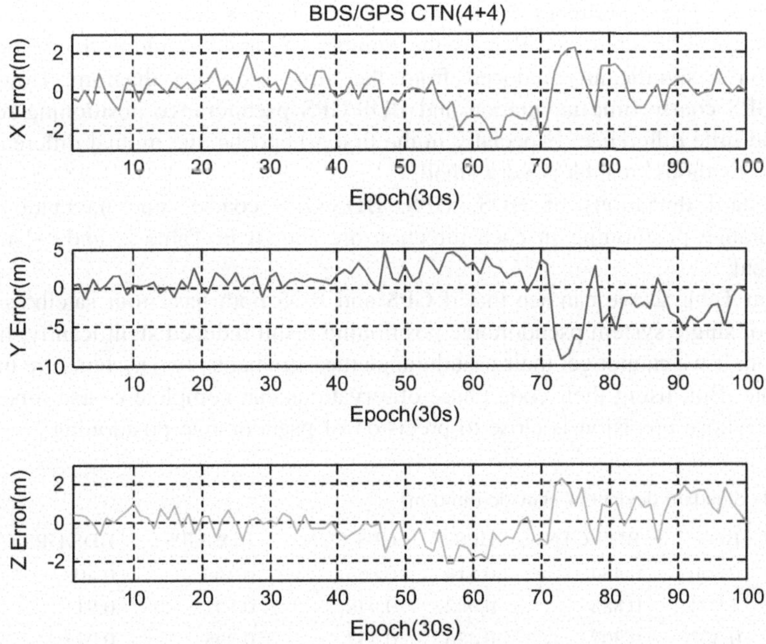

Fig. 8 BDS/GPS coarse time navigation error sequence diagram

diagrams, and Figs. 3, 5, and 7 are the BDS, GPS, BDS/GPS coarse time navigation results and their corresponding pseudorange positioning results mutual difference sequence diagrams.

From Figs. 3, 5, and 7, we can see that BDS, GPS, BDS/GPS coarse time navigation and their corresponding pseudorange positioning results have little difference and most of the mutual differences are within 1 m. So we can conclude that BDS, GPS, BDS/GPS coarse time navigation and their corresponding pseudorange positioning precision are approximate. From Figs. 2, 4, and 6, we can see that BDS, GPS, BDS/GPS coarse time navigation results (the earth-fixed coordinate system) under the *Y*, *X*, *Z* three direction errors are basically within 2 m, which is enough in the daily navigation and positioning.

Standard deviations of BDS, GPS, BDS/GPS coarse time navigation and pseudorange positioning in each direction are shown in Table 1 (*XYZ* is just three-dimensional).

As can be seen from Table 1, the BDS, GPS, BDS/GPS coarse time navigation and their corresponding pseudorange positioning's precision is close, BDS and coarse time navigation's precision is even superior to that of BDS pseudorange positioning. We think there is time difference below 1 s besides 14 s and coarse time navigation can correct the time difference below 1 s.

Considering that receiver is hard to be observed more than 20 satellites under the condition of weak signal. So this paper randomly selected four satellites' observation from BDS and four satellites' observation from GPS to conduct combined coarse time navigation. In this situation, BDS or GPS coarse time navigation is impossible. The experiment results are shown in Figs. 8 and 9.

It is not difficult to see that as the number of satellites reduced, positioning precision *r* significantly reduced from the error sequence diagram. However, BDS/GPS coarse time navigation and BDS/GPS pseudorange positioning results still had little difference. Especially in the first 50 epochs, the mutual difference of the two methods' results is very small.

Standard deviations of BDS, GPS, BDS/GPS coarse time navigation and pseudorange positioning in each direction are shown in Table 2, and '–' means inexistent.

From Table 2, we can see that if GPS and BDS both have four satellites, precision of single system pseudorange positioning result reduced significantly. In this situation, if we cannot get their pseudorange measurements, two systems are unable to locate. But, using their code phase observations can complete coarse time navigation whose precision is close to precision of pseudorange positioning.

Table 1 Standard deviations statistic (unit: m)

	BDS	BDS CTN	GPS	GPS CTN	BDS/GPS	BDS/GPS CTN
X	0.450	0.484	0.514	0.666	0.366	0.408
Y	1.038	0.888	0.942	0.948	0.892	0.911
Z	0.302	0.307	0.436	0.437	0.295	0.297
XYZ	1.171	1.057	1.158	1.238	1.008	1.041

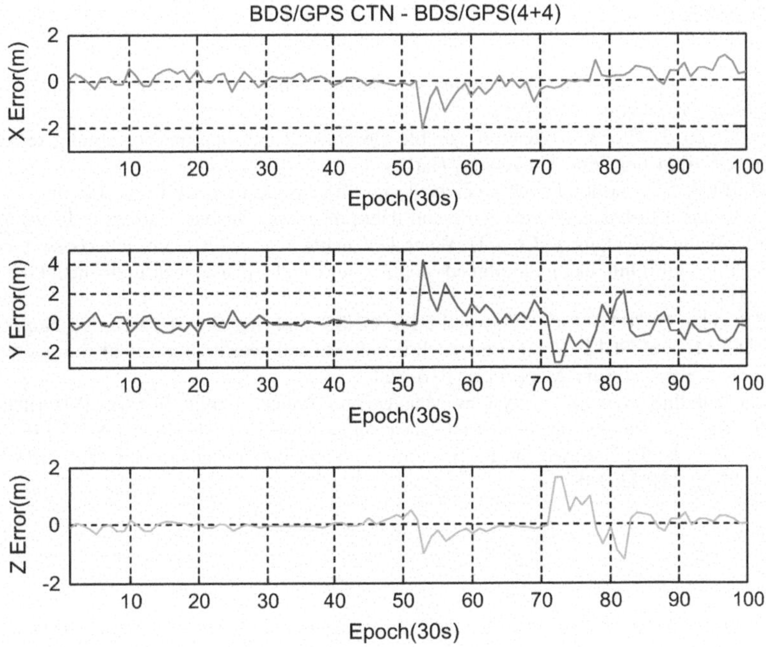

Fig. 9 BDS/GPS coarse time navigation and pseudorange positioning mutual difference sequence diagram

Table 2 Standard deviations statistic (unit: m)

	BDS	BDS CTN	GPS	GPS CTN	BDS/GPS	BDS/GPS CTN
X	1.141	–	3.198	–	0.946	0.949
Y	6.957	–	6.313	–	2.397	2.837
Z	1.945	–	4.936	–	0.838	0.952
XYZ	7.313	–	8.628	–	2.709	3.139

4 Conclusion

For the condition that there are only code phase measurements and conventional methods cannot locate, this paper realized BDS/GPS coarse time navigation algorithm, which effectively extends the scope of application of satellite positioning. And the conclusions are drawn that the precision of coarse time navigation is close to the precision of pseudorange positioning and that coarse time navigation can be used for daily navigation and location-based services. Because visible satellites of the Australian CUT0 station are so many and the signal strength is so strong, so the data is limited and coarse time navigation under weak signal conditions still need further study.

References

1. Song C (2009) Assisted GPS positioning system key technology research. National University of Defense Technology, Changsha
2. Zhang G (2013) Global navigation satellite system and enhance the positioning technology research. Harbin Industrial University, Harbin
3. Xie G (2009) GPS principle and receiver design. Electronic Industry Press, Beijing
4. Xie Z, Huang J (2010) GPS surveying and data processing. Wuhan University Press, Wuhan
5. Chen H, Wang H, Chiang Y, Chen H, Wang H, Chiang Y, et al (2013) A new coarse time GPS positioning algorithm using combined doppler and code phase measurements. GPS Solut 541–551
6. Van Diggelen F (2009) A-GPS assisted GPS, GNSS, and SBAS. Artech House, Norwood
7. Li X, Wu M, et al (2012) Navigation satellite velocity and acceleration calculation method and precision analysis. J Surv Mapp Pract 816–824
8. China's Satellite Navigation System Management Office. Public Service Performance of BeiDou Satellite Navigation System specification (version 1.0). The BDS-OS-PS-1.0
9. Gao W, Guo J (2012) Based on the space-time system BeiDou positioning with GPS fusion. J Surv Mapp 12:743–748

Spacecrafts Navigation Demand Analysis and Simulation on MEO Constellation

Peng Li, Zhonggui Chen, Xiangjun Wu, Wenyu Hao and Lu Wang

Abstract Space navigation users of GNSS can be classified into four categories: low orbit users, medium orbit users, high orbit users, and deep space users. The GNSS space navigation capability can be analyzed by visibility, GDOP, and received EIRP. The transmit antenna pattern and space receiver antenna orientation are important factors which should be considered carefully. In this paper the navigation requirements are put forward, respectively, for different types of space user. And the related factors are studied theoretically. At last, the navigation capability for space users of one MEO constellation is simulated with "satellite navigation mission analysis simulator (SNMAS)." The simulation illustrates that: the constellation can provide almost the same navigation service for the low orbit users and the land users; the "continuous positioning" requirement of medium orbit users can be reached; the "intermittent positioning" and "continuous timing" requirement of high orbit users can also be realized; but the "continuous timing" requirement of deep space users cannot be realized such as lunar orbit users.

Keywords Space navigation · Mission analysis · Navigation satellite constellation · SNMAS

1 Introduction

Navigation satellite constellation is one of the essential parts of PNT infrastructure, which provides orbit determination, attitude determination, timing, and navigation services for land users, ships, aircrafts, and spacecrafts. The American GPS and Chinese Beidou system are designed mainly for the navigation and positioning requirement of land users and aircrafts, and the "narrow beam" design of antenna pattern will lead to bad visibility, high GDOP, and low receiving power [1]. A lot of

P. Li (✉) · Z. Chen · X. Wu · W. Hao · L. Wang
Institute of Spacecraft System Engineering, China Academy of Space Technology, P.O BOX 5142-370 Youyi Road 102, Beijing 10094, People's Republic of China
e-mail: richardlee0914@gmail.com

J. Sun et al. (eds.), *China Satellite Navigation Conference (CSNC) 2016 Proceedings: Volume I*, Lecture Notes in Electrical Engineering 388,
DOI 10.1007/978-981-10-0934-1_22

space navigation experiments have been carried out based on the GPS. In 1982, the satellite “Landsat-4” is the first spacecraft which was equipped with GPS receivers and had got the navigation service [2]. In the middle of 2012, the first test satellite of Galileo, GIOVE-A, had become the first high orbit civilian satellite which conducted space navigation service with GPS after retirement. Although the GIOVE-A orbit is higher about 1000 km than GPS, the satellite can measure its position, velocity, and time with GPS. As a result, many tests and simulation were applied quickly in the high orbit spacecraft navigation [3–9].

The spacecraft navigation performance of GNSS constellation should be studied carefully, because the spacecraft navigation is the important extended function for the GNSS design but the existing GPS and Beidou have a lot of problems when providing services for spacecraft users. The GPS-III mission analysis team sets up different groups in order to carry out demand analysis for different orbit spacecrafts, these groups are named as land service group, medium service group, and high service group [10].

This paper analyzes the navigation requirement for different types of space user, and studies the related factors in space navigation including visibility, GDOP, received EIRP, transmit antenna pattern, and space receiver antenna orientation. At last, the navigation capability for space users of one MEO constellation is simulated with “satellite navigation mission analysis simulator (SNMAS).”

2 Spacecraft Navigation Demand Analysis

The space navigation users of GNSS, according to the height of their orbit, can be classified into four categories: low orbit users, medium orbit users, high orbit users, and deep space users. The low orbit space users are below 3000 km such as space station, earth observation satellite. The medium orbit users are from 3000 to 8000 km, and the high orbit users are from 8000 km to earth stationary orbit (about 36,000 km). The deep space users are beyond 36,000 km. The distribution for different types of space navigation services is shown in Fig. 1.

The low orbit users are similar to the land users, and have high requirements on satellite visibility, GDOP, signal stability and integrity, and so on. Because the navigation signal antenna pattern is saddle surface, the low orbit users can get a similar navigation performance with the land users. With the increase of the orbit, the geometry dilution of position will be reduced, and the navigation usability will be reduced. Therefore, the demand of medium orbit users is the navigation usability, that is, the basic requirement of the user is to receive a positioning service with certain accuracy, which requires more than four visible satellites with low GDOP. And the integrity of the signal is not necessary.

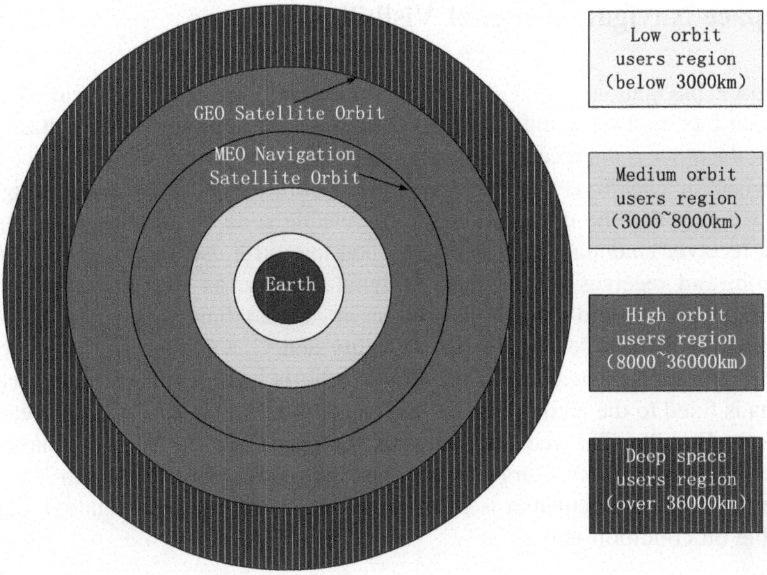

Fig. 1 Space navigation services distribution

The navigation performance will degrade obviously for the high orbit users when the GDOP is getting higher. When the space user is higher than the MEO navigation satellites, the receiving antenna is fixed to the earth in order to receive the signal outside the shadow of the earth, and the visibility will be greatly reduced. The main requirement for the high orbit users is limited positioning and one of the GNSS signals is received properly, which provides accurate timing without expensive space atomic clock [10]. The deep space users will suffer not only the problem of visibility but also the low signal power and high GDOP caused by the long distance. The deep space users do not need position but that one of the GNSS signals is received properly. The requirements for the four types of users are listed in Table 1.

Table 1 The requirements for four types of users

Type of users	Height of orbit (km)	Requirements
Low orbit users	0–3000	Continuous positioning, accuracy is within 20 m
Medium orbit users	3000–8000	Continuous positioning, visibility is over four satellites
High orbit users	8000–36,000	Limited positioning, more than one signal is visible
Deep space users	Above 36,000	More than one signal is visible

3 Space Navigation Signal Visibility Analysis

The space navigation signal's visibility, different from land scenario, has two important factors. One is the navigation satellite's transmit antenna orientation and pattern; the other is the receiver antenna orientation and pattern. For the space users, the navigation satellite is "visible" when the satellites are within the receiving zone of the receiver and the users are within the visible zone of the navigation signal.

The receiver antenna of the land users and low orbit users is fixed backward to the earth, and receives the navigation signal directly, as shown in Fig. 2a. The medium orbit users and some high orbit users can also have their receiver antenna fixed backward to the earth, but the visibility and GDOP will degrade when the height of the users increase. When the user orbit is increased further, the receiver antenna is fixed to the earth receiving the signal outside the shadow of the earth, as shown in Fig. 2b. The receiver antenna for users above the MEO navigation satellites and deep space users can only be fixed backward to the earth.

When the receiver antenna is fixed to the earth, the space navigation signal is available on condition that

$$\begin{cases} \gamma \leq \alpha \leq \beta \\ \mathrm{el}_{\mathrm{te}} \leq \varphi \end{cases} \tag{1}$$

where γ is the earth shade angle, β is the transmit antenna beam half angle, φ is the receiver antenna half angle. α is transmit angle, which is calculated as

$$\alpha = \arccos\left(\frac{\overrightarrow{\mathrm{CS}} \cdot \vec{S}}{\left|\overrightarrow{\mathrm{CS}}\right| \cdot \left|\vec{S}\right|}\right) \tag{2}$$

where $\vec{S}$ is satellite position vector, $\overrightarrow{\mathrm{CS}}$ is the vector from spacecraft to the satellite. $\mathrm{el}_{\mathrm{te}}$ is received signal inclination, and can be obtained as

$$\mathrm{el}_{\mathrm{te}} = \arccos\left(\frac{-\overrightarrow{\mathrm{CS}} \cdot \vec{C}}{\left|\overrightarrow{\mathrm{CS}}\right| \cdot \left|\vec{C}\right|}\right) \tag{3}$$

where $\vec{C}$ is the spacecraft position vector.

When the receiver antenna is fixed backward to the earth, the space navigation signal is available on condition that

$$\begin{cases} \alpha \leq \beta \\ \mathrm{el}_{\mathrm{be}} \leq \varphi \end{cases} \tag{4}$$

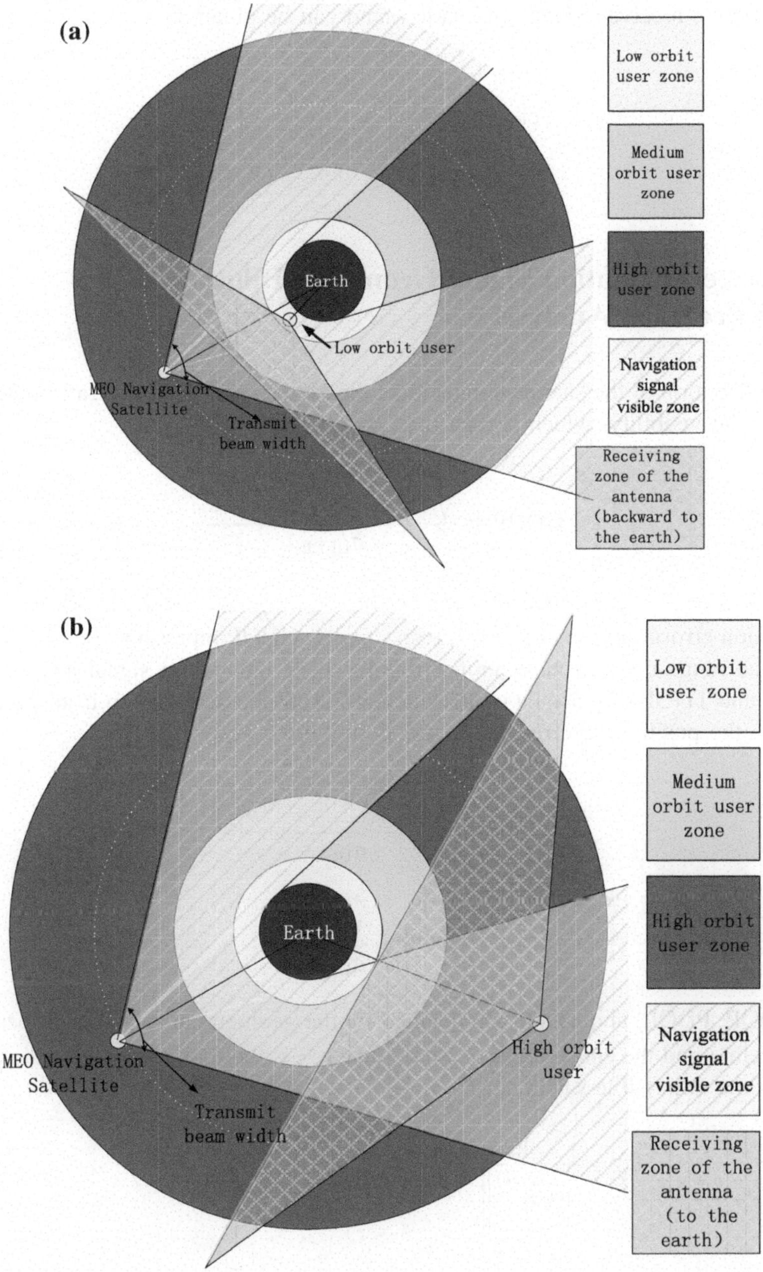

Fig. 2 Space navigation signal visibility diagram. **a** Receiver antenna backward to the earth. **b** Receiver antenna to the earth

where $\mathrm{el_{be}}$ is received signal inclination, and can be obtained as

$$\mathrm{el_{be}} = \arccos\left(\frac{\overrightarrow{\mathrm{CS}} \cdot \vec{C}}{\left|\overrightarrow{\mathrm{CS}}\right| \cdot \left|\vec{C}\right|}\right) \tag{5}$$

4 Space Navigation Signal Geometric Dilution of Precision Analysis

The distribution of the satellites must meet the requirement of spacecraft navigation besides the visibility, which is measured by GDOP [10].

$$\mathrm{GDOP} = \frac{\sqrt{\sigma_{x_u}^2 + \sigma_{y_u}^2 + \sigma_{z_u}^2 + \sigma_{ct_b}^2}}{\sigma_{\mathrm{UERE}}} \tag{6}$$

where σ_{UERE} is use equivalent range error, $\sigma_{x_u}^2$, $\sigma_{y_u}^2$, $\sigma_{z_u}^2$, and $\sigma_{ct_b}^2$ are the three dimension errors and the time error, respectively. GDOP represents the degradation from the range error to navigation error. In space navigation signal analysis, the PDOP and TDOP can also be used, which represent the degradation from the range error to the positioning error and to the timing error, respectively.

$$\mathrm{PDOP} = \frac{\sqrt{\sigma_{x_u}^2 + \sigma_{y_u}^2 + \sigma_{z_u}^2}}{\sigma_{\mathrm{UERE}}} \tag{7}$$

$$\mathrm{TDOP} = \frac{\sqrt{\sigma_{ct_b}^2}}{\sigma_{\mathrm{UERE}}} \tag{8}$$

GDOP, PDOP, and TDOP are decided by the geometrical relationship between the satellite and the space users. Assume that $1^{(k)}$ is the unit vector from the space user to the kth visible satellite,

$$1^{(k)} = \frac{1}{\left\|\overrightarrow{\mathrm{CS}}^{(k)}\right\|}\left(S_x^{(k)} - C_x, S_y^{(k)} - C_y, S_z^{(k)} - C_z\right) \tag{9}$$

where $\overrightarrow{\mathrm{CS}}^{(k)}$ is the vector from the space user to the kth visible satellite, $S_x^{(k)}, S_y^{(k)}, S_z^{(k)}$ is the position of the kth visible satellite, C_x, C_y, C_z is the space user position. GDOP, PDOP, and TDOP can be calculated as

$$\begin{cases} \mathrm{GDOP} = \sqrt{H_{11} + H_{22} + H_{33} + H_{44}} \\ \mathrm{PDOP} = \sqrt{H_{11} + H_{22} + H_{33}} \\ \mathrm{TDOP} = \sqrt{H_{44}} \end{cases} \tag{10}$$

where H is error accuracy matrix, and G is geometry matrix.

$$H = (G^{\mathrm{T}} \cdot G)^{-1} \tag{11}$$

$$G = \begin{bmatrix} \left(-1^{(1)}\right)^{\mathrm{T}} & 1 \\ \left(-1^{(2)}\right)^{\mathrm{T}} & 1 \\ \vdots & \vdots \\ \left(-1^{(K)}\right)^{\mathrm{T}} & 1 \end{bmatrix} \tag{12}$$

5 Space Navigation Signal Link Analysis

The formula of satellite received power is

$$P_{\mathrm{r}} = P_{\mathrm{t}} + G_{\mathrm{t}} - L_{\mathrm{S}} - L_{\mathrm{a}} + G_{\mathrm{r}} - L_{\mathrm{p}} \tag{13}$$

where P_{t} is transmitting power, G_{t} is transmitting antenna gain, L_{S} is free space attenuation, L_{a} is atmospheric loss, G_{r} is receiving antenna gain, L_{p} is antenna mismatch loss. As the space navigation signals do not pass through the atmosphere, the atmospheric loss $L_{\mathrm{a}} = 0\,\mathrm{dB}$.

$$L_{\mathrm{S}} = 20\log\left(\frac{4\pi R_{\mathrm{d}} f}{3 \times 10^{8}}\right) \tag{14}$$

where R_{d} is the distance between navigation satellite and space user, and is decided by the geometrical relationship between the satellite and the space user, which is illustrated by Fig. 3. The transmit loss L_{S} is in direct ratio to the square of R_{d}, as a

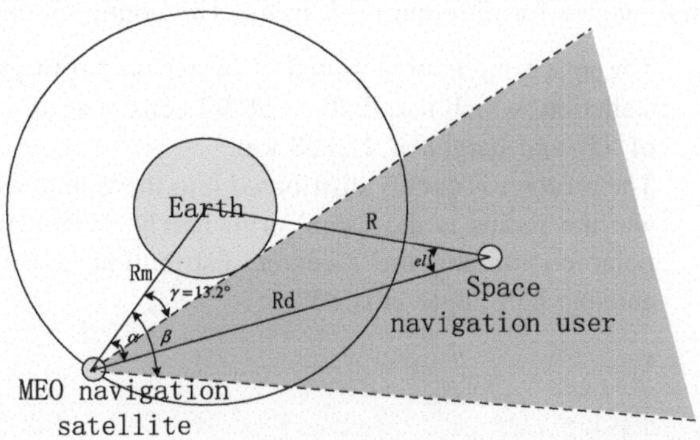

Fig. 3 Space navigation signal geometrical relationship

result, the optimized design is that the transmit antenna gain $G_T(\alpha)$ is also in direct ratio to the square of R_d in the range between earth shade angle γ and transmit antenna beam half angle β. The antenna pattern should be designed as the gain from 0 to γ which is reduced and increased from γ to β.

$$R_d = R_m \cdot \cos\alpha + \sqrt{R^2 - (R_m \sin\alpha)^2} \tag{15}$$

where R_m is MEO satellite orbit radius, R is space user orbit radius, α is transmit angle. Space user received signal inclination el is

$$\mathrm{el} = \sin^{-1}\left[\frac{R_m \sin(\alpha)}{R_g}\right] \tag{16}$$

If the user antenna is fixed to the earth, the distance between the satellite and the space user will be maxim when the line between them touches the earth.

$$R_{max} = R_m \cdot \cos(\gamma) + R_g \cdot \cos(\mathrm{el}) \tag{17}$$

6 Space Navigation Signal Design and Simulation

6.1 Simulation Configuration

According to the requirements of different space navigation users, the space navigation signal transmit EIRP and beam width should be designed carefully, and the visibility, GDOP, received power are the important factors. This paper simulates the navigation capability for space users of one MEO constellation with "satellite navigation mission analysis simulator (SNMAS)," which is a satellite navigation system mission analysis tool, and provides functions such as navigation satellite constellation configuration and generation, users setup and generation, visibility calculation, DOP analysis, transmit/received antenna setup and simulation, link calculation, and received power calculation.

This simulation calculates the visibility and GDOP in different transmit antenna half angles for different orbit users. The configuration is

1. The space navigation signal is based on MEO global navigation system constellation, which use Walker 24/3/2 constellation configuration, with inclination of 55° and height of 21,528 km;
2. The receiver is evenly distributed into the sphere whose center is the earth's core and the radius is the users' orbit height. When the density is 10° in spherical polar coordinate, the receiver distribution is shown in Fig. 4. The receiver antenna half angle φ is 85°;

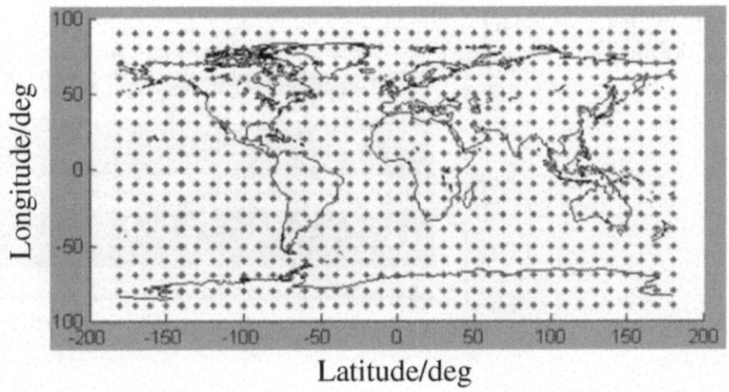

Fig. 4 Space receiver distribution

3. The transmit beam is in B2, the pattern is designed as the antenna gain is over 12, 10, 5, 0, and −2 dBi when the transmit antenna beam half angle β is within ±20°, ±22°, ±30°, ± 35°, and ±40°, respectively (Fig. 5).

6.2 *Visibility and DOP Simulation of Low Orbit Users*

The half angle β is 20° for the low orbit users, the receiver density is 10°, and the user orbit height is 3000 km. The available satellites number and 2σ DOP (95 %) are recorded for the sphere with 3000 km height during a simulation period, and the simulation interval is 10 min. As shown in Fig. 6, the average visible satellites number is 5.7 (95 %), and the DOP is 3.3 (95 %).

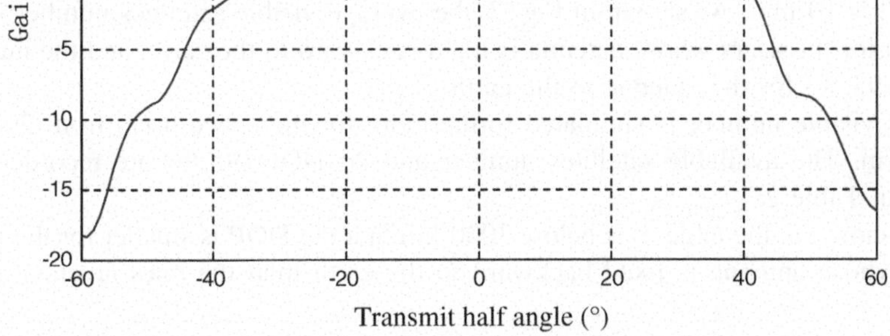

Fig. 5 Space navigation satellite transmit antenna pattern

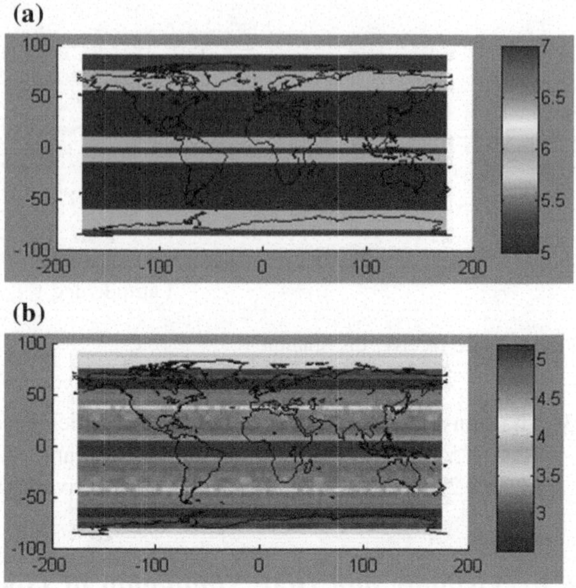

Fig. 6 Simulation result for the low orbit users (3000 km). **a** Visibility. **b** DOP

The simulation of low orbit user illustrates that the MEO navigation constellation provides the low orbit users' continuous and accurate positioning service when the transmit antenna pattern is designed as in Fig. 5 and the user receiver antenna is fixed backward to the earth.

6.3 *Effect Simulation of Receiver Antenna Orientation*

The receiver antenna orientation should be trade-off for the users above medium orbit to decide whether to receive the signal outside the shadow of the earth. The half angle β is 20° for the medium orbit users, the receiver density is 10°, and the user orbit height is 4000 km. The available satellites number (95 %) are recorded for the sphere with 4000 km height during a simulation period, and the simulation interval is 10 min. As shown in Fig. 7, the average visible satellites number is 3.1 (95 %) for the users whose antenna is fixed backward to the earth, and the number is 5.0 (95 %) for the antenna to the earth.

The visible number is simulated furthermore for the space users from 2000 to 4000 km. The available satellites number and 2σ DOP (95 %) are recorded and listed in Table 2.

As shown in the table, it is below 3000 km that the DOP is smaller for the space users whose antenna is fixed backward to the earth than the ones antenna to the

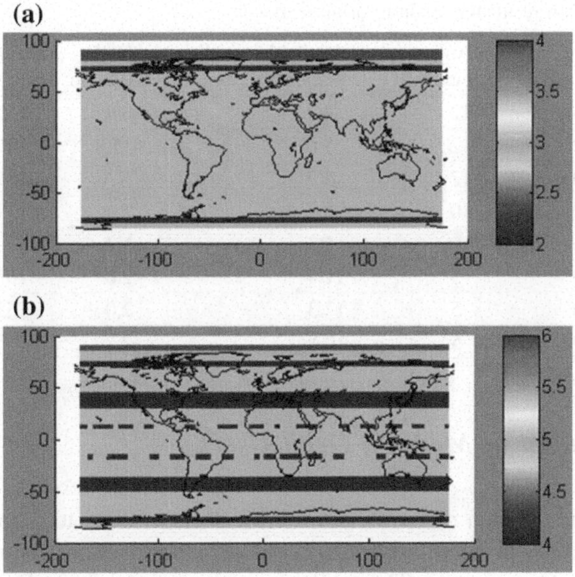

Fig. 7 Visible satellites number simulation for the medium orbit user (4000 km). **a** Receiver antenna backward to the earth. **b** Receiver antenna to the earth

Table 2 Simulation of available number and DOP (2000–4000 km)

User height (km)	Receiver antenna backward to the earth		Receiver antenna to the earth	
	Visible numbers (95 %)	DOP (95 %)	Visible numbers (95 %)	DOP (95 %)
2000	6.6	2.8	6	6.7
2250	6.5	2.9	6.5	4.5
2500	6.0	3.0	6.9	4.0
2750	5.9	3.2	7.2	3.7
3000	**5.7**	**3.3**	**7.7**	3.4
3250	5.2	6.5	7.2	3.6
3500	4.3	24.3	6.3	7.9
3750	3.6	Inf	5.9	8.3
4000	3.1	Inf	5.0	15.9

earth. However, the DOP degrades quickly for the space users backward to the earth above 3000 km, and the degradation is relatively smooth for the ones to the earth. As a result, the low orbit space users (below 3000 km) should fix their antenna backward to the earth, and the medium orbit, high orbit, and deep space users should fix their antenna to the earth.

Table 3 Simulation result for medium orbit users

User height (km)	Half angle β (°)	Antenna gain (dB)	Available satellite number (95 %)	DOP (95 %)	Minimum received power (dBW)
3000	20	12	7.7	3.4	−157.2
4000	20	12	5.0	15.9	−157.6
5000	20	12	3.7	NA	−157.9
5000	22	10	6.4	7.4	−159.9
6000	22	10	5.2	NA	−160.3
6000	30	5	11.4	2.0	−165.3
7000	30	5	12.3	2.0	−165.6
8000	30	5	10.8	2.7	−165.9

6.4 *Simulation of Medium Orbit Users*

Assume that the antenna of the medium orbit users is fixed to the earth, the user height is from 3000 to 8000 km, and the visible satellite number is over 4 which is the medium user requirement. The available satellites number, 2σ DOP (95 %), and minimum received power are recorded during a simulation period. The simulation interval is 10 min, and the result is recorded in Table 3.

The simulation illustrates that the user below 4000 km can receive the signal to get positioning service where the half angle is less than 20°, the user below 6000 km need half angle less than 22°, and the user below 8000 km need half angle less than 30°. The minimum received power for medium orbit user is about −165.9 dBW.

6.5 *Simulation for High Orbit and Deep Space Users*

Assume that the antenna is fixed to the earth, the high orbit user height is from 10,000 to 360,000 km, and the lunar orbit user height is 380,000 km, which is the typical deep space height, the visible satellite number is over 1 which is the user requirement. The available satellites number, 2σ DOP (95 %), and minimum received power are recorded during a simulation period. The simulation interval is 10 min, and the result is recorded in Table 4.

The simulation illustrates that the user below 36,000 km can receive the signal to get timing service where the half angle is less than 30°, the lunar orbit user about 38,000 km must receive the signal where the half angle is less than 40°.

Table 4 Simulation result for high orbit and deep space users (10,000–36,000 km)

User height (km)	Half angle β (°)	Antenna gain (dB)	Available satellite number (95 %)	DOP (95 %)	Minimum received power (dBW)
10,000	30	5	8.2	6.9	−166.4
15,000	30	5	4.8	NA	−167.6
20,000	30	5	3.6	NA	−168.6
25,000	30	5	3.2	NA	−169.6
30,000	30	5	2.8	NA	−170.4
36,000	30	5	2.4	NA	−171.3
380,000	30	5	0.2	NA	−187.7
380,000	35	0	0.5	NA	−192.7
380,000	40	−2	1.6	NA	−194.7

6.6 Simulation Conclusion

1. In the low orbit area below 3000 km, the navigation receiver of spacecrafts should be fixed backward to the earth as the land users. The visibility, DOP, and minimum received power are stable.
2. The available number and DOP degrade quickly for the space users backward to the earth above 3000 km, as shown in Fig. 8. As a result, the low orbit space users (below 3000 km) should fix their antenna backward to the earth in order to receive the navigation signal outside the shadow of the earth.
3. In the medium orbit area from 3000 to 8000 km, the user needs to receive the signal where the half angle is from 20° to 30° in order to get continuous positioning. The minimum antenna gain is 5 dB, and the minimum received

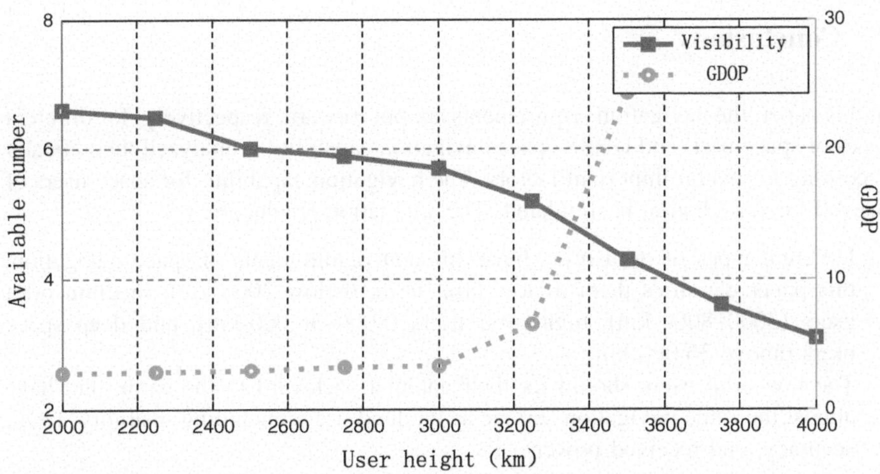

Fig. 8 Simulation of low orbit users (antenna backward to the earth)

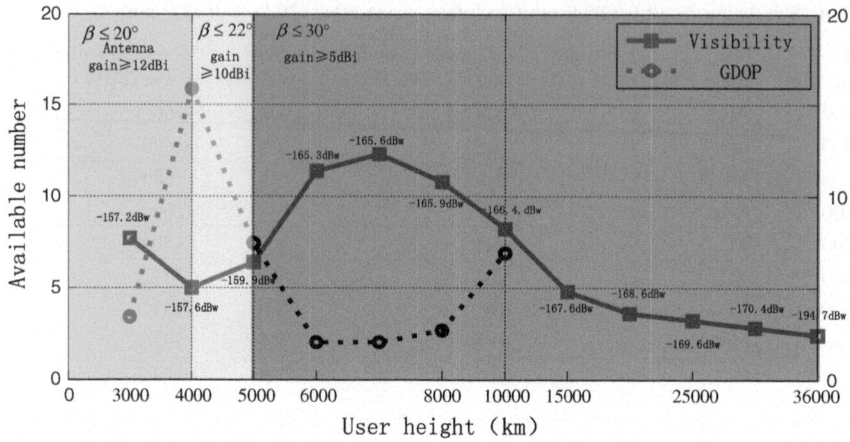

Fig. 9 Simulation of medium and high orbit users (antenna to the earth)

power is −166 dBW, as shown in Fig. 9. The high sensitivity receiver is in medium orbit area.

4. In the high orbit area from 8000 to 36,000 km, the high sensitivity receiver is needed. And continuous positioning cannot be realized above 15,000 km for the degradation of visibility and DOP. However, the available number is above 2, and the timing service is provided.
5. For the lunar orbit user, the signal outside the antenna main lobe must be received and the received power is low as −194.7 dBW. As a result, the timing demand for the lunar orbit user is not met.

7 Conclusions

In this paper, the navigation requirements are put forward, respectively, for different types of space user, and GNSS space navigation capability is analyzed theoretically according to several important factors. The navigation capability for space users of one MEO constellation is simulated. The simulation results are:

1. Different types of orbit users have different requirements of space navigation, this paper classifies them as low orbit users (below 3000 km), medium orbit users (3000–8000 km), high orbit users (8000–36,000 km), and deep space users (above 36,000 km);
2. The low orbit users should fix their antenna backward to the earth, and have almost the same navigation service as the land users, including visibility, DOP, accuracy, and received power;

3. The medium orbit users should fix their antenna to the earth, adopt high sensitivity receiver, and the "continuous positioning" requirement can be reached;
4. The high orbit users should fix their antenna to the earth, adopt high sensitivity receiver, and the "intermittent positioning" and "continuous timing" requirement can be realized;
5. The "continuous timing" requirement of deep space users as lunar orbit user cannot be realized for the low received power.

References

1. Li P, Chen ZG, Gu YN, Si Y (2013) Spacecrafts navigation signal research based on GNSS constellation. In: Proceedings of 4th China satellite navigation conference, vol 2. Springer Press, Berlin, pp 35–48
2. Birmingham WP, Miller BL, Stein WL (1983) Experimental results of using the GPS for Landsat-4 onboard navigation. Navig J Instit Navig 30(3):244–251
3. Kronman JD, McElroy T (1994) Considerations for the application of GPS in satellites: GPS system trade study. In: Proceedings of ION GPS-94, pp 1622–1626
4. Wu S (1996) MicroGPS for orbit determination of earth satellites. Proc ION NTM 1996 509–516
5. Cohen CE (1994) Space flight tests of attitude determination using GPS. Int J Satell Commun 12: 427–433
6. Kronman JD (2000) Experience using GPS for orbit determination of geosynchronous satellite. Proc ION GPS 2000:1622–1626
7. Michael C (2002) Results from the GPS flight experiment on the high earth orbit AMST OSCAR-40 spacecraft. Proc ION GPS 2002, 122–133
8. Qin H, Liang M (2008) Research on postitioning of high earth orbital satellite using GNSS. Space Sci 28(4):316–325
9. He Q, Sun Q (2010) A new solution of autonomous navigation for GEO satellites Based on GNSS. J. Spacecr TT&C Technol 29(1):7–11
10. US: Federal Aviation Administration (2011) Global positioning system standard positioning service performance standard

An Improved Location-Based Service Authentication Algorithm with Personalized *K*-Anonymity

Yu-Meng Ye, Chang-Chun Pan and Gen-Ke Yang

Abstract Location-Based Service (LBS) has been an indispensable part of our daily life. However, LBS may cause the risk of revealing the location privacy of the users, and that could be abused illegally. To defend users' location privacy against the threats, we propose a location information authentication system with a revised algorithm based on a trusted third-party architecture. The improved algorithm is built upon the so-called *CliqueCloak* theorem which is supposed to prevent the privacy disclose caused by side information. Our contribution lies in developing a revised top location algorithm to reprocess the points failing in *K*-anonymity process. Using the set of road network top locations on the map, we could assign those points a new anonymous area. Our experiments show that the improved personalized *k*-anonymity location information authentication algorithm outperforms the classical one in terms of the success rate and the computing efficiency.

Keywords Location privacy protection · *K*-anonymity · LBS · Top Location

Y.-M. Ye · C.-C. Pan (✉)
Department of Automation and Shanghai Key Laboratory of Navigation and Location Based Services, Shanghai Jiao Tong University, Shanghai 200240, China
e-mail: pan_cc@sjtu.edu.cn

Y.-M. Ye
e-mail: regen@sjtu.edu.cn

G.-K. Yang
Department of Automation and Key Laboratory of System Control and Information Processing, Ministry of Education of China, Shanghai Jiao Tong University, Shanghai 200240, China
e-mail: gkyang@sjtu.edu.cn

J. Sun et al. (eds.), *China Satellite Navigation Conference (CSNC) 2016 Proceedings: Volume I*, Lecture Notes in Electrical Engineering 388,
DOI 10.1007/978-981-10-0934-1_23

1 Introduction

Location-Based Services (LBS) can obtain location information from clients of users and then return many kinds of effective information to them. In recent years, LBS has been widely used in various fields. However, when users are using LBS with mobile clients, their information including identity, location, and query will be submitted to the LBS providers. It leads to the leakage of users' privacy.

If the attackers mine users' privacy by semantic analysis with the identity (ID), the content of the query, the time (t) and the location (x, y) sending the query, the IP address, we could use pseudonyms instead of actual IDs to disconnect the links between identities and locations of users. However, the attackers will mine privacy in not only such direct ways but also advanced ways. We need to improve the method to protect the privacy.

Assume that the user has used the pseudonym, the attackers who only obtain false ID, precise time, and location coordinates sending the query, can still attack in at least two ways with the side information. One is called restricted space identification and another is observation identification.

Restricted space identification is the attack based on user's own place such as home and work place. When a user U sends messages in his own place L, and the attacker A has known U as the owner of L, then finds that the intercepted messages were sent from the same user, A can infer that the sender of messages is U. Observation identification is the attack aided by observing user's behaviors when sending queries. If A has known that U was visiting the place P during time interval T (e.g., by monitors), and there was only one user sending messages during T in L, then A can infer the messages intercepted during T were send by U.

In view of the above situation, Sweeney proposed the K-anonymity model [4]. Then Gedik proposed a personalized K-anonymity model and *CliqueCloak* algorithm [1] based on Sweeney's work. The model and algorithm allow users to specify the minimum anonymity level K and the inaccuracy tolerances in temporal and spatial dimensions. And the algorithm defends against advanced attacks like restricted space identification and observation identification. The experiments show that personalized K-anonymity protection method can effectively anonymize messages sent from the users. However, the success rate of anonymity is low due to many users who defined low temporal tolerance cannot be anonymized on time.

We propose the *Top Location* algorithm to improve the *CliqueCloak* algorithm. Using the set of road network top locations, the trusted anonymity server searches for the top location in the constraint box of the user. And the top location does not overlap user's actual location. Then the anonymity server submits several anonymous messages with the location areas including the top location and dummy locations instead of user's actual location. The algorithm raises the success rate and efficiency of the classical K-anonymity algorithm and keeps the effect of anonymity.

2 Related Work

The privacy protection methods can be separated into two categories according to the core concepts [3], the methods using *K*-anonymity concept and the method without *K*-anonymity concept.

2.1 The Methods Without K-Anonymity Concept

Kido H et al. proposed an anonymity method with dummies [2, 3] to protect the location privacy. The user's client generates several dummy locations to send to LBS providers with the actual location of the user. After the client receives a plurality of query results, it will choose the result corresponding to actual location and discard results of dummy query. As service providers cannot distinguish the true location, user's location privacy is protected. Yiu ML et al. proposed a method called *SpaceTwist* [3, 5]. The user sends actual location to the anonymity server. Then the server creates a location different from the user's actual location as the anchor, nearest neighbors are retrieved incrementally until the query is answered correctly by the user's client. These methods only cope with direct privacy attacks without side information and cause more data redundancy.

2.2 The Methods with K-Anonymity Concept

Sweeney et al. proposed a protection model named *k*-anonymity model [4] in 2003. *K*-anonymity protect a release if the information for each user contained in it cannot be distinguished from at least $K - 1$ other users whose information also appears in the release. The attackers cannot distinguish the privacy information when the messages have the same identifier, so as to protect the privacy in the data set. Based on the *K*-anonymity model, Gedik et al. described the personalized *K*-anonymity model and *CliqueCloak* algorithm [1]. It enables each user to specify the minimum anonymity level *K* and the maximum temporal–spatial tolerances when the server anonymizes users' queries. Each of the cliques contains K messages that satisfy the condition of *K*-anonymity.

The methods with *K*-anonymity concept prevent privacy threats fundamentally, including privacy attacks aided with side information such as restricted space identification and observation identification. However, these methods will sacrifice the precision of query and the users may wait for too long time obtaining results of query. If we combine *K*-anonymity methods and non-*K*-anonymity methods, we can obtain the advantages of both kinds of methods for better effect of anonymity.

3 Personalized *K*-Anonymity Location Information Authentication System

In this paper, we propose a personalized *K*-anonymity location information authentication system based on the trusted third-party architecture. We build a trusted third-party server between the mobile clients of users and the semi-honest LBS providers. The third-party server receives users' queries including precise location and the actual identity, anonymizes the queries, and sends them to LBS providers. Obtaining results from LBS providers, the server distributes results to every user. Compared with other architectures such as the distributed architecture and the mobile point-to-point architecture, the trusted third-party architecture gives less data burden to the mobile clients and is easy to maintain and update. Therefore, the trusted third-party architecture be appropriate for the current location information authentication system.

3.1 *Personalized* K-*Anonymity Model*

For reference convenience, Table 1 summarizes some notations used in the paper.

We define the input of system model as follows:

$$m_s \in S : \langle u_{\mathrm{id}}, r_{\mathrm{no}}, \{x, y, t\}, K, \{\mathrm{d}x, \mathrm{d}y, \mathrm{d}t\}, C\rangle \tag{3.1}$$

Table 1 Notation table of model [1]

Notation	Meaning
S	Source message set
T	Transformed message set
m_s	A message in set S
m_t	A message in set T
u_{id} and r_{no}	Sender ID and message number
dx, dy, dt	Temporal and spatial tolerances
$L(m_s) = (x, y, t)$	Spatiotemporal point of m_s
$B_{\mathrm{cn}}(m_s)$	Constraint box of m_s
$\Phi(x, \mathrm{d}x)$, $\Phi(y, \mathrm{d}y)$, $\Phi(t, \mathrm{d}t)$	$[x - \mathrm{d}x, x + \mathrm{d}x]$, $[y - \mathrm{d}y, y + \mathrm{d}y]$ $[t - \mathrm{d}t, t + \mathrm{d}t]$
$B_{\mathrm{cl}}(m_t)$	Cloaking box of m_t
X:$[x_s, x_e]$, Y:$[y_s, y_e]$, I:$[t_s, t_e]$	Maximum and minimum values of cloaking box
$B_m(S')$	MBR of a set of source messages
$G_m(S, E)$	Constraint graph
C	Content of the message

We define the output of system model as follows:

$$m_t \in T : \langle u_{\text{id}}, r_{\text{no}}, \{X : [x_s, x_e], Y : [y_s, y_e], I : [t_s, t_e]\}, C\rangle \tag{3.2}$$

The source message m_s before the conversion and the anonymous message m_t going through the conversion must be in line with the following constraints:

1. *Spatial containment*:

$$m_s . x \in m_t . X, \quad m_s . y \in m_t . Y \tag{3.3}$$

2. *Spatial resolution*:

$$m_t . X \subset \Phi(m_s . x, m_s . \mathrm{d}x), m_t . Y \subset \Phi(m_s . y, m_s . \mathrm{d}y) \tag{3.4}$$

3. *Temporal containment*:

$$m_s . t \in m_t . I \tag{3.5}$$

4. *Temporal resolution*:

$$m_t . I \subset \Phi(m_s . t, m_s . \mathrm{d}t) \tag{3.6}$$

5. *Content preservation*:

$$m_s . C = m_t . C \tag{3.7}$$

6. *Location K-anonymity*:

$$\begin{aligned} &\exists T' \subset T, \text{such that } m_t \in T', |T'| \geq m_s . K, \\ &\forall \{m_{t_i}, m_{t_j}\} \subset T', m_{t_i} . u_{id} \neq m_{t_j} . u_{id}; \\ &\text{and } \forall m_{t_i} \in T', B_{cl}(m_{t_i}) = B_{cl}(m_t) \end{aligned} \tag{3.8}$$

The model enables each user to specify the anonymity level K and the maximum temporal and spatial inaccuracy tolerances as the constraint box $\{\Phi(x, \mathrm{d}x), \Phi(y, \mathrm{d}y), \Phi(t, \mathrm{d}t)\}$. A larger K implies higher degrees of privacy, because it is difficult for attackers to link the user to the query in anonymous messages of K users. The smaller the constraint box is, the more accurate the query results are, because the query location area is closer to user's precise location.

K-anonymity model can be represented with the constraint graph and clustering analysis method. There is a set M which consists of the messages in the constraint graph called $G_m(S, E)$ and the messages in M can be anonymized together. $G_m(S, E)$ is an undirected graph. S is the set of vertices, each representing a message received from a client. And E is the set of edges. There exists an edge $e = (m_{si}, m_{sj})$ in E between two vertices m_{si} and m_{sj} only if they have not come from the same client and their actual locations are in each other's constraint boxes. Given a message m_s in S, the model is requested to find the set M including m_s and then find a subset of M which forms a spatial–temporal cloaking box.

The *CliqueCloak* theorem [1] is shown as follows:

In $G_m(S, E)$, M = $\{m_{s1}, m_{s2}, m_{s3}..., m_{sl}\} \subset S$, and for $\forall 1 \leq i \leq l$, $m_{ti} = \langle m_{si} . u_{id}, m_{si} . r_{no}, B_m(M), m_{si} . C \rangle$. Then, $\forall 1 \leq i \leq l$, m_{ti} is a valid anonymous format of m_{si}, i.e., $R(m_{si}) = m_{ti}$, if and only if the set M forms an l-Clique in the $G_m(S, E)$ and $\forall 1 \leq i \leq l$, $l \geq m_{si}$. K.

From the theorem we know that the system should find an l-Clique in the constraint graph, in which core message m_{sc} and $K - 1$ other messages will be seen as one clique being subject to six constraints.

3.2 *Personalized* K-*Anonymity Algorithm*

We briefly describe the four data structures [1] used in the algorithm of the system.

- The *Message Queue* Q_m collects the messages sent from clients in the order that they are received, then pops messages to the anonymity server one by one to process them.
- The *Constraint Graph* G_m is a dynamic in-memory graph, which contains the messages that are not yet anonymized and not yet dropped due to expiration.
- The Expiration Heap H_m is the priority queue based on the deadline of the messages. Each message in H_m that is not yet anonymized and is not yet dropped due to expiration will be sorted with ascending order. $m_s.t + m_s . dt$ is the sorting keyword of each message m_s.
- The *Multidimensional Index* I_m is used to search spatiotemporal location coordinates of the messages efficiently on the map.

The personalized *K*-anonymity algorithm so-called *CliqueCloak* algorithm [1] processes the messages in four steps as follows:

1. *Zoom-in*: The target message m_{sc} pops from Q_m and joins into G_m, H_m and I_m. Then the server searches for the messages that have edges with m_{sc} in the $G_m(S, E)$, and adds those messages into the subset of G_m, say, G'_m.
2. *Detection*: The server drops m_s in G'_m whose $m_s . K > m_{sc} . K$. The rest (more than $m_{sc} . K - 1$) of messages in G_m' form a subset U. Then it drops m_s whose neighbors (except m_{sc}) are less than $m_{sc} . K - 2$. Finally, the server selects $K - 1$ elements in the U to form the spatial–temporal cloaking box with m_{sc}.
3. *Perturbation*: The server submits the anonymous K-Clique messages with random ID and same location coordinates to the LBS providers. Then these messages will be removed from the system.
4. *Expiration*: A message is considered to be expired if the current time is beyond the upper limit of the message's temporal constraint box. In this step, the server checks the top elements of H_m for any messages that have expired. Then they will be removed from G_m, H_m, Q_m, and I_m.

3.3 Improved Algorithms

The server removes expired messages from system in Step 4 of the *K*-anonymity algorithm, which means that the queries of users fail to be anonymized. In practical application, the user cannot get the results of the query when it is failed in anonymity. The user has to resend the query, which leads to the data redundancy and the waste of time. Worse yet, if the density of users is so low that even though users has been waiting for a long time, there is still no chance to anonymize queries. In the extreme case, it seriously influences the quality of LBS.

In order to solve the problem, we propose the algorithm *Top Location* (Table 2) as the fifth step of the personalized *K*-anonymity algorithm to improve it. We reserve the message after it is expired to anonymize the message in another way.

At the top locations, the density of users could be relatively high. Therefore, we add index of top locations (e.g., the intersections, the bus stations) into the system. The server searches for the top location h in the set of road network top location whose location coordinates are in the constraint box of m_c and are closest to but does not overlap the location of m_c. Then the server uses h as an anchor to generate m_c. $K - 1$ dummy locations around h and form a spatial cloaking box including h and dummy locations. Then the server could send a group of anonymous messages with the spatial cloaking box to the LBS providers. Finally, the server removes m_c from G_m, H_m, Q_m, and I_m whether it is anonymized.

In this algorithm, the coordinates of top location are in the constraint box of the target message, which ensure that the location area submitted does not exceed the maximum inaccuracy tolerance of the users. Because of the constraint that the top location cannot overlap user's precise location coordinates, it is impossible for attackers to know the user staying at location L in the time interval T (aided with side information) and intercept the message including location L at the same time. So we could defend against attacks aided with side information. In addition, as this method and *K*-anonymity method are in the same system, the attackers will not know whether the server submit messages of real users or dummy users. Therefore,

Table 2 Pseudocode of *Top Location* algorithm

Line number	Pseudocode
1	Obtain set C of expired messages and pop the message m_c in C one by one
2	Search for the top locations in road network top location set whose location coordinates are in the constraint box of m_c and join them into the subset A
3	Remove the top location whose distance from the m_c is less than 10 m from A
4	Search A for the top location h closest to m_c as the anchor
5	Generate m_c.K − 1 dummy location coordinates around h
6	Form a spatial cloaking box including h and dummy locations and send messages with the cloaking box to LBS providers
7	Remove m_c from G_m, H_m, and I_m whether it is anonymized

Top Location algorithm can raise the success rate of anonymity without sacrificing the effect of *K*-anonymity. And if the set includes more top locations, it will be easier for expired messages to be anonymized.

4 Simulation Experiment

In order to evaluate the performance of the algorithm, we build the simulation platform to measure several evaluation metrics about the *CliqueCloak* algorithm and the *Top Location* algorithm. The experimental data set randomly generated 10,000 data in the set representing 10,000 user messages. Each of the message has ID, *K*, latitude and longitude, spatial and time constraints. In the experiment, samples about *K* are $K = \{2\}, \{2, 3\}, \{2, 3, 4\}, \{2, 3, 4, 5\}$. Temporal tolerances are randomly generated among 3, 5, 7, 9 s. And the spatial tolerances are generated randomly among 0.0007° (654 m), 0.0008° (747 m), 0.0009° (840 m), and 0.001° (934 m). There are three experimental groups about three different variables. The range of *K*, spatial tolerance, and temporal tolerance has been controlled, respectively. We measured five kinds of evaluation metrics [1] as follows:

1. *Success rate* (SR): the percentage of messages convert from m_s to m_t.
2. *Relative anonymity level* (RAL): the average ratio between number of m_t in one cloaking box and K of each m_t.
3. *Relative spatial resolution* (RSR): the average square root of the ratio between area of m_s's spatial constraint box and m_t's spatial cloaking box.
4. *Relative temporal resolution* (RTR): the average ratio between the temporal constraint of m_s and actual waiting time of m_t.
5. *Message processing time* (MPT): the time for processing each message.

The experimental results show that the improved algorithm raises the success rate of anonymity greatly. As shown in Fig. 1, the success rate of classical personalized *K*-anonymity algorithm decreases with the increase in the range of *K*. And when the

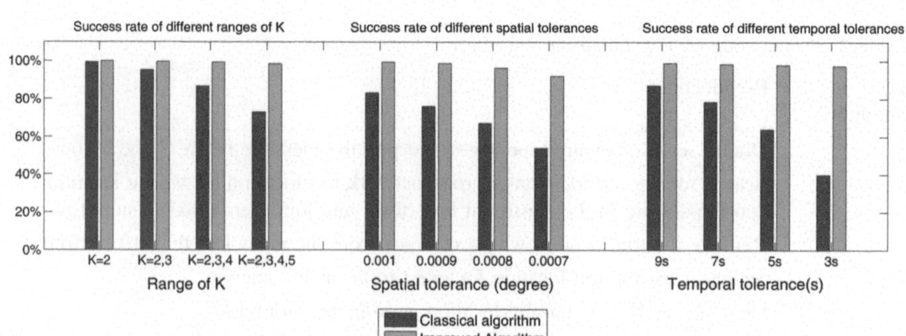

Fig. 1 Comparison under different settings of the success rate for different *K* levels, spatial tolerance, and temporal tolerance

temporal and spatial tolerances decrease, the success rate of classical algorithm decreases significantly low to 39.88 %. In contrast, the average success rate of the *Top Location* algorithm is 98 %, the lowest success rate is 91.97 %. The main reason is that the *Top Location* algorithm tries to reprocess the messages failing to be anonymized in the classical algorithm. It leads to the increase of the success rate of anonymity.

The results of measuring the message processing time for different algorithms are shown in Fig. 2. In the same experimental condition, the processing time of the *Top Location* algorithm at the later stage is lower than classical K-anonymity algorithm. The classical algorithm has five times of sudden rise, which is more than improved algorithm. It means the *Top Location* algorithm has higher efficiency and it is relatively stable in the system.

In addition, we measure the relative anonymity level, relative temporal resolution, and relative spatial resolution. The results are shown in Table 3. According to

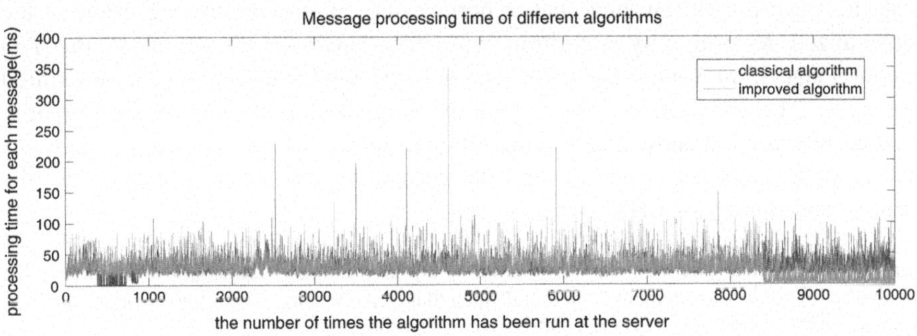

Fig. 2 Message processing time of different algorithms

Table 3 Average measurement results

Variable	Relative anonymity level		Relative temporal resolution		Relative spatial resolution	
	Classical	Improved	Classical	Improved	Classical	Improved
$K = 2$	1	1	9.61	9.61	9.08	9.08
$K = 2, 3$	1	1	7.78	7.78	4.50	4.50
$K = 2, 3, 4$	1.09	1.09	6.84	6.84	3.64	3.64
$K = 2, 3, 4, 5$	1.06	1.06	5.99	5.99	3.40	3.40
0.0007°	1.07	1.07	5.58	5.58	3.44	3.44
0.0008°	1.06	1.06	5.80	5.80	3.29	3.29
0.0009°	1.07	1.07	6.27	6.30	3.14	3.13
0.001°	1.07	1.07	6.63	6.65	3.13	3.08
3 s	1.06	1.06	3.80	3.80	3.82	3.82
5 s	1.06	1.06	5.01	5.01	3.44	3.44
7 s	1.07	1.07	6.15	6.15	3.21	3.21
9 s	1.07	1.07	7.29	7.29	3.18	3.18

the evaluation metrics, the relative anonymity level, relative temporal resolution, and relative spatial resolution of the top location algorithm are almost the same as those of classical *K*-anonymity algorithm. And in both algorithms, the relative anonymity level increases with the enlargement of the range of *K*. The relative temporal resolution increases with the increase of waiting time or the narrowing of the range of *K*. And the relative spatial resolution increases with the enlargement of the area of constraint box and the range of *K*.

5 Conclusion

We propose a secure location information authentication system in this article, which includes the model and the revised algorithm. The model of system is built mainly based on personalized *K*-anonymity model and *CliqueCloak* theorem. In order to make the system have better performance in anonymity, we improve the personalized *K*-anonymity algorithm with *Top Location* algorithm. In the *Top Location* algorithm, we use the set of road network top location to reprocess expired messages. The experiments show that the improved personalized *k*-anonymity location information authentication algorithm outperforms the classical one in terms of the success rate, the stability, and the computing efficiency, which ensure the security and quality of LBS.

Acknowledgements The research is supported by a joint foundation launched by China Satellite Navigation System Management Office and Shanghai Science and Technology Committee with grant No. BDZX0005 and 13511501302.

References

1. Gedik B, Liu L (2008) Protecting location privacy with personalized k-anonymity: Architecture and algorithms. IEEE Trans Mob Comput 7(1):1–18
2. Kido H, Yanagisawa Y, Satoh T (2005) An anonymous communication technique using dummies for location-based services. In: Proceedings of the international conference on pervasive services, 2005, ICPS'05. IEEE, New York, pp 88–97
3. Jia J, Zhang F (2013) Overview of location privacy protection technology. Appl Res Comput 30 (3):641–646
4. Sweeney L (2002) k-anonymity: a model for protecting privacy. Int J Uncertain Fuzziness Knowl Syst 10(05):557–570
5. Yiu ML, Jensen CS, Huang X et al (2008) Spacetwist: managing the trade-offs among location privacy, query performance, and query accuracy in mobile services. In: ICDE, pp 366–375

Indoor Pseudolite Positioning Based on Interferometric Measurement of Carrier Phase

Peng Zhang, Fuyu Sun, Yaming Xu and Jinling Wang

Abstract In indoor environments, GNSS signals endure serious attenuation, hence a GNSS receiver could not output accurate, reliable positioning solutions continuously. Pseudolite positioning system was proposed as a suitable alternative solution for indoor positioning applications. In this research, a new indoor pseudolite positioning system based on carrier phase interferometry technique is presented. This pseudolite-based positioning system can achieve decimeter accuracy without the support of a base station, and without the need of synchronization between pseudolites. In addition, in the proposed new positioning system the integer ambiguity resolution is unnecessary as well. In this pseudolite system, the main part of interferometric measurement is a dual-antenna software receiver. By using the principle of interferometric measurement, dual-antenna software receiver could deduce the spatial relationship between the dual-antenna and pseudolites. As a result, dual-antenna's position can be determined uniquely. Experiment results demonstrate that a horizontal positioning accuracy of decimeter-level accuracy can be obtained.

Keywords Pseudolite · Indoor positioning · Interferometric measurement · Dual-antenna · Software receiver

P. Zhang (✉) · F. Sun · Y. Xu
School of Geodesy and Geomatics, Wuhan University, Wuhan 430079, China
e-mail: pzhang@sgg.whu.edu.cn

P. Zhang · Y. Xu
Key Laboratory of Precise Engineering and Industry Surveying,
National Administration of Surveying, Mapping and Geoinformation,
Wuhan 430079, China

J. Wang
School of Civil and Environmental Engineering, UNSW,
Sydney, NSW 2052, Australia
e-mail: Jinling.wang@unsw.edu.au

J. Sun et al. (eds.), *China Satellite Navigation Conference (CSNC) 2016 Proceedings: Volume I*, Lecture Notes in Electrical Engineering 388,
DOI 10.1007/978-981-10-0934-1_24

1 Introduction

The development of Global Navigation Satellite Systems (GNSS) has revolutionized the traditional surveying, geodesy and navigation for the past decades, and it has been widely applied nowadays to achieve reliable and accurate positioning results. However, in indoor environments, GNSS has a limited ability to provide accurate, reliable and continuous positioning solutions. There are some indoor positioning techniques, like WiFi, RFID and WSN [1], but none of these techniques can provide the accuracy under 1 m.

Pseudolite positioning system was proposed as a suitable alternative solution for indoor positioning applications [2]. In a typical pseudolite system, receivers can achieve cm-level accurate positions by relative positioning method with the measurements from a base station. However, in a traditional pseudolite system all the pseudolites should be kept synchronized all the time. In addition, in indoor environment, multipath effects are serious, leading to signal blockage frequently, which bring huge challenges into integer ambiguity resolution.

In recent years, several pseudolite systems have been developed, such as Locata [3], i-GOing [4] and IMES [5]. The pseudolites in Locata system transmit their own proprietary signal structure in 2.4 GHz ISM band and centimetre-level positioning accuracy can be achieved without the use of a base station. i-GOing system uses continuous GNSS like signals to facilitate future integration in standard GNSS receivers. i-GOing positioning system is able to provide sub-meter level 2-D positioning service even in harsh indoor environments. In Locata and Going, stringent time synchronization requirements should be guaranteed, which may lead to significant deployment costs. In IMES system [4], the receiver position is determined as the position of the pseudolite broadcasting the strongest received signal. Although time synchronization is no longer necessary, the positioning accuracy of IMES is limited to 10–20 m which can't meet the requirements in many indoor applications.

In this research, an indoor pseudolite positioning system based on carrier phase interferometry technique is presented. This new pseudolite-based positioning system can achieve decimeter accuracy without the support of a base station and without the need of synchronization between pseudolites. Although this system is based on carrier phase measurements, the integer ambiguity resolution is unnecessary. The main part of this proposed pseudolite positioning system, based on interferometric measurement is a dual-antenna software receiver. The distance between these two antennas is fixed and shorter than the half of carrier wavelength. By using the principle of interferometric measurements of carrier phase, dual-antenna software receiver could deduce the spatial relationship between the dual-antenna and the pseudolite by extracting the carrier phase difference between these two antennas. With the number of tracked pseudolites increasing, dual-antenna's position can be determined uniquely, and the positioning accuracy can be enhanced accordingly.

This paper is organized as follows. Section 2 introduces the measurement model of this pseudolite positioning system. Section 3 describes the detail of the proposed interferometric measurement method. Section 4 illustrates and discuss the experimental results. The main conclusions of this research are drawn in the last section.

2 Measurement Model

A 2-D geometric relations between pseudolite i and the dual-antenna is depicted in Fig. 1. Antennas a and b are located at the fixed interval less than a half-wavelength of a GPS L1 carrier wave. The coordinates of these two antennas are (x, y) and $(x + \Delta x, y + \Delta y)$ respectively, where (x, y) are unknown values to be estimated. The spatial arrangement of these two antennas composites a baseline. Since the length of this baseline d_{ab} is fixed which can be precisely measured in advance and the angle of the baseline θ_{ab} can also be output by a digital compass in the real time, the value of $(\Delta x, \Delta y)$ can be calculated by $\Delta x = d_{ab} \cdot \cos\theta_{ab}$ and $\Delta y = d_{ab} \cdot \sin\theta_{ab}$. Therefore, $(\Delta x, \Delta y)$ can be regarded as known values. As long as we get antenna a's coordinates (x, y), antenna b's coordinates $(x + \Delta x, y + \Delta y)$ can also be regarded as known. Hence, in the following analysis work, we only focus on the calculation of the coordinates of antenna a.

Assuming the coordinates of pseudolite i's antenna are (x^i, y^i) which can be regarded as known values, then the distance between pseudolite i and antenna a can be expressed as:

$$r_a^i = \sqrt{(x - x^i)^2 + (y - y^i)^2} \tag{1}$$

Fig. 1 Geometric relations between pseudolite and antennas

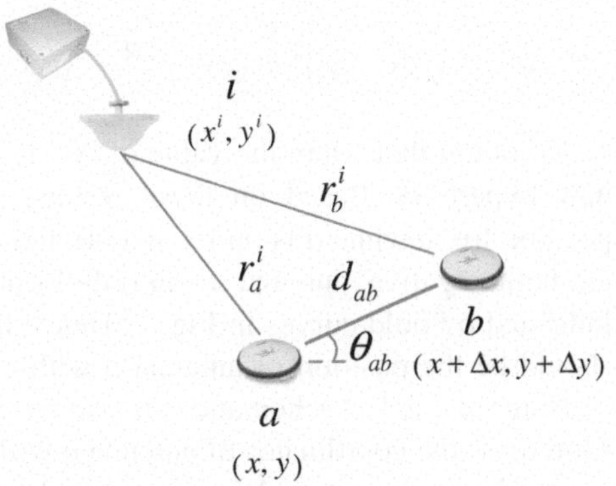

Similarly, r_b^i, the distance between pseudolite i and antenna b can also be calculated as

$$r_b^i = \sqrt{(x + \Delta x - x^i)^2 + (y + \Delta y - y^i)^2} \tag{2}$$

The difference between r_a^i and r_b^i can be presented as:

$$\begin{aligned} \Delta r^i &= r_a^i - r_b^i \\ &= \sqrt{(x - x^i)^2 + (y - y^i)^2} - \sqrt{(x + \Delta x - x^i)^2 + (y + \Delta y - y^i)^2} \end{aligned} \tag{3}$$

3 Interferometric Measurement Method

We transform Eq. (3) as follows

$$\Delta r^i = \sqrt{(x - x^i)^2 + (y - y^i)^2} - \sqrt{[x - (x^i - \Delta x)]^2 + [y - (y^i - \Delta y)]^2} \tag{4}$$

Assume pseudolite i has a virtual mirror point i' with coordinates $(x^i - \Delta x, y^i - \Delta y)$. As explained in Sect. 2, (x^i, y^i) and $(\Delta x, \Delta y)$ are all known. From Eq. (4), we can see that if the value of Δr^i is given, the distance difference from the Point (x, y) to the other two Points (x^i, y^i) and $(x^i - \Delta x, y^i - \Delta y)$ is constant. Consequently, Eq. (4) is a typical hyperbola equation with focal points i and i', and the Point (x, y) is on this hyperbola. More specifically, if $\Delta r^i < 0$, antenna a locates on the right curve of the hyperbola; if $\Delta r^i > 0$, the possible position of antenna a can only be on the left curve of the hyperbola.

If there exists another pseudolite j in this 2-D coordinate system and its coordinates are indicated as (x^j, y^j), another hyperbola equation with focal points j and j' can be achieved as well:

$$\Delta r^j = \sqrt{(x - x^j)^2 + (y - y^j)^2} - \sqrt{[x - (x^j - \Delta x)]^2 + [y - (y^j - \Delta y)]^2} \tag{5}$$

No doubt that when the value of Δr^j is given, antenna a will also locate on this new hyperbola. Based on these results, we can deduce that the only possible position for antenna a is on the intersection of these two hyperbolas. If Δr^i and Δr^j are both negative, antenna a can only locate on the right curve of each hyperbola, indicated by bold curves in Fig. 2. Hence the intersection of these two bold curves should be the position of antenna a with coordinates (x, y).

Naturally, a 3-D schematic can also be deduced with more pseudolites involved. Moreover, the coordinates of antenna a will be determined uniquely by method like trilateral measurements. We ignore the 3-D schematic for brevity in this discussion here.

Fig. 2 Concept of hyperbolic positioning

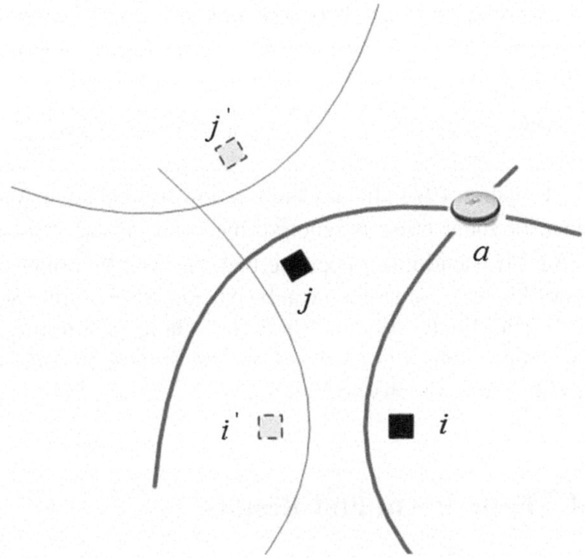

By jointly solving Eqs. (4) and (5), the location of antenna a can be uniquely estimated given the values of Δr^i and Δr^j. Then how to achieve Δr^i and Δr^j needs to be considered. Normally, the carrier waves of pseudolite i will be captured by antenna a and b. With the help of phase-locked loop (PLL), the dual-antenna software receiver will keep tracking these two input carrier waves and output corresponding carrier phase measurements indicated by ϕ_a^i and ϕ_b^i, which are modeled as follows:

$$\phi_a^i = \lambda^{-1} r_a^i + \frac{c}{\lambda}(\delta t - \delta t^i) + N_a^i + \varepsilon_{\phi,a}^i \tag{6}$$

$$\phi_b^i = \lambda^{-1} r_b^i + \frac{c}{\lambda}(\delta t - \delta t^i) + N_b^i + \varepsilon_{\phi,b}^i \tag{7}$$

where δt and δt^i stands for clock bias of dual-antenna receiver and pseudolite i respectively. N_a^i and N_b^i are unknown carrier phase ambiguities in cycles from pseudolite i to antenna a and b respectively. $\varepsilon_{\phi,a}^i$ and $\varepsilon_{\phi,b}^i$ are measurement noise. The difference between ϕ_a^i and ϕ_b^i can be calculated by Eq. (8), where δt and δt^i have been removed.

$$\Delta\phi^i \approx \phi_a^i - \phi_b^i = \lambda^{-1}(r_a^i - r_b^i) + (N_a^i - N_b^i) + \left(\varepsilon_{\phi,a}^i - \varepsilon_{\phi,b}^i\right) \tag{8}$$

Since the distance between antenna a and b is limited to less than half of carrier wavelength, $\left(N_a^i - N_b^i\right)$ is always zero. Equation (8) is therefore reduced to:

$$\Delta\phi^i \approx \phi_a^i - \phi_b^i = \lambda^{-1}\left(r_a^i - r_b^i\right) + \left(\varepsilon_{\phi,a}^i - \varepsilon_{\phi,b}^i\right) = \lambda^{-1}\Delta r^i + \left(\varepsilon_{\phi,a}^i - \varepsilon_{\phi,b}^i\right) \quad (9)$$

Equation (9) indicates the relationship between $\Delta\phi^i$ and Δr^i. In addition, if the measurement noise is ignored, the value of Δr^i can be directly achieved, which is $\lambda\Delta\phi^i$. If pseudolite j is concerned, Δr^j can be obtained as well. As the results, the coordinates of antenna a can be calculated by jointly solving Eqs. (4) and (5) in this 2-D coordinate system. Since the whole positioning calculation is based on the difference values of carrier phase measurements, this technique method is regarded as an interferometry technique.

4 Experiment and Results

In this experiment, the model of pseudolite is GSG-L1. It generates a single L1 carrier and modulates it with the C/A code and the reduced navigation message. Universal Software Radio Peripheral (USRP) was chosen as the front end of dual-antenna software receiver. USRP is intended to be a hardware platform for software radio. With appropriate daughter board configuration, USRP has the potential to capture the entire GNSS family of signals. In this experiment, two DBSRX daughterboards which can handle signals from 800 MHz to 2.4 GHz were assembled in one USRP.

Experiments were conducted in a lab room of $10 \times 7\ \text{m}^2$. Five pseudolite antennas were installed on the roof as shown in Fig. 3. The 3-D coordinates of these five pseudolite antennas were precisely surveyed by a total station.

To verify the proposed carrier phase interferometry technique, the dual-antenna was deliberately placed on four test points in sequence. At each test point, carrier phase measurements of 60 epochs were utilized to calculate the position of antenna a with the proposed interferometry technique. Generally, three pseudolites are

Fig. 3 Distribution of pseudolite antennas

Fig. 4 Results of positioning at each test point (Color figure online)

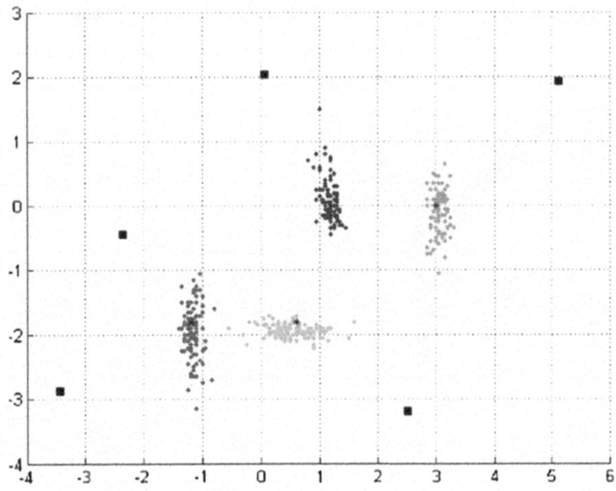

enough for positioning calculation in 3-D coordinate. But in this test, all five pseudolites's carrier phase differences were utilized by dual-antenna software receiver, which brings more chances to demonstrate the performance of this interferometry technique.

Positioning results are illustrated in Fig. 4. The location of each pseudolite is indicated by small black square. The positioning solutions of test points 1, 2, 3 and 4 are indicated by red, blue, green and cyan colors respectively. From Fig. 4, it emerges that at each test point, the minimum positioning error is below 2 dm and the maximum positioning error can be controlled within 1 m. Apparently, if an optimal estimation method is implemented, overall geometric strength for the positioning can be enhanced and a higher horizontal positioning accuracy can be achieved.

At test points 1, 2 and 3, the baseline is parallel with the x axis; and at test point 4, baseline is parallel with the y axis. According to the results, we can further conclude that a higher positioning accuracy can be achieved at the parallel direction of the baseline; On the contrary, at the direction which is perpendicular to the baseline, a lower positioning accuracy will be obtained. This phenomenon can be explained as follows. When the incoming carrier wave is perpendicular to the baseline of two antennas, the carrier phase measurements of the two antennas are almost the same, which means the positioning resolution is very low. On the contrary, if the incoming carrier wave is parallel to the baseline, the difference between carrier phase measurements from two antennas is the biggest which brings higher positioning resolution.

The details of the intersection by five hyperbolas at each test point are show in Fig. 5 separately. The width of each hyperbola means ±0.01 period accuracy of difference of carrier phase measurements can be achieved in these area. The wider of a hyperbolic curve width is, the higher quality of the pseudolite signal is, and as the consequence, the better performance of Interferometric measurement will get.

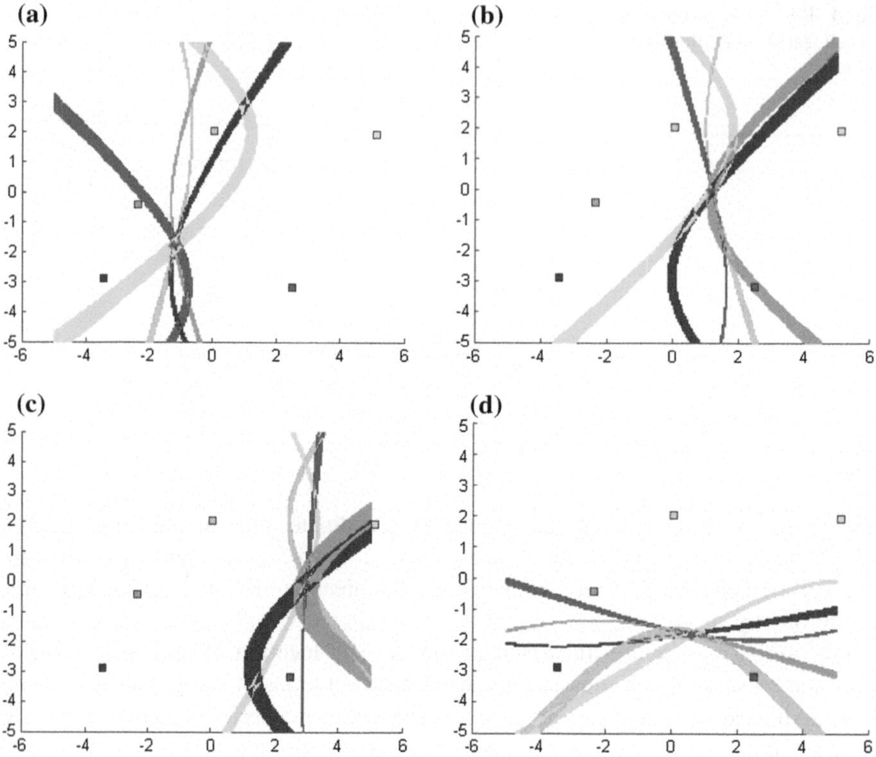

Fig. 5 Interferometry measurement at four test points

5 Conclusion

In this paper, a new indoor pseudolite positioning system based on interferometric measurement of carrier phase has been proposed and investigated. The experimental results have demonstrated that this new pseudolite positioning system can achieve decimeter positioning accuracy without the needs of a base station and synchronization between pseudolites. In addition, integer ambiguity resolution is also unnecessary in the new positioning system. Experiment results with five pseudolites have verified the effectiveness of this interferometry technique. This pseudolite positioning system can be a potential solution for many indoor positioning applications.

Acknowledgements The research is supported by National Natural Science Foundation of China (No. 41374011 and 41474005), National Natural Science Foundation of Hubei Province (No. 2014CFB717).

References

1. Gu Y, Lo A, Niemegeers I (2009) A survey of indoor positioning systems for wireless personal networks. IEEE Commun Surv Tutor 11(1):13–32
2. Wang J (2002) Pseudolite applications in positioning and navigation progress and problems. J Glob Position Syst 1(1):48–56
3. Rizos C, Roberts G, Barnes J, Gambale N (2010) Experimental results of Locata: a high accuracy indoor positioning system. In: International conference on indoor positioning and indoor navigation
4. Kurz O, Mongredien C, Rohmer G (2013) A new multipath detection and mitigation approach for pseudolite systems. In: Proceeding of the 26th international technical meeting of the ION satellite division, Nashville, Tennessee
5. Kohtake N, Morimoto S, Kogure S, Manandhar D (2011) Indoor and outdoor seamless positioning using indoor messaging system and GPS. In: International conference on indoor positioning and indoor navigation (IPIN), Guimarães, Portugal

References

Research and Implementation on BDS/GNSS Real-Time Positioning for Urban High-Precision Location-Based Services

Liang Wang, Zishen Li, Hong Yuan, Jiaojiao Zhao and Kai Zhou

Abstract With the deepening development of mobile Internet technology and Global Navigation Satellite System (GNSS), realizing high-precision positioning and navigation on mobile intelligent terminals is not only an important way to achieve the urban high-precision location-based services (LBS) in the future, but is also one of the development trends and major application fields of the satellite navigation industry. Although those mobile intelligent terminals such as smart phones basically have integrated a GNSS module, their positioning accuracies are generally low which cannot meet the requirements of high-precision location-based services. In this work, a system design scheme for realizing high-precision location-based services was studied and proposed, and a set of BDS/GNSS real-time high-precision positioning software for mobile intelligent terminals is developed based on Android platform. By collecting the observation data stream from an external GNSS module through the built-in Bluetooth module on those intelligent devices and receiving the differential correction data from a nearby reference station via 3G/4G or Wi-Fi wireless network, respectively, users' high-precision position then can be calculated in real-time by the software, and the positioning result will be displayed on Google maps simultaneously. At present, this software supports real time single point positioning and real-time pseudorange differential positioning for BDS/GPS/GLONASS standalone and combined system. The experimental results of combined BDS/GPS/GLONASS real-time positioning show that the accuracies of single point positioning in horizontal and vertical are better than 2.5 and 3.5 m, respectively, and the accuracies of real-time differential positioning in horizontal and vertical are better than 0.5 and 1.0 m, respectively. As a consequence, this contribution can provide some technical references for the realization of urban high-precision location-based services.

L. Wang (✉) · Z. Li (✉) · H. Yuan · J. Zhao · K. Zhou
Academy of Opto-Electronics, Chinese Academy of Sciences, Beijing, China
e-mail: wangliang@aoe.ac.cn

Z. Li
e-mail: lizishen@aoe.ac.cn

L. Wang
University of Chinese Academy of Sciences, Beijing, China

J. Sun et al. (eds.), *China Satellite Navigation Conference (CSNC) 2016 Proceedings: Volume I*, Lecture Notes in Electrical Engineering 388,
DOI 10.1007/978-981-10-0934-1_25

Keywords Location-based services (LBS) · Global Navigation Satellite System (GNSS) · BDS/GNSS · Real-time differential positioning · Android platform

1 Introduction

Navigation and location-based services industry has become one of the most rapidly developing field of emerging information industry after the Internet and mobile communication industry. At the same time, with the development of modern city, the requirements for location-based services (LBS) are become higher and higher, such as vehicle monitoring and scheduling, etc., in which high-precision positioning is the core of location-based services. Meanwhile, with the rapid development of mobile Internet technology, mobile intelligent terminals that are based on the platforms like Android are widely used which play an important role in LBS [1].

Although those mobile intelligent terminals such as smart phones basically have an integrated GNSS module which can provide navigation and positioning services for users, its positioning accuracy is generally low (about 10–20 m) which cannot meet the requirements of high-precision location-based services since they all apply the standard positioning services which provided by Global Navigation Satellite System (GNSS) for positioning [2]. Realizing high-precision positioning and navigation on mobile intelligent terminals is not only an important way to achieve the urban high-precision location-based services in the future, but also one of the development trend and major application field of the satellite navigation industry. In addition, it is also an important form of promoting the industrialization of China's BeiDou Navigation Satellite System (BDS) [3, 4].

GNSS augmentation positioning that is based on the idea of differential positioning is an effective method to realize high-precision positioning which is widely used in practice. The basic idea is using another receiver that is installed on a reference station whose position is pre-known precisely to observe the satellites simultaneously. Thus, the common errors between user and reference station can be eliminated or reduced. As a result, user's positioning accuracy can be improved remarkably. In addition, multi-GNSS combined positioning [5–10] has become an inevitable trend in GNSS-based navigation and positioning and it can significantly increase the number of visible satellites and improve the dilution of precision (DOP), thereby improving the positioning availability, continuity, and accuracy. Especially, with the development of China's BDS and its regional operation in the Asia-Pacific region, it provides excellent opportunities and good foundations for multi-GNSS combined positioning and high-precision navigation services.

Faced with the urgent demand of urban high-precision LBS in the future, a system design scheme for realizing high-precision LBS was studied and proposed in this work, and a set of BDS/GNSS real-time high-precision positioning software for mobile intelligent terminals which aimed at providing users real-time positioning and navigation services is developed based on Android platform.

2 System Design Scheme

Currently, the built-in satellite navigation module in smart devices (like smart phones) usually cannot output the real-time pseudorange and carrier phase measurements, as a result, it is unable to provide differential users positioning solutions, that is to say, it can only output single point positioning results instead whose accuracy is generally low (about 10–20 m) that cannot meet the requirements of high-precision location-based services. In addition, the positioning accuracy is usually poor due to the influence caused by the complex environment in urban.

In view of this, a system design scheme was proposed for realizing high-precision location-based services. By collecting the observation data stream from an external GNSS module through the built-in Bluetooth module on those intelligent devices and receiving the differential correction data from a nearby reference station via 3G/4G or Wi-Fi wireless network, respectively, user's high-precision position then can be calculated in real-time. The designed system mainly consists of a ground network of reference stations, data processing center, broadcast network for differential information, GNSS module, and mobile intelligent terminal. The main system design scheme and implementation framework is shown in Fig. 1 and the main data processing flow is shown in Fig. 2.

Where, the ground network of reference stations is responsible for receiving BDS, GPS and GLONASS satellite signals and getting the real-time pseudorange and carrier phase measurements and ephemeris, and then sending them to the data processing center through a data communication link (via Internet); the data processing center is used to receive the data stream from each reference station in

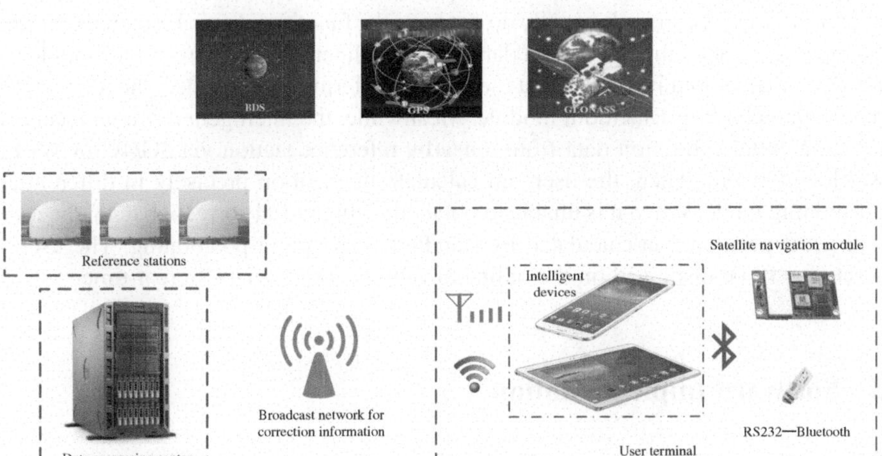

Fig. 1 The design scheme and framework of the BDS/GNSS real-time high-precision positioning system

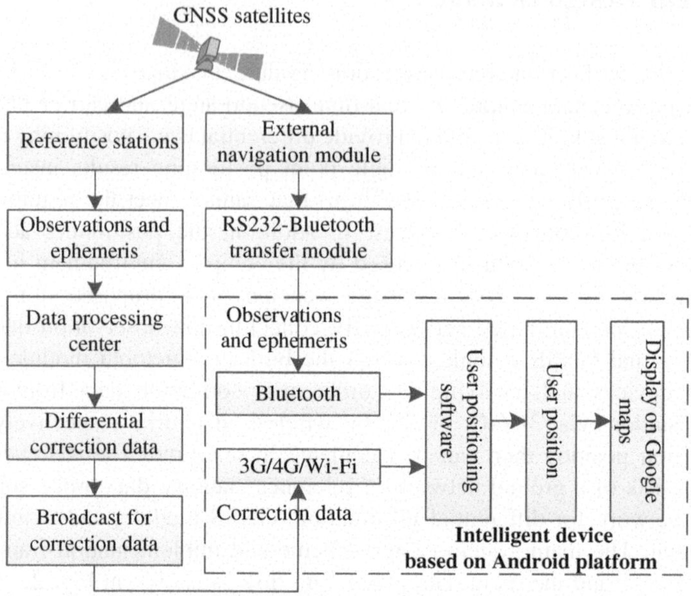

Fig. 2 Flow chart of the main data processing

real-time and calculate the differential correction information and broadcast them to users through a wireless network.

For the user terminals, they are equipped with an external satellite navigation module for receiving satellite signals to get the observations and ephemeris. The satellite navigation receivers usually output its data through a serial port (RS-232), but the intelligent terminals usually do not have the function of serial communication. So, in order to send the navigation data to an intelligent device, a transfer module of RS-232 to Bluetooth is used so that the intelligent terminal can collect the navigation data with its built-in Bluetooth module. Meanwhile, the intelligent terminal receives the differential correction data from a nearby reference station via 3G/4G or Wi-Fi wireless network. Then, the user can calculate its position precisely in differential positioning mode. When it is unable to obtain the differential correction data, the user will get it position that calculated by standard single point positioning. The user's position will be displayed on the online or offline Google maps in real-time.

3 Software Implementation

Based on the scheme designed above, a software run on intelligent devices for user to realize BDS/GNSS real-time high-precision positioning is developed based on Android platform. Currently, the software supports real-time GPS, BDS and

Fig. 3 The operation interface of the software (**a**) The interface for positioning parameter configuration (**b**)

GLONASS standalone and combined single point positioning and pseudorange differential positioning. When the user is in real-time differential positioning mode, it needs to receive the real-time differential data that broadcasted by the data processing center.

The software's running interface and its main operation options are shown in Fig. 3a, the configuration interface for positioning parameters is shown in Fig. 3b, where users can set the corresponding positioning settings based on their own situations. For example, users can choose any combination of GPS, BDS and GLONASS in positioning; at the same time, users can also choose whether to use the dual-frequency observation data. If users are equipped with a satellite navigation module which can output dual-frequency observation data, they are suggested to use the dual-frequency data to get more accurate positioning results.

After positioning successfully, the user's position then can be displayed on maps in real-time. Like the example shown in Fig. 4a, the pink marker in the figure is user's current position. At the same time, the user can see the current sky plot of visible satellites though the option of 'display', as shown in Fig. 4b.

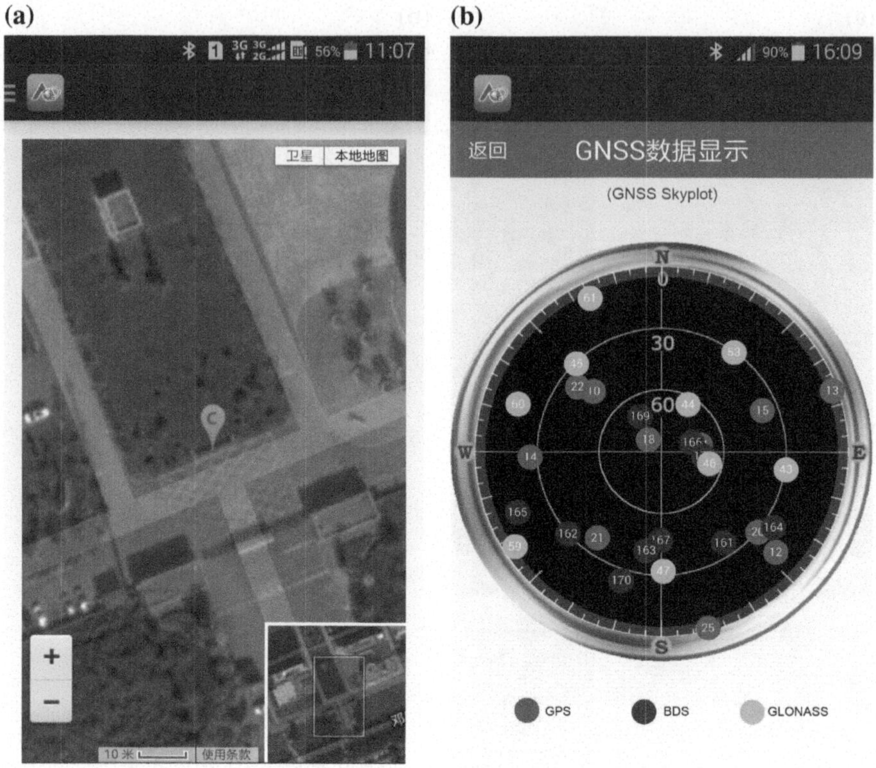

Fig. 4 Display user's position on map (**a**) The sky plot of current visible satellites (**b**) (Color figure online)

4 Experimental Results

In order to validate the positioning performance in static and dynamic situations, a single point positioning test and a real-time differential positioning test are carried out respectively. The reference station (named as AOE) was installed on the roof of the main building of Academy of Opto-Electronics (AOE), Chinese Academy of Sciences in Beijing, China. The observation condition at AOE is very good, i.e., no obstruction in the sky.

In the experiments, the differential positioning were carried out in real-time, while the single point positioning results were obtained by post-processing with the stored data which collected by the software while in real-time differential positioning. All the positioning results were update in 1 Hz. The satellite navigation receiver module in user terminal is NovAtel OEM615 navigation module with the signal processing ability of BDS single frequency (B1) and GPS/GLONASS dual frequency (L1/L2).

Fig. 5 The experimental scene in static test

In addition, in order to analyze the user's positioning accuracy, we used another GNSS receiver as rover station to collect the real-time observation data. It was connected to the same antenna that for the tested user terminal via a power divider. After the experiment, the raw observation data collected by the reference station and the rover station were converted to RINEX format. Then the user's "true" positions were obtained by post-processing RTK positioning via the open source software RTKLIB. What's more, we only use the ambiguity fixed solutions for comparison when analyzing the experimental results.

The static test was carried out in Zhongguancun Park which nearby AOE on January 10, 2016 with an experiment period of about 30 min. The experimental scene is shown in Fig. 5 with good observation conditions around the user. The dynamic test was carried out on the Beiqing Road in Haidian District, Beijing on January 25, 2016 with an experiment period of about 35 min. The GNSS antenna was installed on a car, and the user trajectory was shown in Fig. 6.

The errors in the directions of East (E), North (N), and Up (U) for BDS/GPS/GLONASS combined real-time differential positioning in the static and dynamic test are shown in Figs. 7 and 8, respectively. The statistical positioning accuracies are shown in Table 1, in which the results refer to the statistical accuracy for the probability of 2σ (95.44 %).

Fig. 6 The user trajectory in dynamic test

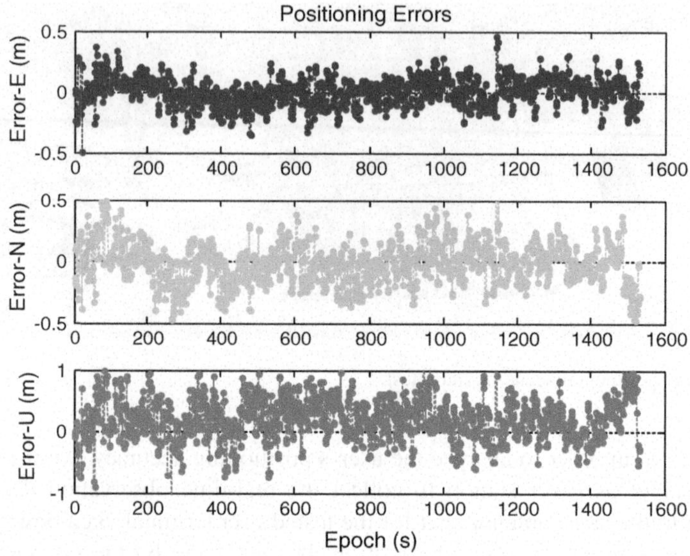

Fig. 7 Real-time differential positioning errors in static test

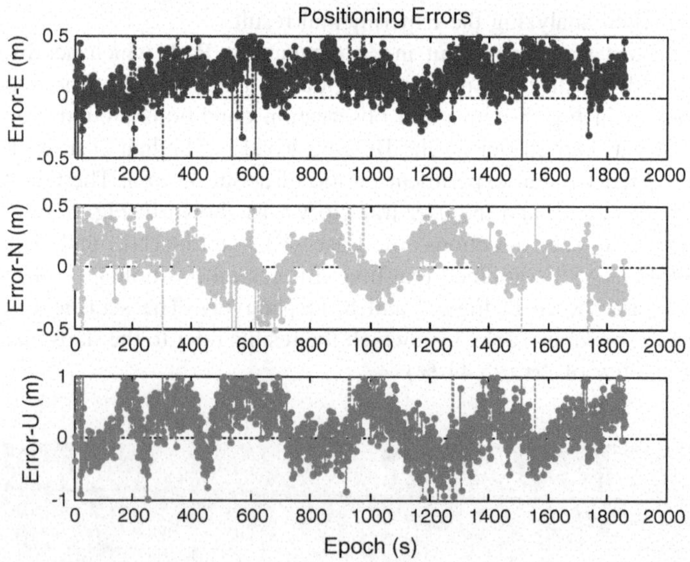

Fig. 8 Real-time differential positioning errors in dynamic test

As it can be seen from the experimental results, the single point positioning accuracies in horizontal are about 1.5 m for static test and 2.5 m for dynamic test, respectively. When it was in differential positioning mode by receiving differential

Table 1 Statistical performances for user positioning

No.	Tests	2σ (95.44 %) accuracy/(m)	
		Horizontal	Vertical
1	Single point positioning in static	1.74	2.68
2	Real-time differential positioning in static	0.37	0.82
3	Single point positioning in dynamic	2.31	3.42
4	Real-time differential positioning in dynamic	0.48	0.94

correction data from reference stations, the positioning accuracies are better than 0.5 m in horizontal and 1.0 m in vertical, respectively, for both static and dynamic test. As a result, the acquired positioning accuracy in this work can meet the needs of urban high-precision location-based service. It should be pointed out that the results presented here are in the experimental conditions of this work. Since the satellite visibility and the surrounding environment in different areas of the city are different, the corresponding results may be different in other areas.

5 Conclusion

Realizing high-precision positioning and navigation on mobile intelligent terminals is an important way to achieve the urban high-precision location-based services in the future. A system design scheme for realizing high-precision location-based services was studied and proposed in this work, and a set of BDS/GNSS real-time high-precision positioning software for mobile intelligent terminals is developed based on Android platform. By collecting the navigation data stream from an external GNSS module through the built-in Bluetooth module on those intelligent devices and receiving the differential correction data from a nearby reference station via 3G/4G or Wi-Fi wireless network meanwhile, respectively, user's high-precision position then can be calculated in real-time by the software. The results show that its positioning performance can meet the requirements of high-precision location-based service in urban. As a consequence, this contribution can provide some technical references for the realization of urban high-precision location-based services.

References

1. Fu Y (2014) The opportunities and challenges of applications in LBS for BeiDou navigation satellite system. Satell Netw 6:28–32
2. Ji S, Gao Z, Wang W (2015) M-DGPS: mobile devices supported differential global positioning system algorithm. Arabian J Geosci 8:6667–6675
3. China Satellite Navigation Office (2013) BeiDou navigation satellite system signal in space interface control document open service signal (version 2.0), 2013

4. Yang Y, Li J, Wang A et al (2014) Preliminary assessment of the navigation and positioning performance of BeiDou regional navigation satellite system. Scientia Sinica Terrae 44:72–81
5. Wang L, Li Z, Yuan H et al (2015) Validation and analysis of the performance of dual-frequency single-epoch BDS/GPS/GLONASS relative positioning. Chin Sci Bull 60:857–868
6. Li H, Dang Y, Bei J et al (2014) Research on model and performance of BDS/GPS/GLONASS multi-mode fusion positioning. Bull Surv Mapp 9:1–5
7. Teunissen PJG, Odolinski R, Odijk D (2014) Instantaneous BeiDou+GPS RTK positioning with high cut-off elevation angles. J Geod 88:335–350
8. He H, Li J, Yang Y et al (2014) Performance assessment of single- and dual-frequency BeiDou/GPS single-epoch kinematic positioning. GPS Solutions 18:393–403
9. Odolinski R, Teunissen PJG, Odijk D (2015) Combined BDS, Galileo, QZSS and GPS single-frequency RTK. GPS Solut 19:151–167
10. Deng C, Tang W, Liu J et al (2014) Reliable single-epoch ambiguity resolution for short baselines using combined GPS/BeiDou system. GPS Solut 18:375–386

The Analysis of Pedestrian's Motion State Based on the Entropy of sEMG

Min Li and Long Zhao

Abstract In order to solve the problem that traditional eigenvalue analysis of time domain and frequency domain cannot classify pedestrian's motion state correctly in pedestrian navigation because of the changing of stride frequency and length, etc., a method for motion state analysis in combination with time domain characteristics and entropy of sEMG was presented. This method analyzed the complexity and amplitude of pedestrian's motion state, estimated the stride frequency, and classified four movement state include walking, running, upstairs, and downstairs according to variation feature of sEMG. The experimental results show that the average classification accuracy went up at least by 8.25 % with presented method, it gets more reliable and stronger anti-interference capacity compared with the characteristics of time domain.

Keywords Entropy · sEMG · Feature extraction · Pedestrian's motion state

1 Introduction

Surface electromyogram (sEMG) is a voltage variation when brain governs muscle to active [1]. As the development of methods to handling sEMG, it is used in many fields such as disease diagnosis [2], artificial limb control [3], gesture language recognition [4], etc. Indoor pedestrian navigation is a relatively accurate navigation system. The current study of indoor navigation system mainly focused on radio frequency signal and got breakthrough in some application fields [5, 6]. However, it

M. Li
Digital Navigation Center, Beihang University, Beijing, China

L. Zhao (✉)
Science and Technology on Aircraft Control Laboratory,
Beihang University, Beijing, China
e-mail: buaa_dnc@buaa.edu.cn

J. Sun et al. (eds.), *China Satellite Navigation Conference (CSNC) 2016 Proceedings: Volume I*, Lecture Notes in Electrical Engineering 388,
DOI 10.1007/978-981-10-0934-1_26

needs fixed hardware devices and specific training, which is high in cost and low in applicability. Pedestrian navigation based on self-contained sensor and realizes navigation through pedestrian dead reckoning (PDR). Commonly used methods are based on the principle inertia but its error is divergent caused by acceleration's twice integration. As for PDR of smart phone its navigation device is inaccurate and the sway of phone will influence the result. Literature [6] proposed a method to analysis motion state and pedestrian navigation, which greatly reduced the error because the step length is limited. Different motion state needs to select different step length model [6], so motion state analysis is precondition in pedestrian navigation. The classification of motion state is based on the eigenvalues, and commonly used eigenvalues include time domain features [3, 6] (mean absolute value (MAV), mean square root, the number of passing zero, the slope of the signal change and wave length), frequency features [3] (average frequency and median frequency), and time-frequency characteristics [7] (wavelet transform, etc.). Most of them classify running from other state by using stride frequency as an important feature, which is not an essential difference between running and other states. Entropy is a nonlinear eigenvalue represented by signal complexity; it reveals the characteristics and mechanism of physiology system because of nonlinearity in human beings [8, 9]. Literature [10] studied the classification of muscle damage by using the sample entropy of sEMG and got good results.

In this paper, we analyzed motion state by using entropy features combined with time eigenvalue, classified different state by model of linear discriminant analysis (LDA) and support vector machine (SVM). This method increased anti-interference ability and is useful when stride frequency and other factors change.

2 Pedestrian's Motion State Analysis Algorithm

2.1 Extraction of Active Segment

The extraction of active segment is to distinguish the segment with and without signal. Commonly used methods include moving average method, short time Fourier transform method, etc. [4]. In this experiment, we chose moving average method, value of energy as standard, N as fixed window length. Energy in every moment is as follows:

$$S_N(i) = \frac{1}{N}\sum_{j=1}^{i+N-1} S(j)^2 \tag{1}$$

where $S(j)$ is the sEMG energy amplitude at time j.

When the value of energy is larger than threshold, the section is active, otherwise it is inactive.

$$S_{\text{rectified}}(i) = \begin{cases} S_N(i) & \text{if} \quad S_N(i) > \text{TH} \\ 0 & \text{if} \quad S_N(i) \leq \text{TH} \end{cases} \tag{2}$$

where TH is the threshold.

2.2 *Extraction of Eigenvalue*

2.2.1 Time Domain Eigenvalue

The time domain eigenvalue describe the characteristic that signal varies with time. Approaches which have been used frequently include MAV, amplitude sum of squares (ASS), integral EMG (IEMG), root mean square (RMS) and frequency, etc. Many literatures [6] used frequency to take running apart from the rest state, which cannot accurately distinguish them while running frequency reduce and walking frequency increase. But running consumes more energy than walking, upstairs consume more than downstairs for each step regardless of frequency, which is the value of amplitude in time domain eigenvalue. The most common and effective amplitude characteristic is MAV and RMS, as follows:

$$\text{MAV} = \frac{1}{N}\sum_{i=1}^{N} |\text{EMG}_i| \tag{3}$$

$$\text{RMS} = \sqrt{\frac{1}{N}\sum_{i=1}^{N} \text{EMG}_\text{i}^2} \tag{4}$$

where N is the length of active segment sequence in each step.

2.2.2 Theory of Sample Entropy

Entropy is a feature which describes the complexity of signal sequence. Greater the signal complexity is, bigger the entropy value is. It is bigger in still state than moving state because the movement of next situation is uncertain in stationary and the complexity is bigger. Sample entropy and approximate entropy are two kinds of popular entropy. Sample entropy is more useful because it needs less data and avoids the biasness caused by vector match itself. The process to get sample entropy is as follows [9]:

(1) For a given one-dimensional time series whose length is N $\{u(i), i = 1, \ldots, N\}$, we construct m-dimension vector $X_i = \{u(i), \ldots, u(i + m - 1)\}$, $i = 1, 2, \ldots, N - m$;

(2) Calculate the distance between arbitrary vector X_i and the rest vectors $X_j (j \neq i)$

$$d_{ij} = \max|u(i+k) - u(j+k)|$$

(3) Count the number of vectors that meet the condition $d_{ij} \leq r$ for vector X_i and the ratio with total vector number denoted as $C_i^m(r)$ when the threshold r is given.
(4) Take the mean of $C_i^m(r)$ and obtain $B^m(r)$ as follows:

$$B^m(r) = \frac{\sum_{i=1}^{N-m} C_i^m(r)}{N-m}$$

(5) Take m plus 1, repeat the process (1)–(4), we can obtain $B^{m+1}(r)$;
(6) Calculate sample entropy as follows:

$$\mathrm{SampEn}(m, r, N) = -\ln[B^{m+1}(r)/B^m(r)]$$

2.3 *Classification*

LDA is a classical algorithm in model recognition. The basic idea is getting a translation to make the new subspace projection has the largest distance between classes and smallest distance in class. But this method has certain request for dimension in case of too few dimensions cause over fitting.

SVM is a model recognition method based on statistics study theory. It can avoid the problems in traditional classification method such as over fitting, dimension disaster, and local minimum; and has good generalization ability under the condition of small sample.

3 Experimental Test and Results Analysis

3.1 *Collection of sEMG*

This experiment used the sEMG measuring device that is ZJE-II model and its sampling frequency is 1000 Hz. We used software provided by the manufacturer to collect data and used calf as the source of signal, the specific acquisition process is shown in Fig. 1.

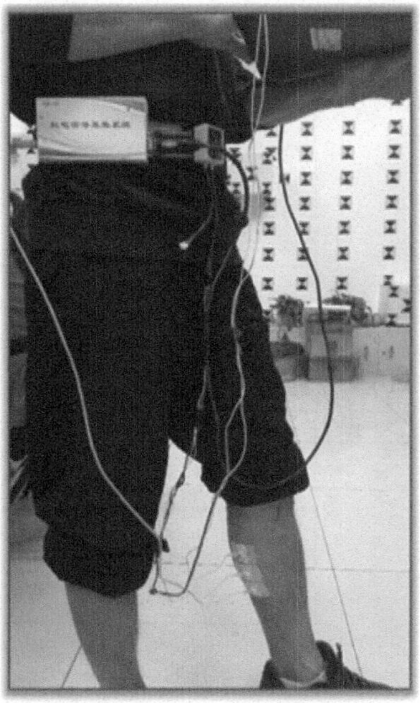

Fig. 1 Process of sEMG

3.2 *The Pre-process of sEMG*

The voltage amplitude of sEMG is less than 150 μV and the frequency is less than 500 Hz. In the actual collection, however, the voltage amplitude is more than 150 μV and frequency is instable because of body movement and interference of power frequency. Therefore, we should remove some unrelated disturbance signal and extract useful signal by pre-processing. The pretreatment methods commonly used are bandpass filter and extraction of active segment.

3.2.1 Bandpass Filter

Bandpass filter can remove noise including interference of power frequency, background noise and baseline drift, etc. In this test, we used the signal processing toolbox in MATLAB for filter processing, chose Chebyshev type I filter with passband frequency of 50–450 Hz, and the passband ripple coefficient is Rp = 1, stop-band attenuation coefficient is Rs = 20. The time domain waveform, before and after processing, is shown in Fig. 2. After eliminating some interference, the wave can be more reliable.

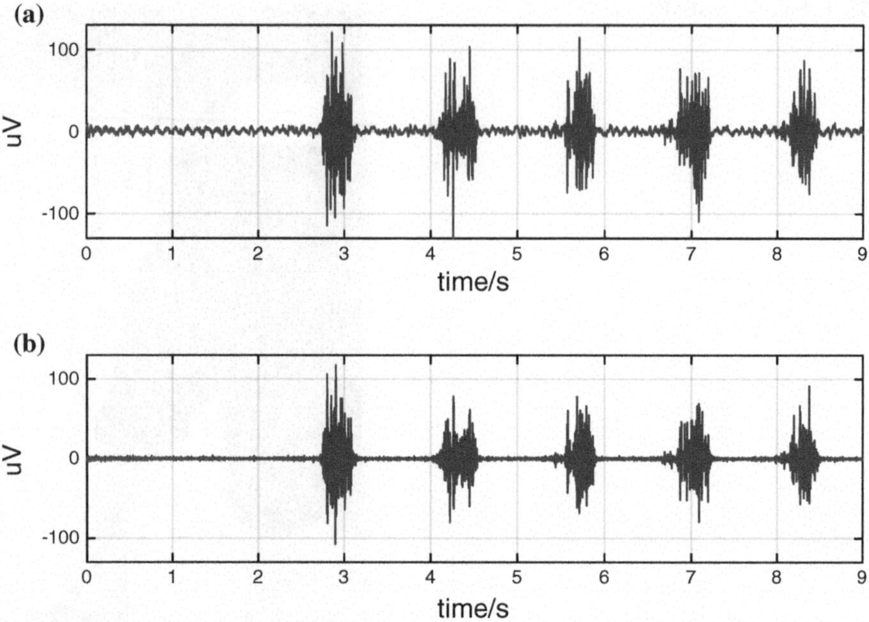

Fig. 2 The time domain waveform before and after processing. **a** Before. **b** After

3.2.2 Eigenvalue Extraction

In the experiment, we chose fixed window length $N = 200$ to calculate energy and hence, can get the energy diagram as illustrated in Fig. 3. We chose left threshold greater than right threshold considering that signal is moving left when calculating the power sequence. The specific choice is according to the value of EMG. Here, the thresholds are selected as follows: the left threshold is TH1 = 100 and the right threshold is TH2 = 20. The number of active segment is same as the number of steps and each active segment is different.

3.3 Extraction of Eigenvalue

Feature extraction is a process that extracts eigenvalues in every step, constructs classifier trained by eigenvalues of number of steps (we chose around 100 steps in this test), then classifies different motion states through these eigenvalues.

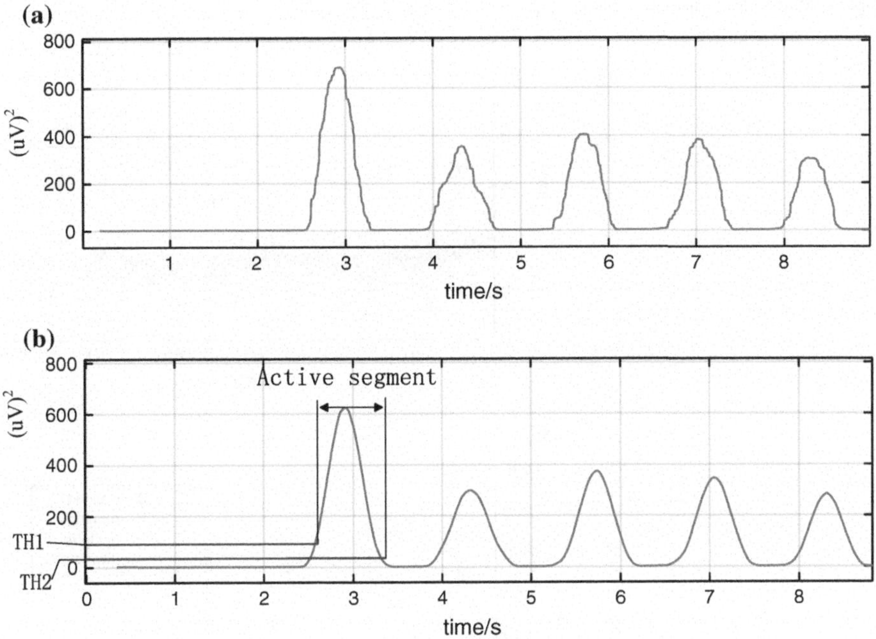

Fig. 3 The extraction of active segment. **a** Energy signal envelop. **b** Smoothing result

3.3.1 Extraction and Analysis of Time Domain Eigenvalue

Choose 100–120 steps from each various motion as sample points and extract MAV and RMS of each step, the scatter diagrams are shown in Fig. 4a, b. From Fig. 4a, we can know that the MAV of walking and downstairs concentrate between 5 and 10 while running and upstairs concentrate between 13 and 20. Figure 4b shows that the MAV of walking and downstairs concentrate between 10 and 15 while running and upstairs concentrate between 24 and 32. These two kinds of time domain characteristic are different on value but basically consistent on the distribution of motion state, that is to say, value of walking and go downstairs is smaller than running and go upstairs. However, MAV is higher in polymerization degree in same state than RMS.

3.3.2 Extraction and Analysis of Sample Entropy

In the extraction of sample entropy, this experiment chose parameters as follows: $m = 2$, $r = 0.2 \times \text{SD}$ (SD is standard deviation of signal sequence), $S = 200$ (definite the value of N to calculate entropy value as length of entropy sequence S), $Z = 20$ (frame length). Every active section contains a number of entropies and entropy eigenvalue of the step is mean of these entropies. The scatter diagram of sample entropy in different motion state is shown in Fig. 5.

Fig. 4 The scatter plot of time domain characteristics. **a** MAV. **b** RMS

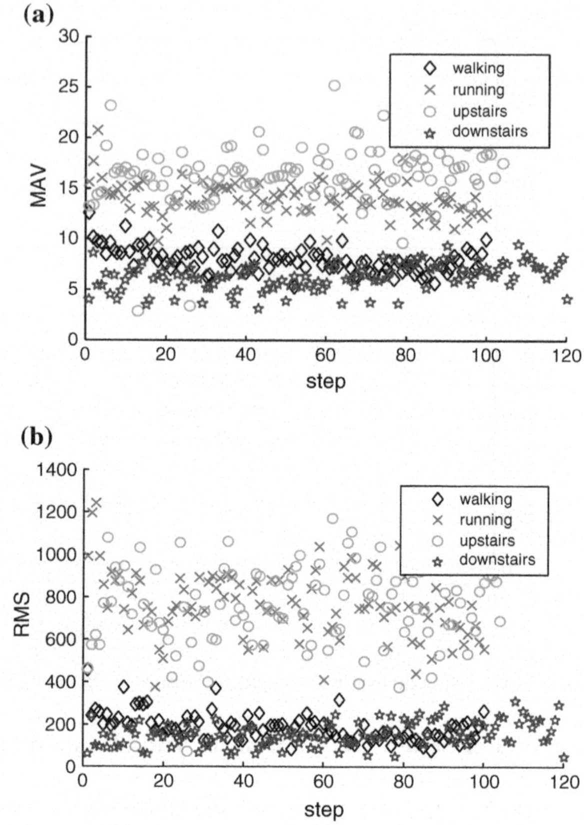

Fig. 5 The scatter plot of entropy

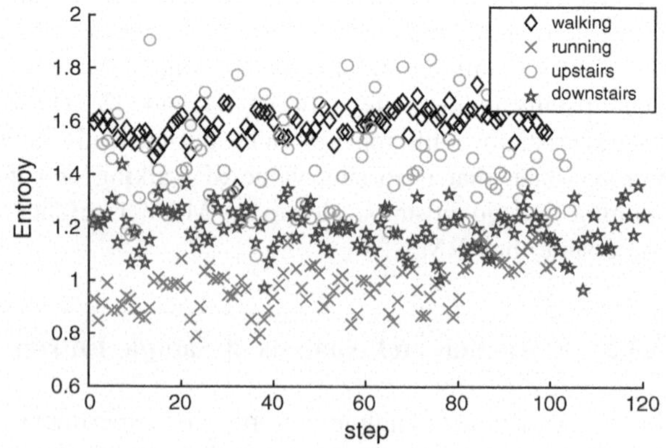

Figure 5 illustrated that entropy of running is concentrated under 1, entropy of downstairs is concentrate about 1.2, entropy of walking is bigger and about 1.6 while entropy of upstairs is disperse and centered on 1.4. It means that entropy

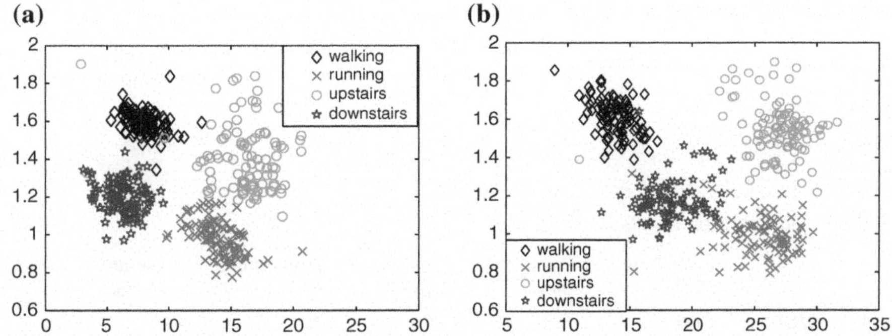

Fig. 6 The scatter plot of entropy combining time domain feature. **a** Object 1. **b** Object 2

eigenvalue can distinguish running from upstairs and other state, walking from running or downstairs. But we cannot identify upstairs from walking and downstairs from this figure.

3.3.3 Combined with the Sample Entropy and Time Domain Features

From the analysis above we conclude that time domain characteristics mainly describe amplitude while entropy describe mean complexity of every step. These two kinds of eigenvalues cannot distinguish all four kinds of motion states, but done when combine them. Draw a scatter diagram with MAV as *X*-axis and entropy as *Y*-axis, the result is shown in Fig. 6a. From this scatter diagram we can get that different motion state with different distribution and classification can be done using these two kinds of eigenvalues: MAV and entropy. In this test, polymerism is large in the same state and coherence is small in different state.

Change the object and stride frequency, we got the scatter diagram by using the same method shown in Fig. 6b. Although every distribution center has changed on the numerical, the distribution and basic position does not change and this method can be a good classification. Therefore, we can conclude that entropy and MAV can be a set of eigenvalues to analysis and classify motion state.

3.3.4 Classification Results

In the analysis of pedestrian's motion state based on the entropy of sEMG, different characteristics represent different aspects of property and has different classification features. In this experiment, we selected the representative characteristic in time domain-MAV and nonlinear characteristic value entropy as classification characteristics, chose 80 % of the experimental sample as the training sample set to construct the classifier and 20 % as the test sample set to test the classifier. Here we

Table 1 Classification results of LDA

	Walking (%)	Running (%)	Upstairs (%)	Downstairs (%)	Average accuracy (%)
MAV + RMS	90	85	80	65	80
MAV + Entropy	95	85	83	90	88.25

Table 2 Classification results of SVM

	Walking (%)	Running (%)	Upstairs (%)	Downstairs (%)	Average accuracy (%)
MAV + RMS	90	85	85	70	82.5
MAV + Entropy	95	90	85	95	92.5

used MATLAB to construct classifier, used LDA and SVM to classify and test four types of motion states, the test results are shown in Tables 1 and 2.

As seen from Tables 1 and 2, after using the method combining MAV and sample entropy to construct classifier, the classification accuracy of walking is as high as 95 % and average accuracy has improved by 8.25 and 10 %, respectively, than by using only two kinds of time domain characteristics, MAV and RMS. Results show that the method presented in this paper is better on classification result than traditional method of time domain features, and the recognition rate will not be affected by the change of stride frequency.

4 Discussion and Conclusion

The analysis of pedestrian's motion state is a basic and important step in pedestrian navigation system, which establishes a good foundation for the choosing of stride length model and pedestrian navigation. This paper proposed nonlinear eigenvalue sample entropy on the basis of time domain characteristics and analyzed the internal complexity of sEMG. First of all gathered the sEMG of several movement state from single channel, then made preprocessing, feature extraction, and classification for these sEMGs of different state; next, constructed classifier use the LDA model and SVM model; finally classified four kinds of pedestrian motion state, namely, walking, running, upstairs, and downstairs. The results of two objects show that the method combined with the entropy eigenvalue and time domain eigenvalue get greater classification result, and antijamming capability is stronger when object change, so that can satisfy the requirement for analysis of pedestrian's motion state in pedestrian navigation.

Acknowledgments Project supported by the National Natural Science Foundation of China (Grant No. 41274038, 41574024), the Aeronautical Science Foundation of China (Gratis No. 2013ZC51027), and the Fundamental Research Funds for the Central Universities.

References

1. Wu D, Sun X et al (2010) Feature collection and analysis of surface electromyography signals. J Clin Rehabil Tissue Eng Res 14(43):8073–8076
2. Prosser LA, Lee SCK et al (2010) Trunk and hip muscle activity in early walkers with and without cerebral palsy-A frequency analysis. J Electromyogr Kinesiol 20:851–859
3. Jahani Fariman H, Ahmad SA et al (2015) Simple and computationally efficient movement classification approach for EMG-controlled prosthetic hand: antificial neural network. Intell Autom Soft Comput 21(4):559–573
4. Zhang X (2010) Human action recognition and interaction based on sEMG. University of Science and Technology of China, Anhui
5. Yang L, Chen Y, Li XY, et al (2014) Tagoram: real-time tracking of mobile RFID tags to high precision using COTS devices. In: 20th annual international conference on mobile computing and networking, pp 237–248
6. Wang Q (2013) Surface electromyography based human gait analysis and its applications. University of Science and Technology of China, Anhui
7. Napoli NJ et al (2015) An EMG comparative analysis of quadriceps during isoinertial strength training using nonlinear scaled wavelets. Hum Mov Sci 40:134–153
8. Chen W (2008) A study of feature extraction from sEMG singal based on entropy. Shanghai Jiao Tong University, ShangHai
9. Zhang Y (2009) The theory of approximate entropy and its application. Chin J Med Phys 1543–1546
10. Yang X, Li Z et al (2014) The research of electromyographic signal based on sample entropy. China Sci Technol Inf 34–36

FOG De-noising Method Based on Empirical Mode Decomposition and Allan Variance

Shanshan Gu, Qinghua Zeng, Jianye Liu and Weina Chen

Abstract To overcome the negative influences of exceptional fiber optic gyro (FOG) signal on the performances of INS/GPS integrated navigation system, a FOG signal de-noising method in the time–frequency domain is proposed based on the empirical mode decomposition and Allan variance (EMDAV) in this paper. Initially, the FOG output signal is decomposed with empirical mode decomposition (EMD), and then the decomposed signal is analyzed by Allan variance. The noise is removed based on the error characteristics analysis of FOG signal in time–frequency domain. Real FOG data is processed by the de-noising method based on EMDAV, and the results show that, the mean and variance of the signal de-noised by proposed method can be reduced with 2 orders of magnitude than the wavelet method. The experiment of INS/GPS integrated navigation system under GPS failure indicates that, DEMAV can improve the navigation accuracy effectively.

Keywords Proceedings fiber optic gyro · Empirical mode decomposition · Allan variance · De-noising

1 Introduction

As a key device for angular velocity detection, gyro plays an important role in inertial navigation system. In order to improve the navigation accuracy in static situation, it is important to analyze the characteristic of gyro signal and to choose a certain method for gyro signal de-noising [1].

The gyro errors commonly contain the deterministic error which can be compensated by calibration or other methods and the random error which is a complex time-varying stochastic process and is difficult to be compensated.

S. Gu · Q. Zeng (✉) · J. Liu · W. Chen
Navigation Research Center, Nanjing University of Aeronautics and Astronautics, Nanjing, China Satellite Communication and Navigation Collaborative Innovation Center, Nanjing, China
e-mail: zengqh@nuaa.edu.cn

J. Sun et al. (eds.), *China Satellite Navigation Conference (CSNC) 2016 Proceedings: Volume I*, Lecture Notes in Electrical Engineering 388,
DOI 10.1007/978-981-10-0934-1_27

Due to the merits of multi-resolution analysis, the wavelet transform has widely applied for gyro signal de-noising. Nonetheless, the problems of frequency overlapping, threshold instability, decomposition, and reconstruction algorithm errors existed in the wavelet transform bring some limitations for the gyro signal processing [2]. Furthermore, when the useful signal is drowned by the exceptional noise, the results of wavelet de-noising are not satisfactory. While being an adaptive local time–frequency analysis method [3], empirical mode decomposition (EMD) does not possess the above disadvantages. The signal can be decomposed by its characteristic to a finite number of empirical mode function by EMD, which is commonly applied in signal de-noising [4], prediction [5], and so on. In [6], EMD is used to remove the MEMS gyro noise. In [2], a bounded noise assist analysis method is introduced to improve the decomposition accuracy of EMD based on the noise spread model of EMD. The noises with low magnitude and high frequency are compressed into early intrinsic mode functions (IMF). However, the consistency and accuracy of fiber optic gyro (FOG) signal model in different application environments need a further investigation. In this paper, a FOG signal de-noising method based on EMD and Allan variance is proposed, which employs Allan variance to analyze the decomposed gyro signal of EMD. Meanwhile, the exceptional signal is de-noised and the proposed method is compared with wavelet method with real gyro data.

The proposed method can be a general method for gyro error performance analysis, which aims at analyzing the characteristics of gyro signal in both the time domain and the frequency domain. It also can provide a theoretical guidance for the EMD-based de-noising methods.

2 De-noising Method Based on EMDAV

The traditional AV analysis method which presents the gyro noise characteristics in the only time domain cannot reflect its distributions among different frequency bands. However, the noise characteristics vary with the frequency. To reveal more details of the gyro noise characteristic and provide the basis for gyro de-noising, the de-noising method based on EMDAV is proposed in this paper.

2.1 The Basic Principle of EMDAV

As a method utilized to analyze the nonlinear and nonstationary signal, EMD decomposes the signal on the basis of time-scale characteristics. There is no need to set the basic functions in advance, and can be applied to decompose any type of signal. Thus, EMD has obvious advantages in processing nonlinear and nonstationary signal.

The EMDAV algorithm proposed in the paper is based on the EMD and AV. After decomposing the gyro signal with EMD, the AV can be obtained with the help of intrinsic mode function. Lastly, the AV logarithmic curve over time and frequency can be depicted. The steps of the EMDAV-based gyro signal characteristics analysis method can be summarized as follows:

1. Find all the local maxima and minima of the original signal $x(t)$.
2. Fit the estimated points of the first step using natural cubic spline to form an upper envelope of $e_{\max}(t)$ and a lower envelope of $e_{\min}(t)$.
3. Compute the mean of the two envelopes,

$$m_i(t) = \frac{e_{\max}(t) + e_{\min}(t)}{2} \tag{1}$$

   where $i = 1, 2, \ldots, n, n$ is number of decomposition level.
4. Refine the estimate $h_i(t)$ of the IMF by subtracting the mean solved in the previous step from the current signal.

$$h_i(t) = x_i(t) - m_i(t) \tag{2}$$

   where

$$x_i(t) = \begin{cases} x(t) & i = 1 \\ x(t) - \sum_{j=1}^{i-1} h_j(t) & i \geq 2 \end{cases} \tag{3}$$

5. Proceed from Step (1) again unless a stopping criterion has been fulfilled. The stopping criterion is [7]:

$$\mathrm{SD} = \frac{\sum_{t=0}^{T} |h_{n-1}(t) - h_n(t)|^2}{\sum_{t=0}^{T} h_{n-1}^2(t)} \tag{4}$$

   where SD is screening threshold. The iteration stops when SD is smaller than a constant, whose value is often set as 0.2–0.3 [8].
   Eventually, the IMF components of $h_1(t), h_2(t), \ldots, h_n(t)$ and the remainder of $r_n(t)$ can be obtained. Then, the original signal can be expressed as:

$$x(t) = \sum_{i=1}^{n} h_i(t) + r_n(t) \tag{5}$$

6. Calculate the AV coefficients correspond to the IMF components and analyze the signal characteristics in different frequency bands.

From the above EMDAV procedures, we can see that the IMF of $h_i(t)$ indicates the different frequency components. With the increase of decomposition level, the frequency of decomposed signal decreases gradually, namely, the first IMF represents the highest frequency component of the original signal. The frequencies of subsequent IMFs reduce almost in the form of the negative power of 2 [9] until the frequency of remainder is very small. Consequently, each IMF corresponds to the signal component of different frequencies.

Compared with traditional AV analysis method, the characteristics and distributions of gyro signal in different frequency bands can be depicted in details with EMDAV algorithm. One more dimensional information can be shown in three-dimensional (3D) logarithmic graph of EMDAV than that in 2D curve of traditional AV.

2.2 *Exceptional Noise Suppression Based on EMDAV*

Exceptional noise is regarded to be an interfered gyro signal with abnormal amplitude which is generally generated by the abnormal changes of the external environment, such as the change of the environmental temperature, the platform vibration, and so on. As a part of inertial measurement units, FOG has been employed for airplane, ship, and so on, where the exceptional noise will degrade the instrument's performance and ability seriously. So, it is necessary to eliminate the influence of this kind of exceptional noise.

Considering that a signal would be decomposed into IMFs in the different frequency domains by EMD, the exceptional noise which is distributed among certain frequency bands can be detected by the EMDAV [10]. The proposed suppression method based on EMDAV in this paper follows the steps below:

1. Analyze the gyro signal with EMDAV;
2. Determine the IMF components which are polluted by the exceptional noise. When the EMDAV coefficient of the ith level IMF is much less than that of the $(i-1)$th level from beginning (i is from the beginning of 1), the IMF components with level from 1 to $(i-1)$ contain exceptional noise;
3. Process the IMF components corresponding to the exceptional noise, which is decided by EMDAV variances;
4. Set a certain threshold for remainder IMF components to reduce other noises determined by EMDAV [11];

$$T_k = \sigma_k \sqrt{2 \ln N} \tag{6}$$

Fig. 1 SPAN-CPT system

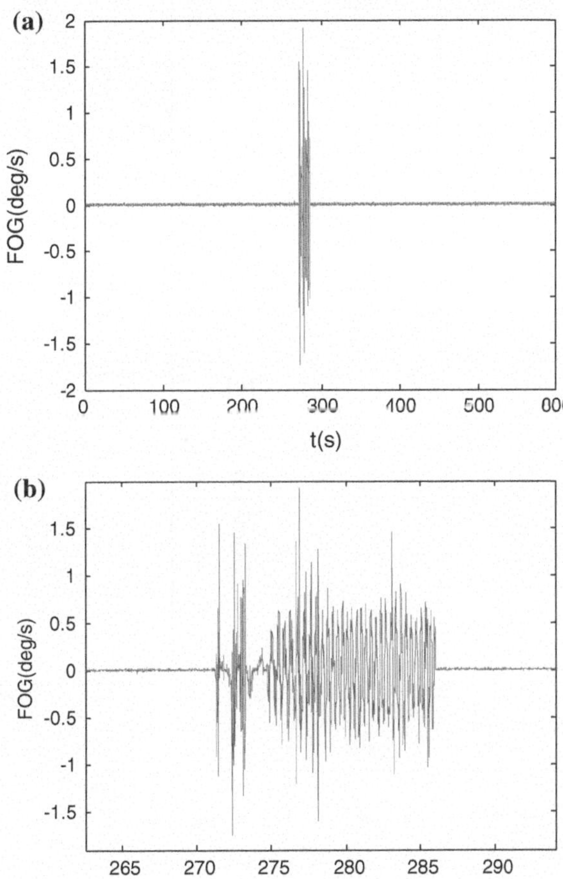

Fig. 2 The origin signal of FOG and details of exceptional noise. **a** Origin signal **b** details of exceptional noise

Fig. 3 Comparison of the signal before and after de-noising. **a** FOG signal **b** Details of common signal **c** Details of exceptional noise

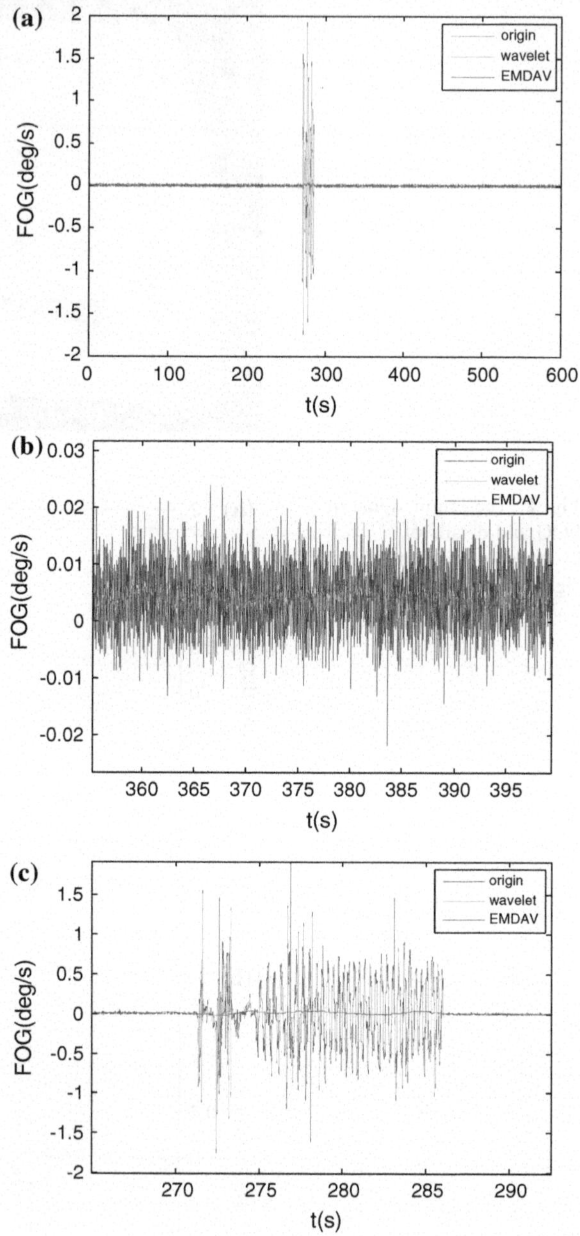

Fig. 4 Comparisons of navigation system errors. **a** Position errors **b** Velocity errors **c** Attitude errors

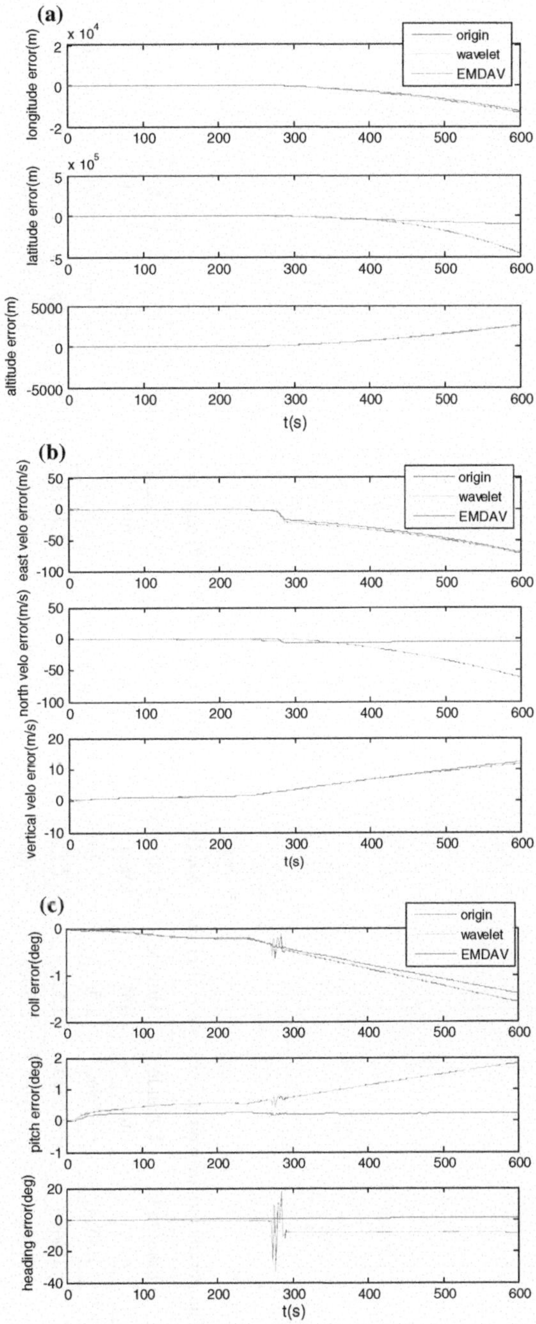

Table 1 The parameters of navigation system after 600 s for the FOG signal de-noised by different algorithms

Algorithm	Longitude (m)	Latitude (m)	Altitude (m)	East velocity (m/s)	North velocity (m/s)	Vertical velocity (m/s)	Roll (deg)	Pitch (deg)	Heading (deg)
Origin	−1.35e4	−4.58e5	2.53e3	−7.19e1	−6.19e1	1.16e1	−1.58e0	1.82e0	−8.39e0
Wavelet	−1.34e4	−4.55e5	2.52e3	−7.17e1	−6.16e1	1.16e1	−1.58e0	1.82e0	−8.37e0
EMDAV	−1.26e4	−1.07e5	2.60e3	−7.02e1	−4.41e0	1.23e1	−1.40e0	2.23e-1	1.20e0

where σ_k is the mean square error of IMF, and N is the decomposition level. When the signal of each IMF component is larger than the threshold of T_k, its value is set to be zero otherwise the original values are reserved;

5. Reconstruct the signal.

3 Experiment Results and Analysis

In actual application, FOG signal is affected by the exceptional noise with large amplitude in local time, which affects the output performance of the signal [8]. In order to simulate the FOG signal polluted by the exceptional noise, practical FOG static signal of SPAN-CPT (as shown in Fig. 1) is acquired. The signal is disturbed artificially during 270–286 s. The signal sampling rate is 100 Hz, the sampling time is 600 s, and one of the signals is shown in Fig. 2.

To verify the performance of the presented noise suppression method, the results of wavelet soft threshold with db4 and the de-noising method based on MEDAV are given in Fig. 3. It can be seen that the processing results of the wavelet and the proposed method are similar for the high-frequency noise, whereas the exceptional noise cannot be suppressed by the wavelet.

The FOG signals before and after filtering are put into the inertial/GPS integrated navigation system, and GPS is failed after 240 s. 360 s later the position errors, velocity errors, and attitude errors are shown in Fig. 4. The parameters of integrated navigation system are listed in Table 1. In Fig. 4, it can be seen that the results of navigation system for the exceptional FOG signal de-noised by wavelet method are similar to that for the origin signal. Nevertheless, it can be seen in Table 1, the results for the exceptional FOG signal de-noised by the proposed method can be improved greatly except the vertical velocity and the altitude.

4 Conclusion

A de-noising method based on EMDAV for FOG is proposed in this paper. The Allan variance is calculated by the signal decomposed based on the characteristic timescale, and the signal characteristics of gyro error in different frequency bands are revealed. Theoretical analysis and experiment of practical FOG gyro data indicate that: the proposed method can separate the exceptional noise and useful signal, and can reduce the disturbance of exceptional noise and high-frequency noise at the same time to improve the accuracy of navigation system.

In addition, the proposed method can be used for the analysis of gyro signal. Compared with the characteristics of gyro error in the sole time domain given by the traditional AV, the EMDAV algorithm can provide more detailed information in the time–frequency domain.

Acknowledgments This work is supported by the National Natural Science Foundation of China (No. 61533008, No. 61374115, No. 61328301, No. 61273057) and by Foundation of Jiangsu Key Laboratory "Internet of Things and Control Technologies", and partly supported by the Project Funded by the Priority Academic Program Development of Jiangsu Higher Education Institutions.

References

1. Lv P, Liu J, Lai J, Qin G (2014) Research on the performance evaluation methods of fiber optical gyro stochastic errors. Chin J Sci Instrum 35(2):412–418
2. Cui B, Chen X, Song R (2015) Application of EMD threshold filtering for fiber optical gyro drift signal de-noising. Acta Optica Sin 2:53–58
3. Gan Y, Sui L, Xiao G et al (2013) EMD de-noising theory considering static and dynamic conditions and its applications in INS. In: Proceedings of China satellite navigation conference (CSNC). Springer, Berlin, pp 507–515
4. Li L, Chai X, Zheng S et al (2014) A de-noising method for track state detection signal based on the statistical characteristic of noise. J Transp Technol 4(04):50741. doi:10.4236/jtts.2014.44029
5. Goel G, Hatzinakos D (2014) Ensemble empirical mode decomposition for time series prediction in wireless sensor networks. In: International conference on IEEE computing, networking and communications (ICNC), pp 594–598
6. Wang L, Cheng X, Jia Z (2013) A novel denoising method and its application to MEMS gyro signal. In: 25th Chinese Control and decision conference (CCDC). IEEE, pp 3248–3253
7. Guo C, Wen Y, Li P, Wen J (2014) Enhancement of leak signals using EMD in pipeline. Chin J Sci Instrum 35(2):412–418
8. Huang NE, Shen Z, Long SR, et al (1998) The empirical mode decomposition and the Hilbert spectrum for nonlinear and non-stationary time series analysis. Proc R Soc London A: Math Phys Eng Sci 454(1971):903–995
9. Gan Y, Sui L, Wu J et al (2014) An EMD threshold de-noising method for inertial sensors. Measurement 49:34–41
10. Gan Y, Sui L (2011) De-noising method for gyro signal based on EMD. Acta Geod Cartogr Sin 40(6):745–750
11. Gan Y, Sui L (2012) EMD threshold de-noising and its applications in INST data processing. Acta Geod Cartogr Sin 41(4):504–509

A Method of X-ray Pulsar-Based Navigation for Constellation in Libration Points

Lu Zhang, Yidi Wang, Wei Zheng and Dapeng Zhang

Abstract A method of X-ray pulsar-based navigation for satellites constellation in Halo orbit is given in this article. The dynamics of libration point orbits offer many opportunities for flexible, low-energy trajectories, and missions around libration points are arising. Autonomous navigation is necessary for spacecraft autonomy in libration point, as well as significantly benefits China's deep space missions. In this paper, the state equation is established according to the dynamic characteristic in the libration point orbit. The difference of pulse arrival time to different satellites can be regarded as the measurements in the measurement equation. Unscented Kalman filter is applied to estimate the state of the system. Simulations demonstrate that this method is feasible for autonomous navigation for satellites constellation in the Sun–Earth libration point orbit. Comparing X-ray pulsar-based method with the traditional inter-satellite links navigation principle, satellites constellation rotation can be solved. Combining inter-satellite links and pulsar measurements in the UKF filter, the state of the system can be estimated more precisely and in a more quick way.

Keywords X-ray pulsar · Libration point · Constellation autonomous navigation · Unscented Kalman filter

1 Introduction

Libration points were first discovered by Eular and Lagrange in their study of the restricted three-body problem. Among the five equilibrium points, the two collinear points are stable whereas the three noncollinear ones are unstable. The dynamics of libration points offer great opportunities for deep space missions. The Sun–Earth L_2

L. Zhang (✉) · Y. Wang · W. Zheng (✉) · D. Zhang
College of Aerospace Science and Engineering, National University of Defense Technology, Changsha, China
e-mail: zhanglu.evergreen@163.com

W. Zheng
e-mail: zhengwei@nudt.edu.cn

J. Sun et al. (eds.), *China Satellite Navigation Conference (CSNC) 2016 Proceedings: Volume I*, Lecture Notes in Electrical Engineering 388,
DOI 10.1007/978-981-10-0934-1_28

point is collinear with the Sun and Earth, and locates in the far side of the Earth about 1.50 million km away. For its superiority of heat, magnetism, and force, L_2 point is an ideal location for cosmic observation and astronomical research. For instance, International Sun–Earth Explorer (ISEE-3) and Terrestrial Planet Finder (TPF) are all representative missions that fully utilize the advantages of the libration points [1].

Navigations for recent missions to libration points are all based on ground tracking and control. Increased libration point missions will lay heavy burden on the ground-based network [2]. Long distance from the Earth and long period of the mission lead a great time delay from ground station to in-orbit spacecraft, which will greatly limit the control ability of ground station network. Thus, autonomous control system is of great importance to libration point missions and autonomous navigation is the premise on which autonomy is based [3]. Navigation satellites constellation in Sun–Earth libration points would provide navigation service for interplanetary detector, as well as establishing the communication between Earth and mars [4]. For the great potential of libration points, the establishment of navigation satellites constellation in libration point orbits is of great significance and could greatly benefit the future deep space exploration in China [5]. Folta proposed a method based on celestial observation, while the measurement vector is composed of Doppler shift, which restricts the autonomy of spacecraft [6]. Hill shows the method named Liaison Navigation, which is based on the strong asymmetry of the three-body force field, and satellite-to-satellite tracking (SST) is the only observation [7]. Different from the two-body problem, the observability matrix will not be ranked deficient in three-body problem.

A pulsar is a neutron star rotating at high velocity, with excellent long-term stability of the rotation period. As one type of celestial navigation, X-ray pulsar-based navigation has the same characteristics as celestial navigation: strong autonomy and interference immunity [8]. In addition, X-ray pulsar-based navigation has unique advantages such as good reliability and high accuracy thus can be used as an excellent beacon.

When a same pulse arrives at two spacecrafts, the time difference of the two arrival time reflects the projection of their relative position on the direction to the pulsar. Take the time difference as the observation data, the influence of ephemeris error and clock error to the precision of navigation can be greatly eliminated [9]. In this paper, the dynamic equation of libration points is given, and the observation equation is established upon the time difference of the same pulse arriving at two spacecraft. The position of the spacecraft is estimated and compared with the navigation method based on satellite-to-satellite track.

2 Dynamic Model of Libration Point Orbits

In circular restricted three-body problems (CR3BP), there are two massive bodies in orbit about their mutual barycenter, and the spacecraft with infinitesimal mass has little impact on the massive bodies. For small eccentricity orbits, the simplified circular orbit can be used to describe the motion of each body.

In the Sun–Earth system, the motion of a spacecraft can be depicted in the Sun–Earth rotation coordinate frame. This coordinate frame has its origin at the barycenter, with the positive x axis direction going from Sun to Earth. The positive z direction is parallel to the direction of the angular momentum of Sun and Earth, and the y axis composes the right-hand coordinate system with x axis and z axis. The coordinate system is duplicated in Fig. 1.

For a constellation which is composed of N satellites, $[\boldsymbol{x}_1 \quad \boldsymbol{x}_2 \quad \ldots \quad \boldsymbol{x}_N]^{\mathrm{T}}$ is chosen as the state vector, among which the state vector $\boldsymbol{x}_{\mathrm{i}}$ of the i-th satellite is $[x_i \quad y_i \quad z_i \quad v_{ix} \quad v_{iy} \quad v_{iz}]$. The state equation $\dot{\mathbf{x}} = f(\mathbf{x}) + \mathbf{w}$ can be established upon the orbital dynamic model of satellite, where $f(\mathbf{x})$ is a combination of $[f_1^{\mathrm{T}}(\mathbf{x}) \quad f_2^{\mathrm{T}}(\mathbf{x}) \quad \cdots \quad f_{\mathrm{N}}^{\mathrm{T}}(\mathbf{x})]^{\mathrm{T}}$.

Where $f_i(\mathbf{x})$ is as follows:

$$\begin{cases} \frac{\mathrm{d}x_i}{\mathrm{d}t} = v_{xi} + w_{xi} \\ \frac{\mathrm{d}y_i}{\mathrm{d}t} = v_{yi} + w_{yi} \\ \frac{\mathrm{d}z_i}{\mathrm{d}t} = v_{zi} + w_{zi} \\ \frac{\mathrm{d}v_{xi}}{\mathrm{d}t} = 2v_{yi} + x_i + f_{xi}(x, y, z) + w_{v_{xi}} \\ \frac{\mathrm{d}v_{yi}}{\mathrm{d}t} = -2v_{xi} + y + f_{yi}(x, y, z) + w_{v_{yi}} \\ \frac{\mathrm{d}v_{zi}}{\mathrm{d}t} = f_{yi}(x, y, z) + w_{v_{zi}} \end{cases} \tag{1}$$

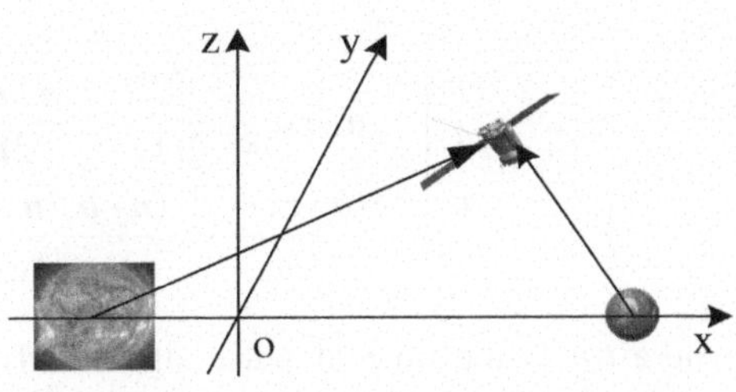

Fig. 1 Coordinate system of Sun–Earth rotation frame

where

$$
\begin{aligned}
f_{xi}(x,y,z) &= -\frac{1-\mu}{r_{1,i}^3}(x_i+\mu) - \frac{\mu}{r_2^3}(x_i-1+\mu) \\
f_{yi}(x,y,z) &= -\left(\frac{1-\mu}{r_{1i}^3}+\frac{\mu}{r_{2i}^3}\right)y_i \\
f_{zi}(x,y,z) &= -\left(\frac{1-\mu}{r_{1i}^3}+\frac{\mu}{r_{2i}^3}\right)z_i
\end{aligned}
\tag{2}
$$

$r_{1i}=\sqrt{(x_i+\mu)^2+y_i^2+z_i^2}$ is the distance from the spacecraft to Sun, and $r_{2i}=\sqrt{(x_i+\mu-1)^2+y_i^2+z_i^2}$ is the distance from the spacecraft to Earth. m_S, m_E are the mass of Sun and Earth, respectively, the three-body gravitational parameter is $\mu = m_E/(m_S+m_E)$. $\boldsymbol{w}$ is the process noise and is assumed to be zero-mean Gaussian noise. The equation of motion can be written as $\dot{\mathbf{x}} = f(\mathbf{x})+\mathbf{w}$.

3 Observation Model Based on Pulsars and Inter-satellite Links

3.1 Observation Equation Based on Pulsars

The theory of X-ray pulsar-based navigation is currently well developed. When photons from the pulsar are received by detector on the spacecraft, the arrival time of photons can be recorded. Based on the pulsar timing model, the arrival time to spacecraft can be compared with the time taken by the photon arriving at solar system barycenter.

The time transfer model of the pulse arriving time to spacecraft and SSB is denoted as follows:

$$
t_{SSB} = t_{SC} + \frac{1}{c}\begin{bmatrix} \boldsymbol{n}\cdot\boldsymbol{r}_{SC} - \dfrac{r_{SC}^2}{2D_0} + \dfrac{(\boldsymbol{n}\cdot\boldsymbol{r}_{SC})^2}{2D_0} \\ -\dfrac{(\boldsymbol{b}\cdot\boldsymbol{r}_{SC})}{D_0} + \dfrac{(\boldsymbol{n}\cdot\boldsymbol{b})(\boldsymbol{n}\cdot\boldsymbol{r}_{SC})}{D_0} \end{bmatrix} + \frac{2\mu_s}{c^3}\mathrm{In}\left|\frac{\boldsymbol{n}\cdot\boldsymbol{r}_{SC}+r_{SC}}{\boldsymbol{n}\cdot\boldsymbol{b}+b}+1\right| \tag{3}
$$

where t_{SC} is the time of pulse signal arriving at the spacecraft; $\boldsymbol{n}$ is the position vector of pulsar; c is the speed of light; D_0 is the position of pulsar at the standard propagation time T_0; $\boldsymbol{b}$ is the relative position of from SSB to the barycenter of Sun and μ_s is the gravitational constant [10].

For constellation in libration points, when two satellites in constellation receive the same pulse signal, the difference time of arrival (DTOA) δt reflects the projection distance of the two satellites in the direction of pulsar. The observation vector can be established based upon the relative position of the two satellites and pulsar [11].

From Eq. (3), the time of arrival (TOA) to the i-th and the j-th satellites can be transferred to SSB, respectively, as follows:

$$t_{\mathrm{SSB}}^{i} = t_{\mathrm{SC}}^{i} + \frac{1}{c}\left[\begin{array}{l} \boldsymbol{n}\cdot\boldsymbol{r}_{\mathrm{SC}}^{i} - \dfrac{\left(r_{\mathrm{SC}}^{i}\right)^2}{2D_0} + \dfrac{\left(\boldsymbol{n}\cdot\boldsymbol{r}_{\mathrm{SC}}^{i}\right)^2}{2D_0} \\ -\dfrac{\left(\boldsymbol{b}\cdot\boldsymbol{r}_{\mathrm{SC}}^{i}\right)}{D_0} + \dfrac{\left(\boldsymbol{n}\cdot\boldsymbol{b}\right)\left(\boldsymbol{n}\cdot\boldsymbol{r}_{\mathrm{SC}}^{i}\right)}{D_0} \end{array}\right] + \frac{2\mu_s}{c^3}\mathrm{In}\left|\frac{\boldsymbol{n}\cdot\boldsymbol{r}_{\mathrm{SC}}^{i} + \boldsymbol{r}_{\mathrm{SC}}^{i}}{\boldsymbol{n}\cdot\boldsymbol{b} + \boldsymbol{b}} + 1\right| \tag{4}$$

$$t_{\mathrm{SSB}}^{j} = t_{\mathrm{SC}}^{j} + \frac{1}{c}\left[\begin{array}{l} \boldsymbol{n}\cdot\boldsymbol{r}_{\mathrm{SC}}^{j} - \dfrac{\left(r_{\mathrm{SC}}^{j}\right)^2}{2D_0} + \dfrac{\left(\boldsymbol{n}\cdot\boldsymbol{r}_{\mathrm{SC}}^{j}\right)^2}{2D_0} \\ -\dfrac{\left(\boldsymbol{b}\cdot\boldsymbol{r}_{\mathrm{SC}}^{j}\right)}{D_0} + \dfrac{\left(\boldsymbol{n}\cdot\boldsymbol{b}\right)\left(\boldsymbol{n}\cdot\boldsymbol{r}_{\mathrm{SC}}^{j}\right)}{D_0} \end{array}\right] + \frac{2\mu_s}{c^3}\mathrm{In}\left|\frac{\boldsymbol{n}\cdot\boldsymbol{r}_{\mathrm{SC}}^{j} + \boldsymbol{r}_{\mathrm{SC}}^{j}}{\boldsymbol{n}\cdot\boldsymbol{b} + \boldsymbol{b}} + 1\right| \tag{5}$$

According to Eqs. (4) and (5), the relative observation model of two satellites receiving the same pulse signal can be simplified to first order, which is

$$n\cdot\Delta r_{ij} = c\cdot\delta t^{ij} \tag{6}$$

δt^{ij} is the time delay of the two satellites receiving the same pulse signal. In the navigation method based upon relative DTOA observation, the position determination error exerted by the ephemeris error and clock error will be obviously eliminated.

Based upon the difference of pulse arrival time, the observation model would be

$$\boldsymbol{y} = h(\boldsymbol{x}) + \boldsymbol{v} \tag{7}$$

where $h_i(\boldsymbol{x}) = \begin{bmatrix} n\cdot\Delta r_{1j} & n\cdot\Delta r_{2j} & \ldots & n\cdot\Delta r_{nj}\end{bmatrix}^{\mathrm{T}}$ is the observation matrix of the i-th satellite, and $\boldsymbol{v}$ is the observation noise.

When the libration point constellation consists of two satellites and observe three pulsars, the observation matrix can be related to the state with the $h(\boldsymbol{x}_t)$ matrix, which is

$$h(2x_t) = \begin{bmatrix} n_{1x} & n_{1y} & n_{1z} & & -n_{1x} & -n_{1y} & -n_{1z} & \\ n_{2x} & n_{2y} & n_{2z} & \mathbf{0}_{3\times3} & -n_{2x} & -n_{2y} & -n_{2z} & \mathbf{0}_{3\times3} \\ n_{3x} & n_{3y} & n_{3z} & & -n_{3x} & -n_{3y} & -n_{3z} & \end{bmatrix} \times \begin{bmatrix} x_A & y_A & z_A & v_{Ax} & v_{Ay} & v_{Az} & x_B & y_B & z_B & v_{Bx} & v_{By} & v_{Bz}\end{bmatrix}^{\mathrm{T}} \tag{8}$$

n_x, n_y, n_z are the position vectors of three pulsars.

Take pulsar ephemeris error as an example, the impact of DTOA observation on navigation precision will be discussed. $\boldsymbol{n}$ is assumed to be the real direction vector of a pulsar, and $\tilde{\boldsymbol{n}}$ is the direction of pulsar with an error of $\Delta\boldsymbol{n}$

$$\tilde{\boldsymbol{n}}=\boldsymbol{n}+\Delta\boldsymbol{n} \tag{9}$$

According to Eq. (3), the pulse arrival time at SSB can be simplified as:

$$\tilde{t}_{\mathrm{SSB}} = t_{\mathrm{SC}} + \frac{1}{c}\tilde{\boldsymbol{n}} \cdot \boldsymbol{r}_{\mathrm{SC}} \tag{10}$$

Δt_{SSB} is the deviation of t_{SSB} caused by pulsar ephemeris error $\Delta\boldsymbol{n}$

$$\Delta t_{\mathrm{SSB}}=\frac{1}{c} \cdot \Delta\boldsymbol{n} \cdot \boldsymbol{r}_{\mathrm{SC}} \tag{11}$$

According to Eq. (6), when the measurement is DTOA instead of TOA, the deviation of the pulse arrival time can be written as:

$$\Delta t^{ij}_{\mathrm{SSB}}=\frac{1}{c} \cdot \Delta\boldsymbol{n} \cdot \boldsymbol{r}^{ij}_{\mathrm{SC}} \tag{12}$$

$\boldsymbol{r}_{\mathrm{SC}}$ is the distance between the spacecraft and SSB, and $\boldsymbol{r}^{ij}_{\mathrm{SC}}$ is the relative position of the i-th and j-th spacecraft. Compared with $\boldsymbol{r}_{\mathrm{SC}}$, $\boldsymbol{r}^{ij}_{\mathrm{SC}}$ is reduced by orders of magnitude. Thus, the position error caused by 1 mas' pulsar ephemeris error can be reduced from hundreds of meters to several meters.

3.2 Observation Equation Based on Inter-Satellite Links

The inter-satellite links observation can be represented by:

$$z = g(\boldsymbol{x}) + \boldsymbol{v} \tag{13}$$

where for constellation which consists of n spacecrafts, $g(\boldsymbol{x}) = [g_{12}(\boldsymbol{x})\ldots g_{ij}(\boldsymbol{x})\ldots g_{n-1,n}(\boldsymbol{x})]\ i,j = 1,2,\ldots,n,$

$$g(\boldsymbol{x}_{ij}) = \left|\boldsymbol{r}_i - \boldsymbol{r}_j\right| + c\left(\delta t_i - \delta t_j\right) \tag{14}$$

The simulation of inter-satellite links in this paper is simplified, and each satellite is considered to be visible to other satellites.

4 Filtering Algorithm

The unscented Kalman filter (UKF) is used to estimate the states of satellites in libration point constellation. The steps of UKF are as follows:

1. Initialization of the state $\hat{\mathbf{x}}_0$ and the covariance matrix $\mathbf{P}_0$.

$$\hat{\mathbf{x}}_0 = E(\mathbf{x}_0) \quad \mathbf{P}_0 = E\left[(\mathbf{x}_0 - \hat{\mathbf{x}}_0)(\mathbf{x}_0 - \hat{\mathbf{x}}_0)^{\mathrm{T}}\right] \tag{15}$$

2. Calculation of the sigma points and weight.

When the libration point constellation consists of two satellites, the dimension of the state vector is 12×1, thus the amount of sigma points is 25. Sigma points $\chi_{0,k-1}, \chi_{1,k-1}, \ldots, \chi_{25,k-1}$ can be calculated according to $\hat{\mathbf{x}}_{k-1}$ and $\mathbf{P}_{k-1}$. $\omega_0^m, \omega_1^m, \ldots, \omega_{25}^m$ are the weight of mean and $\omega_1^c, \ldots, \omega_{25}^c$ are the weight of covariance;

$$\begin{cases} \chi_{0,k-1} = \hat{\mathbf{x}}_{k-1} \\ \chi_{i,k-1} = \hat{\mathbf{x}}_{k-1} + \sqrt{n+\lambda} \cdot \left(\sqrt{\mathbf{P}_{k-1}}\right)_i \quad i = 1, 2, \ldots, n \\ \chi_{i+n,k-1} = \hat{\mathbf{x}}_{k-1} - \sqrt{n+\lambda} \cdot \left(\sqrt{\mathbf{P}_{k-1}}\right)_i \quad i = n+1, \ldots, 2n \end{cases} \tag{16}$$

The weight coefficients for calculating main value and covariance after the unscented transformation are

$$\begin{cases} \omega_0^m = \frac{\lambda}{n+\lambda} \\ \omega_0^c = \frac{\lambda}{n+\lambda} + (1 - \alpha^2 + \beta) \\ \omega_i^m = \omega_i^c = \frac{1}{2(n+\lambda)}, \quad i = 1, \ldots, 2n \end{cases} \tag{17}$$

where n is the dimension of state vector, $i = 1, \ldots, 2n$. $\lambda = \alpha^2(n+k) - n$ is a filtering parameter and α is used to control the distribution of sigma points $(0 \le \alpha \le 1)$; For a multi-variable system, k is generally taken as $k = 3 - n$. β is a parameter related to the prior distribution of state. For Gaussian distribution, β is generally taken as 2.

3. Time update

$$\begin{aligned} \chi_{k|k-1} &= f(\chi_{k-1}) \\ \mathbf{x}_k^- &= \sum_{i=0}^{2n} \omega_i^m \chi_{i,k|k-1} \\ \mathbf{P}_k^- &= \sum_{i=0}^{2n} \omega_i^c \left[\chi_{i,k|k-1} - \mathbf{x}_k^-\right]\left[\chi_{i,k|k-1} - \mathbf{x}_k^-\right]^{\mathrm{T}} + \mathbf{Q}_k \\ \mathbf{Z}_{i,k|k-1} &= \boldsymbol{H}\left(\chi_{i,k|k-1}, \boldsymbol{U}_k, k\right) \\ \mathbf{Z}_k^- &= \sum_{i=0}^{2n} \omega_i^c \mathbf{Z}_{i,k|k-1} \end{aligned} \tag{18}$$

4. Measurement update`

$$
\begin{aligned}
\mathbf{P}_{\tilde{\mathbf{Z}}_k\tilde{\mathbf{Z}}_k} &= \sum_{i=0}^{2n} \omega_i^c \left[\mathbf{Z}_{i,k|k-1} - \hat{\mathbf{Z}}_k^-\right]\left[\mathbf{Z}_{i,k|k-1} - \hat{\mathbf{Z}}_k^-\right]^{\mathrm{T}} + \mathbf{R}_k \\
\mathbf{P}_{\tilde{\mathbf{X}}_k\tilde{\mathbf{Z}}_k} &= \sum_{i=0}^{2n} \omega_i^c \left[\chi_{i,k|k-1} - \mathbf{X}_k^-\right]\left[\mathbf{Z}_{i,k|k-1} - \hat{\mathbf{Z}}_k^-\right]^{\mathrm{T}} + \mathbf{R}_k \\
\mathbf{K}_k &= \mathbf{P}_{\tilde{\mathbf{X}}_k\tilde{\mathbf{Z}}_k}\mathbf{P}_{\tilde{\mathbf{Z}}_k\tilde{\mathbf{Z}}_k}^{-1} \\
\mathbf{K}_k &= \mathbf{P}_{\tilde{\mathbf{X}}_k\tilde{\mathbf{Z}}_k}\mathbf{P}_{\tilde{\mathbf{Z}}_k\tilde{\mathbf{Z}}_k}^{-1} \\
\hat{\mathbf{X}}_k &= \hat{\mathbf{X}}_k^- + \mathbf{K}_k\left(\mathbf{Z}_k - \hat{\mathbf{Z}}_k^-\right) \\
\mathbf{P}_k &= \mathbf{P}_k^- - \mathbf{K}_k\mathbf{P}_{\tilde{\mathbf{Z}}_k\tilde{\mathbf{Z}}_k}\mathbf{K}_k^{\mathrm{T}}
\end{aligned}
\tag{19}
$$

The position and velocity of a satellite can be estimated by the UKF, thus the autonomous navigation of libration point orbits can be realized.

5 Simulation Results

For Halo orbits, in the Sun–Earth rotation frame, the initial states of the spacecrafts in the Sun–Earth rotation are (146,407,196.6, 0, −800,000 km) and (0, 356.83, 0 m/s). The interval of pulsar observation time is one hour, and the total simulation time is 150 days. Pulsar ephemeris error is 0.1 mas, with the initial position error (10, 10, 10 km) and velocity error (10, 10, 10 m/s). The noise of the position and velocity in the dynamic model are respectively 10^{-8} m and 10^{-8} m/s. The measurement noise of TOA is 1 μs. The figure of standard orbits of two spacecrafts in the Sun–Earth rotation frame is depicted in Fig. 2.

PSR B1937+21, PSR B1821−24 as well as PSR B0531+21 are chosen as the navigation sources, with an ephemeris error of 0.1mas. Table 1 lists the parameters of the corresponding pulsars.

Figures 3 and 4 show the simulation results when there are two satellites in the halo orbit constellation.

As the estimation results shown in Figs. 3 and 4, X-ray pulsar-based navigation is a feasible way for autonomous navigation of libration point constellation. Compared with the navigation method based on inter-satellite links, it is obvious to see the relative pulsar-based method can decrease the estimation error in a faster speed. The precision getting from inter-satellite links observation is lower than that from DTOA observation, however, the combination of both DTOA and inter-satellite links can obviously increase the navigation precision.

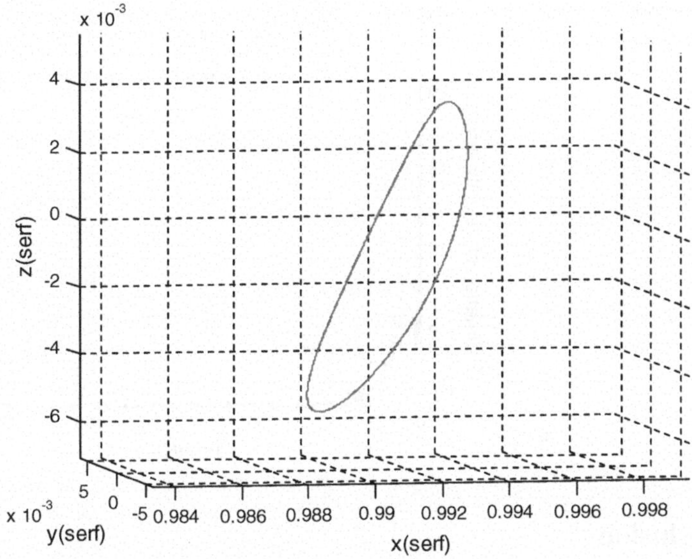

Fig. 2 Standard orbit of constellation in Sun–Earth rotation frame

Table 1 List of X-ray pulsar parameters

	Frequency	Right ascension	Declination
PSR B1937+21	1/0.00156	294.91164	21.58386
PSR B1821−24	1/0.00305	276.13729	−24.86793
PSR B0531+21	1/0.0334	83.63852	22.01469

Fig. 3 *Curve* position error in three directions

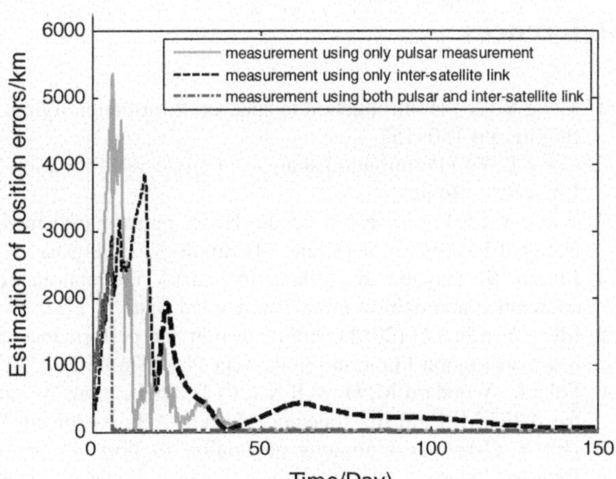

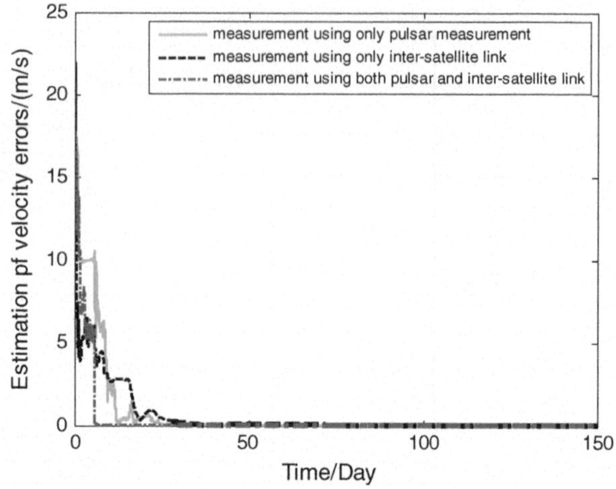

Fig. 4 *Curve* velocity error in three directions

6 Conclusion

The theory of constellation in libration point orbit is proposed in this paper, and simulations are done to prove its feasibility. The relative pulsar-based navigation method can efficiently reduce the effect of pulsar ephemeris error and clock error. Compared with the conventional navigation method based on inter-satellite links, relative pulsar-based navigation can capture the position of constellation effectively. The combination of inter-satellite links and DTOA is a feasible way for position and velocity determination of libration point constellation.

References

1. Meng Y (2014) Introduction to spacecraft formation flying. National Defense Industrial Press, Beijing, pp 150–152
2. Che Z (2008) Perturbation analysis of spacecraft formation flying around Halo orbit. Tsinghua University, Beijing
3. Wang Y (2011) Research on the X-ray pulsar-based navigation in deep space exploration. National University of Defense Technology, Changsha
4. Bhasin K, Hayden JL (2004) In: AIAA international communications satellite systems conference and exhibit monterey. United State, p 1180, 9–12 May 2004
5. Meng Y, Chen Q (2014) Outline design and performance analysis of navigation constellation near earth-moon libration point. Acta Phys Sin 63(24):1–11
6. Folta D, Woodard M, Howell K (2012) Applications of multi-body dynamical environments: the ARTEMIS transfer trajectory design. Acta Astronaut 73:237–249
7. Hill K (2007) Autonomous navigation in libration point orbits. University of Colorado, Boulder
8. Zheng W, Wang Y, Tang G, Liu L (2015) X-ray pulsar-based navigation: theory and applications. Science Press, Beijing, pp 1–3

9. Liu L (2014) Orientation parameters determination of the navigation satellites constellation via X-ray pulsars measurement. National University of Defense Technology, Changsha
10. Chester TJ, Butman SA (1981) Navigation using X-ray pulsar. NASA, Washington. NASA TR N81-27129, pp 22–25
11. Xiong K, Wei C, Liu L (2008) Research on autonomous navigation of satellite constellation based on X-ray pulsars. J Astronaut 29(2):545–549

Indoor Map Aiding/Map Matching Smartphone Navigation Using Auxiliary Particle Filter

Chunyang Yu, Haiyu Lan, Zhenbo Liu, Naser El-Sheimy and Fei Yu

Abstract A map-based pedestrian navigation method is presented in this research, integrating indoor map information and MEMS-based inertial sensors to enhance the performance of pedestrian indoor navigation. This paper demonstrates how Map Matching (MM) and Map Aiding (MA) methods constrain the navigation solution through an auxiliary particle filter. Previous studies examined each method individually, but the combination provides a more accurate position estimation. Additionally, this research introduces a novel cascade structure filter algorithm to reduce the computational burden and improve the estimation speed of APF by decreasing the update frequency of APF. The accuracy and availability of the proposed algorithm are tested and validated by performing experiments in various practical scenarios.

Keywords Map aiding · Map matching · Auxiliary particle filter · Cascade structure algorithm · Indoor pedestrian navigation

---

C. Yu (✉) · F. Yu (✉)
Harbin Engineering University, Harbin, China
e-mail: chunyang.yu@ucalgary.ca

F. Yu
e-mail: feiyu.heu@ucalgary.ca

C. Yu · H. Lan (✉) · Z. Liu (✉) · N. El-Sheimy (✉)
University of Calgary, Calgary, Canada
e-mail: hlan@ucalgary.ca

Z. Liu
e-mail: zhenbo.liu2@ucalgary.ca

N. El-Sheimy
e-mail: elsheimy@ucalgary.ca

J. Sun et al. (eds.), *China Satellite Navigation Conference (CSNC) 2016 Proceedings: Volume I*, Lecture Notes in Electrical Engineering 388,
DOI 10.1007/978-981-10-0934-1_29

1 Introduction

Indoor pedestrian navigation produces a convenient solution to manoeuvring through large indoor environments. The market demand for this system is high because it reduces the time spent locating the user and finding a specific location.

The indoor positioning technologies can be broken down into RF-based technologies and non-RF-based technologies [1]. Compared with RF-based technologies, non-RF-based technologies, including Pedestrian Dead Reckoning [2], Inertial Navigation, Signage and Maps, etc. [1], do not need to instal, maintain, or update. They are less time, money, and energy consuming. An inertial navigation system using inertial sensors is a totally self-contained navigation technology, which is not susceptible to external environments and does not need any preinstalled information [3]. In addition, Micro-electromechanical Systems (MEMS) sensors require lower power consumption, and are smaller in size, and lighter in weight making them readily portable [4]. However, the stand-alone INS system suffers from error drift, which has a significant influence on the final estimated solutions [5].

Currently, map-aided algorithms which do not need additional infrastructure such as power, devices or processing, have been widely used for pedestrian or vehicle navigation. Meanwhile, electronic maps of many public areas, such as airports, hospitals and universities, can be easily downloaded through location-service companies, such as Google™, Openstreet™, or through scanning Quick Response codes accessed on smart-devices. At present, almost all current indoor navigation products, such as Indoors, Meridian, Polestar, and so on, do not employ indoor maps information to calculate position; they merely utilize map information to plot the position solution or the destination in "presentation layer" of the basic algorithm. However, maps can be further used in calculation process through data fusion techniques [6–8]. Previous studies implemented Map Matching (MM) and Map Aiding (MA) methods individually [2, 9, 10], in this paper, considering the characteristics of each method, both methods are applied to complement each other. Therefore, the pedestrian navigation system can make full use of indoor maps and provide a more accurate position estimation.

Additionally, APF is selected in this algorithm because that PF can directly use nonlinear systems and measurement models while also eliminating the need for system linearization. However, it suffers from computational burden and sample impoverishment problems. This research introduces a novel cascade structure filter algorithm to reduce the computational burden and improve the estimation speed of APF by decreasing its update frequency [11]. The accuracy and availability of the proposed algorithm are tested by performing experiments in various practical scenarios. Detail of this algorithm will be described in the following section.

2 Objective and Methodology

The ultimate aim of this research is to provide a reliable low-cost indoor pedestrian navigation solution that enables robust, continuous, and accurate positioning information in indoor environments with minimum expenditure.

2.1 APF/KF Cascade Structure Algorithm

The APF uses the Monte Carlo Approach to approximate the real posterior density through a set of particles, and it is flexible in exploiting aided information [12]. However, because it needs a large number of particles to perform the calculation, it suffers from computational burden which will significantly affect the estimation speed of the system. Compared with the PF, the Kalman Filter (KF) is relatively less complex mathematically and easily applicable. Pure PF pedestrian navigation iteration rely on the update frequency of user's walking speed to update the APF, a process which loses large amounts of data; the introduction of the KF allows the system to make full use of the data from the inertial sensors. Therefore, in this paper, a two-layer cascade structure algorithm is proposed Fig. 1 [11].

In Fig. 1, the cascade structure algorithm contains a lower KF whose update frequency is determined by the high-output frequency of inertial sensors and upper APF which has a relative lower frequency according to the step detection of the pedestrian. In the lower layer, non-holonomic constraints (NHC) or the zero velocity update (ZUPT) is used to correct the preliminary solution of INS depending on the MEMS sensors' condition. The pedestrian step length and heading derived from the lower filter solution are combined with the indoor map

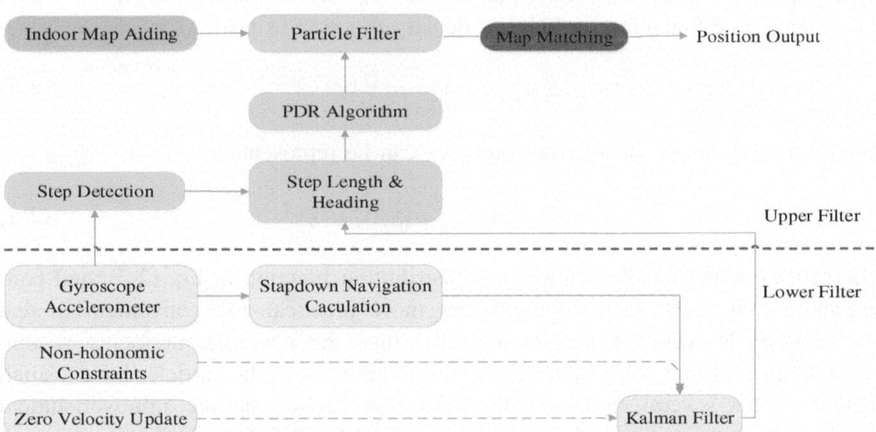

Fig. 1 Two-layer cascade structure algorithm using KF and PF

information through the APF in the upper filter. Finally, when it is suitable, MM method will be used to re-correct the APF estimated position. The detail of the NHC and the lower filter method used in this paper refers to [13, 14].

2.2 *The Upper Particle Filter*

In the upper layer, the indoor map information is used to correct the lower filter solution through APF. MA method is used to reset the weights of particles in the APF method, and MM method is employed to re-correct APF estimation solution.

2.2.1 Auxiliary Particle Filter

In this paper, the estimation of the state sequence is user's position, which is represented as $X_k = \{x_0, \ldots, x_k\}$; and the measurements can be represented as $Y_k = \{y_0, \ldots, y_k\}$. As an implementation of the Bayes method, PF uses a set of weighted particles to represent the posterior probability function $p(x_k|Y_k)$:

$$p(x_k|Y_k) \approx \sum_{i=1}^{N} w_k^i \delta\left(x_k - x_k^i\right) \tag{2.1}$$

where $\delta(\cdot)$ is the Dirac function, and the weight of particle x_k^i is expressed as [15]:

$$w_k^i \propto w_{k-1}^i \frac{p\left(y_k|x_k^i\right) p\left(x_k^i|x_{k-1}^i\right)}{q\left(x_k^i|x_{k-1}^i, y_k\right)} \tag{2.2}$$

where $q(\cdot)$ is the importance distribution function. The traditional SIR method uses the system state transition probability density function as [16]:

$$q\left(x_k^i|x_{k-1}^i, y_k\right) = p\left(x_k^i|x_{k-1}^i\right) \tag{2.3}$$

Therefore, the weight of different particles can be represented as

$$w_k^i \propto w_{k-1}^i p\left(y_k|x_k^i\right) \tag{2.4}$$

SIR method simplifies the importance distribution function in Eq. (2.2), and consequently makes the estimation process more practical and convenient in real applications. However, Eq. (2.4) does not take the new measurements y_k into consideration, which will lead to the situation that most of the particles have a small weight. After several iterations this situation causes sample impoverishment problem. Therefore, the APF method is proposed, which considers the measurements regarding the importance distribution function. APF performs pretests to screen out effective particles from the current particles. Specifically, the particles that can generate effective particles after system propagation in the next step will be kept. Otherwise, the corresponding particle will be deleted [17].

Additionally, the PDR method is used to update the pedestrian's position, and the system model of the upper filter is

$$\begin{cases} \mathrm{E}_{k+1}^{i} = \mathrm{E}_{k+1} + S_{k}^{i} \sin(\psi_{k}^{i}) \\ \mathrm{N}_{k+1}^{i} = \mathrm{N}_{k+1} + S_{k}^{i} \cos(\psi_{k}^{i}) \end{cases} \tag{2.5}$$

where E_{k+1} and N_{k+1} are the position estimation along local East and North at step $k + 1$; S_k^i is the estimated pedestrian step length from step k to $k + 1$; ψ_k^i is the estimated heading at time step k, detail about the PDR method can refer to [18].

2.2.2 Map Matching and Map Aiding Algorithm

MM and MA are two commonly used methods to combine map information with the INS solution. MM is widely used for fixed trajectory applications, such as outdoor vehicle navigation [10], and MA is more flexible and does not require any assumptions about the trajectory, but it is not as powerful as MM [19]. Given this background, both methods are used in this paper to complement each other. The MA algorithm is used to perform the measurement update process of APF through the "cross-wall" method. To be specific, after system propagation, if the new particle passes the wall of the building without passing through the door, then this particle is invalid and its weight will be assigned as zero, which means this particle is deleted from the filter. If a wall is not intersected by the particle during the propagation step, the particle will be remained in the next step.

MM will be added on the MA-based APF solution, if a pedestrian walks in a corridor where his trajectory is fixed. The indoor map date used in this paper was in shape file, and provided by the University of Calgary (U of C) as part of the Arc Geographic Information System (ArcGIS) software. Using the Mapping Toolbox of Matlab, this shape file map data could be easily loaded and converted into x and y coordinates in meters, and saved as vectors in matrixes. Depending on the format of indoor map, the point-to-curve matching algorithm is used in this paper. The algorithm projects the estimated location, P, to the closest link in the network using the following distance form equation [10, 20]:

$$\text{Distance} = \frac{|X_e(Y_1 - Y_2) - Y_e(X_1 - X_2) + (X_1Y_2 - X_2Y_1)|}{\sqrt{(X_1 - X_2)^2 + (Y_1 - Y_2)^2}} \tag{2.6}$$

$$X_P = \frac{[X_e(X_2 - X_1) + Y_e(Y_2 - Y_1)] + (Y_2 - Y_1)(X_1Y_2 - X_2Y_1)}{[(X_1 - X_2)^2 + (Y_1 - Y_2)^2](X_2 - X_1)^{-1}} \tag{2.7}$$

$$Y_P = \frac{[X_e(X_2 - X_1) + Y_e(Y_2 - Y_1)] + (X_2 - X_1)(X_1Y_2 - X_2Y_1)}{[(X_1 - X_2)^2 + (Y_1 - Y_2)^2](Y_2 - Y_1)^{-1}} \quad (2.8)$$

in which (X_e, Y_e) is the position estimated from the MA-based APF algorithm, (X_1, Y_1) and (X_2, Y_2) are the starting points and ending points of a line segment in the digital map. Equation (2.6) is used to detect which line segment the estimated position should be projected to; Eqs. (2.7) and (2.8) are used to project coordinates (X_e, Y_e) to obtain the foot of a perpendicular (X_P, Y_P).

3 Experiment and Results

The validity and feasibility of the proposed algorithm were validated through conducting ground experiments in different scenarios (the third floor of the Calgary Centre for Information Technology (CCIT) building at the UC; the second scenario is on the first floor of the Energy Environment Experiential Learning (EEEL) building at the UC). The sensor data was obtained from Android sensor framework of Xiaomi which has the following MEMS sensors (Table 1) [21]:

To quantitatively show the performance of the proposed algorithm, we locked the user's walking heading for the PDR update at a specific value according to the indoor map to get a reference trajectory [14]. In this way, a reference trajectory for the user can be obtained to compute the estimation errors of the proposed algorithm. Figure 2 is the designed trajectory for test 1, and, Fig. 3 is its INS-derived trajectory. Figure 4 is the MA-based APF method derived estimation results for test 1 in the CCIT building from the proposed two-layer algorithm. Figure 5 is the MA\MM based APF solution. Figure 6 shows the designed walking trajectories for scenario 2. Figure 7 is the lower filter derived solution for this test. Figure 8 is the MA-based APF method derived estimation results from the proposed two-layer algorithm. Figure 9 is the MA\MM based APF solution.

Compare Fig. 3 with Figs. 4 and 8 with Fig. 7, it is easy to see that Map aiding APF method can significantly improve the estimation accuracy of the estimated position. Figures 5 and 9 show that in the corridor area, after adding MM method, the calculated position can be further increased compared with the pure MA method.

Figures 10 and 11 shows the MM\MA estimated solutions presented on the EEEL digital map. Comparing the estimated trajectory with the reference trajectory

Table 1 Sensors of Xiaomi 3

Sensor	Manufacturer	Model
Accelerometer	InvenSense	MPU6050
Gyroscope	InvenSense	MPU6050
Compass	AKM	AK8963
Barometer	Bosch	BMP180

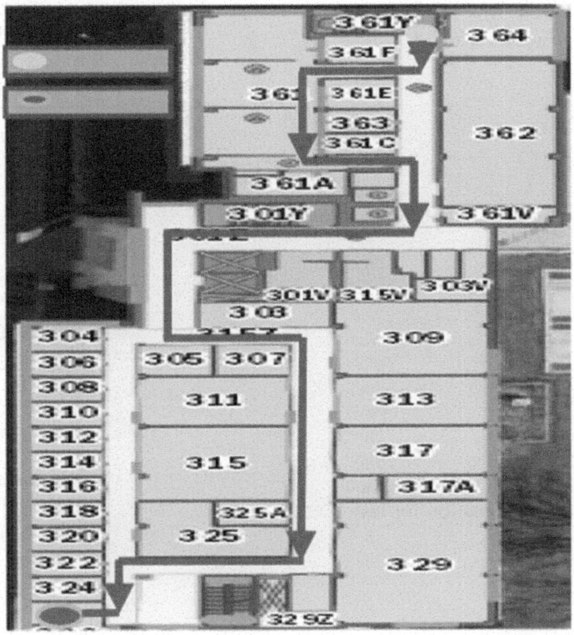

Fig. 2 Designed trajectory of experiments in CCIT building

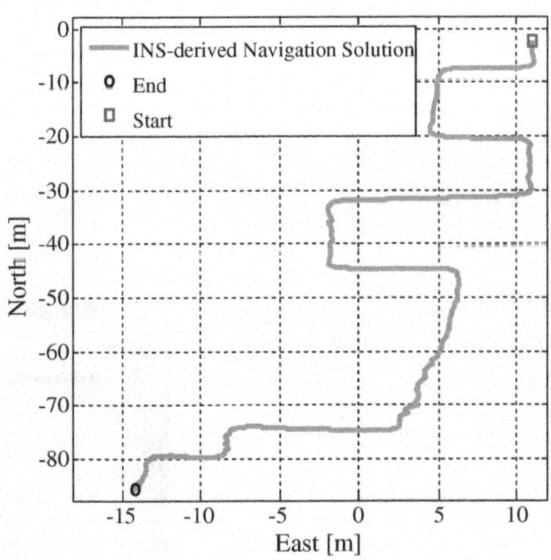

Fig. 3 INS derived position solution for test 1

designed using the method in [18], the RMS of the proposed method is less than 3 meters, which is an high accurate method according to classification of indoor navigation accuracy.

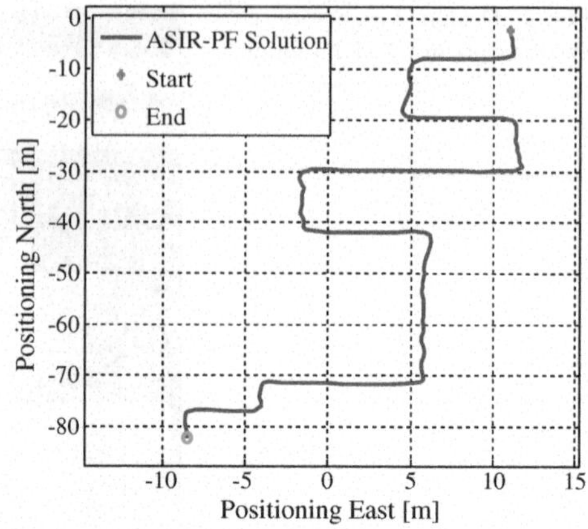

Fig. 4 MA methods derived position for test 1

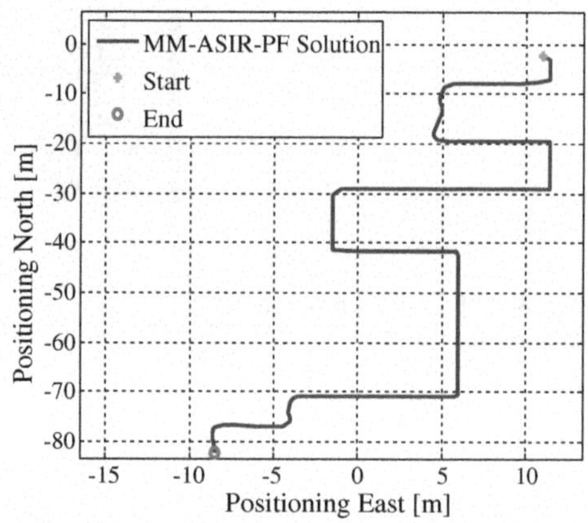

Fig. 5 MA-MM methods derived solution for test 1

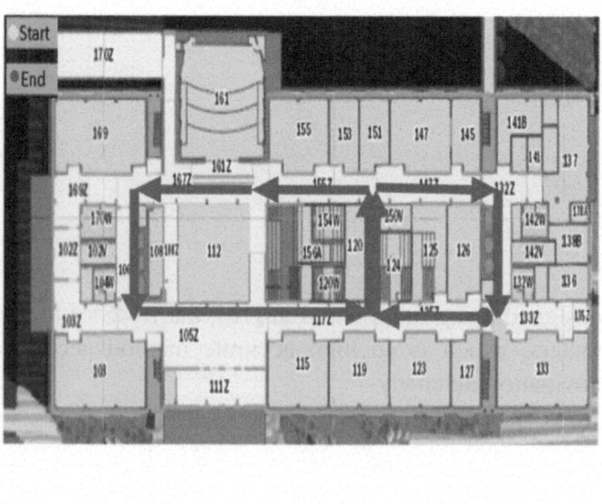

Fig. 6 Designed trajectory of experiments in EEEL building

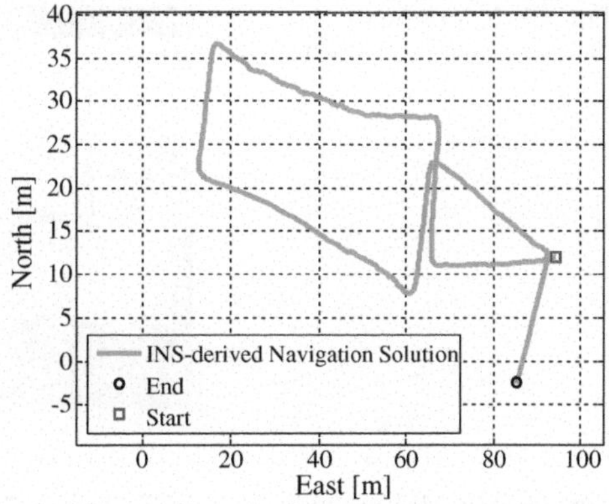

Fig. 7 INS derived position solution for test 2

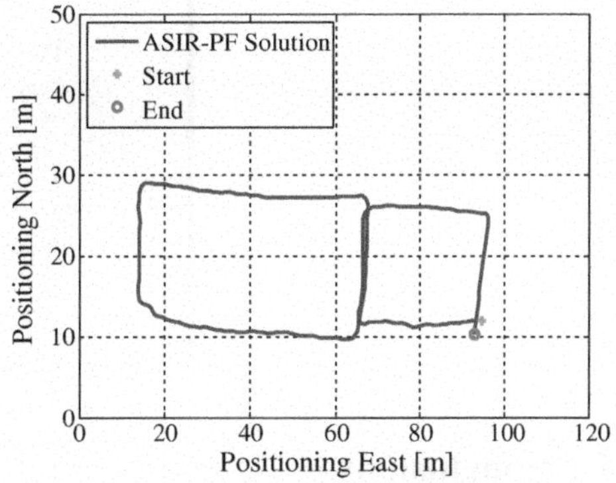

Fig. 8 MA methods derived position for test 2

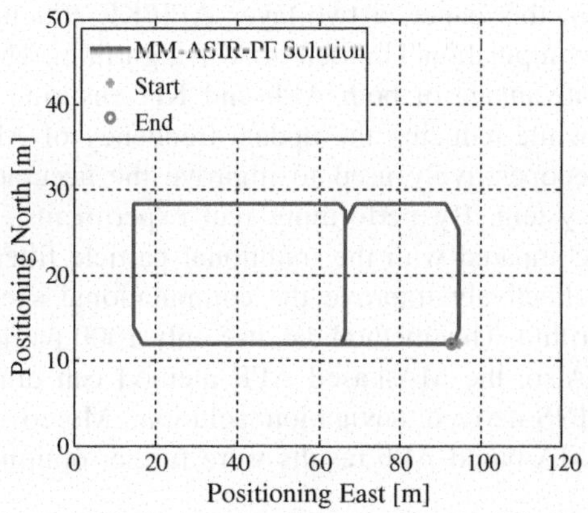

Fig. 9 MA-MM methods derived solution for test 2

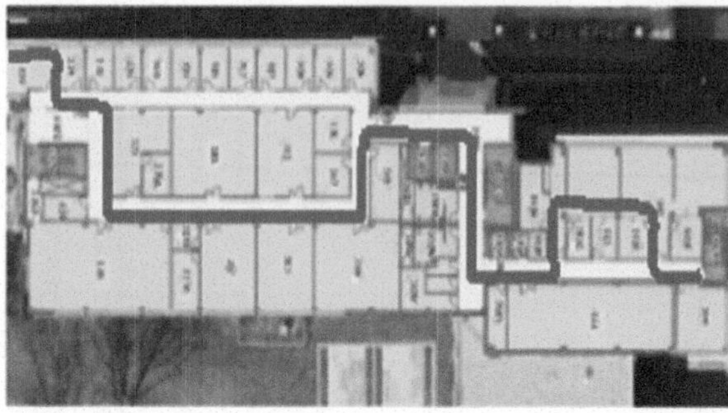

Fig. 10 MA methods derived position for test 1

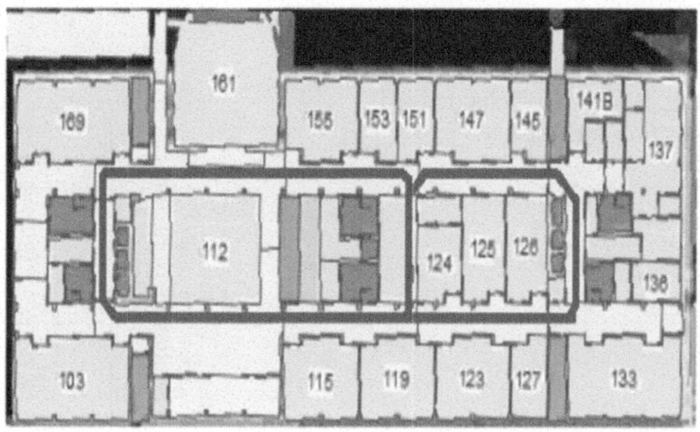

Fig. 11 MA-MM methods derived position for test 2

4 Conclusions

In this paper, a two-layer APF/KF structure algorithm is proposed to solve the computational burden of APF algorithm. The two-layer structure algorithm can take advantage of both APF and KF, ensuring the inertial sensor data utilization rate while reducing the update frequency of APF. Besides, MM and MA method are cooperatively used to improve the accuracy of the indoor pedestrian navigation system. By performing real experiments, we can easily obtain the conclusions: compared with the traditional particle filter, the proposed two-layer structure can effectively improve the computational speed of the APF-based navigation algorithm. This method can use only 1000 particles to complete the estimation process. Also, the MA-based APF method can dramatically improve the accuracy of the INS-derived navigation solution. Moreover by adding the MM algorithm, the MA-based APF results were further optimized.

References

1. Karimi HA (2015) Indoor wayfinding and navigation
2. Bojja J, Collin J, Särkkä S, Takala J (eds) (2015) Pedestrian localization in moving platforms using dead reckoning, particle filtering and map matching. In: Proceedings of IEEE international conference on acoustics, speech and signal processing
3. Titterton D, Weston JL (2004) Strapdown inertial navigation technology. IET
4. Noureldin A, Karamat TB, Georgy J (2012) Fundamentals of INS, GPS and their integration. Springer Science & Business Media, Heidelberg
5. Skog I, Händel P, Nilsson J-O, Rantakokko J (2010) Zero-velocity detection—an algorithm evaluation. IEEE Trans Biomed Eng 57(11):2657–2666
6. Link JAB, Smith P, Viol N, Wehrle K (eds) (2011) Footpath: accurate map-based indoor navigation using smartphones. In: 2011 international conference indoor positioning and indoor navigation (IPIN). IEEE
7. Klepal M, Beauregard S (eds) (2008) A novel backtracking particle filter for pattern matching indoor localization. In: Proceedings of the first ACM international workshop on mobile entity localization and tracking in GPS-less environments. ACM
8. Beauregard S, Klepal M (eds) (2008) Indoor PDR performance enhancement using minimal map information and particle filters. In: Position, location and navigation symposium, 2008 IEEE/ION. IEEE
9. White CE, Bernstein D, Kornhauser AL (2000) Some map matching algorithms for personal navigation assistants. Transp Res C Emerg Technol 8(1):91–108
10. Cossaboom M, Georgy J, Karamat T, Noureldin A (2012) Augmented Kalman filter and map matching for 3D RISS/GPS integration for land vehicles. Int J Navig Obs
11. Krach B, Roberston P (eds) (2008) Cascaded estimation architecture for integration of foot-mounted inertial sensors. In: Position, location and navigation symposium, 2008 IEEE/ION. IEEE
12. Simon D (2006) Optimal state estimation: Kalman, H infinity, and nonlinear approaches. Wiley, New York
13. Lan H, Yu C, Zhuang Y, Li Y, El-Sheimy N (2015) A novel Kalman Filter with state constraint approach for the integration of multiple pedestrian navigation systems. Micromachines 6(7):926–952
14. Lan H, El-Sheimy N (eds) (2014) A state constraint Kalman filter for pedestrian navigation with low cost MEMS inertial sensors. In: Proceedings of the 27th international technical meeting of the satellite division of the institute of navigation (ION GNSS + 2014), Tampa, FL, USA
15. Arulampalam MS, Maskell S, Gordon N, Clapp T (2002) A tutorial on particle filters for online nonlinear/non-Gaussian Bayesian tracking. IEEE Trans Signal Process 50(2):174–188
16. Gustafsson F, Gunnarsson F, Bergman N, Forssell U, Jansson J, Karlsson R et al (2002) Particle filters for positioning, navigation, and tracking. IEEE Trans Signal Process 50(2): 425–437
17. Pitt MK, Shephard N (1999) Filtering via simulation: auxiliary particle filters. J Am Stat Assoc 94(446):590–599
18. Lan H, Yu C, El-Sheimy N (eds) (2015) An integrated PDR/GNSS pedestrian navigation system. In: China satellite navigation conference (CSNC) 2015 Proceedings, vol III. Springer
19. Quddus MA, Ochieng WY, Zhao L, Noland RB (2003) A general map matching algorithm for transport telematics applications. GPS Solutions 7(3):157–167
20. Deng Z, Ruan F, Lu S, Zheng R, Zeng H, Fang Y (2015) A method of map matching based on particle filter in indoor positioning
21. Technology Y (2014) The inertial sensors and environmental sensors in Xiaomi 2014. Available from: http://www.ewisetech.com/Pages/Article/9.aspx. [cited 21 Jan 2014]

A Research of Targets Tracking and Positioning Algorithm Based on Multi-feature Fusion

Pengfeng Chen and Long Zhao

Abstract In order to improve the accuracy and robustness of targets tracking and positioning, this paper proposes a particle filter algorithm based on multi-feature fusion. According to the diversity of character information, a multi-feature fusion strategy based on color, texture, and edge character has been developed, which can realize the comprehensive utilization of various visual features information. A feature criterion function has been addressed to the weighted strategy of multi-feature fusion, which can adjust adaptively the weight of feature and enhance the reliability of target tracking. Combined with binocular stereo vision technology, this algorithm can locate the target by calculating the geometrical relationship between the correspondence pixel points and spatial points. The test results show that the algorithm can realize the target tracking and positioning more accurately.

Keywords Multi-feature fusion · Adaptive weights · Particle filter · Target tracking · Binocular stereo vision · Target positioning

1 Introduction

Target tracking and locating is a prosperous research filed in computer vision which can be applied to many fields such as the pedestrian tracking [1], security surveillance, intelligent transportation, military, etc. However, restricted by the complex environment, positioning accuracy and device factor, the target tracking and locating methods still remain to be further improved.

P. Chen
Digital Navigation Center, Beihang University, Beijing, China

L. Zhao (✉)
Science and Technology on Aircraft Control Laboratory, Beihang University, Beijing, China
e-mail: buaa_dnc@buaa.edu.cn

J. Sun et al. (eds.), *China Satellite Navigation Conference (CSNC) 2016 Proceedings: Volume I*, Lecture Notes in Electrical Engineering 388,
DOI 10.1007/978-981-10-0934-1_30

Tracking location research can be divided into two parts: target tracking and target positioning. In target tracking, there is no single visual feature that can describe the target information completely. Tracking algorithms based on single feature, therefore, are often limited by the practical application environment. According to the intrinsic differences between different features, a feasible solution can be obtained by multi-feature fusion [2] to improve the accuracy and robustness of tracking algorithm. In the research of target positioning, the commonly used methods include global positioning system (GPS), Bluetooth, radio frequency identification (RFID), infrared, computer vision, etc. GPS signal is easily weakened in indoor or high dense-constructed area; Infrared positioning technology needs to launch modulation infrared ray, it can only support the line-of-sight transmission because the ray cannot through the obstacles. In addition, most position technologies need expensive reception terminal devices and relevant communication sensors for targets, so that they are limited to be widely used. Traditional positioning technology [3] based on computer vision can be divided into three sub-processes, i.e., stereo matching, three dimensional reconstruction, and calculating spatial coordinates. Because the stereo vision matching needs to spend much time to extract the future points and has mismatching feature points, real-time performance and robustness of the algorithm are reduced.

In order to solve this problem, an algorithm combined the adaptive weighting particle filter algorithm and multi-feature fusion is presented according to the observation probability density fusion of the particle filter algorithm. In tracking model, color histograms [4], local binary pattern (LBP) and edge character were chosen to describe the target area. Color histogram, LBP, and edge character describe the color distribution of target area, texture information, and the association between the surrounding pixels and target edge information, respectively. In the fusion strategy, the adaptive weighting can be obtained through calculating the probability density distribution of corresponding characteristics for the target area and background area. The adaptive weighting particle filter algorithm based on multi-feature fusion, not only focuses on the similarity between the candidate area and the target area, but also considers distinction between target region and background region. The proposed tracking algorithms can enhance the target tracking precision, and estimate the centroid position of motion target in the image, which can avoid the complex stereo vision matching process in visual positioning. For this reason, the spatial location of target can be calculated by using binocular vision technology and the performance of the algorithm presented in this paper is verified by a practical experimental test.

The paper is organized as follows. The feature extraction is presented in Sect. 2. Section 3 focuses on the target tracking with multi-feature fusion. Section 4 states how to calculate the spatial location. Section 5 gives the experiment analysis. Finally, we draw some conclusions and shed light on future work in Sect. 6.

2 Target Feature Extraction

In the image sequences, there are many typical characteristics that can be used for target tracking. Because each character has its own special properties and different sensitivity to the environment, the choice of characteristics is very significant problem. Taking the distinguishability, stability, and independence of the features into account, the color histogram, LBP, and edge character are chosen as target visual features in this paper.

2.1 *Color Histogram*

Color histogram describes the color distribution in the form of histogram by computing the value of each pixel in the target area. In order to increase the reliability of the distribution, a kernel function is used to compute the weights for every pixel of the target. Then the target model are described as follows:

$$q_{\text{color}}(u) = f\frac{1}{N}\sum_{i=1}^{N} k\left(\left\|\frac{x_i - x_0}{h}\right\|\right)\delta[b(x_i) - u], \quad u = 1, 2, \ldots, m \tag{2.1}$$

$$k(x) = \begin{cases} 1 - x^2 & x \leq 1 \\ 0 & \text{others} \end{cases} \tag{2.2}$$

$$f = 1 \Bigg/ \sum_{i=1}^{N} k\left(\left\|\frac{x_i - x_0}{h}\right\|\right) \tag{2.3}$$

where $k(x)$ is the weighted kernel function, δ is the Kronecker delta function, f is the normalization factor, and h is defined by $h = \sqrt{h_x^2 + h_y^2}$, in which h_x and h_y is the half width, and half height of the target area, respectively.

2.2 *Local Binary Pattern*

The LBP operator [5], shown as a powerful measure of image, can extract the texture feature of target easily. The LBP operator labels the pixel in an image by comparing its neighborhood with the center value and considering the results as a binary number (binary pattern). The LBP operator is described by

$$\text{LBP}(x_c, y_c) = \sum_{p=0}^{P-1} 2^p s(i_p - i_c) \tag{2.4}$$

where i_c denotes the gray value of the center pixel (x_c, y_c), i_p denotes the gray value of the neighborhood, and P is the number of its neighborhood. The function $s(x)$ is defined as follow

$$s(x) = \begin{cases} 1 & \text{if} \quad x \geq 0 \\ 0 & \text{if} \quad x < 0 \end{cases} \tag{2.5}$$

Calculate LBP value of each pixel according to Eq. (2.5), then the texture histogram of LBP character is normalized as

$$q_{\text{color}}(u) = \frac{1}{N} \sum_{i=1}^{N} \delta[b(x_i - u)] \tag{2.6}$$

where $b(x_i)$ is the index function, returning the histogram serial number of pixel x_i, N is the total number of pixels in target area.

2.3 Edge Feature

The edge feature [6] can be described by histogram of oriented gradient. In gray image, the horizontal edge $G_x(x_i)$ and vertical edge $G_y(x_i)$ can be acquired by using edge detection operators to calculate the edge information in X and Y direction, respectively. The gradient $G(x_i)$ and orientation θ of each pixel can be defined as

$$\begin{aligned} G(x_i) &= \sqrt{G_x^2(x_i) + G_y^2(x_i)} \\ \theta &= \arctan\left(G_y(x_i)/G_x(x_i)\right) \end{aligned} \tag{2.7}$$

The weighted oriented gradient histogram is calculated by dividing the orientation into m bins. The histogram of oriented gradient is given by

$$q_{\text{edge}}(u) = C \sum_{i=1}^{N} k\left(\left\|\frac{x_i - x_0}{h}\right\|\right) G(x_i) \delta[b(x_i) - u] \tag{2.8}$$

where kernel function $k(x)$ is shown as Eq. (2.2), the normalization factor C is defined as

$$C = 1 \Big/ \sum_{i=1}^{N} G(x_i) \tag{2.9}$$

3 Target Tracking with Adaptively Multi-feature Fusion

3.1 Multi-feature Fusion

After extracting the target features, the multi-feature fusion is realized by the Gaussian weighted strategy. Given the target state x_k and time k, the entire observation likelihood can be calculated by

$$p(z_k|x_k) = \exp\left\{-\frac{1}{2}\left(\alpha\frac{d_{\mathrm{color}}^2}{\sigma_{\mathrm{c}}^2} + \beta\frac{d_{\mathrm{texture}}^2}{\sigma_{\mathrm{t}}^2} + \gamma\frac{d_{\mathrm{edge}}^2}{\sigma_{\mathrm{e}}^2}\right)\right\} \tag{3.10}$$

where $\sigma_{\mathrm{c}}^2, \sigma_{\mathrm{t}}^2$ and σ_{e}^2 denote the noise variance of the color histogram, LBP character, edge feature, respectively; $d_{\mathrm{color}}, d_{\mathrm{texture}}$ and d_{edge} denote the similarity distance of the color histogram, LBP character, edge feature between target area, and candidate area; α, β and γ are the weights of the color, texture, and edge features, respectively.

The similarity distance d_{f} can be calculated by Bhattacharyya coefficient. During the current frame, assuming that the likelihood of candidate area is p_{f} while the target area is q_{f}, the similarity between candidate area and target area is described as follows

$$\rho_{\mathrm{f}}(q_{\mathrm{f}}, p_{\mathrm{f}}) = \sum_{u=1}^{m} \sqrt{q_{\mathrm{f}}(u)p_{\mathrm{f}}(u)}, \quad \mathrm{f} \in \{\mathrm{color}, \quad \mathrm{texture}, \quad \mathrm{edge}\} \tag{3.11}$$

where

$$d_{\mathrm{f}} = \sqrt{1 - \rho_{\mathrm{f}}(q_{\mathrm{f}}, p_{\mathrm{f}})}, \quad \mathrm{f} \in \{\mathrm{color}, \quad \mathrm{texture}, \quad \mathrm{edge}\} \tag{3.12}$$

The smaller value of d_{f} indicates the higher similarity of character between candidate area and target area, then particle gets bigger weighting in Eq. (3.10).

3.2 Adaptive Weights

If multiple features are characterized by equal or fixed weighting without discrimination in multi-feature fusion, the target tracking would be seriously disturbed by some features, which is sensitive on the occasion of varying environment. In this paper, an evaluating function, used to reflect the ability of identification, has been addressed to calculate the weights of features [7].

Based on the sum of intraclass and interclass variance, the evaluation function for the effectiveness of characteristic is defined as follow:

$$\mathrm{var}\left(q_{\mathrm{f}};\ \mu_{\mathrm{f}}^{\mathrm{t}}, \mu_{\mathrm{f}}^{\mathrm{b}}\right) = \frac{w_{\mathrm{f}}^{\mathrm{t}} w_{\mathrm{f}}^{b}\left(\mu_{\mathrm{f}}^{\mathrm{t}} - \mu_{\mathrm{f}}^{\mathrm{b}}\right)^2}{w_{\mathrm{f}}^{\mathrm{t}}\sigma_{\mathrm{ft}}^2 + w_{\mathrm{f}}^{\mathrm{b}}\sigma_{\mathrm{fb}}^2}, \quad \mathrm{f} \in \{\mathrm{color}, \quad \mathrm{texture}, \quad \mathrm{edge}\} \tag{3.13}$$

where $\mu_{\mathrm{f}}^{\mathrm{t}}$ and $\mu_{\mathrm{f}}^{\mathrm{b}}$ denote the average of target and background area's correspondence characteristic value; $w_{\mathrm{f}}^{\mathrm{t}}$ and $w_{\mathrm{f}}^{\mathrm{b}}$ denote the weights of target and background area's correspondence characteristic value; σ_{ft}^2 and σ_{fb}^2 denote the variance of target and background area's correspondence characteristic value. In Eq. (3.13), the molecules are the sum of interclass, the denominators is the sum of intraclass variance. According to pattern recognition theory, the larger variance in interclass, while the smaller variance in intraclass, the discernment for target and background is more robust.

As mentioned above, the weights can be calculated by

$$\begin{cases} \alpha = \dfrac{\mathrm{var}\left(q_{\mathrm{color}};\ \mu_{\mathrm{color}}^{\mathrm{t}}, \mu_{\mathrm{color}}^{\mathrm{b}}\right)}{\sum \mathrm{var}\left(q_{\mathrm{f}};\ \mu_{\mathrm{f}}^{\mathrm{t}}, \mu_{\mathrm{f}}^{\mathrm{b}}\right)} \\ \beta = \dfrac{\mathrm{var}\left(q_{\mathrm{texture}};\ \mu_{\mathrm{texture}}^{\mathrm{t}}, \mu_{\mathrm{texture}}^{\mathrm{b}}\right)}{\sum \mathrm{var}\left(q_{\mathrm{f}};\ \mu_{\mathrm{f}}^{\mathrm{t}}, \mu_{\mathrm{f}}^{\mathrm{b}}\right)} \\ \gamma = \dfrac{\mathrm{var}\left(q_{\mathrm{edge}};\ \mu_{\mathrm{edge}}^{\mathrm{t}}, \mu_{\mathrm{edge}}^{\mathrm{b}}\right)}{\sum \mathrm{var}\left(q_{\mathrm{f}};\ \mu_{\mathrm{f}}^{\mathrm{t}}, \mu_{\mathrm{f}}^{\mathrm{b}}\right)} \end{cases}, \quad f = \{\mathrm{color}, \quad \mathrm{texture}, \quad \mathrm{edge}\} \tag{3.14}$$

3.3 *Tracking Algorithm Based on Particle Filter*

This paper uses the color histogram, LBP texture, and histogram of oriented gradient to describe the target information and embeds it into the particle filter tracking.

The diagram of the tracking algorithm is shown in Fig. 1.

4 Target Positioning Based on Binocular Vision

After robust tracking, centroid coordinates of targets in binocular video sequences can be acquired. On the basis, the spatial location of target can be calculated with the use of binocular vision technology [8].

In stereo vision system, an arbitrary point $P(X_{\mathrm{w}}, Y_{\mathrm{w}}, Z_{\mathrm{w}})$ is fallen the left and right image on each $P_{\mathrm{L}}(u_{\mathrm{l}}, v_{\mathrm{l}})$ and $P_{\mathrm{R}}(u_{\mathrm{r}}, v_{\mathrm{r}})$. Assume that projection matrix M_{L} and

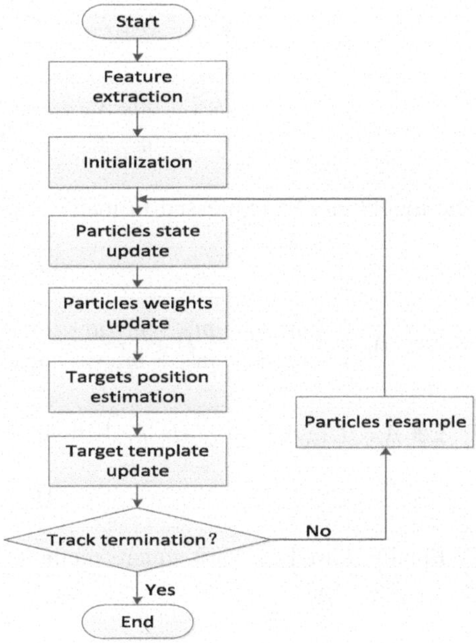

Fig. 1 The structure of tracking algorithm based on particle filter

M_R denote the geometric relationship from P to P_L, from P to P_R, respectively, then model is shown as follows

$$z_{cl}\begin{bmatrix} u_l \\ v_l \\ 1 \end{bmatrix} = \begin{bmatrix} m_{11}^l & m_{12}^l & m_{13}^l & m_{14}^l \\ m_{21}^l & m_{22}^l & m_{23}^l & m_{24}^l \\ m_{31}^l & m_{32}^l & m_{33}^l & m_{34}^l \end{bmatrix} \begin{bmatrix} X_w \\ Y_w \\ Z_w \\ 1 \end{bmatrix} = M_1 \begin{bmatrix} X_w \\ Y_w \\ Z_w \\ 1 \end{bmatrix} \tag{4.15}$$

$$z_{cr}\begin{bmatrix} u_r \\ v_r \\ 1 \end{bmatrix} = \begin{bmatrix} m_{11}^r & m_{12}^r & m_{13}^r & m_{14}^r \\ m_{21}^r & m_{22}^r & m_{23}^r & m_{24}^r \\ m_{31}^r & m_{32}^r & m_{33}^r & m_{34}^r \end{bmatrix} \begin{bmatrix} X_w \\ Y_w \\ Z_w \\ 1 \end{bmatrix} = M_2 \begin{bmatrix} X_w \\ Y_w \\ Z_w \\ 1 \end{bmatrix} \tag{4.16}$$

where M_i, $i = L, R$ is a 3×4 matrix, Z_{cl} and Z_{cr} denote the coordinates of the Z axis of P in the correspondence camera coordinate system.

To further simplify the formula, the model be calculated as follows

$$\begin{cases} u_l = \dfrac{m_{11}^l X_w + m_{12}^l Y_w + m_{13}^l Z_w + m_{14}^l}{m_{31}^l X_w + m_{32}^l Y_w + m_{33}^l Z_w + m_{34}^l} \\[2ex] v_l = \dfrac{m_{21}^l X_w + m_{22}^l Y_w + m_{23}^l Z_w + m_{24}^l}{m_{31}^l X_w + m_{32}^l Y_w + m_{33}^l Z_w + m_{34}^l} \end{cases} \tag{4.17}$$

$$\begin{cases} u_{\mathrm{r}} = \dfrac{m_{11}^{\mathrm{r}}X_{\mathrm{w}} + m_{12}^{\mathrm{r}}Y_{\mathrm{w}} + m_{13}^{\mathrm{r}}Z_{\mathrm{w}} + m_{14}^{\mathrm{r}}}{m_{31}^{\mathrm{r}}X_{\mathrm{w}} + m_{32}^{\mathrm{r}}Y_{\mathrm{w}} + m_{33}^{\mathrm{r}}Z_{\mathrm{w}} + m_{34}^{\mathrm{r}}} \\ v_{\mathrm{r}} = \dfrac{m_{21}^{\mathrm{r}}X_{\mathrm{w}} + m_{22}^{\mathrm{r}}Y_{\mathrm{w}} + m_{23}^{\mathrm{r}}Z_{\mathrm{w}} + m_{24}^{\mathrm{r}}}{m_{31}^{\mathrm{r}}X_{\mathrm{w}} + m_{32}^{\mathrm{r}}Y_{\mathrm{w}} + m_{33}^{\mathrm{r}}Z_{\mathrm{w}} + m_{34}^{\mathrm{r}}} \end{cases} \tag{4.18}$$

The model can be expressed in matrix equation as

$$A_{\mathrm{i}}P = b_{\mathrm{i}} \tag{4.19}$$

where $A_{\mathrm{i}} = \begin{bmatrix} m_{31}^{\mathrm{i}}u_{\mathrm{i}} - m_{11}^{\mathrm{i}} & m_{32}^{\mathrm{i}}u_i - m_{12}^{\mathrm{i}} & m_{33}^{\mathrm{i}}u_{\mathrm{i}} - m_{13}^{\mathrm{i}} \\ m_{31}^{\mathrm{i}}v_{\mathrm{i}} - m_{21}^{\mathrm{i}} & m_{32}^{\mathrm{i}}v_i - m_{22}^{\mathrm{i}} & m_{33}^{\mathrm{i}}u_{\mathrm{i}} - m_{23}^{\mathrm{i}} \end{bmatrix}$, $b_{\mathrm{i}} = \begin{bmatrix} m_{14}^{\mathrm{i}} - m_{34}^{\mathrm{i}}u_{\mathrm{i}} \\ m_{24}^{\mathrm{i}} - m_{34}^{\mathrm{i}}v_{\mathrm{i}} \end{bmatrix}$, $\mathrm{i} = \mathrm{l}, \mathrm{r}$.

Let $A = (A_{\mathrm{l}}, A_{\mathrm{r}})^{\mathrm{T}}, b = (b_{\mathrm{l}}, b_{\mathrm{r}})^{\mathrm{T}}$, then

$$AP = b \tag{4.20}$$

Finally, based on least square method, the coordinate of P can be obtained by $P = (A^{\mathrm{T}}A)^{-1}A^{\mathrm{T}}b$.

5 Experiment

To evaluate the performance of the proposed algorithm, a series of experiments are implemented. The experiments are divided into two parts. The former aims to prove that the accuracy and robustness of particle filter tracking, and the latter is to verify the effectiveness of the positioning.

The test video 1 which has complicated environment is from the Visual Tracker Benchmark (video is available at http://www.visual-tracking.net), and its resolution is 320×240 pixel. The test result is shown in Fig. 2.

As can be seen from Fig. 2, the particle filter based on color histogram is easily disturbed when the target's color is similar to background, and the algorithm presented in this paper can track the target accurately.

The test video 2 which is used to test the location accuracy is from Digital Navigation Center of Beihang University, and its resolution is 1280×960 pixel. The test result is shown in Fig. 3.

Figure 3 gives the results of human tracking indoor. After tracking the target accurately, we can estimate the geometric center coordinate. Then the target location information can be calculated by Eq. (4.20).

In this experiment, we chose front-left corner of room as the origin of coordinate in the binocular positioning in general. The related parameters were measured as follows: camera installation height is 2 m, it's lens optical axis is inclined downward and the angle with vertical plane is $\pi/3$, two camera's axes are parallel with the horizontal space 45 cm, and the calibrated focus length is 3.6 mm. The calculated result of target location is shown in Table 1.

Fig. 2 The results of the targets tracking, **a** particle filter based on color histogram in literature [3], **b** particle filter based on multi-feature fusion

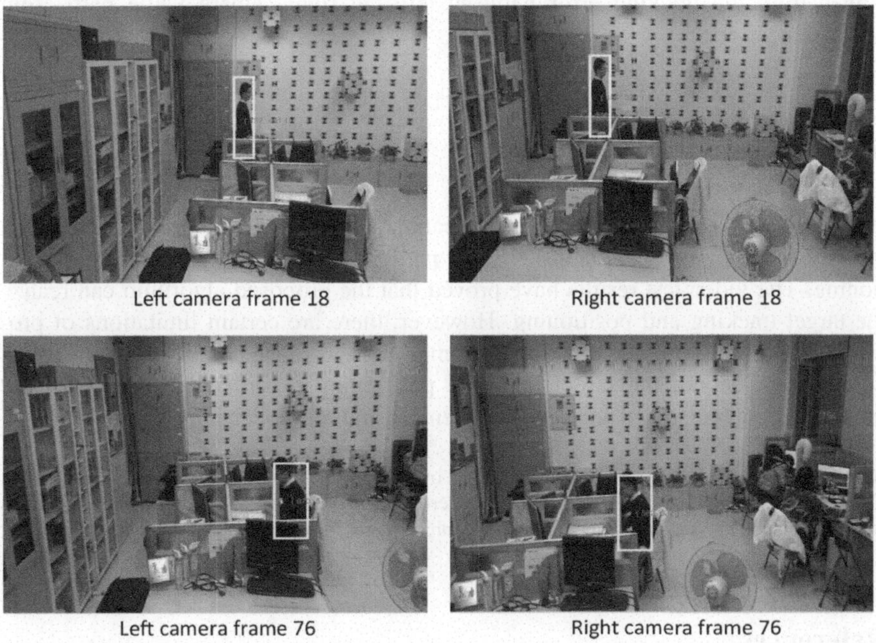

Fig. 3 The tracking results in left and right camera

Some samples of final positioning results are demonstrated in Table 1. The frame indexes are 18, 25, 34, 53, 58, 76, 83, and 97. The calculated mean absolute error of the top 100 frame is (13.32, 12.76, 8.15). The reason is that the estimation of targets' geometric center coordinate may not completely match in left and right

Table 1 The tracking results in left and right camera

Frame number	World coordinate (cm)	Calculating coordinate (cm)
Frame 18	(137, 219, 110)	(131, 212, 108)
Frame 25	(142, 245, 110)	(134, 237, 106)
Frame 34	(148, 250, 110)	(156, 258, 117)
Frame 53	(275, 330, 110)	(269, 337, 103)
Frame 58	(278, 362, 65)	(287, 373, 73)
Frame 76	(405, 382, 65)	(398, 371, 58)
Frame 83	(527, 416, 65)	(543, 407, 70)
Frame 97	(599, 471, 110)	(608, 479, 102)
Mean absolute error of the top 100 frame (13.32, 12.76, 8.15)		

video sequences. Then this will lead to calculating error during binocular vision positioning. According to Table 1, we can see that the proposed positioning method can realize target positioning in the range of permitted error. The error mainly comes from the tracking error between left and right camera while parts from measurement error.

6 Conclusion

In this paper, a target tracking and locating algorithm was reported, including feature extraction, multi-feature fusion, particle filter tracking, and target positioning. The video test results have proved that the proposed algorithm can realize the target tracking and positioning. However, there are certain limitations of proposed algorithm. First, the tracking error may cause estimated error of targets' pixel coordinate. Second, the target real-time positioning is still challenging. All the mentioned contents need further improvement and are under investigation.

Acknowledgements This project is supported by the National Natural Science Foundation of China (Grant No. 41274038, 41574024), the Aeronautical Science Foundation of China (Gratis No. 2013ZC51027), and the Fundamental Research Funds for the Central Universities.

References

1. Pai CJ, Tyan HR, Liang YM et al (2004) Pedestrian detection and tracking at crossroads. Pattern Recogn 37(5):1025–1034
2. Perez P, Vermak J, Blake A (2004) Data fusion for visual tracking with particles. Proc IEEE 92(3):495–513
3. Martins HA, Birk JR, Kelley RB (1981) Camera models based on data from two calibration planes. Comput Graphics Imaging Process 17:173–180

4. Nummiaro K, Koller-Meier E, Gool LV (2003) An adaptive color-based particle filter. Image Vis Comput 21(1):99–110
5. Ojala P, Harwood DM (1996) A comparative study of texture measures with classification based on feature distributions. Pattern Recogn 29(1):51–59
6. Guo S (2012) Approach of image edge detection based on wavelet scale multiplication. Appl Mech Mater 130–134:4282–4285
7. Shoushtarian B, Bez HE (2005) A practical adaptive approach for dynamic background subtraction using an invariant color model and object tracking. Pattern Recogn Lett 26(1):5–26
8. Zhang ZY (1997) Motion and structure from two perspective views: from essential parameters to euclidean motion via fundamental matrix. J Opt Soc Am A 14(11):2938–2950

The Application of Assisted-BDS Positioning in Indoor Environment of Weak Signal

Zhiyong Huang, Dongqing Zhao, Shaolei Peng, Anran Zu and Guojun Li

Abstract In view of the present problems of BDS receiver in the indoor weak signal environments faced with strong signal attenuation, the NLOS propagation, cross-correlation effects, the A-BDS coarse-time navigation, and RAIM algorithm to solve the problems of indoor positioning and the use of BDS data to verify the reliability of above-mentioned method are proposed in this paper. The experiment shows that the coarse-time navigation algorithm is able to provide consecutive and reliable positioning results and also it can be used in the weak signal environments. RAIM algorithm based on the combination of FDE can increase the availability of the positioning result, this, to some extent, solve the problems of NLOS propagation and cross-correlation effects existing in indoor satellite signals.

Keywords Assisted-BDS · Coarse-time navigation · RAIM · Indoor positioning

1 Introduction

The appearance of Assisted-GNSS (A-GNSS) enlarges the range of the conventional receiver applications, making it still achieve positioning results in weak signal conditions, including urban canyons, dense forests, and indoor environments [1]. A series of A-GNSS receiver sensitivity tests shows that the A-GNSS has the ability to solve indoor positioning [2]. However, it is restricted by various factors in practice, especially causing many positioning problems in deep indoor environments, that is, the receiver faces strong signal attenuation, large multipath, NLOS

Z. Huang (✉)
China Aerospace Surveying and Mapping Center, Beijing 102102, China
e-mail: lanruo_hzy@qq.com

D. Zhao · S. Peng
Information Engineering University, Zhengzhou 450001, China

A. Zu · G. Li
Beijing Satellite Navigation Center, Beijing 100094, China

J. Sun et al. (eds.), *China Satellite Navigation Conference (CSNC) 2016 Proceedings: Volume I*, Lecture Notes in Electrical Engineering 388,
DOI 10.1007/978-981-10-0934-1_31

signal propagation, or cross-correlation effect (also called near–far effect) between the strong signal and weak signal generated by the satellite [3].

In view of the positioning problems of BDS receiver faced in indoor environments, this paper proposes some methods that we can acquire positioning results through the use of coarse-time navigation algorithms operate on weak signals and improve the reliability of positioning results by the use of receiver autonomous integrity monitoring (RAIM) algorithm, so as to implement the application of A-BDS positioning in weak signal environments.

2 A-BDS Coarse-Time Navigation

The introduction of A-GNSS precisely aims to enlarge the range of the conventional receiver applications. When the receiver is under a weak signal environment, the traditional receiver does not work well, while A-GNSS receiver in such environment can still capture the signals and get the positioning result because of its high-sensitivity characteristics. However, when the signal is extremely weak to the extent that the A-GNSS receiver cannot obtain the time of week (TOW) [4], because because decoding navigation message bit error rate (BER) is too high, at that time, it also unable to obtain accurate satellite signal transmission time, and the navigation message cannot complete decoding, ultimately, A-GNSS receivers will not be able to complete the positioning. We can provide assisted ephemeris and assisted time through A-GNSS location server, because most of the reference network cannot provide fine time assistance (such as GSM, UMTS, WCDMA time accuracy of 1–2 s, CDMA time accuracy is microsecond), therefore we can only use coarse-time navigation algorithms for positioning. The algorithm is also called snapshot positioning, single shot positioning [5], time-free positioning, and code phase positioning. If take the A-GNSS coarse-navigation algorithm and other key technology into the BDS, we can solve the positioning problems of A-BDS in weak signal environments. We can use different A-BDS positioning mode under different signal strengths. For ease of understanding, the A-BDS positioning algorithm can be roughly divided into the following three categories according to the different signal strengths of the receiver (Table 1).

The A-BDS receiver is able to complete the process of capturing, tracking, bit synchronizing, frame synchronizing, and decoding the navigation message in the

Table 1 Positioning mode of A-BSD receiver under different signal strengths

Signal strength (dBW)	Signal processing	Positioning mode
About >−175	Complete capture and decode TOW	SPP (or coarse-time navigation)
About >−190	Complete capture, cannot decode TOW	Coarse-time navigation
About <−190	Cannot capture signal	Cannot positioning

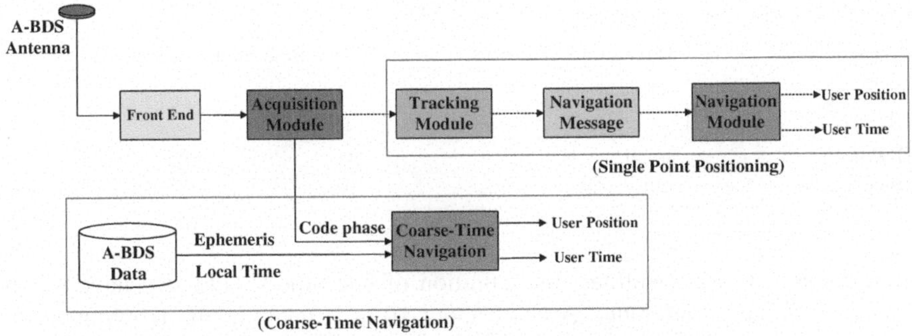

Fig. 1 Coarse-time navigation algorithms and SPP algorithms solver process

normal or a little bit weak signal environments, we can use single point positioning (SPP) mode for calculation on this occasion. Of course, we can also ignore the decoded TOW and use coarse-time navigation algorithm for positioning because of the proposed assisted information, we can get positioning results without having to decode the navigation message and achieve faster time of first fix (TTFF). We can only use coarse-time navigation algorithms to get positioning results when the receiver is in a weak signal environment. The A-BDS receiver cannot complete signal capture in the extremely weak signal conditions, likewise, positioning results cannot be obtained at this time. The two navigation algorithm positioning process is shown in Fig. 1.

As shown above, the code phase value obtained in acquisition module is directly used for coarse-time navigation. We can get two aspects of advantages while estimating TOW for coarse-time navigation. A faster TTFF, because without the implementation of tracking and decoding navigation message and other processes behind. With a higher sensitivity, we still get the positioning results in the case of the signal below the required navigation message that decoding signal strength needed.

In order to complete coarse-time navigation, there we will need a priori information by A-BDS server as follows:

1. Ephemeris information is available (To calculate the satellite position and estimate the predicted pseudoranges).
2. Approximate user location accuracy of tens of kilometers.
3. Seconds precision assisted time.
4. Sub-millisecond pseudorange measurements, i.e., code phase value.

Wherein the assisted time information is critical, however, of how much superior precision assisted time can be regarded as fine time? In A-GPS, 10 ms is generally defined as the time demarcation point of coarse-time navigation [6]. Because the GPS satellites' maximum pseudoranges rate is 800 m/s, satellite position error can generate the maximum of 8 m or so within 10 ms, this level of error range is acceptable for navigation receiver. The maximum pseudoranges rate is different

Table 2 The fine-time requirements of GPS and BDS satellites

The satellites	Maximum speed (m/s)	Fine-time requirements (ms)
GPS	800	10
BDS MEO	700	15
BDS GEO	Small	∞
BDS IGSO	380	25

from different GNSS satellites. Its definition of fine time should also have minor deviation. If the pseudorange error of about 10 m is taken as the requirement of fine-time, the maximum pseudoranges rate of GPS and BDS satellites and their requirements of fine-time is in Table 2.

For BDS, its maximum pseudorange rate of about 700 m/s, the accuracy of assisted time 15 ms can be defined as the dividing line of coarse-time and fine-time (while the observation satellites do not contain MEO satellites, it is 25 ms, and only the GEO satellite is theoretically infinite). When the assisted time is better than 15 ms, the traditional 4-state navigation equations will be able to provide a solution. When the assisted time is worse than 15 ms, then we need to get a solution with 5-state coarse-time navigation equations.

In 5-state coarse-time navigation equations, $\rho^{(k)}$ represents the measured pseudoranges, $\hat{\rho}^{(k)}$ represents the prediction pseudoranges calculated by the approximate location of the user and the satellite position, $\delta\rho^{(k)}$ represent pseudorange residuals. For each satellite, the relationship between $\delta\rho^{(k)}$ and the update values of 5 priori parameters is

$$\delta\rho^{(k)}=\rho^{(k)}-\hat{\rho}^{(k)} = -\mathrm{e}\cdot\delta x_{xyz}+\delta_b+v^{(k)}\cdot\delta_{t_c}+\varepsilon^{(k)} \tag{1}$$

Among them, $v^{(k)}$ is the pseudorange rate, $\dot{\delta}_t^{(k)}$ represents the satellite clock drift, pseudorange rate is obtained as follows:

$$v=\left(\vec{\mathrm{e}}^{(k)}\cdot\vec{v}^{(k)}-\dot{\delta}_t^{(k)}\right) \tag{2}$$

For k satellites, obtained matrix equation is

$$\delta\rho = H\delta x+\varepsilon \tag{3}$$

Among them

$$H=\begin{bmatrix} -\mathrm{e}^{(1)} & 0 & v^{(1)} \\ \vdots & \vdots & \vdots \\ -\mathrm{e}^{(k)} & 0 & v^{(k)} \end{bmatrix} \tag{4}$$

However, the measured pseudoranges $\rho^{(k)}$ cannot be obtained in weak signal environments, what we get is just the measured code phase values from the capture process, the so-called fractional pseudoranges. So how to restore the full pseudoranges from the code phase values is the key to solving the problem of coarse-time navigation. When we present the A-BDS coarse-time navigation equation, the impact of BDS NH code is not considered. According to a given priori position and assisted time, Van Diggelen presents a method to remodel full pseudoranges, which can avoid the problem of millisecond integer ambiguities. The algorithm process is as follows:

1. Select a high elevation angle satellite as the reference satellite, the satellite is specified with integer millisecond ambiguity as $N^{(1)}$, The measured code phase value is $z^{(1)}$, So get the first satellite measurements of pseudorange as

$$\rho^{(1)} = N^{(1)} + z^{(1)} \tag{5}$$

2. The measured pseudoranges and predicted geometric distance link, there are

$$N^{(1)} + z^{(1)} = r^{(1)} - \delta_t^{(1)} + b + \varepsilon^{(1)} = \hat{r}^{(1)} - r_d^{(1)} - \delta_t^{(1)} + b + \varepsilon^{(1)} \tag{6}$$

Among them, $r^{(1)}$ is the true geometric distance, $\hat{r}^{(1)}$ is the obtained rough geometric distance between the satellite at the coarse satellite launch time and the receiver prior position, $\hat{r}^{(1)}$ contains error caused by $r_d^{(1)}$, which brings by priori position and time error, $\delta_t^{(1)}$ is satellite clock error, b is public deviation (receiver clock error), $\varepsilon^{(1)}$ is measurement error. For other satellites, assume its integer millisecond ambiguity as $N^{(k)}$, similarly, the following equation can be obtained:

$$N^{(k)} + z^{(k)} = \hat{r}^{(k)} - r_d^{(k)} - \delta_t^{(k)} + b + \varepsilon^{(k)} \tag{7}$$

3. Place the formula (7) minus (6), erasing the same common errors, the following formula can be obtained

$$\begin{aligned} N^{(k)} = N^{(1)} + z^{(1)} - z^{(k)} + \left(\hat{r}^{(k)} - r_d^{(k)} - \delta_t^{(k)} + b + \varepsilon^{(k)}\right) \\ - \left(\hat{r}^{(1)} - r_d^{(1)} - \delta_t^{(1)} + b + \varepsilon^{(1)}\right) \end{aligned} \tag{8}$$

And in order to get the correct $N^{(k)}$ value, there need to make the priori position and time error be less than 0.5 light-ms in the rounding operation(light-milliseconds, the distance of light propagation within a millisecond), that satisfies the following formula:

$$\left| -r_d^{(k)} + \varepsilon^{(k)} + r_d^{(1)} - \varepsilon^{(1)} \right| < 150\ \text{km} \tag{9}$$

Such expression of $N^{(k)}$ can be obtained as follows:

$$N^{(k)} = \text{round}\left(N^{(1)} + z^{(1)} - z^{(k)} + \left(\hat{r}^{(k)} - \delta_t^{(k)}\right) - \left(\hat{r}^{(1)} - \delta_t^{(1)}\right)\right) \tag{10}$$

This will restore the other satellite pseudoranges by the code phase values, we obtained pseudorange residuals equation of coarse-time navigation as follows:

$$\delta\rho = N + z - \hat{\rho} \tag{11}$$

3 Problems of Indoor GNSS Signal and RAIM Algorithm

3.1 Problems of Indoor GNSS Signal

The indoor GNSS signal faces a strong signal attenuation, the attenuation when in the deep indoor is 1000 times or more than outdoor. However, since the A-GNSS receiver get acquisition assisted data by A-GNSS server, making the receiver can accurately find the correlation peak position in the course of searching satellite signals. In this case, the receiver can have a longer residence time for signal accumulation in the frequency search range, the acquisition sensitivity can be also greatly improved. This makes the A-GNSS receiver acquisition indoor weak signals with possible, and many sensitivity experiments have verified this [7]. However, because the signal is extremely weak, it impossible to decode the navigation message, we can only use the above-mentioned coarse-time navigation algorithm.

In the indoor environment, direct signal of line of sight (LOS) propagation is often subject to strong attenuation, and the reflected non line of sight (NLOS) signal due to bypassing the walls or ceilings has a higher sensitivity. However, because NLOS propagation creates additional paths, there is a big error with the measured pseudoranges, i.e., it is easy to reach 100 m magnitude. If left it untreated, it will produce 100 m magnitude positioning errors. The conditions of the received signal in typical indoor environment [8] are shown in Fig. 2.

When entering the room, since blocked by the ceiling or wall, the GNSS signal produce a series of phenomena like reflection, refraction, and scattering. The direct signal received by the user through the wall block has the correct path, but it has been greatly attenuated. The signal bypassed the wall from the windows or doors into the interior of have a strong power, the path the signal passes through is the path of least resistance [8]. Nevertheless, the path contains a lot of pseudorange errors. A better method is to use RAIM technology for the reflected signals with large errors.

In the indoor environment by the window, it is prone to receive strong signal coming from the window, while other refracted propagation signals that pass

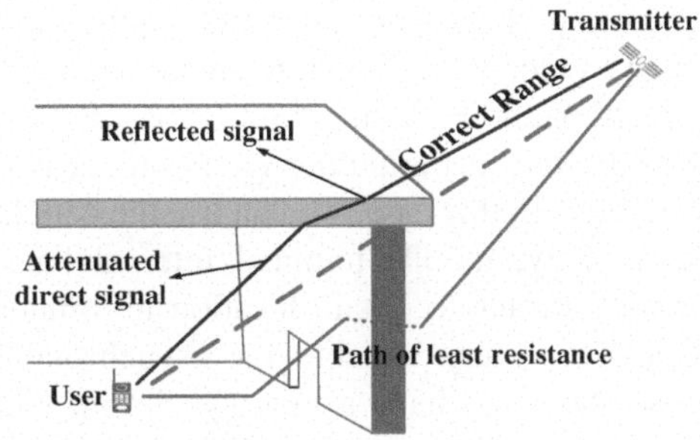

Fig. 2 Signal propagation path of typical indoor environment

through the ceiling and walls will produce a greater attenuation. An error cross-correlation peak may occur between strong and weak signals, the cross-correlation peak is higher than auto-correlation peak of weak signals, this effect is called the cross-correlation effects. The cross-correlation peak of cross-correlation peak is −24 dB, common GNSS receiver experiences this problem less, because its acquisition weak signals are weak. However, A-GNSS and high-sensitivity GNSS receiver have a large signal dynamic range, which is −123 to −160 dBm. It is easy to generate the cross-correlation effects.

Cross-correlation effects will have a great pseudorange error. The receiver captured the wrong cross-correlation peak will lead to any error of a sub-millisecond magnitude pseudorange, the maximum reach is 300 km. In the A-GNSS receiver, the search space is reduced due to the supply of the acquisition assistance data. If we adopt a reasonable order to search satellite, it can greatly reduce the probability of the occurrence of the cross-correlation effects. In addition, after the capture of satellite, the RAIM technology can be used to detect and eliminate the measurements with large error.

3.2 Combinational FDE Algorithm

Because the received direct signal in indoor environments has significant multipath, it can only receive echo-only signal in some cases, the measured pseudoranges contain more errors at this time. When the cross-correlation effect occurs, it will produce the maximum pseudorange error 300 km. In this case, if a sufficient number of satellites available, it is observed that 6 satellites can detect (Fault Detection, FD) the measurement contain a larger error, it is observed that more than seven satellites can detect and exclude (fault detection and exclusion, FDE) a large measurement of error. But before proceeding to RAIM, we should calculate horizontal protection level (HPL) at first, and then compared with the given horizontal alarm level (HAL). If HPL $>$ HAL, it means that satellite constellation have a poor geometry. It does not meet the requirements for the implementation of RAIM, we need to receive the satellite signal anew.

Although the total number of satellites in the constellation BDS increased rapidly, but in the indoor environment it is very difficult to require seven visible satellites. In this case, we can use the cellular network enhanced cell-ID positioning technology, mobile terminal internal barometer, magnetometer, and other positioning means to obtain location information and increase redundancy. Enhanced Cell-ID is very suitable for asynchronous time network, it is not needed the base station to have a time synchronization, we can obtain the terminal location by measuring the received signal strength. And enhanced cell-ID at urban area where dense base have a higher precision, it can reach 100 m. In the suburb where we have rare base, positioning accuracy can reach 550 m [9]. Users who initiate positioning request are mostly located in urban areas, 100 m positioning accuracy is sufficient to exclude outliers as a redundant condition. The barometer has a relatively high elevation positioning accuracy, it can be used to exclude the satellite observations from which produce large elevation errors. Combination FDE algorithm process shown in Fig. 3.

Therefore, the use of coarse-time navigation algorithms can resolve the problem of weak satellite signals positioning. When combinational FDE algorithm is used, to a certain extent, we can solve the big pseudoranges error problems created by cross-correlation effects and nondirect signal, and reduce the number of satellites which are not needed. The combination of two algorithms to solve indoor weak signal positioning provides a feasible idea.

Fig. 3 RAIM algorithm based on a combination of FDE

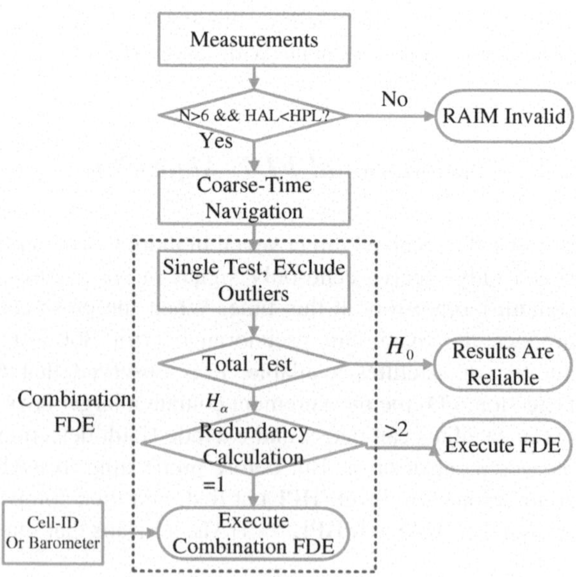

4 Data Simulation

Select a BDS/GPS Quad band dual system receiver, obtain GNSS data of 2650 epochs in Shanghai, the sampling interval is 30 s. Select one of the BDS data carry on coarse-time navigation algorithm simulation experiments.

4.1 Outdoor A-BDS Coarse-Time Navigation Algorithm Verification

Coarse-time navigation algorithm uses code phase measurements and the experimental data obtained for complete pseudorange measurements. Therefore, we use pseudorange measurements modulo 1 ms to get the code phase measurements (without considering the impact of NH code). In order to make the algorithm proceed in coarse-time navigation mode, the real/known transmit time is ignored, and we take the coarse-time error 1 s to verify the effects of different coarse-time errors on the positioning accuracy. For assisted data such as assisted ephemeris A-BDS required in the process of operation, it is administered directly. First of all, we compare the HDOP of 4-state navigation (i.e., SPP) with the 5-state coarse-time navigation, the following comparison chart can be obtained.

Figure 4 shows that, the HDOP of 5-state coarse-time navigation is always greater than 4-state navigation, and in some cases, their difference is approximately zero. And the HDOP of 5-state coarse-time navigation are all less than 3, it indicates that we can provide reliable positioning results for the user.

When the coarse-time error is 1 s, we get three directions (East, North, Up) error sequence diagram is shown in Fig. 5.

When the coarse-time error is 1 s, coarse-time navigation can provide positioning accuracy in the horizontal direction within 15 m, elevation direction within 20 m. It shows that 5-state coarse-time navigation can provide relatively high precision positioning results.

Fig. 4 HDOP of 4-state navigation and 5-state coarse-time navigation

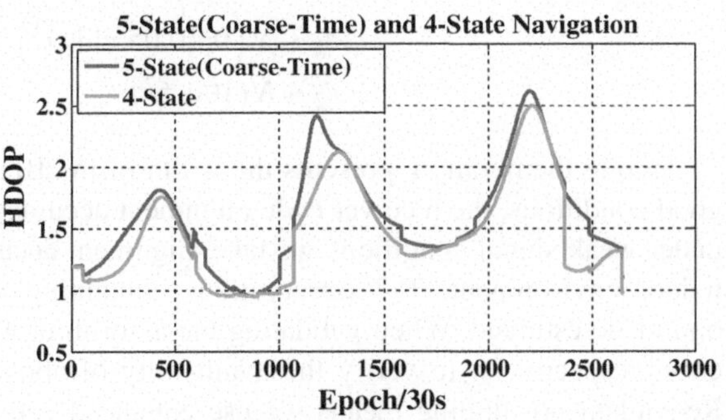

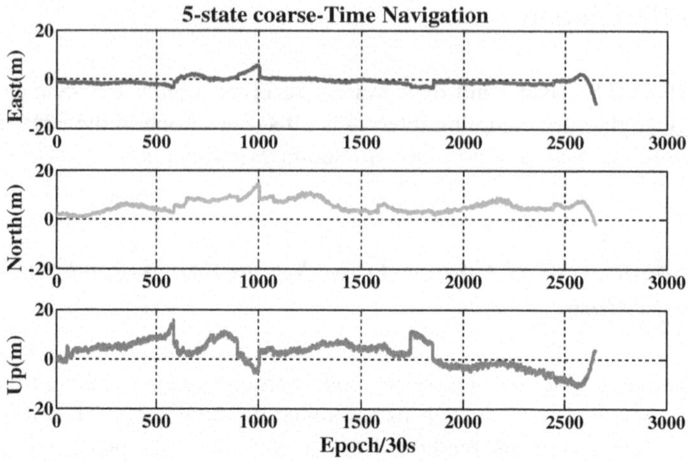

Fig. 5 Error sequence diagram of 5-state coarse-time navigation

4.2 Indoor A-BDS Coarse-Time Navigation Algorithm Simulation Verification

In the indoor environment, GNSS signal facing a great attenuation. Although A-GNSS/HS-GNSS receiver can capture weak signals, thermal noise greatly affects the signal. In addition, indoor environment is faced with significant signal NLOS propagation, which may led to receive the echo-only signal. At this time, the signal contains large gross errors, it is very easy to reach 100 m. Under the circumstances of containing multipath signal, it also easy to cause larger pseudorange error. In order to simulate the thermal noise, we add normally distributed random thermal noise error ε_1 to all observations in each epoch. To simulate the echo-only signal and multipath signal, we add the normally-distributed NLOS propagation errors ε_2, ε_1 and ε_2 to 1–3 observations in each epoch. The distribution of obedience is as follows:

$$\left.\begin{array}{l}\varepsilon_1 \sim N\left(0, (0.05\ \text{chip})^2\right)\\ \varepsilon_2 \sim N(100, 5^2)\end{array}\right\} \tag{12}$$

Among them, chip represents the width of the BDS chip, about 146.5 m, under good conditions, the receiver code alignment accuracy is better than 1/100. So here under weak signal conditions we take alignment accuracy of 1/20. As for the deep indoor environment, 1–3 observations containing a big gross error is a more optimistic estimate. When validating the algorithm, we need to add 1, 2, or 3 gross error, respectively, to verify the availability of above-proposed RAIM algorithm. Redundant positioning means we use enhanced cell-ID, with accuracy of 100 m, which is considered available. The use of A-BDS coarse-time navigation and

Table 3 A-BDS coarse-time navigation positioning results

Number of big errors	Availability (%)	East (m)	North (m)	Up (m)
1	66.3	1.921	29.191	48.225
2	51.9	2.201	41.575	58.313
3	37.0	2.458	47.286	68.130

Table 4 A-BDS coarse-time navigation based on RAIM positioning results

Number of big errors	Availability (%)	East (m)	North (m)	Up (m)
1	92.8	0.806	8.449	11.380
2	61.8	1.861	7.506	37.618
3	33.7	2.4448	7.027	61.028

A-BDS coarse-time navigation based on RAIM algorithm is respectively, positioning calculated. Taken confidence level of 0.95, the positioning results of E, N direction error which is greater than 50 m and the final global test results which not passed is considered unavailable. We can obtain the following three directions RMS results and positioning results availability statistics (Tables 3 and 4).

As can be seen from the above charts, when we add three larger pseudorange errors, the latter can still provide a good level positioning accuracy, while the elevation direction error is large, if we can provide a reference elevation values (such as barometers), it is possible to exclude the height of poor positioning result. The availability of the positioning result is reduced, the reason is that when the observations contained many gross errors, single constellation limited due to the observed measurements, RAIM algorithm cannot exactly exclude observations with errors, which is leading to reduce availability. In this case, we must have more observation information in order to ensure RAIM algorithm to be effective. In theory, multiple systems with more observations can effectively improve positioning availability. The A-GNSS with multi-constellation will have a more impressive application prospects in indoor positioning in the future.

5 Conclusion and Future Work

In this paper, the method of using A-BDS coarse-time navigation to realize the positioning of indoor weak signal environments is proposed, the calculation process is given in detail, and the reliability of the algorithm is verified by the simulation data. The outdoor positioning results show that coarse-time navigation can provide continuous reliable positioning results. Indoor positioning experiments prove that RAIM algorithm can increase availability and improve positioning results accuracy. However, when the gross error is large, the data redundancy is small, and the RAIM algorithm is invalid, we need more observations of satellites or other observations.

There will be more GNSS satellites in our future, and the satellite signal will has a better design [10]. Moreover, the use of other positioning technology to improve GNSS in increasing the availability of A-GNSS indoor applications will appear. The next step we will further study multi-system A-GNSS coarse-time navigation algorithm, and by providing acquisition assistance data to A-GNSS software receiver to get code phase measurements, so as to validate the indoor coarse-time navigation algorithm.

Acknowledgments This study is funded by State Key Laboratory of Geo-information Engineering, NO. SKLGIE2015-M-2-1.

References

1. van Diggelen F (2010) Down and deep. GPS World 2010(12):24–26
2. Singh S (2006) Comparison of assisted GPS and high sensitivity GPS in weak signal conditions. University of Calgary
3. Seco-Granados G, Lopez-Salcedo JA, Jimenez-Banos D et al (2012) Challenges in indoor global navigation satellite systems: unveiling its core features in signal processing. Sig Process Mag IEEE 29(2):108–131
4. Song C (2009) Research on the key techniques of assisted GPS positioning system. National University of Defense Technology, Changsha
5. Schmid A (2007) Single shot positioning. In: IGNSS Symposium, Sydney
6. van Diggelen F (2009) A-GPS assisted GPS, GNSS, and SBAS. Artech House
7. Bryant R, Dougan S, Glennon E (2001) GPS receiver algorithms and systems for weak signal operation. Salt Lake, UT
8. Lachapelle G (2004) GNSS indoor location technologies. J Global Positioning Syst 3(1–2):2–11
9. Zhongliang D, Dejun Z, Jianming H et al (2009) The assisted GNSS boomed up location based services
10. Hu X, Tang Z, Zhou H et al (2009) Analysis on design principles of GPS and Galileo signal structure. Syst Eng Electron 31(10):2286–2293

BeiDou Smart Scenic LBS Network Fusion Research and Application

Baoguo Yu, Yao Wang, Jun Li and Jianhua Ran

Abstract Based on an analysis of the connotation of smart tourism and the perspective of smart scenic construction, a concept of BeiDou Satellite Navigation System (BDS) smart tourism is proposed. Later on, some related key issues are discussed such as system architecture, wide-area continuous coverage positioning network technology, and wide-area continuous coverage communication network technology. Taking Jiuzhai Valley as experimental construction spot, the BDS smart scenic location-based service (LBS) system was developed and deployed. With such technical features such as both way, teamwork, information push, and remote interaction, this system improves the tourist LBS, fine management, and unexpected event handling capacity of the resort. Finally, the paper recommends the BDS smart tourism LBS system to become an important part of the one belt and one road (OBOR) strategy, for it is believed that this system will significantly speeds up the intelligent tourism construction of the neighboring countries, promote the cultural exchange and cooperation, and drives up the economic development of the concerned regions.

Keywords BDS · Smart scenic · LBS · Pseudolite · Mobile communication

---

B. Yu · Y. Wang (✉) · J. Li
State Key Laboratory of Satellite Navigation System and Equipment Technology, Shijiazhuang, Hebei Province, China
e-mail: wy_hrbeu@163.com

B. Yu · Y. Wang · J. Li
The 54th Institute of China Electronic Technology Corporation, Shijiazhuang, Hebei Province, China

J. Ran
Jiuzhai Valley Administration, Aba Zhou, Sichuan Province, China

J. Sun et al. (eds.), *China Satellite Navigation Conference (CSNC) 2016 Proceedings: Volume I*, Lecture Notes in Electrical Engineering 388,
DOI 10.1007/978-981-10-0934-1_32

1 Introduction

As a key infrastructure of spatial information in China, BeiDou Satellites Navigation System was set up regionally and started to provide regional services in 2012. Since then, BDS has been running in a reliable way, with the improved performance enough to satisfy the 10 m-resolution technical requirement. In some regions, the positioning performance is even better than the commitment. According to the BDS globalization plan, BDS is going to provide the basic navigation services for the countries along the OBOR route. The BDS LBS system has already been applied in such sectors as traffic, farming and fishery in China. In 2014, the output value of satellite navigation industry reached 200 billion RMB [1], among which the construction of tourism, especially the "smart tourism" serves as an important branch of BeiDou LBS application and also a linking business for the countries along the route of OBOR.

2 Connotation of Smart Tourism

Smart tourism is a new term derived from the term of smart city. As a product of digital tourism developed in the advanced stage, smart tourism makes use of the new generation of information and communication technology to initiatively collect and bulletin the information about tourism resources, tourism economy, tourism activities, and tourists and let people know about these information through Internet/Mobile Internet and portable Internet access terminals, thereby to arrange their business and travel plans more reasonably, and to perceive the tourism information and utilize it in an easy way. With smart tourism, it is possible to integrate some tourism resources together and thus to customize them for individual tourists and provide personalized services for their safe and enjoyable tours. The development process of tourism informatization may be divided into four stages, i.e., traditional tourism, electronic tourism, digital tourism, and smart tourism (Fig. 1).

The construction and evolution of smart tourism is reflected in three respects: smart services, smart management, and smart marketing.

1. Smart services

From the perspective of tourists, information technology is utilized to upgrade tourist experience and tourism quality. Let tourists enjoy the brand new services and experience brought about by smart tourism throughout the entire process of travel, e.g., tourism information acquisition, tourism planning and decision making, booking and payment for tourism products, tourism enjoyment as well as review

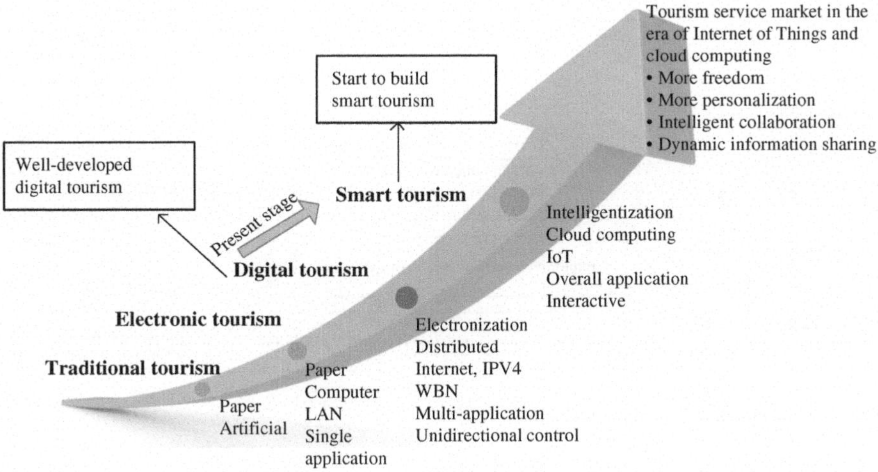

Fig. 1 Evolution of smart tourism

and comments afterwards. Specifically, it is reflected in four respects: navigation, tourist guide, tour map, and shopping.

2. Smart management

Try to push the traditional tourism management towards modern management. With information technology, it is possible to precisely know the information of tourist activities and of tourist business operations, so as to transform the tourism business supervision and administration to process management and real-time management from the traditional passive response and afterwards management.

3. Smart marketing

Based on tourist feedback and data analyses, try to locate tourist attractions and tourist interests, thereby to guide the tourist enterprises to supply the corresponding tourism products and prepare the corresponding marketing subjects, thus to promote innovation of tourist products and marketing approaches.

No matter it is smart services, smart management, or smart marketing, they all take the position information of the objects as basic data. In other words, smart tourism relies on the position information acquired from GNSS-based high-precision continuous positioning in combination with big data processing and exploration as well as the information interaction based on Mobile Internet, so as to satisfy the various demands of tourists, as illustrated in Fig. 2.

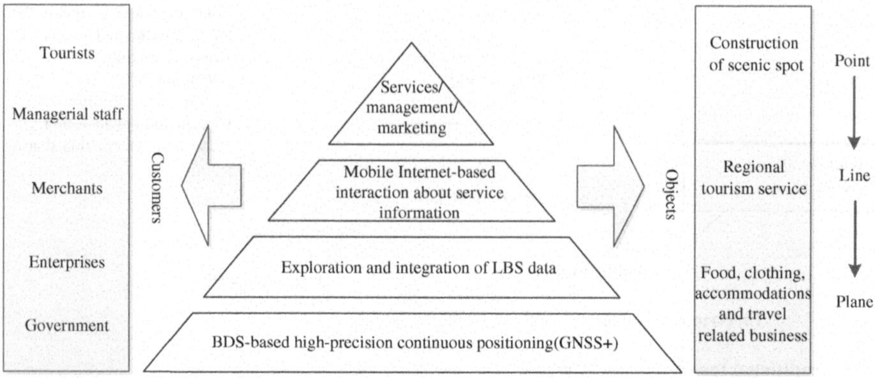

Fig. 2 Concept of BDS smart tourism

3 Key Issues of BDS Smart Scenic

3.1 System Architecture

Based on an open, "cloud plus terminal-supported" network, the architecture design of BDS Smart Scenic is characterized by multi-network integration, multiservice mixture, both-way interaction, configurability, and expandability, as shown in Fig. 3.

Its substrate is composed of wide-area positioning network and wide-area communication network, designed for positioning of and communication with people, vehicles, and objects. Specifically, it integrates Internet, Mobile Internet and BDS RDSS short message communication technology together for communication. In addition, it utilizes BDS/PLS dual-mode compatible positioning technology and RTK high-precision positioning technology to improve the system availability in forest and mountainous areas, so as to satisfy the requirement of high-precision positioning.

In terms of background tourism service platform, the system is planned to build cloud storage, big data exploration, and forecasting, LBS as well as search and rescue service platforms, which are connected to various E-commerce platforms, map service providers, social media, and other business operators through data exchange gateway for providing personalized LBS information. From the viewpoint of customers, the terminals may be classified into two categories: public terminals and dedicated terminals. The former type is generally based on personal smart phones or pad devices by which users may enjoy one-stand registration, mobile payment, smart push, and other services.

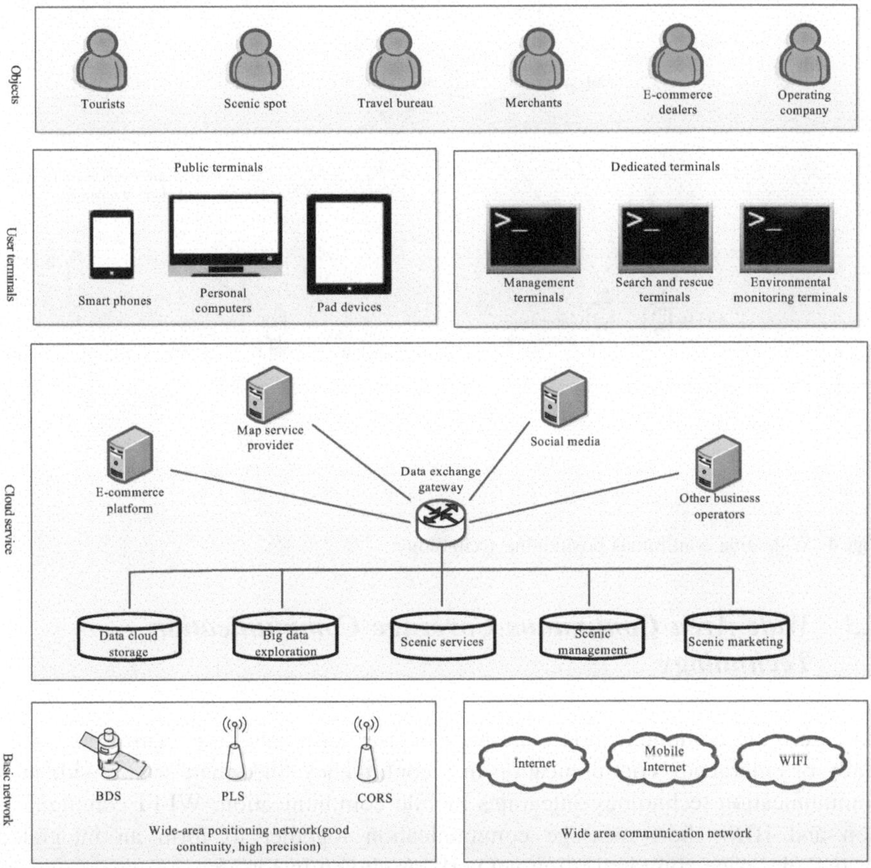

Fig. 3 Architecture of BDS-based smart scenic system

3.2 Wide-Area Continuous Coverage Positioning Technology

The wide-area continuous coverage positioning relies on BDS satellite navigation technology in addition to BDS pseudolite technology and BDS differential augmentation technology for the purpose of satisfying the (indoor and outdoor) positioning requirements of different users, as shown in Fig. 4.

Among them, BDS pseudolite is a technique developed particularly for good continuity and positioning accuracy of terminals. Pseudolite is a device that can transmit navigation signals. It is mainly composed of pseudolite synchro receiver and pseudolite signal transmitter. Comparing with satellite in space, pseudolite has similar signal transmitting capability but more module freedom and more flexible controllability in addition to its distinctive features.

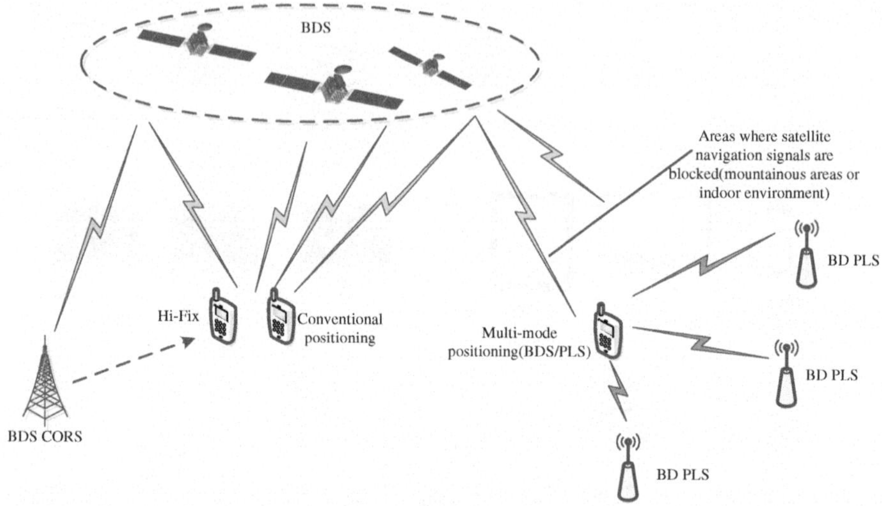

Fig. 4 Wide-area continuous positioning technology

3.3 Wide-Area Continuous Coverage Communication Technology

To assure the communication continuity of user terminals in a scenic spot, especially of emergency communication in a contingency, the smart scenic wide-area communication technology integrates mobile communication, WI-FI communication and BDS short message communication together to form an integrated ground-air-space communication network, as shown in Fig. 5.

Normally, the public terminals carry out information interaction by means of WI-FI, 3G, or 4G while the dedicated terminals do so with the background by means of mobile communication. However, in a contingency, i.e., when the mobile communication network becomes instable or disabled due to different causes, the dedicated terminals will complete information interaction with the background automatically through BDS RDSS. Since no BDS short message communication function is provided with public terminals, they can pair up with the BDS RDSS station nearby by means of Bluetooth and transmit the interactive information to a BDS GEO satellite through the station. If the Scenic Control Center has no the communication capability of receiving BDS short messages, the BDS Scenic Emergency Service Center may forward the communication contents reported by all terminals to the Scenic Control Center through the wire communication network, thereby to guarantee the communication availability between terminals and the background.

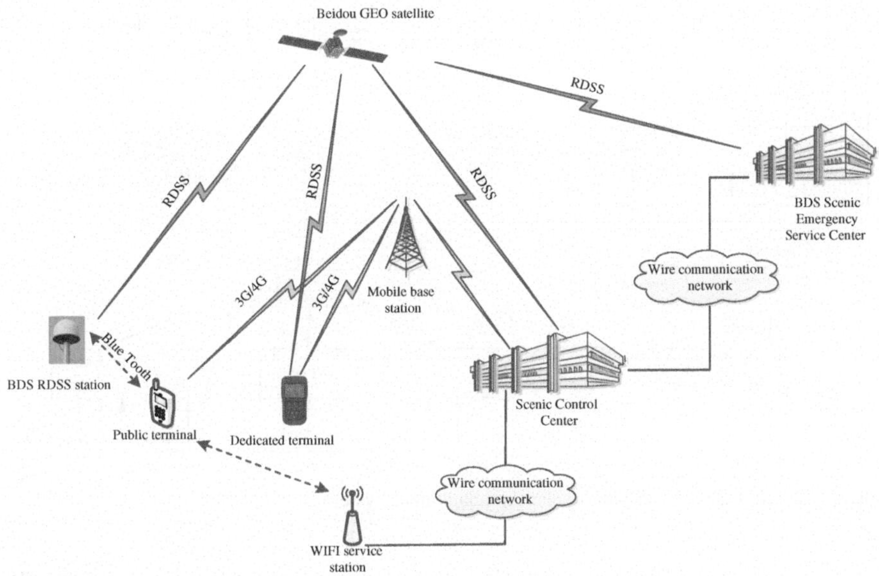

Fig. 5 Scenic wide-area communication network

4 Implementation Practice of BDS Smart Scenic

4.1 System Composition

Considering the requirements of Jiuzhai Valley Scenic Resort for location-based service (LBS), intelligent management and emergency search and rescue, the 54th Research Institute of CETC, taking the strength of "China Technological Innovation Union of Navigational Positioning Chips and Terminal Industry", established an LBS network in combination with the scenic wireless communication network as well as the intelligent management and search and rescue platform, which satisfies the requirements of people and vehicle management and dispatching as well as response to unexpected events. This system is mainly composed of four parts: LBS network, integrated data processing center (IDPC), augmentation system, and user terminals as shown in Fig. 6.

1. LBS network

The LBS network is mainly composed of BDS network, pseudolite network, RFID network, and communication network. The pseudolite network is used to provide satellite augmentation signal and improve the precision and coverage of BDS. The RFID network is composed of many node LBS terminals which can acquire the position information of RF cards in addition to data upload/download function to provide users with multimedia service information. The mobile communication network is composed of the existing 3G mobile communication

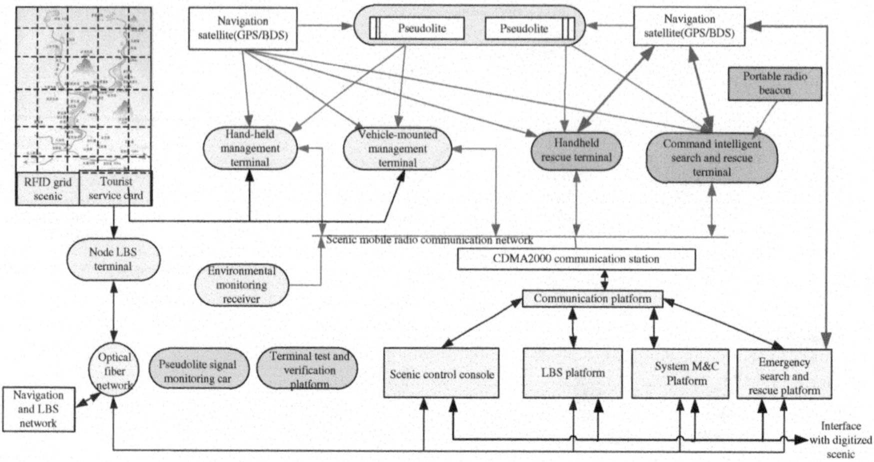

Fig. 6 Jiuzhai Valley BDS smart scenic management system

network and optical fiber network. It connects the pseudolites and the node LBS terminals with the IDPC.

2. IDPC

The IDPC is composed of LBS platform, emergency search and rescue platform, system M&C platform, scenic control platform, as well as, information interaction and communication platform. The LBS platform is used to store the space-time position information of the terminals in the scenic spot and to perform data exploration of the retrieval information from the other platforms. The emergency search and rescue platform is used to respond all the alarms in the scenic area and coordinate the search and rescue terminal holders. The system M&C platform is designed to monitor, control, and display the status of different system components. The scenic control platform is used to manage and coordinate the day-to-day management activities in the scenic spot. The information interaction and communication platform, serving as the communication hub of the entire IDPC, is used to support the information interaction among different platforms and between platforms and terminals.

3. Augmentation system

The augmentation system consists of local augmentation system and wide-area precision point positioning augmentation. The local augmentation system is composed of differential reference stations, which provide differential reference information to the IDPC via the mobile communication network for local augmentation. The wide-area precision point positioning augmentation can provide such information as precision orbit and precision clock error to complete wide-area augmentation service.

4. User terminals

User terminals include handheld management terminal, vehicle-mounted management terminal, hand-held search and rescue terminal, vehicle-mounted search and rescue terminal, high-precision environmental monitoring terminal, tourist RFID card, and smart mobile phone. These terminals work together to jointly complete smart navigation, scenic management, as well as, emergency search and rescue within the scenic area.

4.2 Services Provided by System

4.2.1 Tourist-Oriented LBS

1. One-card services

Each tourist will get an RFID card at the scenic tourist center, which is bound with the personal information of the tourist. Using an interactive node LBS terminal, it is possible to inquire the position information and introduction of scenic spots in the entire scenic area, the position information of sightseeing team members and sightseeing vehicles, visitor density, as well as, the sightseeing routes recommended automatically in addition to emergency alarm.

When a tourist is waiting for sightseeing vehicle at a stop, the local node LBS terminal will read the RFID card of the tourist, report the current position information of the tourist to the LBS platform and push the position information of other team members that is tailored to the tourist's mobile phone. Besides, the node LBS terminal may also display the visitor density of the entire scenic area, the to-be arrival information of sightseeing vehicles, nearby scenic spots information, bulletin information, and sightseeing tips to facilitate tourists to plan their sightseeing routes (Fig. 7).

2. Services for smart mobile phone

By scanning two-dimension code, as shown in Fig. 8, tourists can download the APP of Smart Jiuzhai Valley Tourism into their smart mobile phones. The contents provided by the APP can be divided into five categories: impression, tips, guide, and help. The APP supports a number of family members or friends to make a team with the binding service, so that every member can inquire the position information of others to avoid getting lost. The APP allows visitors to inquire the density distribution information of the entire scenic area and the position information of sightseeing vehicles, thereby to enjoy intelligent LBS through their sightseeing tours (Fig. 8).

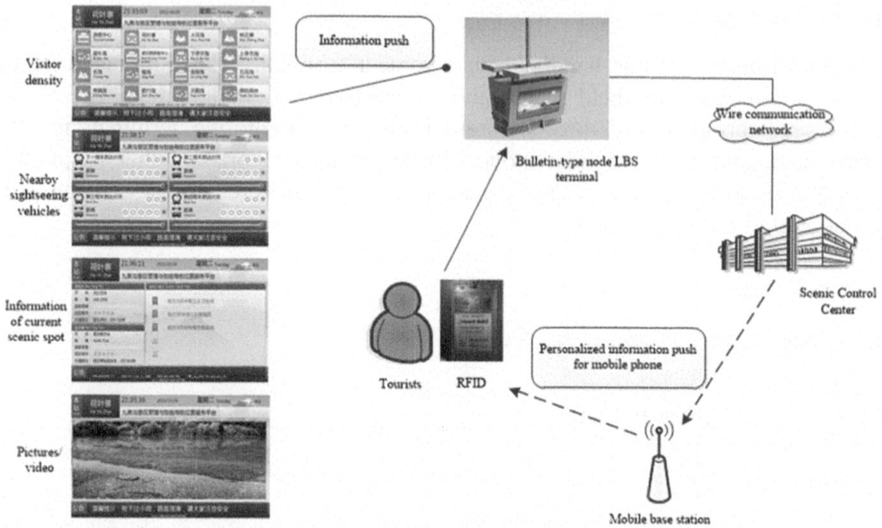

Fig. 7 Location-based service provided by bulletin type node LBS terminal

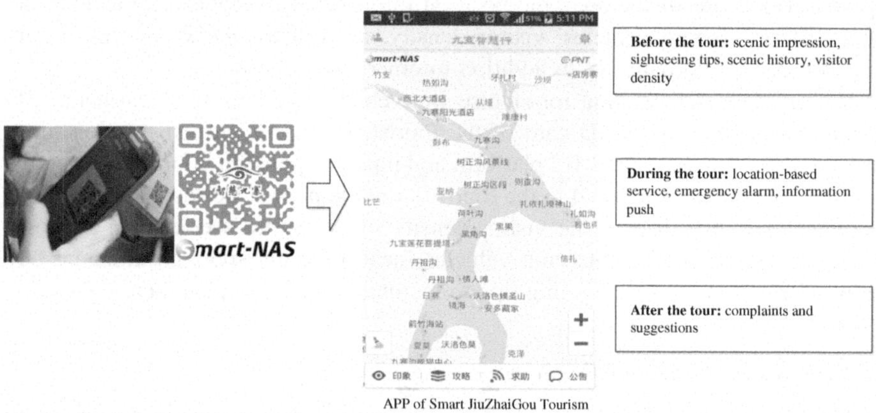

Fig. 8 Location-based service through APP of mobile phone

4.2.2 Scenic-Oriented Intelligent Management

The intelligent management of the scenic area is achievable with the integrated data processing (IDP) platform, including statistic of tourist quantity, intelligent environmental monitoring and data statistics, alarm and emergency search and rescue, bulletin of scenic information, vehicle dispatching, and staff attendance management.

Fig. 9 The visitor density of the scenic area

1. Statistic of tourist quantity and distribution

With the aid of the ticketing system and the access control system of Jiuzhai Valley Scenic Resort, it is possible to count the quantity of tourists who enter the scenic area. Using node LBS terminals, mobile base stations, and video surveillance system the data from multiple systems can be integrated together to calculate the visitor density of the scenic area, as shown in Fig. 9, thereby to provide reference for making diverting and traffic control of tourists in the scenic area.

2. Intelligent environmental monitoring and data statistics

One high-precision environmental monitoring terminal is provided for the staff of the scenic spot, which can be used to report environmental parameters in a real-time manner such as the hydrology, insect attack, biodiversity, geologic, and terrestrial changes. The scenic control platform receives and stores the environmental monitoring information. The data statistic based on long-term collected data may provide reference for the scenic managerial staff to conserve the ecological environment in the scenic area (Fig. 10).

3. Vehicle monitoring and dispatching

By means of real-time surveillance of the position information of sightseeing vehicles and operation vehicles in the scenic area, it is possible to dispatch those vehicles by voice. In addition, by further processing the position data of vehicles, it is possible to sum up the total daily mileage and the times of round trips of every sightseeing vehicle, thereby, to provide basic data for working out a reasonable dispatching strategy (Figs. 11 and 12).

Fig. 10 Collection of hydrographic information (of high precision)

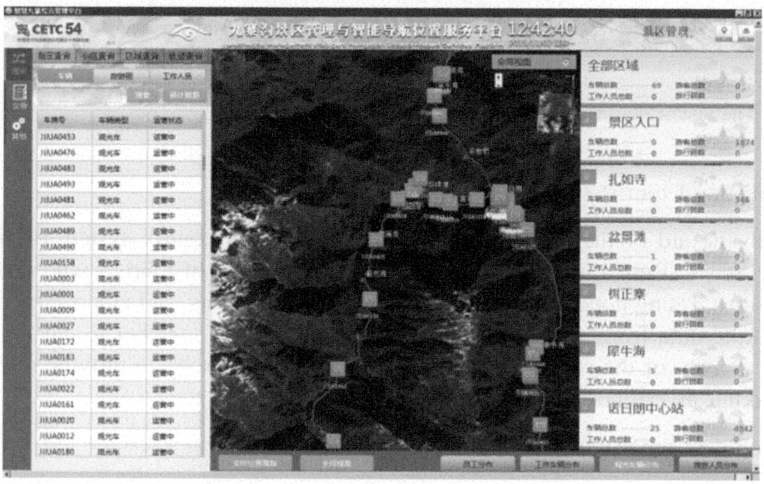

Fig. 11 Real-time position monitoring of vehicles

4. Alarm and emergency search and rescue

Working personnel can use any terminal that logs in the system to give an alarm. After the alarm message is received and verified, the emergency search and rescue platform will launch the contingency plan and work out a search and rescue plan with reference to the position information of the working personnel near the site. The integrated information processing platform and the search and rescue team may conduct real-time and two-way communication with command/handheld search and rescue terminals. When 3G is unavailable, the BDS RDSS short message can be used. In any way, the search and rescue team can reach the alarm location and take rescue action.

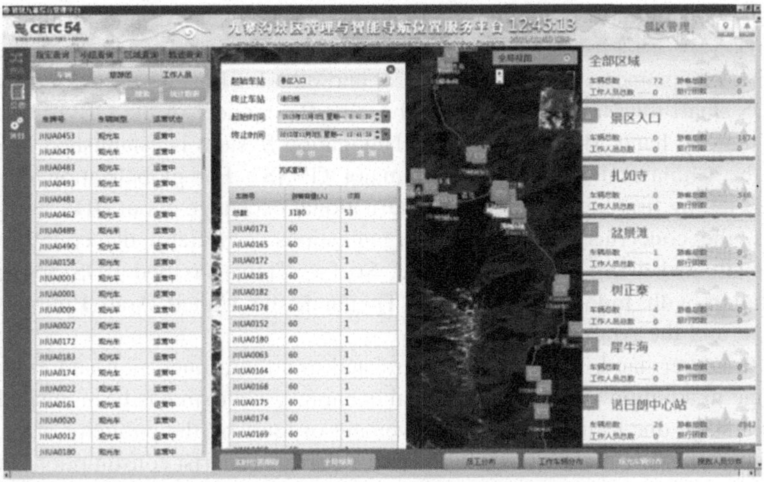

Fig. 12 Data statistic of sightseeing vehicles

5. Staff attendance management

The staff carries handheld management terminals including cleaning workers and vehicle service persons. Before starting to work, they must get their handheld management terminals to log in the demonstration system, which will report their position at a certain time interval. At the end of work hours, they have to log out the network. The scenic control platform will count their workloads with reference to the reported position information.

5 Application Prospect in the OBOR Strategy

This investigation revealed that dual-frequency operation, a rapid acquisition function, and attitude determination are required for miniature multifunctional GPS receivers. The specifications for these functions are summarized in Table 2. It is necessary to increase the total number of channels to receive dual-frequency signals and attitude determination. At most, each antenna views 12 satellites. Forty-eight performance channels are thus required, and the receiver weight should be less than 2.5 kg. However, it is difficult to increase the number of channels because conventional GPS receivers process the signals in one microprocessor. To resolve this computational complexity, signal tracking for use in combination with firmware has to be developed. We can realize up to 48-channel processing by using integrated circuits and digital signal processors (DSP).

OBOR is a term short for the New Silk Road Economic Belt and the 21st Century Maritime Silk Road. This strategy is purposed to promote China's collaboration with the countries along the route in economic policy, to implement regional cooperation at a higher level and in a comprehensive and profound way and to jointly establish a regional economic cooperation framework under the guideline of openness, inclusion, balance, and all-win. For the following reasons, the BDS smart tourism will absolutely become the pioneer of the OBOR strategy.

On the one hand, BeiDou Satellites Navigation System is established independently by China and among the four major GNSSs. For globalization of BDS, China is taking firm steps to perform international exchange about BDS with other GNSSs, the concerned international organizations, especially the countries and regions situated along the OBOR route. So far, China has tied BDS application cooperation with Indonesia, Pakistan, Thailand, the United Arab Emirates, the ASEAN, and the Arab League.

On the other hand, tourism is more related to cultural inclusion and emotional exchange in addition to consumption and economy. Tourism has been pushed onto the front stage due to the cultural involvement of tourism, the publicity of tourists, the affinity of tourism activities and the particularity of tourist consumption. For the same reasons, tourism will inevitably become the lead and guide of the OBOR strategy.

As the product of BeiDou Satellites Navigation System applied to tourism, BDS Smart Tourism has tremendous development potential. Following the point-line-plane development trend, a megaregion smart tourism ecosystem will come into being. By combining it with smart city network, high precision and continuous LBS will be provided. Along with the integration with car networking, the tourist experience of mass tourism will be upgraded. Ultimately, a series of BDS smart scenic standard systems will be established. These systems will be popularized into the countries covered by the OBOR strategy, powerfully drive up the level of smart tourism in these countries, boost their economic development, and facilitate the multicultural exchange and cooperation.

References

1. Chinese Beidou navigation industry development blue book in 2014–2015
2. Zhang L, Li N, Liu M (2012) On the basic concept of smarter tourism and its theoretical system. Tourism Tribune 27(5):66–73
3. Yu B, Ye H, Gan X (2011) The research on multi-mode pseudolite network and its enhancement technology. In: China satellite navigation conference 2011
4. Song Q, Zhang B, Li S (2013) Study of configuration technology of ground pseudolite. Comput Measur Control 21(3):743–746
5. Xu Y, Chen X, Huang L et al (2012) Research on service pattern and platform design of location based common service on the basis of multi-application system. Geogr Geo-Inf Sci 28 (3):40–43, 48

Extract Common-Mode Error in Middle-Scale GPS Network Using Principal Component Analysis

Hao Zhang, Cuilin Kuang and Chenlong Lu

Abstract Based on the correlation coefficient between the time series of the GPS network, extracting the common model error (CME) using principal component analysis (PCA) in Crustal Movement Observation Network of China (CMONOC). This method does not need the assumption of spatially uniform distribution like stacking. Compared with the traditional regional stacking (RS), correlation weighted regional stacking (CWRS), and traditional PCA, this method extracts CME more accurately and reduces the influence of local effects during extracting the CME.

Keywords GPS · Correlation · Common model error · Spatiotemporal filtering

1 Introduction

With the improvement of observation technology and accuracy of positioning, continuous GPS observation becomes the main measure of deformation monitoring and is widely used in analysis of regional crustal deformation, study of slow earthquake event, analysis of global plate motion, and many other observation fields. China launched the scientific state key projects "Crustal Movement Observation Network of China (CMONOC)" officially in 1997. In the first phase of the project, they built 25 cors in China's mainland, including the Yongxing Island in Hainan. When the second phase project of CMONOC was launched, the number of cors has increased to 260. Meanwhile, the number of cors built by local government and other departments has been over 400 and it is expected to be over 1000 in the next 10 years.

There are some spatial-correlated public noises in GPS observation time series, which we used to call it common-mode error (CME) [1–3]. The methods we mainly take to eliminate the common-mode errors in coordinate are regional stacking

H. Zhang (✉) · C. Kuang · C. Lu
Department of Surveying and Remote Sensing Science, Central South University, Changsha 410083, China
e-mail: 15200831421@163.com

J. Sun et al. (eds.), *China Satellite Navigation Conference (CSNC) 2016 Proceedings: Volume I*, Lecture Notes in Electrical Engineering 388,
DOI 10.1007/978-981-10-0934-1_33

(RS) [4], principal component analysis (PCA), and Karhunen–Loeve expand (which the range of the area is less than 500 km, normally). Recent studies have indicated that the correlation of GPS coordinate time series will reduce with the increase of the distance [6]. Therefore, the CME of GPS observation network with big space scale will not be uniformed any more. The solution used to take is dividing the area into subrange. For example, SOPAC has divided its whole PBO network into several subdomains [4]. But this method is limited. On the one hand, we lack a unified standard for the divisions. On the other hand, the results after filtering of the edges interlaced are not unified. In this paper, we study the correlation of residual time series and extract the CME of CMONOC using correlation-based PCA.

2 Spatiotemporal Filtering

2.1 *Regional Stacking*

For a regional network daily station coordinate residual time series with n stations and spanning m days (north, east, or vertical), the CME is computed for days i = 1, 2, …, m by

$$\varepsilon_i = \frac{\sum_{k=1}^{n} \frac{x_{i,k}}{\sigma_{i,k}^2}}{\sum_{k=1}^{n} \frac{1}{\sigma_{i,k}^2}} \tag{1}$$

where $x_{i,k}$ is the residual value of the kth station at epoch i and $\sigma_{i,k}$ is its standard error.

2.2 *Correlation Weighted Regional Stacking*

Previous studies have indicated that the correlation besides the station's can well represent the similarity of their CME inside. Therefore, Tian and Shen [7] puts out correlation weighted stacking, which adds correlation coefficient as weight to the traditional RS. The CME is computed by

$$\varepsilon_{i,j} = \frac{\sum_{k=1}^{n} \frac{x_{j,k}}{\sigma_{j,k}^2} \times \frac{r_{k,j}}{\max\{|r_{k,j}|\}}}{\sum_{k=1}^{n} \frac{1}{\sigma_{j,k}^2} \times \frac{r_{k,i}}{\max\{|r_{k,j}|\}}} \tag{2}$$

where $x_{i,k}$, $\sigma_{i,k}$ and $\sigma_{i,k}$ is defined in (7). $r_{k,j}$ is the correlation coefficient besides the kth and jth station.

2.3 Principal Component Analysis

PCA is an orthogonal function analysis. It has a certain correlated matrix decomposed into unrelated vector (called principal component). The principal components represent the temporal variations, and the eigenvectors represent the spatial responses to the principal components. It is a statistical method for dimension reduction.

For the residual value time series X, the element of covariance matrix B used in PCA is defined as

$$b_{i,j} = \frac{1}{m-1}\sum_{k=1}^{m} x_{k,i}\, x_{k,j} \tag{3}$$

where the $x_{k,i}$, $x_{k,j}$ is defined in (1) $(i,j = 1, 2, 3, \ldots, n)$. The $(n \times n)$ symmetric matrix B can be decomposed as

$$B = V\Lambda V^{\mathrm{T}} \tag{4}$$

where the Λ is a diagonal matrix ranked n consisting of nonzero eigenvalues. And V is a $n \times n$ matrix with orthonormal rows. The data matrix X can be expanded as

$$X = AV^{\mathrm{T}} \tag{5}$$

which means

$$x_{i,j} = \sum_{k=1}^{n} a_{i,k}\, v_{j,k} \tag{6}$$

And the A is a $m \times n$ matrix. The element of it can be computed by

$$a_{i,j} = \sum_{k=1}^{n} x_{i,k}\, v_{k,j} \tag{7}$$

where $v_{j,k}$ is the jth element of kth principal component. $a_{i,k}$ and $v_{j,k}$ is called pattern, corresponding to pattern's time characteristics and spatial response, respectively. If we arrange the eigenvectors so that the eigenvalues are in descending order, the first few principal components represent most of the contributors to the variance of the residual time series. The principal component can get from

$$\varepsilon_{i,j} = \sum_{k=1}^{p} a_{i,k} v_{j,k} \tag{8}$$

where p is the number of principal components which have the bigger eigenvalues. Due to Dong's [3] research, the principal components which has the consistent spatial responses will be considered as CME.

2.4 Principal Component Analysis Based on Correlation

The PCA cannot identify the stations which strongly affected by the local loading and the results will be contaminated because of this. In order to extract CME more accurately, the paper proposes a method combined PCA and correlation coefficient.

The elements of correlation coefficient matrix r_{xy} is computed by

$$r_{xy} = \frac{\sum_{k=1}^{N}(x_k - \bar{x})(y_k - \bar{y})}{\sqrt{\sum_{k=1}^{N}(x_k - \bar{x})^2 \sum_{k=1}^{N}(y_k - \bar{y})^2}} \tag{9}$$

where N is the total number of public epoch; $\bar{x}$ and $\bar{y}$ is the residual values` mean of the two GPS stations. The process of our experiment is (1) traversal to one stations of CMONOC; (2) computing the correlation coefficients with other stations; (3) choose the stations which its correlation coefficient with the traversal station is above 0.8; (4) extract CME from the network constituted of the chose stations.

3 Experimental Results and Analysis

3.1 Data Preprocessing

The experiment needs to extract the residual time series by removing the structural information and the nonstructural information. Assume that the coefficient a is for the coordinate of the station; coefficient b is for the velocity; coefficients c, d represent annual periodicity; e, f represent half annual periodicity; g_j represent the changing velocity during the earthquake; h_j is the amplitude logarithmic decreased; H is a step function; v_i is the noise. This paper uses the models as follows:

$$\begin{aligned} y(t_i) &= a + bt_i + c\sin(2\pi t_i) + d\cos(2\pi t_i) \\ &\quad + e\sin(4\pi t_i) + f\cos(2\pi t_i) + \sum_{j=1}^{N_{\text{quake}}} g_j H(t_i - T_j^{\text{quake}})t_i \\ &\quad + \sum_{j=1}^{\text{Nquake}} h_j \ln[1 + (t_i - T_j^{\text{quake}}/\tau_j)H(t_i - T_j^{\text{quake}}) + v_i \end{aligned} \tag{10}$$

There are gross errors in time series of observations due to many external or internal factors and these errors must be excluded before the analysis. Because the methods mentioned in this paper need the time series do not have any missing data, this paper interpolate the time series with the Kriged–Kalman filter [8]. And we correct the phase step due to the antenna replacement.

3.2 Assumption

The simulation experiment selects 12 stations of CMONOC which have the high observation quality. We added amplitude of 4 and 8 annual periodicity noises on the site No. 5 and site No. 12 as local effects, respectively. Using the PCA and PCA based on correlations, we obtained the same results in *N*, *E*, *U* directions. Due to space limitations, here only make the illustration on *E* direction. After filtering, the spatial responses is like the picture below.

Figure 1 shows that PCA has been unable to distinguish CME from the principal component due to the influences of the local loading; and spatial response also shows huge difference before and after. Figure 2 improves that the PCA based on correlation can filter the unusual station. And the spatial response of the first

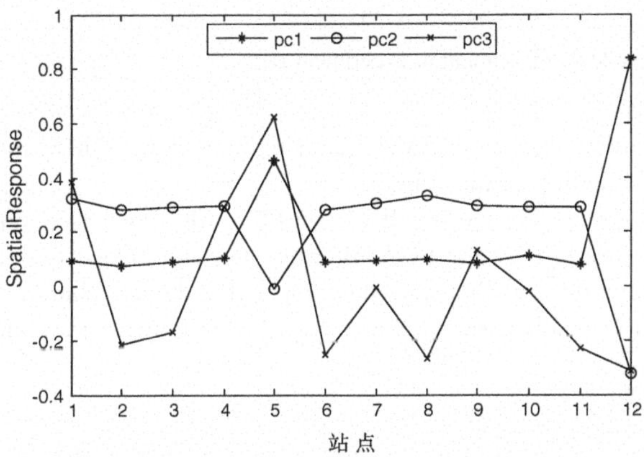

Fig. 1 Spatial response of PCA

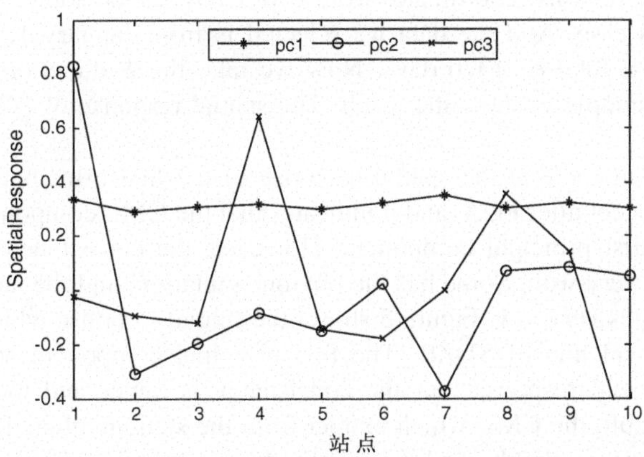

Fig. 2 Spatial response of PCA based on correlation

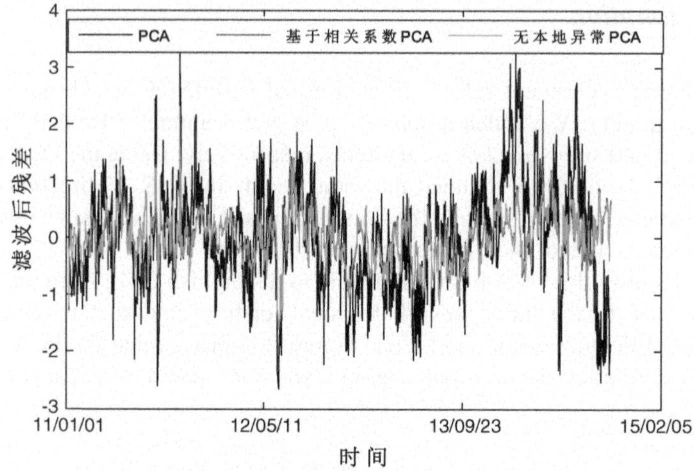

Fig. 3 Filtering residuals of simulation experiment (E)

principal component is more uniform. After extracting the CME, the residual time series of the first station is like Fig. 3.

As you can see above, the result of the PCA filtering based on correlation is better than which only uses the PCA filtering, and is similar to which did not add the local loading. All these characteristics imply that the correlation coefficients can filter those stations with a strong local error, which overcomes the weakness of PCA.

3.3 *Case Analysis*

We choose 208 stations' coordinates from CMONOC as the source of data for the analysis of this case. And the data we selected is from January 1, 2011–2014 in October 1 and a total of 1370 days. Now we take the *E* direction of the SDZB station as an example to show the result. The spatial response of PCA is shown in Fig. 4.

Figure 4 improves that the spatial response of the first principal component is generally consistent after PCA, and it indicates that the CME component we need to extract is the first principal component. However, the consistency of the spatial response is not very well. Some has the obvious mutation, and the amplitude of the spatial response is very low. Figure 5 shows the spatial response which takes 0.8 as correlation critical line of SDZB. The first principal component still has a concordance in spatial response, but the consistency is better and the amplitude is higher. As a result, the CME which extract from the stations filtered by correlation coefficient are more reliable, and the amplitude is more accurate.

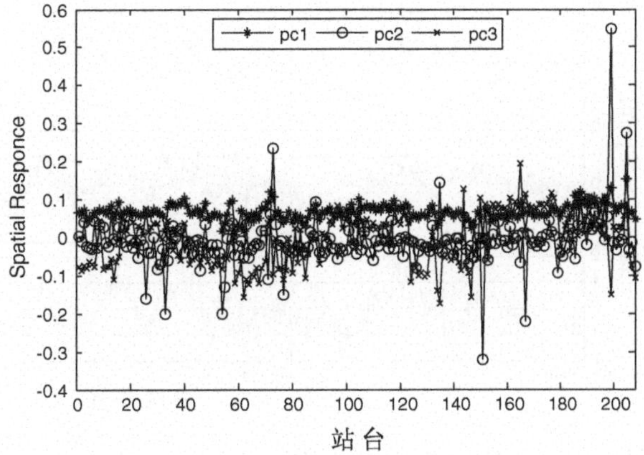

Fig. 4 Spatial response of PCA(SDZB)

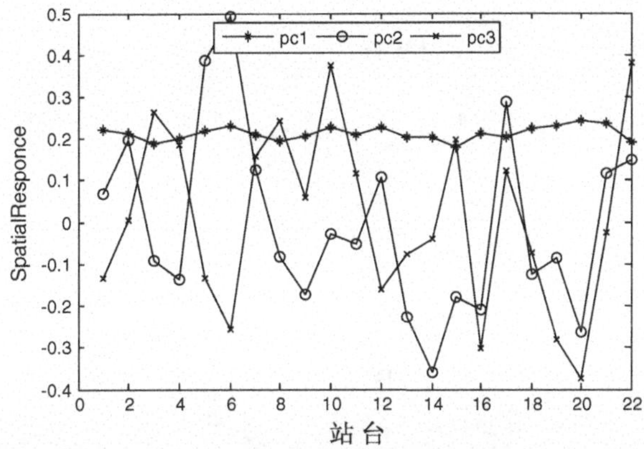

Fig. 5 Spatial response of PCA based on correlation (SDZB)

Figure 6 shows the results using different filtering methods to extract the CME. And we can get the similar result using three common filtering methods, but the residual error is significantly smaller using the method that we mentioned in this paper. The effects have been improved especially in the second half of 2011 to the beginning of 2012, the beginning of 2013 to the middle of 2013, and the end of 2014.

Figure 7 is the overall distribution of RMS of CMONOC after filtering. We can see the RMS of residual error is minimal in most station using PCA based on correlations coefficient, and the effects are remarkable in the area where the stations are distributed densely. Due to space of paper, we just list the distribution map of *E* direction, and the results of *N* direction and *U* direction are similar. As we can see from the distribution, the RMS of the method put forward this paper is much smaller than other three methods in the area where distribution of the stations is dense. But the effects are not obvious in the northeast and southwest region where distribution of the stations is sparse. Meanwhile, some of the stations cannot meet

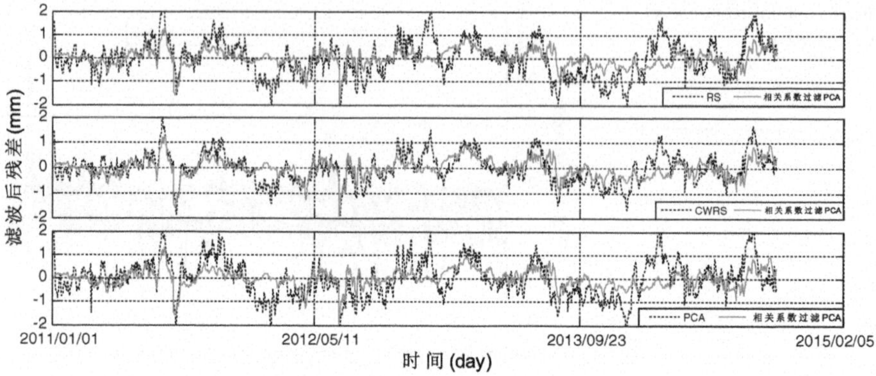

Fig. 6 Filtering residuals of SDZB(E)

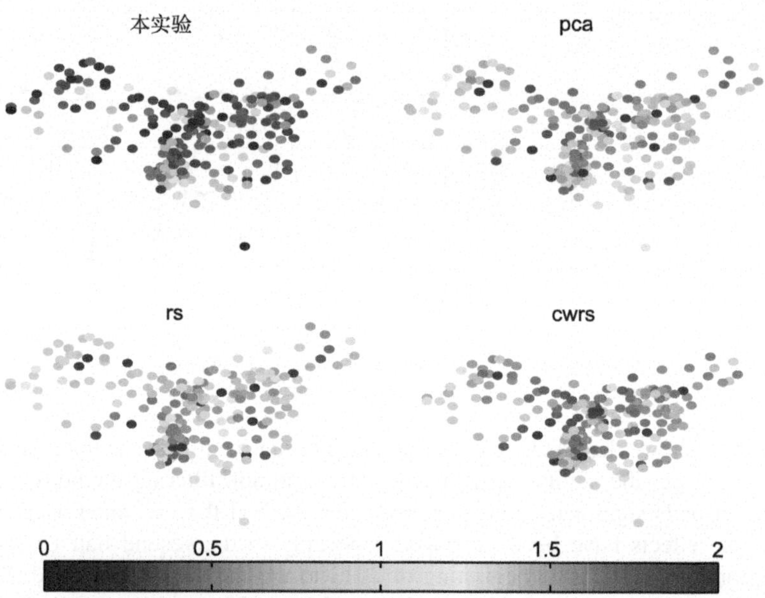

Fig. 7 Distribution of RMS in CMONOC(E)

two kinds of conditions in the experiment, so these stations cannot adopt the method mentioned in this paper to improve.

At last, we count the RMS of all the stations with all directions. The improvement rate is similar at direction *N* and *E*. And improvement rate of our means is highest among all the filtering methods (Table 1).

Table 1 The RMS lifting rate of residual series

Filtering	Lifting rate of RMS		
	N (%)	*E* (%)	*U* (%)
PCA based on correlation	47	42	39
PCA	29	28	18
CWRS	42	38	35
RS	21	20	25

4 Conclusion and Discussion

This paper proposes a spatial filtering method based on correlations. It overcomes the limitation of the traditional technique, such as the disunion of the standard for the area division, and rejects the hypothesis of space uniform distribution. It is more tallying with the actual situation than RS. Simultaneously, it also avoids the stations which affected by local loadings during the filtering and promote the result. Nonetheless, this method is more effective for regions with intensive stations. For distant regions with fewer stations, we need to do something one more step.

References

1. Wdowinski S, Bock Y, Zhang J et al (1997) Southern California permanent GPS geodetic array: spatial filtering of daily positions for estimating coseismic and postseismic displacements induced by the 1992 Landers earthquake. J Geophys Res Part B Solid Earth 102(8):18057–18070 (Printed Edition)
2. Nikolaidis R (2002) Observation of geodetic and seismic deformation with the global positioning system
3. Dong D, Fang P, Bock Y et al (2006) Spatiotemporal filtering using principal component analysis and Karhunen-Loeve expansion approaches for regional GPS network analysis. J Geophys Res Solid Earth (1978–2012) 111(B3)
4. Zhang J (1996) Continuous GPS measurements of crustal deformation in southern California. University of California, San Diego
5. Hu JC, Wu JC, Sun YF (2009) Comparison among three spatiotemporal filtering methods for regional GPS networks analysis. J Geod Geodyn 29(3):95–99
6. Williams SDP, Bock Y, Fang P et al (2004) Error analysis of continuous GPS position time series. J Geophys Res Solid Earth (1978–2012) 109(B3)
7. Tian YF, Shen YK. Correlation weighted stacking filtering of common-mode component in GPS observation network. Acta Seismologica Sinica 33(2):198–208
8. Mardia KV, Goodall C, Redfern EJ et al (1998) The kriged Kalman filter. Test 7(2):217–282

The BDS Multipath Hemispherical Map Based on Double Difference Residuals and Its Application Analysis

Qiang Shi, Wujiao Dai, Fanhe Zeng and Cuilin Kuang

Abstract Sidereal filtering is effective to mitigate multipath errors from observations or coordinates by taking advantage of the sidereal repeatability of GPS multipath signals. However, the characteristics of BDS multipath effect are different from that of GPS because the constellation and orbit of BDS are more complicated than that of GPS. According to the repeatability of BDS multipath signals, the multipath hemispherical map (MHM) is established for each satellite pair based on double difference residuals, and then the validity of the model is tested using real field observation data. The results show that the BDS MHM can reduce the root mean square (RMS) of double difference residuals by 83.6 % on average and improve the coordinate accuracy by 20.0, 52.2, and 16.5 % in *E*, *N*, and *U* directions, respectively. While observation sidereal filtering method improves the coordinate accuracy by 17.5, 50.0, and 13.0 %, and coordinate sidereal filtering method only improves the coordinate accuracy by 12.5, 42.2, and 7.8 % since it does not consider the repeat periods of MEO satellites multipath signals, which are about 7 days. Compared to the sidereal filtering methods, MHM is not only more effective to multipath mitigation, but also easier to implement in real-time applications since it can avoid calculating the accurate repeat period of multipath signals for each satellite.

Keywords BDS · Multipath effect · Sidereal filtering · Multipath hemispherical map

1 Introduction

In short-baseline GNSS measurements, various error sources, such as satellite and receiver clock errors, can be eliminated by double difference processing. Ionospheric and tropospheric delays can also be almost mitigated by difference technique.

Q. Shi (✉) · W. Dai · F. Zeng · C. Kuang
Department of Surveying and Remote Sensing Science,
Central South University, Changsha 410083, China
e-mail: promise_shi@163.com

J. Sun et al. (eds.), *China Satellite Navigation Conference (CSNC) 2016 Proceedings: Volume I*, Lecture Notes in Electrical Engineering 388,
DOI 10.1007/978-981-10-0934-1_34

Nevertheless, site-dependent multipath errors cannot be mitigated by the double difference technique, which becomes the dominant error source in GNSS measurements [1].

Currently, multipath can be mitigated by using three processing strategies [2]: receiver antenna design, receiver signal filtering and post-processing techniques. Although the special receiver antenna and receiver-internal correlation techniques can mitigate some multipath signals, the remaining errors are still the dominant error source in GNSS short-baseline measurements. The post-processing techniques are used to mitigate the multipath errors from the observation or coordinate domain, and then improve positioning accuracy, such as antenna array method [3], weighting GNSS observation based on signal noise ratio (SNR) [4], sidereal filtering techniques, [5] and so on. Due to the nearly exact repetition of GPS satellite geometry above a site every sidereal day, multipath signals are highly correlated across subsequent days under the same antenna and reflector environment, and it is possible to use the sidereal repeat periods to mitigate these errors. However, obtaining the accurate repeat periods of multipath signals is the key to sidereal filtering techniques, Choi et al. [6], Axelrad et al. [7], Ragheb et al. [8] and Agnew and Larson [9] researched the repeat periods of multipath signals and the results showed different satellites have different orbital repeat periods, therefore, it is not suitable to take a constant time as the repeat period of multipath signals. The sidereal filtering techniques include two methods, one of which is based on coordinate domain, which extracts the multipath errors from the coordinate sequences, and then timeshifted with a repeatable period to subtract the filter value from subsequent measurements, and the other is based on observation domain extracting multipath errors from observations, and mitigated from single or double differenced residuals. However, the multipath extracted from the coordinate sequences are the composite errors of all satellites, and the repeat periods calculated based on the coordinate sequences are not the accurate repeat periods of all satellites. In addition, this method needs a high-quality filtering method to denoise from the coordinate sequences. In contrary, the observation sidereal filtering method considers the repeat period of every satellite pair, which is more reasonable than coordinate sidereal filtering method. However, it is not obvious which shift period to use when different satellites are visible over the observation time period. Therefore, Zhong et al. [10] proposed a new sidereal filtering method based on GPS single differences, which allows the single differences of each satellite to be shifted individually to take account of the different orbital repeat periods of GPS satellites, and the results showed the sidereal filtering method based on single differences was better than that based on double differences. Although the sidereal filtering can weaken multipath signals in some extend, the degree of repeatability decreases as the interval between first day and subsequent days' increases. When the repeatability reduces to a low degree, the multipath model will not be effective to correct the multipath errors [11]. Therefore, Dong et al. [12] built the multipath hemispherical map (MHM) using a common receiver clock for the two antennas based on multipath spatiotemporal repeatability. This method can mitigate the multipath effectively and it is suitable for real-time processing, which avoids frequently calculating the accurate repeat period of multipath signals for each satellite. However, Dong et al. [12] built the MHM based on a common receiver clock for the two antennas, which is not suitable for conventional high-precision data processing.

Currently, some researches have focused on BDS multipath signals. Wu et al. [13] analyzed the multipath characteristics of GEO/IGSO satellites, and mitigated the pseudorange multipath errors by CNMC algorithm. Shi et al. [14] analyzed the pseudorange multipath characteristics of BDS and GPS, and the results showed that the noise level of BDS GEO and IGSO code measurements are higher than that of the GPS satellites as a whole. Ma and Shen [15] extracted the multipath signals of pseudorange and phase based on the combinations of triple frequency measurements, and researched the multipath periods of GEO, IGSO, and MEO satellites using the foregoing spectrum analysis methods. Ye et al. [16] investigated the characteristics of BDS carrier phase multipath signals and improved the positioning results by mitigating BDS multipath errors based on coordinate and observation domain. Wang et al. [17] researched the characteristics of GEO multipath signals in code range using spectral analysis and correlation analysis, and improved the precision of SPP by mitigating the extracted multipath.

In summary, the current researches on BDS multipath signals are mainly focused on pseudorange multipath or preliminary study on carrier phase multipath. Although Ye et al. [16] improved the positioning accuracy by mitigating the carrier phase multipath signals using sidereal filtering based on coordinate and observation domain, the delay time of the adjacent days is not constant, therefore, the sidereal filtering needs to calculate the accurate multipath delay time of each satellite for everyday, which increases the complexity of data processing. In addition, the sidereal filtering based on single difference can mitigate GPS multipath effectively, but Ye et al. [16] recovered the single differenced residuals from double differenced residuals using an independent constraint based on the assumption that the sum of observed satellites' single differenced residuals will be zero instead of using single difference model to adjust directly. However, the carrier phase multipath errors of GEO satellites are systematic, therefore, it is not suitable to recover the single differenced residuals from double differenced residuals using a simple "zero mean" assumption, which will add new errors to BDS positioning results. For this reason, drawing lessons from GPS MHM [12], we propose a new method based on double difference model, which is suitable to mitigate BDS multipath signals effectively and real timely named "BDS multipath hemispherical map", and comparative analysis of BDS MHM and sidereal filtering is carried out using real field observation data.

2 Multipath Basic Principle and the Establishment of BDS MHM

2.1 Multipath Basic Principle

The receiver only receives the direct signal from satellites under the ideal condition. However, the real signal received by receiver is a superposition of the direct and reflected signal due to the presence of reflecting objects around the antenna. The superposed signal has a time delay with a longer path than the original signal causing the large propagation delay, and that is the multipath effect. The principle of multipath signals is shown in Fig. 1.

When the reflector is known, the carrier phase errors caused by multipath signals can be calculated as follows:

$$\begin{cases} S_M = \dfrac{\Delta_c}{2\pi}\lambda = \dfrac{\lambda}{2\pi}\arctan\left(\dfrac{\alpha\sin\Delta\varphi}{1+\alpha\cos\Delta\varphi}\right) \\ \Delta\varphi = 4\pi\cdot s\cdot\sin\beta/\lambda \end{cases} \quad (1)$$

where s is the distance between antenna and reflector; β is the incident angle of reflection signal; λ is the signal wavelength. It can be seen from Eq. (1) that when we know the reflector and satellites position, the multipath errors can be calculated, and the maximum of the multipath error is 0.25 λ while $\Delta_c = \pi/2$.

2.2 The Establishment of BDS MHM

The MHM is proposed based on the spatiotemporal repeatability of multipath signals [12]. It can be seen from Fig. 1 and Eq. (1) that if the multipath environment is keept unchanged, the multipath error will be a constant for the reflected signal which has the

Fig. 1 Principle of multipath signals

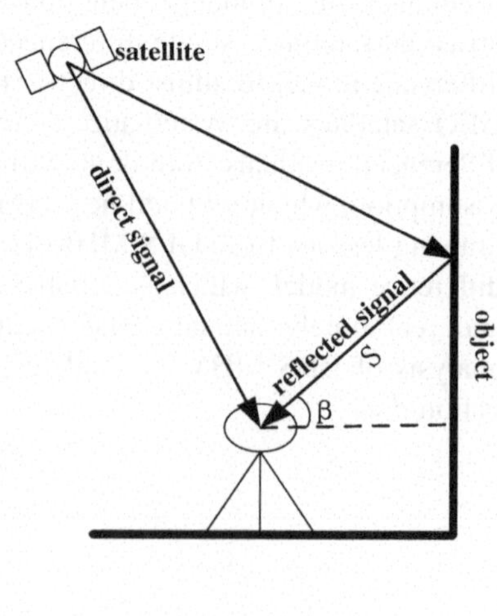

same elevation and azimuth. Based on this feature, the multipath signals are divided into spatial grid network related to the elevation and azimuth, taking the average of satellites' multipath errors located in the grid as the multipath error value of this grid.

As BDS has three different constellations which operate more complex than GPS, the characteristics of BDS multipath signals are different from that of GPS. Dong et al. [12] used a common receiver clock for the two antennas to build MHM based on single difference model, which did not have universal applicability. Therefore, we build BDS MHM based double difference model to mitigate the BDS multipath signals, which can also remove the receiver and satellite clock errors. As GEO satellites are basically stationary and the five GEO satellites are visible all day in most areas of China, we take C03 satellite as the fixed reference satellite to avoid the frequent change of reference satellite, which has a high elevation angle (about 57°) with little multipath signals. The modeling process is as follows:

First, taking C03 satellite as the reference satellite to calculate the double difference residuals of each satellite pair based on double difference model. Because of the low frequency characteristics of multipath signals, the low frequency wavelet filter is used to recover the high-precision low frequency multipath signals [18]. In this paper, the db8 wavelet is used to filter the high frequency noise, which has better denoising function. Second, building the MHM for GEO, IGSO, and MEO satellites respectively, and 1° by 1° grid MHM models are constructed for IGSO and MEO satellites, while we construct 0.1° by 0.1° grid MHM models for GEO satellites doing repetitive movement in small areas to mitigate GEO multipath signals effectively, which change slowly with their movements. Finally, in order to consider the repeat periods of all satellites especially for MEO satellites, we use seven days' data to build the MHM model. Each satellite multipath error for every grid is calculated within seven days, and the average of seven days for every grid is taken as its multipath error value.

3 Data Collection and Processing

Data were collected at sampling rate of 1 Hz with a 0° elevation mask angle from November 18, 2015 to December 01, 2015 on the rooftop of Mining Building at Central South University using Trimble Net R9 for 14 days. Simultaneous observation was taken by two Trimble receivers in this experiment, one of which was base station (Fig. 2a) located in the open environment without any shelter, and the other was rove station (Fig. 2b) located in strong multipath environment. In this experiment, one reflector was mounted on the north of rove station to create strong multipath environment. In order to completely eliminate the ionospheric and tropospheric delays and orbital errors using double difference model, the design length of this baseline was 6.5 m.

We processed BDS observations using our in-house GNSSDMS software with a 10° elevation mask angle, which is designed for short-baseline deformation monitoring based on Visual Studio 2010, operating in both static and kinematic modes.

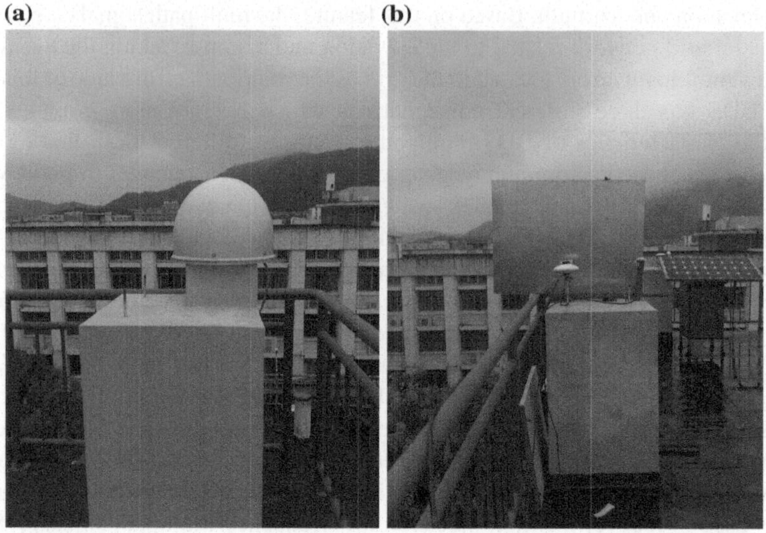

Fig. 2 **a** Base station. **b** Rove station

Here only the B_1 phase observations were processed using single-epoch kinematic module in this software. First, we processed the observations of first seven days to obtain the multipath errors of each satellite pair, and then built the BDS MHM following the step in Sect. 2. Secondly, the observations of last seven days were processed using the BDS MHM as well as sidereal filtering based on coordinate and observation domain to test the validity of BDS MHM by making a comparative analysis of these results.

4 Analysis of Experiment Result

In order to visually observe the variations of BDS multipath errors with the motion of satellites, we related the multipath errors of GEO, IGSO, and MEO with their movements, and showed the spatial variations of multipath errors in graphical

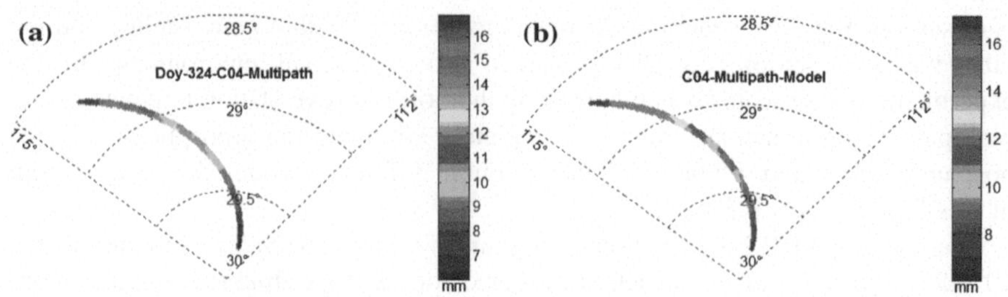

Fig. 3 **a** The sky map of C04 multipath errors in Doy-324. **b** The sky map of C04 MHM

representation (Figs. 3, 4, and 5). Due to space limitation we selected one satellite of each kind constellation to analyze.

Figure 3a shows the variations of C04 multipath errors with its motion in Doy 324, and Fig. 3b represents the C04 MHM based on the observations of first seven days. The value of the color represents the multipath error value. Owing to GEO satellites do repetitive movement in small areas, the sky map of C04 multipath errors is drawn in Fig. 3 using an enlarged scale. It shows that the multipath signals from GEO satellites are not a constant but are changing slowly. Moreover, the sign of C04 multipath error value is accordant, which indicates the multipath errors of GEO satellites are systematic. In addition, the sky map of multipath errors in Fig. 3a is similar to that in Fig. 3b, which indicates that, the MHM of C04 satellite is basically accurate.

The variations of C10 multipath errors with its motion in Doy 324 and the MHM of C10 satellite based on the observations of first seven days are visible in Fig. 4a and b. The two figures show that when satellite moves to the north, the color in the sky map will become deepened, which means, the multipath errors become large, which is in accordance with the actual observation environment. We mounted one reflector on the north of rove station to create strong multipath environment in this experiment, and therefore the multipath errors will become large when satellite moves toward the reflector. We can also see from Fig. 4 that the multipath errors are large when satellite has a low elevation, which is more vulnerable to multipath effect. In addition, the sky map of multipath errors in Fig. 4a is similar to that in Fig. 4b, which indicates the MHM of C10 satellite is basically accurate.

Figure. 5 shows the variations of C14 multipath errors with their motions over two consecutive days (Doy 324 and 325). Since MEO satellites for BDS are designed for global coverage, the range of MEO satellites movement is larger than that of GEO and IGSO, which can be seen from Figs. 3, 4, and 5. When satellite moves toward the reflector, the multipath errors will become large, which is similar to IGSO satellites. Due to the orbital periods of MEO satellites are about seven days (see [16]), multipath errors for MEO satellites have no relationship over two consecutive days, which can be seen in Fig. 5.

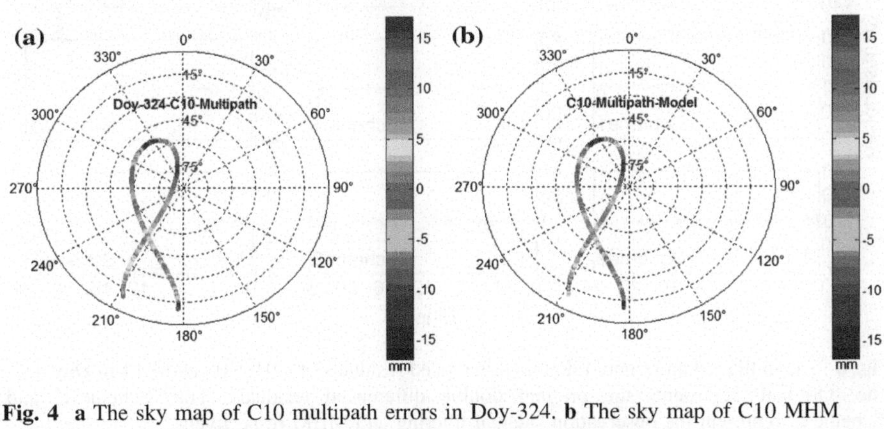

Fig. 4 **a** The sky map of C10 multipath errors in Doy-324. **b** The sky map of C10 MHM

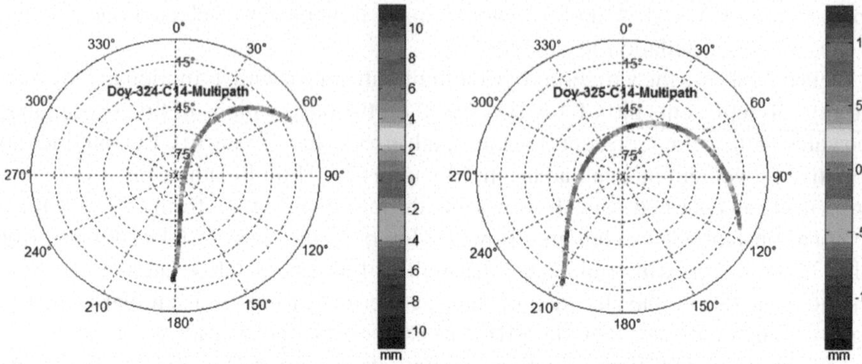

Fig. 5 The sky map of C14 multipath errors in Doy-324 and Doy-325

In order to test the validity of BDS MHM, the observations of last seven days were processed using the BDS MHM as well as sidereal filtering based on coordinate and observation domain by making a comparative analysis of these results. Owing to the MHM and observation sidereal filtering eliminate the multipath errors from observations, the improvements on observations for these two methods are compared to analyze the validity of BDS MHM. Due to space limitation, one satellite of each kind constellation was selected to analyze its double difference residuals in Doy 332 (see Fig. 6). The double difference residuals were denoised to eliminate high frequency noise using wavelet filtering method.

The original and filtered residuals of C04, C08, and C14 in Doy 332 are shown in Fig. 6. The upper part in Fig. 6 represents the original double differenced

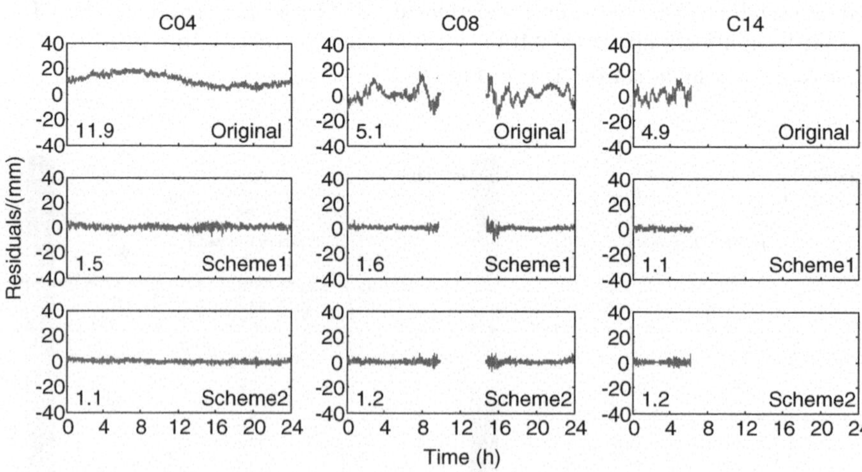

Fig. 6 The unfiltered and filtered double differenced residuals of C04, C08 and C14 in Doy 332. The 'Original' represents the original double differenced residuals. The 'Scheme 1' and 'Scheme 2' represent the observation sidereal filtering and MHM respectively

Table 1 The RMS statistics of double differenced residuals (unit: mm)

PRN	Original	Scheme 1	Improvement %	Scheme 2	Improvement %
C04	11.9	1.5	87.4	1.1	90.7
C08	5.1	1.6	68.6	1.2	76.5
C14	4.9	1.1	77.5	1.2	75.5
Ave	7.3	1.4	80.8	1.2	83.6

residuals, and the middle of that represents the filtered double differenced residuals based on observation sidereal filtering, and the bottom part of Fig. 6 represents the filtered double differenced residuals based on MHM. The number in the bottom left corner of each subplot indicates the RMS value of the double differenced residuals, and the unit is millimeter. As shown in Fig. 6, the improved double differenced residuals tend to be white noise. In order to quantitatively analyze the improvements of these two methods on observations, the RMS values of remaining residuals are calculated in Table 1. From this table, after sidereal filtering and MHM are applied, the improvements for double differenced observation residuals are evident. The average improvement of observation sidereal filtering is 80.8 % while that of MHM is 83.6%, which indicates that the improvement effect of MHM is better than that of observation sidereal filtering.

To further test the validity of BDS MHM, the positioning results before and after applying the BDS MHM are analyzed as well as comparing to the sidereal filtering results. The coordinate residuals sequences of one day are shown in Fig. 7.

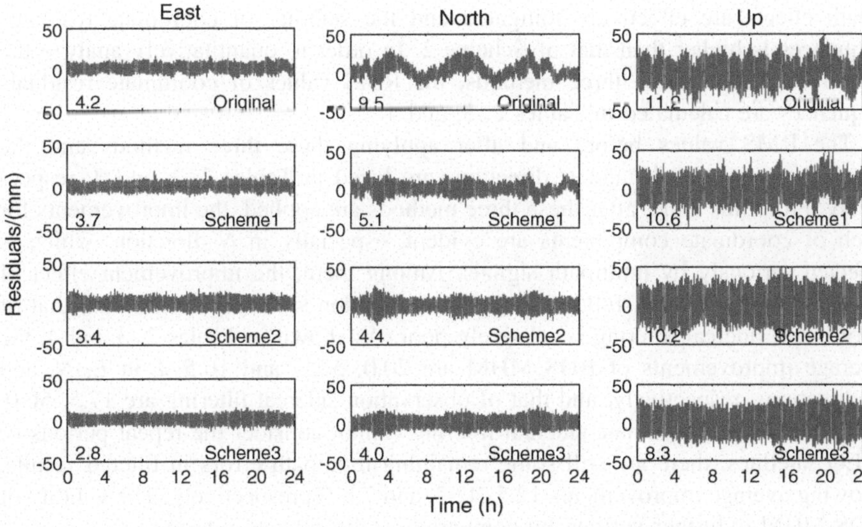

Fig. 7 The unfiltered and filtered residuals of *E*, *N*, *U* directions in Doy-332. The number in the *bottom left corner* represents the RMS value of the coordinate residuals. The 'Original' represents the original coordinate residuals. The 'Scheme 1', 'Scheme 2' and 'Scheme 3' represent the filtered coordinate residuals based on coordinate domain, observation domain and MHM

The original and filtered residuals of *E*, *N*, and *U* directions in Doy 332 are shown in Fig. 7. The upper part in Fig. 7 represents the original coordinate residuals, which are not white noise but periodic errors affected by multipath signals, especially in *N* direction, and that is in accordance with the actual observation environment. As shown in Fig. 2b, due to the reflector was located in the north of rove station, the north direction would be affected seriously by multipath signals. Scheme 1 represents coordinate sidereal filtering, which effectively mitigates the multipath signals and the coordinate residuals sequences become more stable. However, there are still some large fluctuations in the residuals of *N* and *U* directions, such as 2–3 h and 19–24 h in *N* direction, which indicates that there are still some remaining multipath errors which are not mitigated by coordinate sidereal filtering. Owing to the repeat periods of MEO satellites are seven says, the multipath signals of MEO satellites have no repetition over two consecutive days, which are related to their orbital movements. Therefore, coordinate sidereal filtering cannot completely mitigate multipath errors, which does not consider the repeat periods of MEO multipath signals.

Scheme 2 represents sidereal filtering based on observation domain. The multipath repeat periods of GEO, IGSO, and MEO are considered respectively, of GEO and IGSO satellites are approximately a sidereal day, while around seven sidereal days for the MEO satellites. Due to observation sidereal filtering considers the orbital repeat periods of MEO satellites, the coordinate residuals sequences are more stable than that of coordinate sidereal filtering, which tend to be white noise. Scheme 3 represents the BDS MHM, and this model also consider the orbital repeat periods of MEO satellites by taking the average of seven days' multipath errors to establish BDS MHM. Intuitively, as shown in the bottom part of Fig. 7, the multipath effects are effectively mitigated and the stability of coordinate residuals sequences is higher than that of Scheme 2. In order to quantitatively analyze the improvements of these three methods, the RMS values of coordinate residuals sequences are calculated in Tables 2, 3, and 4.

The RMS values before and after applying these three methods and the improvements in *E*, *N*, and *U* directions are listed in Tables 2, 3, and 4, respectively. From this table, after these three methods are applied, the improvements for each of coordinate components are evident, especially in *N* direction, which is affected seriously by multipath signals. Among them, the improvement effect of BDS MHM is the best, followed by the observation sidereal filtering, and that of coordinate sidereal filtering is relatively poor. As shown in Tables 2, 3 and 4, the average improvements of BDS MHM are 20.0, 52.2, and 16.5 % in *E*, *N*, and *U* directions, respectively, and that of observation sidereal filtering are 17.5, 50.0, and 13.0 %. As coordinate sidereal filtering cannot consider the repeat periods of MEO satellites, there are still some remaining multipath errors in filtered results, showing average improvements 12.5, 42.2, and 7.8 %, respectively. The validity of BDS MHM is further verified by comparing to the sidereal filtering.

Table 2 The RMS statistics of coordinate residuals in *E* direction (unit: mm)

Doy	Originall	Scheme 1	Improvement %	Scheme 2	Improvement %	Scheme 3	Improvement %
329	3.2	2.8	12.5	2.7	15.6	2.7	15.6
330	3.9	3.2	17.9	2.9	25.6	2.8	28.2
331	4.3	3.7	14.0	3.4	20.9	3.4	20.9
332	4.2	3.7	11.9	3.4	19.0	2.8	33.3
333	4.5	4.1	8.9	3.9	13.3	4.0	11.1
334	4.1	3.7	9.8	3.4	17.1	3.6	12.2
335	3.9	3.6	7.7	3.5	10.3	3.4	12.8
Ave	4.0	3.5	12.5	3.3	17.5	3.2	20.0

Table 3 The RMS statistics of coordinate residuals in *N* direction (unit: mm)

Doy	Originall	Scheme 1	Improvement %	Scheme 2	Improvement %	Scheme 3	Improvement %
329	8.2	4.9	40.2	4.1	50.0	3.9	52.4
330	9.3	5.0	46.2	4.3	53.8	4.1	55.9
331	9.5	5.3	44.2	4.5	52.6	4.5	52.6
332	9.5	5.1	46.3	4.4	53.7	4.0	57.9
333	9.6	5.5	42.7	4.7	51.0	4.4	54.2
334	8.8	5.1	42.0	4.6	47.7	4.2	52.3
335	8.1	5.3	34.6	4.8	40.7	4.9	39.5
Ave	9.0	5.2	42.2	4.5	50.0	4.3	52.2

Table 4 The RMS statistics of coordinate residuals in *U* direction (unit: mm)

Doy	Originall	Scheme 1	Improvement %	Scheme 2	Improvement %	Scheme 3	Improvement %
329	9.9	9.1	8.1	8.3	16.2	8.1	18.2
330	10.1	9.4	6.9	8.8	12.9	8.7	13.9
331	11.5	10.6	7.8	9.9	13.9	10.2	11.3
332	11.2	10.6	5.4	10.2	8.9	8.3	25.9
333	12.4	12.1	2.4	11.3	8.9	11.4	8.1
334	12.3	11.3	8.1	10.4	15.4	10.6	13.8
335	12.9	11.4	11.6	10.9	15.5	10.2	20.9
Ave	11.5	10.6	7.8	10.0	13.0	9.6	16.5

5 Conclusion and Discussion

According to the special design of BDS constellation, we improved the GPS MHM using double difference model by taking C03 satellite as the fixed reference satellite, and the BDS MHM was proposed in this paper, which is suitable to mitigate BDS multipath signals real timely. The real data were used to test the validity of BDS MHM by comparing to sidereal filtering, and some conclusions can be drawn as follows:

Given that the movements of different kind satellites are different, we built the model for each pair of satellites to increase the effectiveness of BDS MHM. From the sky map of BDS multipath effects, we can conclude that the multipath effects of GEO satellites have a systematic shift which are systematic errors while that of IGSO and MEO satellites are random errors. The comparative analysis of BDS MHM and observation sidereal filtering based on double difference residuals shows that the multipath signals are effectively mitigated by BDS MHM, showing average improvement 83.6 %, and the filtered double difference residuals tend to be white noise.

BDS multipath signals can be effectively mitigated by both MHM and sidereal filtering. As the multipath model is shifted as a whole in coordinate sidereal filtering, which does not consider the orbital repeat periods of MEO satellites, the improvement effect of that is relatively poor, while that of observation sidereal filtering is better, which considers the orbital repeat periods of each satellite pair. However, observation sidereal filtering needs to calculate the accurate repeat period of multipath signals for each satellite over two (for GEO and IGSO satellites) or seven (for MEO satellites) days, which increases complexity to data processing. The BDS MHM based on the observations of seven consecutive days not only considers the orbital repeat periods of each satellite pair, but also avoids the complexity to calculate the accurate delay times for each satellite. In addition, this model is easier to implement in real-time multipath mitigation, and the average improvements of positioning results are 20.0, 52.2, and 16.5 % in *E*, *N*, and *U* directions, respectively.

Although the BDS MHM proposed in this paper can effectively mitigate the multipath errors, there are still some deficiencies. First, despite its high elevation, the C03 satellite may be blocked by obstacles in actual measurement. When C03 satellite is invisible or unhealthy, the BDS MHM needs to be reestablished. Second, if there is a big change in surrounding environment, the BDS MHM needs to be rebuilt, which is established based on the assumption that the multipath environment keeps unchanged, and that is a common drawback of all filtering method. Due to the particularity of GEO multipath effects, it is not suitable to recover the single differenced residuals from double differenced residuals using a simple "zero mean" assumption. Therefore, the conversion method deserves further study, and then to test the validity of BDS MHM based on single difference residuals.

References

1. Leick A (2004) GPS satellite surveying. Wiley, Hoboken
2. Dai WJ, Huang DW, Cai CS (2014) Multipath mitigation via component analysis methods for GPS dynamic deformation monitoring. GPS Solutions 18(3):417–428
3. Ray JK (2000) Mitigation of GPS Code and carrier phase multipath effects using a multi-antenna system. University of Calgary, Canada
4. Lau L, Moke E (1999) Improvement of GPS relative positioning accuracy by using SNR. J Survey Eng 125(4):185–202
5. Ge L, Han S, Rizo C (2000) Multipath mitigation of continuous GPS measurements using an adaptive filter. GPS Solutions 4(2):19–30
6. Choi K, Bilich A, Larson KM et al (2004) Modified sidereal filtering: implications for high rate GPS positioning. Geophys Res Lett 31(22). doi:10.1029/2004GL021621
7. Axelrad P, Larson K, Jones B (2005) Use of the correct satellite repeat period to characterize and reduce site-specific multipath errors. In: Proceedings of ION GNSS, pp 2638–2648
8. Ragheb AE, Clarke PJ, Edwards SJ (2007) GPS sidereal filtering: coordinate- and carrier-phase-level strategies. J Geodesy 81(5):325–335
9. Agnew DC, Larson KM (2007) Finding the repeat times of the GPS constellation. GPS Solutions 11(1):71–76
10. Zhong P, Ding X, Yuan L, Xu Y et al (2010) Sidereal filtering based on single differences for mitigating GPS multipath effects on short baselines. J Geodesy 84(2):145–158
11. Dai WJ, Wu XX, Luo FX (2012) GPS multipath effect processing method based on augmented parameters Kalman filtering. Geomatics Inf Sci Wuhan Univ 37(4):423–427 (in Chinese)
12. Dong D, Wang M, Chen W et al (2015) Mitigation of multipath effect in GNSS short baseline positioning by the multipath hemispherical map. J Geod. doi:10.1007/s00190-015-0870-9
13. Wu X, Zhou J, Wang G et al (2012) Multipath error detection and correction for GEO/IGSO satellites. Sci China Phys Mech Astron 55(7):1297–1306
14. Shi C, Zhao Q, Hu Z, Liu J (2013) Precise relative positioning using real tracking data from COMPASS GEO and IGSO satellites. GPS Solutions 17(1):103–119
15. Ma XY, Shen YZ (2014) Multipath error analysis of COMPASS triple frequency observations. Sci Res 5:12–21
16. Ye SR, Chen DZ, Liu YY, Jiang P, Tang WM, Xia PF (2015) Carrier phase multipath mitigation for BeiDou navigation satellite system. GPS Solutions 19(4):545–557
17. Wang GX, de Jong Kees, Zhao QL, Hu ZG, Guo J (2015) Multipath analysis of code measurements for BeiDou geostationary satellites. GPS Solutions 19(1):129–139
18. Satirapod C, Khoonphool R, Rizos C (2003) Multipath mitigation of permanent GPS stations using wavelets. In: Proceedings of 2003 international symposium on GPS/GNSS. Tokyo, Japan, pp 15–18

Part III
BDS/GNSS Test and Assessment Technology

BeiDou Compatible Indoor Positioning System Architecture Design and Test Evaluation

Chao Ma, Jianyun Chen, Jun Yang, Dexiang Ming and Jinling Wang

Abstract In recent years, indoor space of human activities has been more and more huge and complex. Point of interest (POI) has been increasingly abundant. There is an intense trend of positioning and navigation in parking lot, shopping mall, and airport. Meanwhile, it is a matter for computer to recognize the specific position in the room on exact-marketing, intelligent manufacturing, robot and no one care etc. These requirements bring in a huge opportunity for IPS (indoor positioning system). Although GNSS has made great success in the aspect of positioning in recent years, traditional positioning (satellite positioning, base station location) can not be able to meet the indoor positioning because of technical restriction. At present, using Wi-Fi or Bluetooth beacon positioning technology that it acquires user's position by detecting the RSSI power and matching beacon position faces huge challenges on the account of laying a number of beacon station on Wi-Fi or bluetooth positioning system, poor stability in the results of positioning and the terrible noise interference. The paper comes up with a positioning system of BeiDou compatible system, which is prominent to reduce the change of intelligent mobile terminal and the complexity of infrastructure construction on indoor positioning, realizes indoor–outdoor positioning compatibility and seamless connection. The core of the system is high performance multi-output satellite navigation signal simulator. Every output port of the simulator is provided with independent output function of single satellite signal and produces the pseudolite signal that can be compatible with BeiDou system air interface and design special broadcast content. The receiver combines carrier-phase smoothing with different positioning technology and achieves indoor–outdoor seamless positioning by BeiDou positioning terminal. This paper builds BeiDou compatible indoor–outdoor positioning testing system and provides the test data and conclusion by test evaluation.

C. Ma (✉) · J. Chen · J. Yang · D. Ming
College of Mechatronic Engineering and Automation, National University of Defense Technology, Changsha, China
e-mail: 512639054@qq.com

J. Wang
School of Surveying and Spatial Information Systems, The University of New South Wales, Sydney, Australia

J. Sun et al. (eds.), *China Satellite Navigation Conference (CSNC) 2016 Proceedings: Volume I*, Lecture Notes in Electrical Engineering 388,
DOI 10.1007/978-981-10-0934-1_35

Keywords Indoor positioning · Pseudolites · Signal compatibility · Navigation signal simulation

1 Introduction

Indoor positioning system (IPS) is a framework that acquires the position at location of people and objects mainly adopting the wireless communication, base station location, and inertial navigation in the indoor environment [1, 2]. Location-based services (LBS) have been widely applied in people's lives. LBS became an important part of national defense, economic construction, and social life. With the modernization of the cities, the number of large buildings is rapidly increasing. More than 80 % of people's daily lives are in indoor environment. The requirement of indoor LBS is dramatically increasing. Indoor LBS is needed in fields such as special care, key building management, Internet of Things, and personal service [3–5]. Traditional positioning (satellite positioning, base station location) cannot be able to meet the indoor positioning because of technical restriction. GPS cannot realize positioning due to the obstruction of wall. There is a low distribution density of mobile base station so that the precision of positioning results is thousands meters by neighbor joining method. Even though increasing the number of base stations and using TOA (time of arrival) method, the precision of positioning results is 200 m. In the face of broad market space for LBS in the indoor environment, it cannot be too much to develop IPS. The current indoor wireless location technologies have Wi-Fi, Bluetooth, infrared, UWB, RFID, and ultrasonic etc. Table 1 [6] explains the positioning precision, merits and demerits about all kinds of positioning technologies.

We can find the common problems about the incompatibility of indoor–outdoor positioning technology, difficulty to integrate intelligent mobile terminal and complexity to construct infrastructure are the bottleneck that restricts the mass commercial application in Table 1. Therefore, the paper comes up with a BeiDou

Table 1 Indoor positioning technology comparison

Technology	Precision	Merits	Demerits
Wi-Fi	2–50 m	Wide network, low cost, strong communication	Vulnerable to environmental interference
Bluetooth	2–10 m	Small in size, easily integrating, easily popularize	Short communication distance, poor stability
Infrared	5–10 m	High location precision	Short transmission distance, vulnerable to interference line of sight
UWB	6–10 cm	High precision, strong penetrability	High cost, small coverage
RFID	5 cm–5 m	Low cost, high precision	Without the ability of communication, short distance
Ultrasonic	1–10 cm	High location precision	Short transmission distance affected by environmental temperature and humidity

compatible IPS, which realizes the compatibility of indoor–outdoor positioning decreases the change of intelligent mobile terminal and reduces the construction of infrastructure.

2 The Framework of Pseudolite-Based Indoor Positioning System

The core of the system is high performance multi-output satellite navigation signal simulator that is provided with the function of channel simulation. Every output port of the simulator is equipped with independent output function of single satellite signal and produces the pseudolite signal that can be compatible with BeiDou system air interface and design special broadcast content.

BeiDou IPS includes BeiDou time receiver, multi-output satellite navigation signal simulator, pseudolite antennas, and link wire in Fig. 1. BeiDou time receiver receives air BeiDou signals timing for multi-output satellite navigation signal simulator in order to synchronize the time system of indoor–outdoor. Multi-output satellite navigation signal simulator broadcasts the pseudolite signals whose time uses BeiDou time and whose coordinate system adopts CGCS2000. The navigation message is the ephemeris parameter of the fixed pseudolite.

The innovation of the system is all the pseudolite signals generated by multi-output satellite navigation signal simulator so that the hard problem about the

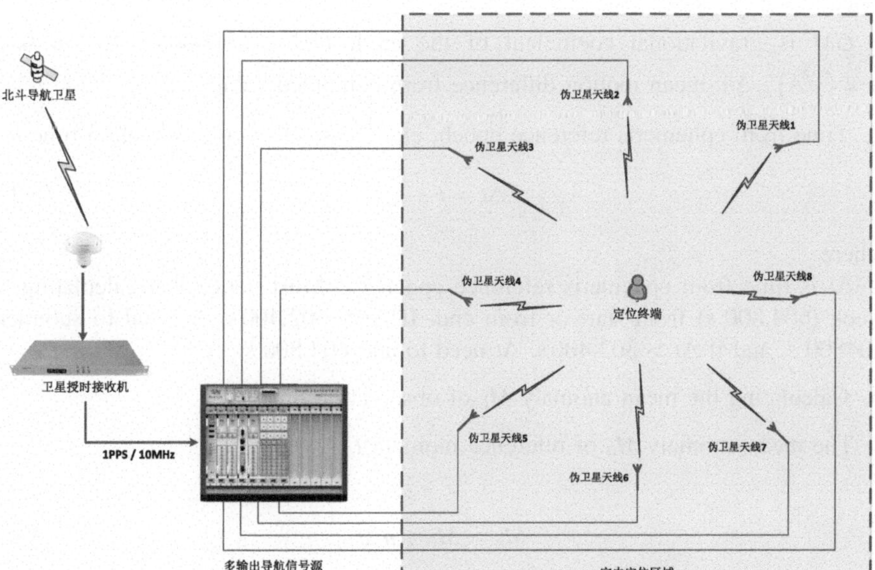

Fig. 1 Framework of Beidou indoor positioning system

high precise time synchronization of pseudolite system and the estimation of clock bias is completely solved. Otherwise, application of simulator also simplifies the infrastructure construction.

3 The Key Technical Analysis of Pseudolite System

3.1 Fixed Satellite Position Message Broadcast Technology

In order to realize indoor–outdoor seamless connection, the system needs to broadcast the same navigation signals as BeiDou. Traditional positions of satellites are calculated by 6 track number and correct number based on the Kepler's three laws. It is an appropriate method to set zero for the time-related parameter related to satellites angular velocity so that the location of the satellite do not change over time under any circumstances. For MEO/IGSO satellites, traditional navigation position calculation process is as follows [7, 8]:

1. Corrected mean motion n

Computed mean motion n_0, use the equation

$$n_0 = \sqrt{GM} \Big/ a^{\frac{3}{2}} \tag{3.1}$$

$$n = n_0 + \Delta n \tag{3.2}$$

where

GM is gravitational coefficient of the earth $GM = 3.986005 \times 10^{14}\,\mathrm{m}^3/\mathrm{s}^2$, $a = \left(\sqrt{A}\right)^2$, Δn: mean motion difference from computed value.

2. Time from ephemeris reference epoch, clock correction of observation time t

$$\Delta t = t - t_{oe} \tag{3.3}$$

where

Δt is time from ephemeris reference epoch t_{oe}. Must notice that calculating a week (604,800 s) from start or from end. If $\Delta t > 302{,}400\,\mathrm{s}$, Δt need to subtract 604800 s, and if $\Delta t > 302{,}400\,\mathrm{s}$, Δt need to add 604,800 s.

3. Calculating the mean anomaly M_k of observation time

The mean anomaly M_0 of reference moment t_{oe}

$$M_k = M_0 + n\Delta t \tag{3.4}$$

4. Eccentric anomaly E_k

$$E_k = M_k + e \sin E_k \tag{3.5}$$

where

e: orbital eccentricity. The equation can be solved by iterative method.

5. True anomaly f_k

$$\cos f_k = \frac{\cos E_k - e}{1 - e \cos E_k} \tag{3.6}$$

$$\sin f_k = \frac{\sqrt{1 - e^2} \sin E_k}{1 - e \cos E_k} \tag{3.7}$$

$$f_k = \arctan \frac{\sqrt{1 - e^2} \sin E_k}{\cos E_k - e} \tag{3.8}$$

Inverse trigonometric function, and can get true anomaly f_k.

6. Argument of latitude ϕ_k

$$\phi_k = \omega_0 + f_k \tag{3.9}$$

where

ω_0: argument of perigee.

7. Orbital perturbation correction

$$\left. \begin{aligned} \delta u &= C_{us} \sin 2\phi_k + C_{uc} \cos 2\phi_k \\ \delta r &= C_{rs} \sin 2\phi_k + C_{rc} \cos 2\phi_k \\ \delta i &= C_{is} \sin 2\phi_k + C_{ic} \cos 2\phi_k \end{aligned} \right\} \tag{3.10}$$

where

C_{uc}, C_{us}, C_{rc}, C_{rs}, C_{ic}, C_{is}: perturbation parameters. δu: argument of latitude Correction.δr: radius correction. δi: inclination correction.

8. Corrected argument of latitude u_t, Corrected r_t, corrected inclination i_t

$$u_k = \phi_k + \delta u \tag{3.11}$$

$$r_k = a(1 - e \cos E_k) + \delta r \tag{3.12}$$

$$i_k = i_0 + \delta i + \dot{i} \cdot \Delta t \tag{3.13}$$

9. Positions in orbital plane

Time t, GPS positions in orbital plane:

$$\left.\begin{array}{l} x_p = r_k \cos u_k \\ y_p = r_k \sin u_k \end{array}\right\} \quad (3.14)$$

10. Corrected longitude of ascending node Ω_k

$$\Omega_k = \Omega_0 + (\dot{\Omega} - \omega_e)\Delta t - \omega_e t_{oe} \quad (3.15)$$

where

Ω_0: longitude of ascending node of orbit plane at weekly epoch, $\omega_e = 7.29211567 \times 10^{-5}\,\text{rad/s}$, $\dot{\Omega}$: rate of inclination angle.

11. Earth-fixed coordinates

$$\begin{bmatrix} x_s \\ y_s \\ z_s \end{bmatrix}_{TS} = \begin{bmatrix} x_p \cos \Omega_k - y_p \cos i_k \sin \Omega_k \\ x_p \sin \Omega_k + y_p \cos i_k \cos \Omega_k \\ y_p \sin i_k \end{bmatrix} \quad (3.16)$$

In order to get the fixed satellites positioning when calculating the navigation message, it is necessary to make the following revision of the parameters in the above process:

1. Satellite average angular velocity $n = 0$

   Making the value associated with time to zero

$$\Delta n = -\sqrt{\frac{\mu}{A^3}} \quad (3.17)$$

thus $n = n_0 + \Delta n = 0$, then we have the mean anomaly $M_k = M_0 + nt_k = M_0$.

2. Orbital eccentricity $e = 0$

   Orbital eccentricity $e = 0$, so $E_k = M_k - e \sin E_k = M_k$,

$$\left\{\begin{array}{l} \sin v_k = \frac{\sqrt{1-e^2} \sin E_k}{1-e \cos E_k} = \sin E_k = \sin M_k \\ \cos v_k = \frac{\cos E_k - e}{1-e \cos E_k} = \cos E_k = \cos M_k \end{array}\right. \Rightarrow v_k = M_k \quad (3.18)$$

   Thus

$$\phi_k = v_k + \varpi = M_0 + \varpi \quad (3.19)$$

$$\delta u_k = 0 \Rightarrow u_k = \phi_k + \delta u_k = M_0 + \varpi \quad (3.20)$$

3. Satellite ephemeris correction setting zero

Perturbation parameters C_{uc}, C_{us}, C_{rc}, C_{rs}, C_{ic}, C_{is} set zero, thus

$$\begin{cases} e = 0 \\ \delta r_k = 0 \end{cases} \Rightarrow r_k = A(1 - e\cos E_k) + \delta r_k = A \tag{3.21}$$

4. Rate of inclination angle IDOT = 0

$$\begin{cases} \text{IDOT} = 0 \\ \delta i_k = 0 \end{cases} \Rightarrow i_k = i_0 + IDOT \cdot t_k + \delta i_k = i_0 \tag{3.22}$$

5. Rate of right ascension $\dot{\Omega} = \dot{\Omega}_e$

$$\dot{\Omega} = \dot{\Omega}_e \Rightarrow \Omega_k = \Omega_0 + (\dot{\Omega} - \dot{\Omega}_e)t_k - \dot{\Omega}_e t_{oe} = \Omega_0 - \dot{\Omega}_e t_{oe} \tag{3.23}$$

After the above parameter assignment, positions in orbital plane are as follows:

$$\begin{cases} x_k = r_k \cos u_k = A\cos(M_0 + \varpi) \\ y_k = r_k \sin u_k = A\sin(M_0 + \varpi) \end{cases} \tag{3.24}$$

MEO/IGSO satellites' coordinates in CGCS2000 coordinate system

$$\begin{cases} X_k = A\cos(M_0 + \varpi)\cos(\Omega_0 - \dot{\Omega}_e t_{oe}) - A\sin(M_0 + \varpi)\cos i_0 \sin(\Omega_0 - \dot{\Omega}_e t_{oe}) \\ Y_k = A\cos(M_0 + \varpi)\sin(\Omega_0 - \dot{\Omega}_e t_{oe}) + A\sin(M_0 + \varpi)\cos i_0 \cos(\Omega_0 - \dot{\Omega}_e t_{oe}) \\ Z_k = A\sin(M_0 + \varpi)\sin i_0 \end{cases} \tag{3.25}$$

When $M_0 + \varpi = \alpha$, $\Omega_0 - \dot{\Omega}_e t_{oe} = \beta$

$$\begin{cases} X_k = A\cos\alpha\cos\beta - A\sin\alpha\cos i_0 \sin\beta \\ Y_k = A\cos\alpha\sin\beta + A\sin\alpha\cos i_0 \cos\beta \\ Z_k = A\sin\alpha\sin i_0 \end{cases} \tag{3.26}$$

To control α, β and i_0, pseudolite arbitrary coordinates can be realized.

- Situation 1

Select $i_0 = \frac{\pi}{2}$, positioning formulas is

$$\begin{cases} X_k = A\cos\alpha\cos\beta \\ Y_k = A\cos\alpha\sin\beta \\ Z_k = A\sin\alpha \end{cases} \tag{3.27}$$

Thus,

$$\begin{cases} M_0 + \varpi = \alpha = \sin^{-1}\left(\frac{Z_k}{A}\right) \\ \Omega_0 - \dot{\Omega}_e t_{oe} = \beta = \sin^{-1}\left(\frac{Y_k}{A\cos\alpha}\right) \end{cases} \quad (3.28)$$

When $M_0 = \frac{\pi}{2}$

$$\varpi = \sin^{-1}\left(\frac{Z_k}{A}\right) - M_0 \quad (3.29)$$

$$\Omega_0 = \sin^{-1}\left(\frac{Y_k}{A\cos\alpha}\right) + \dot{\Omega}_e t_{oe} \quad (3.30)$$

Ephemeris is updated for every hour or for two hours at a time, thus, with the increasing of t_{oe}, Ω_0 will eventually be more than $\pm\pi$ and does not meet the requirements of the ephemeris choreography. So if Ω_0 more than $\pm\pi$, it is necessary to change Ω_0 into $\pm\pi$ by modulus method.

Another problem of the above methods is unable to restore the correct phase about β, which may lead to positioning failure because of the sign flip of X_k or Y_k.

- Situation 2

  First make sure $\Omega_0 - \dot{\Omega}_e t_{oe} = \beta$, when $\Omega_0 = \frac{\pi}{2}$, $\beta = \frac{\pi}{2} - \dot{\Omega}_e t_{oe}$

  When

$$\begin{cases} A_1 = A\cos\beta \\ A_2 = A\sin\beta \end{cases} \quad (3.31)$$

Thus

$$\begin{cases} X_k = A_1\cos\alpha - A_2\sin\alpha\cos i_0 \\ Y_k = A_2\cos\alpha + A_1\sin\alpha\cos i_0 \\ Z_k = A\sin\alpha\sin i_0 \end{cases} \quad (3.32)$$

$$-\frac{X_k - A_1\cos\alpha}{Y_k - A_2\cos\alpha} = \frac{A_2}{A_1}$$

$$A_1^2\cos\alpha - A_1X_k = A_2Y_k - A_2^2\cos\alpha$$

$$(A_1^2 + A_2^2)\cos\alpha = A_1X_k + A_2Y_k$$

$$\cos\alpha = \frac{A_1X_k + A_2Y_k}{A_1^2 + A_2^2} \quad (3.33)$$

$$\sin i_0 = \frac{Z_k}{A\sqrt{1 - (\cos\alpha)^2}} \quad (3.34)$$

This method avoids the case 1, so it is appropriate to adopt situation 2.

However, GEO satellite in computing the longitude of ascending node cannot offset the value associated with time, so GEO satellite cannot regard as a pseudolite.

3.2 Ephemeris Parameters Scaling Factor Revision

Although the above methods to realize the pseudolite position do not change satellite calculation method and the navigation message structure, it is necessary to revise some ephemeris parameters scaling factor because of ephemeris choreography of word length limit.

1. Satellite average angular velocity correction Δn

Number of bits Δn are 16 bit in the ephemeris choreography, the scaling factor is 2^{-43}, thus the effective range is $\pm 3.73 \times 10^{-9}$. However, the requirement of pseudolite ephemeris design needs Eq. (3.17), even though $\sqrt{A}$ has a maximum 8192 to calculate, $\Delta n = 3.6316 \times 10^{-5} - 1.2 \times 10^{-3}$, it is obvious that ephemeris choreography cannot be valid. The solution to solve the problem is to change the scaling factor

$$F_{\Delta n} = \left\lceil \log_2 \left(\frac{\Delta n_{\max}}{2^{15}} \right) \right\rceil = 24 \tag{3.35}$$

2. Rate of Right Ascension $\dot{\Omega}$

Number of bits $\dot{\Omega}$ are 24 bit in the ephemeris choreography, the scaling factor is 2^{-43}, thus the effective range is $\pm 9.54 \times 10^{-7}$. The requirement of pseudolite ephemeris design needs $\dot{\Omega} = \dot{\Omega}_e = 7.2921150 \times 10^{-5}$. The same solution should be done

$$F_{\dot{\Omega}} = \left\lceil \log_2 \left(\frac{\dot{\Omega}_e}{2^{23}} \right) \right\rceil = 36 \tag{3.36}$$

3.3 Square Root of the Semi-major Axis Revision

Another problem of part is the precision of $\sqrt{A}$. The scaling factor of $\sqrt{A}$ is 2^{-19}, thus the precision of $\sqrt{A}$ is 1.9073e−06 m. The precision is sufficient for regular satellite, nevertheless, it is difficult to fit fixed pseudolite because of the accumulation of pseudolite location changing along with time.

As stated before, Eq. (3.17) fixes the pseudolite location. But due to the truncation error by the scale factor, the semi-major axis is

$$\sqrt{\tilde{A}} = \sqrt{A} + \delta_A \tag{3.37}$$

δ_A range from 0 to 1.9073e−06 m. Thus, satellite average angular velocity is

$$n = \frac{\sqrt{\mu}}{\left(\sqrt{A} + \delta_A\right)^3} - \frac{\sqrt{\mu}}{\left(\sqrt{A}\right)^3} \neq 0 \tag{3.38}$$

According to the simulation, the largest satellite coordinates in three-dimensional direction may produce the drift of 15.6 m every two hours.

The solution can be solved using the characteristics of high precision of Δn in navigation messages to revise the error of equation (3.37), the average angular velocity of the receiver is n_0

$$\tilde{n}_0 = \sqrt{\frac{\mu}{\tilde{A}^3}} = \frac{\sqrt{\mu}}{\left(\sqrt{\tilde{A}}\right)^3} = \frac{\sqrt{\mu}}{\left(\sqrt{A} + \delta_A\right)^3} \tag{3.39}$$

Revised Δn

$$\Delta n = -\frac{\sqrt{\mu}}{\left(\sqrt{A} + \delta_A\right)^3} \tag{3.40}$$

among

$$\delta_A = \text{Round}\left(\sqrt{A} \times 2^{19}\right) \times 2^{-19} \tag{3.41}$$

From the above, the satellite positions can be perfectly still.

4 Simulation and Result Analysis

Through the above analysis, experimental scene is built in (Fig. 2).

By changing the relative ephemeris file, we can see that the output simulation control interface shows the constant coordinates of satellites and zero of satellites velocity in Fig. 3.

Sampling the signals of multi-output navigation signal simulator, software receiver calculates the user location in Fig. 4. The precision of positioning is 3 or 4 m rough in Fig. 5.

Simulation with MATLAB in Fig. 6.

Fig. 2 Beidou pseudolite system test environment

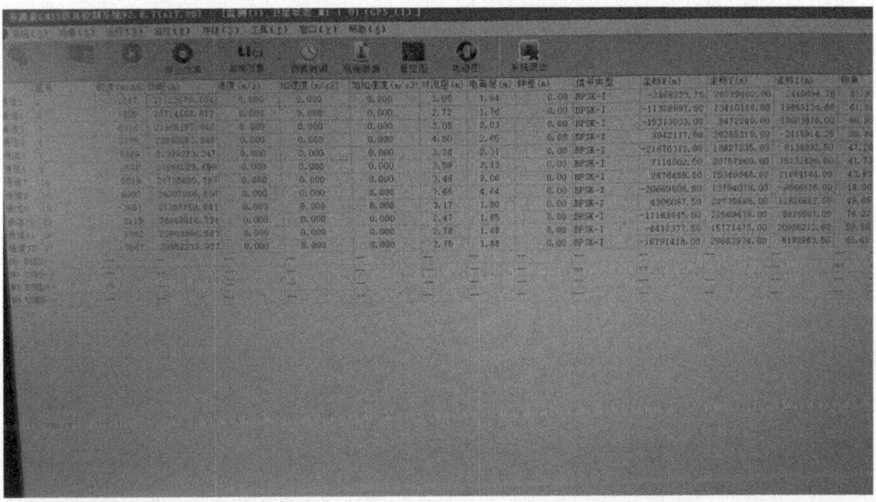

Fig. 3 Multi-output simulation control interface

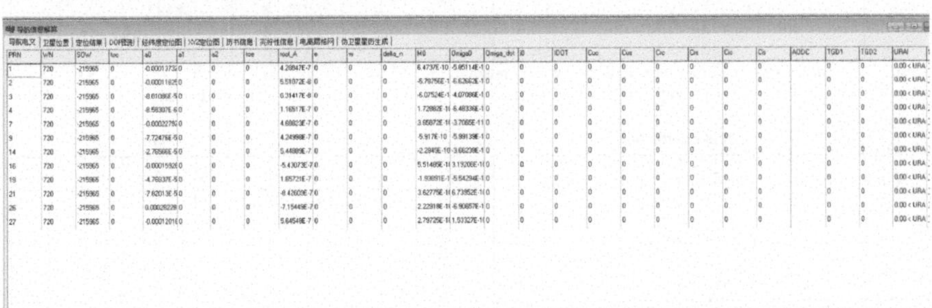

Fig. 4 The navigation message parsing interface

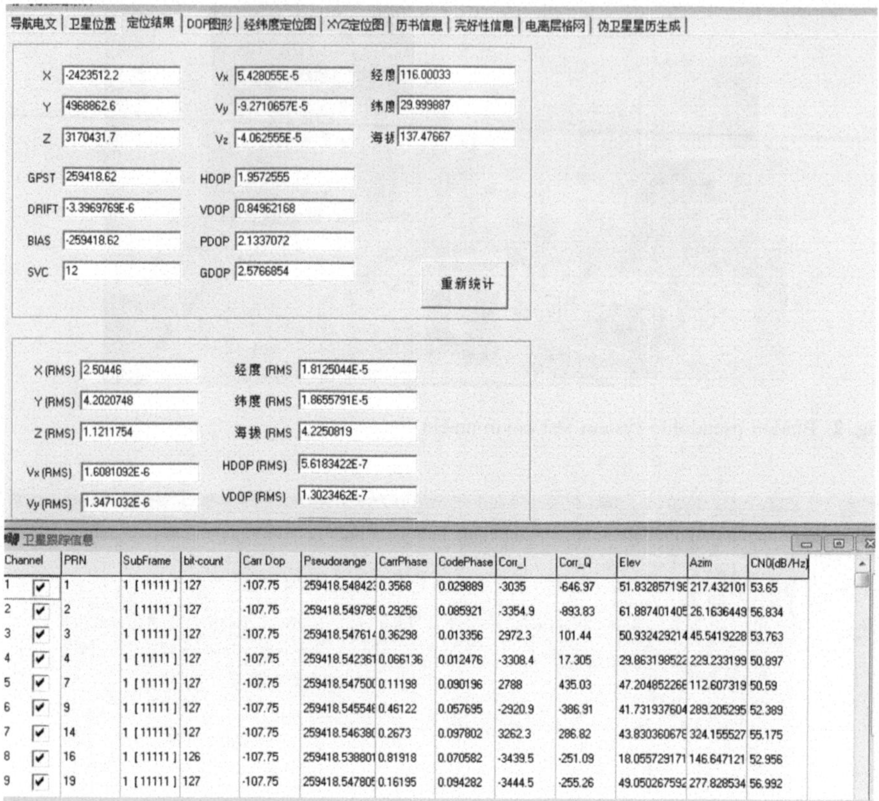

Fig. 5 Position calculating interface

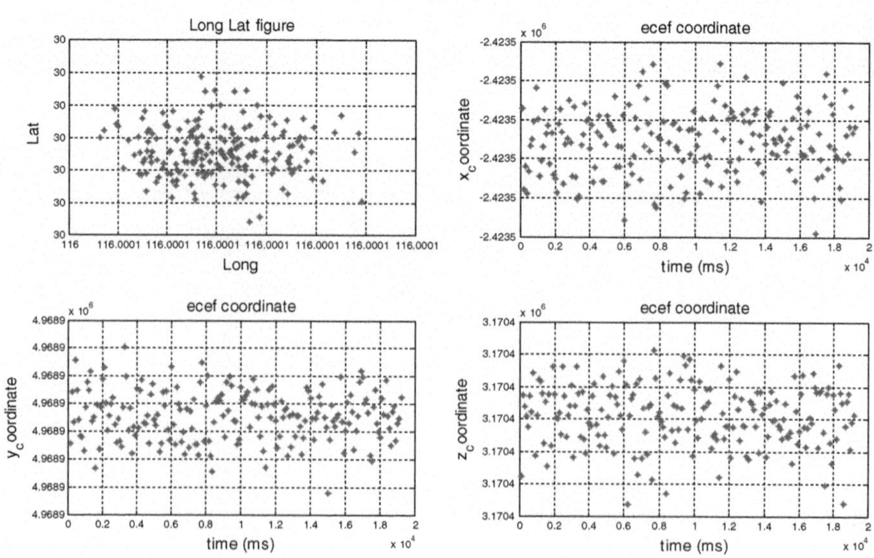

Fig. 6 Positioning results simulation

The process of positioning in the 19.3 s (positioning per 100 ms), the precision of *X*, *Y*, and *Z* coordinate is 1.6538, 1.7931, and 0.9212 m.

5 Conclusion

The result of simulation shows the feasibility and rationality of the IPS with simple structure and the desired requirements.

References

1. Deng Z et al (2013) Situation and development tendency of indoor positioning. 中国通信 10 (3):42–55
2. Yang Q, Pan, SJ, Zheng VW (2008) Estimating location using Wi-Fi. Intell Syst IEEE 23(1):8–13
3. Lin CJ et al (2010) Application of intelligent agent and RFID technology of indoor position: safety of kindergarten as example. In: 2010 international conference on IEEE machine learning and cybernetics (ICMLC), pp 2571–2576
4. Zhou S, Pollard JK (2006) Position measurement using Bluetooth. IEEE Trans Consum Electron 52(2):555–558
5. Wang X (2011) Research on pseudolite positioning technique. Shanghai Jiao Tong University, Shanghai
6. Wan L, Zhang L, Xu Y, Zheng X (2015) Overview of indoor positioning: classification, method and application. Geomatics World 8–14 + 30
7. CD (2000) Interface control document—Navstar GPS space segment/Navigation user interfaces, ICD-GPS-200C
8. BeiDou navigation satellite system signal in space interface control document (Version 2.0)

Test System for BDS User Terminal Based on RF Replay Apparatus

Deyong Xian, Peirong Fan, Hailing Wu and Qian Wang

Abstract RF signal of traditional real road test system for BDS user terminal is usually unrepeatable while the system is complex and expensive. In this paper, an economic and signal-replicable BDS user terminal test system, constructed with RF signal record and playback apparatus, is described and tested. Compared with commercial satellite navigation signal simulators, experiment results show that the RF signal provided by record and playback apparatus has a maximum carrier-to-noise ratio attenuation of 0.8 dB, which is similar to that of signal simulators. However, the latter device is able to provide much more critical and useful live noises of the ambient. In addition, to characterize the test system, two kinds of test scenario (suburban road and city canyon) are proposed, and experimental data demonstrate a high positioning accuracy (<17.04 m) and usability (>87 %) performance of BDS user terminals under replayed signals.

Keywords Record and playback · Beidou · Terminal test · Carrier-to-noise ratio

National Natural Science Foundation (41474027).

D. Xian (✉) · P. Fan · Q. Wang
Beijing Satellite Navigation Center, Beijing 100094, China
e-mail: xi_an_yong@126.com

P. Fan
Department of Precision Instrument, Tsinghua University, Beijing 100084, China

H. Wu
Beijing Institute of Tracking and Telecommunication Technology (BITTT),
Beijing 100094, China

J. Sun et al. (eds.), *China Satellite Navigation Conference (CSNC) 2016 Proceedings: Volume I*, Lecture Notes in Electrical Engineering 388,
DOI 10.1007/978-981-10-0934-1_36

1 Introduction

In recent years, signal simulators for satellite navigation have been playing an important role in BDS user terminal performance test, because of their capability of generating multiple scenarios with high sensitivity and great accuracy which provides great convenience for quantitative evaluation of BDS user terminals. However, under some challenging environments (e.g., urban canyons), the navigation signal is blocked significantly by buildings and trees, power strength and multipath effect of the signal vary a lot, making it quite difficult for these satellite signals to be simulated precisely.

The appearance of RF signal recording and playback apparatus provide developers and researchers with an option of not only processing live and repeatable signals, but also an alternative substitute of the complicated and expensive live test system to improve user experience.

This paper describes a BDS user terminal test system based on RF replay apparatus and is organized as follows. The system composition is illustrated and characterized in Chaps 2 and 3 respectively. Chapter 4 introduces an example of terminal test with the system above while the conclusion is drawn in Chap 5.

2 BDS Navigation Terminal Test System

As shown in Fig. 1, the BDS user terminal test system consists of a RF replay device, a GPS/INS integrated navigation subsystem, a referential BDS receiver, and a post-processing software. To be more specific, the navigation subsystem is designed to provide referential position for BDS terminals, while the high sensitivity GNSS receiver is used for evaluation of the replayed navigation signal. Post-processing software helps to analyse positioning and velocity accuracy of the system.

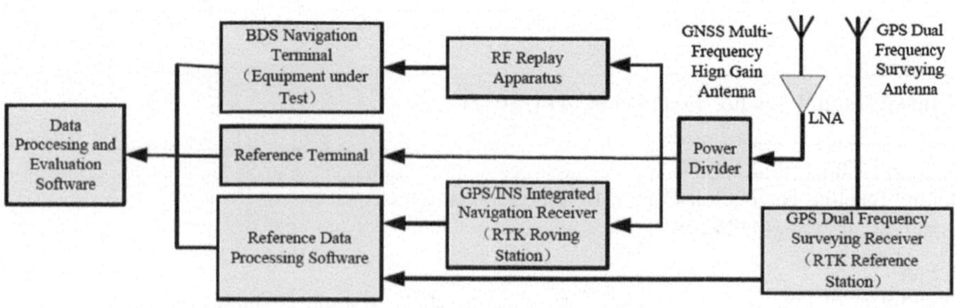

Fig. 1 BDS navigation terminal system based on RF reply apparatus

2.1 RF Replay Apparatus

There are two modes of operation for the RF replay apparatus: recording mode and playback mode. During the process of signal recording, complicated space signal is collected and stored lively to a disk, and in replay mode, the recorded signal is backward converted to navigation signal to simulate the realistic working environment. Through this method, a set of stable, reliable, and repeatable test signal is available for GNSS user terminal developers.

2.1.1 Signal Recording

The process of signal recording [1] is illustrated in Fig. 2 where an external multimode and multifrequency GNSS antenna is equipped to receive space satellite navigation signals. Practically, a high-performance antenna with 2–3 dB higher gain than the commercial one is usually chosen to compensate the signal power loss caused by data processing. The navigation signal is then amplified, filtered, and downconverted before input a high-speed analog-to-digital converter (ADC).

2.1.2 Data Storage

In this paper, the acquisition and storage of BDS B1 frequency (signal frequency: 1561.098 MHz, bandwidth: 4 MHz) signal is analyzed. When use an 8-bit ADC, under 62 MHz sample frequency and 62 MB/s data speed, the bus transfer rate required reaches up to 496 Mbps, which brings out a stringent demanding of RF replay apparatus in data transfer, storage, and reading [2]. In view of the above reason, digital downconversion and extraction are carried out on the recorded data to lower its transfer rate and facilitate date storage. Meanwhile, the data storage and control process (shown in Fig. 3) are realized based on embedded high-performance processor and PCI Express bus, as well as running a Linux operating system to respond for high-speed storage/read, PC interconnections (such as USB2.0, Gigabit Eth, etc.) and human–machine interface control.

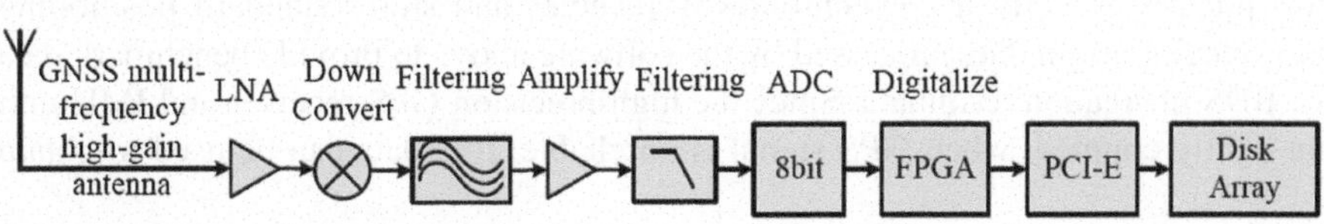

Fig. 2 Recording of navigation signal

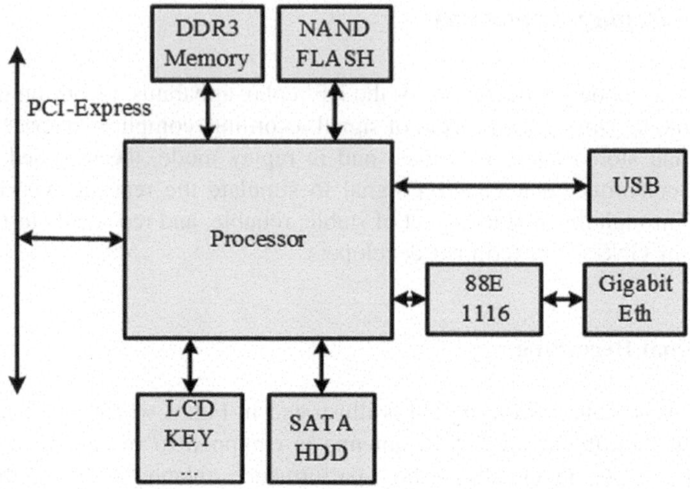

Fig. 3 Data storage and control

2.1.3 Signal Replay

Functionality of the signal playback section is demonstrated in Fig. 4. Essentially, it works as an inverted receiver front end where the data stored in disk is converted to analog signal through a matching digital-to-analog converter (DAC). A low-pass reconstruction filter is used to remove high-frequency components from the conversion process. In the mixing stage, frequency of the navigation signal is moved up to the correct carrier value (in this case, 1561.098 MHz). After band-pass filter and attenuation, the signal is output to BDS user navigation terminals.

2.2 GPS/INS Integrated Navigation Subsystem

The GPS/INS integrated navigation system, consisting of a GPS dual-frequency surveying antenna, a RTK roving station, a RTK referential station and post-processing software, synchronously receives and stores standard positioning and velocity origin data processed by the software above to provide benchmark data for BDS navigation terminals. Since the high-precision GPS receiver and IMU unit are tightly coupled, when GPS signal is unreliable, INS data can be used to obtain

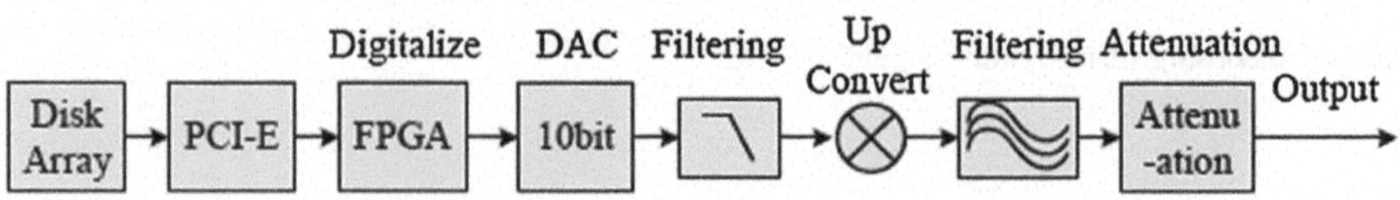

Fig. 4 Playback of navigation signal

high precision location and attitude information. Conversely, when INS system error gets larger, GPS data can be used for calibration. Through this technology, reliability and sustainability of the test system is guaranteed.

2.3 *Data Processing and Evaluation Software*

Based on the database system, the software adopts an offline post-processing method for positioning data evaluation. Under control of the data acquisition and storage module, location and raw observation information, which are prepared according to the NMEA-0183 protocol, are transferred through serial ports. In the next stage, unqualified parts of the data are rejected by data pre-procedure module. At last, positioning and velocity measurement result are analyzed and assessed in the evaluation module with an evaluation report created.

3 Performance Analysis of RF Replay Apparatus

3.1 *Main Performance Index*

Performance index of the RF replay apparatus is mainly included in the data acquisition section. Table 1 shows the regular values of these parameters.

3.2 *Noise Characteristic*

According to system model of the BDS user terminal, carrier-to-noise ratio (C/N_0) is a key factor of signal quality. Idealistically, there is no CNR change between

Table 1 Main performance index of RF replay apparatus

No.	Type	Parameter	Index
1	RF front end	LNA gain	30 dB
2		LNA noise figure	0.8 dB
3		Dynamic range	>70 dB
4	Recording ADC	Sample resolution	8 bit
5		Spurious dynamic range	75 dB
6	Playback ADC	Resolution	10 bit
7		Spurious dynamic range	75 dB
8	Data Processing	Sample storage rage	64 MB/s
9		Readout rate	50 MB/s
10	Noise figure from signal acquisition to replay		1 dB

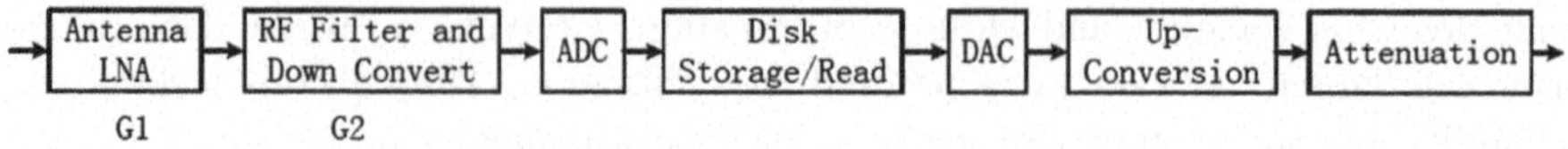

Fig. 5 Signal processing of RF replay apparatus

acquisition and replay process of navigation signal [3]. However, in practice thermal noise and ADC quantization noise of digital devices will unavoidably lead to CNR deterioration. Therefore, this paper characterizes noise influence of the signal acquisition, storage, and replay to verify the consistency between playback signal and the live one.

In this paper, the equivalent noise factor of BDS B1 frequency is estimated. A simplified signal processing flowchart is displayed in Fig. 5.

(1) Acquisition of the signal can be considered as a successive N-stage receiver, thus according to the Friis formula, the total noise figure F_{tot} can be expressed as following.

$$F_{\text{tot}} = 1 + \sum_{n=1}^{N} \frac{F_n - 1}{\prod_{k=0}^{n-1} G_k}, \quad G_0 = 1 \tag{1}$$

where G_n is the power gain of the nth stage, F_n is the noise figure. According to the formula, F_{tot} is dominated by noise performance of the device right after the antenna. In the RF replay apparatus described in this paper, a high power gain (30 dB) and low noise (0.8 dB) amplifier is applied for the first stage signal processing to prevent CNR deterioration caused by large noise figure.

(2) ADC Quantization Noise Analysis

For finite RF front-end bandwidth, the signal attenuation due to 1-bit quantization error is 1.96 dB and the attenuation of 2-bit ADC quantization is 0.55 dB, while it becomes 0.16 dB if a 3-bit ADC is used. When it comes to 4 bit (or higher) ADC, there will be basically no help to reduce the quantization loss. In the signal acquisition system, an 8-bit ADC is used to quantify the samples, therefore the actual impact of quantization noise on navigation signal CNR is negligible.

(3) The thermal noise power of DAC and frequency conversion process is considered:

$$P_R = 10 \log \text{KTB} = -138\,\text{dBW} \tag{2}$$

(4) For the process of data storage and read, there is only pure digital process which leads to neither signal distortion nor CNR change.

(5) The signal attenuation procedure has only passive device, where noise figure equals to the loss and there are no more thermal noise.

According to the analysis above, equivalent noise figure of a RF replay apparatus is about 1 dB, thus will cause 1 dB CNR deterioration during the signal analysis.

3.3 CNR Analysis

In this section, the CNR of live and replayed signal was compared and analyzed based on a high-sensitivity BDS terminal. The BDS space vehicle (SV) 01, 07, and 11 were chosen to demonstrate the CNR difference under suburban roads with no obstruction, results are displayed in Figs. 6, 7 and 8.

Mean value of CNN difference between the live and replayed signal is shown in Table 2 with a maximum value of 0.8 dB.

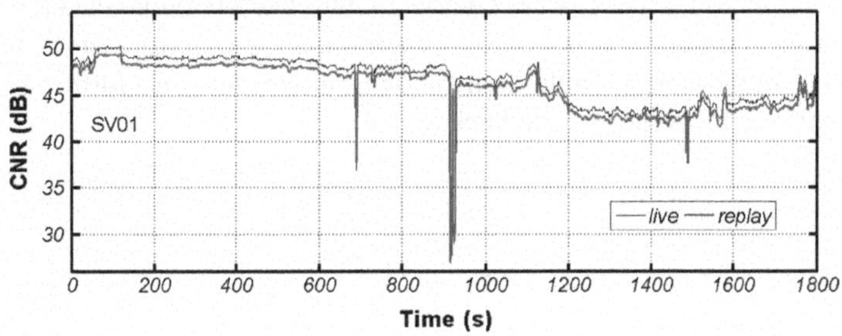

Fig. 6 C/N_0 for BDS SV01 (GEO)

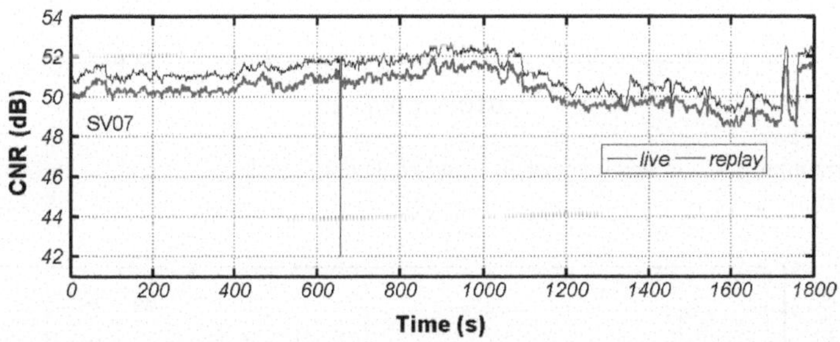

Fig. 7 C/N_0 for BDS SV07 (IGSO)

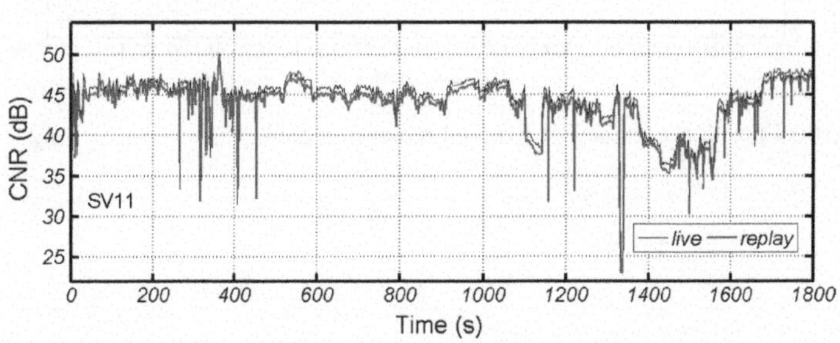

Fig. 8 C/N_0 for BDS SV11 (MEO)

Table 2 Average of C/N0 error

Parameter	SV01	SV07	SV11
Average CNR (dB)	0.79	0.8	0.51

3.4 Positioning Performance Evaluation

Positioning performance of navigation terminals is significantly depended on the playback signal quality, therefore a high sensitivity GNSS referential terminal was used to position simultaneously when recording the navigation signal. Afterward, the same terminal was applied under the playback signal to position again, then the data generated in both process were compared with the referential data respectively. Results can be seen in Figs. 9 and 10. Where positioning error curves of the live and replayed signal share a similar trend and absolute value, thus further verified the usability and reliability of the playback signal.

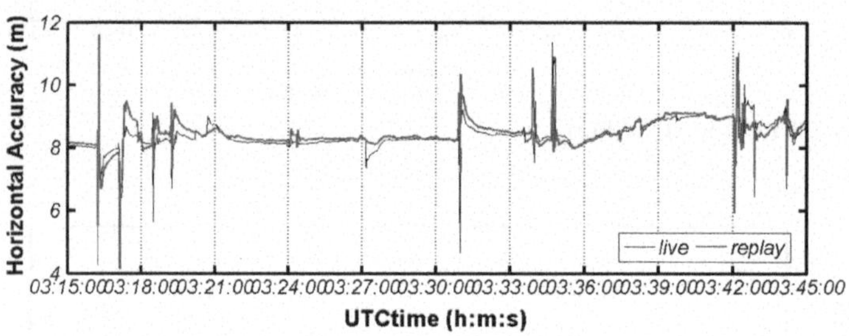

Fig. 9 Horizontal position accuracy of live and replayed signal

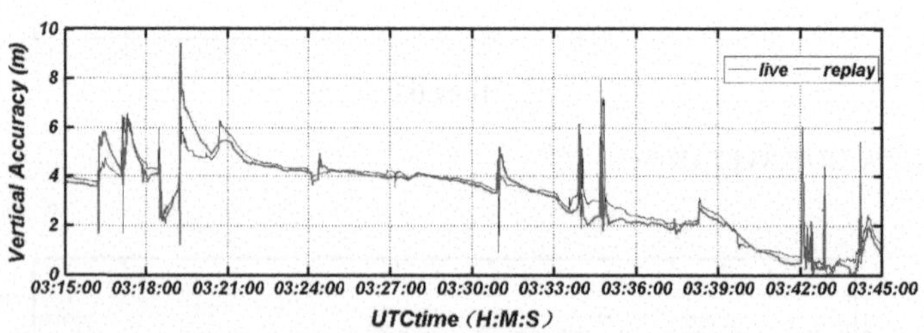

Fig. 10 Vertical position accuracy of live and replayed signal

4 BDS Navigation Terminal Test

Two testing scenarios, including urban canyons and suburb roads (shown in Figs. 11 and 12) respectively, are designed to evaluate performance of the BDS terminal test system described in this paper with a variety of BDS terminals.

Positioning accuracy is defined in the following manner: truncating 3600 s positioning data output from the terminal under test, if the effective result number is no less than 3420 (3600 × 95 %), the [n × 95 %]th largest positioning error is defined as positioning accuracy of the terminal where n is the number of samples collected.

Fig. 11 Test scene in city canyon

Fig. 12 Test scene in suburban route

Table 3 Testing result of 5BDS user terminals

Terminal No.	Positioning accuracy (m)	Positioning usability (%)
1	17.04	97.97
2	11.42	87.36
3	10.69	98.75
4	16.62	97.89
5	11.96	1.81

Positioning usability is determined as follows: Among all positioning errors of the 3600 points, the number of points with an error smaller than 30 m is defined as the terminal's effective positioning numbers.

$$\text{Positioning usability} = \text{effective anchor points/total points.}$$

According to the scheme above, multiple BDS receivers were evaluated by the terminal test system, and results of positioning and usability calculation are shown in Table 3.

It can be seen from the table that based on the signal acquisition and replay system, researchers can not only compare the performance of different terminals, but also be able be choose a more appropriate algorithm to make improvements.

5 Conclusion

This paper describes a BDS terminal test system based on a RF replay apparatus, and with experimental tests, CNR difference between live and playback signals is smaller than 0.8 dB. In terms of positioning error, both curves have a consistent trend and absolute values. At last, experiments on various BDS user terminals demonstrated feasibility, repeatability, and economic of the system.

References

1. Backén S, Akos DM, Wilson SG (2011) RF replay system for narrowband GNSS if signals. IEEE Trans Aerosp Electron Syst 47(1):586–595
2. Zhang T, Zhang H, Yan K, Liu J, Niu X (2012) Design and verification of built-in if data record and playback function in GNSS receivers. In: 25th International technical meeting of the satellite division of The Institute of Navigation, Nashville, TN
3. Ilie I, Hini R, Cardinal JS, Blood P, Fortin D (2011) Record and play-back system for GNSS: what you need to know for successful testing. In: 24th International technical meeting of the satellite division of The Institute of Navigation, Portland OR

Study on Satellite Navigation Service Availability Concept and Simulation Analysis

Tianqiao Zhang, Hongbing Wang and Liuchen Chen

Abstract Service availability for accuracy is a basic element in the performance frame of the GNSS services. In the system designing phase, service availability is a key link to the satellite constellation and maintenance strategy, the distribution of internal reliability performance, and the decision of service accuracy. In the system operation monitoring phase, service availability is one of the emphases of system service performance assessment and evaluation. In this paper, the understanding for the concept of service availability in the field of satellite navigation is introduced first. Then, some key problems in the design of satellite navigation service availability are studied, a calculation formula of service availability is proposed based on multiple constraints such as single satellite availability, ranging error and user operating conditions, and the accuracy threshold for the service availability is established which is different from the GPS SPS. Finally, according to two configuration alternatives of the BeiDou global constellation, simulation analysis of the service availability of the Asia-Pacific region is made, pointing out the difference between the two alternatives, providing a reference for the future BeiDou global system optimization.

Keywords Satellite navigation · Service availability · Reliability design · Test evaluation

1 Introduction

Service performance of global navigation satellite system includes accuracy, availability, integrity, continuity, each element of the four indicators depth cross-linking, where availability is a reliable guarantee for the remaining three indicators, is perceived by the user of satellite navigation services reliability, robustness direct factor. The satellite navigation service accuracy, integrity, conti-

T. Zhang (✉) · H. Wang · L. Chen
Beijing Satellite Navigation Center, Beijing 100094, China
e-mail: gary_85@163.com

J. Sun et al. (eds.), *China Satellite Navigation Conference (CSNC) 2016 Proceedings: Volume I*, Lecture Notes in Electrical Engineering 388,
DOI 10.1007/978-981-10-0934-1_37

nuity, these we usually refer to is all based on certain availability guarantee. Throughout the promises made by GPS, GLONASS, Galileo promises to ICAO, IMO [1], the availability indicators of satellite navigation service are generally better than the 99 %, in order to meet wide variety of user needs. Since the range accuracy performance of signal in space, broadcasting from the four major global navigation satellite systems, is relatively close, so the service availability depends mainly on the availability of navigation satellite constellations.

Concerned to the global navigation satellite constellation, NAVSTAR GPS, as the same as GLONASS, uses a constellation of 21 + 3 configuration in the early design. With the global satellite navigation applications continuously increasing, the United States completed the adjustment of the original GPS constellation from January 2010 to June 2011, replacing to 24 + 3 constellation layout. As on December 2015, the actual number of GPS satellites is more than 33 in orbit. For the GLONASS system, with the recovery of full operation ability, the constellation upgrades plan also puts on the agenda by the Russian Central Machinery Manufacturing Institute Governing Council in December 2011. In the future, almost all GLONASS satellites in the space will be adjusted orbit phase, with a great supplement of satellites in orbit. A six orbital plane-30 satellite program has been recognized by the vast majority of Russian experts [2]. Beidou global navigation satellite system intends to provide a world-class satellite navigation service, and expect superior performance than other systems in the key service areas; the optimization design and continuous improvement for constellation configuration are inevitable, in the foreseeable future.

The following section in this paper, conducted in-depth analysis to the main factors affecting the availability of satellite navigation services, so the calculation formula of service availability is proposed based on multiple constraints such as single satellite availability, ranging error and user operating conditions, and a concept of accuracy threshold of service availability is established which is different from GPS SPS (4-th Edition, 2008) [3]. Concerning to two alternatives for BeiDou global constellation configuration, a comparative analysis for service availability was given in the Asia-Pacific region. The results show that, 3GEO + 3IGSO + 24MEO program in shielding angle of 10 degrees cannot guarantee service availability as 99 % in Asia-Pacific region, while 5GEO + 5IGSO + 24MEO program at shielding angle of 10 degrees can still achieve the performance. With the shielding angle increasing, the advantages of 5GEO + 5IGSO + 24MEO program are obvious in the Asia-Pacific region.

2 Service Availability and Accuracy Threshold

Satellite navigation service availability refers to a specified percentage of time, when the service accuracy and integrity provided by the navigation system meets to the user require [3, 4], is a measure of the probability of the navigation system performance. Service availability of satellite navigation depends on two aspects of factors: constellation value and single satellite availability. In where, constellation value is related to constellation geometry, with single satellite availability relating to satellite short-term planned interruption, short-term unplanned interruption, and long-term interruption.

As an important indicator of coverage performance in the designated service area, constellation value mainly characterized the availability of constellation geometric dilution of precision at a certain threshold requirement. The ranging error from all satellites are usually assumed identical and statistically independent, when calculating the constellation geometric dilution of precision. But in actual situation, ranging error from different satellites vary obviously, especially from satellites in different orbits. Some literatures [5, 6] have proposed to use weighted DOP when measuring constellation availability, considering the constraint of user observation angle, constellation value is defined as follows:

$$\mathrm{CV} = \left. \frac{\sum_{t=t_0}^{t_0+kT} \sum_{i=1}^{L} \mathrm{bool}(\mathrm{WPDOP}_i(t) < Th_{\mathrm{DOP}}) \cdot \mathrm{area}_i}{k * \sum_{i=1}^{L} \mathrm{area}_i} \right|_{ev} \tag{1}$$

where t_0 is the initial time point, T is the time step, k is the time length in statistic; $\mathrm{WPDOP}_i(t)$ is weighted DOP value in the i-th earth surface grid area at time t, Th_{DOP} is the threshold of weighted DOP, bool() is a Boolean function, L is the total number of the earth's surface grid area, area_i is the i-th surface grid area, ev is the user shielding angle.

For a large global navigation satellite constellation, no satellite failure is almost impossible. Therefore, in the assessment of service availability for a satellite navigation system, different satellite-fault conditions must be taken into account, counting the constellation availability in various probability of satellite failure, whereby the proposed service availability formula as follows, based on multiple constraints such as single satellite availability, ranging error, user shielding angle.

$$A = \sum_{m=0}^{M} P_m \cdot CV_m = \sum_{m=0}^{M} P_m \cdot \left. \frac{\sum\limits_{t=t_0}^{t_0+kT} \sum_{i=1}^{L} \mathrm{bool}(\mathrm{WPDOP}_{i,m}(t) < Th_{\mathrm{DOP}}) \cdot \mathrm{area}_i}{k * \sum_{i=1}^{L} \mathrm{area}_i} \right|_{ev} \tag{2}$$

where m represents the m-th satellites fault state, M represents the total number of failure status in the satellite constellation; P_m describes the probability of a single fault condition, which is determined by the number of fault satellites, satellite type, satellite availability of different type. CV_m describes the constellation value in the m-th fault condition.

From formula (2), there are five influential factors to the service availability of satellite navigation: (1) User Constraint, related to constellation observation conditions and the number of visible satellites; (2) Satellite Constellation Configuration and Range Accuracy, related to constellation weighted geometric dilution of precision; (3) Service Availability Threshold, related to user requirement for service performance; (4) Single Satellite Availability (there are differences of different satellites orbits), related to the probability of each constellation fault condition; (5) Examined Area, related to the regional average results of service availability.

The availability threshold above is a threshold for weighted geometric dilution of precision, that is, efficacious in system design stage. In order to test and evaluate the system performance, service availability must be assessed in position domain, it should use position accuracy threshold directly, thereby giving accuracy threshold of service availability as follows:

$$Th = \sum_{m=0}^{M} P_m \cdot \text{WPDOP}_{99\,\%,m} = \sum_{m=0}^{M} P_m \cdot \left. \frac{\sum_{i=1}^{L} \text{sort}_{99\%}(\text{WPDOP}_{i,m}(t)) \cdot \text{area}_i}{\sum_{i=1}^{L} \text{area}_i} \right|_{ev} \tag{3}$$

where $\text{sort}_{99\,\%}()$ represents 99 % of the time probability.

The accuracy threshold of service availability of the above is obviously different from GPS SPS (4-th Edition, 2008). In the service specification for GPS SPS, the threshold accuracy of the global service availability is selected as 95 % probability positioning accuracy at the worst observation point, where the concept is not rigorous. But the given accuracy threshold calculation method of service availability in this paper, can stand statistical results testing.

3 Constellation Simulation

According to two constellation alternatives for BeiDou system already mentioned in open literatures, the comparison of service availability simulation is given below, one 3GEO + 3IGSO + 24MEO program used is the literature [4], another is 5GEO + 5IGSO + 24MEO program expanding based on current Beidou region system used is the literature [7]. For the single satellite availability, referring to the literature [8], GEO value 93.84 %, IGSO value 94.77 %, MEO value 96.61 %. For the weighting amount of ranging accuracy, referring to [4], UERE (GEO)/UERE (IGSO, MEO) = 1.375. Since both the global basic constellations used Walker (24/3/1) program, the service performance was mainly compared in the Asia-Pacific region. According to the principle that slightly better than GPS global performance whose 99 % PDOP is 3.3 [3], the WPDOP availability threshold assigned 2.8 in Asia-Pacific region.

In accordance with the single satellite availability above, the statistical probability of all fault conditions for two constellations are shown in Tables 1, 2, where fault satellite number is within five. Table 1 includes 331,212 kinds of satellite-fault state, the statistical coverage reached 99.46. Table 2 includes 174,437 kinds of satellite-fault state, the statistical coverage reached 99.76. There are the whole statistical probabilities of constellation fault sate approximated in Tables 1, 2.

With the probability of each constellation fault state, WPDOP value of the service area above, calculating the Eq. (2), service availability of satellite navigation can be obtained. Table 3 gives the different service performance in the

Table 1 5GEO + 5IGSO + 24MEO constellation fault state statistics

Total number of fault satellites	Number of fault states	Combined probability
0	1	0.1831
1	34	0.3192
2	561	0.2695
3	5984	0.1469
4	46,376	0.0581
5	278,256	0.0178
Total	331,212	0.9946

Table 2 3GEO + 3IGSO + 24MEO constellation fault state statistics

Total number of fault satellites	Number of fault states	Combined probability
0	1	0.2366
1	30	0.3497
2	435	0.2495
3	4060	0.1144
4	27,405	0.0379
5	142,506	0.0097
Total	174,437	0.9976

Table 3 Performance comparison of the two constellations in the Asia-Pacific region

Shielding angle	Full probability performance statistics	5GEO + 5IGSO + 24MEO	3GEO + 3IGSO + 24MEO
5	Availability of WPDOP (≤2.8)	99.87 %	99.14 %
	Average of WPDOP	1.30	1.51
10	Availability of WPDOP (≤2.8)	99.21 %	96.38 %
	Average of WPDOP	1.51	1.77
15	Availability of WPDOP (≤2.8)	93.13 %	82.81 %
	Average of WPDOP	1.82	2.22

Asia-Pacific region, on the different condition of user shielding angles: 5-degree, 10-degree, 15-degree. The results show that, at high user shielding angle condition, the more GEO, IGSO satellites the better service performance for the Asia-Pacific region.

The following figures show that, under 15-degree shielding angle condition, the geometric dilution of precision distribution for 5GEO + 5IGSO + 24MEO, 3GEO + 3IGSO + 24MEO (Figs. 1 and 2).

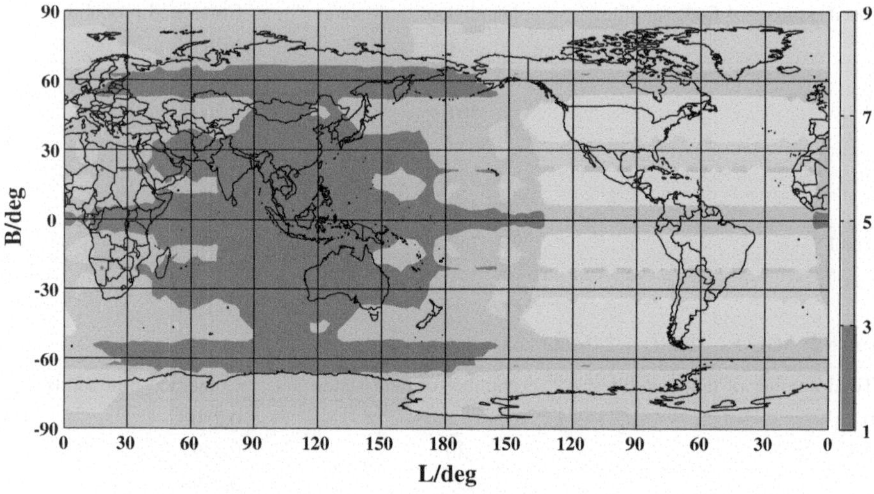

Fig. 1 99 % of WPDOP for 5GEO + 5IGSO + 24MEO constellation under 15-degree shielding angle

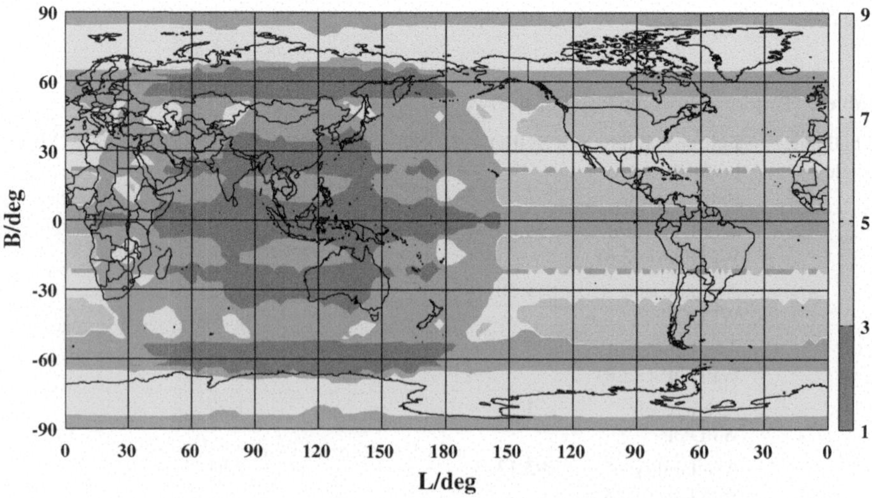

Fig. 2 99 % of WPDOP for 3GEO + 3IGSO + 24MEO constellation under 15-degree shielding angle

4 Conclusion

When the user shielding angle increased to 10-degree, navigation constellation configuration 3GEO + 3IGSO + 24MEO program cannot guarantee service availability as 99 % in Asia-Pacific, while 5GEO + 5IGSO + 24MEO program can still achieve the performance. With the increase in shielding angle , the advantages of 5GEO + 5IGSO + 24MEO program are obvious in the Asia-Pacific region. On the background that the actual satellite number of GPS, GLONASS is already or will be more than 30. Beidou global navigation satellite system intends to provide a world-class satellite navigation service, and expect superior performance than other systems in the key service areas, the optimization design and continuous improvement for constellation configuration are inevitable, in the foreseeable future.

References

1. International Civil Aviation Organization (2006) Aeronautical telecommunications volume I radio navigation aids (sixth edition)
2. Congress of the United States Congressional Budget Office (2011) The global positioning system for military user: current modernization plans and alternatives
3. Department of Defense of United States of America (2008) Global Positioning system standard positioning service performance standard. 4th Edition September 2008. http://pnt.gov/public/docs/2008/sp-sps2008.pdf
4. Falin LYZYW (2010) Analysis of global availability for beidou 2nd generation navigation system. Electron Meas Technol 33(7):15–18
5. Sairo H, Akopian D, Takala J (2003) Weighted dilution of precision as quality measure in satellite positioning. IEEE Radar Sonar Navig 150 (6):430–436 (2003)
6. Feixue WMSGW, Shaowei Y (2007) Weighted Geometric dilution of precision's analysis for mixed constellation navigation system. Chin Space Sci Technol 10(5):50–56
7. Cai H, Li X (2015) Weakness analysis of navigation constellation based on service availability, CSNC2015
8. Li G, Li J, Jiao W, Gui Q (2010) Analysis of PDOP availability of navigation constellation based on satellite service intermittence. Geomatics Inf Sci Wuhan Univ 35(7):841–845
9. Zheng H, Li H, Yang Z (2011) The analysis of constellation availability for satellite navigation system. Aerosp Control 29(3):87–94

Study on Integrated Testing and Evaluation Method of GNSS Timing Performance

Feng Zhu, Xiao-hui Li, Hui-jun Zhang and Yin-hua Liu

Abstract GNSS timing is the process for users to obtain the UTC time from the satellite navigation signal. At present, each GNSS has the timing function, but timing performance characterization of different navigation systems is not unified as well as the published data not comparable, which makes the users great inconvenience. According to the users' using characteristics, the paper proposes the conception and method to test the bias between UTC broadcasted by the satellites and actual UTC, which studies the timing testing methods of single satellite and multiple satellites, and referring to the timing performance of GPS, GLONASS, and BDS we develop the integrated testing and evaluation experiments. The result shows GPS timing error is better than 20 ns, and GLONASS timing performance is less than GPS with timing error relatively higher fluctuation; compared with GPS and GLONASS, timing evaluation of BDS has still a certain distance that sometimes the maximum timing error can be up to 120 ns, thus from the users' perspective the UTC precision obtained by the BeiDou satellites is less than GPS and GLONASS.

Keywords Global navigation satellite system · Coordinated universal time · Testing and evaluation methods · Uncertainty · Timing error

1 Introduction

As GNSS becomes fast increasing and gradually perfecting, multimode navigation application has been entering a new period. In order to avoid "blocking" among different navigation systems and provide better timing service for achieving compatibility and interoperability, it is necessary to consider many complicated problems such as time system offset, coordinate transformation, satellite identification,

F. Zhu (✉) · X. Li · H. Zhang · Y. Liu
Key Laboratory of Precision Navigation Positioning and Timing, CAS,
National Time Service Center CAS, Lintong District, China
e-mail: zhufeng@ntsc.ac.cn

J. Sun et al. (eds.), *China Satellite Navigation Conference (CSNC) 2016 Proceedings: Volume I*, Lecture Notes in Electrical Engineering 388,
DOI 10.1007/978-981-10-0934-1_38

and so on. Consequently, the time reference has become one of the first importantly solved problems in multimode navigation. The world recognized time standard of Coordinated Universal Time (UTC) is the basic time reference of GNSS, which provides traceability with each navigation system for timing service [1].

As the hysteretic nature of UTC, the traceability data in navigation message has just broadcasted with the bias between system time and UTC (*K*), and the bias between UTC (*K*) and UTC is lagging broadcasted by BIPM circular T [2]. However, some users ignore the influence that is feasible for the users in the international area. But for the users across different regions and countries in GNSS compatibility and interoperability, it demands for UTC time as the reference uniformly, in order to modify the user time to UTC. According to ITU standard broadcasting requirement, the time bias between GNSS time and UTC must be controlled within 100 ns, thus it requires the bias between GNSS broadcasting UTC time and actual UTC time less than 100 ns; so it is necessary to analyze and test the satellites broadcasting UTC time which evaluates the timing performance of navigation system [3].

The paper utilizes the standard time and frequence resource of UTC (NTSC), which proposes the method of testing the bias between satellites broadcasting UTC and actual UTC, and it studies on integrated testing and evaluating the timing performance of GPS, GLONASS, and BDS. First, the paper introduces the performance characterization system of testing and evaluation method. Secondly, it analyzes the timing performance of each satellite, and with data of Circular T and CV comparison it studies the timing performance of each navigation system through integrated multiple satellites test. Then it compares and analyzes the timing performance of three navigation system in 2015 with the conclusion given at last.

2 Testing and Evaluation Method

The international testing method is generally through two-way satellites or CV comparison methods, which establishes the comparison link between navigation system control station and traceability reference time UTC (*k*), thus can calculate the bias between navigation system time and UTC [4]. In this way, although the user can obtain a high timing precision, it cannot reflect the real UTC precision from satellite signal in the practical condition, which is also unable to evaluate the timing performance of each satellite.

Referring to the characteristic and requirement of satellite timing performance, the paper analyzes the testing and evaluation function, and from the users' characteristic it proposes the concept and method for testing the bias between satellites broadcasting UTC time and actual UTC, with the basic principle showed in Fig. 1.

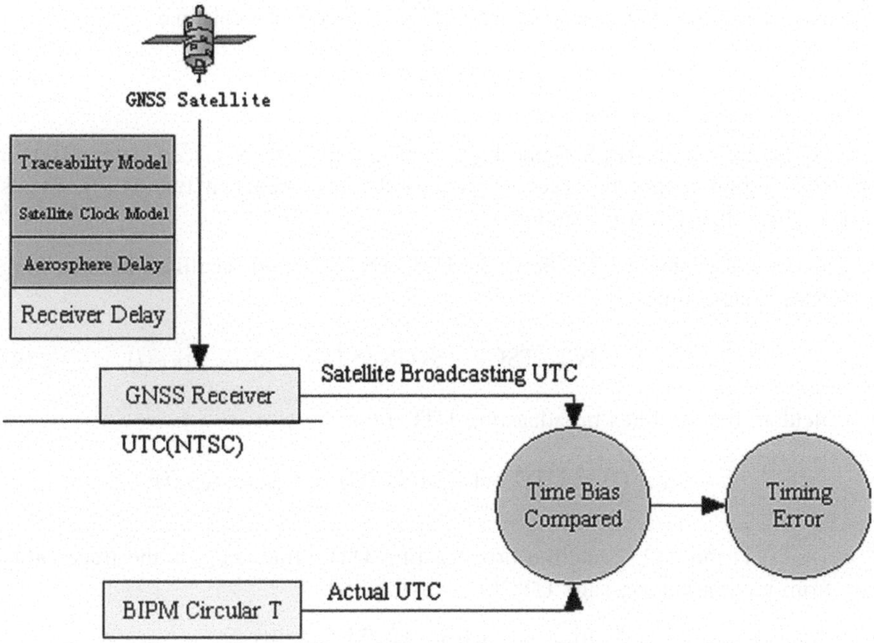

Fig. 1 Testing and evaluation principle of GNSS timing performance

- Calculate bias $\tau_{\mathrm{S,R}}(i)$ between satellites broadcasting GNSS system time and receiver time by pseudorange modification

$$\begin{aligned}\tau_{\mathrm{S,R}}(i) &= \mathrm{SvGNSST}(i) - \mathrm{REV} \\ &= \rho/c - \rho_0/c - \tau_{\mathrm{clk}} - \tau_{\mathrm{rela}} - \tau_{\mathrm{iono}} - \tau_{\mathrm{tropo}} - \tau_{\mathrm{sagnac}} - \tau_{\mathrm{chan}} - \tau_{\mathrm{other}}\end{aligned} \tag{1}$$

SvGNSST(i) is the NO. i satellite broadcasting system time, ρ is the pseudorange from the receiver, τ_{clk} is the satellite clock bias, τ_{rela} is the relativistic effect calibration, τ_{iono} and τ_{tropo} are relatively the ionosphere delay and troposphere delay, τ_{sagnac} is the Sagnac modification, τ_{chan} is the receiver delay, τ_{other} is the other delay including multipath effect, ρ_0/c is the geometry path delay from satellite to receiver showed as below

$$\rho_0/c = \sqrt{(x_{\mathrm{S}} - x_{\mathrm{R}})^2 + (y_{\mathrm{S}} - y_{\mathrm{R}})^2 + (z_{\mathrm{S}} - z_{\mathrm{R}})^2}/c \tag{2}$$

$x_{\mathrm{S}}, y_{\mathrm{S}}, z_{\mathrm{S}}$ is the satellite position calculated from navigation message; $x_{\mathrm{R}}, y_{\mathrm{R}}, z_{\mathrm{R}}$ is the receiver position measured from survey department;

- Measure the bias between UTC (NTSC) and receiver local time

$$\tau_{\mathrm{C,R}} = \mathrm{UTC(NTSC)} - \mathrm{REV} \tag{3}$$

UTC(NTSC) is the 1PPS signal kept by National Time Service Center, REV is the 1PPS signal output by receiver, $\tau_{\mathrm{C,R}}$ is the bias between two 1PPS signals measured by time interval counter;

- Calculate the bias $\tau_{\mathrm{C,S}}(i)$ between UTC (NTSC) and satellites broadcasting GNSS system time

$$\tau_{\mathrm{C,S}}(i) = \mathrm{UTC(NTSC)} - \mathrm{SvGNSST}(i) = \tau_{\mathrm{C,R}} - \tau_{\mathrm{S,R}}(i) \tag{4}$$

- Calculate the satellites broadcasting UTC time

$$\tau_{\mathrm{S,UTC}}(i) = \mathrm{UTC} - \mathrm{UTC(NTSC)} = \tau_{\mathrm{trace}} - \tau_{\mathrm{C,S}}(i) \tag{5}$$

$\tau_{\mathrm{S,UTC}}(i)$ is the NO. i satellite broadcasting UTC time, τ_{trace} is the traceability data from navigation message UTC(k) – SvGNSST;

- Obtain the actual UTC time τ_{BIPM} from BIPM Circular T:

$$\tau_{\mathrm{BIPM}} = \mathrm{UTC} - \mathrm{UTC(NTSC)} \tag{6}$$

- Calculate the satellite timing error $\sigma_{\mathrm{S,UTC}}(i)$:

$$\sigma_{\mathrm{S,UTC}}(i) = \tau_{\mathrm{BIPM}} - \tau_{\mathrm{S,UTC}}(i) \tag{7}$$

In conclusion, this testing and evaluation method includes some errors from the signal propagation path, but it can reflect the real UTC precision obtained by users. If we share the evaluation result to the users, it will make them easier to master and apply.

3 Performance Characterization

Testing and evaluation method is mainly performance characterized by the uncertainty, and the uncertainty contains random error and system error among timing performance, which mainly exists in the middle parts including pseudorange measurement, mathematical model, receiver calibration, Circular T data, and time interval counter [5]. Analyzing with the uncertainties of these variables, we can get the combined uncertainty of testing and evaluation method by standard deviation.

Table 1 Uncertainty analysis of testing and evaluation methods of GNSS timing performance (*Unit* ns)

Uncertainty source (ε)	Uncertainty
Pseudorange measurement	1
Mathematical model	2
Receiver calibration	2
Circular T data	5
Time interval counter	0.1
Combined uncertainty ($\sqrt{\Sigma\varepsilon_i^2}$)	5.83

$$\varepsilon = \sqrt{\varepsilon_\rho^2 + \varepsilon_{\text{Math}}^2 + \varepsilon_{\text{Rev}}^2 + \varepsilon_{\text{BIPM}}^2 + \varepsilon_{\text{TIC}}^2} \quad (8)$$

Among them, ε_ρ is the uncertainty of pseudorange measurement, $\varepsilon_{\text{Math}}$ is the uncertainty of mathematical model, ε_{Rev} is the uncertainty of receiver calibration [6], $\varepsilon_{\text{BIPM}}$ is the uncertainty of UTC-UTC (NTSC) published by Circular T, ε_{TIC} is the uncertainty of time interval counter. According to this evaluation standard, it reflects the precision of satellites timing performance, which shows the degree of closeness between satellites broadcasting time and UTC time (Table 1).

We can see that the error is mainly reflected on the uncertainty of Circular T data, with the total combined uncertainty is 5.83 ns.

Testing and evaluation result is mainly characterized through maximum, minimum, average, and standard deviation. The maximum and minimum characterizes the range between satellites broadcasting time and UTC; the average characterizes the accuracy between satellites broadcasting time and UTC which reflects the timing precision of navigation system; the standard deviation characterizes the degree of dispersion in satellites broadcasting UTC time which reflects the distribution of timing error.

4 Testing and Evaluation Results

4.1 Single Satellite Testing Result

The following testing results are the part satellites timing performance of GPS, GLONASS, and BDS in which the NO. 13 BeiDou satellite is not visible, and the testing time is from January 1st, 2015 to November 30th, 2015 (Figs. 2, 3, 4, 5, and 6; Table 2).

As can be seen, the timing errors of GPS satellites ranges within ±20 ns in 2015, and the timing errors of GLONASS satellites is a little higher ranging within ±40 ns, which are both better than BeiDou satellites. The timing error of BeiDou MEO satellites have some gross error mainly due to the lower satellites elevation which causes the larger errors in ionospheric and tropospheric [7]. The testing result can be inferred that the BDS traceability model exists a large error.

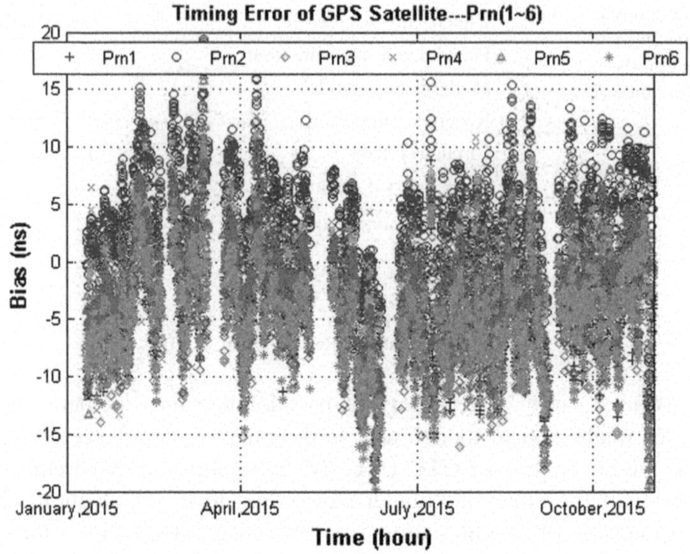

Fig. 2 Timing error of GPS (1–6)

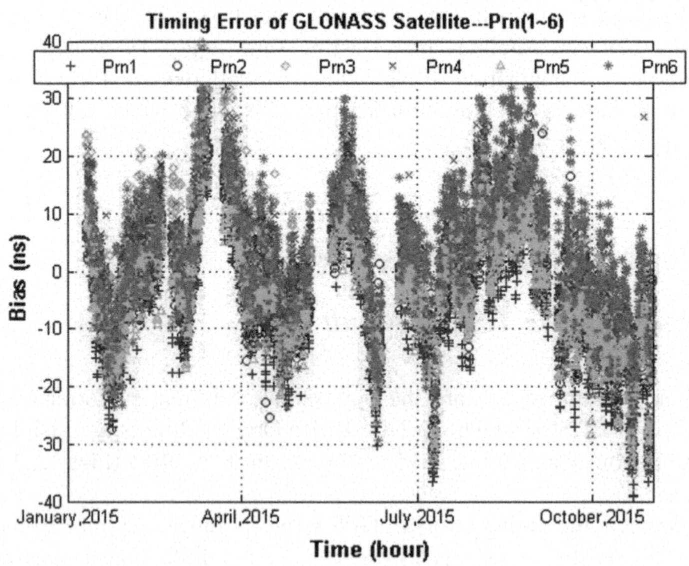

Fig. 3 Timing error of GLONASS (1–6)

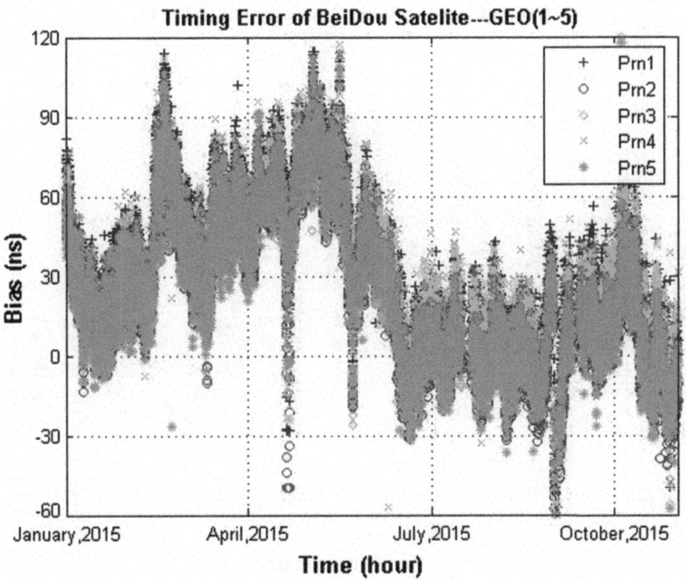

Fig. 4 Timing error of BDS (1–5)

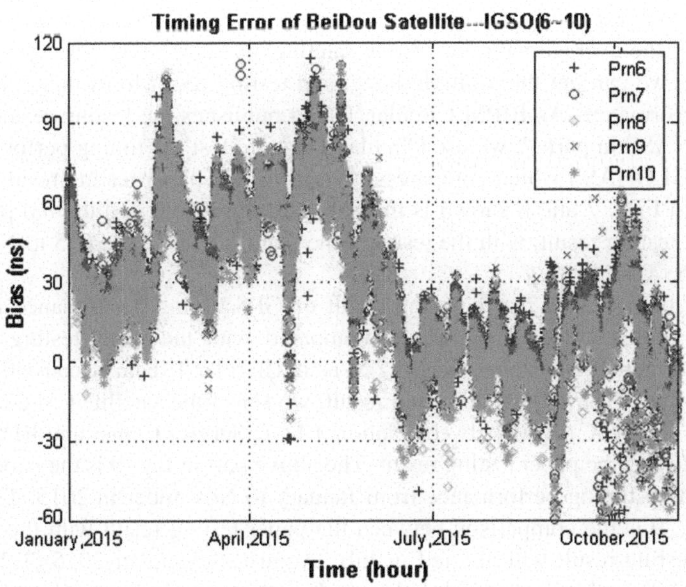

Fig. 5 Timing error of BDS (6–10)

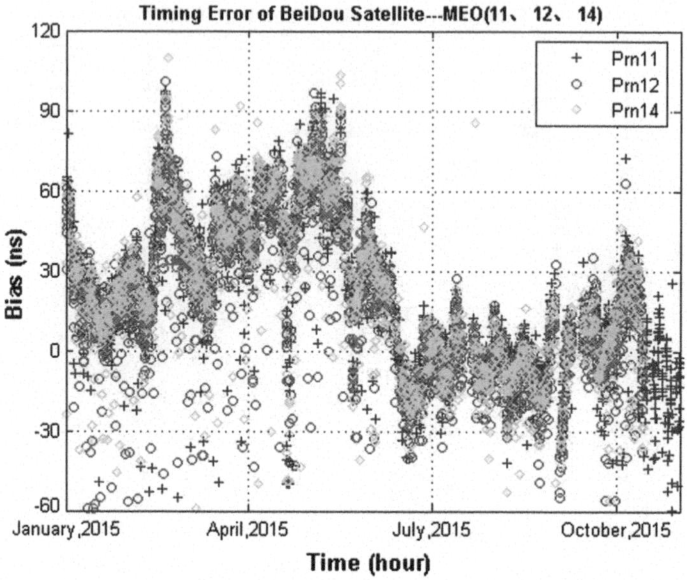

Fig. 6 Timing error of BDS (11, 12, 14)

4.2 Multiple Satellites Integrated Testing Result

Calculating the average value of visible satellites in each navigation system at the same time, we can get the multiple integrated testing result to evaluate the GNSS timing performance. As BIPM Circular T also publishes the testing result of GPS and GLONASS in part V, we use Circular T data to test the timing performance of GPS and GLONASS which compares and testifies the paper testing result [8]. The blue part in Figs. 7 and 8 shown is the paper testing result, and the red part is the Circular T testing result, with the testing time from January 1st, 2015 to November 30th, 2015 (Table 3).

As can be seen, the paper testing result of GPS timing performance is almost consistent with Circular T. However, compared with the paper testing result of GLONASS there is a system offset of 220 ns in Circular T data. Because Circular T has not included the BDS monitoring result, we select the satellite CV comparison data of UTC (NTSC)-BDT between National Time Service Center and BDS control station to testify the paper testing result. The above part in Fig. 9 is the paper testing result of BDS timing performance from January to November in 2015. The below part in Fig. 9 is the comparison between the paper testing result with the blue and CV data testing result with the red in July, August, September 2015 (Table 4).

The three month testing result below in Fig. 9 shows the paper testing result is almost consistent with CV comparison data, which both reflects BDS exists a

Table 2 Timing error of GNSS satellites statistics (*Unit* ns)

Prn	Maximum	Minimum	Average	Standard deviation
GPS 1	10.54	−21.33	−4.23	4.52
GPS 2	19.34	−14.31	3.88	4.48
GPS 3	10.83	−20.78	−5.1	4.66
GPS 4	18.97	−20.69	−1.86	4.77
GPS 5	16.23	−21.06	−2.01	4.54
GPS 6	10.85	−20.07	−5.29	4.41
GLONASS 1	35.53	−41.96	−4.75	10.99
GLONASS 2	31.49	−34.14	−0.63	10.55
GLONASS 3	43.26	−35.41	2.93	11.53
GLONASS 4	41.20	−35.10	2.74	11.26
GLONASS 5	32.63	−33.83	−2.82	10.48
GLONASS 6	38.22	−29.33	3.99	11.35
BDS 1	114.56	−56.86	37.41	27.26
BDS 2	110.15	−57.56	21.13	27.45
BDS 3	119.75	−56.91	24.60	26.73
BDS 4	116.99	−59.50	35.81	26.97
BDS 5	117.69	−59.56	25.26	28.51
BDS 6	115.80	−58.18	29.43	27.33
BDS 7	111.69	−59.92	27.37	27.63
BDS 8	107.77	−59.22	24.16	27.33
BDS 9	115.16	−54.07	27.23	27.75
BDS 10	111.11	−59.51	24.83	28.02
BDS 11	97.16	−59.79	21.91	28.49
BDS 12	100.59	−59.16	18.65	27.92
BDS 14	109.76	−59.31	21.95	28.83

certain timing error. Thus the paper testing result above in Fig. 9 indicates the maximum BDS timing error can be up to 120 ns in 2015, sometimes it exists a large bias between BDS broadcasting UTC and the actual UTC.

4.3 Comparison Result of GNSS Timing Performance

The following result is monthly timing performance comparison of GPS, GLONASS, and BDS in 2015. As it shown, the blue part represents GPS, the green part represents GLONASS, the red part represents BDS (Figs. 10 and 11).

As it can be seen, GPS timing error range from −2 to 6 ns which approaches nearly zero after April, with the standard deviation of each month better than 5 ns; GLONASS timing error range from −13 to 10 ns, and the standard deviation slightly higher than GPS; BDS timing error range from 0 to 70 ns, but the standard

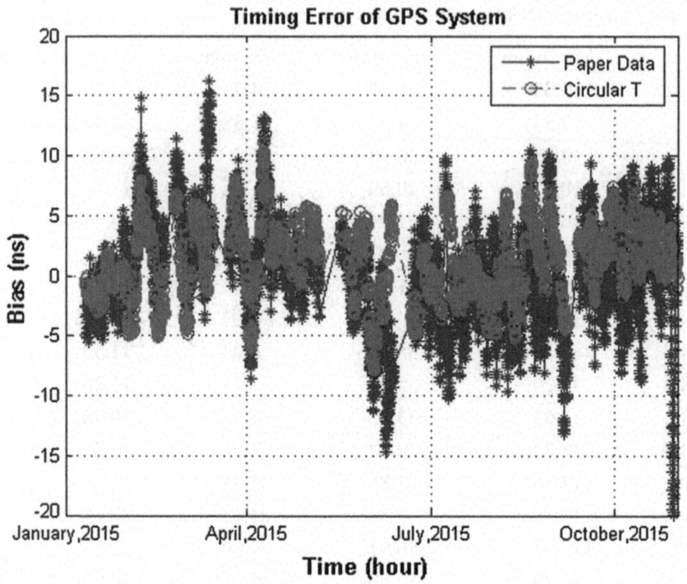

Fig. 7 Timing error of GPS (Color figure online)

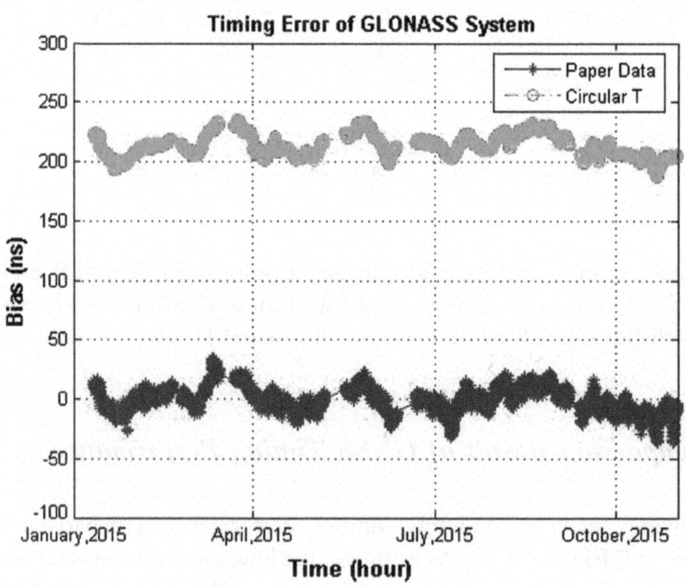

Fig. 8 Timing error of GLONASS (Color figure online)

Table 3 Timing error of GPS and GLONASS statistics (*Unit* ns)

GNSS	Data source	Maximum	Minimum	Average	Standard deviation
GPS	Paper	16.21	−22.80	0.27	4.47
	Circular T	11.8	−7.9	1.15	3.18
GLONASS	Paper	34.43	−36.42	−0.80	10.46
	Circular T	235.20	186.80	213.47	9.48

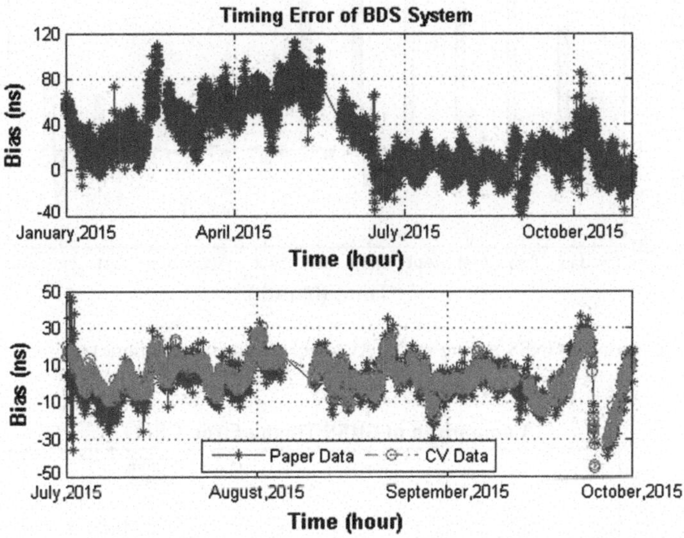

Fig. 9 Timing error of BDS (Color figure online)

Table 4 Timing error of BDS statistics (*Unit* ns)

Testing time	Data source	Maximum	Minimum	Average	Standard deviation
January–November	Paper	112.56	−38.95	27.37	27.14
July, August, September	Paper	66.30	−38.95	2.01	10.83
	CV comparison	31.02	−45.53	3.93	9.42

deviation is generally higher than 10 ns. It indicates that the BeiDou system time fluctuates relatively higher than GPS and GLONASS [9], while the traceability data from BDS navigation message still has some errors, thus from the users' perspective the UTC precision obtained by the BeiDou satellites is less than GPS and GLONASS.

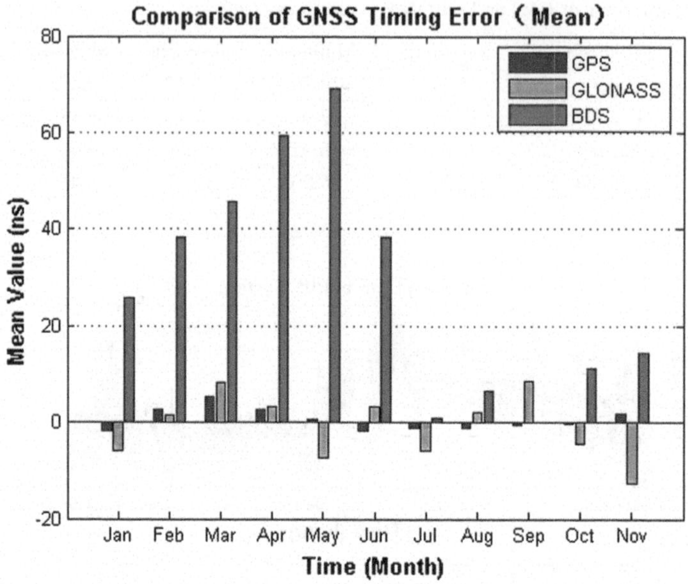

Fig. 10 Comparison of GNSS timing error (Average) (Color figure online)

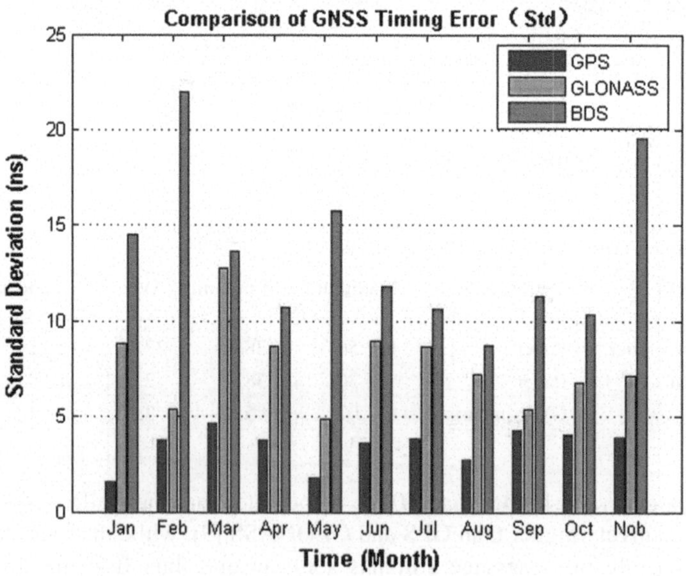

Fig. 11 Comparison of GNSS timing error (Standard deviation) (Color figure online)

5 Conclusion

From the system perspective and according to the users' characteristics, the paper evaluates the UTC precision obtained by the users instead of satellites broadcasting, and this conversion method clears the testing object which not only makes the testing process easy operating but also reflects the users' actual timing errors. From the testing result, although the BDS timing error basically satisfies the ITU demanding within 100 ns, its timing performance still exists a certain distance compared with GPS and GLONASS.

The system officially operating since April 2014, which regularly releases out the evaluation results through the FTP files. Thus, it can not only penetrate into positioning and timing performance of GNSS for advancing it widely used in defense application,. but also lay the foundation for the further optimization of system design and improvement of BDS timing performance, which makes the users have been able to select.

Acknowledgments Funded by National Natural Science Foundation of China (Grant No. 61127901).

References

1. Zhu F, Li X, Fan D (2013) Evaluation of GNSS timing performance based on circular T and analysis of system time offset. In: 2013 IEEE 11th international conference on IEEE electronic measurement and instruments (ICEMI), pp 776–780
2. Petit G, Arias F, Harmegnies A et al (2012) UTCr: a rapid realization of UTC. In: European frequency and time forum (EFTF). IEEE, New York, pp 24–27
3. Sha H, Li P, Zhan J et al (2013) Comparsion between timing methods in BeiDou satellite navigation system. European frequency and time forum and international frequency control symposium, pp 90–92
4. Jiang Z, Lewandowski W (2012) Inter-comparison of the UTC time transfer links. European frequency and time forum (EFTF). IEEE, New York, pp 126–132
5. Zhu F (2015) The time parameter of satellite navigation and its test method. University of Chinese Academy of Sciences
6. Zhong X, Tang Z, Meng Z et al (2013) A study on high precision calibration of zero value for navigation signal simulator. In: China satellite navigation conference (CSNC) 2013 Proceedings. Springer, Berlin, pp 469–477
7. Wu X, Hu X, Wang G et al (2013) Evaluation of COMPASS ionospheric model in GNSS positioning. Adv Space Res 51(6):959–968
8. Jiang Z, Lewandowski W (2012) Use of GLONASS for UTC time transfer. Metrologia 49 (1):57
9. Gao W, Jiao W, Xiao Y et al (2011) An evaluation of the Beidou time system (BDT). J Navig 64(S1):S31–S39
10. Han C, Liu L, Zhao J (2009) Conception, definition and evaluation of pseudo-range measurement. J Astronaut 30(6):2412–2425

Performance Assessment of Galileo In-Orbit Satellite Clocks

Biyun Yu, Huijun Zhang and Xiaohui Li

Abstract At present, the Galileo in-orbit satellites include four in-orbit validation (IOV) satellites and eight full operational capability (FOC) satellites. The accuracy of Galileo positioning is better than 2 m. Thirty FOC satellites are expected to be launched completely by 2019, and a complete constellation of full operational capability will be achieved by the time. The quality of Galileo satellite clock affects the performance of navigation and reliability. Based on this, a solution of satellite clock assessment is given, the standard time and frequency signal of National time service center of Chinese Academy of Sciences are taken as references. The performance of all Galileo in-orbit satellites clocks is assessed by receiving the signal in space and using the way of dual-frequency ionosphere-free combination, which includes frequency deviation, frequency drift, frequency stability, and the consistency of the frequency stability over time. At the same time, the uncertainty of relative frequency, frequency stability, and time offset of Galileo satellites clocks is analyzed. The evaluation results show that the orbit satellite clock frequency stability is (τ = 60 s and τ = 300 s) at e−13 level, the uncertainty of frequency stability is in the order of e−14, and the stability is consistent over time. When the satellite clock time is corrected to the Galileo system time (GST), the time offset of GST relative to UTC (NTSC) is between −40 and −70 ns, and the satellite clock stability after clock correction is slightly improved. The uncertainty of measurement of system time offset is 3.37 ns. Research methods and achievements of this paper can be the significant references with respect to Galileo stand-alone positioning, multi-GNSS combination positioning, integrity monitoring, the Galileo system time interoperability, etc.

Keywords Galileo · Satellite clock · Assessment · Stability · Uncertainty

---

B. Yu (✉) · H. Zhang · X. Li
Key Laboratory of Precision Navigation Positioning and Timing, CAS,
National Time Service Center, CAS, Lintong District, China
e-mail: yubiyun13@mails.ucas.ac.cn

J. Sun et al. (eds.), *China Satellite Navigation Conference (CSNC) 2016 Proceedings: Volume I*, Lecture Notes in Electrical Engineering 388,
DOI 10.1007/978-981-10-0934-1_39

1 Introduction

Nowadays, there exist four major satellite navigation systems in the world, including American GPS, Russian GLONASS, and two systems under construction —European Galileo satellite navigation system and Chinese Beidou satellite navigation system. Multi-GNSS system navigation, positioning and timing have become an increasingly important trend in the field of satellite navigation, which have the advantages of improving the availability of satellite navigation system, accuracy of satellite navigation, at the same time, enhancing the measurement accuracy, reliability and efficiency, etc. [1].

Among them, European satellite navigation system was initiated and provided with main financial support by European Space Agency and the European Union, in order to establish a civil global satellite navigation system which is independent, and has excellent performance which is better than that of GPS and good interoperability between it and existing satellite navigation systems (GPS, GLONASS) [2]. Currently, Galileo satellite navigation system is in the phase of constellation deployment, which includes manufacturing satellites, launching satellites, constructing ground facilities, and putting them into use. Until March 28th, 2015, Galileo system has four IOV satellites and eight FOC satellites and it is expected that 30 satellites will be launched by 2019 to complete Galileo system.

With the development of the Galileo satellite navigation system, its performance of atomic clock is being highly valued, because as time reference of navigation systems ranging and one of the core payload satellite navigation system, the performance of atomic clock directly determines the precision of navigation and time–frequency transmission. Two passive hydrogen maser (PHM, 1000 s stability is better than 1e−14) and two rubidium atomic frequency sources (RAFS, 100 s stability is better than 5e−13) are loaded on each satellite of the Galileo satellite system. Among them, a hydrogen atomic clock is used as the master clock, and another atomic hydrogen and rubidium atomic clock are used as a hot spare clock. In order to achieve seamless handover, four atomic clocks are synchronized with each other in real time and monitored by ground stations and control units. The measurement value of the time of atomic clock is transmitted to 29 ground control stations through the unique communication network. After calculation, the deviation between the time of atomic clock and the time of ground system GST will be written into the broadcast message [3]. Galileo system time is produced by the precision time facility (PTF), the function of PTF is the produce of GST time and the time transfer, making the precise time synchronization between the Galileo clock time and the Galileo system time GST [4].

The time and frequency of satellite navigation system are provided by loaded atomic clock, and time system is one of the core elements of satellite navigation and positioning. Therefore, the performance of atomic clock of satellite navigation system and system time synchronization directly determines the accuracy of satellite navigation and positioning. In order to study the positioning accuracy and the monitoring precision of time difference of GNSS system, it is necessary to evaluate the performance of Galileo atomic clock. Atomic clock performance and system time synchronization are mainly composed of frequency accuracy, drift rate, stability, and other indicators. To evaluate the performance of Galileo atomic clock has a significant effect on Galileo single system positioning, multi-system positioning, integrity monitoring and Galileo system time interoperability.

2 The Test Method of Galileo Satellite Clock Assessment

In order to assess the Galileo satellite clock, the test system should be built, it can get the satellite clock offset of Galileo satellite using the clock offset data to assess the performance of satellite clock. Coordinated Universal Time UTC (NTSC) held by the National Time Service Center of Chinese Academy of Sciences is used as the reference time scale. It obtains time offset between satellite clock and UTC (NTSC) by receiving signal in the space, because parameters between Galileo clock time and system time GST are broadcasted by the satellite ephemeris, so as to get the difference between Galileo system GST and UTC (NTSC), then it can realize the assessment of Galileo satellite clock. The principle is shown in Fig. 1.

Figure 1 shows the principle of assessing Galileo satellite clock. First, the standard time scale UTC (NTSC) 1PPS held by the National Time Service Center is used as the reference time signal, and UTC (NTSC) 10 MHz is used as reference frequency signal. Secondly, GPS/Galileo receiver drives system time to GPST to keep the receiver clock and the external input UTC (NTSC) 1PPS consistent.

Through receiving GPS and Galileo system signals in the space, GNSS receiver uses dual ionosphere combination and the Quasi-Zenith model to eliminate the influence of the ionosphere and troposphere under the condition that coordinates of receiver are accurately calibrate. By the GNSS data acquisition and processing IPC, we obtain the time deviation between UTC (NTSC) and the Galileo satellite clock through pseudorange correction method as shown (1):

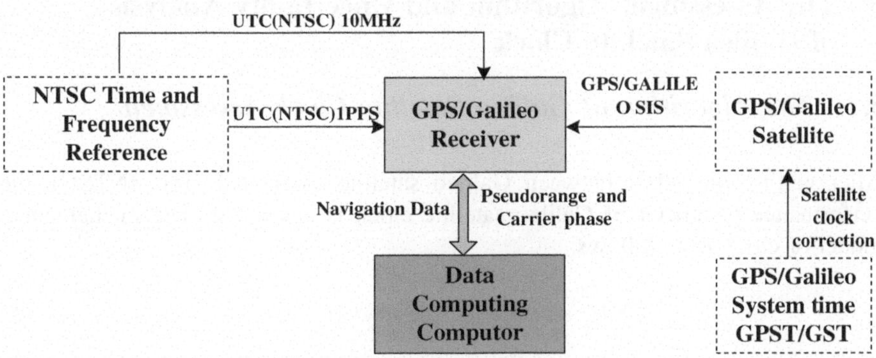

Fig. 1 Assessing principle of Galileo satellite clock

$$\begin{aligned} \mathrm{UTC(NTSC)} - \mathrm{SVT} = {} & (P - \rho - d_{\mathrm{trop}} - d_{\mathrm{ion}})/c \\ & + \tau_{\mathrm{ref}} - (\tau_{\mathrm{rev}} + \tau_{\mathrm{cable}} + \tau_{\mathrm{antenna}}) \end{aligned} \tag{1}$$

After clock correction, the offset between UTC (NTSC) and GST is obtained as shown (2):

$$\begin{aligned} \mathrm{UTC(NTSC)} - \mathrm{GST} = {} & (P - \rho - d_{\mathrm{trop}} - d_{\mathrm{ion}} + \mathrm{d}T)/c \\ & + \tau_{\mathrm{ref}} - (\tau_{\mathrm{rev}} + \tau_{\mathrm{cable}} + \tau_{\mathrm{antenna}}) \end{aligned} \tag{2}$$

where P is pseudorange, ρ is geometrical distance, c is light speed, $\mathrm{d}T$ is satellite time correction, d_{trop} is tropospheric delay, d_{ion} is ionosphere delay, τ_{ref} is reference delay, τ_{rev} is receiver internal delay, τ_{cable} is receiver cable delay, τ_{antenna} is receiver antenna delay.

Galileo system removes the ionosphere error by dual-frequency correction model using the E1 and E5a frequencies, the model is as follows [5]:

$$d_{\mathrm{ion}} = \frac{f_2^2}{f_1^1 - f_2^2} \times (P_2 - P_1) \tag{3}$$

where f_1 (E1 carrier frequency is 1575.42 MHz) and f_2 (E5a carrier frequency is 1176.45 MHz) is the carrier frequency of Galileo system, P_1 and P_2 are the pseudoranges of the two frequencies.

The simple model is used in GNSS time offset monitoring system to remove the tropospheric error. The model is as follows:

$$d_{\mathrm{trop}} = \frac{F1}{\mathrm{Sin}(\theta) + F2} \tag{4}$$

where θ is the satellite elevation, $F1$ = 2.47, $F2$ = 0.0121.

3 The Assessment Algorithm and Uncertainty Analysis of Galileo Satellite Clock

3.1 The Algorithm of Galileo Satellite Clock Assessment

After getting the offset between Galileo satellite clock and UTC (NTSC), the performance parameter of Galileo satellite clock is abstracted. The mathematical model of clock is as follows:

$$x(t) = a_0 + a_1(t - t_0) + a_2(t - t_0)^2 \tag{5}$$

where $x(t)$ is the satellite clock time offset to UTC (NTSC) at time t, a_0 is initial phase, a_1 is the frequency deviation, a_2 is the frequency drift rate, and t_0 is the starting point of model.

Using the satellite clock offset data to calculate the frequency stability; the model is as follows [6]:

$$\sigma_y^2(\tau) = \frac{1}{2(N-2)\tau^2} \sum_{i=1}^{N-2} [x_{i+2-2x_{i+1}+x_i}]^2 \tag{6}$$

where $x(i)(i = 1, 2, \ldots N)$ is the measurement of phase, the unit is time unit; N is the number of the measurement data in interval of τ, τ is the interval of sampling data.

3.2 *The Uncertainty Analysis of Galileo Satellite Clock*

The measurement of uncertainty refers to the change of the measurement result that is not sure, it is an estimated value of the measured true values with a certain range. It is a parameter of measurement results, it indicates the dispersion of the measured values. The evaluation method of measurement uncertainty depending on the different valuation methods is divided into class A and class B [7].

3.2.1 The Uncertainty Analysis of Galileo Satellite Clock Relative Frequency

Using $\sigma_y(\tau)$[8] ($k = 1$) as the Type A uncertainty of Galileo relative frequency, since there is no significant Type B component for frequency, the combined uncertainty U_c can be considered as the Type A uncertainty by this equation:

$$U_c = \sqrt{U_a^2 + U_b^2} = U_a = \sigma_y(\tau) \tag{7}$$

It can be seen that the uncertainty of Galileo satellite clock relative frequency can be expressed with the Galileo satellite clock frequency stability when we use a coverage factor of $k = 1$.

3.2.2 The Uncertainty Analysis of Galileo Satellite Clock Frequency Stability

Coordinated Universal Time UTC (NTSC) held by the National Time Service Center of Chinese Academy of Sciences is used as the reference time scale, we obtain the clock data by receiving signals in the space to evaluate the Galileo satellite clock

Table 1 Noise type and K_n noise factor

Noise type	K_n
Random walk FM	0.75
Flicker FM	0.77
White FM	0.87
Flicker PM	0.99
White PM	0.99

frequency stability; the factors of frequency stability uncertainty are the observation noise of tracks, transport links, and receivers. In general, we determine the dominant noise type of measuring data. The formula that represents Allan variance method, which computes uncertainty of frequency stability, is as follows [9]:

$$\text{ErrorBars} = K_n \sigma_y(\tau)/\sqrt{N} \tag{8}$$

where ErrorBars is the uncertainty of frequency stability, K_n is noise factor, N is the number of the measurement data point, $\sigma_y(\tau)$ is frequency stability by Allan Deviation (Table 1).

4 Galileo Satellite Clock Assessment Result

4.1 The Source Data of Galileo Satellite Clock Assessment

Currently, GPS/Galileo receivers of National Time Service Center can track eight Galileo satellites, each satellite can be seen 5 to 6 h. Receiving the follow data through the GPS/Galileo receiver:

- The pseudorange and the carrier phase of E1 and E5a, the sample interval is 60 s;
- The satellites position and elevation, the sample interval is 60 s;
- The IGS precise orbit and clock data;

The Galileo system is developing, and not all the navigation data of Galileo satellite can be received, so the IGS precise orbit and clock data are used to assess the performance of Galileo satellite clock. The result of June 2015 and November 2015 is as follows.

4.2 Galileo Satellite Clock Frequency Stability Result

The various indicators of performance can be computed using the data of Galileo satellite clock data, the average time is 60 s and 300 s. The result is shown in

Tables 2 and 3, at the same time, the uncertainty of frequency stability is analyzed. According to Barnes first partial function, we determine that dominant noise is, respectively, Flicker FM noise and white FM noise during the measurement process of the three IOV satellites (11, 12 and 19) and two FOC satellites (22 and 26) of Galileo systems when (τ = 60 s) and (τ = 300 s). The dominant noise is random frequency noise during the measurement process of other two FOC satellites (14 and 18) of Galileo systems when (τ = 60 s) and (τ = 300 s).

Table 2 shows the assessment results of the initial phase a_0, frequency deviation a_1, frequency drift rate a_2, and frequency stability (ADEV) of the Galileo every satellite average interval of 60 s; the uncertainty of the frequency stability is estimated, at the same time it gives the result of various indicators after clock correct to GST.

Table 2 Initial phase, frequency offset, frequency drift, and frequency stability of Galileo satellites (τ = 60 s)

PRN	Type	a_0Unit: s	a_1Unit: s/s	a_2Unit: s/s^2	(ADEV) (τ = 60)	Uncertainty (τ = 60)	Note
11	IOV	−2.48e−04	−6.49e−10	−3.82e−14	1.00e−12	3.48e−14	Clock uncorrected
		−4.70e−08	9.50e−12	−4.16e−14	1.00e−12	3.48e−14	Clock corrected
12	IOV	3.10e−03	9.46e−09	−2.17e−14	8.56e−13	3.46e−14	Clock uncorrected
		−4.64e−08	−5.68e−12	−1.91e−12	8.55e−13	2.97e−14	Clock corrected
14	FOC	−1.88e−03	−2.04e−08	−8.12e−12	1.29e−10	5.88e−12	Clock uncorrected
		−6.25e−08	2.68e−09	8.10e−12	1.29e−10	5.84e−12	Clock corrected
18	FOC	−4.11e−04	−2.05e−08	−1.01e−11	9.88e−13	4.01e−14	Clock uncorrected
		5.32e−08	3.25e−09	−1.01e−11	9.88e−13	4.01e−14	Clock corrected
19	IOV	3.60e−06	3.63e−11	4.74e−14	9.73e−13	3.72e−14	Clock uncorrected
		−4.25e−08	−2.61e−11	3.39e−14	9.73e−13	3.72e−14	Clock corrected
22	FOC	−8.06e−04	−1.06e−08	1.06e−14	8.85e−13	3.74e−14	Clock uncorrected
		−4.86e−08	1.24e−12	1.35e−14	8.84e−13	3.74e−14	Clock corrected
26	FOC	−5.65e−04	−2.41e−08	2.27e−15	8.63e−13	3.28e−14	Clock uncorrected
		−4.56e−08	5.60e−13	2.57e−15	8.62e−13	3.28e−14	Clock corrected

Table 3 Initial phase, frequency offset, frequency drift, and frequency stability of Galileo satellites (τ = 300 s)

PRN	Type	a_0 Unit: s	a_1 Unit: s/s	a_2 Unit: s/s^2	(ADEV) (τ = 300)	Uncertainty (τ = 300)	Note
11	IOV	−2.48e−04	−3.24e−09	−9.85e−13	9.63e−13	8.51e−14	Clock uncorrected
		−4.71e−08	5.17e−11	−1.07e−12	9.61e−13	8.49e−14	Clock corrected
12	IOV	1.33e−03	3.93e−08	−6.15e−14	7.54e−13	6.60e−14	Clock uncorrected
		−4.79e−08	−2.65e−11	2.09e−14	7.49e−13	6.55e−14	Clock corrected
14	FOC	−2.59e−03	−1.04e−07	−2.72e−10	2.48e−12	2.84e−13	Clock uncorrected
		2.13e−07	1.11e−08	−2.72e−10	2.49e−12	2.85e−13	Clock corrected
18	FOC	−5.61e−03	−1.08e−07	−2.54e−10	1.49e−12	1.63e−13	Clock uncorrected
		2.10e−07	1.10e−08	−2.57e−10	1.60e−12	1.73e−13	Clock corrected
19	IOV	8.72e−06	3.51e−10	1.01e−14	8.41e−13	7.71e−14	Clock uncorrected
		−4.71e−08	3.20e−11	−5.99e−14	8.29e−13	7.60e−14	Clock corrected
22	FOC	9.53e−05	−1.19e−07	3.06e−13	7.42e−13	6.52e−14	Clock uncorrected
		−4.31e−08	−5.04e−11	3.69e−13	7.28e−13	6.39e−14	Clock corrected
26	FOC	6.45e−05	−1.20e−07	−3.59e−13	7.96e−13	6.99e−14	Clock uncorrected
		−4.65e−08	2.52e−11	−2.15e−13	8.06e−13	7.09e−14	Clock corrected

Table 3 shows the average interval of 300 s Galileo every single satellite clock indicator of the initial phase, frequency error, frequency drift rate, and frequency stability (ADEV) sampling interval of 300 s. The uncertainty of the frequency stability (τ = 300) is also shown. Finally, the result after correcting to GST is given.

Tables 2 and 3 show:

1. When Galileo three IOV satellites (E11, E12 and E19) at (τ = 60 s) and (τ = 300 s), the frequency stability is at e−13. Four Galileo FOC satellites, E22 and E26 satellites at (τ = 60 s) and (τ = 300 s), the frequency stability is the order of e−13.
2. Galileo navigation system in August 2015 had sent the two satellites to the wrong orbit which is E14 and E18 FOC satellite. Their frequency stability of

less than e−13 orders of magnitude, the frequency offset, and frequency drift rate are significantly larger than the other two satellites.

3. When Galileo 3 IOV satellites and two FOC satellites (E22, E26) at the ($\tau = 60$ s) and ($\tau = 300$ s), the uncertainty of frequency stability caused by noise is at the e−14 middleweight, the uncertainty of E14 and E18 satellite frequency stability is greater than other satellites.

4.3 ADEV Consistency Over Time

We do a study about the consistency of frequency stability of Galileo in-orbit satellite from September 12, 2015 to October 12, 2015. However, there are no Orbit and clock difference data of FOC satellites in IGS precise orbit clock error products. Therefore, we only studied the consistency of 3 IOV satellites' frequency stability.

Figure 2 shows that the Allan deviation consistency which is fixed by satellite clock error of the 3 IOV satellites (E11, E12 and E19) which is consistent.

1. Over the time, the frequency stability of E11 is changed from 7.924e−13 to 2.207e−12, the frequency stability of E12 is changed from 8.508e−13 to 1.945e −12, the frequency stability of E19 is changed from 7.449e−13 to 1.415e−12.
2. The standard deviation of frequency stability of three IOV satellite is 3.6274e −13, 2.7501e−13, and 1.8960e−13, the Allan of the IOV satellite is consistent over time, the values are about e−12 middleweight.

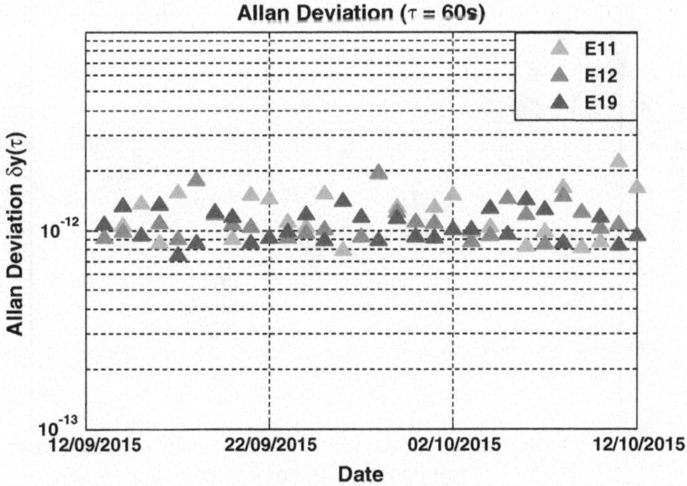

Fig. 2 Allan deviation (at 60 s)

4.4 *The Monitor Result and Uncertainty Analysis of Galileo System Time Offset*

4.4.1 Galileo System Time Offset Monitor Result

Take UTC (NTSC) of National time service center of Chinese Academy of Science as a reference and calculate the value of UTC (NTSC)-GST of the Galileo in-orbit satellites by receiving space signals. Galileo satellite data is collected by the multiple GPS/Galileo receivers of National time service center. The Galileo visual period (NTSC)-GST value within the October 25, 2015 to October 31, 2015 of National time service center is analyzed by using the data of pseudorange, ephemeris, and precise orbit clock error released by IGS.

From the figure:

1. From Figs. 3 and 4, the offset of UTC (NTSC) and GST of the 3 IOV satellites and 2 FOC satellites (22, 26) is variable between −70 and −30 ns, and is consistent over time.
2. The two Galileo FOC satellites (14, 18) were first launched into low elliptical orbit inclination in August 2015. One of them was gradually pushed to the satellite by low-orbit maneuver point in November 2015, so that it tracks more round. It reaches the new target track at November 29, 2015 and transmits the first navigation signal. As can be seen from Fig. 5, the offset of UTC (NTSC)

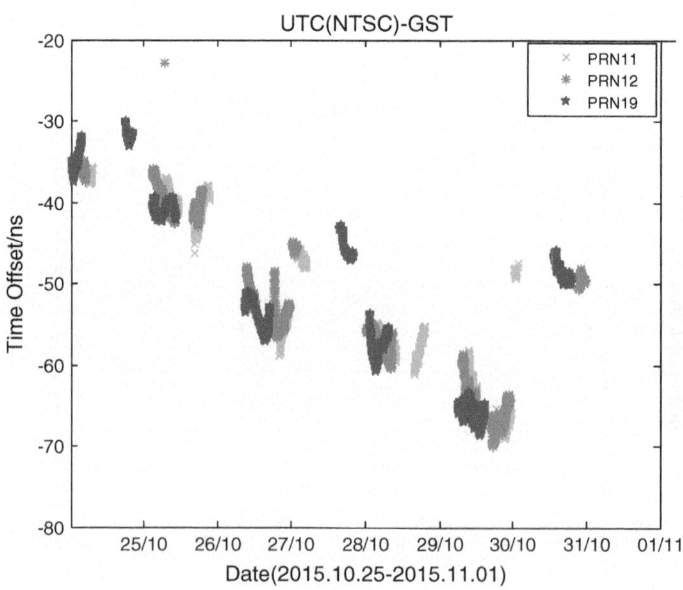

Fig. 3 Galileo 3 IOV satellites time offset relative to UTC (NTSC) after clock correction to GST

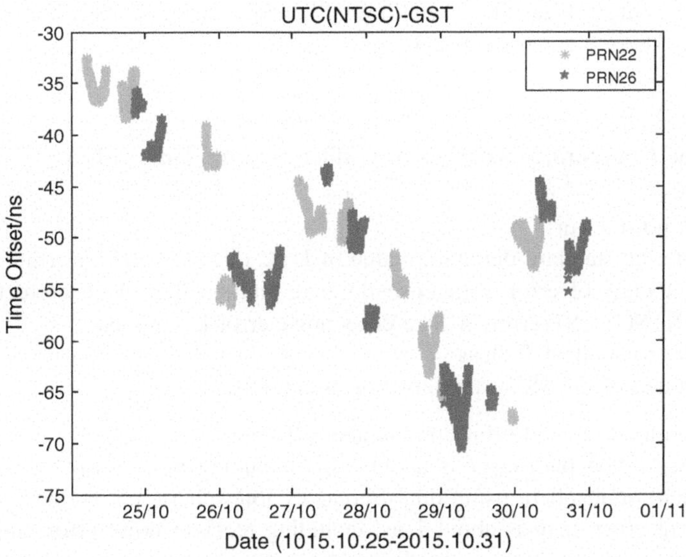

Fig. 4 Galileo 22 and 26 satellites time offset relative to UTC (NTSC) after clock correction to GST

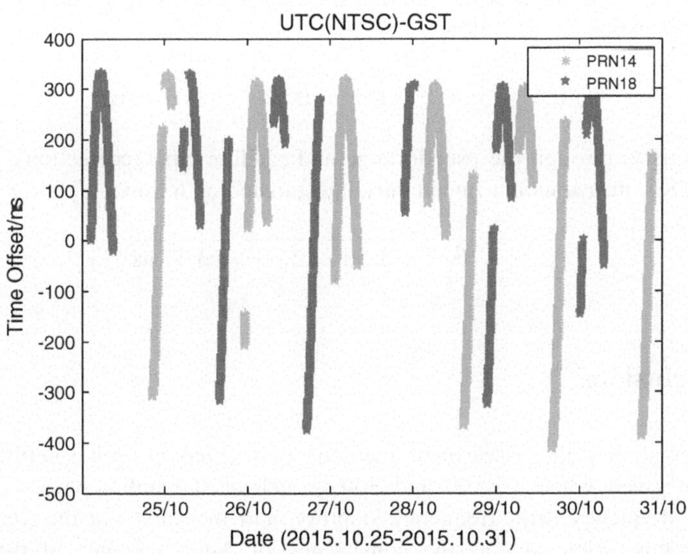

Fig. 5 Galileo 14 and 18 satellites time offset relative to UTC (NTSC) after clock correction to GST

and GST of Galileo FOC satellites (14, 18) is variable between −400 and 400 ns. There is a certain trend.

4.4.2 The Uncertainty Analysis of Galileo System Time Offset

1. *A* class noise error:
   Estimate the standard deviation value of UTC (NTSC)-GST determine the size of the *A* class of error. After pseudorange and satellite clock corrections, get UTC (NTSC)-GST error, *A* Size class noise error is 1.56 ns.
2. Error calculation of *B* class
   The budget of GNSS system error accuracy [10]:
   (a) satellite clock and orbit error is about 0.1 ns;
   (b) transmission path error is about 3 ns, including the ionosphere and troposphere, ionosphere using dual-frequency ionospheric corrections;
   (c) the receiver error is about 1 ns, including receiver noise error, the receiver delay error, the receiver hardware delay deviation;
   (d) multipath error is about 0.5 ns;
   (e) antenna delay error is about 0.5 ns;
   (f) reference clock error is about 1 ns;

Above six single monitoring satellite, is an error caused by a Class *B* error, and total error is estimated as follows:

$$\sqrt{0.1^2+3^2+1^2+0.5^2+0.5^2+1^2} \approx 3.39\,\text{ns}$$

In summary, through the pseudorange and satellite clock corrections, get UTC (NTSC)-GST, measurement uncertainty is estimated as follows:

$$\sqrt{A^2+B^2} = \sqrt{1.56^2+3.39^2} = 3.37\,\text{ns}$$

5 Conclusions

This paper studies the assessment methods of Galileo in-orbit satellites. Some values are given when τ is 60 and 300 s, such as the initial phase, frequency deviation, frequency drift, frequency stability, and measurement uncertainty, etc. Secondly, this article studies the consistency of Allan variance of the Galileo in-orbit satellite clock. Finally, the article analyzes the UTC (NTSC)-GST time offset of 3 IOV satellites and 4 FOC satellites.

The results showed that:

1. The frequency stability is ($\tau = 60$ s and $\tau = 300$ s) at e−13 level of the five satellites (11, 12, 19, 22, 26), which is less than e−13 for the other two FOC satellites (14, 18) of the Galileo in-orbit satellites.
2. Allan variance which is fixed by satellite clock error of the 3 IOV satellites is consistent over time.
3. The offset of UTC (NTSC) and GST of the 3 IOV (11, 12, 19) satellites and 2 FOC satellites (22, 26) is between −40 and −70 ns, and is consistent over time. However, for the other two FOC satellites (14, 18), the offset of UTC (NTSC) and GST is various between −400 and 400 ns, there is certain tendency.
4. The uncertainty of Galileo system time measurement is 3.37 ns.

Acknowledgments Funded by GNSS system time offset monitor and prediction supported by the "West Light" No. Y507YR0501.

References

1. Vanschoenbeek I, Bonhoure B, Boschetti M et al (2007) GNSS time offset: effects on GPS-Galileo interoperability performance. Inside GNSS 60–70
2. Luo SM (2003) Galileo systems—Development of European global navigation satellite system (GNSS) program. Professional Committee of China Institute of Navigation beacon radio navigation study group "in 2003 annual meeting" Proceedings of 2003
3. Wu H, Li X, Lu X (2011) The basic of satellite navigation system time. Science Press, Beijing
4. Yue D-M, Luo Q-M, Li Q et al (2011) Introducing and analysis of Galileo time system. In: 2011 national time and frequency conference
5. Kou Y-H (2007) GPS principles and applications, 2nd edn. Electronic Industry Press, Beijing
6. Li X, Yang X (2010) Precision measuring of the time and frequency signal. Science Press, Beijing
7. Fei Y (2010) Accuracy theory and data analysis. China Machine Press, Beijing
8. IEEE Standard 1139-1999 (1999) IEEE Standard Definitions of Physical Quantities for Fundamental Frequency and Time Metrology—Random Instabilities, prepared by IEEE Standards Coordinating Committee 27 on Time and Frequency, 1 Mar 1999
9. Stable32 Frequency Stability Analysis Version 1.3 User Manual
10. Zhang X (2013) 2014 research on the improvement of GNSS system time offset monitoring methods. University of Chinese Academy of Sciences, Beijing

A Precision Evaluation Method for Inter-satellite Link Measurement of Ka Band Based on SLR

Zhenguo Xiao, Wende Huang, Jun Yang, Shiqian Song and Hao Xie

Abstract Improving the autonomous operation ability has become a main goal of satellite navigation system in the new century. The technology to apply inter-satellite link (ISL) to perform high precision measuring between satellites is a technology to implementing autonomous operation of navigation satellite constellation, and Dual-One-Way Ranging (DOWR) of Ka band is an effective way to carry out high precision measurement between satellites. Ka DOWR data of satellite-ground could be achieved by setting up the measuring link between orbiting satellite and pseudolite on the ground, analyzing precision it combined with precision analysis of troposphere and ionosphere correcting models, we can further infer the ranging precision of satellite–satellite. In order to evaluate the precision of satellite–ground Ka DOWR data, a method of using SLR data as exterior checking for comparison is proposed. The raw data of Ka and SLR contain environment effect, and their measuring time and spatial reference do not coincide. To achieve high precision evaluation of Ka DOWR data by SLR, it is needed to go through the environment error correction, time, and spatial reference formalization and some other steps. The errors in the steps will have impact on the final precision evaluation result. In order to verify the feasibility of this evaluation method, this paper starts from the principal of Ka DOWR and SLR, then analyzes the error sources that impact the evaluation result and expounds the principle and process of this evaluation method. By setting up the ground test simulation platform of ISL, the Ka satellite-ground DOWR and SLR data are obtained and analyzed. The simulation results show that the evaluation precision of this proposed method is in the magnitude of centimeters, which can provide reference for the analysis of the satellite–ground test data of ISL.

Keywords Autonomous operation · Inter-satellite link (ISL) · Dual-One-Way ranging (DOWR) of Ka band · Satellite laser ranging (SLR) · Ranging precision evaluation · SLR comparison method · Evaluation precision

Z. Xiao (✉) · W. Huang · J. Yang · S. Song · H. Xie
College of Mechanical Engineering and Automation, National University of Defense Technology, Changsha 410073, China
e-mail: xzgdemail@126.com

J. Sun et al. (eds.), *China Satellite Navigation Conference (CSNC) 2016 Proceedings: Volume I*, Lecture Notes in Electrical Engineering 388,
DOI 10.1007/978-981-10-0934-1_40

1 Introduction

Satellite navigation and positioning systems are playing more and more important roles in the national production and the construction of national defense. In order to ensure the operation of the navigation satellite and the ability to provide navigation service when the ground nets are destroyed by war or natural disasters, building a new generation of satellite navigation system with capacity for autonomous operation is becoming a hot spot of research and the development direction [1]. Inter-satellite link (ISL) technology is a key technology for autonomous operation of navigation constellation; the main navigation systems have all researched and demonstrated it. At present, the US has applied ISL of UHF band in the GPS system, and plans to use Ka band for replacement in its next generation Block III navigation satellite [2]. New satellite in Beidou system also introduced Ka ISL technology to realize autonomous navigation function, and is now in the stage of on-orbit test.

In order to evaluate the precision of Ka ISL measurement, we can launch a satellite with the ISL devices, and establish the measurement link with the pseudolite on the ground. Compared with the whole constellation on-orbit test, the ground test verification has the advantages of low cost, multiple assessment methods, etc. If we only use orbit data to assess the Ka DOWR data, it requires precision orbit below the centimeter level to achieve centimeter level precision evaluation result, it is difficult to get orbit data with that high precision, and this also affects the real time evaluation. SLR is launched by NASA in the early 1960s, aiming at using space technology to study the earth dynamics, geodesy, the earth physics and astronomy, and other areas of the discipline of technology [3]. SLR is highly precise and without ambiguity or clock error, besides, it is not affected by the ionosphere [4]. Thus it is widely used in satellite clock error calculating and satellite ephemeris, satellite precise orbit determination, and precision evaluation [5–8].

In view of the above background, this paper analyses the feasibility of using SLR data to evaluate the precision of the satellite–ground observation data under the Ka satellite link system. Starting from the principle of Ka band ISL and SLR, it discusses the principle and process of precision evaluation for Ka ISL measurement based on SLR data, analyses the error sources that impact the precision of evaluation result in data processing. By building the ground test simulation platform of the ISL, the Ka satellite–ground DOWR and SLR data are obtained, which verify the feasibility and precision of this evaluation method.

2 Basic Principles

The principle of using satellite–ground test platform to evaluate precision of Ka DOWR ranging is shown in Fig. 1, which is carried out by colocation observation of Ka ground pseudolite and SLR laser station.

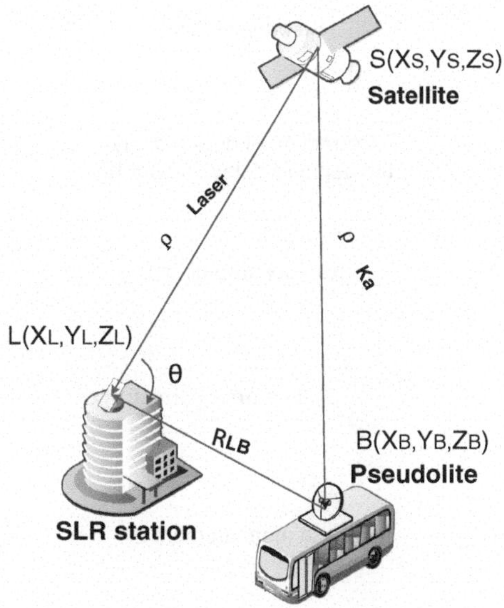

Fig. 1 Colocation observation of Ka and SLR

The principle of SLR technology is to measure the time interval τ of laser pulses traveling from the laser station to satellite, thus the distance ρ_{SLR} between station and satellite is calculated, and the observation equation can be presented by

$$\rho_{\mathrm{SLR}} = c\tau/2 = \rho + \varepsilon_{\mathrm{SLR}}^{\rho} \tag{1}$$

where c is the speed of light, ρ is the geometric distance, and $\varepsilon_{\mathrm{SLR}}^{\rho}$ is ranging error of SLR.

DOWR can get two observations in a measuring cycle: one is the sending observation and the other as the receiving. Therefore, observation equations are as follows

$$\rho^{ij}(t_0, t_1) = R^{ij}(t_0, t_1) + c\left[\delta t^{j}(t_1) - \delta t^{i}(t_0)\right] + \varepsilon^{ij} \tag{2}$$

$$\rho^{ji}(t_2, t_3) = R^{ji}(t_2, t_3) + c\left[\delta t^{i}(t_3) - \delta t^{j}(t_2)\right] + \varepsilon^{ji} \tag{3}$$

where ρ is the pseudo range, i and j are satellites numbers, t_0 and t_2 are the sending time of measurement signal while t_1 and t_3 means the receiving time; R is the geometric distance, δt is the satellite clock error, ε means other measure errors.

The original observation data of Ka and SLR contain environmental impact, and are not consistent with the measurement time and spatial reference point. To achieve the high precision comparison of the two kinds of data, it is needed to go through the environmental error correction, time and space formalization and some other steps, Fig. 2 shows the detailed data processing and assessment process.

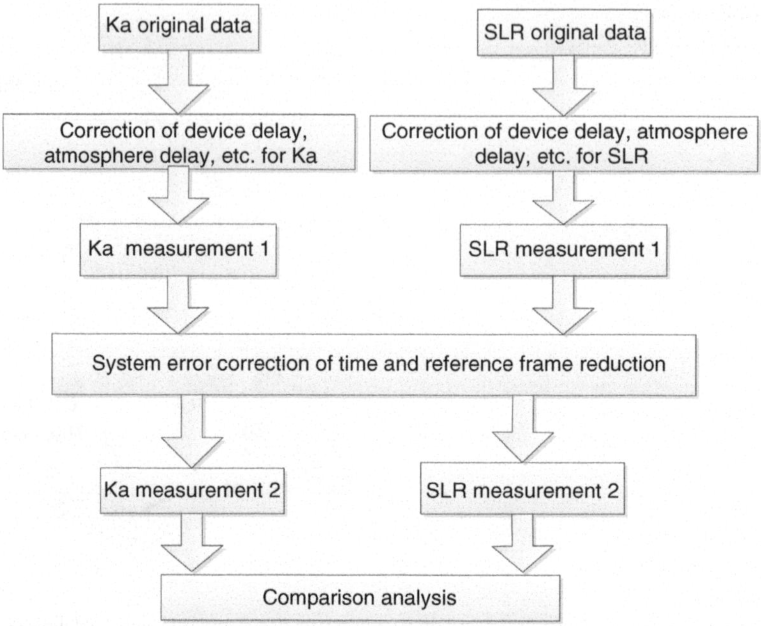

Fig. 2 Flowchart of raw Ka and SLR data processing

To achieve the comparison of the two, it is needed to go through the error correction, time and space formalization and some other steps. The errors in these steps will have impact on the final result of precision evaluation.

2.1 Error Correction Residual

The original data of SLR contains tropospheric delay, device delay, relativistic effects, etc., and Ka observation data contain ionospheric delay, device delay, tropospheric delay, antenna phase center deviation, etc., it is required to amend these errors.

In Ref. [8], the method of preprocessing the SLR data is given, and the residual error of SLR observation data is about 0.03 m, which is mainly noise and troposphere correction residual error.

Compared with satellite–satellite measurements, satellite–ground measurements have extra influence of ionosphere and troposphere.

When the microwave signals pass through the ionosphere, the ionospheric delay is [9]

$$I = \pm 0.4028 \frac{\text{TEC}}{f^2} \tag{4}$$

where *TEC* means the total number of electrons in the signal propagation path, *f* is the signal frequency with the unit of GHz. The sign is positive for pseudorange and negative for phase observation. The equation indicates, with the higher frequency, the less ionosphere influence. For the Ka band, the ionospheric delay is about centimeter scale, then the ionospheric delay residual error can be ignored when using the IGS ionospheric data for high precision correction.

In the standard atmospheric state, the relationship between the atmospheric refraction coefficient n_0 and the signal wavelength λ is as follows [10]

$$(n_0 - 1) \times 10^6 = 287.604 + 4.8864\lambda^{-2} + 0.068\lambda^{-4} \tag{5}$$

where the unit of λ is micron. The changing ranging of refraction coefficient for signal frequencies varying from 1 to 100 GHz is 1.000287604000000 to 1.000287604000544, the difference is very small that can be ignored. Currently, IGS released tropospheric delay product (ZPD) within the precision of 5 mm [11]. Therefore, the influence of tropospheric correction residual on evaluation precision can be ignored.

2.2 *Time Formalization*

SLR system and Ka ranging system use different clocks, and the interval between the observation data are also different for two systems, causing the observation time of data from two systems are not aligned. Besides, the observations are not instantaneous value, making it necessary to calculate the data to instantaneous value. Therefore, the time formalization is mainly divided into two parts: first, formalization of the observation value to make them instantaneous values; secondly, formalization of SLR and Ka DOWR data to make them time aligned. Specific steps are

(1) Unify Ka measurement data and SLR data into the same time frame, such as BeiDou time.
(2) Due to the influence of light time, each observation data corresponds to a time interval. For the ease of comparison, it is necessary to normalize the data to instantaneous value. In this paper, the preprocessed Ka one-way observation data are normalized to receiving time which removes the influence of light time. The SLR data are normalized to the time the laser reaches the laser reflector of satellite.
(3) To eliminate the clock error impact on evaluation result, it is required to decouple the Ka DOWR data. This means to add the one-way observations of uplink and downlink at the same time obtained in step 2 to get the value without clock error.

(4) In order to eliminate the inconsistency between the SLR data and the Ka DOWR data, the method of polynomial interpolation or dynamic estimation can be used to normalize the two data to the same time.

2.3 Space Formalization

The essence of distance measurement is to get the distance information between two points in space. Therefore, different reference points result in different measurement results. The Ka observation data obtained after error correction and time formalization can be regarded as the instantaneous distance between the mass point of the satellite and the ground pseudolite. The corresponding SLR data is the distance from the mass point of the satellite to the mass point of SLR station.

As shown in Fig. 1, when evaluating the precision of Ka measurement by comparison with SLR data, the location of the satellite $\vec{S}$ is provided by the satellite precise ephemeris, the ground Ka pseudolite position $\vec{B}$ and laser station position $\vec{L}$ are known precisely, through the triangle of them in the same coordinate system, using cosine theorem

$$\cos\theta = \frac{\overrightarrow{\mathrm{LS}} \cdot \overrightarrow{\mathrm{LB}}}{\left|\overrightarrow{\mathrm{LS}} \cdot \overrightarrow{\mathrm{LB}}\right|} \tag{6}$$

We can solve angle θ; through angle θ, SLR ranging value ρ_{Laser} and the distance R_{LB} between pseudolite and laser station on the ground, using the theory of cosine theorem can get Ka geometric distance ρ_{Ka}^{C}:

$$\rho_{\mathrm{Ka}}^{C} = \sqrt{\rho_{\mathrm{SLR}}^{2} + R_{\mathrm{LB}}^{2} - 2\rho_{\mathrm{SLR}} \cdot R_{\mathrm{LB}}\cos\theta} \tag{7}$$

Finally, compare ρ_{Ka}^{C} with the normalized Ka measurement ρ_{Ka}^{O} and do the statistical analysis to assess the precision of Ka DOWR measurement.

3 Data Simulation and Analysis

3.1 Simulation Data Generation

According to the principle of Ka ISL DOWR, the simulation environment is built, and based on the mechanism of Ka ISL, observation data between satellite and ground are generated. Using the precise satellite orbit and ground position information to generate SLR data; when generating Ka observation data, it is subject to

Table 1 Initial orbit data of satellite

	X	*Y*	*Z*
Position (m)	−36,031,725.2928148	21,816,922.7212969	0.00000000
Speed (m/s)	−914.338824312	−1510.075722549	2521.133980296

Table 2 Position data of Ka ground station and laser station

	X	*Y*	*Z*
Ka station (m)	−2,177,527.700641	4,388,901.473554	4,070,001.155546
Laser station (m)	−2,177,524.700641	4,388,904.473554	4,070,004.155546

receiving time, one-way pseudorange is calculated through the reverse iteration of signal propagation delay time to calculate the location of the emitter.

In the simulation of the generation of Ka satellite–ground two-way ranging data, we ignore the influence of device delay, the troposphere and ionosphere delay, only add random error with the mean value of 0 m and the standard deviation of 0.1 m. The time difference between the clock error and the two systems is not considered. Basic parameters of the simulation are set as follows.

The satellite initial orbit: orbital epoch is 2,456,976.5 (Julian day); satellite initial position and velocity of *X*, *Y*, *Z* direction in ECI are shown in Table 1.

X, *Y*, and *Z* coordinates of Ka and SLR station location (mass point location) in ECF are shown in Table 2.

3.2 Simulation Data Analysis

Simulation generates 2 h data (2400 data), the two-way ranging data of Ka is as shown in Fig. 3 (to facilitate observation, show only a part of the data).

After the Ka DOWR data are normalized to the mean time of downlink receiving time and uplink receiving time, results in ρ_{Ka}^{O}, compare the formalization results with theory geometric distance calculated from SLR data, Fig. 4 shows the difference between the value ρ_{Ka}^{O} and ρ_{Ka}^{C}.

As seen in Fig. 4, the two-way observation formalization ρ_{Ka}^{O} of Ka and the distance calculated from SLR data are identical, the difference between the two is random noise. Under the simulation conditions, the results show that the mean error is about 0.02 m, the standard deviation is about 0.09 m, which have the difference of mean error 0.02 m, and standard error 0.01 m to the added random error. Considering that the SLR data itself contains error, the final assessment precision can be maintained at the centimeter level.

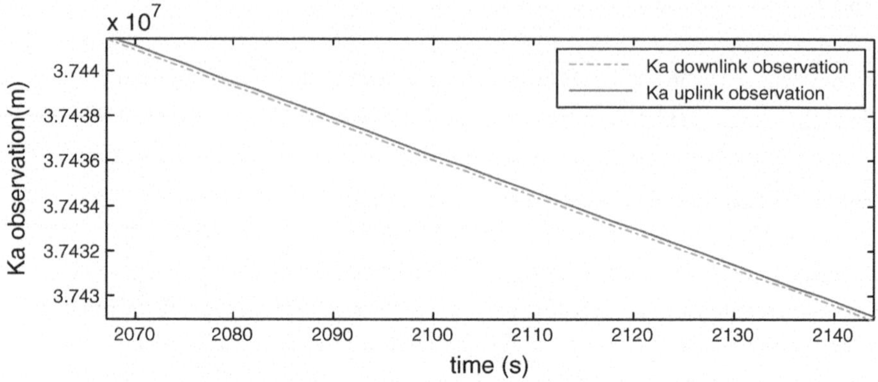

Fig. 3 Up and down observation data of Ka (color figure online)

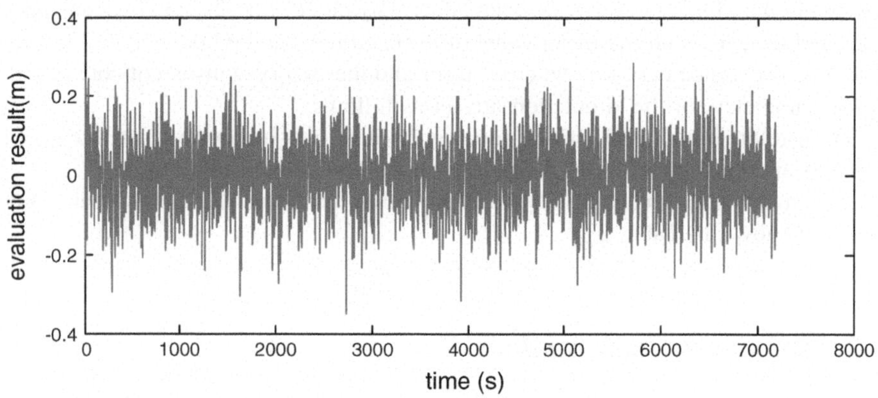

Fig. 4 Difference between observation and calculated range for Ka

4 Conclusion

The following conclusions can be drawn from the above discussion and analysis

(1) Due to the high precision and non-clock error feature of SLR data, it can provide an independent external check method for the Ka satellite–ground two-way ranging data of the satellite link system, and realize the high precision evaluation of the Ka DOWR data.

(2) In the process of evaluation, the error correction of observation data and the formalization of space and time are required. The evaluation results are affected by the error correction precision of the observed data and the precision of the data space and time formalization.

(3) Considering the influence of error correction precision and space and time formalization error of two kinds of observation data, in this paper, the precision of the SLR-based Ka DOWR data precision evaluation method is in the centimeter level.

This paper puts forward the SLR comparison method for precision evaluation of Ka inter-satellite link observation data, discusses the basic principle and specific process of the method, and analyses the error terms. It can provide some reference for data processing under the inter-satellite link technology test.

References

1. Chen Z (2012) Key technologies of autonomous operation for navigation satellite constellation using inter-satellite tracking data. National University of Defense Technology, Changsha
2. Qin W, Yang X, Wei P, Li Z (2014) A calibration method for system error of inter-satellite equipment. J Astronaut 35(10)
3. Han G, Qu F, Guo J, Cao L (2012) Analysis and realization of day-time ranging of SLR. Chin J Sci Instrum 33(4):885–889
4. Sheng CZ, Gan WJ, Zhao CM et al (2013) Precise orbit determination of JASON-2 satellite: analysis of GPS, SLR and DORIS. Sci Sin-Phys Mech Astron 43(2):219–224
5. Qin X, Yang Y, Jiao W, Wang G (2004) Determination of navigation satellite clock bias using SLR and pseudorange data. Acta Geodaetica et Cartogr Sinica 33(3):205–209
6. Shen X (2011) Using SLR to determine satellite LAGEO's precisely orbit. Shandong University, Shandong
7. Fu C (2012) The Algorithm of compass precise orbit determination based on SLR and pseudorange observation. Tsinghua University, Beijing
8. Mao Y, Jia X, Sun F (2007) Validating satellite broadcast ephemeris by SLR data. J Zhengzhou Instit Surv Mapp 24(4):263–266
9. Xie G (2009) Principles of GPS and receiver design. Publishing House of Electronics Industry, Beijing
10. Li Z, Haung J (2005) GPS surveying and data processing. Wuhan University Press, Wuhan
11. Zhao Z (2014) Several kinds of troposphere delay correction model impact on GPS precise point positioning. Chang'an University, Xi'an

A BDS Constellation URE Evaluation Method Based on the Probability Statistics Characteristics Under AOD

Tongyu Che, Chuanding Zhang, Li Yang and Wei Feng

Abstract BeiDou Navigation Satellite System (BDS) can provide regional services at present. Its ground segments are mostly within the territory of china which leads to the lack of whole arc monitoring for some satellites. In order to reflect the influence of BDS's mixed constellation (GEO + IGSO + MEO) and the exploration of broadcast ephemeris due to lack of whole arc monitoring, the paper conducted the URE evaluation of BDS constellation based on the constrained condition of Age of Data (AOD). The paper got the URE features of single satellite by analyzing the probability statistics characteristics under AOD, and then evaluated the constellation's URE based on the weight conducted from the above. The verification of the method was done by the real data.

Keywords BDS · Mixed constellation · AOD · URE · Probability statistics · Weighting algorithm

1 Introduction

BeiDou Navigation Satellite System (BDS) is with the abbreviation as BDS. It is Chinese global satellite navigation system with independent construction, independent operation, and the compatibility with world's other satellite navigation systems [1]. The current BDS space segment consists of five GEO satellites

T. Che (✉) · L. Yang
Information Engineering University, Beidou Navigation Technology Collaborative Innovation Center of Henan, Zhengzhou, China
e-mail: gnssche@foxmail.com

C. Zhang · W. Feng
Beijing Satellite Navigation Center, Beijing, China
e-mail: 13607665382@163.net

W. Feng
Institute of Geodesy and Geophysics, Chinese Academy of Sciences, Wuhan, China

J. Sun et al. (eds.), *China Satellite Navigation Conference (CSNC) 2016 Proceedings: Volume I*, Lecture Notes in Electrical Engineering 388,
DOI 10.1007/978-981-10-0934-1_41

(C01-C05), five IGSO satellites (C06-C10), and four MEO satellites (C11-C14) [2] (up to October 2015).

The broadcast ephemeris of BDS was broadcasted every hour. The basic navigation information includes: whole week number (WN), user range accuracy index (URAI), independent health identification (SatH1), satellite equipment time delay (TGD1), ephemeris parameters and age of data ephemeris (AODE), clock error parameters, and age of data clock (AODC).

Age of Data (AOD) consists of AODE (or issue of data ephemeris, IODE) and AODC (or issue of data clock, IODC). According to ICD, AODE is the extrapolated interval of ephemeris parameters. It indicates the time difference between the reference epoch of ephemeris parameters and the last observation epoch for extrapolating ephemeris parameters. Similarly, AODC is the extrapolated interval of clock correction parameters. It indicates the time difference between the reference epoch of clock correction parameters and the last observation epoch for extrapolating clock correction parameters. AODE and AODC are updated at the start of each hour in BDT, and it is 5 bits long with definitions in Table 1.

Because of BDS's mixed constellation and its ground segment stations locate mostly within the territory of China, the lack of whole arc monitoring of BDS IGSO/MEO often occurs which leads to the increase in IODE/IODC.

Figures 1 and 2 are variation curves of IODE/IODC of all three kinds of satellites (GEO/IGSO/GEO) within BDT week 456. The horizontal axis is time, expressed in day of week. The longitudinal axis is the value of IODE/IODC. C11 satellite is MEO, C06 satellite is IGSO and C01 satellite is GEO.

From Figs. 1 and 2, it can be concluded that the GEO satellite's IODE is 1 (occasionally is 2) and IODC is 0 (occasionally is 1). Besides, IODE and IODC of IGSO and MEO satellites' have significantly changes. IODE of IGSO can reach 6 and of MEO can reach 25. IODC of IGSO can reach 5 and of MEO can reach 20. Particular attention should be given to the condition that IODC of MEO does not increase accordingly with IODC when IODC is over 20, but restarts from 0.

This is because BDS GEO satellites are full arc visible by system ground stations with real-time measurement data. The control center can make use of the latest data to upload broadcast ephemeris. But BDS IGSO and MEO satellites are lack of full arc monitoring sometimes with no observation data during that time. The control segment cannot make use of the latest data to upload broadcast ephemeris

Table 1 The definition of IODE/IODC

IODE	IODC	Definition
<25	<25	Age of the ephemeris/clock correction parameters in hours
25	25	Age of the ephemeris/clock correction parameters is two days
26	26	Age of the ephemeris/clock correction parameters is three days
…	…	…
31	31	Age of the ephemeris/clock correction parameters is over seven days

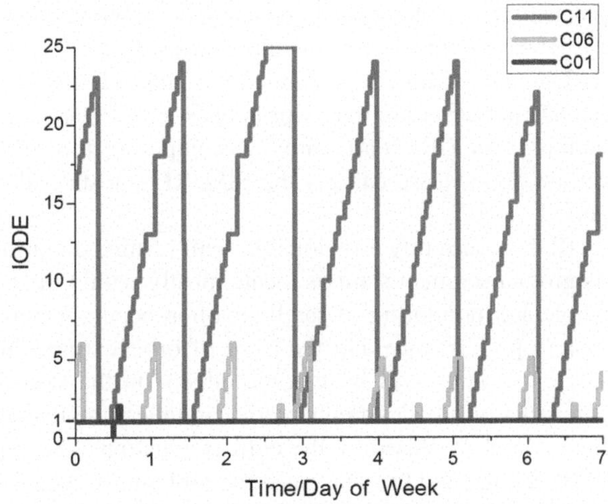

Fig. 1 IODE graph of BDS satellites in BDT week 456

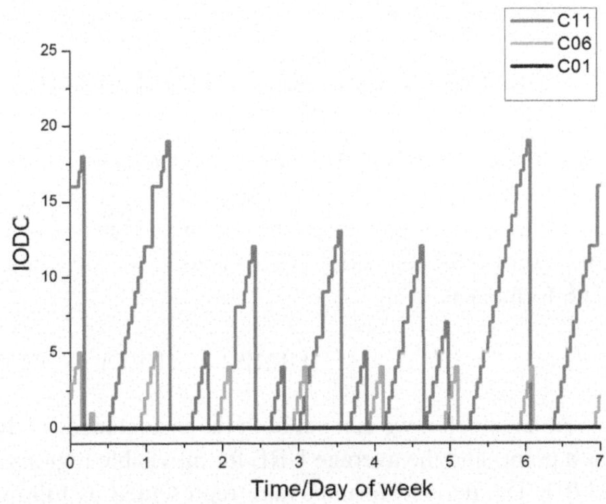

Fig. 2 IODC graph of BDS satellites in BDT week 456

but relies on the extrapolated orbit. So, it will result in significant increase in the AOD.

Many researchers have conducted the system performance evaluation of BDS. Yang assessed the evaluation of regional satellite navigation system and put forward some basic theories and methods of navigation system evaluation [3]. Su discussed the URE algorithm in navigation constellation and the computing model based on weighted method [4]. Pan evaluated the BDS broadcast ephemeris and came to the conclusion that its 3D error is within 5 m [5]. Guo did simulation and analysis of the service accuracy of BDS with pseudo satellites [6]. Geng conducted BDS precision ephemeris evaluation and the analysis of broadcast ephemeris precision [7]. P. Steigenberger and Oliver Montenbruck did some researches about the BDS system precision evaluation [8, 9].

The GPS-SPS-PS document published by US Department of Defense (DOD) defines some basic principles of GPS system evaluation [10]. There is a definition in URE Part: only by computing the SISURE of each satellite under certain periods, we can get the SISURE statistic value of each satellite under that period.

BDS is currently a regional system of mixed constellation and the situation that its ground segment stations locate mostly within the territory of China. The lack of whole arc monitoring of satellites often occurs which leads to the dropping of the system performance and the need of orbit extrapolation for broadcast ephemeris. The exploration will have some effects on the statistic value of URE. The AOD parameter of broadcast ephemeris can reflect the feature of exploration (as shown in Figs. 1 and 2). Based on the current situations, the paper conducted the evaluation of BDS URE and the analysis of its probability statistics characteristics under AOD. At last, a method of BDS constellation URE evaluation was put out based on that characteristic.

2 The URE Evaluation Method of BDS

User Range Error (URE) only considers about the effect of space segment and ignores the effect in propagation path. So, it is important for the navigation precision evaluation in system level. For navigation constellation, the index of URE and User Equipment Error (UEE) is used for the evaluation of system performance. The formula is

$$(\mathrm{UERE})^2 = (\mathrm{URE})^2 + (\mathrm{UEE})^2 \tag{1}$$

At an epoch *t*, we can get both the instantaneous URE for a single user (regarded as a point) and the average URE for all visible regions (the cutoff angle to a satellite is 0°). The derived formulas are represented as follows.

2.1 *The Formula of Instantaneous URE*

We define a space rectangular coordinate system for the analysis (Fig. 3). The earth center is *o*. *X*-axis is parallel to the tangential direction of the movement of the satellite. *Z*-axis is the vector from the earth center to the mass of the satellite. *Y*-axis is perpendicular to the *XZ* plane and the *XYZ* constitutes the right coordinate system.

Fig. 3 The sketch of IURE

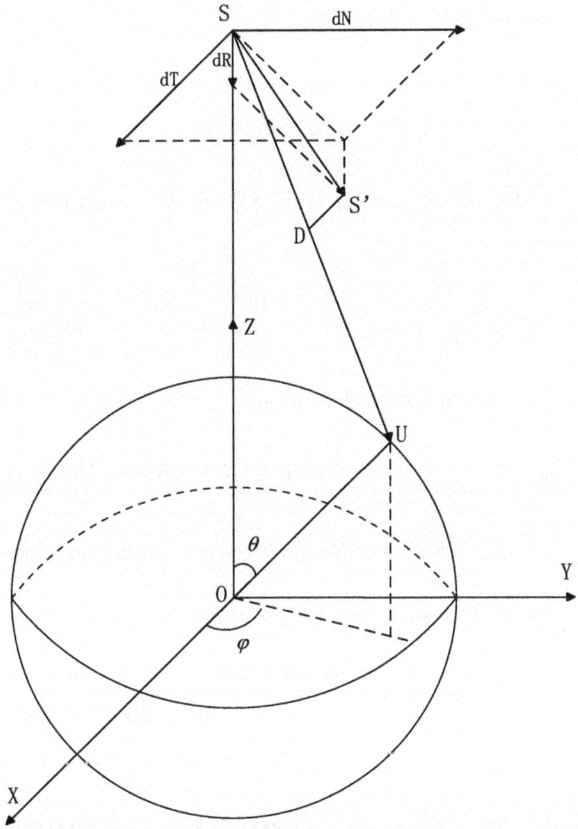

At an epoch t, the calculation of instantaneous URE formula of a user U on the earth to a satellite S is as follows. Assume that the earth is a sphere, the radius of earth is $r = \left|\overrightarrow{OU}\right|$, the value is Re. The distance between the mass of satellite to the center of earth is $r' = \left|\overrightarrow{OS}\right|$, the value is Rs. $U(\varphi, \theta, 0)$ is a point on the earth with geodetic height of zero.

Assume that the cutoff angle of satellite S to user U is 0°, so $\angle SUO = 0°$. S′ is the position of satellite calculated by broadcast ephemeris. $\left|\overrightarrow{SD}\right|$ is the IURE according to the definition above.

$S(0, 0, Rs)$ and $S'(\mathrm{d}t, \mathrm{d}n, Rs + \mathrm{d}r)$ are two points.

According to polar coordinate system definition,we can get a point u:

$$u(Re\ \sin\theta\ \cos\varphi, Re\ \sin\theta\ \sin\varphi, Re\ \cos\theta).$$

So we get vectors

$$\begin{cases} \overrightarrow{SU} = (Re\ \sin\theta\ \cos\varphi, Re\ \sin\theta\ \sin\varphi, Re\ \cos\theta - Rs) \\ \overrightarrow{SS'} = \{\mathrm{d}t,\ \mathrm{d}n,\ \mathrm{d}r\} \end{cases} \tag{2}$$

$\left|\overrightarrow{SD}\right|$ is the projection of $\overrightarrow{SS'}$ on $\overrightarrow{SU}$. According to vector projection formula:

$$\mathrm{Proj}_{\overrightarrow{SU}}\overrightarrow{SS'} = \frac{\overrightarrow{SU}\cdot\overrightarrow{SS'}}{\left|\overrightarrow{SU}\right|} \tag{3}$$

According to the definition:

$$\left|\overrightarrow{SU}\right| = (Re^2 + Rs^2 - 2ReRs\ \cos\theta)^{1/2} \tag{4}$$

$$\overrightarrow{SU}\cdot\overrightarrow{SS'} = \mathrm{d}t\,Re\ \sin\theta\ \cos\varphi + \mathrm{d}n\,Re\ \sin\theta\ \sin\varphi + \mathrm{d}r(Re\ \cos\theta - Rs) \tag{5}$$

Bring formulas (5) and (4) to (3), we get

$$\mathrm{URE}_{\mathrm{instant}} = \frac{\mathrm{d}t\,Re\ \sin\theta\ \cos\varphi + \mathrm{d}n\,Re\ \sin\theta\ \sin\varphi + \mathrm{d}r(Re\ \cos\theta - Rs)}{(Re^2 + Rs^2 - 2Re\,Rs\ \cos\theta)^{1/2}} \tag{6}$$

2.2 *The Formula of BDS Average URE*

The calculation formula of IURE is discussed above. It can calculate the URE of each position and do the average to calculate the average URE in the region. Subsection method can also be used. The accuracy of the tangential, radial, and normal position error and the accuracy of the prediction of clock are obtained together by rules of weight.

BDS MEO is of a height of 21,525 km above the earth surface. The radius of earth is 6400 km. So, the max angle $\angle oss' = \tan^{-1}(ss'/r') \approx 13.88°$ (see the definition in Fig. 3). The average URE under the cutoff angle of 0° in the area is [11]:

$$\mathrm{URE}_{\mathrm{MEO}} = \sqrt{(0.98R - T)^2 + 1/54(T^2 + N^2)} \tag{7}$$

BDS GEO/IGSO is of a height of 35,786 km. The radius of earth is 6400 km.,so the max angle $\angle oss' = \tan^{-1}(ss'/r') \approx 8.692^\circ$ (see the definition in Fig. 3). The average URE under the cutoff angle of 0° in the area is:

$$\mathrm{URE}_{\mathrm{GEO/IGSO}} = \sqrt{(0.99R - T)^2 + 1/127(T^2 + N^2)} \quad (8)$$

3 Experimental Plan

The experimental plan of the evaluation is as follows:

1. Pre-Preparing for the broadcasted.
   BDS broadcast its ephemeris with no time delay. Because of the accuracy of precise ephemeris is of one order magnitude to the broadcast ephemeris, precise ephemeris is regarded as actual position of satellites. We collected BDS broadcast ephemeris of 28 days (from 14th September 2014 to 12th October 2014) (from BDT week 454 to BDT week 457). MGEX of IGS provided precise ephemeris of mixed constellation. We collected precise ephemeris of BDS from GFZ (GeoForschungsZentrum, Potsdam) analysis center (filename: gbm18100.sp3—gbm18136.sp3). BDS Satellite C13 was removed from the experiment because of its fault.
2. Setting the time period.
   BDS has 14 satellites in service. But thc data has to be removed when there is orbit maneuver, attitude maneuver, the attitude control model from yaw-steering to orbit-normal and SatH1 parameter of broadcast ephemeris is marked as 1 (unhealthy).
   BDS broadcast its ephemeris every hour. In theory, the difference between any epoch T and TOE (Time of ephemeris) should be lower than half an hour (1800 s) ($|\mathrm{d}t| = |T - TOE| \leq 1800\,\mathrm{s}$). But in practice, there may be lack of broadcast ephemeris for some reasons. So, we dropped the epoch of $|\mathrm{d}t| = |T - TOE| > 1800\,\mathrm{s}$ to ensure the $|\mathrm{d}t| = |T - TOE|$ value was not the main factor in evaluation.
3. Computing the difference between the broadcast and precise ephemeris.
   The time system of precise ephemeris is GPST and the coordinate system is IGB08. The time system of BDS broadcast ephemeris is BDT and the coordinate system is CGCS2000. According to ICD, the BDT/TAI/GPST need transformation. In the experiment, the unity of time and coordinate system is necessary. In the same system, after computing the position of satellites form broadcast ephemeris R_{brod} and from the precise ephemeris R_{precise}, the 3D position error $r = |R_{\mathrm{broc}} - R_{\mathrm{precise}}|$ was estimated. After projecting that on satellite body fixed coordinate, the errors in tangential, radial and normal positions $r(R, T, N)$ were estimated.

The algorithm of the calculation of BDS broadcast ephemeris can be obtained in ICD. Especially, being different from the usage of GPS broadcast ephemeris, BDS broadcast ephemeris provides the mass center as the center calculated from the broadcast ephemeris but not the center of antenna. The precise ephemeris also provides the mass center. So, there is no need to compute the difference between the mass center and antenna center as usually done in experiment with GPS ephemeris.

4. Computing the URE.
   We computed the URE of each satellite with the data above in formulas (7) and (8), after setting the time interval d*T*, we got a series of URE according to time.
5. Doing the statistic of URE based on the condition of AOD
   After classifying the URE based on AOD, we did the corresponding statistic.
6. Setting up the function of URE with AOD.
7. Real data verification.

4 Experimental Analysis

4.1 The Error of Broadcast Ephemeris in RTN Directions

We set the example of satellite C01(GEO)/C06(IGSO)/C14(MEO) to analyze the error of broadcast ephemeris in RTN direction. As we can see from Figs. 4, 5 and 6, the *X*-axis is the day of week. The *Y*-axis is the error between the broadcast ephemeris and precise ephemeris (the unit is meters).

From the figures above, we can see: the error in *R* direction is much smaller than that in *T* and *N* directions; the precision of GEO ephemeris is worst especially in

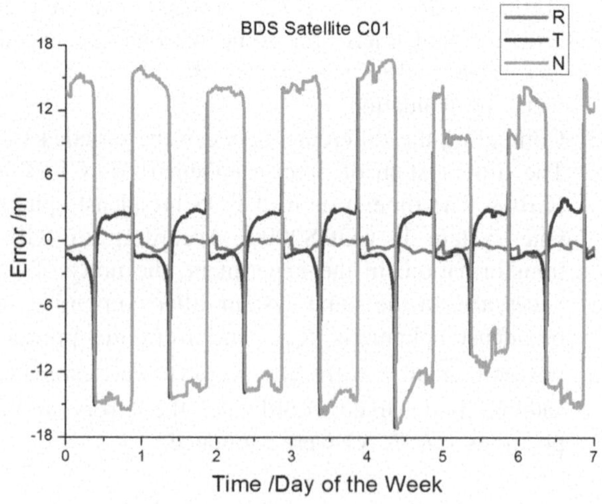

Fig. 4 The error of C01 (GEO) in RTN directions in BDT week 456

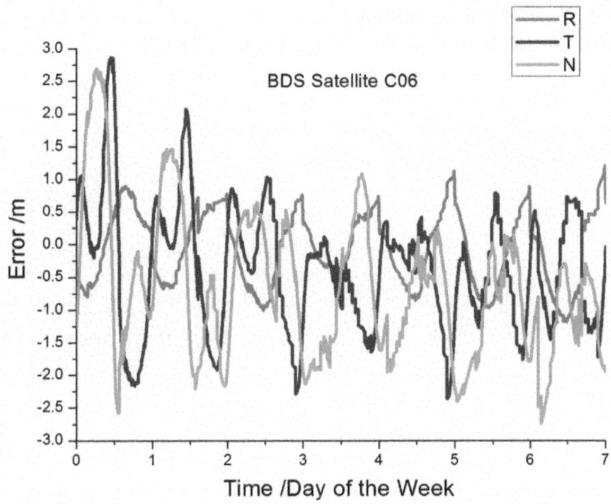

Fig. 5 The error of C06 (IGSO) in RTN directions in BDT week 456

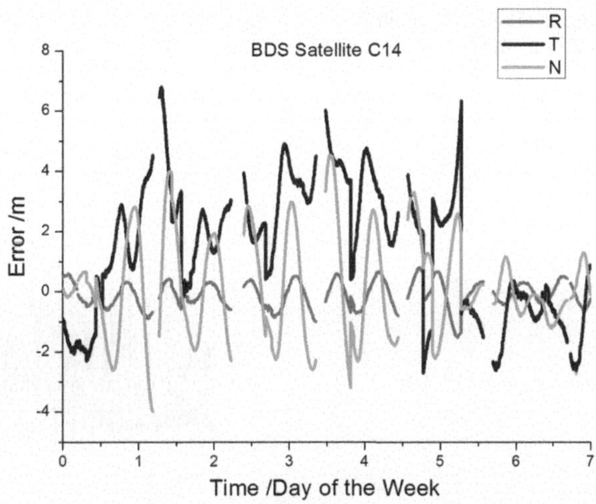

Fig. 6 The error of C14 (MEO) in RTN directions in week 456

N direction. Because BDS broadcast ephemeris provides the mass center as the center but not the center of antenna, the expectation of error in *R* direction is almost 0 without any systematic error.

4.2 The Series of BDS URE

The series of URE were plotted in the follows. The *X*-axis is the day of week. The *Y*-axis is the URE (the unit is meters). C13, C02, C03, C05 satellites were removed because of their frequent maneuvers and bad data quality (Figs. 7, 8 and 9).

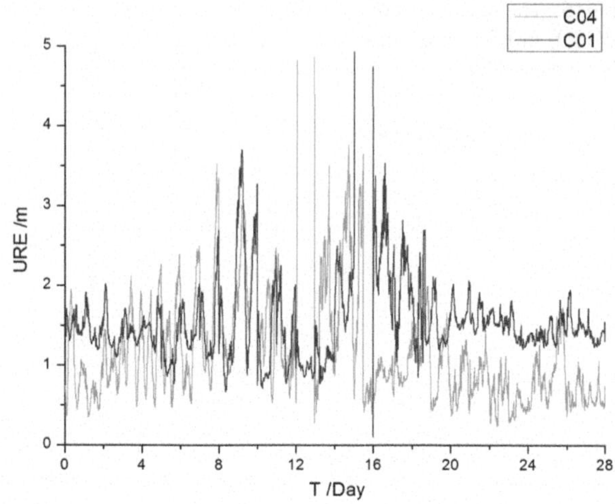

Fig. 7 The URE graph of GEO in BDT week 456

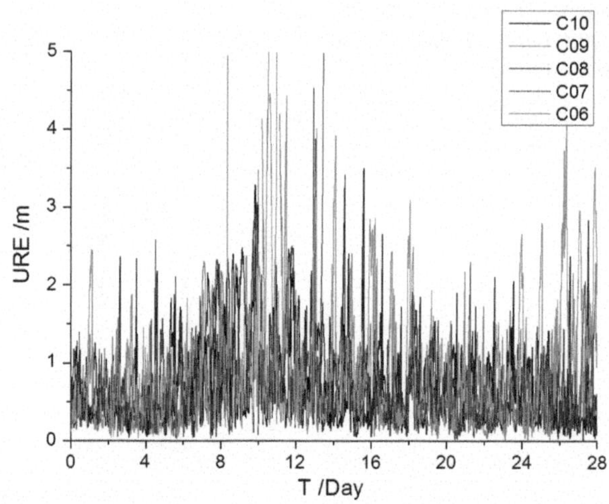

Fig. 8 The URE graph of IGSO in BDT week 456

As we can see: the URE ranges from 0 to 2 m most of the time. But there exist periodic variations. So, the variation is not smooth. The URE of IGSO and MEO is better than that of GEO.

4.3 The Probability Statistics Characteristics of URE Based on AOD

We made the probability density plot of URE under certain AODs to figure its probability statistics characteristics. URE represents the error of broadcast orbit and

Fig. 9 The URE graph of MEO in BDT week 456

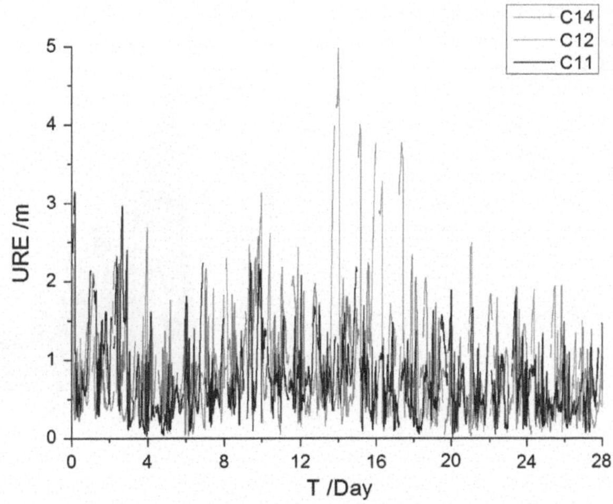

clock. The broadcast ephemeris under the same AOD is conducted from the same raw data. So, they should have similar error characteristic.

Figures 10, 11, and 12 are the probability density plots of three kinds of satellites of URE. The *X*-axis is the value of URE. The *Y*-axis is the number the value appears.

From the figures above, we founded that URE did not appear in a likely normal distribution (or Gaussian distribution). The plot was not symmetry around the center axis of value. Most values appeared in the region toward the zero value. The frequency of large value is quite small. By such characteristic, we founded that it is not strict to represent the BDS URE value under certain AOD with the average value but 95 % dividing point was better.

Fig. 10 The URE probability density plot of C01 (IODE = 1)

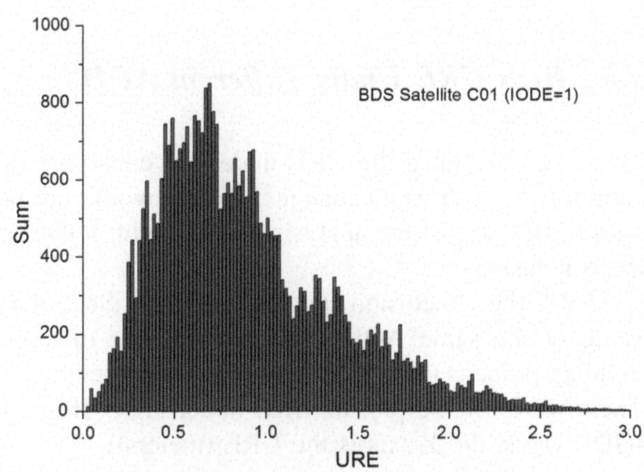

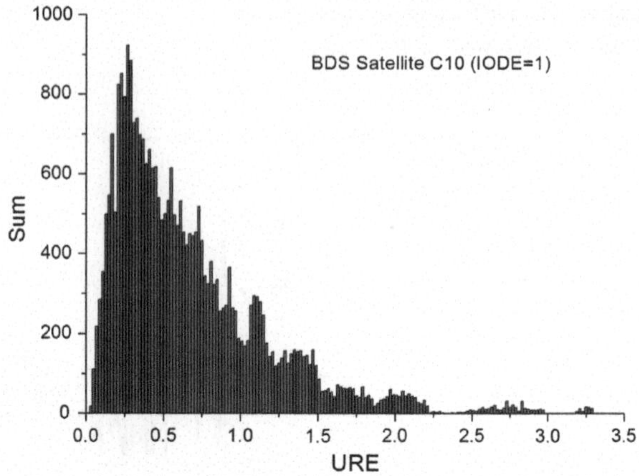

Fig. 11 The URE probability density plot of C10 (IODE = 1)

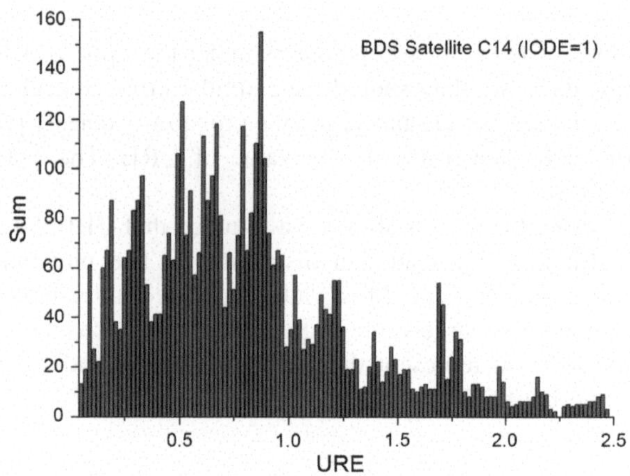

Fig. 12 The URE probability density plot of C14 (IODE = 1)

4.4 BDS URE Under Different AOD

Then, we calculated the URE under the constraint of AOD. GEO did not appear significantly with AOD changes, so they would not be calculated. IODC increased with IODE, so just the analysis of URE with IODE is necessary for us to come to the conclusions.

Under the conclusion that we got from Sect. 4.3. After sequencing the URE value of the same AOD from the smallest to largest, we conducted the 95 % dividing point as the symbolized point to present the URE under that certain AOD. Then, we draw the plot of URE under AOD [Figs. 13 and 14. The *X*-axis is the IODE value, the *Y*-axis is the URE (meters)].

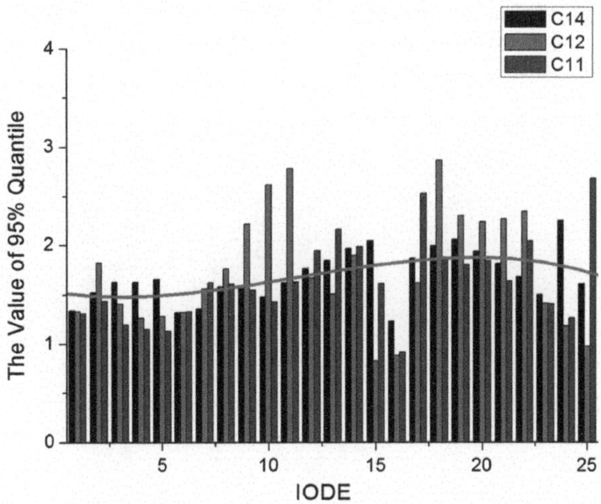

Fig. 13 URE's variety with IODE of BDS MEO

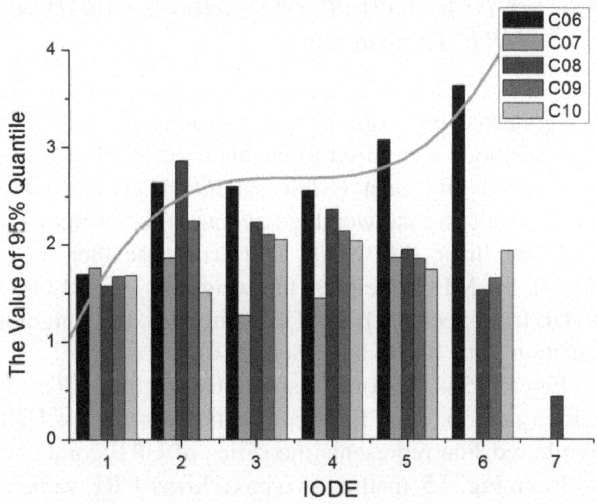

Fig. 14 URE's variety with IODE of BDS IGSO

As we can see, URE of MEO does not raise significantly with IODE, but increase with periodic change. Under IODE of 16, there is a sharp drop down of value. The fitting formula with 3rd orders is $Y(\text{MEO}) = 1.52301 - 0.03153X + 0.00591X^2 - 1.728\text{E}{-4}X^3$.

The variation of IGSO is similar with that of MEO. The fitting formula with 3rd orders is $Y(\text{IGSO}) = 0.06386 + 2.27173X - 0.6556X^2 + 0.06291X^3$.

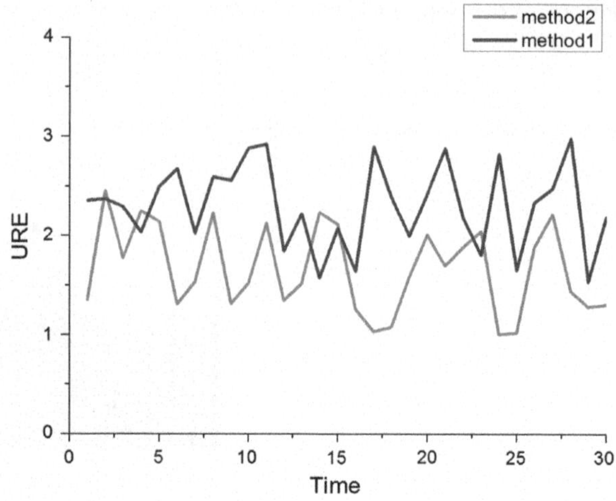

Fig. 15 The comparison figure of two methods of BDS URE (color figure online)

4.5 *A New Method BDS Mixed Constellation URE Evaluation*

We collected BDS observation and navigation data from BJFS station of one day. Two methods were used to evaluate the BDS visible constellation's URE. Method one: computing each visible satellite URE and conducting the average. Method two: conducting the weighted average. The principle of weighting is as follows: for GEO satellites, the weight is 1/2(because their insignificantly appearance under AOD); for MEO satellites, the weight is 1/Y(MEO) (Y(MEO) is the function with IODE from Sect. 4.4); for IGSO satellites, the weight is 1/Y(IGSO) (Y(IGSO) is the function with IODE from Sect. 4.4).

Figure 15 is the comparison of two methods. The *X*-axis is time, the *Y*-axis is the URE (meters). Blue line represents the series of URE conducted from method 1. While red line represents the series of URE conducted from method 2,

From Fig. 15, method two has a lower URE value because satellites of low AOD have more impact on the method's function.

5 Conclusion

1. BDS MEO and IGSO have better value of URE than GEO but the frequency of large URE value of IGSO is significantly big.
2. BDS URE does not appear in a likely normal distribution (or Gaussian distribution). Most values appear in the region towards the zero value. The frequency of large value is quite small. By such characteristic, we find that it is not strict to

represent the BDS URE value under certain AOD with the average value but 95 % dividing point is better.
3. URE of BDS satellites do not rise significantly with AOD, but an increase with periodic change. URE of GEO insignificantly appears with AOD.
4. With the method we suggest, we can get a lower URE value because the satellites of low AOD have more impact on the evaluation. The URE plot will be smoother. Such method can be applied in the system performance evaluation of mixed constellation.

Acknowledgments Thanks for the support of The National Natural Science Fund (41174026), Open fund of State key lab in geodetic and earth dynamics (SKLGED 2014-3-5-E) and the data from MGEX.

References

1. China Satellite Navigation Office. BeiDou navigation satellite system signal in space interface control document open service signals (version 2.0)
2. China Satellite Navigation Office (2013) BeiDou navigation satellite system open service performance standard (version 1.0), pp 1–2
3. Yang Y, Li J, Wang A et al (2014) Preliminary assessment of the navigation and positioning performance of BeiDou regional navigation satellite system. Sci Chin Earth Sci 57:144–152. doi:10.1007/s11430-013-4769-0
4. Guo R, Tang B, Liu L et al (2014) Simulation analysis of BeiDou service performance with pseudolite augment. Bull Survey Mapp 10:7–10 + 32
5. Pan L, Cai C (2014) Accuracy assessment of BeiDou broadcast ephemeris. Bull Survey Mapp 09:16–18
6. Su T, Wen Y, Jingyang FU et al (2014) A discussion about method of calculating URE. Sci Survey Mapp 11:3–5
7. Geng T, Su X, Xu X et al (2015) Research on precise and broadcast ephemeris orbit of BeiDou navigation satellite system. Chin Sci Paper 09:1023–1026 + 1032
8. Steigenberger P et al (2013) Orbit and clock analysis of compass GEO and IGSO satellites[J]. J Geodesy 87(6):515–525
9. Montenbruck O, Hauschild A, Steigenberger P et al (2013) Initial assessment of the COMPASS/BeiDou-2 regional navigation satellite system. GPS Solutions 17(2):211–222
10. US Department of Defense (2008) GPS Standard Positioning Service (SPS) Performance Standard 4th Edition[S]. USA
11. Zhou S, Hu X, Wu B (2010) Analysis of orbit determination and prediction accuracy for a regional tracking network. Sci Chin Phys Mech Astron 06:800–808

Design and Verification of Vehicle-Based GNSS/INS Integrated Navigation Semi-hardware Simulation and Test Platform

Wen-xue Liu, Sheng-qun Geng, Xiang-hong Li and Hong Yuan

Abstract Currently, the test of the GNSS/INS integrated navigation system is primarily a real testing in the vehicle or a computer software simulation test. There are defects of high cost, difficult to repeat the testing scenario, and other issues in the former and the test not comprehensive enough in the latter. This paper proposes a semi-hardware type simulation and testing platform design, which is based on GNSS simulator, high-precision three-axis turntable and master control system, and builds a semi-hardware test platform. The platform has a flexible programmable configuration and repeatable test scenarios and the platform also has good stability and scalability. Based on iFEN NavX-NCS satellite navigation simulator and SGT320E precision three-axis turntable,we built semi-physical simulation test platform for the design implementation and verification scheme. The real test results of NovAtel's SPAN-SE commercial GNSS/INS navigation system verify the availability, effectiveness, and stability of the platform.

Keywords GNSS · INS · Integrated navigation · Semi-hardware type simulation and testing

1 Introduction

Global Navigation Satellite System (GNSS) and inertial navigation system (INS) are the most commonly used two means of navigation, they have their own advantages and disadvantages. GNSS/INS integrated navigation system can learn from each other, to overcome the shortcomings of both respective systems, with

W. Liu (✉) · X. Li · H. Yuan
University of Chinese Academy of Sciences, Beijing 100094, China
e-mail: liuwenxue010@163.com

W. Liu
Academy of Opto-Electronics, Chinese Academy of Sciences, Beijing 100094, China

S. Geng
Beihang University, Beijing 100191, China

J. Sun et al. (eds.), *China Satellite Navigation Conference (CSNC) 2016 Proceedings: Volume I*, Lecture Notes in Electrical Engineering 388,
DOI 10.1007/978-981-10-0934-1_42

high precision, high reliability, and resistance to harsh environments, etc., is one of the hot research field navigation. The test method of GNSS/INS system has been a problem in the art and attracted widespread attention.

Currently, Research on GNSS/INS system testing method is mainly divided into two kinds in which one is software-based simulation test and the other is a real testing in the vehicle. Both test methods have advantages and disadvantages. The former uses computer software to simulate the moving vehicle attitude, acceleration information and digital IF signal of Navigation satellite which the moving vehicle received simultaneously, then using the information to complete the testing of GNSS/INS system [1]. This method is used for GNSS/INS navigation system verification of algorithm level, there is a certain sense in algorithm research, but it completely or partially ignored the actual INS devices such as clock drift, bias, and device noise [2, 3] and other characteristics affect the hardware itself. The latter typically uses high-precision location information (such as differential GNSS positioning data) vehicle platform as a reference, by data processing in real-time way or post-contrast the tested GNSS/INS system and the reference related experimental test data to achieve performance evaluation of the tested GNSS/INS integrated navigation system. This program can undoubtedly test the GNSS/INS integrated navigation system in hardware and software to conduct a comprehensive assessment, but its high cost, test scenarios and test scenarios inflexible, and other problems are difficult to completely duplicate the scene of the GNSS/INS tests.

This paper proposes a GNSS/INS integrated navigation semi-physical simulation test platform solutions, which is based on a common GNSS satellite navigation simulator as GNSS satellite signal source and high-precision three-axis turntable as the carrier's attitude simulator, and an acceleration of the simulated data of auxiliary information and system control simulation system, the semi-physical simulation test platform use them together to form a vehicle GNSS/INS integrated navigation test system. This GNSS/INS integrated navigation test platform is with high fidelity, repeatable test scenarios, test scenarios can be flexibly configured, etc. The real test results of NovAtel's SPAN-SE commercial GNSS/INS navigation system verify the availability, effectiveness of the platform, and test scenarios can be stable reproducibility.

2 Semi-hardware Test Platform General Designs

In the GNSS/INS integrated navigation system testing, it is required to have GNSS RF signal simulation, attitude and acceleration vector motion scene simulation in a unified scenarios. In this paper, GNSS/INS integrated navigation semi-physical simulation and test platform are divided into carrier model simulation module, test platform control module, test hardware module, and data display and analysis module. The carrier model simulation module generates movement state of vehicle, including position and velocity information of the vehicle, inertial navigation, and accelerometers data. The test platform control module fuses the information which is from the carrier model simulation module and sends it to the hardware of this platform, so that each part of the hardware to produce the corresponding motion vector unified scene. The test platform data display and analysis module are responsible for testing the various data platforms which summarizes display and analysis of test results.

The structure diagram of GNSS/INS integrated navigation semi-physical simulation test platform which is proposed in this paper is shown as Fig. 1.

The workflow and test mode of this platform are that, the simulation software simulates to produce motion vectors and motion error of the vehicle, and the test platform control module reads the generated simulation data which are including carrier trajectory, attitude data and accelerometer data, then sent to GNSS satellite

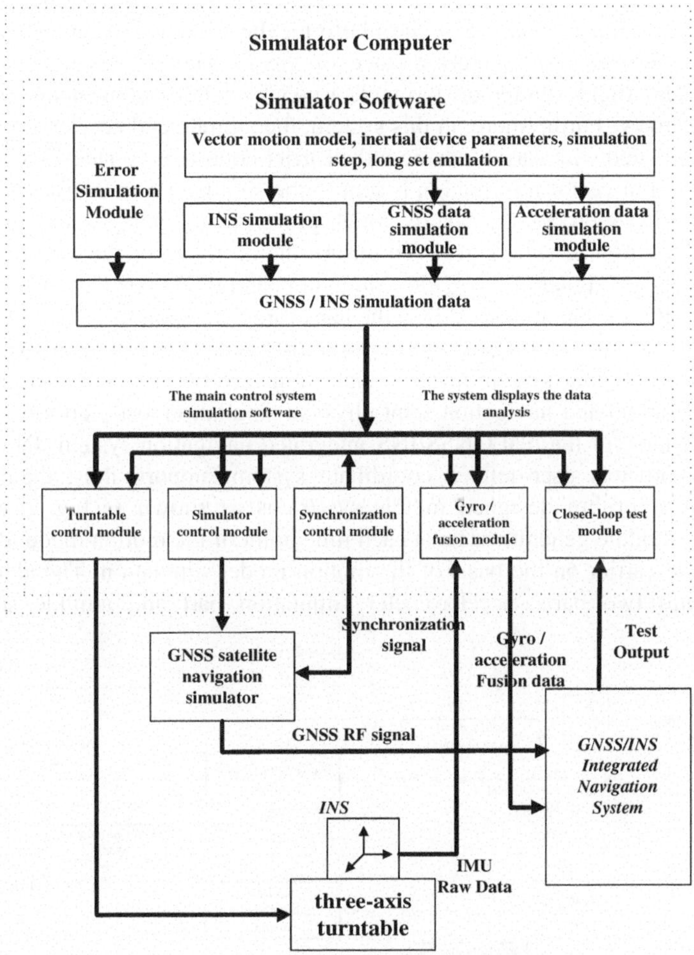

Fig. 1 The platform structure diagram of GNSS/INS integrated navigation semi-physical simulation test platform

navigation simulator, three-axis turntable and gyro/acceleration data fusion module. The GNSS satellite navigation simulator generates matching GNSS RF carrier signal. The tested GNSS/INS integrated navigation system received the GNSS RF signal and fused the INS data generated by the platform, and then completed GNSS/INS integrated navigation solver. The test platform data display and analysis module receive the output data by the tested GNSS/INS integrated navigation system, and the results were compared with the expected data to complete the relevant performance analysis and evaluation.

3 Semi-hardware Test Platform Key Technical Analysis

In platform design of this paper, the platform, the technical specifications of the platform is limited by the performance of GNSS satellite navigation platform simulator and flight simulation turntable triaxial related technical indicators and control accuracy. Furthermore, in this system, the attitude and acceleration data are need to be related with carrier position, the carrier coordinate system, and the initial attitude. Motion vector information is sent to the simulator and three-axis turntable by DSP + FPGA processing boards which processes the delay correction to produce the synchronize sync signal. To ensure the accuracy of the system platform scene, the trigger signal of navigation simulators and three-axis turntable should be corrected with the signal delay due to the hardware corresponding delay. Therefore, the number of simulation software system INS data generation algorithms and control system signal synchronization algorithms are two key technologies in the GNSS/INS integrated navigation semi-physical simulation test platform.

In the course of the real GNSS/INS integrated navigation system, INS data are motion parameters user-defined coordinate system support, thus, the INS data generation algorithm designed in this system is as shown in Fig. 2. INS data generation module generates data at each time point of the motion model download body on the carrier on the basis of the motion model simulation. The algorithm is divided into two parts, accelerometer simulation part and attitude simulation

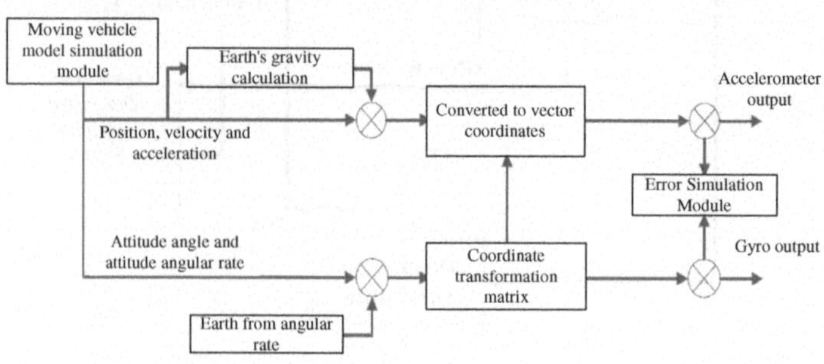

Fig. 2 The structure diagram of INS data generation algorithm

section. Accelerometer simulation part is responsible for calculating the movement of the carrier during movement of Earth's gravitational acceleration and local, sum of the two are converted to the coordinate system and then superimposed error simulation module accelerometer error part is the accelerometer output observation data. The attitude simulation part gets each time attitude trajectory data-based carrier attitude that changes during the motion which will provide a platform for HIL simulation test vectors attitude control system to complete the simulation. Meanwhile, the attitude simulation part is also responsible for calculating portion Attitude transformation matrix from the local level to coordinate vector coordinates.

In the INS data, the accelerometer measures the acceleration with respect to inertial coordinate system acceleration. Accelerometer measurements include three parts, vector motion acceleration, and gravitational acceleration due to the Earth rotation and velocity vector coordinates caused. Acceleration measurement is expressed in the form of vector coordinates, the accelerometer simulated module can be described by Formula (1),

$$f_{ib}^{b} = R_{l}^{b}\left[\dot{V}^{l} + (2\Omega_{ie}^{l} + \Omega_{el}^{l}) \times V^{l} - g^{l}\right] \tag{1}$$

In Formula (1), f_{ib}^{b} represents accelerometer output vector, R_{l}^{b} represents a transformation matrix from the local level to coordinate vector coordinates, $\dot{V}^{l}$ represents the acceleration of the carrier of showing in the local horizontal coordinate system, Ω_{ie}^{l} represents symmetric matrices of angular velocity vector, Ω_{el}^{l} represents symmetric matrices of the ω_{el}^{l}, g^{l} represents the local gravity vector, $g^{l} = [0, 0, -g]^{T}$, g is calculated according to general.

Gyroscope is used for measuring rotation angular velocity of the carrier with respect to the inertial coordinate system. Factors constitute gyroscope measurements including carrier itself and coordinate rotation angular rotational angular velocity caused so on. Similarly, the rotational angular velocity measurements in the form of vector coordinates are expressed. Gyro simulation model represented by the following Formula (2):

$$\omega_{ib}^{b} = R_{l}^{b}(\omega_{ie}^{l} + \omega_{el}^{l}) + \omega_{lb}^{b} \tag{2}$$

In Formula (2), ω_{ib}^{b} represents gyro output vector; R_{l}^{b} represents a transformation matrix from the local level to coordinate vector coordinates; ω_{ie}^{l} secured relative to the inertial rotation angular coordinates expressed in the local coordinate system of the horizontal coordinates; ω_{el}^{l} is the local horizontal coordinate system relative to the fixed coordinate system of representation at the local level rotation angular coordinates; ω_{lb}^{b} is the carrier of the coordinate system relative to the local horizontal coordinate system that represents the vector rotation angular coordinates. It is calculated by the following Eq. 3:

$$\omega_{lb}^{b} = \begin{bmatrix} 0 \\ \dot{r} \\ 0 \end{bmatrix} + R_2^T(r) \begin{bmatrix} \dot{p} \\ 0 \\ 0 \end{bmatrix} + R_2^T(r) R_l^T(p) \begin{bmatrix} 0 \\ 0 \\ \dot{A} \end{bmatrix} = \begin{bmatrix} \cos r & 0 & \cos p \\ 0 & 1 & \sin p \\ \sin r & 0 & \cos p \, \cos r \end{bmatrix} \cdot \begin{bmatrix} \dot{p} \\ \dot{r} \\ \dot{A} \end{bmatrix} \tag{3}$$

In Formula (3), (p, r, A) represents carrier motion attitude angle, $(\dot{p}, \dot{r}, \dot{A})$ represents the rate of change of the attitude angle.

As can be seen from the above accelerometers and gyroscopes simulation model, when the movement of the carrier state of motion (i.e., position, velocity, acceleration, attitude angle change rate) is known, INS original observation data generated in accordance with the above model can simulate.

In the error simulation module of the semi-physical simulation test platform, INS attitude simulation is using three-axis turntable. We need to simulate the errors of INS accelerometer according to different levels of INS. For accelerometers error simulation, we use a random error model which generally use a random number, a random walk process or Order Gauss–Markov process to represent.

The other key of the test platform systems technology is to keep the GNSS RF signal and INS data in a synchronous scene. The accuracy of time synchronization is directly related to the extent of its scope of application of the test platform [4, 5]. In the GNSS/INS integrated navigation system, the Kalman filter algorithm for filtering the estimated timing data is carried out in accordance with certain queued and then into the navigation filter. The error caused great impact on navigation and positioning accuracy of the system. For example, in the loose combination GNSS/INS integrated navigation system, the dynamic model can be presented as Formulas (4) and (5) [5],

$$\dot{e} = Ae + \tilde{\omega} \tag{4}$$

$$Y = He + v \tag{5}$$

Suppose the time synchronization offset difference of GNSS system data and the INS data is Δt, then the measurement equation impact by the time synchronization can be presented as Formula (6),

$$Y(t) = H[X_{GNSS}(t) - X_{INS}(t + \Delta t)] + v \tag{6}$$

At time t, for the Taylor series expansion, the measurement equation can be presented as Formula (7),

$$Y(t) = He + v + \xi \tag{7}$$

In Formula (7), $\xi = \dot{X}_{INS}(t)\Delta t$. It can be seen as time synchronization deviation of the system introduced by the measurement noise.

Seen from the above, In the GNSS/INS integrated navigation system, keeping GNSS and INS data time synchronization is critical. In hardware of the test platform in this paper, major time delay includes satellite navigation simulator hardware delay, response delay of three-axis turntable, and processing delay of INS data fusion. High-speed FPGA + DSP signal and information processing board in control system to achieve high-precision time synchronization and error adjustment time. As seen from Fig. 1, the start of GNSS simulator is controlled by a trigger signal, and the same as the three-axis turntable, the processing delay of the INS data fusion is in the FPGA + DSP board and the start of it is also a flag in the FPGA program. So in this system, the time delay of the three module is marked as following, Δt_1 represents the hardware delay of the simulator, Δt_2 represents response delay of three-axis turntable, Δt_3 represents processing delay of INS data fusion. Each delay specific relationship is as shown in Fig. 3:

From Fig. 3 we can draw that, by modifying the hardware delay guarantee is fed to the GNSS/INS integrated navigation system INS RF signal and data synchronization keep scene. If not, the time synchronous error will be enlarged to varying degrees, thus worsening the GNSS/INS system performance.

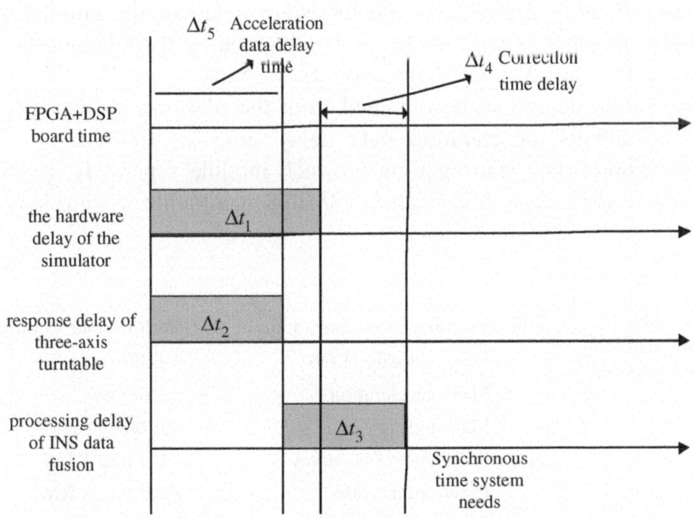

Fig. 3 The synchronize time chart of modules in the Platform

4 Design Platform Implementation

In this paper, we build a car GNSS/INS integrated navigation test system based on the GNSS/INS integrated navigation system test platform design scheme. In the platform, the GNSS simulator which is used in Germany iFEN company NavX-NCS satellite navigation simulators, flight simulator uses a three-axis turntable China Aviation Industry Corporation developed the 33 SGT320E-speed flight simulation turntable. Technical specifications of the simulator and three-axis turntable we used are as following.

Indicators Tables 1 and 2 above can be seen, NavX-NCS satellite navigation simulator signal dynamic range, SGT320E type of flight simulation turntable respective technical indicators are fully able to meet the test requirements of the vehicle-mounted GNSS/INS integrated navigation system. The actual device image is as shown in Fig. 4.

As shown in Fig. 4a is the actual device image of the NavX-NCS simulator, (b) is the actual device image of the SGT 320E three-axis turntable, and (c) is the actual device image of the FPGA + DSP board used to processing the information.

In the GNSS/INS test platform, FPGA + DSP board as a time synchronization system and adjusting the reference to adjust the various modules to meet different trigger delay time required to synchronize the final scene. In the implementation of the platform, the data of each module is discrete in time, which simulators carrier trajectory update rate 10 Hz, update rate Axis Flight Simulator data 200 Hz. The test platform for automotive GNSS/INS integrated navigation system, its system time synchronization accuracy 5 ms scene to meet automotive GNSS/INS integrated navigation requirements. In various delays shown in Fig. 3, the time delay of the processing delay of INS data fusion can be obtained by the algorithm in the FPGA and DSP, it is $\Delta t_3 = 5\,\text{ms}$, the hardware delay of the simulator and the response delay of three-axis turntable can be obtained by the relevant the technical specifications, they are $\Delta t_1 = 268\,\text{ms}$ and $\Delta t_2 = 25\,\text{ms}$.

Correction time delay can be obtained from the platform system $\Delta t_4 = \Delta t_2 + \Delta t_3 - \Delta t_1 = -238\,\text{ms}$, acceleration data delay time is $\Delta t_5 = \Delta t_2 = 25\,\text{ms}$. By selecting the appropriate starting time for each module separately, platform synchronized delay correction can reach GNSS/INS navigation system synchronized scenes.

Table 1 Technical specifications of iFEN NavX-NCS Simulator

Technical indicators Item	Performance parameters
Max. velocity (LOS)	±22800 m/s
Max. acceleration	±1570 m/s^2
Max. jerk	±15600 m/s^3
Pseudorange accuracy	<1.0 mm RMS
Pseudorange rate	<1.0 mm/s RMS
Signal resolution	0.1 dBm

Table 2 Technical specifications of SGT320E three-axis turntable

序号	Item		Performance parameters
1	Wobble error	Outer frame	±5″
		Mid frame	±5″
		Inner frame	±5″
2	Axis verticality of inner frame-mid frame		±5″
3	Axis verticality of mid frame-outer frame		±5″
4	Triaxial intersection		0.5 mm (inner diameter ball)
5	The maximum angle	Three frame	Three frame continuous rotation
6	Angular position accuracy		±5″
7	Angular position repeatability		±3″
8	Angular velocity range	Outer frame	±0.01 to ±250 (°)/s
		Mid frame	±0.01 to ±250 (°)/s
		Inner frame	±0.01 to ±400 (°)/s
9	Triaxial angular velocity accuracy and stability	$\omega < 1\,(^\circ)/s$	$\pm 1 \times 10^{-2}$
		$1 < \omega \leq 10\,(^\circ)/s$	$\pm 1 \times 10^{-3}$
		$\omega \geq 1\,(^\circ)/s$	$\pm 1 \times 10^{-4}$
10	The maximum angular acceleration	Outer frame	$200\,(^\circ)/s^2$
		Mid frame	$300\,(^\circ)/s^2$
		Inner frame	$300\,(^\circ)/s^2$

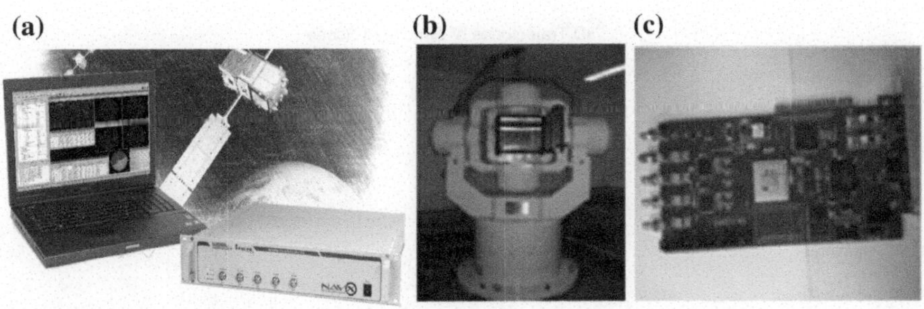

Fig. 4 The actual objects of the platform

5 Platform Test Results and Analysis

In this paper, NovAtel's SPAN-SE model GNSS/INS integrated navigation system as an object the car GNSS/INS integrated navigation semi-physical simulation test platform to test the performance of the platform. The test is divided into two terms as following.

Table 3 Technical specifications of SPAN-SE integrated navigation system

Horizontal position accuracy (RMS)	1.5 m
IMU measurement	100 Hz
INS solution	100 Hz
Time accuracy	50 ns
Gyroscope bias	±20°/h
Gyro bias stability	±1°/h
Gyro angle random walk	$0.0667°/\sqrt{h}$
Accelerometer bias	±50 mg
Accelerometer bias stability	±0.75 mg

1. Test the system performance of the platform in simulation scenarios
2. The use of high-precision differential system NovAtel PP6 car track data with a carrier attitude measurement, acceleration measurement data as a platform to test the platform performance scene

The specifications of the SPAN-SE model GNSS/INS integrated navigation system used in the test performance is shown in Table 3.

1. Test results of the system performance in simulation scenarios
   In the simulation scenarios, we simulate vehicle for movement within a 38 m * 158 m rectangular playground. The initial state of North front ENU coordinates, speed is of 1–2 m/s. Under this scenario, the three-dimensional trajectories are simulated as shown in Fig. 5.

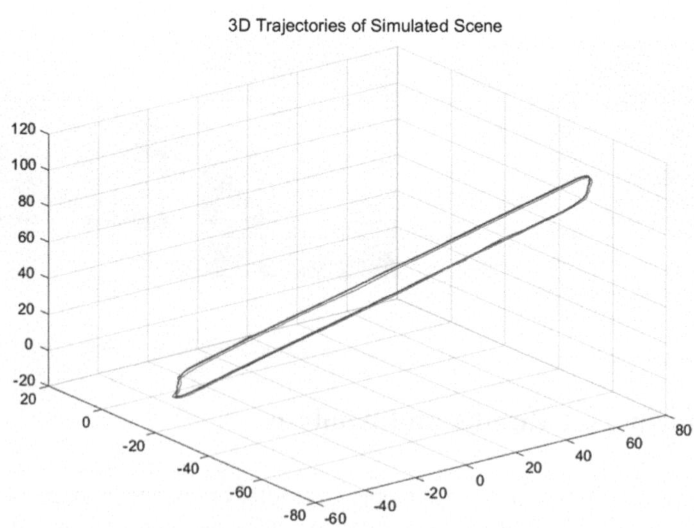

Fig. 5 3-D trajectories of simulated scene

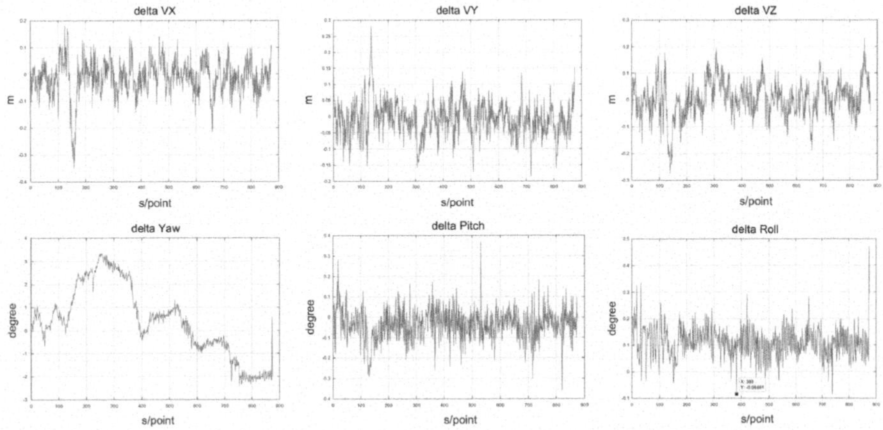

Fig. 6 The velocity and attitude error *curve* (about 15 min)

Table 4 Test results error table

Errors item with expectations	RMS	Minimum error	Maximum error
ECEF X	1.780	−0.400	3.200
ECEF Y	0.808	−0.100	2.200
ECEF Z	0.603	−2.300	0.600
ECEF VX	0.064	−0.340	0.180
ECEF VY	0.053	−0.184	0.280
ECEF VZ	0.065	−0.280	0.220
ROLL	0.124	−0.085	0.473
PITCH	0.085	−0.359	0.368
YAW	1.583	−2.381	3.313

Under the platform speed and attitude error for the test results ECEF coordinate system as shown in Fig. 6, the next platform to test the error data analysis is listed in Table 4.

2. The platform performance test results at real Scene 1
Scene 1 is designed as a rectangular path planning. Speed of 30–50 km/s, in the vicinity of Beijing Astronautic Town. NovAtel's reference board is PP6 precision multimode differential receiver, differential reference station standing CAS Institute, baseline length 5–6 km. Specific planning paths and sports and a three-dimensional trajectory are shown in Figs. 7 and 8.
Under the platform speed and attitude error for the test results ECEF coordinate system is as shown in Fig. 9, the next platform to test the error data analysis is listed in Table 5.

Fig. 7 The road map of Scene 1

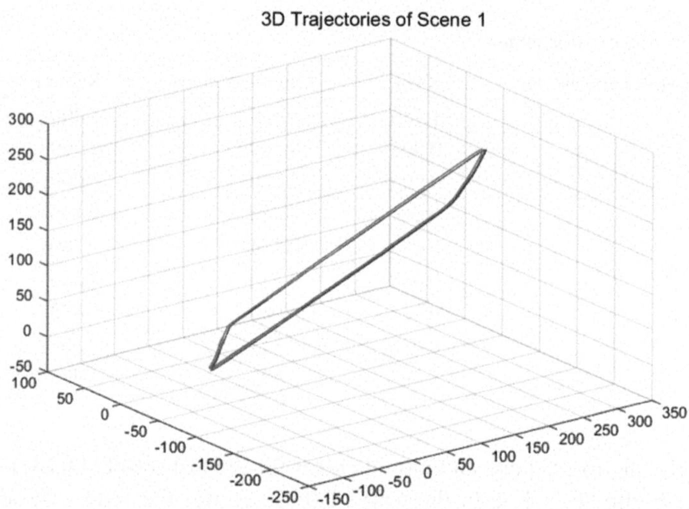

Fig. 8 3-D trajectories of Scene 1

3. The platform performance test results at real scene2
   Scene 2 is designed as path planning for a wide range, speed of 30–50 km/s. Position as a starting point from Beijing Stone Bridge, counterclockwise around the City of Victoria to Paul Temple Bridge. The data acquisition mode is the same as Scene 1, baseline length 5–6 km. Specific planning paths and sports and a three-dimensional trajectory are shown in Figs. 10 and 11.

Under the platform speed and attitude error for the test results ECEF coordinate system as shown in Fig. 12, the next platform to test the error data analysis is listed in Table 6.

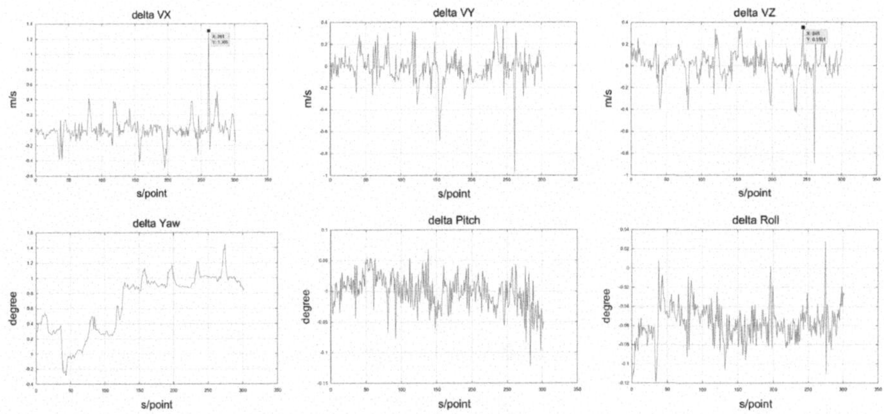

Fig. 9 The velocity and attitude error *curve* (about 5 min)

Table 5 The scene 1 test results error statistics table

Errors item with expectations	RMS	Minimum error	Maximum error
ECEF X	1.489	−0.893	2.191
ECEF Y	1.097	−2.756	0.554
ECEF Z	1.154	−1.677	1.862
ECEF VX	0.146	−0.493	1.305
ECEF VY	0.139	−0.974	0.378
ECEF VZ	0.125	−0.890	0.350
ROLL	0.088	−0.119	0.028
PITCH	0.133	−0.123	0.068
YAW	0.543	−0.286	1.452

By analyzes of the test results of the simulation scene and two actual sports scenes in this platform performance, SPAN-SE receiver test results are in error within the system performance index of its nominal. This verifies the semi-physical simulation test platform of effectiveness and availability.

Error analysis of three consecutive test results is given below under the simulation test platform that utilizes the same NovAtel's SPAN-SE model GNSS/INS navigation system of imitation and real scene Scene 1, respectively, in Tables 7 and 8.

As can be seen from Tables 7 and 8, in three consecutive tests the semi-physical simulation test platform, each time the error integrated navigation systems are provisions within its target range. Thus, the test scene of this test platform is available and repeatable. The test was repeated three times in each scene, the position error of the RMS error of the three horizontal comparison integrated navigation system, speed error, and the attitude error within 0.3, 0.03 and 0.01, respectively, they are so tiny that it can be understood as the system noise error. So the GNSS/INS integrated navigation semi-physical simulation test platform has better system stability.

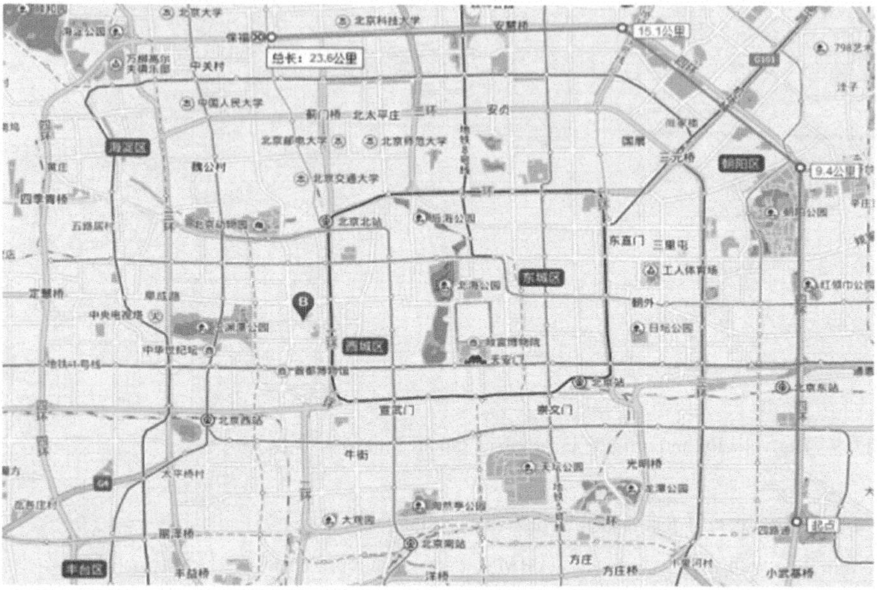

Fig. 10 The road map of Scene 2

Fig. 11 3-D trajectories of Scene 2

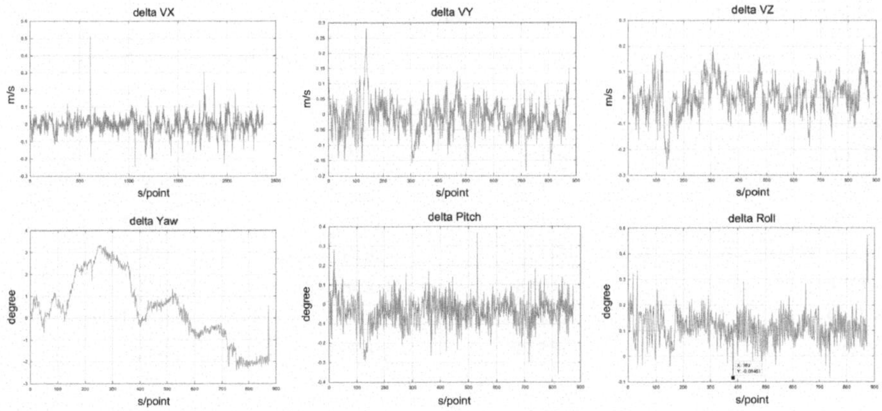

Fig. 12 The velocity and attitude error *curve* (about 45 min)

Table 6 The Scene 2 test results error statistics table

Errors item with expectations	RMS	Minimum error	Maximum error
ECEF X	1.455	−1.413	3.302
ECEF Y	0.951	−2.638	2.386
ECEF Z	1.050	−1.609	3.373
ECEF VX	0.048	−0.245	0.508
ECEF VY	0.056	−0.225	1.592
ECEF VZ	0.058	−0.411	1.414
ROLL	0.089	0.349	−0.351
PITCH	0.059	0.293	−0.418
YAW	0.603	1.737	0.001

Table 7 Error statistics table repeats the simulate scene three times

Errors item with expectations	RMS first time	RMS second time	RMS third time
ECEF X	0.603	0.603	0.599
ECEF Y	0.808	0.840	0.828
ECEF Z	1.780	1.789	1.784
ECEF VX	0.064	0.064	0.067
ECEF VY	0.053	0.053	0.055
ECEF VZ	0.065	0.065	0.068
ROLL	0.124	0.125	0.198
PITCH	0.085	0.086	0.087
YAW	1.583	1.587	1.504

Table 8 Error statistics table repeats the simulate scene three times

Errors item with expectations	RMS first time	RMS second time	RMS third time
ECEF X	1.489	1.340	1.550
ECEF Y	1.097	1.121	1.208
ECEF Z	1.154	1.196	1.201
ECEF VX	0.146	0.176	0.177
ECEF VY	0.139	0.151	0.158
ECEF VZ	0.125	0.132	0.153
ROLL	0.088	0.083	0.090
PITCH	0.133	0.138	0.151
YAW	0.543	0.560	0.624

6 Conclusion and Discussion

This paper proposes a GNSS/INS integrated navigation semi-physical simulation test platform program, then we uses iFEN company NavX-NCS satellite navigation simulator, 33 SGT320E type three-axis flight simulation turntable which is developed by China Aviation Industry Corporation and associated control and synchronization software and hardware to achieve the automotive GNSS/INS integrated navigation semi-physical simulation test platform. Finally, NovAtel's SPAN-SE GNSS/INS integrated navigation system as a DUT test the platform in several simulated scenarios and practical car scenes. Actual test results show that the proposed semi-physical simulation test platform has completed a simulated vehicle movement effectively functional and performance testing GNSS/INS integrated navigation system, the platform has the GNSS/INS test multiple scenarios stable repeatability and flexible programmable scenes.

As can be seen from the paper, the larger error in the system platform is testing several simulation scenarios. In our analysis, this is may be due to the deviation of error simulation and actual carrier INS's error. Therefore, the semi-physical simulation test platform's further work is necessary to conduct further research on this, in order to improve the accuracy of the test platform in the system several imitation scene, and may increase the amount of a typical carrier sports scene, to meet the GNSS/INS navigation system to test different scenarios.

In addition, now time synchronization accuracy of the platform is 5 ms, which can meet the functional and performance testing of general automotive GNSS/INS integrated navigation system. But under high dynamic environment, some tiny time synchronization error will be enlarged to varying degrees, which may cause GNSS/INS solver error, therefore the application range of the platform is limited. By analysis of the platform hardware and software configurations, we can get that satellite navigation simulator and table indexes can meet the high dynamic test requirements within a certain range, thus further research to improve the platform time synchronization accuracy can make it meet high dynamic scene within a certain range of GNSS/INS system testing needs.

The GNSS/INS integrated navigation test platform is proposed in this paper, the data of the platform can be obtained either from the simulation system or obtained from the real test. The platform enables GNSS/INS test in the same scene repeated stability, the test scenarios can be configured according to specific user needs flexibly, thus it will save the cost of testing GNSS/INS system. So, it has broad application prospects in the development and testing GNSS/INS integrated navigation system.

References

1. Petovello MG, Lachapelle G (2006) Comparison of vector-based software receiver implementations with application to ultra-tight GPS/INS integration. In: Proceedings of ION GNSS, p 6
2. Hong S, Chang YS, Ha SK, Lee MH (2002) Estimation of alignment errors in GPS/INS integration. In Proceedings of Institute of Navigation (ION) GPS 2002, Portland, OR, pp 527–534
3. Wei M, Schwarz KP (1990) Testing a decentralized filter for GPS/INS integration. In: Position location and navigation symposium, 1990 (IEEE PLANS'90). Record. The 1990's—a decade of excellence in the navigation sciences, IEEE, pp 429–435
4. Hong S, Lee MH, Kwon SH et al (2004) A car test for the estimation of GPS/INS alignment errors. IEEE Trans Intell Transp Syst 5(3):208–218
5. Ding WD, Wang JL, Mumford P et al (2005) Time synchronization design for integrated positioning and geo referencing systems. In: Proceedings of SSC, pp 1265–1274
6. Kennedy S, Hamilton J, Martell H (2006) Architecture and system performance of SPAN–NovAtel's GPS/INS solution. Proc IEEE/ION PLANS 2006:23–25
7. Ding W, Wang J, Rizos C et al (2007) Improving adaptive Kalman estimation in GPS/INS integration. J Navig 60(03):517–529
8. Hong S, Lee MH, Chun H-H, Kwon S-H, Speyer JL (2005) Observability of error states in GPS/INS integration. IEEE Trans Veh Technol 54(2)
9. Cutright C, Braasch MS (2002) GPS and INS flight test instrumentation of a fully aerobatic turbojet aircraft. In: Aerospace conference proceedings, IEEE, vol 3, pp 1361–1368
10. Snyder S, Schipper B, Vallot L et al (1992) Differential GPS/inertial navigation approach/landing flight test results. Aerosp Electron Syst Mag IEEE 7(5):3–11
11. Lee HK (2002) Calibration of measurement delay in global positioning system/strapdown inertial navigation system. J Guid Contr Dyn 25(2):240–247
12. Childers D, Gelderloos H (2001) Test results for an off the shelf GPS/INS during approach and landing testing of the X-40A. In: 20th conference on digital avionics systems, DASC. IEEE, vol 2, pp 8B7/1–8B7/12
13. Phuong NHQ, Kang HJ, Suh YS et al (2006) A GPS/INS integration system for land vehicle application. In: International joint conference on SICE-ICASE, IEEE, pp 482–487

Analysis of the Influence of Geomagnetic Storms on the Ionospheric Model of Beidou/GPS System

Xing Li, Hongliang Cai, Dong Li, Changjiang Geng, Liang Chen and Jin Xu

Abstract Geomagnetic storm activity will cause a certain degree of disturbance on the global ionosphere; ionospheric model error which is one of the main error sources of satellite navigation system will affect the service performance of the satellite navigation system. In order to analyze the influence of geomagnetic storms on the satellite navigation system, especially the Beidou satellite navigation system, this paper takes a strongly storm which took place from March 9th, 2015 to March 21st as an example, respectively, analyze BDS and GPS ionospheric delay precision when ionosphere is quiet or disturbed. Further, it makes contrast the difference of positioning accuracy between BDS and GPS when ionopheric disturbance happened in March, 2015. The results show that BDS and GPS ionopheric model are affected by the storm, both the ionopheric model correction rate and service performance are decreased.

Keywords Geomagnetic storm · Ionospheric delay · Beidou/GPS · Klobuchar model · Positioning accuracy

---

X. Li
Beijing Institute of Tracking and Telecommunication Technology, Beijing, China

H. Cai (✉)
Beijing Institute of Tracking and Telecommunication Technology, Beijing 100094, China
e-mail: caibanyu@126.com

D. Li · C. Geng · L. Chen
Test and Assessment Research Center, China Satellite Navigation Office, Beijing, China

D. Li · C. Geng · L. Chen
GNSS System Engineering Center, China Academy of Aerospace Electronics Technology, Beijing, China

J. Xu
China Academy of Space Technology, Beijing, China

J. Sun et al. (eds.), *China Satellite Navigation Conference (CSNC) 2016 Proceedings: Volume I*, Lecture Notes in Electrical Engineering 388,
DOI 10.1007/978-981-10-0934-1_43

1 Introduction

The cancelation of SA policy makes the ionospheric delay to be the largest source of error in the satellite navigation positioning, especially for single frequency users, who need to depend on the ionospheric model parameters broadcasted by navigation system to make ionospheric delay correction. BDS and GPS systems are using the Klobuchar model, the difference is that BDS uses a fixed geographic coordinates, while GPS uses a sun-fixed geomagnetic coordinate system [1–4].

March 17 2015 04:30 of universal time (Beijing time at 12:30), G1 geomagnetic storms happened, later geomagnetic storms become strong to G4 geomagnetic storm, becoming the strongest geomagnetic storms in the current solar cycle. Affected by these geomagnetic storms, the global ionospheric is disturbed in different degrees, ionospheric storm appeared in some areas [5]. In order to analyze the influence of geomagnetic activity on Beidou and GPS ionospheric model and its service performance, this paper analyzes the ionosphere correction accuracy of the Beidou system Klobuchar eight-parameter model (hereinafter referred to as "BDS-K8") and GPS Klobuchar eight-parameter model (referred to as GPS-K8) during the ionosphere is quiet or disturbed, and carries a contrast on their system service performance.

2 Analysis of the Influence of Geomagnetic Storms on BDS and GPS Ionospheric Model

2.1 Evaluation Model

The accuracy of the ionospheric model correction can be evaluated by the model correction rate (CorPer) [6–8], which is defined as follows (reference standard as the true value):

$$\mathrm{CorPer} = \sum\nolimits_{i=1}^{n} \left(1 - \frac{|\mathrm{TEC}_{\mathrm{mod}} - \mathrm{TEC}_{\mathrm{ref}}|}{\mathrm{TEC}_{\mathrm{ref}}}\right) / n \quad (1)$$

where, $\mathrm{TEC}_{\mathrm{mod}}$ is to be evaluated for the calculation of the ionospheric model, $\mathrm{TEC}_{\mathrm{ref}}$ is the ionosphere reference, N for statistic numbers. In order to accurately evaluate the accuracy of GPS-K8 model and BDS-K8 model, we must first select a high precision reference. At present, the ionospheric products generated by the CODE ionospheric processing center have high accuracy, and it will be used as the true reference value of the ionosphere [9]. In accordance with the above formula, the BDS-K8 and GPS-K8 ionospheric model correction rate can be obtained, which can be used to compare the accuracy of BDS and GPS ionospheric model during when ionosphere is calm or disturbed.

2.2 Accuracy Evaluation of Ionospheric Model in Service Area

By the end of 2012, BDS started to provide services for the Asia-Pacific region, covering a range of (55°S–55°N, 70°E–150°E) [10, 11]. This paper selects the system broadcasting ionospheric parameter data during when the ionosphere is calm which referred to March 9–13 2015 (day of year 068-072) and the ionosphere is disturbed which refered to March 17–21 2015 (day of year 076-080), the high-precision global ionospheric results calculated by CODE are chosen as the reference, as to evaluate the ionospheric model correction rate of BDS and GPS in their own service area. The results are shown as follows (Fig. 1).

From the above figure we can see that during March 9–13 2015 (day of year 068-072) when ionosphere is calm, ionospheric model correction rate is about 60–70 % in their service area, the two model correction accuracy is similar; when geomagnetic storms happened, their model correction accuracy declined from March 17, 22:00 to March 18, 16:00; the worst accuracy occurred in March 18 UTC 4:00–6:00 (East District 8 12:00–14:00), during which BDS-K8 accuracy is 4.04 % and GPS-K8 accuracy is 4.97 %; after the geomagnetic storms, the two model correction accuracy happens to a temporal fluctuations, the accuracy is about 50 % which is worse than the normal accuracy. As the figure above showed, the correction accuracy of the BDS and GPS ionospheric model is correlated with solar activity, the ionospheric correction accuracy will decrease when the solar activity is active.

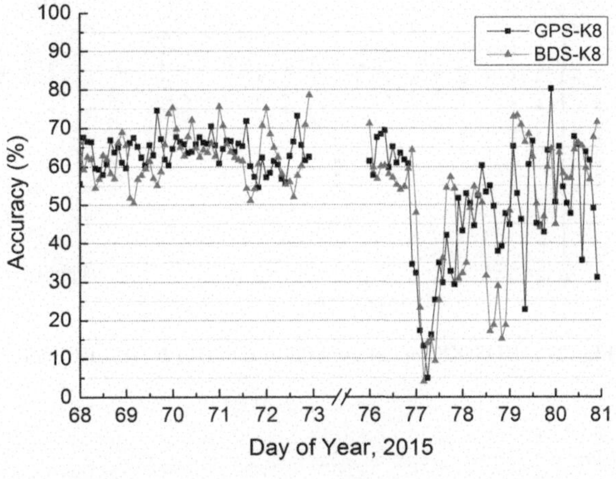

Fig. 1 Ionospheric model correction of BDS-K8 and GPS-K8 in the respective service area

2.3 Accuracy Evaluation of Ionospheric Model Over the Site

In order to further analyze the influence of geomagnetic storms, the absolute ionospheric delay correction on the site over is calculated. Considering the service range of BDS, the northern hemisphere selects the domestic iGMAS sites, including the KunMing Station, Lhasa station, WuHan Station, ShangHai Station, XiAn Station, Beijing Station, Changchun Station, and Urumchi Station, the southern hemisphere selects the same latitude of M-GEX CUT0 and UNX3, geographic latitude as Table 1. The over-site ionospheric correction rate of BDS-K8 model and GPS-K8 model from March 9, 2015 to March 21st 2015 (DOY068-080 days) is shown in Fig. 2a and b.

Shown as Fig. 2, correction rate of BDS-K8, GPS-K8 model are generally equal; their correction rate decrease during geomagnetic storm period, and correction rate of GPS-K8 is less than BDS-K8 in most of the sites.

Table 1 Site location coordinates

Site	Logogram	Latitude	Longitude
KunMing	KUN1	25°N	102°E
LaSha	LHA1	29°N	91°E
WuHan	WHU1	30°N	114°E
ShangHai	SHA1	31°N	121°E
XiAn	XIA1	34°N	109°E
BeiJing	BJF1	39°N	116°E
ChangChun	CHU1	43°N	125°E
Urumchi	GUA1	43°N	87°E
Pace	CUT0	32°S	116°E
Sydney	UNX3	34°S	151°E

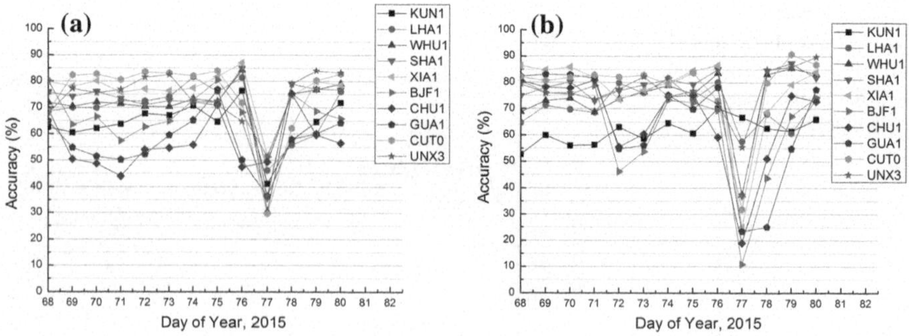

Fig. 2 **a** BDS-K8 model correction rate (%) **b** GPS-K8 model correction rate (%)

3 Analysis of the Influence of Geomagnetic Storms on Service Performance

In order to verify the influence of geomagnetic storm on the service performance of satellite navigation system, single point positioning accuracy which uses different ionospheric delay model parameters is calculated for single frequency users. In the troposphere calculating, the Saastamoinen model is used to calculate the delay correction, and the relativistic correction and the Earth rotation correction are calculated using the model [12–15].

When calculating the ionospheric delay correction, BDS use CODE, BDS-K8 model, GPS use GPS-K8, CODE model, than these results are compared and analyzed.

This paper selects two observation stations: CUT0 station and Wuhan station, focuses on the analysis of BDS B1 positioning error and GPS L1 positioning error based on their own ionospheric parameter correction during March 11, 2015 and March 18, 2015, meanwhile using the results of CODE to compare. Results show as follow (Table 2).

The above positioning results show that:

In the ionosphere quiet period, the standard positioning using ionospheric model is similar to that using CODE model (Figs. 3a, b and 4a, b).

After the occurrence of geomagnetic storms, the horizontal positioning error which is corrected by GPS-K8 ionospheric error was almost not affected compared to CODE (Fig. 4c, d), while horizontal positioning error corrected by BDS-K8 ionospheric error appeared 1–2 m error compared to CODE (Fig. 3c, d).

After the occurrence of geomagnetic storms, vertical positioning error calculated with GPS and BDS ionospheric model appeared big difference with CODE, about 4–9 m (Figs. 3c, d and 4c, d).

Table 2 Average positioning error in March 18, 2015 and March 11, 2015

		Ionosphere model	Quiet period (3–11)		Disturbance period (3–18)	
			Level	Altitude	Level	Altitude
BDS SPP	CUT0	CODE	4.29	5.72	5.10	6.03
		BDS-K8	6.51	7.36	7.61	9.42
	WuHan	CODE	3.00	8.17	5.35	6.30
		BDS-K8	3.04	10.53	6.31	15.54
GPS SPP	CUT0	CODE	2.84	4.14	2.57	3.69
		GPS-K8	3.14	4.01	2.59	7.67
	WuHan	CODE	3.30	5.98	2.03	3.88
		GPS-K8	4.69	7.61	2.77	10.34

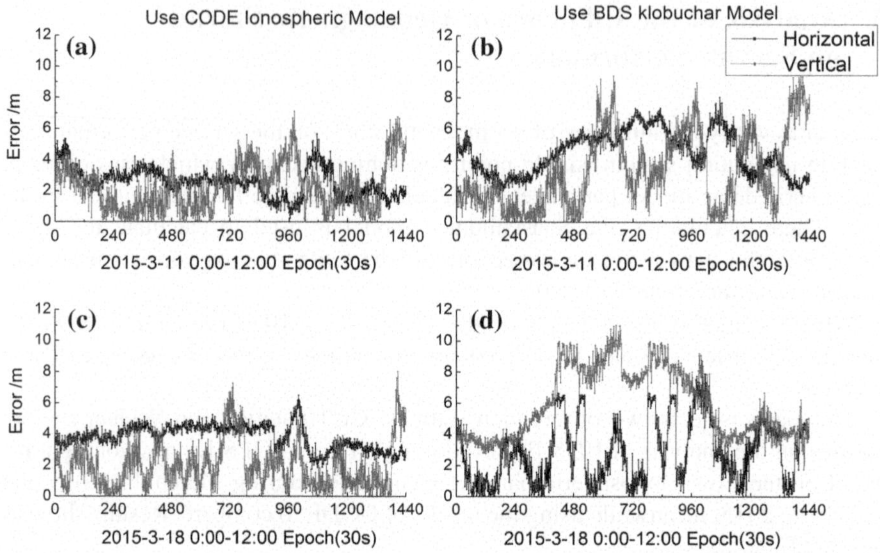

Fig. 3 March 11, 18, compass standard positioning results BDS and K8 CODE contrast

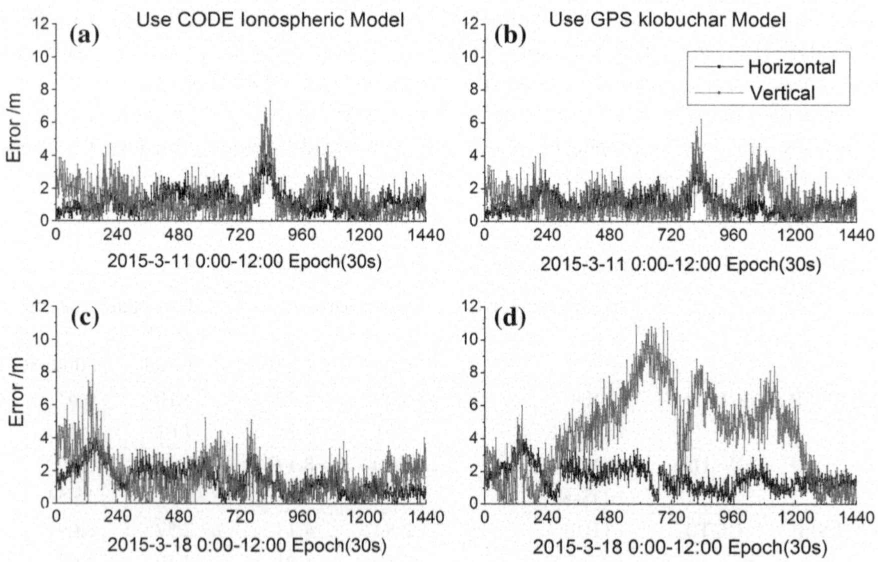

Fig. 4 March 11 18, GPS standard positioning results CODE and K8 GPS contrast

4 Conclusions

In this paper, the relationship between solar activity index and the ionospheric accuracy is analyzed in detail; the correction accuracy of the BDS-K8 and GPS-K8 models are compared in the global and single station at different time periods; the single point positioning accuracy of the two system when ionosphere is calm or disturbed is analyzed in depth. Through experiments, the following conclusions are drawn:

1. In March 2015, ionospheric model correction accuracy of BDS and GPS were decreased due to geomagnetic storms which are caused by solar activity;
2. When ionosphere is calm, BDS-K8 model is equivalent to GPS-K8 model, model correction rate is 60–70 %; while BDS-K8 is little better than GPS-K8 during the ionospheric disturbance;
3. During geomagnetic storms, standard positioning was affected no matter using BDS-K8 or GPS-K8. GPS horizontal direction error change is relatively stable, while error of BDS was about 1–2 m; vertical error calculated using BDS-K8 model and GPS-K8 model has a greater difference with that calculated using CODE model, vertical error was about 4–9 m.

At present, China's Beidou global navigation satellite system construction is being carried out in full swing, the geomagnetic storm events found that BDS ionospheric model is insufficient, further study concerning ionosphere modeling is needed, as to improve navigation system to respond to emergencies.

Acknowledgments This paper is funded by the China National Scholarship Fund.

References

1. Wu X et al (2012) Evaluation of COMPASS ionospheric model in GNSS positioning. Adv Space Res. doi:10.1016/j.asr.2012.09.039
2. Li Z, Huang J (2010) GPS measurement and data processing. Wuhan University Press, Wuhan
3. Zhang J, Ping J, Zhu W et al (2006) Brief review of the ionospheric delay models. Prog Astron (1):16–26
4. Hernandez-Pajares M, Juan JM, Sanz J et al (2009) The IGS VTEC maps: a reliable source of ionospheric information since 1998. J Geodesy 83(11):263–275
5. National Satellite Meteorological Center, space weather monthly report (2015). 3 http://cdc.cma.gov.cn
6. Li W, Li J, Dai W (2009) Study on the methods of updating Klobuchar ionospheric delay correction model. Sci Surv Mapp (5):49–51
7. Elliott DK, Christopher JH (2006) Understanding GPS principles and applications, 2nd edn.[S.l.]. Arteh House, Norwood, pp 224–239
8. Yu M, Guo J, Guo J (2004) Ionospheric delay values from Klobuchar model and dual-frequency measurements: comparison and analysis. Bull Surv Mapp 6:5–8
9. Li Z, Zhang X (2009) New techniques and precise data processing methods of satellite navigation and positioning. Wuhan University Press, Wuhan

10. Wang J, Dang Y, Xue S (2007) The application of NeQuickIonospheric model in China. Sci Surv Mapp (4):38–40
11. China Satellite Navigation office, BeiDou open service performance standard (V 1.0). BDS-OS-PS-1.0, 2013.12
12. Liu J, Ye S (2002) GPS precise point positioning using un-differenced phase observation. Geomatics Inf Sci Wuhan Univ 27(3):234–240
13. Yi C, Chen Y, Zhu J et al (2011) An approach to regional real time precise point positioning based on IGS ultra-rapid orbit. Acta Geodaeticaet Cartographica Sinica 40(2):226–231
14. Wu X, Ping J, Liu L et al (2011) Hardware delay solution of regional satellite navigation system. Geomatics Inf Sci Wuhan Univ 36(10):1218–1221
15. Bei J, Li W, Gu S (2011) RAIM method of Precise point positioning user. Acta Geodaeticaet Cartographica Sinica 40(sup):63–67

Channel Characterization for Beidou Satellite Communications

Xiaoxi Jin, Rendong Ying and Peilin Liu

Abstract With the continuous promotion of Beidou satellite navigation applications, improving the navigation terminal performance is a hot issue under city environment because of its wide application and complex signal propagation. Based on received Beidou satellite signal power shadowing under different environments in city application, the paper proposed a statistical Markov model with three states: line-of-sight (LOS), shadow block, and using a combination of mixture Gaussian functions to approximate the distribution of signal variations. Channel parameters are evaluated from the experimental data in shanghai and utilized to verify the proposed simulated model. Signal waveforms and probability density functions show proposed model close agreement with measurements.

Keywords Beidou satellite · Propagation · Markov statistical modeling · Typical scenes

1 Introduction

In recent years, the applications of satellite navigation technology are getting more and more attentions, especially in urban environment applications. The significant feature of all these scenarios is that the received signal is weaker than general signal power, and influenced by the environment. How to improve the performance of the GNSS terminals in these environments is an important question, which is concerned urgently by users. Now, for this above question there are two research points.

X. Jin (✉)
Beijing Satellite Navigation Center, 100094 Beijing, China
e-mail: 838617547@qq.com

X. Jin · R. Ying · P. Liu
Shanghai Key Laboratory of Beidou Navigation and Location-Based Service,
School of Electronic Information and Electrical Engineering School,
Shanghai Jiao Tong University, Dongchuan Road 800,
Weidianzi Building 309, 200240 Shanghai, China

J. Sun et al. (eds.), *China Satellite Navigation Conference (CSNC) 2016 Proceedings: Volume I*, Lecture Notes in Electrical Engineering 388,
DOI 10.1007/978-981-10-0934-1_44

(1) improving terminal sensitivity to adjust all weak applications; (2) studying on GNSS signals fading features in different environments for adaptive receivers [1]. Now, a lot of study has been carried out for the first point, while study of the second part for BDS signals is almost empty with some research being on for GPS signals. Therefore, it is essential to start some study of BDS signals statistical characteristics.

Now, GNSS receivers are mainly tested by GNSS simulator. GNSS simulator simulates satellite signals propagation by simulation model library which includes satellite ephemeris model, atmospheric model, and clock error model without various environments model [2] which is a close factor for terminal performance. Without environments model, the simulated signals are incomplete and untrue compared with live signals, which is difficult to quantitatively assess receiver performance. The work of signals statistical characteristics is helpful for GNSS simulators to simulate environmental satellite signals. So, the study of signal fading characteristics is meaningful.

There are some universities and research institutions in China start to analyze BDS signals with a pity of without research results. Abroad, fading models for such an outdoor propagation channel have been investigated by German Aerospace Centre (DLR), university of Calgary, and Spirent Company. Briso and Alonso [3] suggest a model consisting of two states, a direct state and a shadowed state. The direct state is modeled by the combination of a lognormal process and a Rayleigh process and shadowed state modeled by a Ricean process. Another model is the urban three-state fade model developed by Akturan and Vogel [4]. The direct state is modeled by Ricean distribution, the second state modeled by Rayleigh distribution, and the block state modeled by Rayleigh distribution.

This paper mainly focuses on statistical properties of the Beidou satellite navigation signal in dynamic city environment. Selecting this application to study is because of its widespread feature. The paper proposes a statistical BDS signal modeling method based on Markov model and Mixed Gaussian distribution (GM). We use software receiver and the method of GM-MM to analyze signals of three typical scenes (urban canyon environment, high-speed environment, and rural shade environment). Finally, the experimental data of the three environments validates the simulated data.

2 Method of BDS Signals Channel Model

There are a lot of signals characteristic parameters for navigation signals including signal envelope, phase, doppler spectra, and satellite constellation, which all contribute to signal statistical modeling. In this paper, we focus on signal envelope, so in the next paper, the feature of signal is specific to signal power.

In different circumstances, the probability of signals is distinct. We assume that there are three parts to describe BDS signals: line-of-sight, shadowed links, and blocked signals. This explains why our models try to classify propagation events according to the degree of shadowing and quantify these events independently.

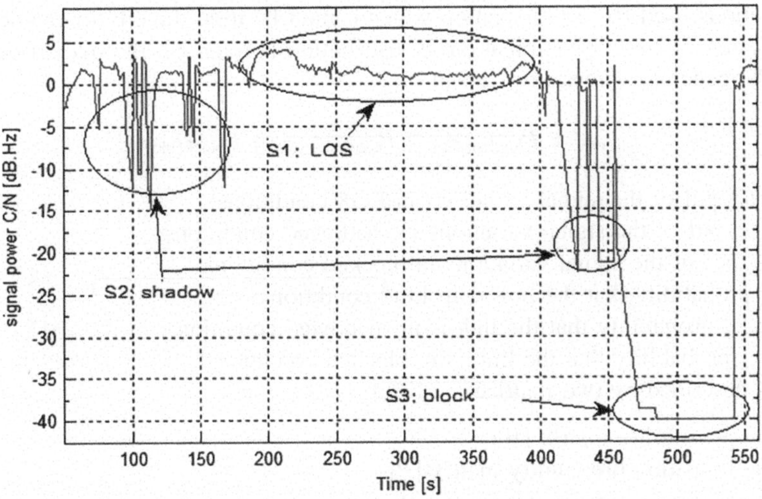

Fig. 1 Received signal power in different shadowing conditions

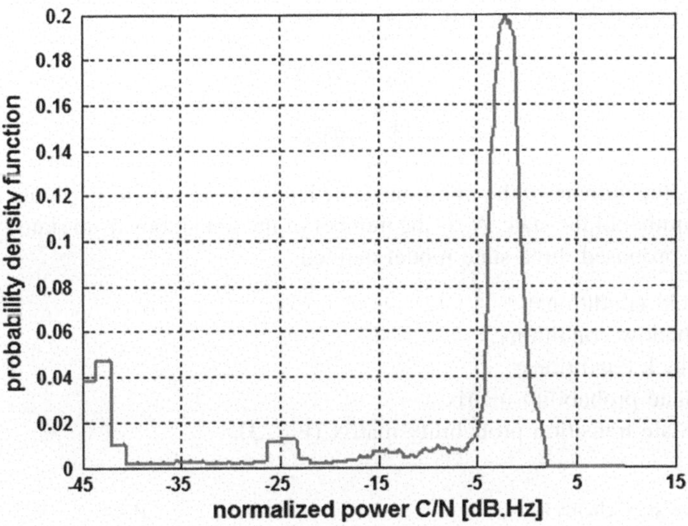

Fig. 2 Probability dense function of received signal in three states

To illustrate signal feature, one example of measured time series is presented. In Fig. 1 the signal power under different conditions is presented. In Fig. 2 the probability density function of the signal is presented.

In Fig. 1 it can be observed how the received signal is subject to very marked variations. In Fig. 2 it can be observed how the probability density function (pdf) is modal for this particular case. These observations clearly suggest that classical distributions used radio propagation which cannot be used directly to characterize the channel. Rather, a combination of distribution is required, the overall pdf follows the general expression given in [5]

$$f_{\text{overall}} = P_{\text{los}} f_{\text{shadow}} + P_{\text{shadow}} f_{\text{block}} + P_{\text{block}} f_{\text{block}} \quad (1)$$

where f_{los} pdf of the signal variations in LOS conditions;
f_{shadow} pdf of the signal variations in shadowed conditions;
f_{block} pdf of the signal variations in blocked conditions;
P_{los} probability that the link is in LOS conditions;
P_{shadow} probability that the link is in shadowed conditions;
P_{block} probability that the link is in blocked conditions;
P is described by two matrices:

- State probability matrix-[W]
- State transition probability matrix-[P].

Each element in matrix [P], P_{ij} represents the probability of change from state I to state j. the overall probability for each state is contained in matix [W] where each element, W_i, represents the total probability of being in state I. The matrix elements W_i, and P_{ij} can be defined as follows:

$$P_{ij} = N_{ij}/N_i \quad (2)$$

$$W_i = N_i/N \quad (3)$$

where N_i is the number of the minimum state length corresponding to state I and N is the total number of the state, N_{ij} is the number of the transitions from state I to state j.

So, we proposed three-state-model defined:

- -S1: LOS conditions;
- -S2: Shadow conditions;
- -S3: Block conditions;
- -[P]: State probability matrix;
- -[W]: State transition probability matrix (Fig. 3).

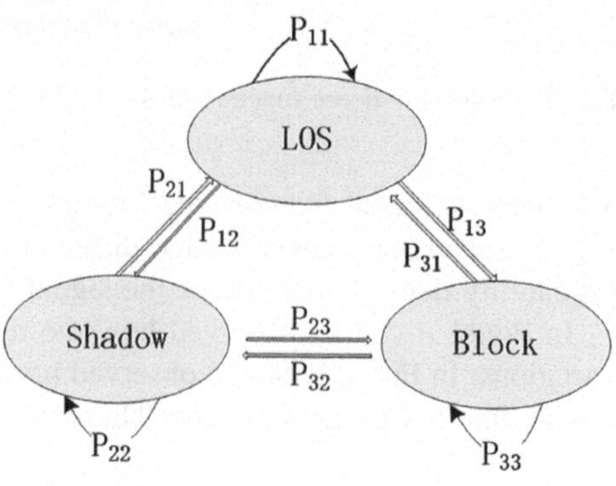

Fig. 3 Three-state channel model of BDS signal

The distribution of the three states is described by mix-Gaussian distribution:

$$f \sim \sum_{i=1}^{k} w_i \mathrm{N}(\mu_i, \sum\nolimits_i) \quad (4)$$

So, the GM-HMM model parameter:

$$\lambda = (P, W, \mu, \sum) \quad (5)$$

3 Typical Scenes Modeling

3.1 *Experiment Platform*

The setup for the measurement campaign is depicted in Fig. 4a. The antenna was mounted on a mechanically steerable platform, which was placed on top of the SJTU measurement van as depicted in Fig. 1. The first intermediate frequency in L-band was down-converted to a second IF of 62 MHz, which were finally sampled with accuracy of 8 bits. The tool of analysis is soft receiver developed by SJTU in Fig. 4b.

3.2 *Environments Definition*

After some elementary tests of calibrating experimental platform, the experiments were performed in Shanghai, summer of 2015, following the itinerary depicted in Fig. 5. More than 20 GB of data were recorded, classified into three types of urban canyon, high way, and suburban by geographical features. In urban canyon environment, satellite signals are usually blocked by large and high buildings, and still with some good visibility. In high way environment, signals are almost LOS except extremely few bridge and truck. In suburban environment, signals are shadowed by small obstacles such as single tree and lower house.

3.3 *Signal Characteristic for Typical Environments*

The paper did analysis on Xuhui shopping district, Xinzhuang residential, A5 urban highway. The signal power figures for these three environments are detailed below:

(a)

(b)

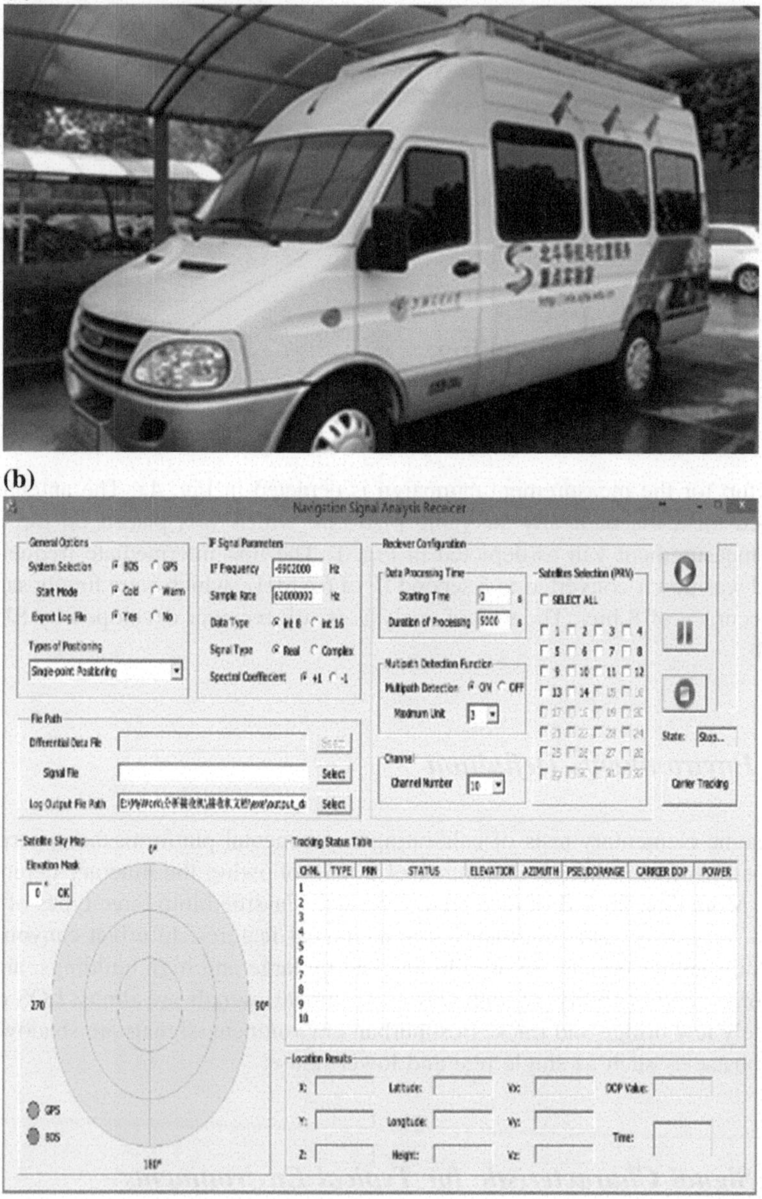

Fig. 4 Measurement analysis system for BDS signal **a** measurement van **b** analysis receiver

Fig. 5 Measurement campaign itinerary

Figure 6a shows an example of the received signal power from a channel recording in xuhui shopping district. The figure shows a high-frequency fading process. Relatively "good", "bad", and "block" channel periods can be distinguished. For instance, there are four areas in 200–400 s, 400–600 s, 1000–1200 s, where the signal is blocked. Figure 6b shows an example of the received signal power in xinzhuang residential. The figure shows signal quality of urban environment which is better than urban environment with smaller fading amplitude and fading frequency without blockage state Fig. 6c shows an example of the received

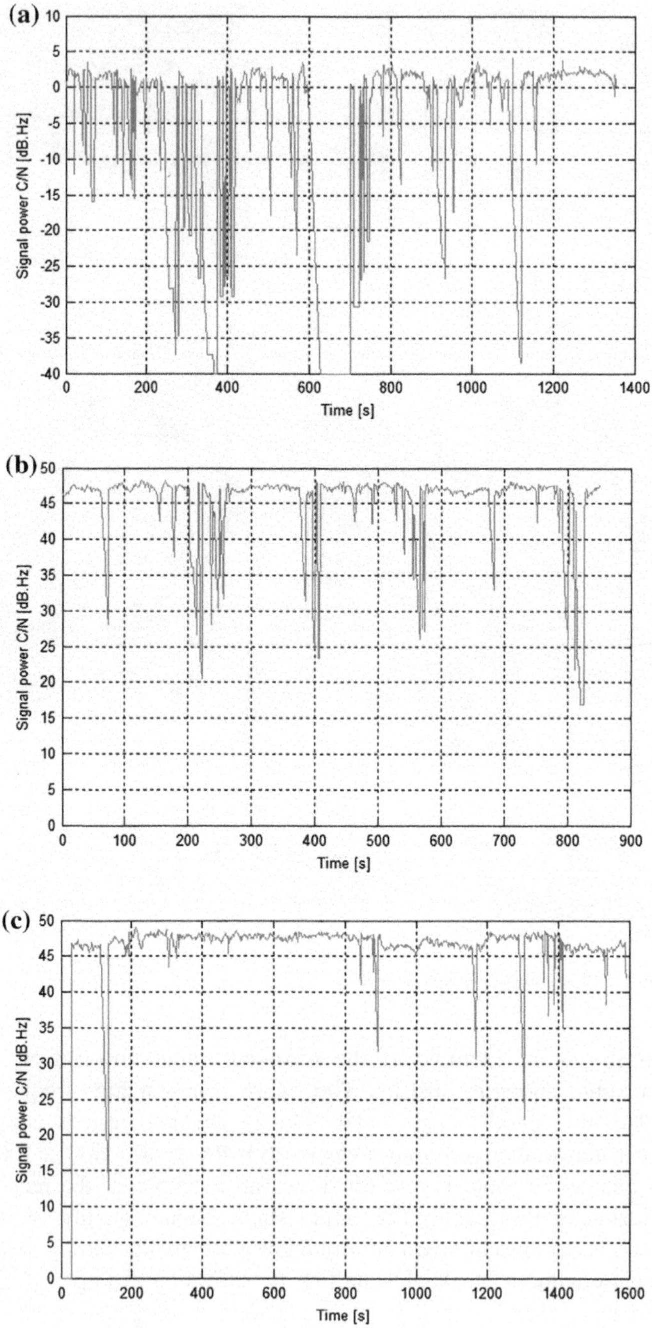

Fig. 6 Signal power for different environments **a** urban **b** suburban **c** highway

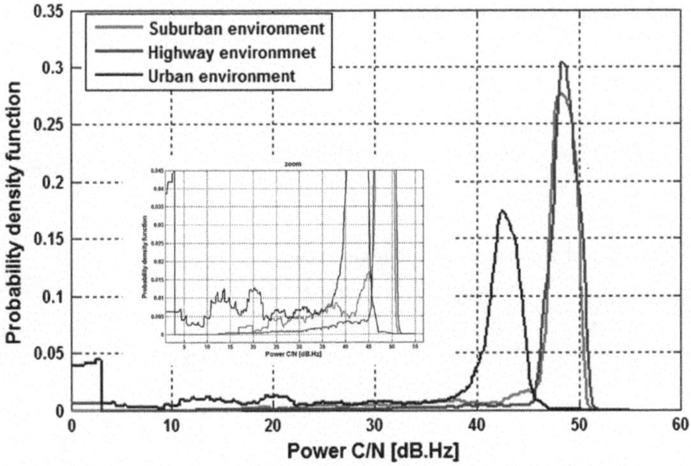

Fig. 7 Probability density function for highway environment

signal power in A5 highway. The figure shows signal in highway environment almost in LOS state, except occasionally fading without blockage state.

Figure 7 shows the probability density function of the received signal power in urban canyon environment and suburban environment and highway environment. The peak in 40–50 dB indicates that signals are in LOS state, and the peak in 0 in urban environment is due to the reception of the direct satellite signal in blocked areas of huge buildings, with the signal between block state and LOS state in the shadow state. From the figure, the pdfs of highway environment and suburban environment are similar, but the zoom figure shows that the pdf of suburban is coarse than highway.

Above all, the signal in urban environment is with "LOS-shadow-block" states; in suburban environment is with "LOS-shadow" states; in highway environment is with " LOS-shadow" states.

3.4 Channel Model for Typical Environments

Tables 1, 2, and 3 summarize the parameters relative to the three-state model earlier presented for the different environments listed in Sect. 2, the first three columns contain the transition probabilities *P*; the fourth and fifth columns contain the parameter μ Σ; the last column represents the probabilities *W*.

Table 1 Channel model parameters of urban scene

	LOS	Shadow	Block	μ	Σ	W
LOS	0.9231	0.037	0.0399	41.1	1.9	56 %
Shadow	0.1148	0.8561	0.0291	18	21	32 %
Block	0.0117	0.0015	0.9868	0	2	12 %

Table 2 Channel model parameters of suburban scene

	LOS	Shadow	Block	μ	Σ	W
LOS	0.9414	0.0586	–	47.1	0.4	68 %
Shadow	0.1127	0.6873	–	42.0	6.0	32 %
Block	–	–	–	–	–	0 %

Table 3 Channel model parameters of highway scene

	LOS	Shadow	Block	μ	Σ	W
LOS	0.9714	0.0286	–	47.3	0.66	87 %
Shadow	0.2179	0.7821	–	44.8	2.1	13 %
Block	–	–	–	–	–	0 %

4 Comparison with Measured Data

Correlation coefficients for typical scenes parameters estimated are shown in Fig. 8 and Table 4.

The results indicate that the simulated data are similar to those measured data with correlation coefficient not less than 0.8. So, the proposed model is suitable.

5 Conclusions

In this paper, the method described here was able to generate fading signals in city applications for BDS satellite signals. An important assumption made is that three state and mix-Gaussian distribution. This method can be also used for multipath signals.

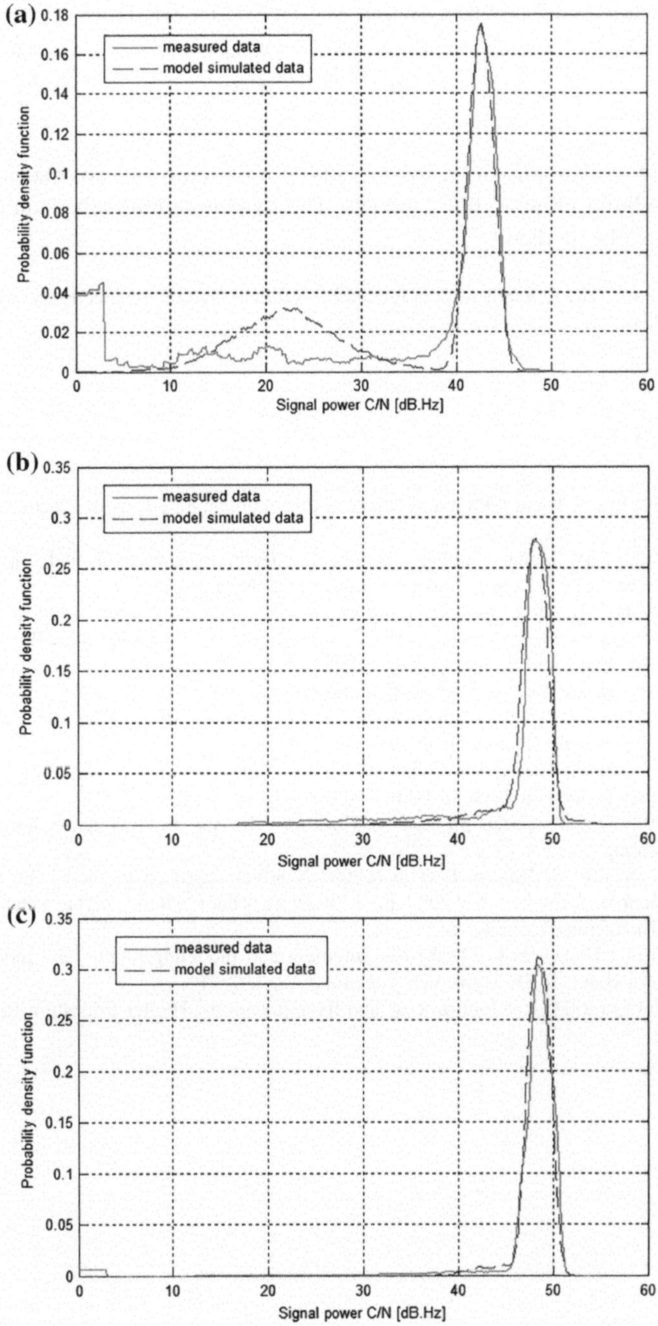

Fig. 8 Simulated and measured pdf for different environments **a** urban **b** suburban **c** highway

Table 4 Correlation coefficient between simulated data and measured data

Correlation coefficient	Urban	Suburban	Highway
ρ	0.8269	0.9485	0.9034

The future work about this research will focus on more extensive scenarios' data analyzing to build more accurate models. The dynamic multipath channel at urban areas will also be studied.

Acknowlegments The research work is funded by National Natural Science Foundation of China (No. 61304225).

References

1. Xie G (2009) Principles of GPS and receiver design. Publishing House of Electronics Industry, Beijing
2. Feng F (2009) GPS signal simulation source test technology research and implementation. Beijing University of Post and Telecommunications, Beijing
3. Brisco C, Alonso JI (1998) Statistical and empirical models for LMSC fading based on GPS measurement. In: First international workshop on radiowave propagation modelling for SatCom services at Ku-Band and above, October 1998, ESTEC, Noordwijk, The Netherlands. http://www.estec.esa.nl/xewww/cost255/ws/98c08/98c08pgm.html
4. Akturan R, Vogel WJ (1997) Path diversity for LEO satellite-PCS in the urban environment. Antennas Propag IEEE Trans 45(7):1107–1116
5. Fontán FP, Vázquez-Castro M, Cabado CE et al (2001) Statistical modeling of the LMS channel. IEEE Trans Veh Technol 50(6):1549–1567
6. Yao T (1992) Digital speech processing. Huazhong University of Science and Technology Press, Huazhong
7. Alasseur C, Scalise S, Husson L et al (2008) A novel approach to model the land mobile satellite channel through reversible jump markov chain monte carlo technique. Wirel Commun IEEE Trans 7(2):532–542
8. Scalise S, Ernst H, Harles G (2008) Measurement and modeling of the land mobile satellite channel at Ku-Band. IEEE Trans Veh Technol 57(2):693–703
9. Chen L, Zhu Q (2003) Random process and its applications. Higher Education Press, Beijing

A Signal Separation Algorithm for GNSS Signal Quality Assessment

Hao Yan, Baowang Lian and Lin Zhang

Abstract With the wide application of multiplexing technology in navigation satellite system, the traditional processing method for one channel signal is no longer suitable for the analysis of composite baseband waveform, which requires signal separation. However, there is a lack of effective navigation signal separation methods currently. First, on the basis of the traditional average algorithm, a signal separation algorithm for signal quality assessment is proposed, and the mathematical model of the algorithm is presented. Utilizing the random of the data bits modulated on different channels, the algorithm can achieve the signal separation by accumulating the multiple code epochs which are selected from the composite signal conditionally. Second, the effect of the carrier phase error, code phase error, and Doppler shift on the separated signal is analyzed. What is more, the difference between the waveforms of the ideal signal and the separated signal is studied, and the estimation error of the separated signal waveform parameters is provided. Finally, the simulated and measured Galileo E1 signals are separated. The baseband waveforms of the composite signal and one channel signal are compared, and the eye diagrams of signal before and after separation are analyzed. The above theoretical analysis and experimental results show that the algorithm is reliable and effective.

Keywords GNSS · Signal separation · Average algorithm · Signal quality assessment

---

H. Yan (✉) · B. Lian · L. Zhang
Department of Electronics Engineering, Northwestern Polytechnical University, 710129 Xi'an, China
e-mail: yanhao311@126.com

J. Sun et al. (eds.), *China Satellite Navigation Conference (CSNC) 2016 Proceedings: Volume I*, Lecture Notes in Electrical Engineering 388,
DOI 10.1007/978-981-10-0934-1_45

1 Introduction

GNSS signal quality assessment plays an important role in the construction of navigation satellite system and the fault detection of satellite. At present, for the traditional single carrier and single component navigation signal, the code chip waveform in time domain can be directly observed and analyzed by using a high gain antenna and the average algorithm.

Owing to the use of multiplexing technology, the modern navigation signals are mixed with each other in time and frequency domain. It is difficult to evaluate the single signal wave by traditional means. Therefore, the signal separation must be done first. Now, the research of signal separation technology mainly focuses on the blind signal separation. The calculation of blind signal separation algorithms is complicated and the separation result cannot meet the requirements of signal evaluation. Moreover, the blind signal separation technology cannot take full advantage of the navigation signal's prior information, so it is not suitable for the separation of navigation signals. Through the research on the navigation signal system, we find that most of the navigation signals can be separated from the frequency domain, time domain, and phase orthogonality. However, there are some multiplexing signals that do not have the separability mentioned above (such as the public signal of Galileo E1). For this kind of signal, a signal separation algorithm for signal quality assessment is proposed in this paper.

This paper introduces the average algorithm [1] first. On this basis, the mathematical model of the signal separation algorithm is proposed, and the detailed algorithm process is represented. In order to ensure that the separation algorithm will not cause significant distortion for the signal, the influence of carrier phase error, code phase error, and Doppler shift on the separation results is analyzed. Taking the BPSK signal as an example, the impact of the separation algorithm on the accuracy of the distortion parameter estimation is analyzed. Finally, the separation algorithm is utilized to separate the simulated and measured signal. The results demonstrate that the proposed algorithm is simple to be used and has an effective performance.

2 The Average Algorithm

The average algorithm for analyzing the time domain signal distortion is suggested by reference [1]. Using the periodic repetition of PN code, the algorithm averages over multiple PN code epochs, so as to improve the SNR (signal-to-noise ratio) of the baseband signal. One of the application conditions of the average algorithm is that the signal must be resampled to ensure an integral number of samples per epoch (or even per chip) [2].

After carrier synchronization, a *PN* code epoch can be written as

$$\begin{aligned} s_j(t) &= a_j c(t/k_d + \Delta\tau_j) \cdot \cos(2\pi \Delta f_j t + \Delta \theta_j) + n_j(t), \\ &0 \le t < T_{pn}, \\ k_d &= 1 - f_{\mathrm{Dop}}/R_c \end{aligned} \tag{1}$$

where, a_j is a navigation data bit; $c(t)$ represents a *PN* sequence and T_{pn} denotes its period; Δf_j, $\Delta\theta_j$ and $\Delta\tau_j$ represent the carrier phase error, carrier frequency error, and code phase error; $n_j(t)$ is Gauss white noise; k_d [1] is used to measure the extension or compression of the code caused by the Doppler effect, while f_{Dop} and R_c represent the carrier Doppler shift and the code rate, respectively.

After the navigation bit modulated on each code epoch is stripped off, the corresponding points of *N* code epochs can be averaged, as shown in formula (2). In the ideal case, the signal power is constant while the noise power is reduced to 1/*N*. Thus, the SNR is improved by *N* times.

$$\begin{aligned} \frac{1}{N}\sum_{j=1}^{N} s_j(t) &= \frac{1}{N}\sum_{j=1}^{N} c(t/k_d + \Delta\tau_j)\cos(2\pi\Delta f_j t + \Delta\theta_j) + \frac{1}{N}\sum_{j=1}^{N} n_j(t), \\ &0 \le t < T_{pn} \end{aligned} \tag{2}$$

3 Separation Algorithm

For composite signal, the navigation bits of different channels are independent of each other, so the average algorithm cannot be used directly. The mathematical model of the traditional average algorithm needs to be improved in order to realize the separation of composite signal. If there are *M* components in the composite signal, the carrier is eliminated completely ($\Delta f_j = 0$, $\Delta\theta_j = 0$), and there is no phase difference between the corresponding points of each epoch ($\Delta\tau_j = 0$), the average result can be shown as

$$\begin{aligned} \frac{1}{N}\sum_{j=1}^{N} s_j(t) &= c_0(t/k_d) + \frac{1}{N}\sum_{i=1}^{M-1} c_i(t/k_d)\cdot\sum_{j=1}^{N} a_{j0}a_{ji} + \frac{1}{N}\sum_{j=1}^{N} n_j(t), \\ &0 \le t < T_{pn} \end{aligned} \tag{3}$$

where $c_i(t)$ represents the *PN* sequence of channel *i*. Before the average in Eq. (3), only the navigation bits of channel 0 were stripped off. Obviously, if $\sum_{j=1}^{N} a_{j0}a_{ji} = 0$, the energy of channel *i* will be zero.

Considering the existing navigation signal system, the non-separable navigation signal usually contains two channels only. So, the Eq. (3) can be simplified as

$$\frac{1}{N}\sum_{j=1}^{N} s_j(t) = c_0(t/k_d) + \frac{1}{N}c_1(t/k_d)\cdot\sum_{j=1}^{N} a_{j0}a_{j1} + \frac{1}{N}\sum_{j=1}^{N} n_j(t) \qquad (4)$$

$$0 \le t < T_{pn}$$

Then, the condition of separating out $c_0(t)$ is

$$\sum_{j=1}^{N} a_{j0}a_{j1} = 0 \qquad (5)$$

If the navigation bits modulated on each epoch are random, the condition (5) can be satisfied when $N \to \infty$. However, in practice, the navigation bits cannot be completely random and N is finite. In order to satisfy the condition, the navigation bits need to be demodulated first, and then N code epochs which meet the condition (5) are selected to be averaged, which requires that N must be even. Therefore, the key to achieve GNSS signal separation is to select out N code epochs which satisfy the condition (5). The signal separation algorithm flow is:

1. Initializes the sign imbalance indicator flag = 0;
2. Read in a code epoch, and assign the bit modulated on $c_0(t)/c_1(t)$ to the variable a_{j0}/a_{j1};
3. If $|\text{flag} + a_{j0}a_{j1}| < 2$, updates the indicator $\text{flag} = \text{flag} + a_{j0}a_{j1}$, and then continues to execute the next step; otherwise, discard this epoch and return to step 2;
4. Resample the raw data;
5. Realize the carrier synchronization and strip the navigation bit modulated on $c_0(t)$;
6. Add the results of Step 5 to the corresponding points of earlier accumulation results, and the new accumulation results are obtained;
7. If the accumulation time has not reached N, return to Step 2; otherwise, the procedure ends.

4 Algorithm Analysis

The purpose of the separation algorithm is to evaluate navigation signal quality, which requires that the separation signal can truly reflect the wave characteristics. In this section, we will investigate the effect of the carrier phase error, code phase error, Doppler shift, and the imbalance of bit sign on separation results.

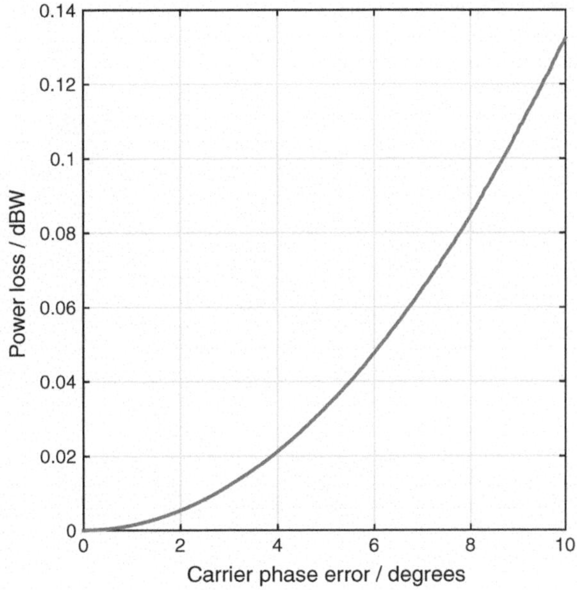

Fig. 1 Relationship between carrier phase error and power loss

4.1 *Carrier Phase Error*

Formula (2) indicates that the carrier phase error can cause the power loss. It is assumed that the carrier phase error within a code epoch is constant, and the carrier phase error is subject to Gauss distribution. The relationship between the carrier phase error and the power loss is given in Fig. 1. Ignoring the effect of mechanical vibration, for the signal with high carrier-noise ratio (such as 60 dB Hz), the tremble of carrier loop can be controlled within 5° [3], the impact of which can be ignored.

4.2 *Code Phase Error*

In the accumulation process, if the code phase of corresponding points is not aligned, the accumulation results will be affected by "average effect", which means the result is the average value around the corresponding point. The average effect will affect the separated signal waveform, mainly for

1. The jump edge of waveforms will be expanded;
2. The position and magnitude of the extrema within a chip will be changed.

Figure 2 shows the average results with different code phase errors. For this example, the signal is affected by analog distortion [4] and the code phase error of each code epoch obeys Gauss distribution. Only one code chip is illustrated in Fig. 2. While the code phase error is smaller, the jump edge of the chip is steeper and the chip waveform is more ideal. When the code phase error is less than 0.01 chips, the

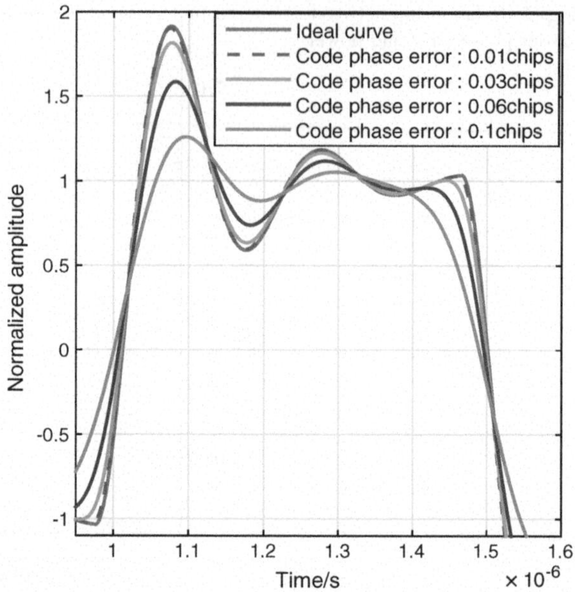

Fig. 2 The effect of code phase error on the accumulation wave (Color figure online)

accumulation curve is very close to the actual one. Therefore, 0.01 chips can be taken as the threshold of the code phase error. For the signal with high noise ratio, by choosing the appropriate code loop parameters, the mean square error of the code phase caused by the thermal noise can be far less than 0.01 chips [3].

4.3 Doppler Shift

Since the satellite is moving, the Doppler shift is changing. Different Doppler shifts can cause different degrees of expansion or compression of *PN* code period, which will cause the phase error between corresponding points. Therefore, the accumulation time is limited. If the Doppler shift is maintained in a code epoch, the code length is c_L, and the code rate of epoch i is R_i (including Doppler shift), the duration of the two code epochs can be expressed as

$$T_1 = \frac{c_L}{R_1}, T_2 = \frac{c_L}{R_2} \tag{6}$$

It is required that the difference between two code epochs' duration must be less than 1 sampling period ($1/f_s$), then there is

$$|T_1 - T_2| < \frac{1}{f_s} \tag{7}$$

The code rate difference between two code epochs is expressed as $\Delta R = R_2 - R_1$. We can obtain

$$|\Delta R| < \frac{R_1^2}{f_s \cdot c_L \pm R_1} \tag{8}$$

When $\Delta R > 0$, in formula (8), the '$\pm$' takes '$-$', otherwise takes '+'. Taking the code rate as 1.023 Mcps, the sampling rate as 250 MHz, and the code length as 10,230, then the change of the code rate caused by the varying of Doppler shift will not exceed 0.4092 cps. For the receiver which is stationary to the ground, the accumulation time can reach several minutes, which can meet the general needs.

4.4 The Imbalance of Bit Sign

The imbalance of bit sign refers to that condition (5) is not satisfied. If condition (5) is not satisfied, $c_1(t)$ will exist in the separation results, which make the amplitude of some chips of $c_0(t)$ to be enhanced while the amplitude of the others is weakened. The power fluctuation of separated signal will occur, which is manifested as "double eyelids" in the eye diagram, as shown in Fig. 3a. Supposing that the power of two signals is P_0 and P_1, the sign imbalance of $c_1(t)$ is defined as

$$u = \frac{1}{N}\left|\sum_{j=1}^{N} a_{j0}a_{j1}\right| \tag{9}$$

Then, the enhanced amplitude is

$$A_+ = \sqrt{P_0} + u\sqrt{P_1} \tag{10}$$

The reduced amplitude is

$$A_- = \sqrt{P_0} - u\sqrt{P_1} \tag{11}$$

The distortion of separated signal can be measured by the amplitude ratio

$$A_R = \frac{A_-}{A_+} = \frac{\sqrt{P_0/P_1} - u}{\sqrt{P_0/P_1} + u} \tag{12}$$

Figure 3a, b give the eye diagrams [7] of separated signal E1b with different u. Figure 3a does not eliminate the imbalance of bit sign (u = 0.05), so there is a phenomenon of "double eyelids". According to the Eq. (12), the amplitude ratio is

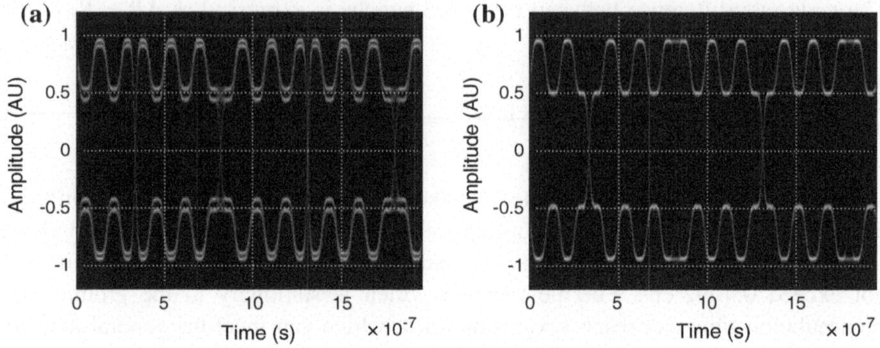

Fig. 3 The eye diagram of separated signal E1b **a** $u = 0.05$ **b** $u = 0$

$$\begin{cases} A_{R-} = 0.8295 \\ A_{R+} = 0.9477 \end{cases}$$

In comparison, the simulation results match the calculated results. Using the algorithm flow proposed in this paper, Fig. 3b eliminates the imbalance of bit sign and the eye diagram is ideal.

5 Analyses of Simulation and Measured Data

In this section, the effect of the separation algorithm on signal waveform is analyzed by a signal accumulation experiment first, ensuring that the algorithm will not introduce an additional distortion to the signal. Then, in a signal separation experiment, the simulated and the measured Galileo E1 signal are separated, and the effectiveness of the separation algorithm is verified.

5.1 The Signal Accumulation Experiment

In the signal accumulation experiment, the BPSK modulation adopted by BeiDou B1 is used. In the signal, the TMC [4, 8] distortions are simulated, whose digital duty cycle is 0.504, $Fd = 5$ MHz and $\sigma = 7$ MNep/s. The CNR of the simulation signal is 80 dB Hz. Under different accumulation times, the average results are presented in Fig. 4. In order to prevent the overlap between different curves, the 10 times, 100 times, and 1000 times accumulated curves move 0.1, 0.2, and 0.3 down, respectively. It is worth noting that with the increase of the accumulation time, the waveform of average signal is more smooth and its difference from the ideal signal is smaller.

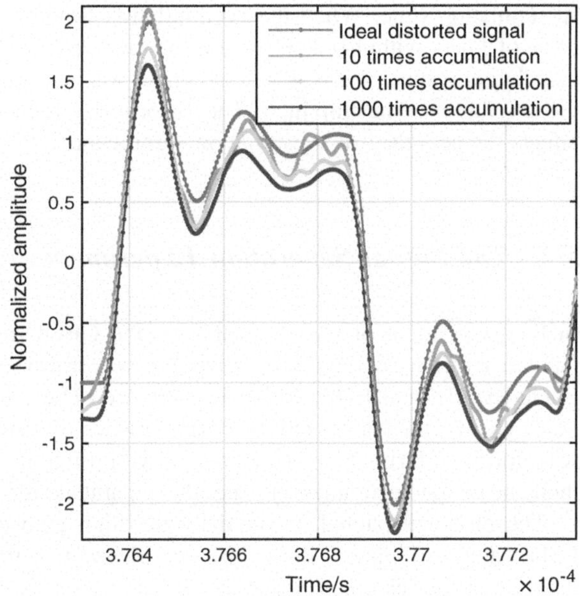

Fig. 4 The chip waveforms under different accumulation times (Color figure online)

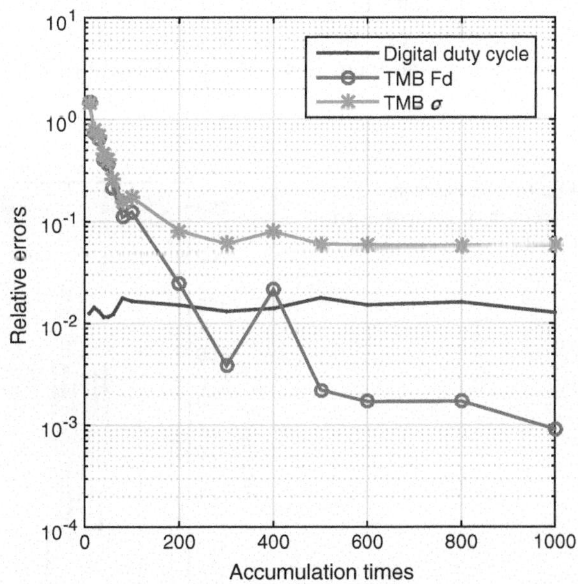

Fig. 5 The parameter estimation errors of the distortion in time domain (Color figure online)

Figure 5 shows the relative error of the time domain distortion parameters under different accumulation times. The calculation results of the digital duty cycle are affected by accumulation time less, whose relative error is within 2 %. The relative errors of the other two parameters generally decrease with the increasing of the accumulation time. When the accumulation time is greater than 500, the relative error of Fd is within 0.2 % and the relative error of σ is around 6 % stably. The calculation results of σ have a fixed error, which is mainly caused by the decline in the amplitude of the chip jitter. The distortion of the waveform is not obvious, but it affects the accuracy of σ estimation significantly.

5.2 The Signal Separation Experiment

In the experiment of signal separation, the CBOC modulation [5, 6] adopted by Galileo E1 is used. Figure 6 shows the waveform of composite baseband signal (includes E1c and E1b), which is untreated. The composite signal is affected by noise greatly and cannot tell the characteristics of a single chip. Figure 7 shows the waveforms of E1b which are separated with different accumulation times. With the increase of the accumulation time, the separated signal waveform is more ideal.

The measured signal acquired by the high gain antenna is from Galileo-IOV FM4 satellite. Figure 8 shows the eye diagram of the composite signal. The eye diagram of the separated signals is shown in Fig. 9a, b, respectively. The eyelids of composite signal are thick, and there are two kinds of "eye" in the eye diagram, which cannot reflect the characteristics of a single channel. For the separated signal, the trace of the eye diagram is clear and neat, which indicates that the noise of the signal is low and there is no obvious distortion in time domain.

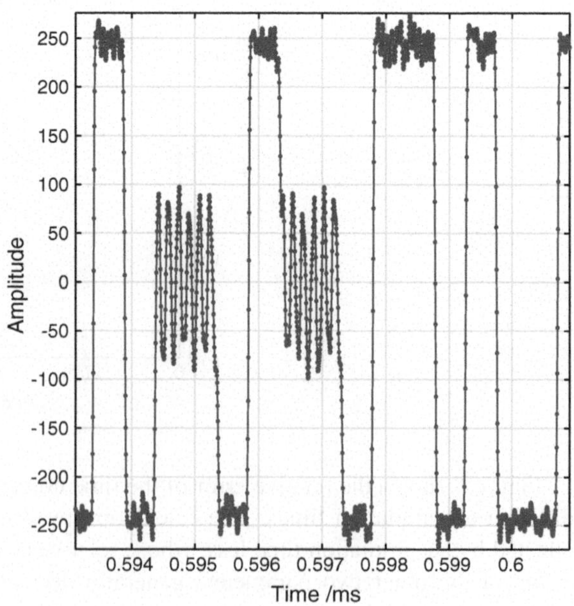

Fig. 6 The baseband waveform of composite E1 signal

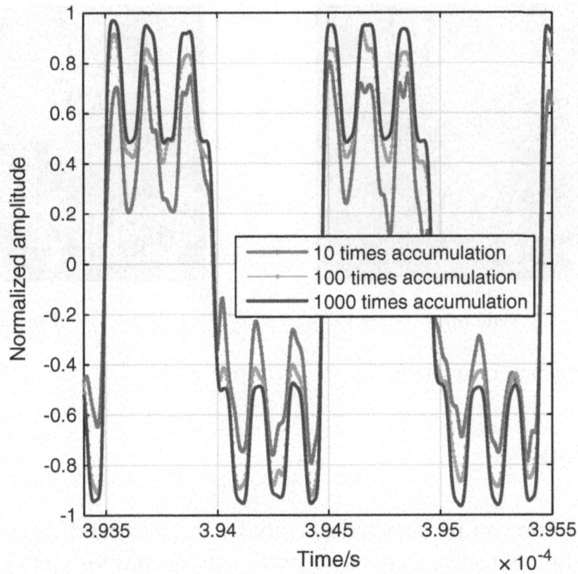

Fig. 7 The waveforms of separated E1b under different accumulation times (Color figure online)

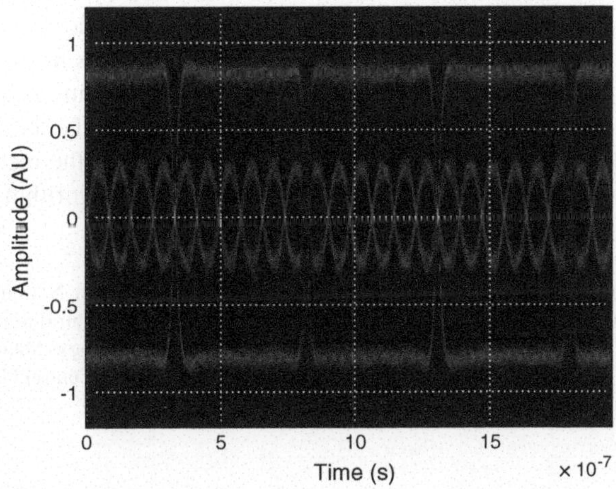

Fig. 8 The eye diagram of composite signal E1

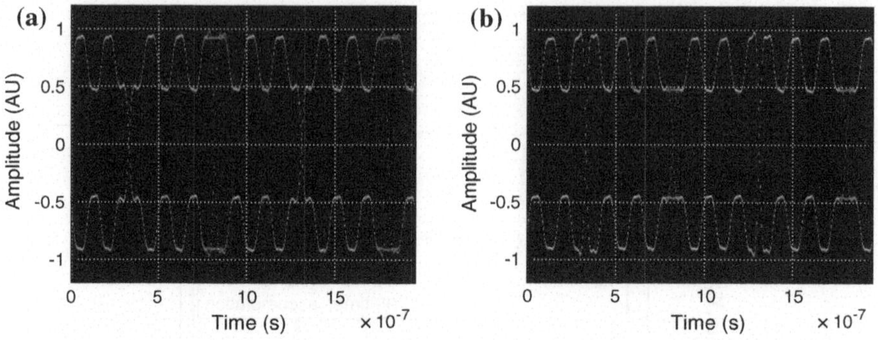

Fig. 9 The eye diagrams of separated signals **a** E1b **b** E1c

6 Conclusion

The signal distortion can be detected directly by the baseband waveform of single channel signal. In this paper, a signal separation algorithm for GNSS signal quality assessment is proposed, which can realize the composite signal separation and improve the separated signal's SNR. The separation results can be affected by the poor tracking results, Doppler shift, accumulation time, and the imbalance of bit sign. However, the additional distortion can be controlled by choosing appropriate processing parameters. According to the mathematical model of the separation algorithm, the condition for signal separation is given. What is more, the algorithm flow of separating two signals is summarized. The signal accumulation experiment and the signal separation experiment were conducted to verify the reliability and validity of the separation algorithm. The separation results of the measured Galileo E1 signals show that the proposed signal separation algorithm has a good performance.

Acknowledgments Thanks to Dr. Xue Wang and Dr. Chengyan He in National Time Service Center for their inspiration. This work was supported by the National Natural Science Foundation of China under Grants 61301094, the National Science and Technology Major Project under Grants GFZX0301040117 and the Fundamental Research Funds for the Central Universities under Grant 3102015ZY040.

References

1. Mitelman AM, Akos D M, Pullen SP et al (2002) Estimation of ICAO threat model parameters for operational GPS satellites. In: Proceedings of international technical meeting of the satellite division of the institute of navigation, pp 12–19
2. He C, Lu X (2009) New coherent accumulation algorithm for further property analysis of GNSS baseband signals. J Data Acquisition Proces 1

3. Xie G (2009) Principles of GPS and receiver design. Publishing House of Electronics Industry, Beijing, pp 292–295, 317–319
4. Wong G, Phelts RE, Walter T et al (2010) Characterization of signal deformations for GPS and WAAS satellites. In: Proceedings of the 23rd international technical meeting of the satellite division of the institute of navigation (ION GNSS 2010), Portland, OR, pp 3143–3151
5. Galileo OS (2016) SIS ICD. Available: http://www.gsc-europa.eu/system/files/galileo_documents/Galileo_OS_SIS_ICD.pdf. Accessed 21 Jan 2016
6. Borre K, Akos DM, Bertelsen N et al (2007) A software-defined GPS and Galileo receiver: a single-frequency approach. Springer Science & Business Media, New York
7. Soellner M, Kurzhals C, Hechenblaikner G et al (2008) GNSS offline signal quality assessment. ION GNSS, Florida
8. XianZhi L, Lei C, Jian X, et al (2015) Research of modeling technique for waveform distortion of navigation signal in time domain, 2015

Analysis on Pseudorange Biases Between GNSS Navigation Satellites

Xiaochao Feng, HengLin Chu, JianCheng Liu, KuiXing Liu, Hao Gao and Qian Wang

Abstract Under ideal conditions, the correlation peak of navigation signals emitted by the navigation satellites is completely symmetrical, and the pseudoranges obtained by the receiver to different satellites are consistent. However, live signals differ from this ideal due to characteristic difference of satellites' navigation overload, the troposphere and the ionosphere, multipath effects and other factors, the signals emitted by each satellite reaching the ground show nonideal characteristics and differences on each other resulting in the biases of pseudorange measured by receiver to the different satellites. In this paper, based on the signal correlation domain analysis and observations data domain analysis, we analyzed GNSS navigation signals, respectively with navigation signal quality monitoring equipment and monitoring receiver. The analysis results show that the pseudorange between different satellite navigation signals occur 2–3 ns biases, and the ideality and inconsistency of navigation signals emitted by different satellites might be the main factor of pseudorange biases, and receiver channel design, loop design, and other factors can also affect pseudorange bias value.

Keywords Beidou system · Pseudorange biases between satellites · Signal quality · Zero-line

1 Introduction

Whether precise orbit determination, clock error monitoring of navigation satellites, ionosphere and troposphere monitoring, or other business processing in the GNSS Ground Operating and Controlling System, or positioning, navigation and timing in

Supported by nation natural fund (41474027).

X. Feng (✉) · H. Chu · J. Liu · K. Liu · H. Gao · Q. Wang
Beijing Satellite Navigation Center, Beijing 100094, China
e-mail: fengxc_bd2@sina.com

J. Sun et al. (eds.), *China Satellite Navigation Conference (CSNC) 2016 Proceedings: Volume I*, Lecture Notes in Electrical Engineering 388,
DOI 10.1007/978-981-10-0934-1_46

the terminal user application, we usually assume that the pseudorange values for all satellites is consistent, the ignorance to pseudorange biases between satellites influences the further improvement of the processing precision. In the field of high precision measurement, the mitigation of the pseudorange biases to process precision is mainly how to calculating code deviation between differential code biases (DCB) and carrier phase deviation fractional carrier phase biases (FCB) parameters, and focus on the validity and stability of the DCB and FCB parameter values. From the GPS SV19 number satellite events, ideality and differences of navigation signals emitting by satellites which are in contribution to the pseudorange bias should arouse our attention and concern.

Based on the signal correlation domain and observation data domain analysis, by respectively using GNSS system signal quality monitoring equipment and monitoring receiver, the paper analyzes pseudorange biases phenomenon of GNSS navigation signals between satellites.

2 Signal Quality and Pseudorange Bias

The correlation peak of ideal navigation signal is symmetrical, the symmetry has nothing to do with the correlation spacing; when the signal is not ideal, such as abnormal signal distortion, the correlation peak will no longer be symmetrical, the most typical example is the abnormal GPS SV19 signal in 1993. To analyze the signal of SV19 anomalies, researchers have proposed various abnormal models of navigation signals, the international civil aviation organization (ICAO) recommended 2OS model, this model comprehensively reflects the satellite failures of all situations. The general mathematical model of navigation satellite payload is shown in Fig. 1 [2]. Among them, the navigation signal generating unit generates navigation baseband signal, the bandwidth of the digital filter for signal emitting

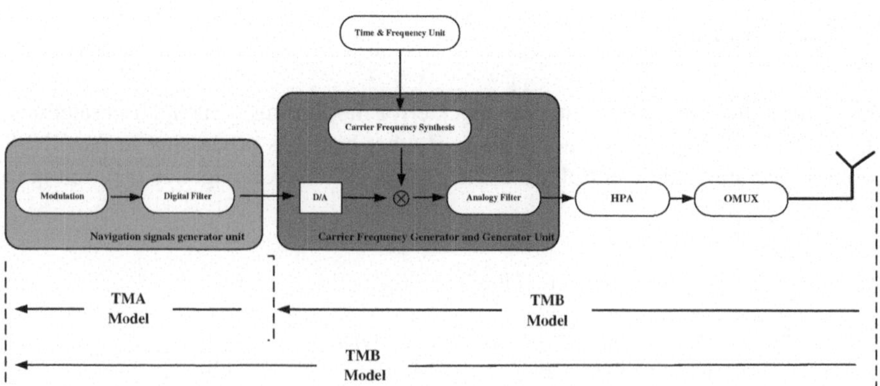

Fig. 1 Payload model

bandwidth; frequency generation and modulation unit convert digital signals to analog signals, and through a few level frequency modulation signal to the carrier frequency; Complete high power amplifier to amplify signal power; output multiplexer (OMUX) combines all the navigation signals, coupled to the same antenna output. Any component's abnormality of the satellite payload can make the signal distortion.

According to component of the navigation satellite payload, 2OS model divides the abnormal signal into three categories: digital circuit anomalies (Threat Model A), analog circuit anomalies (Threat Model B) and the combination of both (Threat Model C), the three models almost cover all possible abnormal signal failures. The abnormal signal model is considered by ICAO as standard, also known as ICAO model, there are three key parameters in the model chips in early/delay time (chips), damping coefficient (MNerpers/s), and the damped oscillation frequency (MHz). The meaning of each parameter is shown in Fig. 2a. The scope of parameters is shown in Fig. 2b.

1. TMA Model

TMA model represents that the digital circuit of satellite payload is abnormal and is modeled as early/delay Δ chip for a rising or falling edge of a pseudocode. The time domain waveform and correlation peak are shown in Fig. 3, correlation peak appears ceiling phenomenon.

2. TMB Model

TMB model represents that the analog circuit of satellite payload is abnormal and represented by a second-order system response. The time domain waveform and the correlation peak is shown in Fig. 4, correlation peak appears deformation and asymmetric phenomenon.

3. TMC

TMC model is the combination of the TMA and TMB model. The time domain waveform and the correlation peak are shown in Fig. 5. Due to the attenuation

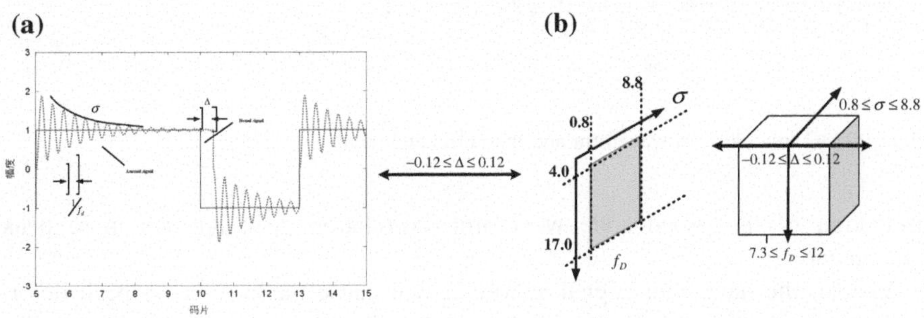

Fig. 2 ICAO model parameters

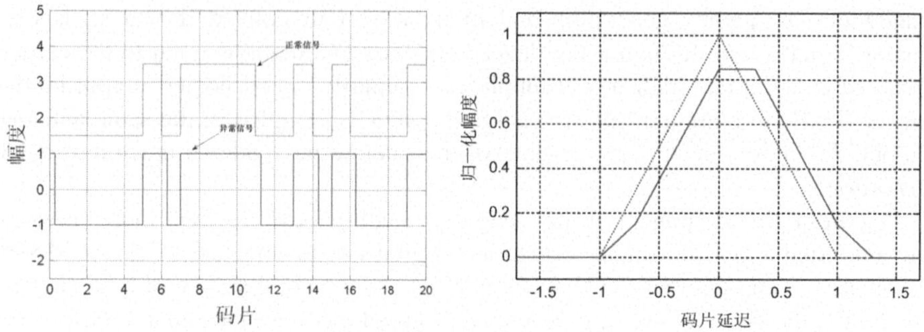

Fig. 3 TMA time domain waveform and the correlation peak

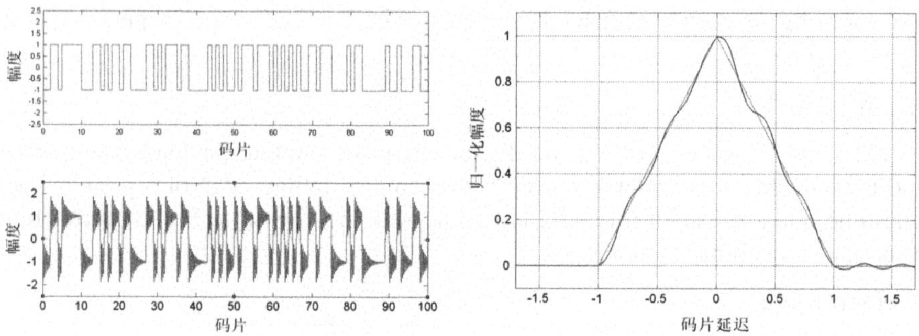

Fig. 4 TMB time domain waveform and the correlation peak

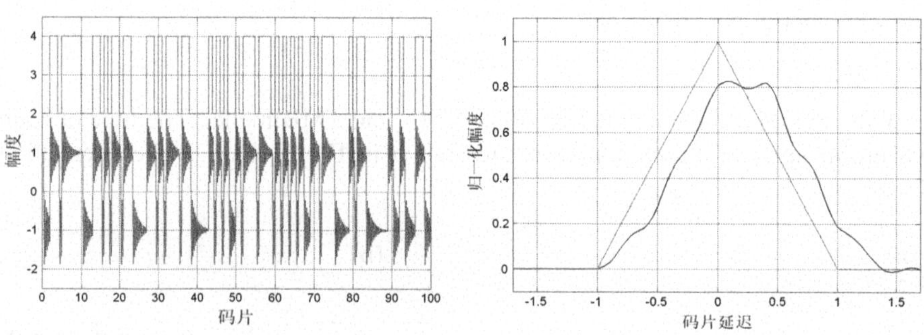

Fig. 5 TMC time domain waveform and the correlation peak

oscillation of the pseudocode waveform, correlation peak appears more peak phenomenon.

Visible, the navigation signal abnormal will cause the correlation peak deformation, causing the pseudorange measurement biases.

3 Correlation Domain Analysis

In the correlation domain analysis, we generally use the correlation peak symmetry, correlation loss and phase discrimination curve (S-curve) passing zero biases such as monitoring indexes to analyze the signal quality. The accurate analysis of signal quality usually adopts the S-curve passing zero biases, characterizing S-curve passing zero biases value under different correlation spacing. Ideally, the S-curve should be zero (or zero biases value is very small). When the signal is abnormal and different design conditions of GNSS receiver biases by adding the zero biases, will cause the pseudorange fixed biases.

Based on the navigation signal quality monitoring equipment of the BeiDou ground operating and controlling system, we carried out the navigation signal correlation peak monitoring test; the equipment is shown in Fig. 6. By high gain directional antenna and omnidirectional anti-multipath antenna, we received and analyzed the space live navigation signal of BeiDou system. At the same time, we also carried out comparative analysis to emulating navigation signal by the standard signal source. The RF sampling equipment can realize the sampling of two channel signal with 50 MHz/s bandwidth (I/Q way). Signal processing and analysis subsystem adopts high performance workstation, which is connected to the sampling equipment, data storage, and playback equipment with Ethernet, receiving and processing correlation peak of digital sampling signal. Data storage unit flow rate of 600 MB/S with storage capacity is up to 3T, which completely meets the requirements of continuous 24 h monitoring.

On January 18, 2011, we carried out the correlation peak monitoring tests to the BeiDou space live navigation signal, and analyzed the correlation peak detection curve and phase discrimination curve (S-curve) of B1I and B3I signals by reprocessing the sampling data, the results are shown in Figs. 7 and 8.

Figure 7 shows that, correlation peak detection curve reflects nearly symmetric for the B1I and B3I signal, but the peak around the curve shows a small deformation which is similar to the TMB model of ICAO. In Fig 8, the horizontal axis denotes the correlation spacing with the unit for chip, and ordinate denotes S-curve

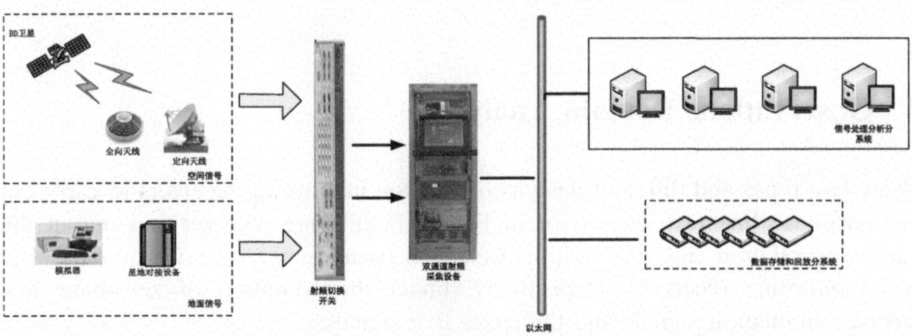

Fig. 6 Signal quality monitoring test equipment composition block diagram

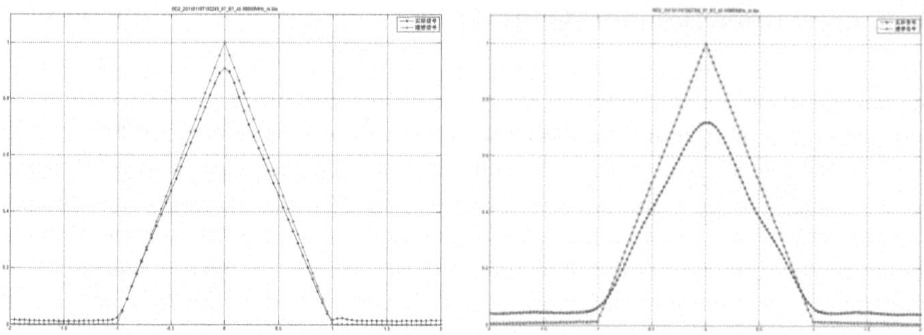

Fig. 7 Correlation peak detection curve of B1I and B3I signal

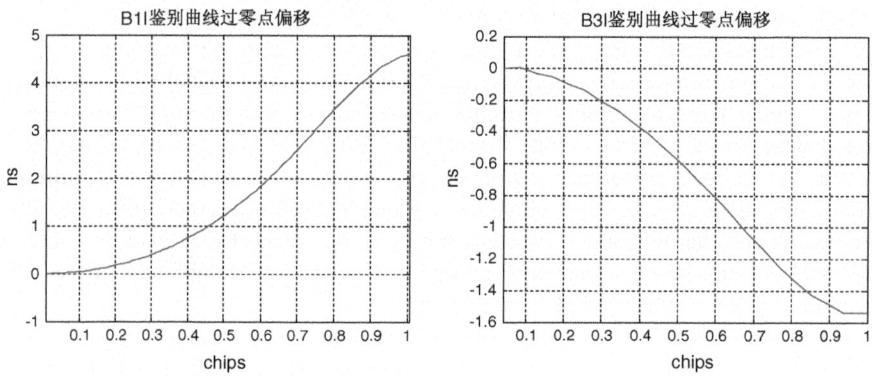

Fig. 8 Phase-discrimination curve of B1I and B3I signal

zero biases under different correlation spacing with the unit for ns under the whole matching process conditions. Ideally, the S-curve passing zero biases is usually very small. It can be seen that the correlation spacing for a chip, B1I phase S-curve passing zero biases reached around 4 ns, B3I around zero offset 1.5 ns, this creates a pseudorange measurement biases.

4 Observations Domain Analysis

Using two types and different designs of BeiDou monitoring receivers to carry out the zero-base line test, as shown in Fig. 9, monitoring receiver can output the narrow correlation and anti-multipath channel pseudorange observation data. The two monitoring receivers, respectively, under the condition of zero-base line receives simulation signals and the space live signals.

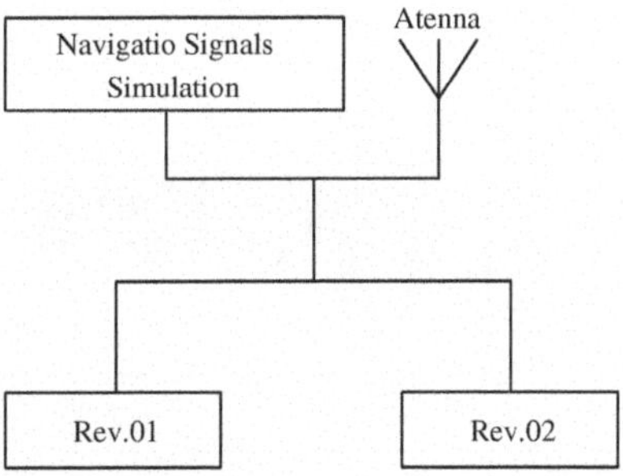

Fig. 9 Zero-base line test

We processed observations data of two zero-base line monitoring receiver by double difference processing, which can eliminate satellite orbit error, satellite clock error, ionosphere error, troposphere, and so on.

Pseudorange observation equation is as follows:

$$P = \rho + \delta\rho + \delta t_r - \delta t_s + \delta t_{\text{ion}} + \delta t_{\text{trop}} + \varepsilon \tag{4.1}$$

P for the pseudorange observation, ρ for satellite and station distance, δt_r, δt_s, δt_{ion}, δt_{trop}, respectively, for the receiver clock error, satellite clock error, the ionosphere delay and tropospheric delay, ε for diameter and other noise error.

We can get double difference pseudorange observation equation to the above observation equation between two receivers 01 and 02, between two satellites j and k, respectively.

$$\Delta\nabla P_{12}^{jk} = \Delta\nabla\delta\rho_{12}^{jk} + \Delta\nabla\varepsilon_{12}^{jk} \tag{4.2}$$

As a long time observation, we can take the $\Delta\nabla\varepsilon_{12}^{jk}$ mean value close to zero. Double difference processing in the satellite pseudorange values for reference, the average double difference results reflect two receivers on the same satellite biases relative to SV01.

As seen from the Figs. 10 and 11, under the simulation signal conditions, two receivers B1I/B3I zero-base line pseudorange biases are less than 1 ns; under space live signal conditions, the B1I zero-base line pseudorange biases is up to 4–6 ns, B3I biases within 1 ns.

We also set the partial 10 channel correlation spacing values of the one monitoring receiver to 0.1–1.0 chip, and adjust the B1I signal digital filter bandwidth values to 6, 10 and 12 MHz for analyzing pseudorange biases. Figure 12 gives pseudorange biases changing by different correlation spacing and filter bandwidth. The figure shows that the receiver correlation space and the filter bandwidth parameter will affect the pseudorange biases value.

At the same time, we analyzed pseudorange observation data obtained by four monitoring receivers setting at Beijing, Harbin, ShanTou, and Urumqi and made difference of narrow correlation and anti-multipath pseudorange, then pseudorange observations of the rest satellites make difference to SV01 benchmark,

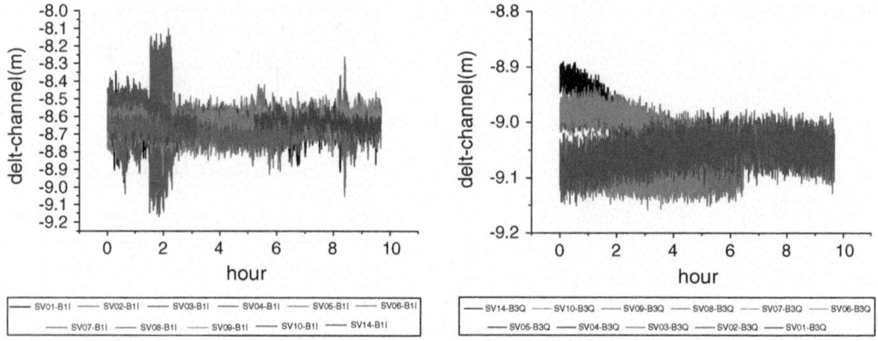

Fig. 10 B1I/B3I zero-base pseudorange with simulation navigatio signals (Color figure online)

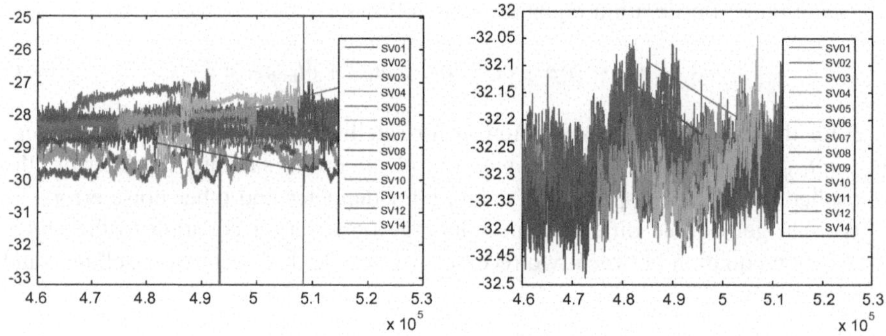

Fig. 11 B1I/B3I zero-base pseudorange with space live navigation signals (Color figure online)

Fig. 12 B1I/B3I zero-base pseudorange with different correlation space and the filter bandwidth parameter (Color figure online)

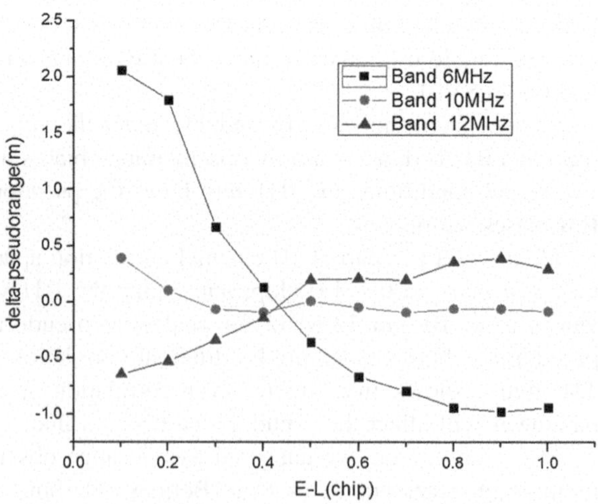

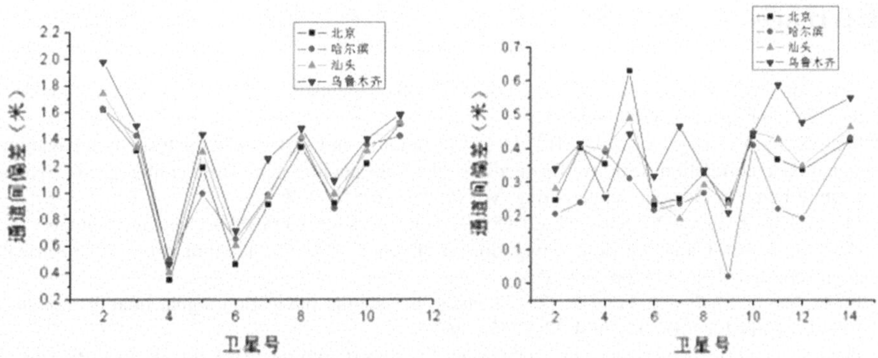

Fig. 13 B1I/B3I zero-base pseudorange with different observation stations by space live navigation signals (Color figure online)

the observatory to different satellite pseudorange biases analysis results is as shown in Fig. 13. Seen from the figure, the pseudorange biases between narrow correlation and multipath channel distance biases at different sites on the same satellite signal exist strong correlation, which reflect that pseudorange biases mainly relate to the satellite signal characteristics.

5 Conclusion

Comprehensive to the above analysis result, BeiDou receiver's pseudorange biases between satellites mainly appeared on the B1 pseudorange observations, each satellite pseudorange biases value relatively fixed, and the S-curve passing zero biases of B1 are basically identical. The B3 pseudorange biases signal is relatively small. The satellite signal abnormality is the main factor of pseudorange biases, and the receiver channel design factors such as the parameters of the loop also affect the pseudorange biases value.

Two technical measures can be taken to weaken the impact on the business process by the satellite pseudorange biases. First, we can carry out a large number of parameter adjustment tests under space live signals conditions, selecting the optimum parameters of the signal processing circuits, and making each satellite pseudorange biases down to a minimum; Second, according to different satellite pseudorange biases value is relatively fixed, the pseudorange value of different satellite appends delay correction to correct biases because of the satellite signal's un-ideality. Although the two technical measures can weaken the pseudorange biases between satellites to the effect for the system's business process, but the existence of pseudorange biases will cause difficulty to further improve the precision of system business and meet the index to achieve higher precision.

References

1. Hauschild A, Montenbruck O (2014) A study on the dependency of GNSS pseudorange biases on correlator spacing. GPS Solut 11
2. Feng X, Fu Y, Gong L, Liu J (2015) Detect techniques and experiment analysis on navigation signal code correlation peak. In: Proceedings of China satellite navigation conference (CSNC). Lecture notes in electrical engineering, vol I(340), pp 517–527
3. Liu J, Fan J, Feng X, Sang H (2015) Ranging bias of COMPASS satellite signals. In: Proceedings of China satellite navigation conference (CSNC). Lecture notes in electrical engineering, vol II(341), pp 347–356
4. Jin G, Wang M, Fan J, Feng X (2012) Image of satellite navigation signal quality monitor. J Guilin Univ Electron Technol 32(5):358–363
5. Phelts RE (2001) Multicorrelator techniques for robust mitigation of threats to GPS signal quality. Stanford University, Palo Alto
6. Phelts RE, Akos DM, Enge P (2000) Robust signal quality monitoring and detection of evil waveforms. ION, pp 1180–1190
7. Jock RIC, Philip BB, John WC (2004) GPS signal quality monitoring system. In: ION GNSS 17th international technical meeting of the satellite division. Long Beach, CA, pp 2239–2245
8. Mitelman AM (2004) Signal quality monitoring for GPS augmentation systems. Ph.D. thesis, Standford University, California, p 12
9. Mitelman AM, Phelts RE et al (2004) Signal deformations on nominally healthy GPS satellites. ION

Part IV
BDS/GNSS User Terminal Technology

Weak Signal Acquisition Algorithm for Indoor Positioning Receiver

Jun Mo, Zhongliang Deng, Jichao Jiao, Shu Jiang, Shengchang Yu and Fuhai Xu

Abstract Positioning receivers are required to be capable of receiving weak signals to satisfy indoor applications. When searching weak signals, the receiver may unsuccessfully be acquired due to the noise. This paper is focused on the problem of weak signal acquisition in Time and Code Division-Orthogonal Frequency Division Multiplexing (TC-OFDM) system which is based on the mobile broadcast networks. We provide theoretical analysis of the TC-OFDM signal, and propose a new acquisition algorithm based on coherent/noncoherent integral for weak signal acquisition. This algorithm compensates noncoherent integral loss for considering the effect of decision threshold that caused by noncoherent integral loss, which can increase the relative acquisition peak and achieve weak signal acquisition effectively. The experimental results indicate that the proposed algorithm can significantly improve the TC-OFDM receiver's ability of weak signal acquisition by 1–3 dB, and reduce the probability of weak signal acquisition leakage.

Keywords Weak signal · Noncoherent integral loss · TC-OFDM

1 Introduction

With the rapid development of the wireless networks, LBS is gradually into our daily lives [1]. Currently, GNSS can provide accurate position information for our outdoor activities. With the gradual increase of large buildings in the city, our activity space has changed from outdoor to indoor environment, according to the report "In the future, 80 % living-time will be in the indoor environment" [2]. Therefore, the precise location of the indoor environment is becoming more and more important. The TC-OFDM system discussed in this paper is developed for the precise location

J. Mo (✉) · Z. Deng · J. Jiao · S. Jiang · S. Yu · F. Xu
School of Electronic Engineering, Beijing University of Posts and Telecommunications, Beijing, China
e-mail: mojun@bupt.edu.cn

J. Sun et al. (eds.), *China Satellite Navigation Conference (CSNC) 2016 Proceedings: Volume I*, Lecture Notes in Electrical Engineering 388,
DOI 10.1007/978-981-10-0934-1_47

of the indoor environment, which can provide wide area high-precision indoor and outdoor location service.

Today, most high-precision positioning systems are based on CDMA (code division multiple access, CDMA) [3], TC-OFDM is based on CDMA, the CDMA signal is superimposed on a mobile broadcast signal, without affecting the user under the premise of receiving mobile broadcast signal, to provide users with precise location services. In the TC-OFDM signal transmission process, due to obstructions and co-channel interference and other factors, the TC-OFDM signal strength greatly reduced. When the signal carrier to noise ratio (CNR) is below a certain value, common parallel code acquisition algorithm cannot detect the faint TC-OFDM signals. Therefore, the TC-OFDM weak signal acquisition is the key point of the TC-OFDM system applications.

2 Algorithm Description

2.1 *TC-OFDM Signal*

For discussion, assume that the positioning receiver only receives a TC-OFDM signal, this signal through down-conversion RF front-end and low-pass filter, the complex signal can be received:

$$r(t) = aD(t-\tau)S(t-\tau)\exp\{j[2\pi(f_{\mathrm{IF}}+f_d)(t-\tau)+\theta_{\mathrm{IF}}]\}+n(t) \tag{1}$$

where a denotes the amplitude of the intermediate frequency signals from the base station, $D(t)$ is the data code, $S(t)$ is GOLD code, τ denotes propagation delay, f_d is the incoming Doppler shift, f_{IF} is intermediate frequency, and $n(t)$ is the noise signal. Complex signal after A/D conversion, the discrete digital IF signal can be processed by the baseband receiver, the signal can be then expressed as

$$i(n) = aD(n-\tau)S(n-\tau)\sin(2\pi(f_{\mathrm{IF}}+f_d)t(n)+\theta_{\mathrm{IF}})+n_i(n) \tag{2}$$

$$q(n) = aD(n-\tau)S(n-\tau)\cos(2\pi(f_{\mathrm{IF}}+f_d)t(n)+\theta_{\mathrm{IF}})+n_q(n) \tag{3}$$

where n_i is the I path noise signal and n_q is the Q path noise signal; the noise signal stands for the additive Gaussian white noise.

2.2 *Coherent/Noncoherent Integration for the New Acquisition Algorithm*

In the TC-OFDM system, the acquisition is designed to establish the conditions for tracking loop, which is a rough acquisition of the pseudocode phase and residual carrier frequency of the location signal. Acquisition algorithm for GNSS receivers is commonly used in two categories, which is time domain serial acquisition algorithm and parallel acquisition algorithm based on FFT [4]. The advantage of time domain serial acquisition algorithm [5] is that the search can be the code phase and carrier frequency that are taken into account; the disadvantage is that in the worst case to all the code phase and frequency should be searched again, long time consumption. Parallel acquisition algorithm based on FFT [6] is divided into two processes: parallel frequency search and parallel code phase search. The advantage of Fourier transform and inverse Fourier transform is faster than time domain serial acquisition algorithm, but that needs bigger storage space. In addition, the Fourier transform and inverse Fourier transform will produce delay, and affected by the data bit transitions.

Time domain serial acquisition algorithm and parallel acquisition algorithm based on FFT has its own advantages and disadvantages, according to the characteristics and particularity of the TC-OFDM system, combined with the acquisition algorithm of the system is based on FPGA hardware implementation, using improved parallel correlator for signal acquisition, this method can effectively improve the acquisition speed.

The first step of the baseband signal processing is for code acquisition. When determining the Gold code information contained in the IF signal, the phase of the Gold code is given. Then the local code generator can reproduce same phase with the Gold code. The structure of code acquisition is shown in Fig. 1.

After A/D conversion processing discrete digital IF signal *i*(*n*) and *q*(*n*), through parallel-to-serial conversion module local codes are then multiplied by the sum, coherent integration results *I*'(*n*) and *Q*'(*n*), is expressed as follows:

$$I'(n) = \sum_{k=0}^{N-1} i(k)C(k) = aND(n)R(\tau)\text{sinc}(f_e T_{\text{coh}})\cos(\phi_e) + n_i = I(n) + n_I \quad (4)$$

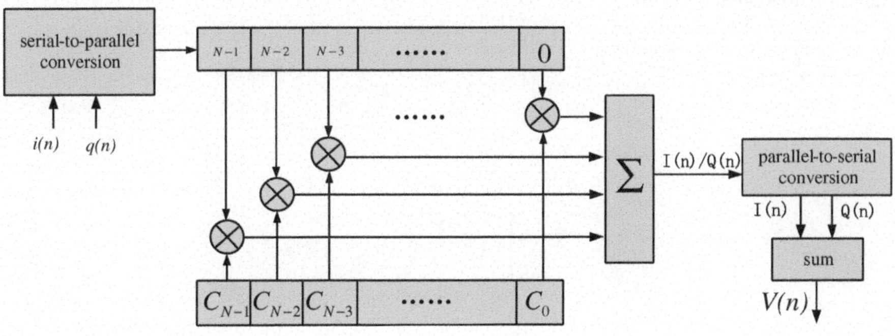

Fig. 1 Acquisition schematic

$$Q'(n) = \sum_{k=0}^{N-1} q(k)C(k) = aND(n)R(\tau)\mathrm{sinc}(f_e T_{\mathrm{coh}}) \cdot \sin(\phi_e) + n_Q = Q(n) + n_Q \quad (5)$$

where $R(\tau)$where $R(\tau)$ is corresponding autocorrelation function, f_e is the frequency error, T_{coh} is integration time, ϕ_e is the phase error, n_I is *I* path noise, n_Q is *Q* path noise. *I*(*n*) and *Q*(*n*) square after serial-parallel conversion module, then acquire output power *V*(*n*):

$$V(n) = (I(n) + n_I)^2 + (Q(n) + n_Q)^2 = (I^2(n) + Q^2(n)) + (2I(n)n_I + 2Q(n)n_Q) + (n_I^2 + n_Q^2) \quad (6)$$

where n_I and n_Q are all the normal noise which has zero mean and variance equal. On the right side of second sign in Eq. (6), the first part is signal autocorrelation power value, the mean of the second part is zero, but the third does not mean zero. So *V*(*n*) can be also expressed:

$$V(n) = (I^2(n) + Q^2(n)) + (n_I^2 + n_Q^2) = a^2N^2R^2(\tau)\mathrm{sinc}^2(f_e T_{\mathrm{coh}}) + (n_I^2 + n_Q^2) \quad (7)$$

When the slide of the local code and the input signal code phase-coherent are same, then $\tau = 0$ and $R^2(\tau)$ have the maximum. Function sinc value is shown in Fig. 2, when $f_e = 0$, *V*(*n*) has a max value; when $|f_e| \leq 1.7$ kHz, the impact on the integral value of the frequency error does not exceed 10 %. Therefore, during the signal acquisition in the TC-OFDM system, it cannot necessarily consider the influence of frequency error on the integral result, can directly do codes acquisition.

Due to the noncoherent integration is not affected by the data bit transition and not affected by the frequency error, it is a simple and easy way to improve the acquisition sensitivity of the receiver [7]. But there is noncoherent integration

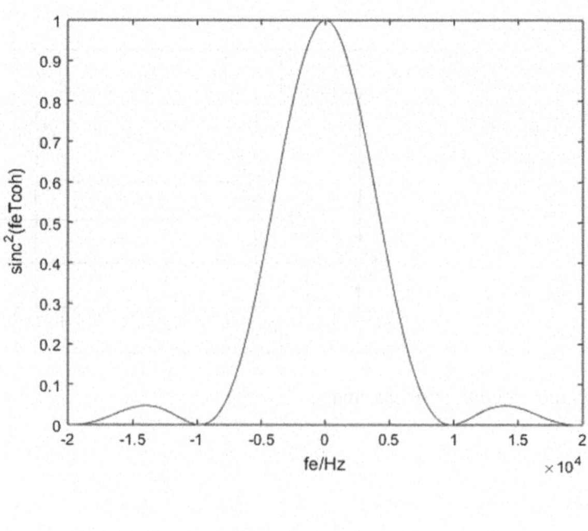

Fig. 2 Function *curve* of sinc

square loss, the signal-to-noise ratio (SNR) of the weak signal is suppressed. Noncoherent integration is accumulated by the following equation:

$$Z(n) = \sum_{l=0}^{x-1} V_l(n) = \sum_{l=0}^{x-1} (I_l^2(n) + Q_l^2(n)) + \sum_{l=0}^{x-1} (n_{I,l}^2 + n_{Q,l}^2) \tag{8}$$

From Eq. (8), $\sum_{l=0}^{x-1} (n_{I,l}^2 + n_{Q,l}^2)$ is the white noise distribution. The noncoherent integration is the energy power amplitude of the noise, which can cause the so-called noncoherent loss. Therefore, to achieve the above-mentioned coherent/noncoherent integration acquisition algorithm, noncoherent integration loss must be compensated.

Noncoherent integration value contains the true signal and the noise amplitude power energy:

$$P_{\text{nc}} = \max(Z(n)) = P_m + P_{\text{noise}} \tag{9}$$

where P_{nc} denotes effective signal power in noncoherent integration, P_m is real effective signal power, P_{noise} is the noise amplitude in noncoherent integration. The value of the noise characterization in noncoherent integration can be expressed as:

$$P_{\text{nn}} = \text{second}(Z(\text{n})) = P_s + P_{\text{noise}} \tag{10}$$

where P_{nn} is the value of the noise characterization in noncoherent integration, P_s is the value of the real noise characterization in noncoherent integration. Noncoherent integration threshold is determined as follows:

$$\frac{P_{\text{nc}}}{P_{\text{nn}}} = \frac{P_m + P_{\text{noise}}}{P_s + P_{\text{noise}}} > R_{\text{th}} \tag{11}$$

where R_{th} is acquisition threshold. From Eq. (11), weak signal can be undetected due to P_{noise}. When P_{noise} can be estimated, the noncoherent integration loss can be compensated. The probability of undetected weak signal can be reduced. This paper proposes a method of compensation for noncoherent integration loss. This method can be seen from Eqs. (12) and (13).

$$P'_{\text{noise}} = \frac{\sum_{n=1}^{\text{len}} Z(n) - P_{\text{nc}} - P_{\text{nn}}}{\text{len} - 2} \tag{12}$$

$$P_{\text{noise}} \approx \alpha P'_{\text{noise}} \tag{13}$$

where P'_{noise} is the mean value of noncorrelation integration result excluding the maximum value and the second largest value, len is the code length, α is the weight value. α is determined based on actual measurement. Noncoherent integration threshold determined after noncoherent integration loss compensation is as follows:

$$\frac{P_{\mathrm{nc}} - P_{\mathrm{noise}}}{P_{\mathrm{nn}} - P_{\mathrm{noise}}} \geq R_{\mathrm{th}} \tag{14}$$

When Eq. (14) is satisfied, signal acquisition is successful.

3 Test Results and Analysis

In order to evaluate the effect of the proposed acquisition algorithm based on the combination of coherent/noncoherent integration, we conducted a series of simulations on MATLAB. The coherent integration time is 0.1 ms and the number of noncoherent is 50 times.

Figure 3 is a TC-OFDM signal acquisition renderings used by the proposed algorithm, and SNR is −23 dB (corresponding positioning signal strength is 128 dBm). In the Fig. 3, the horizontal coordinate is represented as the code phase, the vertical coordinate is expressed as the final integral value. Weak signal is undetected due to the cumulative effect of the noise power which is called noncoherent loss. Relative acquisition peak without noncoherent integration loss compensation is $1.5 \times 10^7/8.6 \times 10^6 = 1.8$; relative acquisition peak with noncoherent integration loss compensation is $1.0 \times 10^6/3.5 \times 10^5 = 2.9$, and $\alpha = 1$.

Figure 4 is in the conditions of SNR = −23 dB (corresponding positioning signal strength is 128 dBm), the TC-OFDM signal acquisition is 500 times continuously, compare the changes of relative acquisition peak before and after noncoherent integration loss compensation. From the Fig. 4, the probability of success acquisition increases significantly after noncoherent loss compensation.

Fig. 3 Acquisition renderings

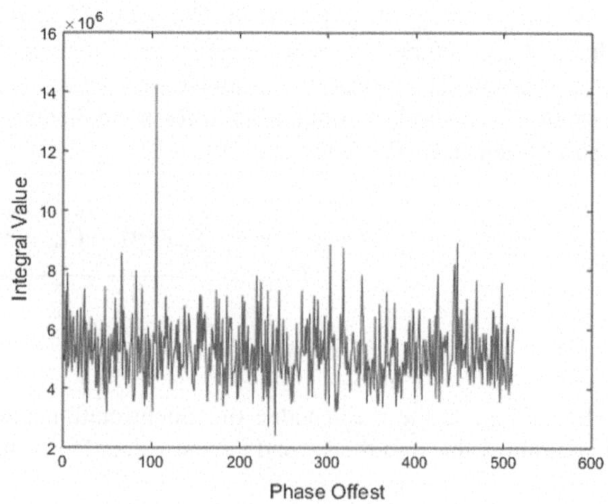

Fig. 4 Changes of relative acquisition peak (Color figure online)

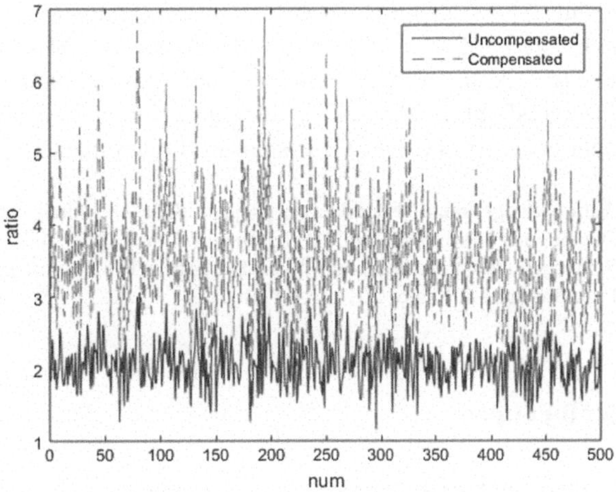

Table 1 Acquisition success rate under different SNR

SNR (dB)	Corresponding positioning signal strength (dBm)	Success rate with uncompensation (%)	Success rate with compensation (%)
−23	−128	96.4	99.8
−24	−129	81.6	96.8
−25	−130	50.4	87.6
−26	−131	23.8	65.8
−27	−132	0.76	46.4

Table 1 shows the success rate of acquisition in different SNR test, the TC-OFDM signal acquisition is 500 times continuously in each case, where $R_{th} = 1.5$ and $\alpha = 1$. It records the percentage of success acquisition before and after noncoherent integration loss compensation. From the Table 1, the success rate of acquisition with noncoherent integration compensation is higher than uncompensated.

4 Conclusion

This paper discusses the problem of weak signal acquisition in the TC-OFDM system. In this paper, TC-OFDM signal is theoretically analyzed, and a new acquisition algorithm based on coherent/noncoherent integral for weak signal acquisition is proposed. This algorithm compensates for noncoherent integration loss, and reduces the cumulative effect of the noise power due to noncoherent integration. It can significantly improve the ability of the weak signal acquisition and reduce the probability of weak signal acquisition leakage. The simulation results show that the success rate of acquisition with noncoherent integration compensation is higher than uncompensated. Moreover, the algorithm is implemented on the FPGA. It shows that this method can improve the TC-OFDM receivers' ability of weak signal acquisition by 1–3 dB, and effectively enhance the application of TC-OFDM system in the complex indoor environment.

Acknowledgements Foundation project: The National High Technology Research and Development Program ("863" Program) of China (No. 2015AA124101). The National Natural Science Foundation of China (No. 61401040). The National Science and Technology Support Program of China (No. 2014BAK12B00).

References

1. Jia D (2012) Thoughts and several suggestions on LBS development in China. Geomat World 06:22–24
2. Deng Z, Yu Y, Yuan X, Wan N, Yang L (2013) Situation and development tendency of indoor positioning. China Commun 10(3):42–55
3. Hu Y, Deng Z, Jiao J, Xia J, Hu Y, Yuan Z (2015) Cross correlation mitigation algorithm for indoor positioning receiver. In: The 6th China satellite navigation conference, p 1
4. Deng Z, Yu Y, Lianming X, Yuan X (2013) Indoor and outdoor wireless location and navigation. Beijing University of Posts and Telecommunications Press, Beijing, pp 159–168
5. Xie N, Zhang F (2012) GPS signal acquisition algorithm based on PN code cross-correlation properties. Inf Electron Eng 10(3):247–251
6. Dong X, Zhao Z, Liu S, Wang B, Liu H, Lin L (2013) A direct detection algorithm for weak GPS Signals with the presence of cross-correlation interference. GNSS World China 38(1):1–7
7. Xie G (2009) Principles of GPS and receiver design. Publishing House of Electronics Industry, Beijing, pp 376–384

Combined BDS and GPS Adaptive Vector Tracking Loop in Challenge Environment

Yulong Song and Baowang Lian

Abstract Satellite navigation receiver would have difficulties in continuous positioning or gets a bad positioning performance in dense urban environment, mostly due to low signal-to-noise-ratio and frequently signal outage. To improve robustness under the challenging environment, a combined BeiDou Satellites System (BDS) and global positioning system (GPS) adaptive vector tracking loop is proposed. An adaptive extended Kalman filter (EKF) based on innovation sequences is employed to tune the tracking loop, and generate navigation solutions. The algorithms are implemented on Matlab software receiver and digital intermediate frequency (IF) signals grabbed by a global navigation satellite system (GNSS) signal sampler under blocked outdoor scenarios are utilized to verify the performance. The tracking and positioning results of vector and scalar tracking loop are compared.

Keywords Challenge environment, BDS · GPS · Adaptive vector tracking loop · EKF

1 Introduction

In dense urban environment, GNSS signals are often weak or even interrupted due to building block, leading to frequently loss of lock and bad positioning results. In contrast to traditional scalar tracking, vector tracking would not tune the carrier and code NCO of each channel independently utilizing discriminators output. The common navigation solution will be employed to estimate the carrier frequency, code frequency and code phase of all channels, which can improve the low signal-to-noise-ratio tracking performance and anti-interference ability. What's more, vector tracking loop can relock the signal once it comes out again after brief outage. Hence, vector tracking loop improves the robustness of receiver.

Y. Song (✉) · B. Lian
School of Electronics and Information, Northwestern Polytechnical University, Xi'an, China
e-mail: yulongsong.nwpu@foxmail.com

J. Sun et al. (eds.), *China Satellite Navigation Conference (CSNC) 2016 Proceedings: Volume I*, Lecture Notes in Electrical Engineering 388,
DOI 10.1007/978-981-10-0934-1_48

Extended Kalman filter is usually employed in a vector tracking loop. It collects the pseudorange and pseudorange rate observations from every tracking loop and estimates navigation solutions directly. The common navigation solutions would then be used to tune parameters of each loop. Signal travel delay, namely code phase, is decided by line-of-sight (LOS) position vectors from receiver to satellites, and the carrier frequency is decided by LOS velocity vectors. Hence, the vector tracking loop would keep locked as long as receiver position and velocity would not mutate. EKF would constrain receiver's state even during signal outage, so vector tracking loop could relock signal once it recovered.

The concept of vector delay lock loop (VDLL) is first proposed by Spiller in the monograph [1], which described its basic structure and advantages. However, few implementation details were given. More detailed discussions of vector tracking loop and deeply integrated algorithms were given by Pany, Kaniuth and Eissfeller [2–4] from University FAF Munich, and ipex software receiver was used as the validation platform. Won, Eissfeller and Dotterbock [5, 6] analyzed performance of vector tracking loop in signal fading environment, and compared different forms of Kalman filter approaches. Lashley, Bevly and Hung [7–11] from Auburn University has done a great deal of work on implementation and performance analysis of VDLL and vector delay frequency lock loop (VDFLL). Zhao and Akos [12] developed vector tracking loop on Matlab software receiver platform, and experiments under actual scenarios were given.

The articles mentioned above do not give detailed derivations of vector loop tuning, and all discussions are for GPS. This paper presents the loop tuning derivations, and describes the implementation of BDS/GPS dual system adaptive EKF vector tracking loop on the open source Matlab GPS software receiver [13]. The tracking and positioning results under actual block scenarios are also given.

2 Mechanism of BDS/GPS Vector Tracking Loop

2.1 Fundamental Principles

In traditional scalar tracking architecture, each loop generates pseudorange/pseudorange rate observations for navigation module and the outputs of PLL/DLL discriminator are utilized to tune local carrier and code NCO independently. In contrast, vector tracking algorithm uses EKF to determine receiver's position, and tunes loop parameters (carrier frequency, code frequency, and code phase) of every channel using LOS position and velocity vector from receiver to satellites. Receiver's position and velocity will not mutate with the help of EKF, so the loop parameters would keep steady. Hence, continuous estimation of loop parameters could be kept even during brief signal outages, and the loop would be relocked as soon as signal recovered.

Fig. 1 Architecture of vector tracking loop

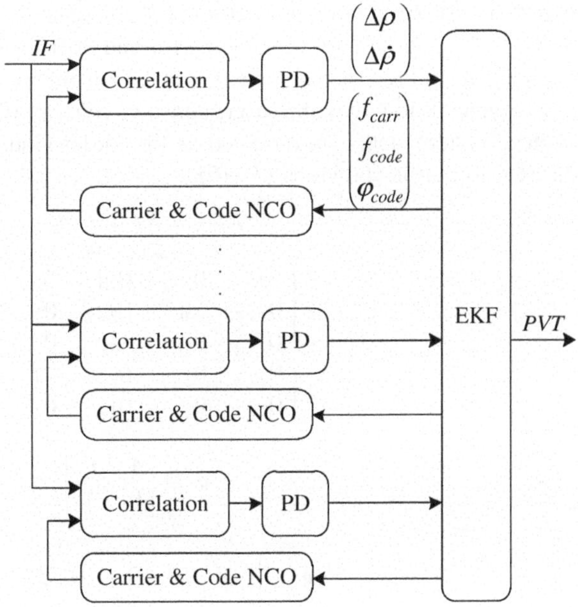

As shown in Fig. 1, intermediate frequency (IF) signal is correlated with local carrier and code, then code phase error $\Delta\tau$ and carrier frequency error Δf are generated by DLL and FLL discriminator. Observations vector z, including pseudorange error $\Delta\rho$ and pseudorange rate error $\Delta\dot{\rho}$, can be obtained as

$$z = \begin{bmatrix} \Delta\rho \\ \Delta\dot{\rho} \end{bmatrix} = \begin{bmatrix} \Delta\tau \cdot c/R_c \\ -\Delta f \cdot c/f_0 \end{bmatrix} \tag{1}$$

where c is light speed; R_c is chip frequency of PRN code; f_0 is GNSS signal nominal frequency. EKF module would make use of z from every channel to solve the navigation state, and determines carrier frequency f_{carr}, code frequency f_{code} and code phase φ_{code} for each.

2.2 Motion Model

EKF uses state transition equations to make priori state estimation. This paper chooses receiver state error as system state, shown as

$$\boldsymbol{X} = \left[\Delta x, \Delta\dot{x}, \Delta y, \Delta\dot{y}, \Delta z, \Delta\dot{z}, \Delta\delta t_u^B, \Delta\delta f_u^B, \Delta\delta t_u^G, \Delta\delta f_u^G\right]^T \tag{2}$$

where $\Delta x, \Delta y, \Delta z$ are position errors in X-, Y-, Z-axis; $\Delta\dot{x}, \Delta\dot{y}, \Delta\dot{z}$ are velocity errors in X-, Y-, Z-axis; $\delta t_u^B, \delta f_u^B$ are clock error and drift for BDS; $\delta t_u^G, \delta f_u^G$ are clock error and drift for GPS. All the states are in ECEF frame.

For general low dynamic user, constant velocity (CV) motion model is the best choice. System states are modeled as first-order Gauss-Markov processes, and the discrete format is shown as (3)–(6).

$$\boldsymbol{X}_k = F\boldsymbol{X}_{k-1} + \boldsymbol{\omega}_{k-1} \tag{3}$$

$$F = \begin{bmatrix} \kappa & 0_{2\times2} & 0_{2\times2} & 0_{2\times2} & 0_{2\times2} \\ 0_{2\times2} & \kappa & 0_{2\times2} & 0_{2\times2} & 0_{2\times2} \\ 0_{2\times2} & 0_{2\times2} & \kappa & 0_{2\times2} & 0_{2\times2} \\ 0_{2\times2} & 0_{2\times2} & 0_{2\times2} & \kappa & 0_{2\times2} \\ 0_{2\times2} & 0_{2\times2} & 0_{2\times2} & 0_{2\times2} & \kappa \end{bmatrix} \tag{4}$$

$$\kappa = \begin{bmatrix} 1 & T_s \\ 0 & 1 \end{bmatrix} \tag{5}$$

$$\boldsymbol{\omega} = \left[0, \varepsilon_{\dot{x}}, 0, \varepsilon_{\dot{y}}, 0, \varepsilon_{\dot{z}}, 0, \varepsilon_f^B, 0, \varepsilon_f^G\right]^T \tag{6}$$

where F is state transition matrix, shown as (4)–(5); T_s is the integration interval; $\boldsymbol{\omega}$ is system noise vector; $\varepsilon_{\dot{x}}$, $\varepsilon_{\dot{y}}$,$\varepsilon_{\dot{z}}$ are velocity noise; ε_f^B,ε_f^G are receiver clock drift noise for BDS and GPS, respectively.

2.3 Measurement Model

Measurement model shown as (7) is used to get EKF posterior state estimation.

$$\boldsymbol{Z} = \boldsymbol{HX} + \upsilon \tag{7}$$

$$\boldsymbol{Z} = \left[\Delta\rho_1^B, \ldots, \Delta\dot{\rho}_1^B, \ldots, \Delta\rho_1^G, \ldots, \Delta\dot{\rho}_1^G, \ldots\right]^T \tag{8}$$

$$\boldsymbol{H} = \begin{bmatrix} -a_{x,1}^B & 0 & -a_{y,1}^B & 0 & -a_{z,1}^B & 0 & 1 & 0 & 0 & 0 \\ \vdots & \vdots & \vdots & \vdots & \vdots & \vdots & \vdots & \vdots & \vdots & \vdots \\ 0 & -a_{x,1}^B & 0 & -a_{y,1}^B & 0 & -a_{z,1}^B & 0 & 1 & 0 & 0 \\ \vdots & \vdots & \vdots & \vdots & \vdots & \vdots & \vdots & \vdots & \vdots & \vdots \\ -a_{x,1}^G & 0 & -a_{y,1}^G & 0 & -a_{z,1}^G & 0 & 0 & 0 & 1 & 0 \\ \vdots & \vdots & \vdots & \vdots & \vdots & \vdots & \vdots & \vdots & \vdots & \vdots \\ 0 & -a_{x,1}^G & 0 & -a_{y,1}^G & 0 & -a_{z,1}^G & 0 & 0 & 0 & 1 \\ \vdots & \vdots & \vdots & \vdots & \vdots & \vdots & \vdots & \vdots & \vdots & \vdots \end{bmatrix} \tag{9}$$

where υ is process noise vector; $\mathbf{Z}$ is observations vector, shown as (8), including pseudorange error $\Delta\rho^B$,$\Delta\rho^G$ and pseudorange rate error $\Delta\dot{\rho}^B$,$\Delta\dot{\rho}^G$; H is measurement matrix, shown as (9), and $a^B_{x,i}$,$a^B_{y,i}$,$a^B_{z,i}$ are LOS unit vector from receiver to ith BDS satellite, $a^G_{x,j}$,$a^G_{y,j}$,$a^G_{z,j}$ from receiver to jth GPS satellite. All superscript B is for BDS and G for GPS.

3 Implementation of BDS/GPS Vector Tracking Loop

3.1 Adaptive EKF

Prediction stage of EKF is shown as (10)–(11).

$$\mathbf{X}_k^- = F\mathbf{X}_{k-1}^+ \tag{10}$$

$$P_k^- = FP_{k-1}^+F^T + Q_k \tag{11}$$

where P is system state error covariance matrix and Q is system noise covariance matrix. Superscript '−' denotes prior estimation and '+' denotes posterior estimation.

Correction stage of EKF is shown as (12)–(14).

$$K_k = P_k^-H^T\left(HP_k^-H^T + R_k\right)^{-1} \tag{12}$$

$$\mathbf{X}_k^+ = \mathbf{X}_k^- + K_k\left(\mathbf{Z} - H\mathbf{X}_k^-\right) \tag{13}$$

$$P_k^+ = (I - K_kH)P_k^- \tag{14}$$

where K is Kalman gain, and R is process noise covariance matrix.

CV motion model is employed, and system error states are modeled as first-order Gauss-Markov processes. Hence, the initialization of Q can be obtained from (15)–(18).

$$Q = \begin{bmatrix} \sigma_x^2\alpha & 0_{2\times2} & 0_{2\times2} & 0_{2\times2} & 0_{2\times2} \\ 0_{2\times2} & \sigma_y^2\alpha & 0_{2\times2} & 0_{2\times2} & 0_{2\times2} \\ 0_{2\times2} & 0_{2\times2} & \sigma_z^2\alpha & 0_{2\times2} & 0_{2\times2} \\ 0_{2\times2} & 0_{2\times2} & 0_{2\times2} & Q_{clk}^B & 0_{2\times2} \\ 0_{2\times2} & 0_{2\times2} & 0_{2\times2} & 0_{2\times2} & Q_{clk}^G \end{bmatrix} \tag{15}$$

$$\alpha = \begin{bmatrix} \frac{T_s^3}{3} & \frac{T_s^2}{2} \\ \frac{T_s^2}{2} & T_s \end{bmatrix} \tag{16}$$

$$Q_{\text{clk}}^{B}=\begin{bmatrix} S_t^B+S_f^B\frac{T_s^3}{3} & S_f^B\frac{T_s^2}{2} \\ S_f^B\frac{T_s^2}{2} & S_f^BT_s \end{bmatrix} \tag{17}$$

$$Q_{\text{clk}}^{G}=\begin{bmatrix} S_t^G+S_f^G\frac{T_s^3}{3} & S_f^G\frac{T_s^2}{2} \\ S_f^G\frac{T_s^2}{2} & S_f^GT_s \end{bmatrix} \tag{18}$$

where σ_x^2,σ_y^2,σ_z^2 are variance of $\varepsilon_{\dot{x}}$,$\varepsilon_{\dot{y}}$,$\varepsilon_{\dot{z}}$, respectively; S_t^B,S_f^B are spectrum density of receiver clock bias and drift for BDS, and S_t^G,S_f^G for GPS.

The initial value of process noise covariance matrix R is often obtained as a rule of thumb. However, improper R will reduce EKF performance, or make it could not converge. Adaptive estimation of R based on innovation sequences can solve the problem. The relationship between process noise vector and innovation sequence can be obtained as (19) according to (7). Then adaptive R could be obtained as (20).

$$\begin{aligned} \boldsymbol{\upsilon}_k &= \boldsymbol{Z}_k - H\boldsymbol{X}_k = \boldsymbol{Z}_k - H\boldsymbol{X}_k^- + H\boldsymbol{X}_k^- - H\boldsymbol{X}_k \\ &= \boldsymbol{v}_k - H\left(\boldsymbol{X}_k - \boldsymbol{X}_k^-\right) = \boldsymbol{v}_k - H\boldsymbol{e}_k^- \end{aligned} \tag{19}$$

$$R_k = E\left[\boldsymbol{\upsilon}_k\boldsymbol{\upsilon}_k^T\right] = E\left[\boldsymbol{v}_k\boldsymbol{v}_k^T\right] - HP_k^-H^T \approx \frac{1}{N}\sum_{j=k-N+1}^{k}\boldsymbol{v}_j\boldsymbol{v}_j^T - HP_k^-H^T \tag{20}$$

where $\boldsymbol{v}_k$ is innovation sequences; $\boldsymbol{e}_k^-$ is prior state error; N is the adaptive estimation update period.

3.2 *Loop Tuning*

3.2.1 Code Loop Tuning

Pseudorange measurement equation is shown as

$$\rho = r + \left(\delta t_u - \delta t^{(s)}\right) + I + T \tag{21}$$

where r is actual range between receiver and satellite s; δt_u,$\delta t^{(s)}$ are clock bias of receiver and satellite s in meters; I, T are ionosphere and troposphere delay in meters.

Omitting I and T, the signal travel time τ from satellite s to receiver can be obtained as

$$\tau = \frac{\rho}{c} = \frac{\|p_s - p_u\| + \left(\delta t_u - \delta t^{(s)}\right)}{c} \tag{22}$$

where c is light speed; p_s, p_u are position vector of satellite s and receiver; $\|\cdot\|$ denotes Euclidean distance.

The travel time variation $\Delta\tau$ between kth and $(k-1)$th loop period is obtained as

$$\begin{aligned}\Delta\tau &= \frac{\left\|p_{s,k}-p_{u,k}\right\| - \left\|p_{s,k-1}-p_{u,k-1}\right\| + \left(\Delta\delta t_u - \Delta\delta t^{(s)}\right)}{c}\\ &= \frac{\boldsymbol{a}\cdot(\Delta p_s - \Delta p_u) + \left(\Delta\delta t_u - \Delta\delta t^{(s)}\right)}{c}\\ &= \frac{1}{c}\begin{bmatrix} a_x & a_y & a_z & -1\end{bmatrix}\begin{bmatrix}\Delta x^{(s)} - \Delta x_u\\ \Delta y^{(s)} - \Delta y_u\\ \Delta z^{(s)} - \Delta z_u\\ \Delta\delta t^{(s)} - \Delta\delta t_u\end{bmatrix}\end{aligned} \tag{23}$$

where Δ denotes variation between kth and $(k-1)$th loop period.

Then the variation of initial code phase of kth loop period compared with $(k-1)$th loop period is shown as

$$\Delta\varphi = f_c \cdot \Delta\tau \tag{24}$$

where f_c is chip rate of PRN code. As a result, variation $\Delta\varphi$ can be used to tune code phase of code loop.

3.2.2 Carrier Loop Tuning

Calculate the derivative of pseudorange measurement equation (21), so as to get pseudorange rate measurement equation (25).

$$\dot{\rho} = \begin{bmatrix} a_x & a_y & a_z & -1\end{bmatrix}\begin{bmatrix}\dot{x}^{(s)} - \dot{x}_u\\ \dot{y}^{(s)} - \dot{y}_u\\ \dot{z}^{(s)} - \dot{z}_u\\ \delta f^{(s)} - \delta f_u\end{bmatrix} \tag{25}$$

Doppler frequency f_d then be estimated as

$$f_d = -\frac{\dot{\rho}}{\lambda} = -\frac{1}{\lambda}\begin{bmatrix} a_x & a_y & a_z & -1\end{bmatrix}\begin{bmatrix}\dot{x}^{(s)} - \dot{x}_u\\ \dot{y}^{(s)} - \dot{y}_u\\ \dot{z}^{(s)} - \dot{z}_u\\ \delta f^{(s)} - \delta f_u\end{bmatrix} \tag{26}$$

Carrier frequency f_{carr} and code frequency f_{code} can be obtained according to f_d, shown as (27)–(28).

$$f_{\mathrm{carr}} = f_{\mathrm{IF}} + f_d \tag{27}$$

$$f_{\mathrm{code}} = f_c + \frac{f_c}{f_0} f_d \tag{28}$$

where f_{IF} is IF; f_0, f_c are carrier and code nominal frequency, respectively.

4 Experiment Results and Analysis

4.1 Facilities

An open source front-end HackRF is employed as digital IF signal sampler. The center frequency of grabbed signal is set as 1567.5 MHz, and the bandwidth of band pass filter (BPF) is 20 MHz, sampling rate is 20 MHz, output format is I/Q 8 bits unsigned integer. Since the clock source of HackRF is a passive crystal oscillator with poor frequency stability, it is hardly to operate GNSS signal. As a result, a 10 MHz rubidium clock is utilized as the external clock source. The experiment facilities are shown as Fig. 2, where two cascaded lead–acid batteries are the power source of rubidium clock, and a multiple GNSS frequency band dish antenna is used.

Fig. 2 HackRF IF signal sampling system

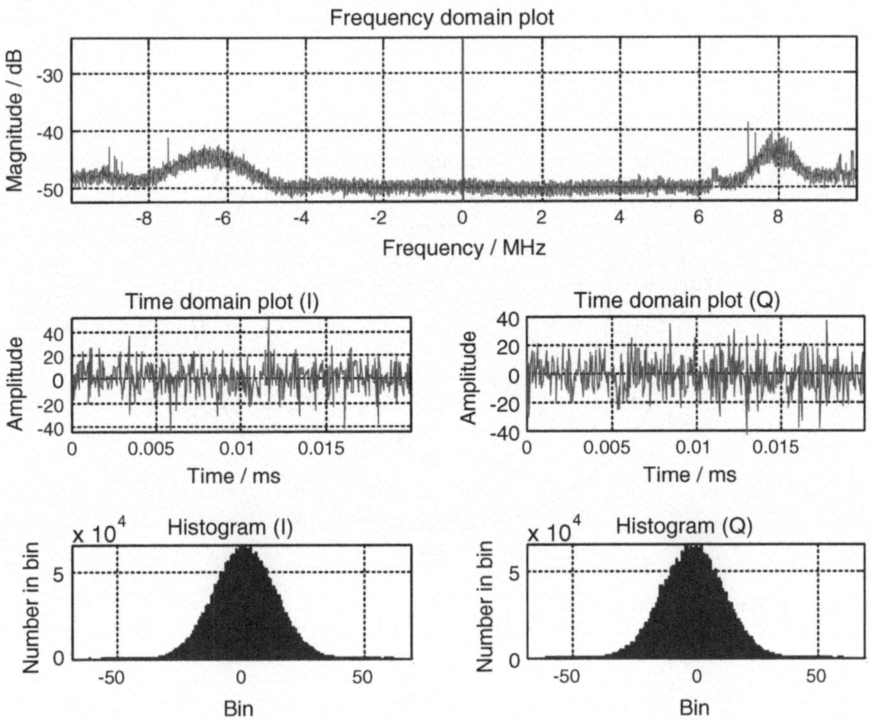

Fig. 3 IF signal analysis in time and frequency domain

The time and frequency domain analysis of the IF signal grabbed by HackRF is shown in Fig. 3. The left spectrum peak is BDS B1 signal, and the right one is GPS L1 signal. Then the signal can be used for BDS/GPS dual system experiments. The histogram shows that the RF gain and dynamic range is proper.

4.2 *Low Dynamic Experiment*

The scenario is chosen as roof of laboratory, and move from west to east along the south wall. At first, let the antenna see the open sky and move for a while. Then put it lower than and closer to the south wall and keep moving for 20 s, so that the BDS GEO satellite signals are blocked. The BDS satellites sky plot without block is shown as Fig. 4.

The update period of adaptive estimation is 100 ms, and Fig. 5 shows the variation of the first diagonal element of matrix *R*, *P*, *K*. It can be seen that at 100 ms *R* and *K* have a mutation, and *P* reduces to near the zero sharply. After that, all of them keep steady. Figure 6 shows the corresponding position and velocity results and root mean square (RMS) errors are labeled. The velocity results are much better since 100 ms.

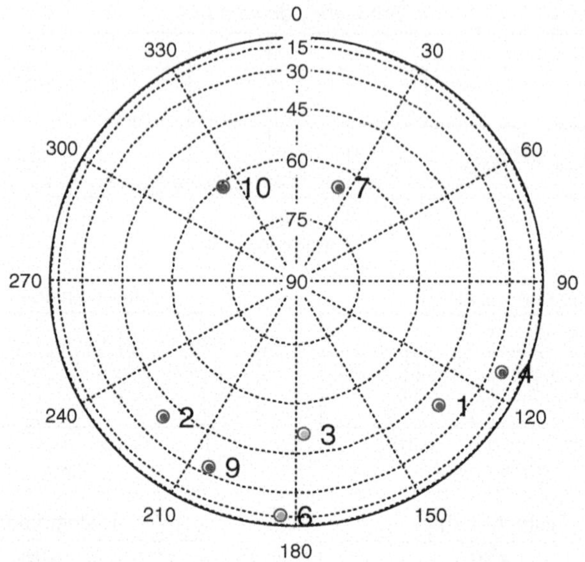

Fig. 4 Sky plot of BeiDou satellites

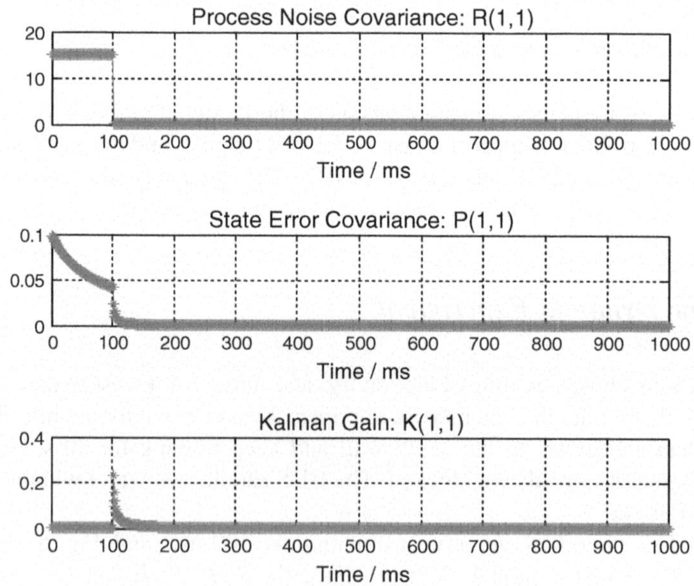

Fig. 5 Filter parameters of adaptive extended kalman filter

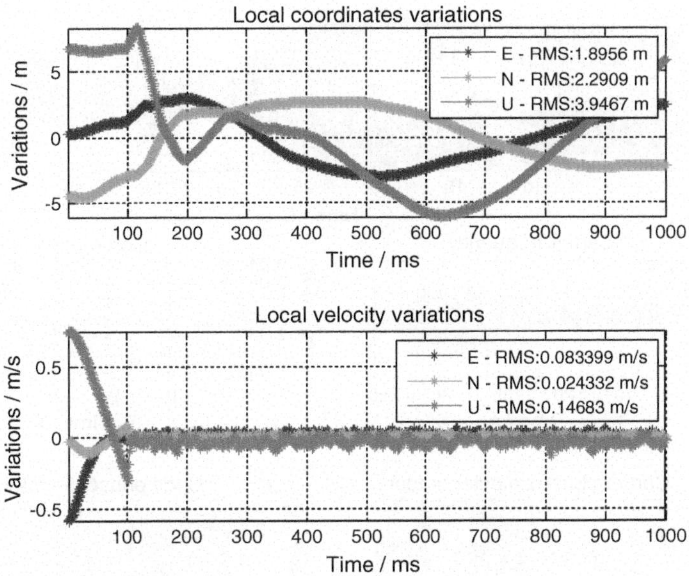

Fig. 6 Position and velocity results utilizing adaptive extended kalman filter

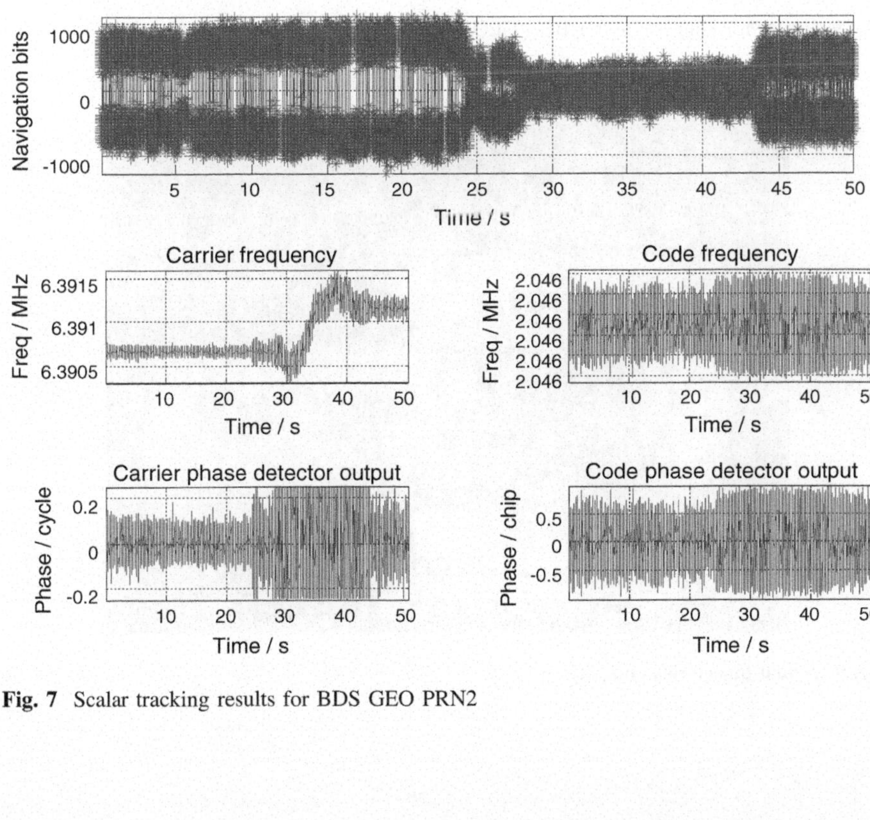

Fig. 7 Scalar tracking results for BDS GEO PRN2

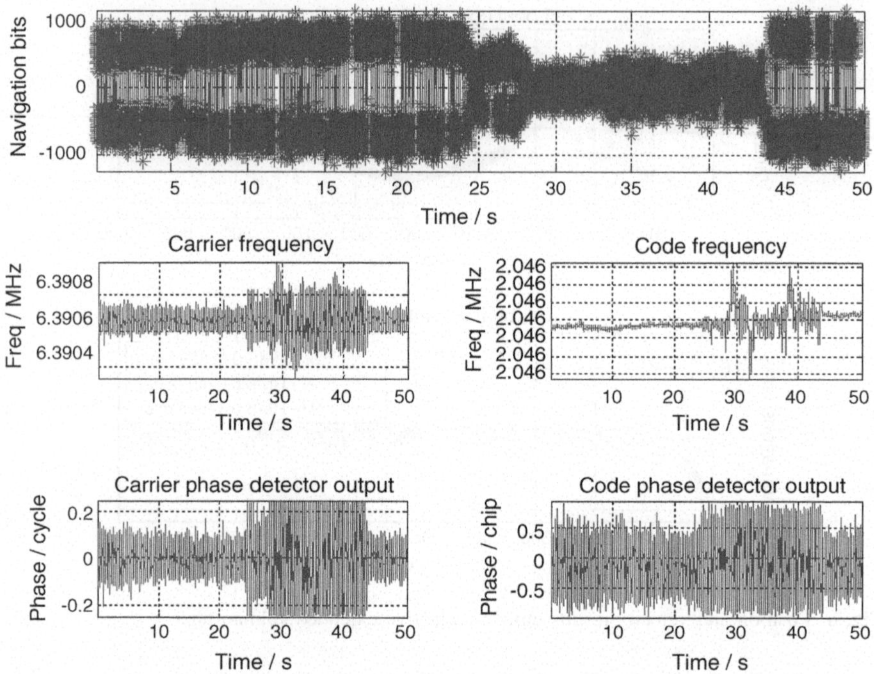

Fig. 8 Vector tracking results for BDS GEO PRN2

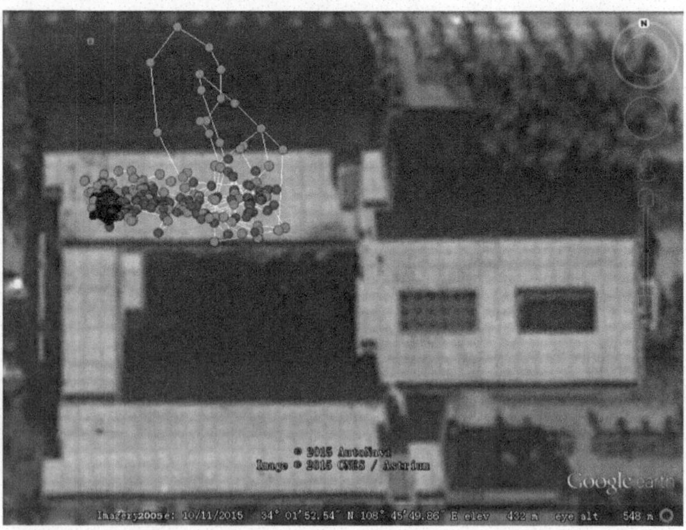

Fig. 9 Actual trajectories comparison

Figures 7 and 8 show the scalar and vector tracking results of BDS GEO PRN2 satellite, respectively, including navigation bits, carrier frequency, code frequency, carrier, and code phase errors. Both tracking loops lose lock after signal outage, and scalar tracking loop could not relock when signal recovers, while vector tracking loop does.

Figure 9 shows the actual trajectories of both scalar and vector loop on Google Earth. The red points are position results at initialization stage. Ephemeris decoding and initial navigation resolving are carried out at this stage. Yellow and green points are scalar and vector tracking trajectories. Both of them did not utilize RAIM algorithm to remove blocked signal. Scalar tracking has bad position results under signal blocked scenario and vector tracking has better robustness.

5 Conclusions

This article gives the exact implementation of adaptive EKF and loop tuning method for BDS/GPS vector tracking loop. The proposed vector tracking loop is implemented on Matlab software receiver platform. With the digital IF signals sampled by HackRF, the performance of the vector tracking loop under actual blocked scenario is validated. Adaptive parameters estimation makes an obviously better state error covariance and velocity result. The proposed algorithm helps receiver to keep continuous positioning during signal outage, and the vector tracking loop would be relocked as soon as the signal recovered. Hence, the proposed vector tracking loop has better robustness than traditional scalar tracking loop.

References

1. Parkinson BW, Spilker JJ, Axelrad P et al (1996) Global positioning system: theory and applications. American Institute of Aeronautics and Astronautics, Washington DC
2. Pany T, Kaniuth R, Eissfeller B (2005) Deep integration of navigation solution and signal processing. In: ION GNSS, pp 1095–1102, 13–16 Sept 2005
3. Pany T, Kaniuth R, Eissfeller B (2005) Testing a vector delay/frequency lock loop implementation with the ipex software receiver. In: Proceedings of GPS/GNSS Symposium, pp 1–9
4. Pany T, Eissfeller B (2006) Use of a vector delay lock loop receiver for GNSS signal power analysis in bad signal conditions. In: IEEE/ION PLANS, pp 893–903, 25–27 May 2006
5. Won J-H, Eissfeller B (2010) Effectiveness analysis of vector-tracking-loop in signal fading environment. In: Satellite navigation technologies and european workshop on GNSS signals and signal processing, pp 1–6
6. Won J-H, Dotterbock D, Eissfeller B (2010) Performance comparison of different forms of kalman filter approaches for a vector-based GNSS signal tracking loop. J Inst Navig 57 (3):185–199

7. Lashley M, Bevly DM (2007) Analysis of discriminator based vector tracking algorithms. In: ION NTM pp 570–576, 22–24 Jan 2007
8. Lashley M, Bevly DM (2007) Comparison of traditional tracking loops and vector based tracking loops for weak GPS signals. In: ION GNSS, pp 789–795, 25–28 Sept 2007
9. Lashley M, Bevly DM (2009) Vector delay/frequency lock loop implementation and analysis. In: ION ITM, pp 1073–1086, 26–28 Jan 2009
10. Lashley M, Bevly DM, Hung JY (2010) A valid comparison of vector and scalar tracking loops. In: IEEE/ION PLANS, pp 464–474
11. Lashley M, Bevly DM, Hung JY (2009) Performance analysis of vector tracking algorithms for weak GPS signals in high dynamics. IEEE J Sel Top Signal Process 3(4):661–673
12. Zhao S, Akos DM (2011) An open source GPS/GNSS vector tracking loop—implementation, filter tuning, and results. In: ION ITM, pp 1293–1305, 24–26 Jan 2011
13. Borre K, Akos DM, Bertelsen N et al (2007) A software-defined GPS and Galileo receiver: a single-frequency approach. Birkhäuser Boston, New York

An Unambiguous Multipath Mitigation Method Based on Double-Delta Correlator for BOC Modulation Signal

Bo Qu, Longlong Li, Lang Bian, Xiaoliang Wang and Yansong Meng

Abstract With the development of global navigation satellite system (GNSS), binary offset carrier (BOC) modulation signal has been widely used for GPS and Galileo. However, because of the effect of the subcarrier, the autocorrelation function (ACF) of BOC signal has a main peak and some side peaks. The traditional delay-locked loop (DLL) may be locked on the side peaks of ACF, which causes an extra tracking error. Dual-estimate loop method is an effective unambiguous tracking method, which makes use of two DLLs to track the code phase and the subcarrier phase in BOC signal. However, the multipath effect is not taken into account for dual-estimate loop method. This paper proposes an unambiguous multipath mitigation method, which employs the dual-estimate loop method to solve the tracking ambiguity problem for BOC signal and makes use of double-delta correlator in the subcarrier tracking loop to mitigate the multipath effect. In addition, a bias code delay estimate value is used in the subcarrier tracking loop to enhance the performance on the multipath mitigation. Simulation results demonstrate that the proposed method can provide the same level of tracking performance as the traditional DLL tracking technique in the case of high carrier-to-noise ratio (CNR) and have the excellent performance on the multipath mitigation.

Keywords GNSS · Dual-estimate loop · Double-delta correlator · Multipath · Delay-locked loop

1 Introduction

BOC modulation signal has been widely used in GNSS systems, such as the modernized GPS [1] and Galileo [2]. The BOC modulation signal has the split power density spectrum and more high frequency component, so it could provide

B. Qu (✉) · L. Li · L. Bian · X. Wang · Y. Meng
China Academy of Space Technology, Xi'an, China
e-mail: qb840528@163.com

J. Sun et al. (eds.), *China Satellite Navigation Conference (CSNC) 2016 Proceedings: Volume I*, Lecture Notes in Electrical Engineering 388,
DOI 10.1007/978-981-10-0934-1_49

the better tracking accuracy than binary phase shift keying (BPSK) modulation which has been used in GPS L1C/A signal.

A traditional receiver employs a delay-locked loop (DLL) to track the BPSK signal. Because the ACF of BPSK modulation signal has only one peak, the receiver could track the C/A code phase accurately. However, BOC signal makes use of the sine-subcarrier or cosine-subcarrier to modulate the pseudorandom noise (PRN) code, and the ACF of BOC signal has one main peak and several side peaks. The traditional receiver, which is used to track the BOC signal, may lock on the side peaks to cause a large tracking error.

Fortunately, some unambiguous tracking methods have been proposed to solve the tracking ambiguity problem mentioned above. The bump-jumping (BJ) method utilizes five correlators to avoid DLL locking on side peaks and has a good tracking accuracy [3]. However, the performance of the BP method gets worse with carrier-to-noise ratio (CNR) becomes low. The BPSK-like method treats the BOC signal as two BPSK signal so as to track BOC signal unambiguously [4]. Nevertheless, this method does not utilize the high frequency component in BOC modulation effectively, which decrease the tracking accuracy. The ASPeCT method combines the ACF of BOC signal with the cross-correlation of BOC signal and BPSK signal to cancel the side peak of ACF of BOC signal [5]. Unfortunately, the ASPeCT method is only suitable for BOC(n, n) signal. The dual-estimate loop method employs DLL and subcarrier-locked loop (SLL) to track the PRN code and subcarrier separately [6, 7]. This method has a good unambiguous tracking accuracy.

Although the dual-estimate loop method has a good tracking performance for BOC modulation signal, the multipath effect is not taken into account. The multipath signals may decrease the performance of the dual-estimate loop method.

This paper proposes the dual-estimate loop method with early code phase based on double delta. This method could avoid the tracking ambiguity problem of BOC signal and mitigate the effect of multipath signals. The proposed method utilizes the DLL and SLL to track BOC signal unambiguously, in which the SLL employs the double-delta correlator to mitigate the effect of multipath signals. In addition, the local PRN code phrase used by the SLL can be earlier than the local prompt PRN code phase, which can improve the multipath mitigation performance.

The paper is organized as follows: In the Sect. 1.2, the dual-estimate loop method is briefly introduced. Section 1.3 introduces the proposed dual-estimate loop method with early code phase based on double delta. Section 1.4 shows the multipath mitigation performance and the tracking accuracy of the proposed method. Finally conclusions are drawn in Sect. 1.5.

2 Dual-Estimate Loop Method

The dual-estimate loop method utilizes the DLL and SLL to track BOC modulation signal. The DLL tracks the PRN code and obtain low accurate unambiguous time delay estimate, while the SLL tracks the subcarrier and obtain high accurate ambiguous time delay estimate [6].

The illustration of dual-estimate loop is shown in Fig. 1.

First, the received BOC signal is downconverted and filtered to get the baseband signal. The DLL and SLL are used to track the PRN code and the subcarrier of BOC signal. The time delay estimate $\hat{\tau}_c$ of DLL is employed to modify its number control oscillator (NCO) to synchronize the PRN code phase in the received BOC signal. The code generator produces the early code $c_E(t)$, the prompt code $c_P(t)$, and the late code $c_L(t)$. The time delay estimate $\hat{\tau}_{sc}$ of SLL is employed to modify its NCO to synchronize the subcarrier phase in the received BOC signal. The subcarrier generator produces the early subcarrier $sc_E(t)$, the prompt subcarrier $sc_P(t)$, and the late subcarrier $sc_L(t)$. The prompt code $c_P(t)$ is multiplied by the early subcarrier $sc_E(t)$, the prompt subcarrier $sc_P(t)$, and the late subcarrier $sc_L(t)$ to obtain the early local reference signal $c_P(t)sc_E(t)$, the prompt local reference signal $c_P(t)sc_P(t)$, and the late local reference signal $c_P(t)sc_L(t)$ for SLL. These reference signals are correlated and integrated with the received signal to obtain the discriminator output of SLL. Similarly, the prompt subcarrier $sc_P(t)$ is multiplied by

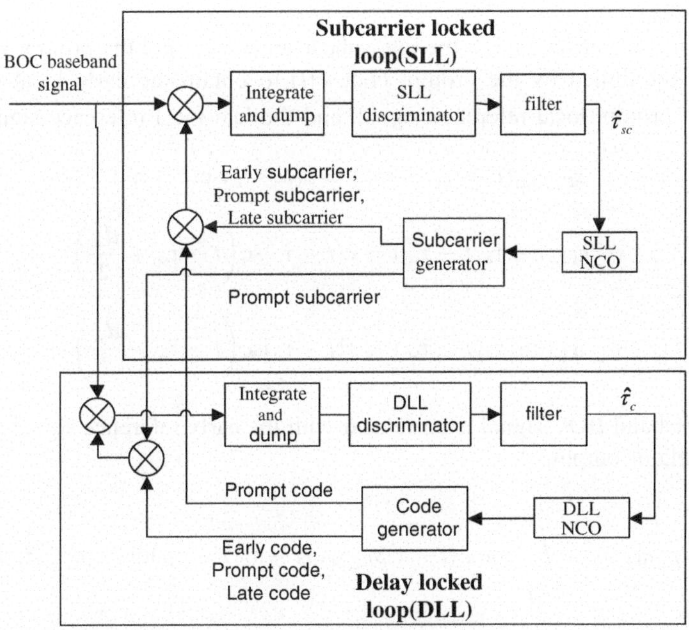

Fig. 1 Dual-estimate loop illustration

the early code $c_E(t)$, the prompt code $c_P(t)$, and the late code $c_L(t)$ to obtain the early local reference signal $c_E(t)\mathrm{sc}_P(t)$, the prompt local reference signal $c_P(t)\mathrm{sc}_P(t)$, and the late local reference signal $c_L(t)\mathrm{sc}_P(t)$ for DLL. These reference signals are correlated and integrated with the received signal to obtain the discriminator output of DLL. The discriminators, such as dot product, can be employed by DLL and SLL.

When the dot product discriminator is utilized in DLL and SLL, the effect of data bit can be neglected. The baseband BOC signal can be expressed as follows:

$$s_{\mathrm{BOC}}(t-\tau) = c(t-\tau)\mathrm{sc}(t-\tau) \tag{1}$$

where τ is the time delay of the received signal.

The early subcarrier, the late subcarrier, and the prompt subcarrier can be represented by

$$\mathrm{sc}_E(t) = \mathrm{sc}\left(t - \hat{\tau}_{\mathrm{sc}} + \frac{d_{\mathrm{sc}}}{2}\right) \tag{2}$$

$$\mathrm{sc}_L(t) = \mathrm{sc}\left(t - \hat{\tau}_{\mathrm{sc}} - \frac{d_{\mathrm{sc}}}{2}\right) \tag{3}$$

$$\mathrm{sc}_P(t) = \mathrm{sc}(t - \hat{\tau}_{\mathrm{sc}}) \tag{4}$$

where d_{sc} represents the correlator spacing of SLL, $\hat{\tau}_{\mathrm{sc}}$ represents the subcarrier time delay estimate.

The early subcarrier $\mathrm{sc}_E(t)$, the late subcarrier $\mathrm{sc}_L(t)$, and the prompt subcarrier $\mathrm{sc}_p(t)$ are multiplied by the prompt code $c(t)$ to obtain the early local reference signal, the prompt local reference signal, and the late local reference signal.

$$s_{P\_\mathrm{SLL}}(t-\hat{\tau}_c, t-\hat{\tau}_{\mathrm{sc}}) = c(t-\hat{\tau}_c)\mathrm{sc}(t-\hat{\tau}_{\mathrm{sc}}) \tag{5}$$

$$s_{E\_\mathrm{SLL}}(t-\hat{\tau}_c, t-\hat{\tau}_{\mathrm{sc}}) = c(t-\hat{\tau}_c)\mathrm{sc}\left(t-\hat{\tau}_{\mathrm{sc}} + \frac{d_{\mathrm{sc}}}{2}\right) \tag{6}$$

$$s_{L\_\mathrm{SLL}}(t-\hat{\tau}_c, t-\hat{\tau}_{\mathrm{sc}}) = c(t-\hat{\tau}_c)\mathrm{sc}\left(t-\hat{\tau}_{\mathrm{sc}} - \frac{d_{\mathrm{sc}}}{2}\right) \tag{7}$$

The baseband BOC signal is correlated with the early reference signal to get the early correlator output.

$$R_{E\_\mathrm{SLL}} = \int_{(k-1)T}^{kT} s_{\mathrm{BOC}}(t-\tau)s_{E\_\mathrm{SLL}}(t-\hat{\tau}_c, t-\hat{\tau}_{\mathrm{sc}})\mathrm{d}t = \chi\left(\hat{\tau}_c^*, \hat{\tau}_{\mathrm{sc}}^*\right) \tag{8}$$

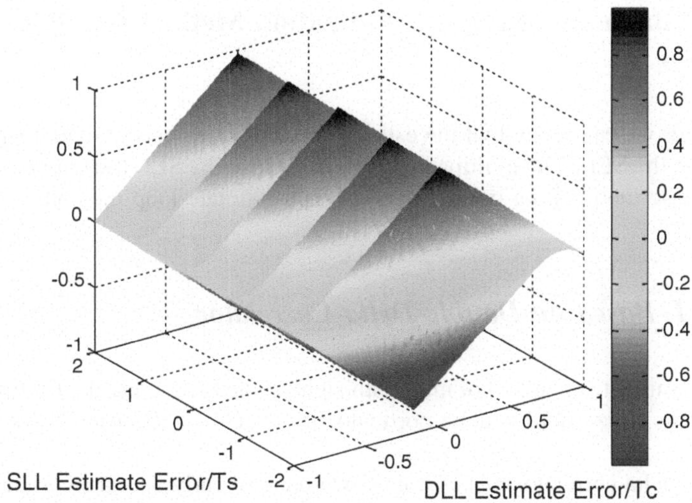

Fig. 2 Two-dimensional function

where the track error of DLL is denoted as $\hat{\tau}_c^* = \hat{\tau}_c - \tau$, $\hat{\tau}_{sc}^* = \hat{\tau}_{sc} - \frac{d_{sc}}{2} - \tau$ represents the time difference between the early local reference signal and the received signal. Similarly, we can obtain the late correlator output between the late local reference signal and the received signal.

As shown in Fig. 2, the cross-correlation function (CCF) $\chi(\hat{\tau}_c^*, \hat{\tau}_{sc}^*)$ is a two-dimensional function. The CCF $\chi(\hat{\tau}_c^*, \hat{\tau}_{sc}^*)$ has one peak in the DLL dimension so that the DLL can track the PRN code unambiguously. The CCF $\chi(\hat{\tau}_c^*, \hat{\tau}_{sc}^*)$ have multiple peaks in the SLL dimension, so SLL tracks the subcarrier ambiguously.

After obtaining the early correlator output, the prompt correlator output, and the late correlator output, the dot product discriminator can be expressed as

$$D_{\text{SLL}} = (R_{E\_\text{SLL}} - R_{L\_\text{SLL}})R_{P\_\text{SLL}} \tag{9}$$

When the discriminator output is close to zero, the tracking loop keeps tracking the received signal steadily. The prompt subcarrier generated by the subcarrier generator has the same phase as the subcarrier of the received BOC signal, and we can get the time delay estimate $\hat{\tau}_{sc}$. Similarly, we can get the PRN code time delay estimate $\hat{\tau}_c$ from DLL.

The unambiguous high accurate time delay estimate $\hat{\tau}$ can be obtained by combining the subcarrier time delay estimate $\hat{\tau}_{sc}$ with the PRN code time delay estimate $\hat{\tau}_c$:

$$\hat{\tau} = \hat{\tau}_{sc} + N * T_s \tag{10}$$

where $\hat{\tau}$ is the combined time delay estimate, T_s is half period of the subcarrier, $N = \text{round}\left(\frac{\hat{\tau}_c - \hat{\tau}_{sc}}{T_s}\right)$, round$(\cdot)$ means get the nearest integer.

3 Unambiguous Multipath Mitigation Method for BOC Signal

In Sect. 1.2, we can know that the estimate accuracy of dual-estimate loop method depends on the SLL. The multipath signal may cause the SLL tracking error, which may decrease the tracking performance of dual-estimate loop method.

3.1 SLL Based on Double-Delta Correlator

In order to mitigate the effect of multipath signals, the SLL in the dual-estimate loop method may utilize double-delta correlator to track the subcarrier in the received signal.

Double-delta correlator makes use of two pairs of early and late correlators with different correlator spacing to correlate with the received signal. As shown in Fig. 3.

E_1 and L_1 are one pair of early and late correlators output with correlator spacing d_1, while E_2 and L_2 are the other pair of early and late correlators output with correlator spacing d_2. The correlator spacing of E_2 and L_2 is two times as much as that of E_1 and L_1.

The following discriminator can be used in SLL with double-delta correlator.

$$D = (E_1 - L_1) - \frac{1}{2}(E_2 - L_2) \tag{11}$$

Then the discriminator can be expressed as follow:

$$D_{\mathrm{SLL}} = (R_{E_1\_\mathrm{SLL}} - R_{L_1\_\mathrm{SLL}})R_{P\_\mathrm{SLL}} - \frac{(R_{E_2\_\mathrm{SLL}} - R_{L_2\_\mathrm{SLL}})}{2}R_{P\_\mathrm{SLL}} \tag{12}$$

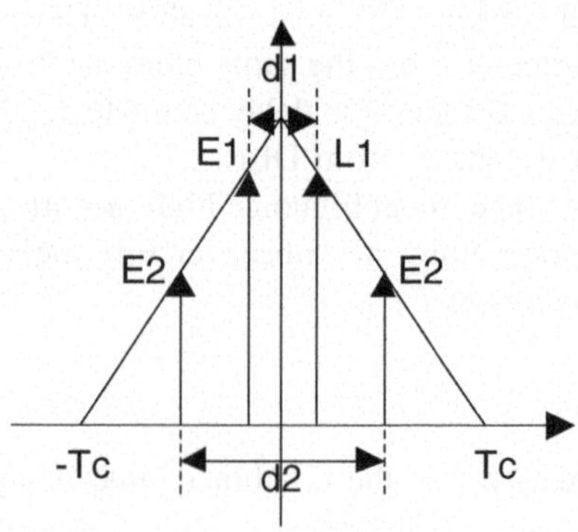

Fig. 3 Double-delta correlator illustration

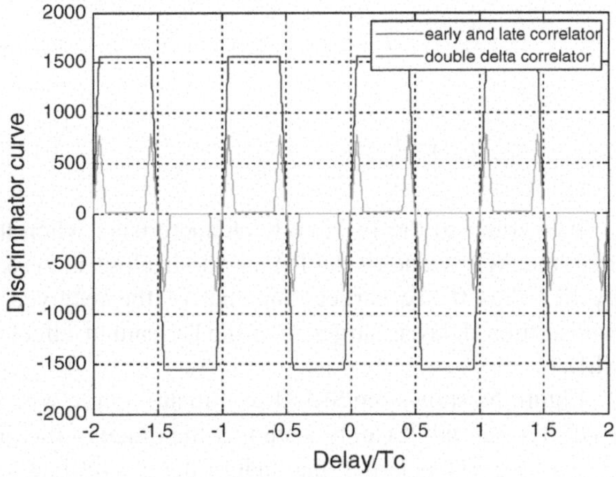

Fig. 4 Discriminator curve for early and late correlator and double-delta correlator (color figure online)

where $R_{E_1\_SLL}$ and $R_{L_1\_SLL}$ represent the early and late correlator outputs whose correlator spacing is d_1. $R_{E_2\_SLL}$ and $R_{L_2\_SLL}$ represent the early and late correlator outputs whose correlator spacing is d_2.

When the correlator spacing d_1 is 0.1Tc (The time of one PRN code chip is denoted as Tc) and d_2 is 0.2Tc, the subcarrier discriminator curve for BOC(1,1) signal is shown in Fig. 4. The range of double-delta correlator output is smaller than that of early and late correaltor. When the signal delay is in the range[−2Tc, 2Tc], the discriminator curve of early and late correaltor has only some zero-crossing point, and its discriminator output is not equal to zero except these zero-crossing points. If the received signal contains multipath signals, the discriminator curve may have a zero-crossing bias, which causes the tracking error for SLL. However, the discriminator curve of double-delta correlator is equal to zero except the range [−0.1Tc, 0.1Tc] around zero-crossing point, which can mitigate the multipath effect on the SLL.

3.1.1 The Early Code Phase for SLL

SLL based on double-delta correlator can mitigate the multipath effect on the dual-estimate loop method. However, if the multipath delay is close to half period of the subcarrier in BOC signal, the tracking performance of SLL may be affected.

In order to further mitigate the multipath effect, SLL can make use of the PRN code with early phase τ_{bias} after that the receiver tracks the received signal steadily. In this condition, the early reference signal, the late reference signal, and the prompt reference signal can be represented as follows:

$$s_{P\_SLL}(t-\hat{\tau}_c, t-\hat{\tau}_{sc}) = c(t-\hat{\tau}_c+\tau_{bias})\mathrm{sc}(t-\hat{\tau}_{sc}) \quad (13)$$

$$s_{E\_SLL}(t-\hat{\tau}_c, t-\hat{\tau}_{sc}) = c(t-\hat{\tau}_c+\tau_{bias})\mathrm{sc}\left(t-\hat{\tau}_{sc}+\frac{d_{sc}}{2}\right) \quad (14)$$

$$s_{L\_SLL}(t-\hat{\tau}_c, t-\hat{\tau}_{sc}) = c^{-}(t-\hat{\tau}_c+\tau_{bias})\mathrm{sc}\left(t-\hat{\tau}_{sc}-\frac{d_{sc}}{2}\right) \quad (15)$$

According to the PRN code characteristic, when the signal delay is larger than 1Tc, the ACF value of the PRN code is close to zero. If the PRN code phase used by SLL has 0.5Tc earlier than that of the received signal, the multipath signal, whose time delay is larger than 0.5Tc, cannot affect the tracking performance for SLL.

Figure 5a shows the SLL discriminator curve with different early code phase for BOC(1,1) signal. It can be seen that the range of the discriminator output with early code phase 0Tc is larger than that of the discriminator output with early code phase 0.5Tc. When CNR is low, the SLL with early code phase 0Tc tracks the subcarrier more stable than the SLL with early code phase 0.5Tc. Therefore, the method increasing the early code phase to improve the performance of multipath mitigation is only suitable for the high CNR condition.

The early code phase may cause the zero-crossing bias to the discriminator curve of SLL. Figure 5b shows the larger figure about the discriminator curve of SLL with early code phase 0.5Tc. We can see that the zero-crossing bias is about 4.2×10^{-3}Tc, which may result in the tracking error for SLL. Therefore, the zero-crossing bias needs to be corrected before receiving the BOC signal.

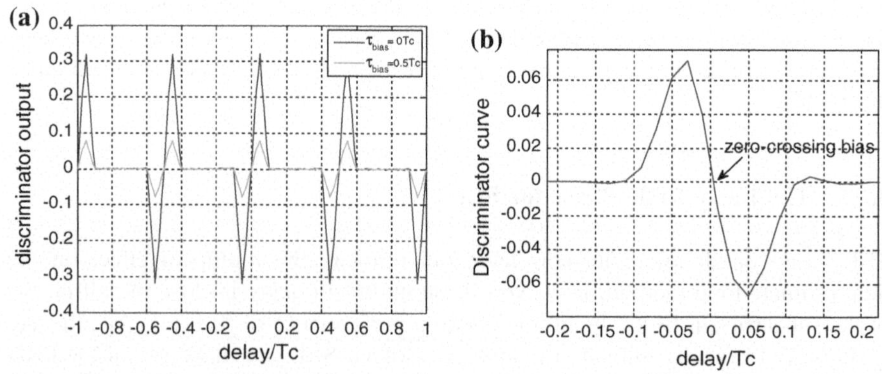

Fig. 5 **a** SLL discriminator curve for different early code phase. **b** Larger figure of the discriminator with early code phase 0.5Tc (color figure online)

3.2 *Two Tracking Phase*

The tracking process of the receiver for BOC signal can be divided into two stages: rough tracking stage and precision tracking stage.

After acquiring BOC signal, the receiver enters the rough tracking stage. In this stage, the traditional dual-estimate loop is used to track BOC signal so that DLL and SLL enters the stable tracking status. In addition, the receiver needs estimate CNR.

If DLL and SLL track the BOC signal steadily, the receiver starts the precise tracking stage. When the CNR estimate is high, the double-delta correlator is used in SLL instead of the early and late correlator, and the PRN code with early phase is also utilized in SLL to mitigate the multipath effect.

4 Simulation Results

The multipath performance and tracking accuracy of the proposed tracking technique are analyzed in this section.

Figure 6 shows the multipath mitigation performance of the proposed method, dual-estimate loop technique (DET) and the traditional DLL with early minus late power (EMLP) discriminator. The simulation employs one direct signal and one multipath signal. The multipath to direct ratio (MDR) of amplitude is −6 dB, the time delay of multipath signal varies from 0 to 1.2Tc. The front-end bandwidth is 20 MHz, the bandwidth of the tracking loop is 1 Hz, and the integration time is 1 ms. The correlator spacing of the traditional DLL with EMLP discriminator is 0.2Tc. The correlator spacing of DLL and SLL in traditional dual-estimate loop are 0.5Tc and 0.2Tc, respectively. The correlator spacing d_1 and d_2 of double-delta

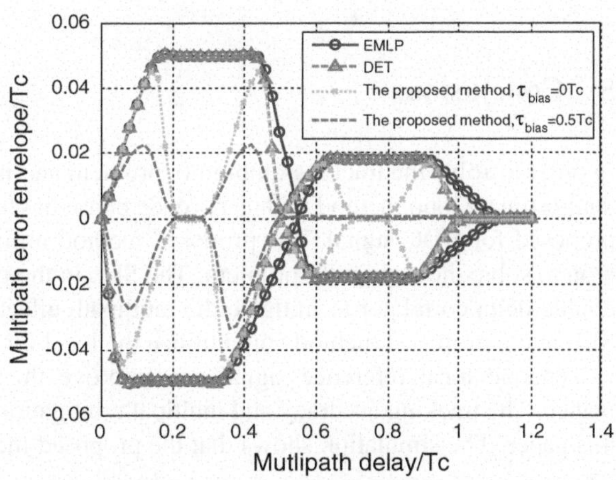

Fig. 6 Multipath error envelope versus multipath delay (color figure online)

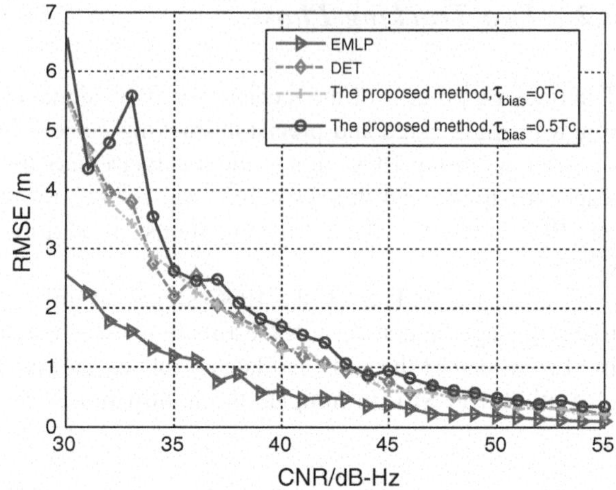

Fig. 7 Code tracking RMSE versus CNR (color figure online)

correlator in the proposed method are 0.2Tc and 0.4Tc, respectively. The early code phase used in the proposed method is 0Tc and 0.5Tc, respectively.

It can be seen that DET has the same the multipath performance as the traditional DLL with EMLP. The multipath performance of the proposed method is better than that of DET and DLL with EMLP. When the early code phase increases from 0Tc to 0.5Tc, the multipath envelop is equal to zero in the multipath delay range [0.5Tc, 1Tc], which can further improve the multipath mitigation performance.

Figure 7 shows the root mean square error (RMSE) of code tracking versus the CNR of the proposed method, DET and the traditional DLL with EMLP. We can see that the proposed method has the same level accuracy as that of the traditional DLL with EMLP and DET under high CNR. When the CNR is low, the tracking accuracy of the proposed method is lower than that of the traditional DLL with EMLP. Therefore, the proposed method is suitable for the situation that the CNR is high and the receiving environment has many multipath signals.

5 Conclusions

In order to solve the tracking ambiguity problem and mitigate the multipath effect, a dual-estimate loop method with early code phase based on double-delta correlator is proposed for BOC signal. The proposed method utilizes the idea on dual-estimate loop to solve the ambiguity problem. The SLL in the proposed method employs the double-delta correlator to mitigate the multipath effect. When the CNR is high, the SLL in the proposed method could utilize the local PRN code with early code phase to generate local reference signal and improve the multipath mitigation performance. The tracking accuracy and multipath mitigation performance are analyzed in this paper. The simulation shows that the proposed method has the better multipath mitigation performance than the traditional DLL with EMLP and DET. When the CNR is high, the proposed method can reach the same level tracking performance as the traditional DLL with EMLP and DET.

References

1. Hudnut K, Titus WC, Captain U (2004) GPS L1 civil signal modernization (L1C). Final report of interagency GPS executive board stewardship project, vol 204
2. Hein GW, Godet J, Issler JL, Martin JC, Lucas-Rodriguez R, Pratt T (2002) The Galileo frequency structure and signal design. In: Proceedings of 2002 ION, pp 1273–1282
3. Fine P, Wilson W (1999) Tracking algorithm for GPS offset carrier signals. In: Proceedings of 1999 ION national technical meeting, pp 671–676
4. Martin N, Leblond V, Guillotel G, Heiries V (2003) BOC(x,y) signal acquisition techniques and performances. In: Proceedings of 2003 ION GPS/GNSS, pp 188–198
5. Julien O, Macabiau C, Cannon ME, Lachapelle G (2007) ASPeCT: unambiguous sine-BOC (n, n) acquisition/tracking technique for navigation applications. IEEE Trans Aerosp Electron Syst 43(1):150–162
6. Hodgart MS, Blunt PD (2007) A dual estimate receiver of binary offset carrier modulated signals for global navigation satellite systems. Electron Lett 43(16):877–878
7. Hodgart MS, Simons E (2012) Improvements and additions to the double estimation technique. In: Proceedings of satellite navigation technologies and European workshop on GNSS signals and signal processing, ESA Workshop on Noordwijk. IEEE, pp 1–7

An Improved Adaptive Kalman Filter Carrier Phase Locking Loop Under Ionospheric Scintillation

Hang Ruan, Birong Xu, Lei Zhang and Feng Liu

Abstract In global navigation satellitesSystem (GNSS) receiver, ionospheric scintillation may lead huge tracking error and frequent cycle clips to the carrier phase tracking loop. When strong scintillation, the phase lock loop (PLL) of receiver may even lose lock. Although the frequency lock loop (FLL) is insensitive to the ionospheric scintillation, the tracking accuracy is poor which is insufficient for most GNSS applications. In this paper, we propose an improved adaptive Kalman filter-based carrier phase locking loop for the scintillated signal tracking. In the loop, the signal amplitude is added in the state vector of the Kalman filter. By calculating the statistic result of the amplitude variance, the amplitude is adaptively estimated. At the same time, the phase discriminator is normalized by the amplitude estimation results. The Kalman filter is redesigned based on the new structure so as to improve the stability of the loop and reduce the effects of amplitude attenuation and phase jitter on the loop. Simulation is carried out based on the ionospheric scintillation data obtained according to the A. J. Stanford model. The results show that the proposed loop can keep working in the continuously moderate and instantaneous strong ionosphere scintillation with less cycle clips.

Keywords GNSS · Kalman filter · Ionospheric scintillation · Phase locking loop

---

H. Ruan · B. Xu · L. Zhang (✉)
School of Information and Electronics, Beijing Institute of Technology,
Beijing, China
e-mail: aerolong@bit.edu.cn

H. Ruan
e-mail: ruanhang@bit.edu.cn

B. Xu
e-mail: xubirong@bit.edu.cn

F. Liu
Beijing Racobit Electronic Information Technology Co., Ltd., Beijing, China
e-mail: liufeng@racobit.com

J. Sun et al. (eds.), *China Satellite Navigation Conference (CSNC) 2016 Proceedings: Volume I*, Lecture Notes in Electrical Engineering 388,
DOI 10.1007/978-981-10-0934-1_50

1 Introduction

When the wireless radio signal broadcasted by Global Navigation Satellite System (GNSS) travels across the ionosphere, the irregularities in the electrons density will cause rapid fluctuations in the signal amplitude and phase. This phenomenon is called ionospheric scintillation [1, 2]. During ionospheric scintillation, the carrier phase tracking loop in GNSS receiver will be seriously disturbed, resulting in frequent cycle slips, high bit error rate, or even losing lock in the severe conditions which means that the receiver cannot work. It will be a great challenge for the GNSS-based applications including aviation security, carrier phase measurement based high-precision position, ionospheric scintillation monitoring, etc. [3]. Therefore, a robust carrier phase lock loop is in great demand for the signal tracking under ionospheric scintillation.

In conventional GNSS receiver, carrier tracking is implemented in a second- or third-order phase-locked loop (PLL) to achieve carrier phase tracking [4]. However, in traditional PLL, the phase jitter and amplitude scintillation caused by ionospheric scintillation often leads to cycle slips, navigation data bit errors or even loss of lock. In order to enhance the tracking performance under ionospheric scintillation, frequency lock loop-assisted phase lock loop (FLL-assisted-PLL) is used to overcome the dynamic errors caused by phase jitter [5–7] utilizes a prefilter on the In-phase correlator to reduce the influence of amplitude scintillation on the discriminator. While the FLL-assisted PLL is a multi-input system, the stability of the loop is poor. Furthermore, the frequency discriminator introduces additional noise which degrades the signal tracking accuracy. To further improve the tracking performance under ionospheric scintillation, [8] proposes a Kalman filter-based PLL, which use a Kalman filter to replace the second- or third-order filter in the traditional PLL to increase the signal tracking stability. According to the signal's characteristics under ionospheric scintillation that the amplitude and phase are changing, [9] proposes a carrier-to-noise ratio estimation based adaptive Kalman filter PLL. But these methods are still unable to overcome the discrimination bias caused by the phase and amplitude jitter due to ionospheric scintillation.

This paper presents an improved adaptive Kalman filter-based PLL for the carrier tracking of GNSS receiver under ionospheric scintillation. In the filter, the signal amplitude state is introduced. By computing the variance of the signal amplitude from the correlator outputs, the scintillated signal amplitude is estimated by an adaptive Kalman filter. Afterwards, the carrier phase discriminator is normalized by the estimated amplitude to improve the tracking stability of the PLL. Based on the A. J. Stamford ionospheric scintillation model, the adaptive Kalman filter-based PLL is simulated. The results inform that, compared with the traditional PLL and ordinary Kalman filter-based PLL, the proposed loop can better keep tracking the signal with less cycle slip under continuous moderate scintillation and instantaneous strong scintillation.

2 Signal Model Under Ionospheric Scintillation

Usually, there are a variety of different scales of inhomogeneous plasma block involved in the normal electron density distribution of ionosphere which is known as the uneven structure of the ionosphere. When traveling across the ionosphere, the electromagnetic wave will be refracted and scattered by the occurrence and development of the uneven gas in it which will cause the signal energy redistribution in space and time. Then the short-term and irregular variations including signal amplitude fading and phase jitter will occur on the propagated electromagnetic wave. This phenomenon is called ionospheric scintillation. Ionospheric scintillation occurs randomly, but in general, the frequency and intensity of its occurrence is related to the time, solar activity, latitude, and the geomagnetic environment.

In order to access the intensity of ionospheric scintillation, the amplitude scintillation indices and phase scintillation standard deviation are proposed. The amplitude scintillation indices is represented by S_4, defined as the signal power normalized standard deviation [10] which is expressed as

$$S_4 = \sqrt{\frac{E(\delta I^2) - E(\delta I)^2}{E(\delta I)^2}} \tag{1}$$

where δI is signal power, $E\{\cdot\}$ is the operator of expectation calculation. In the real implementation, S_4 is calculated from the signal power tracked by the GNSS receiver. The phase-standard deviation σ_ϕ evaluates the phase jitter of signal under ionospheric scintillation. It is usually computed by the detrended 50 Hz phase observations. Based on the amplitude and phase scintillation indices, people define ionospheric scintillation as Table 1 [11].

As the occurrence of ionospheric scintillation is random, the corresponding scintillation model is design so as to analyze the ionospheric scintillation effect on the signal. The typical model is the AJ-Stanford model proposed AJ Van Dierendonck [10]. The expression of signal under ionospheric scintillation is

$$E = Ae^{j\phi} = (A_0\delta A)e^{j(\phi_0 + \delta\phi)} \tag{2}$$

where A_0 is the signal amplitude, δA is the variation of the amplitude, ϕ_0 is the signal carrier phase, and $\delta\phi$ is the phase variation. In usual cases, δA is 1, $\delta\phi$ is 0. When ionospheric scintillation occurs, δA and $\delta\phi$ will both vary randomly.

Table 1 Ionospheric scintillation level

Case	S4	σ_ϕ
Strong	0.9	0.6
Moderate	0.6	0.3
Weak	0.4	0.2

According to the analysis based on a large amount of experiment data, δA obeys Nakagami-m distribution. The probability density function (PDF) is

$$p_n(\delta A) = \frac{2\delta A(1+K)}{\Omega} I_0\left(2\delta A\sqrt{\frac{K+K^2}{\Omega}}\right) \mathrm{e}^{-K-\delta A m \delta A^2/\Omega} \tag{3}$$

where $\delta A > 0$, $m = 1/S_4^2$, $\Gamma(\cdot)$ is the gamma function. In Nakagami-m distribution, $m > 0.5$ so that the amplitude scintillation indices is limited to $S_4 < \sqrt{2}$ in the scintillation model. In fact, it has covered most cases of ionospheric scintillation.

In the model, the phase variation due to ionospheric scintillation has the power spectral density (PSD)

$$S_{\phi_p}(f) = \frac{T}{(f_0^2 + f^2)^{p/2}} \quad \text{radians}^2/\text{Hz} \tag{4}$$

where f_0 is the frequency corresponding to the maximum irregularity size in the ionosphere. T is the spectral strength at 1 Hz, and p is the slope of the PSD. The variance of the phase variations is

$$\sigma_\phi^2 = 2\int_{\tau_c^{-1}}^{\infty} S_{\phi_p}(f)\mathrm{d}f \tag{5}$$

where τ_c is a system parameter related to the phase-stability time of the receiver. It is usually set to 100 s. In the model, the signal phase jitter $\delta\phi$ obeys Gaussian distribution as follows:

$$f(\delta\phi) = \frac{1}{\sigma_{\delta\phi}\sqrt{2\pi}} \mathrm{e}^{-\frac{(\delta\phi)^2}{2\sigma_{\delta\phi}^2}} \tag{6}$$

Based on the statistics characteristics of the signal amplitude and phase, the time sequence of the scintillated signal is generated. Furthermore, a cascade of Butterworth filter is used as a shaping filter to make the time sequence time-dependent so as to gain the PDF similar to the real data. The signal characteristics under different ionospheric scintillation are drawn in Fig. 1. It shows that the signal amplitude and phase vary tempestuously under ionospheric scintillation so that the tracking loop in receiver should be able to deal with problems of high dynamic and low signal power at the same time. Hence, in order to obtain the high response speed and perfect stability, we design the tracking loop based on the Kalman filter.

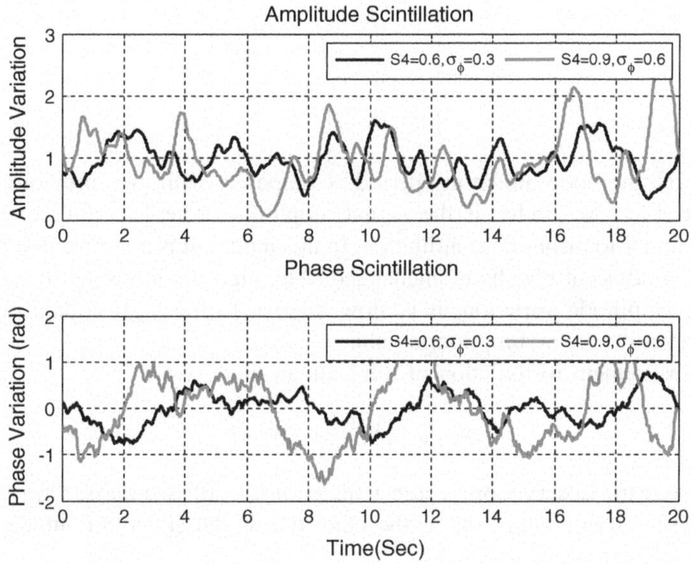

Fig. 1 Amplitude and phase variations

3 Enhanced Adaptive Kalman Filter PLL

3.1 Enhanced Kalman Filter Model

Under ionospheric scintillation, the signal's amplitude and phase both vary tremendously which bring great difficulties to lock the carrier phase of GNSS signal. In order to improve performance of GNSS receiver under ionospheric scintillation, we propose an improved adaptive Kalman filter-based PLL. Its state vector is

$$X_k = \begin{bmatrix} \hat{A}_k & \Delta\hat{\phi}_k & \Delta\hat{\omega}_k & \Delta\hat{\alpha}_k \end{bmatrix}^T \tag{7}$$

where $\hat{A}_k$ is the amplitude estimation, $\Delta\hat{\phi}_k$ is the phase estimation error, $\Delta\hat{\omega}_k$ is the signal frequency estimation error, $\Delta\hat{\alpha}_k$ is the frequency rate estimation error. The state transition equation of the Kalman filter system is

$$X_{k+1} = AX_k + w_k \tag{8}$$

where w_k is the process noise matrix, A is the state transition matrix expressed as

$$A = \begin{bmatrix} 1 & 0 & 0 & 0 \\ 0 & 1 & \Delta t & 0.5\Delta t^2 \\ 0 & 0 & 1 & \Delta t \\ 0 & 0 & 0 & 1 \end{bmatrix} \tag{9}$$

where Δt is the loop updating period. Compared with the previous Kalman filter-based PLL, we induced the signal amplitude state to estimate the signal amplitude under ionospheric scintillation. In this model, it is assumed that the signal amplitude at adjacent epochs is unchanged. Although the ionospheric scintillation will cause amplitude variation, it is slow relative to the high update rate of the loop. As a result, the assumption is valid.

The measurement formulation of the Kalman filter system is

$$y_{k+1} = HX_{k+1} + v_{k+1} \tag{10}$$

where y_{k+1} is the observation vector, which consists of two parts. One of them is carrier phase discriminator output; the other one is the signal amplitude statistics value. y_{k+1} is expressed as

$$y_{k+1} = [\delta\phi_{k+1} \quad C_{k+1}]^T \tag{11}$$

$\delta\phi_{k+1}$ is the carrier phase discriminator output. Considering the presence of the navigation signal data, the Costas loop discriminator is used and donated by

$$\delta\phi_k = I_k \times Q_k \tag{12}$$

I_k and Q_k are the in-phase branch (I) and the quadrature branch (Q) correlation results. C_{k+1} is the signal amplitude estimating value expressed as

$$C_k = \sqrt{I_k^2 + Q_k^2} \tag{13}$$

v_{k+1} is the measurement noise vector. H is the measurement matrix expressed as

$$H = \begin{bmatrix} 1 & 0 & 0 & 0 \\ 0 & 1 & -\Delta t/2 & \Delta t^2/6 \end{bmatrix} \tag{14}$$

Referring to (12), it is clear that the signal power is contained in the Costas discriminator so that it is unavailable in the real implementation. Otherwise, the frequent variation of the signal amplitude caused by the ionospheric scintillation will result in extremely larger discrimination error. Thus, the discriminator should be power normalized. The normalized discriminator is expressed as

$$\delta\phi_k = \frac{I_k \times Q_k}{I_k^2 + Q_k^2} \tag{15}$$

However, the discriminator will also be disturbed by amplitude scintillation, especially when severe amplitude attenuation occurs. In this case, the value of $I_k^2 + Q_k^2$ will remarkably reduce which induces huge carrier phase error discrimination results and breaks the stability of the loop.

For the purpose to eliminate the influence of the huge amplitude variation on the carrier phase discriminator, in the proposed adaptive Kalman filter-based PLL, the estimated signal amplitude is utilized for the normalization of the discriminator which is expressed as

$$\delta\phi_k = \frac{I_k \times Q_k}{\hat{A}_k^2} \tag{16}$$

Using the adaptive Kalman filtering method, the effects of the amplitude attenuation caused by ionospheric scintillation on the phase discriminator is eliminated so that the loop stability is improved. In the following, we will introduce the adaptive processing of the Kalman filtering system.

3.2 Adaptive Processing

Similar to the classical Kalman filtering processing, the processing of the proposed Kalman filtering system with the signal amplitude estimation is summarized as below [12].

In accordance with the previous epoch's status, we can gain the state estimation of the present epoch as

$$\bar{X}_{k+1} = AX_k \tag{17}$$

Supposing the state covariance matrix is P_k, the process covariance matrix is Q_k and the measurement covariance matrix is R_k, the state covariance transition process is

$$\bar{P}_{k+1} = AP_kA^T + Q_k \tag{18}$$

$$K_{k+1} = \bar{P}_{k+1}H^T\left(H\bar{P}_{k+1}H^T + R_k\right) \tag{19}$$

$$P_{k+1} = (I - HK_{k+1})\bar{P}_{k+1} \tag{20}$$

The process of state correction is

$$X_{k+1} = \bar{X}_{k+1} + K_{k+1}(z_k - H\bar{X}_{k+1}) \tag{21}$$

It can be seen that, throughout the Kalman filtering system, the parameters needed to be configured are the process covariance matrix Q_k and observation covariance matrix R_k. According to the structure of Kalman filtering in the paper, the estimation of the signal amplitude and carrier phase can be analyzed separately. Hence, Q_k and R_k are be rewritten as

$$Q_k = \begin{bmatrix} Q_{A,k} & 0 \\ 0 & Q_{X,3\times 3} \end{bmatrix},\ R_k = \begin{bmatrix} R_{A,k} & 0 \\ 0 & R_\phi \end{bmatrix} \tag{22}$$

where Q_A is the signal amplitude estimation process noise covariance matrix, $Q_{X,3\times 3}$ is the process noise covariance matrix of the signal phase error, frequency error and frequency rate error. R_A is the signal amplitude observation error covariance matrix. As the observed value of the amplitude is a scalar, R_A is a scalar. R_ϕ is the signal phase observation error covariance matrix and it is also a scalar.

As the ionospheric scintillation parameters cannot be estimated accurately in real time, the $Q_{X,3\times 3}$ is fixed in the proposed method. It is composed of the frequency rate random walk, the clock's frequency error and the clock's phase error which is expressed as

$$Q_{X,3\times 3} = q \times (2\pi)^2 \times \begin{bmatrix} \frac{\Delta t^5}{20} & \frac{\Delta t^4}{8} & \frac{\Delta t^3}{6} \\ \frac{\Delta t^4}{8} & \frac{\Delta t^3}{3} & \frac{\Delta t^2}{2} \\ \frac{\Delta t^3}{6} & \frac{\Delta t^2}{2} & \Delta t \end{bmatrix} + S_g\omega_{RF}^2 \begin{bmatrix} \frac{\Delta t^3}{3} & \frac{\Delta t^2}{2} & 0 \\ \frac{\Delta t^2}{2} & \Delta t & 0 \\ 0 & 0 & 0 \end{bmatrix} + S_f\omega_{RF}^2 \begin{bmatrix} \Delta t & 0 & 0 \\ 0 & 0 & 0 \\ 0 & 0 & 0 \end{bmatrix} \tag{23}$$

where q is the frequency rate random walk, S_g is phase jitter error of the receiver clock, S_f is the frequency jitter error of the receiver clock. Q_A is calculated from the noise channel outputs. It indicates the signal amplitude variance without consideration of the ionospheric scintillation.

R_ϕ is the variance of the discriminator outputs expressed as [4]

$$R_\phi = \frac{1}{2T \cdot C/N_0}\left(1 + \frac{1}{2T \cdot C/N_0}\right) \tag{24}$$

R_A is computed from the received signal amplitude. By calculating Q_A and R_A, the signal amplitude jitter under different level of ionospheric scintillation can be evaluated. Hereafter, it is used for the adaptive processing of the signal amplitude estimation. The flow diagram of the proposed method is shown in Fig. 2. The process is summarized below.

Step 1: According to the outputs of the noise channel, the amplitude variance is calculated and denoted by $Q_{A,k}$. The signal amplitude variance is computed from the *I*/*Q* branch correlation results which is represented as

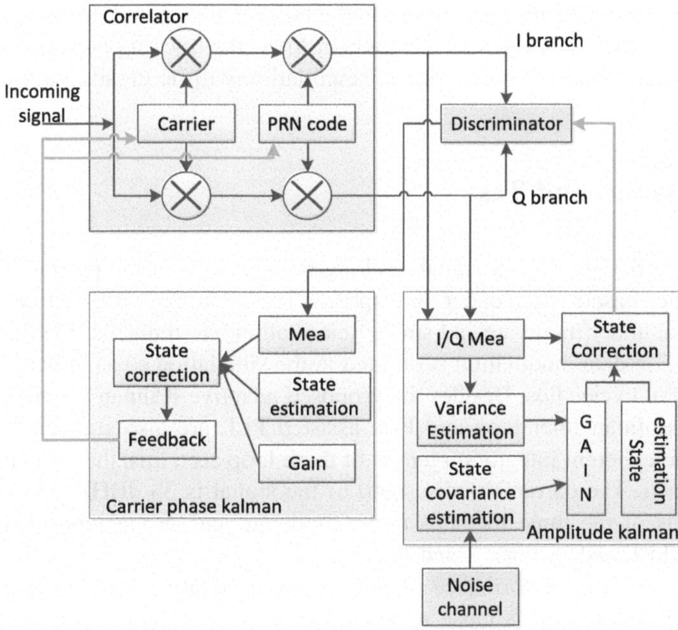

Fig. 2 The process of the adaptive Kalman filter loop

$R_{A,k}$. (As the measurement of each epoch is an instantaneous value, in order to reduce the impact of the noise jitter on the amplitude estimation accuracy, multiple epochs' correlation results are used in the estimation).

Step 2: Depending on the updated state and measurement error covariance results, Kalman filter gain is obtained. Together with the state prediction and measurement results, the state of signal amplitude is corrected.

Step 3: The amplitude estimation results after the correction is sent into the carrier phase discriminator for the normalization to gain the phase measuring results

Step 4: The discriminator output, together with the carrier phase state prediction results and the calculated Kalman filtering gain, is used to update and correct the state vector.

Step 5: According to the updated state, the carrier frequency is estimated and feedback to the correlator for generation of the local carrier.

In the proposed adaptive Kalman filtering system, as the variance of the system Gaussian white noise is unchanged, the Kalman filter gain is mainly decided by the signal amplitude variation. If the signal amplitude scintillation is strong, the signal amplitude variance is large and the Kalman filter gain is decreased to reduce the impact of the amplitude fading on the phase estimation. When the scintillation is weak, the signal amplitude variance is reduced and the Kalman filter gain is increased so as to keep the response speed. Obviously, the adaptive processing will improve the normalization accuracy and enhance the signal tracking stability. Together with the advantages of the Kalman filter, the tracking performance of the signal tracking loop under ionospheric scintillation will be greatly improved.

4 Simulation and Test

In order to verify the GNSS signal tracking performance of the proposed adaptive Kalman filter-based PLL under ionospheric scintillation, we generate the GPS L1CA signal under moderate and strong scintillation based on A.J. Stanford model. The navigation data modulation is omitted in the simulation so as to better evaluate the number of cycle slips. Besides the proposed adaptive Kalman filtering loop, the traditional Kalman filter loop and FLL-assisted PLL are also simulated for comparison. The coherent integration length of these loop are 1 ms, the standard OCXO model is used. The carrier-to-noise ratio of the signal is 38 dBHz. The I/Q correlation results of the three methods under moderate and strong ionosphere scintillation are illustrated in Figs. 3 and 4.

It shows that the performance of the proposed adaptive Kalman loop and the traditional Kalman loop is close under moderate scintillation and both are better than the FLL-assisted PLL. For strong scintillation, the adaptive Kalman loop can

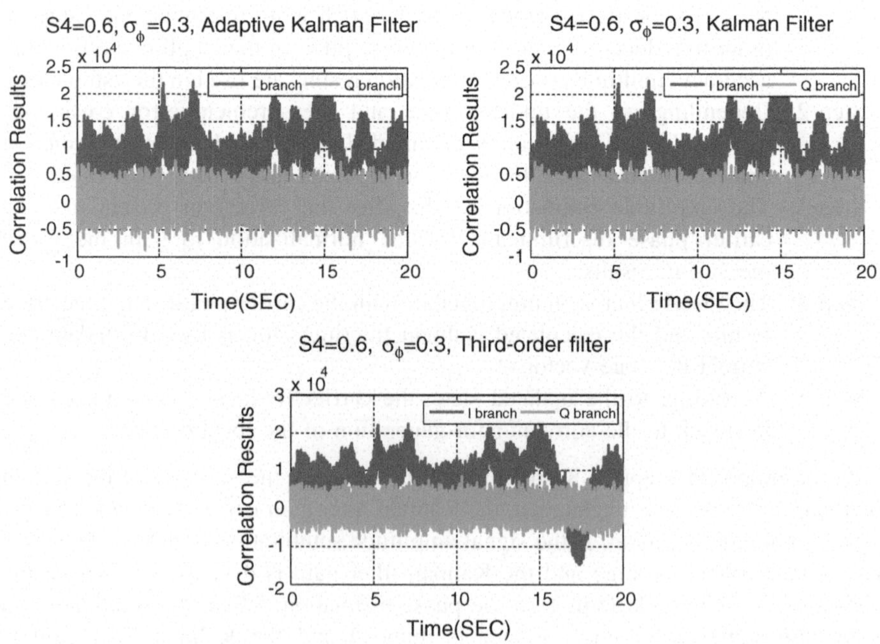

Fig. 3 Loop tracking results under moderate scintillation

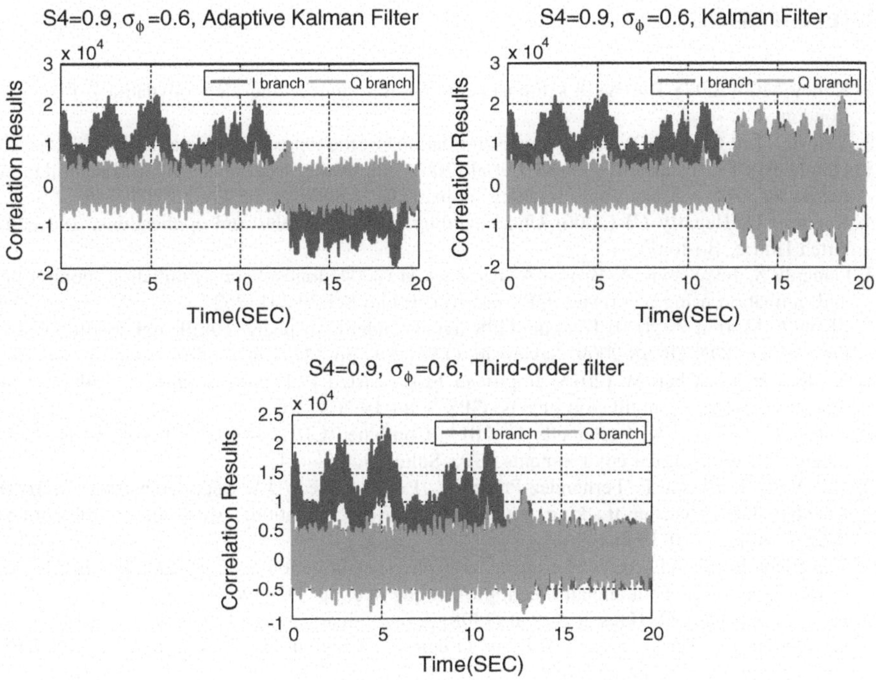

Fig. 4 Loop tracking results under strong scintillation

keep tracking the signal but the other two loops fail to correctly lock the signal. It is validated that the proposed method can improve the performance of GNSS receiver under ionospheric scintillation, especially strong scintillation.

5 Conclusion

This paper presents an adaptive Kalman filter-based PLL for the GNSS signal tracking under ionospheric scintillation. In the Kalman loop, the signal amplitude is estimated adaptively. The estimation results are used for the normalization of the discriminator to improve the accuracy of the carrier phase error discrimination. The simulation results show that the proposed method can significantly improve the signal tracking performance of GNSS receiver under ionospheric scintillation.

References

1. Kelley MC (2009) The Earth's Ionosphere: plasma physics and electrodynamics. Elsevier, London
2. Yeh KC, Liu CH (1982) Radio wave scintillations in the ionosphere. Proc IEEE 70:324–360
3. Humphreys TE, Psiaki ML, Hinks J et al (2009) Simulating ionosphere-induced scintillation for testing GPS receiver phase tracking loops. IEEE J Sel Top Sig Proc 3:707–715
4. Kaplan ED, Hegarty CJ (2006) Understanding GPS. Principles and applications, 2nd edn. Artch House, Boston
5. Ganguly S, Jovancevic A, Brown A, Kirchner M (2004) Ionospheric scintillation monitoring and mitigation using a software GPS receiver. Radio Sci 39:241–262
6. Zhang L, Morton Y (2009) Tracking GPS signals under ionosphere scintillation conditions. In: Proceedings ION-GNSS-2009, Savannah, GA, USA, pp 227–234
7. Xu R, Liu ZZ, Chen W (2015) Improved FLL-assisted PLL with in-phase pre-filtering to mitigate amplitude scintillation effects. GPS Solut 19:263–276
8. Gernot C, O'Keefe K, Lachapelle G (2011) Combined L1/L2 Kalman filter-based tracking scheme for weak signal environments. GPS Solut 15:403–414
9. Vilà-Valls J, Closas P, Fernández-Prades C, López-Salcedo JA, Seco-Granados G (2015) Adaptive GNSS carrier tracking under ionospheric scintillation: estimation vs mitigation. IEEE Commun Lett 19:961–964
10. Van Dierendonck AJ (2001) Measuring ionospheric scintillation effects from GPS signals. In: Proceedings of the 57th annual meeting, Albuquerque, pp 391–396
11. Pullen S, Opshaug G, Hansen A et al (1998) A preliminary study of the effect of ionospheric scintillation on WAAS user availability in equatorial regions, In: Proceedings of ION GPS 1998, Nashville, TN, pp 687–699
12. Won JH, Pany T, Eissfeller B (2012) Characteristics of Kalman Filters for GNSS signal tracking loop. IEEE Trans Aerosp Electron Syst 48:3671–3681

Optimized Design of Vector Frequency Lock Loop Based on Non-coherent Discriminator for BDS Navigation Signal

Honglei Lin, Xiaomei Tang, Yingxiang Liu, Huaming Chen and Gang Ou

Abstract The update rate of vector tracking loop (VTL) is usually the same as the message rate modulated in navigation signal, and BeiDou satellites broadcast D1 NAV message and D2 NAV message which will lead conflict of update rate for VTL. To solve this problem, a novel VTL based on non-coherent discriminator is proposed and a method to optimize its update interval is also introduced. Analysis and simulation show that for low dynamic user, the update interval can be designed longer than 0.1 s and the complexity of VTL can be reduced largely. If the update interval is 0.3 s, this algorithm can track the signal broadcasted by GEO satellites in steady when C/N0 is lower as 20 dBHz.

Keywords BeiDou navigation system · Vector frequency lock loop · Non-coherent discriminator · Optimized design

1 Introduction

Vector tracking loop (VTL) was studied for Global Positioning System (GPS) receivers by many researchers in recent years. VTL has a better tracking performance in sensitivity, signal reacquisition, usability, and so on [1]. A method of vector delay lock loop (VDLL) combining navigation filter and delay lock loop to track all satellite in a vector loop is proposed [2]; it is a first structure of VTL. Afterward, the vector frequency and delay lock loop (VFDLL) is studied in detail [3] and its tracking threshold and dynamic are analyzed and verified by simulations [4]. Vector frequency lock loop (VFLL), VDLL, and VFDLL belong to the concept of VTL. The performance and validity of VTL are studied in different environments,

H. Lin · X. Tang · Y. Liu · H. Chen · G. Ou (✉)
National University of Defense Technology, Deya. 109, Changsha 410073, China
e-mail: ougang_nnc@163.com

H. Lin
e-mail: linhonglei1989@163.com

J. Sun et al. (eds.), *China Satellite Navigation Conference (CSNC) 2016 Proceedings: Volume I*, Lecture Notes in Electrical Engineering 388,
DOI 10.1007/978-981-10-0934-1_51

such as fading channel [5], city and canyon [6] or weak signal environment [7]. Because the message rate of GPS L1CA signal is 50 Hz, the update rate of VTL is usually 50 Hz, which makes computation burden heavily [1].

With the development of BeiDou Navigation System (BDS), B1I signal is providing services to customers in the Asia Pacific Region. The obvious feature of B1I signal is that D1 NAV message (message rate is 50 bps) and D2 NAV message (message rate is 500 bps) are broadcast by different type of satellites at the same time [8]. Because the message rate of D2 NAV signal is 10 times of D1 NAV signal, the coherent time of D2 NAV signal is 1/10 of D1 NAV signal. To scalar tracking loop, the tracking threshold of D2 NAV signal is lower than D1 NAV signal. To VTL, the update interval for D1 NAV signal and D2 NAV signal are conflicted.

To make the update interval of VTL or STL unrelated to message rate, a pre-filter is designed to process the results of discriminator in GPS/INS integration systems, which can reduce the update rate of VTL [9]. Non-coherent square tracking loop is usually used in spread spectrum system; it can derive the unknown message [10]. A memory non-coherent discriminator is used in STL [11], where an exponential weighting function is introduced based on non-coherent square tracking loop, analysis and simulation shows that this discriminator is valid for navigation signal tracking. So, a novel VTL based on the non-coherent discriminator is studied in this paper, where the update interval can be optimized as an important parameter by different optimal model.

This paper is organized as follows. In Sect. 2, a novel VFLL based on non-coherent discriminator is introduced and the non-coherent discriminator is analyzed in detail. In Sect. 3, based on the relationship of update interval and performance of VFLL, a method to optimize the update interval is proposed and its numerical simulation results are described for low and middle dynamic receivers. In Sect. 4, some simulations are carried out to verify those results. Simulation and analysis show that there exists optimal update interval for the novel VFLL based on non-coherent discriminator. And to low dynamic receiver, the update interval of VFLL can be designed longer than 0.1 s, this can reduce the computation burden greatly. If the update interval is 0.3 s, the novel VFLL can also track GEO signal steadily when the C/N0 is lower as 20 dBHz.

2 Model of Vector Frequency Lock Loop

2.1 *Processing of Traditional VFLL*

A type structure of VTL as shown in Fig. 1 [4].

Compared with scalar tracking loop, VFLL can forecast the signal NCO control parameters according to the position and velocity of satellites and receiver, this processing is called the feedback control.

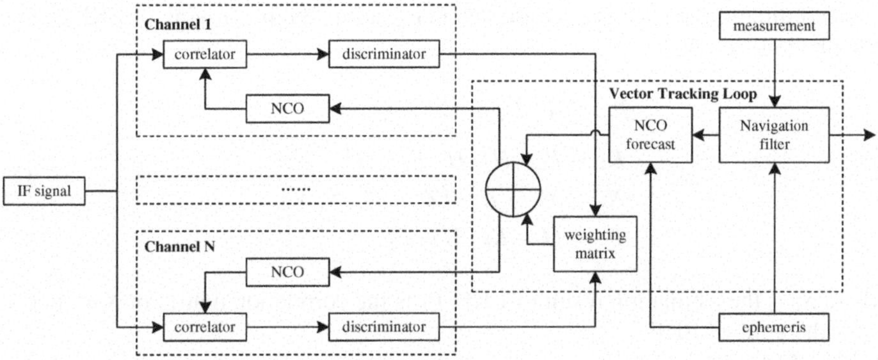

Fig. 1 A type structure of VFLL based on coherent discriminator for signal tracking

When the navigation is an extend Kalman filter, its predicted equation is

$$X_k^- = \Phi \cdot X_{k-1} + \omega_k \tag{1}$$

where $X = [v_x, a_x, v_y, a_y, v_z, a_z, \delta f]$, is the state vector of receiver, v_x, v_y, v_z are its velocity in ECEF coordination, a_x, a_y, a_z are its accelerations, δf is the frequency bias of clock, Φ is transmitting matrix, ω_k is processing noise and its correlation matrix is Q.

The transmitting matrix of dimensionality x, y, z are the same, to x dimensionality, it has

$$\Phi_x = \begin{bmatrix} 1 & 0 \\ 0 & T \end{bmatrix} \tag{2}$$

The processing correlation matrix is

$$Q_x = S_x \begin{bmatrix} T^3/3 & T^2/2 \\ T^2/2 & T \end{bmatrix} \tag{3}$$

To frequency bias, because it is one dimensional, $\Phi_c = 1$ and $Q_c = S_f \cdot T$, where T is update time of VTL, S_x is noise power density of acceleration, S_f is noise power density of frequency. So Φ and Q can be achieved according to Φ_x, Φ_c, Q_x and Q_c.

The measurement equation of EKF is

$$Z_k = H \cdot X_k - H \cdot X_k^s + \varepsilon_k \tag{4}$$

where $Z_k = [f_1, f_2, \ldots, f_N]$ is the observation vector at time k, and R is its correlation matrix. f_i is the doppler of form satellite i to the receiver, N is the number of satellites. H is the measurement matrix and it is determined by the position of satellites and receiver. ε_k is the observation error vector.

According to the state and measurement equation, the processing of EKF can be described as below.

$$\begin{aligned} P_k^- &= \Phi \cdot P_{k-1} \cdot \Phi^{\mathrm{T}} + Q \\ K_k &= P_k^- \cdot H^T \left(H \cdot P_k^- \cdot H^{\mathrm{T}} + R\right)^{-1} \\ X_k &= X_k^- + K_k \cdot \left(Z_k - H \cdot X_k^-\right) \\ P_k &= (I - K_k \cdot H) \cdot P_k^- \end{aligned} \tag{5}$$

where X_k is the estimation result of EKF, P_k is the correlation matrix of X at time k, K_k is the gain of EKF.

To VFLL, the result of estimation of X can be also be written as

$$X_k = X_k^- + K_k \cdot \Delta Z_k \tag{6}$$

where $\Delta Z_k = [\Delta f_1, \Delta f_2, \ldots, \Delta f_N]$, is the result of discriminator, so $K_k \cdot \Delta Z_k$ is the correction term for predicted state vector. According to X_k, VFLL can get the frequency control parameter of channel NCO, it is

$$\hat{Z}_{k+1} = H \cdot X_{k+1}^- - H \cdot X_{k+1}^s + H \cdot K_k \cdot \Delta Z_k \tag{7}$$

where Z_{k+1} is the NCO frequency control words of each channel.

2.2 VFLL Based on Non-coherent Discriminator

The update time of VFLL is designed according to the discriminator which is decided by the message rate of navigation signal. To L1C/A signal, it is no longer than 20 ms. To design VFLL more flexibly, a non-coherent discriminator is adopted instead of the coherent discriminator in Fig. 1. So the update time can be adjusted by the number of accumulation. A typical structure of non-coherent discriminator is based on square operation showed as Fig. 2, where longer accumulation time is carried out after removing navigation message bits through square operation [11].

In Fig. 2 T_{coh} is the coherent time, M is the number of non-coherent, so the total accumulation time before Atan2 discriminator is $T_c = M \cdot T_{\mathrm{coh}}$, the output of Atan2 needs to multiply 1/2 to counteract the doubling frequency caused by square

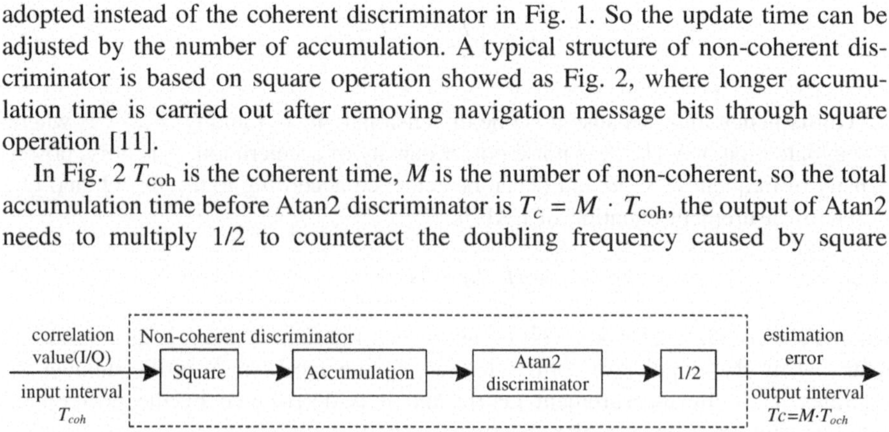

Fig. 2 Design of non-coherent frequency discriminator based on square operation

operation. Through this non-coherent, the data interval changes from T_{coh} to T_c. The square loss brought by square operation is studied in reference [12]. After M times accumulation, the radio of signal to noise power is

$$\text{SNR}_d = \frac{M \cdot T_{\text{coh}}^2 \cdot (C/N_0)^2}{2 + 4 \cdot T_{\text{coh}} \cdot C/N_0} \tag{8}$$

But the result is ignored the Doppler loss caused by integral processing. Doppler loss can be written as [13]

$$L_d \approx \sin c^2(\pi \cdot 2 \cdot \Delta f_d \cdot T_c) \tag{9}$$

where Δf_d is Doppler, coefficient 2 represents the doubling frequency caused by square operation. When the Doppler changes as a linear model at a rate a_f, its loss is

$$L_d = \sin c^2\left(\pi \cdot a_f \cdot T_c^2\right) \tag{10}$$

So the radio of signal to noise power before Atan2 discriminator is

$$\text{SNR}_d = \frac{M \cdot T_{\text{coh}}^2 \cdot (C/N_0)^2}{2 + 4 \cdot T_{\text{coh}} \cdot C/N_0} \cdot \sin c^2\left(\pi \cdot a_f \cdot T_c^2\right) \tag{11}$$

Reference [13] gave the thermo-noise jitter of scalar tracking loop using coherent discriminator

$$\sigma_t = \frac{1}{2\pi \cdot T_{\text{coh}}} \cdot \sqrt{\frac{4 \cdot F \cdot B_L}{C/N_0}\left(1 + \frac{1}{T_{\text{coh}} \cdot C/N_0}\right)} \tag{12}$$

where B_L is its bandwidth, coefficient F is 2 when signal power is nearly tracking threshold and otherwise is 1.

Reference [14] describes the relationships of thermo-noise jitter between scalar tracking loop and Atan2 discriminator, which is

$$\sigma_t^2 = 2 \cdot B_L \cdot T \cdot \sigma_d^2 \tag{13}$$

where T is the loop update interval, to coherent discriminator $T = 2 \cdot T_{\text{coh}}$, according to (12) and (13) the thermo-noise jitter of Atan2 discriminator is

$$\sigma_d = \frac{1}{2\pi \cdot T} \cdot \sqrt{\frac{F}{\text{SNR}}\left(1 + \frac{1}{\text{SNR}}\right)} \tag{14}$$

where T is accumulation time before discriminator, and SNR is its signal to noise power radio. So to non-coherent discriminator based on square operation in Fig. 2, its thermo-noise jitter is

$$\sigma_d = \frac{1}{2} \cdot \frac{1}{2\pi \cdot T_c} \cdot \sqrt{\frac{F}{\mathrm{SNR}_d}\left(1 + \frac{1}{\mathrm{SNR}_d}\right)} \tag{15}$$

3 Tracking Performance Optimal Design

The update interval time of VFLL based on non-coherent discriminator is a flexible parameter which needs to be designed reasonably to reach a better performance. This party mainly studies the problem of optimal design of T_c.

3.1 Tracking Performance Analysis

The tracking threshold of discriminator is usually 1/4 of its scopes in experience [13], so to non-coherent discriminator, the threshold is

$$3 \cdot \sigma_f \leq Th = 1/(4 \cdot T_c) \tag{16}$$

where σ_f is standard error of frequency tracking error, when ignored, the jitter caused by receiver clock mainly includes thermo-noise jitter σ_t and error caused by receiver dynamic f_e, which is

$$3 \cdot \sigma_f = 3 \cdot \sigma_t + f_e \leq 1/(4 \cdot T_c) \tag{17}$$

To σ_t, it needs to solve the steady equation of EKF first [4], which is

$$\Phi P_s^- \Phi^T - \Phi P_s^- H^T \left(H P_s^- H^T + R\right)^{-1} H P_s^- \Phi^T + Q - P_s^- = 0 \tag{18}$$

where P_s^- is the predicted correlation matrix steadily, according to (5), the estimation precision P_s of X is easy to get, so to satellite i, its thermo-noise jitter is

$$\sigma_t^i = \sqrt{\mathrm{diag}(H_i \cdot P_s \cdot H_i^T)} \tag{19}$$

where H_i is the ith row of matrix H. To f_e, according to Ref. [4], it needs to solve the error of X caused by dynamic, which is

$$\Delta X_s = (I - \Phi + K_s \cdot H \cdot \Phi)^{-1} \cdot (I - K_s \cdot H) \cdot B \cdot u \tag{20}$$

So, the frequency jitter caused by dynamic error can be obtained as following

$$f_e^i = H_i \cdot \Delta X_s \cdot f_R / c \tag{21}$$

where f_R is carrier frequency of signal, c is the light speed, u is acceleration dynamic error and $B = [B_x, B_y, B_z]$ is error transmitting matrix, B_x, B_y, and B_z are the same, they are

$$B_x = B_y = B_z = \begin{bmatrix} T_c^2/2 \\ T_c \end{bmatrix} \tag{22}$$

When H, R and S_x is given, σ_t and f_e are the function of T_c.

3.2 Optimal Design of Loop Update Interval

As we know, T_c can affect the σ_t and f_e, it also has been related with the computation of VFLL, so this part will study the optimal choice of T_c.

If the target is to achieve high Doppler tracking precision, this optimal problem becomes how to find the T_c, which makes the tracking jitter minimum under given dynamic U_{th} and signal C/N0, so the optimal model can be described as

$$\begin{cases} \text{find} & T_c \\ \min & \sigma_f(T_c) \\ \text{s.t.} & u = U_{\mathrm{th}} \end{cases} \tag{23}$$

If the target is to achieve high tracking sensitivity, this problem becomes how to find the T_c, which makes the tracking threshold C/N0 is minimum under given dynamic U_{th}, so the optimal model can be described as

$$\begin{cases} \text{find} & T_c \\ \min & C/N_0 \\ \text{s.t.} & u \leq U_{\mathrm{th}} \\ & 3 \cdot \sigma_t(T_c) + f_e(T_c) \leq 1/(4 \cdot T_c) \end{cases} \tag{24}$$

Similarly, the optimal model aimed in other target can also be built. The optimal T_c may be different in different targets.

Those models can be solved by exhaustive search algorithm in numerical simulation. This part will take BDS constellation as examples to solve optimal solution of (23). There are six satellites in the constellation, and PRN are 2, 4, 5, 6, 9, 14. The first three satellites are GEO satellites which broadcast D2 NAV message and the last three satellites are MEO/ISGO satellites which broadcast D1 NAV message.

In the numerical simulation, the maximum acceleration dynamic error Uth in (23) is set as 0.1 g/s, where $g = 9.8\ \mathrm{m/s^2}$, the signal C/N0 is set as 30 dBHz and S_x is 1. The coherent time of GEO satellites is 0.002 s and others are 0.02 s. The linear Doppler loss model is set as (10). To find the optimal T_c with the restriction of integer multiples of 0.02 s, search all possible solution in the scope of T_c = 0.04–0.5 s. The solutions are shown as Figs. 3 and 4.

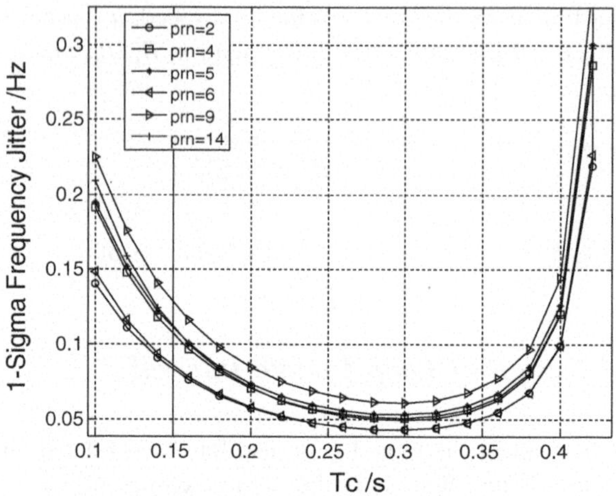

Fig. 3 $1 - \sigma$ jitter when $a_f = 0.1$ g

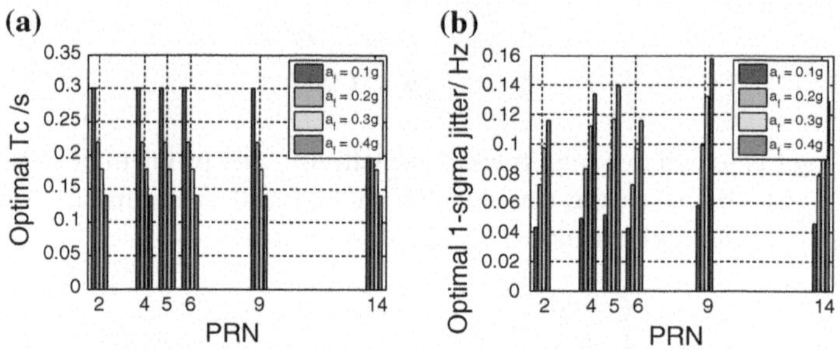

Fig. 4 Optimal solutions in different dynamic. **a** Optimal update interval. **b** Minimum $1 - \sigma$ tracking jitter (Color figure online)

Figure 3 describes the relationship of all satellites between $1 - \sigma$ frequency tracking jitter and update interval T_c when $a_f = 0.1$ g. From the result, it is obvious that there exist optimal T_c which make the tracking jitter is minimum.

Figure 4 describes the optimal update interval T_c and minimum frequency jitter when $a_f = 0.1, 0.2, 0.3, 0.4$ g. According to the solution, it seems that the when the Doppler dynamic is larger, the optimal T_c is smaller, and its minimum σ_f is larger. When $a_f = 0.4$ g, the optimal update interval is 0.14 s.

4 Simulation and Analysis

The simulation is carried out to verify the performance of VFLL based on non-coherent discriminator furthermore. In simulation, the VFLL is replaced by VFDLL without loss of generality. The trajectories of satellites are set as BDS constellation and the trajectory of receiver is shown as Fig. 5. All of trajectories are generated by STK software. Figure 5a described the horizontal motion trajectory where origin of ENU coordinate is its beginning point, Fig. 5b described its velocity of dimension x, y, and z in ECEF coordinate. The total time is 10 min.

In this simulation, C/N0 = 30 dBHz is set for all satellite and $S_x = 1$. The update interval is simulated from 0.1 to 0.6 s and stepped by 0.1 s. To evaluate the tracking sensitivity at the same, the unitary frequency jitter σ_{u} is defined as the multiples of frequency jitter σ_f and update interval T_c,

$$\sigma_u = \sigma_f \cdot T_c \tag{25}$$

According to (16), σ_u is smaller, the tracking sensitive is higher.

Figure 6 is the simulation result. Figure 6a shows the 1σ frequency jitter in different update interval and Fig. 6b is the 95 % precision of velocity result. This figure verifies the conclusion that there exists optimal T_c which makes the frequency jitter minimum. In this condition, the velocity accuracy is also higher. From the result, when $T_c = 0.4$ s, the 1σ frequency jitter is smaller than 0.05 Hz, and the 95 % precision of 3D velocity is about 0.05 m/s.

Figure 6c shows the unitary frequency jitter of all satellites in different T_c. Similarly, there also exists optimal $T_c = 0.3$ s which makes the unitary frequency jitter minimum. So in this condition the tracking threshold is lower.

Figure 6d describes the computation in different update interval. Where EKF represents the time cost by navigation filter and NCO represents the time cost by

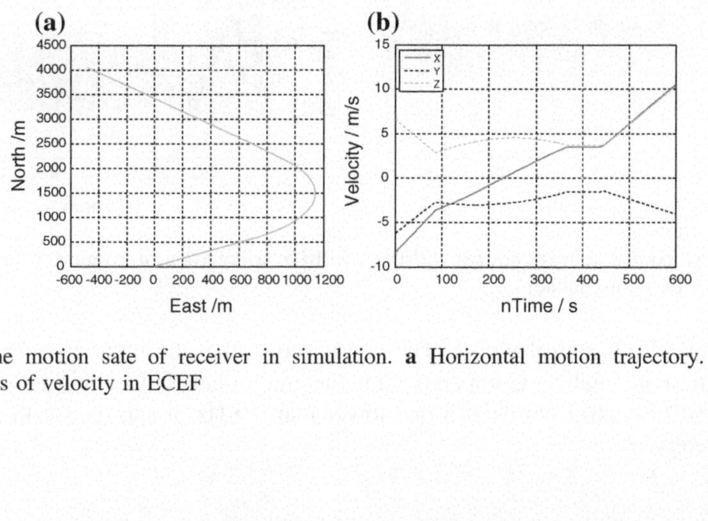

Fig. 5 The motion sate of receiver in simulation. **a** Horizontal motion trajectory. **b** Three dimensions of velocity in ECEF

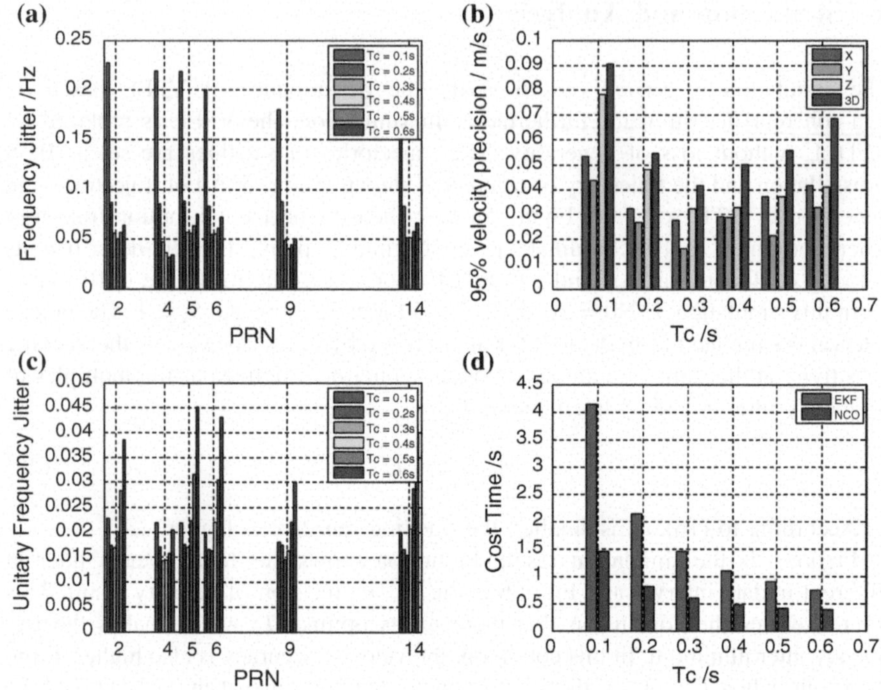

Fig. 6 Tracking performance in different T_c. **a** Frequency jitter. **b** Precision of velocity. **c** Unitary frequency jitter. **d** Computation result (Color figure online)

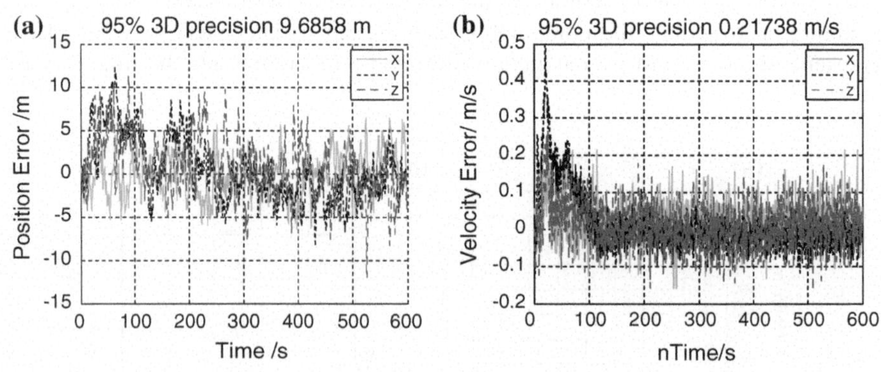

Fig. 7 Position and velocity error in C/N0 = 20 dBHz. **a** Precision of position. **b** Precision of velocity (Color figure online)

calculating NCO control words. The result shows that the interval is longer, the computation is smaller. Compared with the traditional VFLL, this novel VFLL updates NCO control words at a rate lower than 10 Hz, it can reduce the computation greatly.

To verify the tracking sensitivity performance, another simulation is carried out in the condition of $T_c = 0.3$ s and C/N0 = 20 dBHz. Figure 7 is the final position and velocity error and their 95 % precision are 9.68 m and 0.21 m/s. If the noise factor is 1 dB, 20 dBHz means the signal power is −153 dBm. This simulation indicates that this method can track GEO signal at power lower to −153 dBm steadily, but the cost of VFLL based on non-coherent discriminator is depressing the performance of MEO/IGSO signal tracking sensitivity, it also means the disparity performance of the two type signal can be reduced by this method.

5 Conclusion

A novel VFLL based on non-coherent discriminator is studied, and through adjusting the non-coherent accumulation time the update interval of VFLL can be designed more flexibly. Furthermore, based on the relationship between update interval T_c and tracking performance, different optimal models of T_c are built for different optimal targets. Taking BDS constellation as examples, the optimal parameter is analysed by numerical solution and simulation verification. The results show that there exists optimal update interval for VFLL based on non-coherent discriminator. After optimal design, for low dynamic receiver, VFLL can be updated at a rate lower than 10 Hz. It can greatly reduce the computation of VFLL compared with the traditional 50 Hz for L1C/A signal. In the simulation, the VFLL based on non-coherent discriminator can track GEO signal at C/N0 is 20 dBHz in steady when the update interval is 0.3 s.

References

1. Lashley M, Bevly D, Petovello M (2009) What are vector tracking loops and what are their benefits and drawbacks. Inside GNSS 16–21
2. Parkinson B, Sprilker J (1996) Fundamentals of signal tracking theory. In: Global positioning system: theory and applications, vol I. AIAA, Washington, DC
3. Lashley M, Bevly D (2009) Vector delay/frequency lock loop implementation and analysis. In: ION international technical meeting, Anaheim, CA, pp 1073–1086
4. Lashley M, Bevly D (2009) Performance analysis of vector tracking algorithms for weak GPS signals in high dynamics. IEEE J Sel Topics Signal Process 3(4):661–673
5. Won J, Eissfeller B (2010) Effectiveness analysis of vector-tracking-loop in signal fading environment. In: 5th ESA workshop on IEEE satellite navigation technologies and European workshop on GNSS signals and signal processing (NAVITEC), pp 1–6
6. Lashley M, Bevly D (2011) Comparison in the performance of the vector delay/frequency lock loop and equivalent scalar tracking loops in dense foliage and urban canyon. In: 24th ITM of the ION, Portland OR, pp 1786–1803
7. Pany T, Eissfeller B (2006) Use of a vector delay lock loop receiver for GNSS signal power analysis in bad signal conditions. In: Proceedimgs of IEEE/ION PLANS, San Diego, CA, pp 893–903

8. China Satellite Navigation Office (2013) BeiDou navigation satellite system signal in space interface control document open service signal (Version 2.0)
9. Schmidt G, Phillips R (2010) INS/GPS integration architecture performance comparisons. RTO-EN-SET-116, NATO RTO, pp 1–22
10. Holmes J, Chen J (2013) Spread spectrum systems for GNSS and wireless communication. Publishing House of Electronics Industry, Beijing
11. Borio D, Sokolova N, Lachapelle G (2009) Memory discriminators for non-coherent integration in GNSS tracking loops. In: Proceedings of the European navigation conference, ENC'09 Naples, Italy
12. Lin H, Pang J, He L et al (2015) A high sensitive carrier tracking algorithm for BDS GEO navigation signal: In Proceedings of CSNC, Xi'An
13. Xie G (2009) Principles of GPS and receiver design. Publishing House of Electronics Industry, Beijing
14. Betz J, Kolodziejski K (2009) Generalized theory of code tracking with an early-late discriminator part I: lower bound and coherent processing. IEEE Trans Aerosp Electron Syst 45(4):1538–1550

A Software-Defined Approach to STAP Nulling Algorithms for GNSS Anti-jamming

Hailong Xu, Xiaowei Cui and Mingquan Lu

Abstract Spatial Temporal Adaptive Processing (STAP) nulling algorithms have been widely used in GNSS anti-jamming receivers using multi-element antennas. Traditionally, these algorithms are implemented on hardware due to their high computational complexity. This makes it costly and inconvenient to test and verify different algorithms and parameters. In order to address this issue, a GNSS anti-jamming platform for STAP nulling algorithms is implemented using a software approach. On this platform, raw data from eight element channels can be processed real-time with configurable algorithms and parameters. Weights can be calculated by various methods such as SMI, LMS, and RLS. Plenty of intermediate data can be observed immediately or saved for post analysis. To meet the real-time requirement, a GPU is used as an accelerator and a batched method is adopted in programming. Besides, a software receiver is integrated for anti-jamming performance evaluation. Test results show that this platform has a good performance in both computational efficiency and interference mitigation, making it a promising product in research and prototyping, as well as practical applications such as ground stations.

Keywords GNSS anti-jamming · Antenna array · Software radio · Spatial temporal adaptive processing (STAP) · Nulling · GPU · Sample matrix inverse (SMI) · Least mean square (LMS) · Recursive least square (RLS)

1 Introduction

Due to the vulnerability of Global Navigation Satellite System (GNSS) signals and the more and more challenging radio environment the receiver faces, GNSS anti-jamming has become an issue of great importance. Among the most effective countermeasures against jamming in the receiver is using multi-element antenna

H. Xu · X. Cui (✉) · M. Lu
Department of Electronic Engineering, Tsinghua University, Beijing, China
e-mail: cxw2005@mail.tsinghua.edu.cn

J. Sun et al. (eds.), *China Satellite Navigation Conference (CSNC) 2016 Proceedings: Volume I*, Lecture Notes in Electrical Engineering 388,
DOI 10.1007/978-981-10-0934-1_52

and adaptive algorithms, among which Spatial Temporal Adaptive Processing (STAP) nulling algorithms have been widely used due to their frequency resolution and the advantage that no prior knowledge of desired signals is necessary [1]. Traditionally, these algorithms are implemented on hardware such as Field Programmable Gate Arrays (FGPA) and Application-Specific Integrated Circuits (ASIC) because of their high computational complexity. However, this approach suffers from drawbacks including high cost, long development cycle, as well as a low level of flexibility, configurability and extendibility, making it inconvenient and costly to test and verify algorithms during the early stage of development. In order to address this issue, a GNSS array antenna anti-jamming platform with high flexibility, configurability and extendibility is designed and implemented using a software-defined approach in this paper. On this platform, the performance of different algorithms and parameters under varied jamming scenarios can be tested in real environments, making it a test-bed of great value for both research and product prototyping.

Based on the high data acquisition capability offered by professional hardware and software, as well as the rich parallel computational resources provided by a GPU, this platform is able to process raw data from as many as eight antenna elements real-time using STAP nulling algorithms with various weights calculation methods such as Sample Matrix Inverse (SMI), Least Mean Square (LMS) and Recursive Least Square (RLS). Algorithms and parameters can be configured flexibly. Besides, a software receiver is integrated after the STAP module to validate the anti-jamming performance and generate position-velocity-time (PVT) results. Apart from a real-time mode, the platform can also be operated in a data-fetching or post-processing mode, which makes it convenient to compare different algorithms and parameters under exactly the same interference scenario.

2 System Design

Figure 1 shows the system design of the platform. The hardware of this platform comprises an array antenna and its followed RF front-end, as well as an enhanced PC which includes an off-the-shelf multi-channel data acquisition card, a RAID disk array, and a hybrid processor which is made up of a CPU and a GPU. The software which comprises an anti-jamming processing module and a software receiver runs on the hybrid processor. After the 8-element patch antenna array and the RF front-end, the radio signal in L band is transformed to multi-channel intermediate frequency (IF) signals of about 45 MHz. In the real-time mode, the IF signals are digitalized by the data acquisition card with a 14-bit resolution and a sampling rate of 60 MSPS. Then, the multi-channel sampled data are down converted digitally to zero IF and decimated by three in the FPGA of the card. Thus, the raw data transferred to the anti-jamming processing module is 20 MSPS complex data at zero IF. After the anti-jamming processing, the multi-channel raw data turn to single-channel outputs, which are then processed by the software

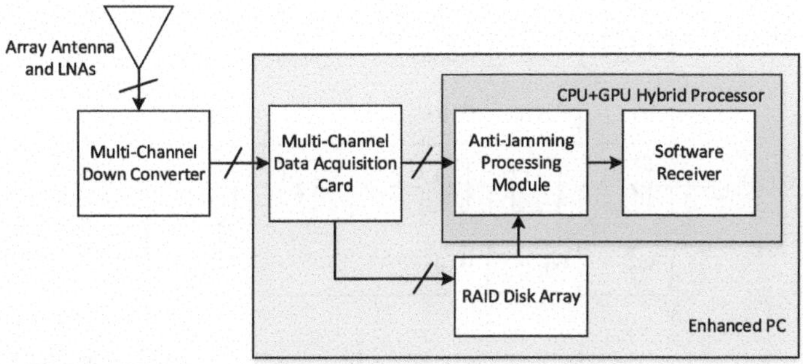

Fig. 1 System architecture and data flow of this platform. The *slash on the line* indicates that the data stream is multi-channel

receiver for navigation. In the post-processing mode, the raw data ready for anti-jamming processing are read from a file on the disk array.

CPU+GPU hybrid processor is the computing core of the whole system. The CPU is Intel's i7 quad-core processor with Single-Instruction-Multiple-Data (SIMD) techniques. The GPU is NVIDIA's Geforce GTX TITAN X, which comprises 3072 cores and 12 GB memory. Data exchange between the CPU, the GPU, the disk array and the data acquisition card is through PCI-E bus. The raw data bandwidth can reach 640 MB/s when the number of antenna elements is eight.

3 Algorithms

STAP extends spatial only processing (SOP) by adding a Finite Impulse Response (FIR) filter behind each antenna element, thus inducing resolution in the temporal (frequency) domain [2]. In nulling, the weights distort the antenna pattern by forming nulls in interference directions, while the information of desired signals is ignored. The weight vector can be calculated by a SMI method, or other iterative methods such as LMS and RLS [3]. Details are introduced below.

3.1 STAP

The diagram of STAP is shown in Fig. 2, in which M denotes the number of antenna elements and N denotes the FIR filter tap length. $x_i(n)$ $(i = 1, 2, \ldots, M)$ represents the complex sample raw data of element i at sampling period n. If the input data vector is defined by

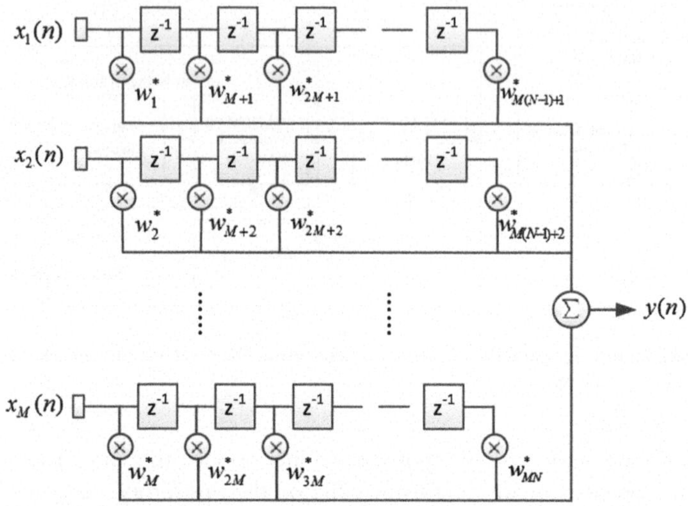

Fig. 2 The diagram of STAP with M antenna elements and N filter taps

$$\begin{aligned}\mathbf{x}(n) = [&x_1(n), x_2(n), \ldots, x_M(n),\\ &x_1(n-1), x_2(n-1), \ldots, x_M(n-1),\\ &\ldots,\\ &x_1(n-N+1), x_2(n-N+1), \ldots, x_M(n-N+1)]^{\mathrm{T}}\end{aligned} \tag{3.1}$$

and the weight vector is defined by

$$\mathbf{w} = [w_1, w_2, \ldots, w_{MN}]^{\mathrm{T}} \tag{3.2}$$

then the output of the anti-jamming process is given by

$$y(n) = \mathbf{w}^H \mathbf{x}(n) \tag{3.3}$$

Weights can be calculated under various criterions. If the Linear Constraint Minimum Variance (LCMV) criterion is used, then the weight vector is given by

$$\mathbf{w}_{\mathrm{opt}} = \frac{\mathbf{R}^{-1}\mathbf{s}}{\mathbf{s}^H \mathbf{R}^{-1}\mathbf{s}} \tag{3.4}$$

in which $\mathbf{R} = E\{\mathbf{x}\mathbf{x}^H\}$ is the covariance matrix, and $\mathbf{s}$ is the constraint vector, whose elements are all zero except for one element which is 1. In this case, the criterion is also called power inversion (PI), meaning that the weights are calculated to get the minimum output power [4].

3.2 SMI

In practical, the covariance matrix is usually estimated by

$$\hat{\mathbf{R}} = \frac{1}{K}\sum_{k=0}^{K-1}\mathbf{x}(n-kN)\mathbf{x}(n-kN)^H \tag{3.5}$$

where K is the number of snapshots taken. k is multiplied by N so that successive input vectors do not overlap with each other and can be regarded as independent. After $\hat{\mathbf{R}}$ is obtained, the weight vector $\mathbf{w}$ can be calculated by (3.4). This method contains matrix inverse calculation so that it is called SMI [4].

3.3 LMS Iteration

The LMS method makes the weight vector vary along the direction in which the output power decreases the fastest, meanwhile satisfy certain linear constraints. Here, the iterative equation is given directly by

$$\mathbf{w}(n+1) = \mathbf{w}(n) - \mu\left(\mathbf{I} - \frac{\mathbf{s}\mathbf{s}^H}{\mathbf{s}^H\mathbf{s}}\right)\mathbf{R}\mathbf{w}(n) \tag{3.6}$$

where $\mathbf{I}$ is the identity matrix and μ is the iterative step. If $\mathbf{R}$ is replaced by the instant estimation of one snapshot, then iterative equation turns to

$$\mathbf{w}(n+1) = \mathbf{w}(n) - \mu\left(\mathbf{I} - \frac{\mathbf{s}\mathbf{s}^H}{\mathbf{s}^H\mathbf{s}}\right)\mathbf{x}(n)\mathbf{x}(n)^H\mathbf{w}(n) \tag{3.7}$$

In LMS, the convergence speed and steady state response is mainly controlled by μ, which is usually calculated by a normalization method given by

$$\mu = \frac{\bar{\mu}}{\mathbf{x}(n)\mathbf{x}(n)^H + \sigma} \tag{3.8}$$

where $\bar{\mu}$ changes between 0 and 1, while σ is a small number to prevent the denominator being too close to zero.

3.4 RLS Iteration

The RLS method uses all the input vectors from the beginning to the present to estimate the covariance matrix, i.e.,

$$\hat{\mathbf{R}}(k) = \sum_{i=0}^{k} \lambda^{k-i}\mathbf{x}(i)\mathbf{x}(i)^H \tag{3.9}$$

where λ is called the "forget factor" which usually ranges in [0.95,1) in practice. $\hat{\mathbf{R}}(k)$ can be calculated iteratively from $\hat{\mathbf{R}}(k-1)$, and based on (3.4) and the matrix inversion lemma, iterative equations of RLS can be derived and given by

$$\mathbf{R}(k)^{-1} = \frac{1}{\lambda}(\mathbf{I} - g(k)\mathbf{x}(k)^H)\mathbf{R}(k-1)^{-1} \tag{3.10}$$

$$g(k) = \frac{\mathbf{R}(k-1)^{-1}\mathbf{x}(k)}{\lambda + \mathbf{x}(k)^H\mathbf{R}(k-1)^{-1}\mathbf{x}(k)} \tag{3.11}$$

$$\mathbf{w}(k) = \frac{\Lambda(k-1)}{\Lambda(k)}\frac{1}{\lambda}(\mathbf{I} - g(k)\mathbf{x}(k)^H)\mathbf{w}(k-1) \tag{3.12}$$

$$\Lambda(k) = \mathbf{s}^H\mathbf{R}(k)^{-1}\mathbf{s} \tag{3.13}$$

4 The Software-Defined Approach

The software needs the most developed labour in this project. It is developed in C++ with Microsoft Visual Studio 2010 and NVIDA CUDA 7.0 on Windows 7. Details are introduced in this section.

4.1 *Software Architecture*

The software architecture comprises of the anti-jamming processing module, the software receiver module and the GUI module; as Fig. 3 shows. The whole software is driven by three CPU threads. The data-fetching thread is responsible for getting multi-channel raw data from the data acquisition card (the real-time mode) or the disk array (the post-processing mode) and then delivering them to the anti-jamming processing module for covariance matrix estimation, weight vector calculation and output calculation in the data processing thread. The output single-channel data are then transferred to the software receiver module for signal acquisition, correlation, tracking and PVT calculation. The GUI thread is responsible for middle status display.

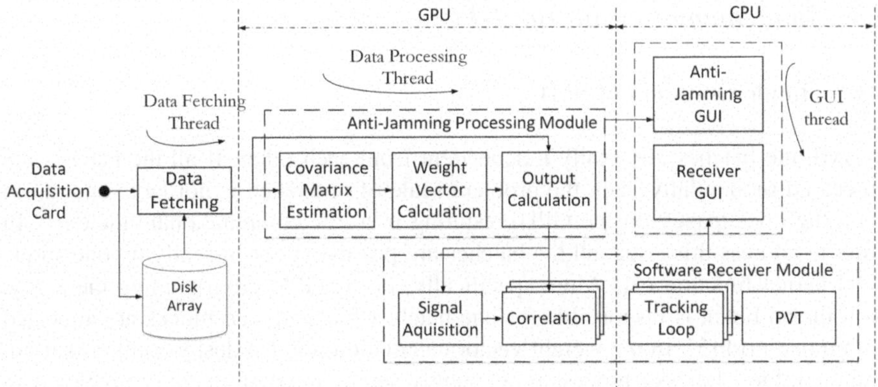

Fig. 3 The software architecture

4.2 Data Partition

Different from the stream processing in a FPGA implementation, the data stream must be divided into chunks before being transferred to the GPU. The data layout of one chunk is shown in Fig. 4, in which rows present antenna elements while columns present sampling periods. The chunk length is set to 50 ms, which is resulted from a balance between memory cost and data copying time expense. Further in SMI, each data chunk is divided into Z batches, each of which comprises K input vectors with no overlap. Each data batch corresponds to one unique weight vector, so the weight updating period is given by

$$T_{\text{update}} = KNT_{\text{s}} \tag{4.14}$$

where T_{s} is the sampling period.

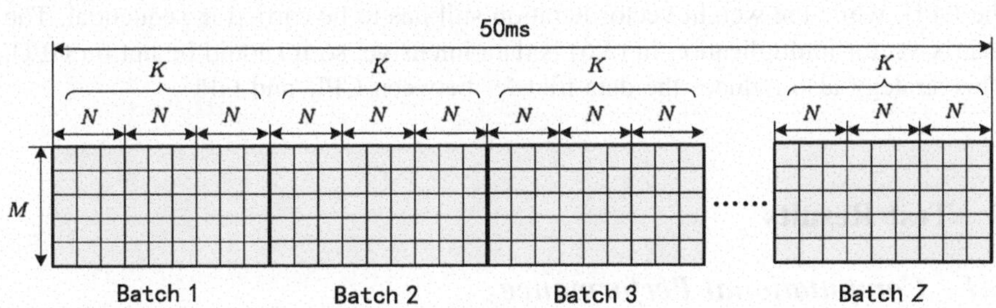

Fig. 4 One chunk of data segmented into blocks

4.3 Implementation on the GPU

4.3.1 Implementation of SMI

In SMI, the batches are totally independent from each other. If all the batches are processed sequentially, then the problem scale of one batch is not large enough to get a high occupancy on the GPU, resulting in a low computational efficiency. In order to address this issue, all the batches are processed concurrently by one single GPU kernel function call. More specifically, in the GPU programming hierarchy, each thread block is responsible for one batch, while Z thread blocks are allocated within one grid [5]. In the weight vector calculation, the Cholesky method is used. Although the Cholesky process is seemingly serial, a certain degree of parallelism can still be achieved by attaching different rows to different threads within one thread block.

4.3.2 Implementation of LMS and RLS

In the LMS or RLS iteration, the latter weight vector is strictly dependent on the former one, resulting complete concurrency of all the batches is unachievable. Thus, the weight vectors of all batches are calculated on the CPU and then transferred to the GPU to calculate the outputs in a batch-concurrent manner. SIMD technique of the CPU is adopted to improve computational efficiency.

If the iteration is conducted in a sample-by-sample manner, then the count of iterations within one chunk is so large that it is hard for the CPU to satisfy the real-time requirement. To reduce computational cost, an "interval-sample" iteration method is adopted, in which one iteration is carried out per K_{interval} sampling periods using the current input vector. By increasing K_{interval}, the computational complexity can be reduced, until the real-time requirement is met.

For LMS, a "block" iteration method can also be adopted, in which the estimated covariance matrix is put into (3.6) for weight vector iteration. Covariance matrix estimation and outputs calculation of all batches can be performed concurrently on the GPU, while the weight vector iteration still has to be carried in sequential. The matrix-vector multiplication in (3.6) is data-intensive, so it should be put on GPU. This strategy also reduces the data transfer between CPU and GPU.

5 Test Results

5.1 Computational Performance

First, the computational performance is given. In the test, M is set to 8 and N is set to 5, while the sample number corresponding to one weight vector varies and the

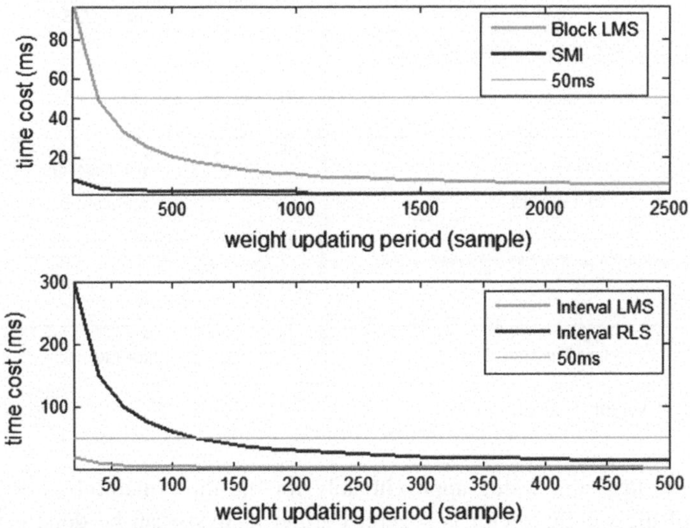

Fig. 5 Computational performance

time cost of processing one chunk of data is measured. Results are given in Fig. 5. By meeting the real-time requirement, we mean that the time cost should be less than 50 ms and leave plenty of idle time for other operations. It can be seen that the SMI implementation has a good computational efficiency, which is attributed to the full parallelism of all the batches. The time cost of block-LMS is far larger than interval-LMS due to the covariance matrix estimation. For interval-RLS, K_{interval} should be large enough to satisfy the real-time requirement.

5.2 Anti-Jamming Performance

As an example, this paper gives the anti-jamming performance of four algorithms using Compass B3 signal in a real scenario which comprises one wideband interference. The interference-signal-ratio (ISR) is about 60 dB, while the interference bandwidth is 10 MHz centred at B3 frequency point. M is set to 8 and N is set to 3. For SMI, $K = 200$. For block-LMS, $\bar{\mu} = 0.2$ and $K = 200$. For interval-LMS, $\bar{\mu} = 0.2$, $\sigma = 10^{-9}$ and $K_{\text{interval}} = 200$. For interval-RLS, $\lambda = 0.9999$ and $K_{\text{interval}} = 200$. As a result, the weight updating period is 30 μs for the first two algorithms and 10 μs for the last two. In order to compare different algorithms, the post-processing mode is used.

The weight vectors can be saved on this platform. Figure 6 gives the variation of the weight vector 2-norm from the beginning to 0.5 s. It can be seen that among the three iterative methods, the convergence speed of interval-RLS is the fastest, while interval-LMS is less stable than the other two methods. The weight vector 2-norm

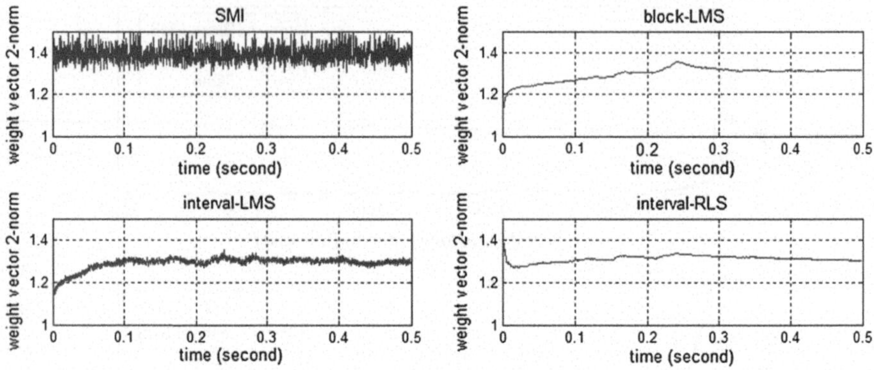

Fig. 6 Weight vector 2-norm variation

converges to the same value approximately for the three methods. For SMI, the variation of the weight vector is severe. Further analysis can be done using these saved weights.

Figure 7 gives the C/N0 results of a specific satellite after the anti-jamming processing. It can be seen that the interference can be mitigated effectively for all the four algorithms, and the anti-jamming performance from high to low is block-LMS, interval-RLS, SMI and interval-LMS, respectively. It is suspected that this is partly attributed to the stability degree of the weight vector.

Fig. 7 C/N0 variation of PRN 9

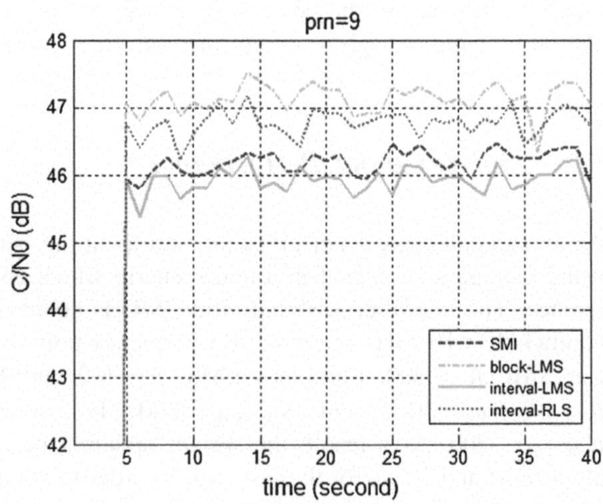

6 Conclusion

The software-defined platform developed in this paper outperforms previous hardware implementations with a high level of flexibility, configurability and extendibility. Various STAP nulling algorithms and parameters are implemented and can be tested on it. The batched method is used in GPU programming. Test results show that this platform has a good performance in both computational efficiency and anti-jamming. Based on its extendibility, the platform can be used as not only a test-bed for research and prototyping, but also as a real product in specific applications such as ground stations.

Acknowledgement This work is supported by the National Science Foundation of China (GrantU1333203) and the Civil Aviation Administration of China (Grant MHRD20140102).

References

1. Sklar JR (2003) Interference mitigation approaches for the global positioning system. Lincoln Lab J 14(2):167–179
2. Fante RL, Vaccaro J (2000) Wideband cancellation of interference in a GPS receive array. IEEE Trans Aerosp Electron Syst 36(2):549–564
3. Van Trees HL (2002) Optimum array processing: part IV of detection, estimation, and modulation. Wiley, New York
4. Moore TD (2002) Analytic study of space-time and space-frequency adaptive processing for radio frequency interference suppression. The Ohio State University
5. NVIDIA C (2015) NVIDIA CUDA C programming guide. NVIDIA Corporation 120:18

A Correlation Interferometer-Based Spoofing Detection Technology Research

Guangwei Fan, Baoguo Yu and Kaiwei Yang

Abstract For detection and identification of GNSS spoofing, this paper proposes to use multi-antenna two-channel direction-finding method and the carrier phase observed quantities that are provided by the receiver to estimate the arrival direction of GNSS signal or spoofing signal, which is then compared with the ephemeris data solved by the receiver. In such a way, it is possible to make high-precision estimation of the arrival direction of spoofing signal. The validity of the proposed algorithm was validated with a simulation test.

Keywords Spoofing · Carrier phase · Detection

1 Introduction

GNSS spoofing interference refers to the act that intentionally generates spurious signals that have parameters identical or similar to those of the genuine GNSS signal and prevent the navigation receiver from acquiring or tracking the genuine navigation signal, thereby to adversely affect the navigation and positioning service quality and performance of the GNSS receiving system within the effective coverage. Therefore, it is necessary to investigate the detection and parameter estimation techniques of GNSS spoofing signal, for it is of great significance in eliminating the effect of spoofing interference to navigation system, supporting the stable service of GNSS and improving the resistance [1, 2] of GNSS to spoofing attack.

With regard to GNSS spoofing interference, Chinese and oversea researchers already proposed some detection techniques [2–6], for example, using the CNR of received signal or the absolute power rate, using the pseudo-range rate of signal, using the carrier phase rate or Doppler frequency rate of received signal, cross-correlation detection at different frequency points and arrival time detection

G. Fan (✉) · B. Yu · K. Yang
State Key Laboratory of Satellite Navigation System and Equipment Technology, Hebei, China
e-mail: fgweihb@163.com

J. Sun et al. (eds.), *China Satellite Navigation Conference (CSNC) 2016 Proceedings: Volume I*, Lecture Notes in Electrical Engineering 388,
DOI 10.1007/978-981-10-0934-1_53

method. However, these techniques are effective only for some particular spoofing types or scenarios, not of universal applicability.

This paper proposes a method to estimate the direction of arrival (DOA) of the received signal with multiple channels and multiple antennas on the basis of real-time computation of the carrier phase of navigation signal by the receiver itself, which is then compared with the results acquired by ephemeris calculation to detect the spoofing interference. The algorithm principle and implementation process are described, and a simulation test was given to evaluate the algorithm validity.

2 Receiving Antenna Model

In radio reconnaissance, five-element two-channel circular array is usually adopted to detect the arrival direction of interference wave. In the detection of GNSS spoofing interference, this method may be borrowed for direction finding of spoofing signal, i.e., to obtain the real-time phases and the phase difference with double channels. The antenna model is as shown in Fig. 1.

M identical omnidirectional arrays are uniformly distributed along the circumference of radius R on the plane xy, as shown in Fig. 1. Spherical coordinates are used to indicate the arrival direction of incident plane wave and the origin point O of the coordinate system is at the center of the array, i.e., the center of circle. The depression angle θ of the signal source refers to the angle between the line from the origin point to the signal source and the axis Z. The direction vector $a(\theta, \varphi)$ represents the array response when the DOA is (θ, φ), which can be expressed as follows.

Fig. 1 Flow pattern of circular array

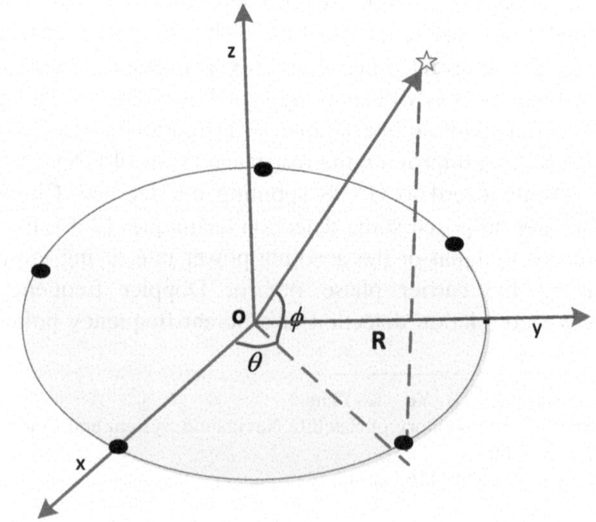

$$a(\theta,\varphi) = \begin{bmatrix} \exp[j2\pi R\sin\theta\cos(\phi-\gamma_0)/\lambda] \\ \exp[j2\pi R\sin\theta\cos(\phi-\gamma_1)/\lambda] \\ \vdots \\ \exp[j2\pi R\sin\theta\cos(\phi-\gamma_{M-1})/\lambda] \end{bmatrix} \tag{2.1}$$

where, $\gamma_m = 2\pi m/M(m = 0, 1, \ldots, M-1)$ and R represents the radius of the array.

3 Estimation of Carrier Phase

The navigation signals broadcasted by a satellite reach the satellite navigation receiver near the earth's surface through spatial transmission over a certain distance. At the receiving end, the GNSS signals can be expressed with the following equation.

$$x(t) = \sum_{i=1}^{L} S_i(t-\tau_i)\cos\left(2\pi\left(f_{D,i}+f_{IF}\right)t\right) \tag{3.2}$$

where, $f_{D,i}$ represents the Doppler shift that arrives at the receiver, L represents the number of received satellite signals, τ_i represents the time propagation delay (TPD) resulting from the electromagnetic wave refraction caused by the ionospheric layer and the tropospheric layer, and $s_i(t)$ represents the signal coming from the ith satellite, $s_i(t) = d_i(t) * C_i(t)$, where $d_i(t)$ represents the data code and $C_i(t)$ represents the spread-spectrum pseudocode.

The spoofing signal has the same spread-spectrum code and carrier frequency as the genuine navigation signal, but different time delay and data code for the purpose of spoofing the normal positioning of the receiver. The spoofing interference at the receiver can be expressed as follows:

$$J(t) = \sum_{j=1}^{K} d'_j(t)C_j(t-\tau_j)\cos\left(2\pi\left(f_{D,j}+f_{IF}\right)t\right) \tag{3.3}$$

where, K represents the number of spoofing signals, $d'_j(t)$ represents the data code of the spoofing signal, which is the same as that of the genuine navigation signal if it is repeater spoofing, but different from that of the genuine navigation signal if it is generator spoofing, and $C_j(t-\tau_j)$ represents the pseudocode of the spoofing signal, which is identical to that of the genuine navigation signal. The time delay of the spoofing signal will be greater than that of the genuine GNSS signal if it is repeater spoofing.

In case of spoofing attack, the signal received at the receiving end can be expressed as follows:

$$r(t) = \sum_{i=1}^{L} x_i(t) + \sum_{j=1}^{K} J_i(t) + n(t) \tag{3.4}$$

where, $n(t)$ represents the system noise.

After coherent integration, the signals coming through Channel I and Channel Q can be expressed as follows, respectively:

$$Y_I(n) = ad(n)R(\tau)\sin c(f_e T_{coh})\cos\phi_e + n_I \tag{3.5}$$

$$Y_Q(n) = ad(n)R(\tau)\sin c(f_e T_{\mathrm{coh}})\sin\phi_e + n_Q \tag{3.6}$$

where, a represents the signal amplitude, τ represents the phase difference between the pseudocode of the received satellite signal and the native code, f_e represents the frequency difference between the frequency of the received carrier and the search frequency, ϕ_e represents the phase difference between two carriers, n_I and n_Q represent the uncorrelated zero-mean Gaussian noises of Channel I and Channel Q, respectively.

If the equidirectional signal $Y_I(n)$ and the orthogonal signal $Y_I(n)$ are combined together, then it can be expressed in plural form as follows:

$$\begin{aligned} r_P(n) &= Y_I(n) + jY_Q(n) \\ &= aD(n)\cos(\omega n + \theta_e) = A_P(n)e^{j\phi(n)} \end{aligned} \tag{3.7}$$

where, the amplitude $A_P(n)$ of the plural vector $r_p(n)$ contains the information of data code, and its phase angle $\phi(n)$ reflects the phase difference and the frequency difference between the original signal and the replicate signal.

It is possible to know the phase difference between the navigation signal and the reference signal with the aid of a phase discriminator. The corresponding calculation formula can be expressed as below:

$$\phi_\varepsilon = \arctan\left(\frac{Q_P}{I_P}\right) \tag{3.8}$$

When the actual phase difference is within the range of −90° through +90°, this phase discriminator will remain working linearly and the phase-discrimination result has nothing to do with the amplitude.

With a number of receivers that have the same native code, it is possible to measure the carrier phases of the same signal arriving at different receiving antennas. And then, by calculating the carrier phase difference between different antennas, it is possible to know the arrival direction of the received signal.

4 Calculation of Two-Channel Phase Difference Between Multiple Antennas

With the direction-finding method based on correlation interferometer, two channels and multiple antennas, it is possible to estimate the arrival direction of GNSS signal or spoofing signal in form of two-dimensional wave using two receivers. Plus, double baselines can provide a wider viewing angle range than single baseline.

The signal direction-finding algorithm of double baselines is based on calculation of the phase difference. Let us assume that there is a four-element antenna (N, S, E, and W), the distances of the antenna away from NS and EW are both d, the azimuthal angle and the elevation of the arrival wave are θ and φ, respectively. If a signal arrives at the direction-finding antennas 1 and 2 at the angle θ with the visual axis of the antennas, then the phase difference ϕ between the two antennas can be calculated with the following formula:

$$\phi = 2\pi \frac{1}{\lambda} \sin\theta \tag{4.9}$$

where, λ represents the signal wavelength and l represents the distance between the two antennas. If the phase responses of the two channels are identical with each other, then the phase difference of the output signal from the receiver is still ϕ The information of the phase difference can be extracted with a phase discriminator.

$$U_c = K\cos\phi \tag{4.10}$$

$$U_s = K\sin\phi \tag{4.11}$$

where, K represents the system gain.

Now, get the azimuth of the incoming signal by means of angular transformation.

$$\phi = \arctan\frac{U_s}{U_C} \tag{4.12}$$

Put it into Eq. (4.10) to obtain

$$\theta = \arcsin\frac{\phi\lambda}{2\pi l} \tag{4.13}$$

5 Application of Double-Baseline Two-Channel Direction-Finding Algorithm to Spoofing Detection and Identification

The double-baseline two-channel receivers adopt the same native code in correlation with the signals received by the two receivers, so as to work out the phase difference between the signal received via two receiving channels and the native code. This phase difference can then be used to derive the phase difference between the signals received via two channels, thereby to apply the direction-finding algorithm of correlation interferometer to direction-finding of spoofing attack. By controlling the two-channel receivers to work with different receiving.

Antennas with the aid of an antenna channel switching unit, it is possible to estimate the arrival direction of the received navigation signal or spoofing signal, which is then compared with the calculated ephemeris to complete the spoofing detection. The entire process is shown in the flow chart (Fig. 2).

Let us assume that the phases of the signal received through the two channels are θ_i and θ_i, respectively, and the phase difference between the two channels is $\varphi_{ij} = \theta_i - \theta_j$. Select N combination modes of antennas and determine N combinations of phase difference. Redefine the phase difference as φ_n, in which $n = 1, 2, \ldots, N$. If the sample of phase difference is written into the form of vector, then $\varphi = \{\varphi_1, \varphi_2, \ldots \varphi_N\}$. Save the phase differences previously measured in different directions as a datasheet that is represented with φ_0 The correlation function can be defined as below

$$\rho_m = \frac{\varphi^{\mathrm{T}} \varphi_{0m}}{\left[\varphi^{\mathrm{T}} \varphi\right]^{1/2} \left[\varphi_{0m}^{\mathrm{T}} \varphi_{0m}\right]^{1/2}} \tag{5.14}$$

where, $m = 1, 2, \ldots, M$ represents the range of search angle and φ_{0m} represents the phase difference array corresponding to the mth direction.

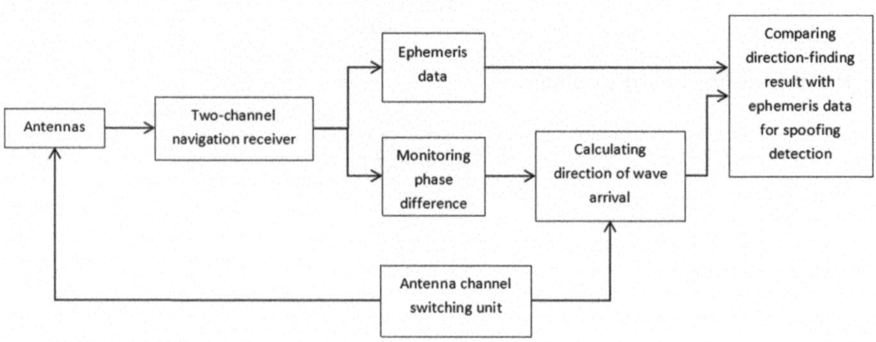

Fig. 2 Principle of spoofing detection

By calculating the real-time phases of the signals received from the same satellite but through different receiving channels with two-channel receivers and then working out the phase difference between them, it is possible to estimate the arrival direction of the received signal based on the principle of correlation interferometer. Since the tracking and positioning process of a receiver is pretty quick and the phase difference is made available in less than 1 s with two-channel receivers, therefore, the two-channel switching process for multiple antennas is very quick too. Let us assume that the satellite or spoofing source remains still relative to the receiver in the process of channel switching for direction finding. A circular multi-element antenna is adopted to estimate the azimuth and elevation of the received signal, which are then compared with the satellite position information solved by the receiver to complete the detection and identification of spoofing signal.

Therefore, it is acceptable to simplify the acquiring and tracking process of receiver. Assuming that the frequency and pseudocode initial phase of the received signal are known, the detection and identification procedure of spoofing attack on the basis of multi-element reception can be described as follows:

1. Determine the baseline selecting sequence for direction finding.
2. The receiver uses the homologous native codes to acquire and track the received navigation signal or spoofing signal, and then work out the carrier phases of the signal received via different satellite tracking channels using Eq. (3.8).
3. Calculate the carrier phase difference of the same signal but received via two channels of the first baseline.
4. Poll different baselines to know the carrier phase differences of the signal received from the same satellite but through different baselines.
5. Use Eq. (5.14) to calculate the correlation values that the carrier phase differences are correlated with the raw data acquired from different directions.
6. Find the maximum value among the correlation values and determine the arrival direction of signal with reference to the subscript corresponding to the maximum value.
7. Compare the elevation angle and azimuthal angle of every incoming signal with the satellite position that is acquired by the receiver by means of ephemeris calculation. If it matches the satellite position obtained by ephemeris calculation, it suggests that the incoming signal is a navigation signal; otherwise, it will be considered as spoofing signal.

6 Performance Simulation

To validate the performance of the proposed detection algorithm, the BDS signals at different frequency points were taken as example for simulation validation. The test environment was so set that a 7-element circular array was used as receiving array with additive white Gaussian noise as ambient noise at the input signal-noise ratio of −20 dB. An antenna channel switching unit was employed for the two-channel receivers to switch into different antenna channels. The circular array was of radius 0.25 m and composed of seven elements. The baseline groups were set to 1/4, 2/5, 3/6, 4/7, 5/1, 6/2, and 7/3, respectively.

We took BDS signal at B3 frequency point as example for detection and identification of GNSS signal and spoofing signal in the simulation test. We assumed that satellite signals entered Channels 1, 2, and 3 of multi-antenna double receivers at the incidence azimuthal angles of 30°, 120°, and 160°, respectively, at the same SNR of −45 dB, thereby to validate the proposed algorithm in direction-finding performance of satellite signal. The direction-finding precisions for different signals are shown in Figs. 3, 4, and 5.

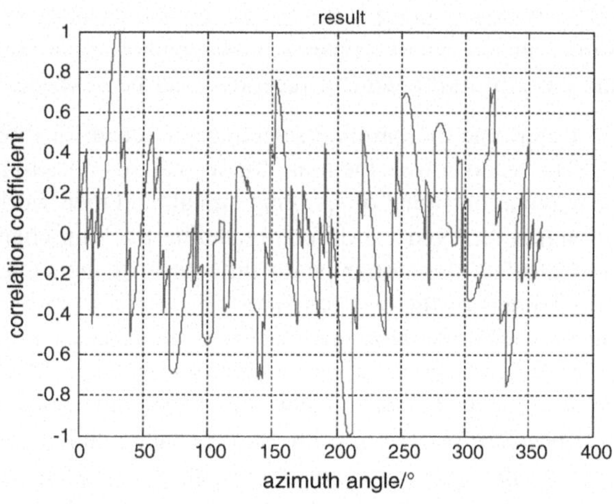

Fig. 3 Direction-finding result via Channel 1

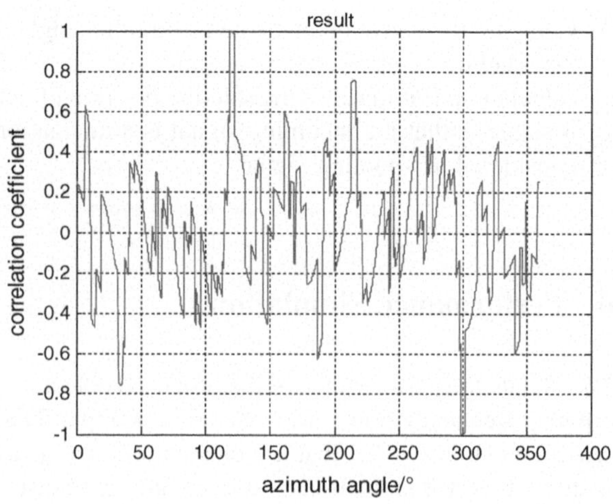

Fig. 4 Direction-finding result via Channel 2

Fig. 5 Direction-finding result via Channel 3

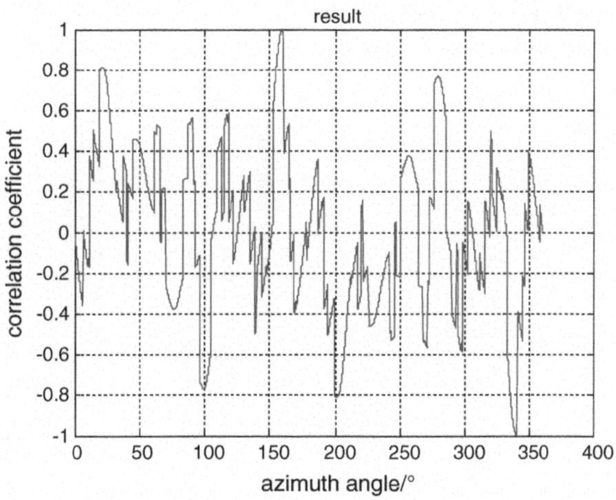

Fig. 6 Direction-finding result for Channel 1

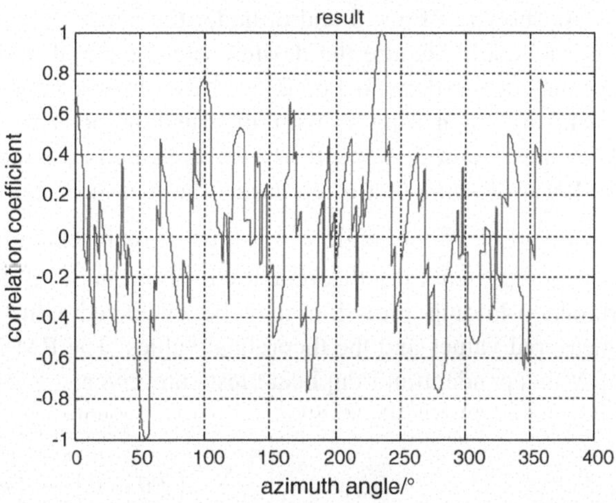

The direction-finding results corresponding to three satellites are shown in Figs. 1, 2, and 3, respectively. By acquiring and tracking the signal of higher correlation gain with receivers, it is possible to estimate the carrier phase of the received signal. Based on the real-time carrier phases of two antennas, we may work out their carrier phase difference. By switching antenna elements, it is possible to know several phase differences and thereby to know the direction of the received satellite signals that are of low power.

When the satellite signal was subject to the attack from a repeater spoofing signal coming in the direction of 235° and of NPR −30 dB, the satellite signals coming through Channels 1 and 2 would be spoofed if the receiver acquired and tracked the signal of stronger power. The signal direction-finding results for Channels 1 and 2 are as shown in Figs. 6 and 7, respectively.

Fig. 7 Direction-finding result for Channel 2

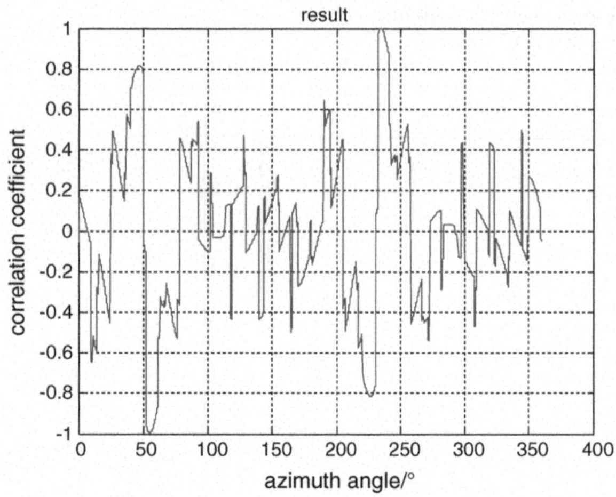

As shown in Figs. 5 and 6, under the attack of a strong spoofing signal, neither channel could acquire the desired satellite signal normally but the spoofing signal coming in the direction of 235°. Now, it was very easy to identify the spoofing signal by comparing it with the satellite position at 30° and 60° obtained in Channels 1 and 2, respectively, by ephemeris calculation.

Let us assume that θ represents the measured azimuthal angle, φ represents the measured elevation angle every time, $\widehat{\theta}$ represents the theoretical azimuthal angle, and $\widehat{\varphi}$ represents the theoretical elevation angle of the desired signal source. The direction-finding precision may be evaluated with the RMS error between the measured values and the theoretical values. The RMS error is expressed as follows if N independent and duplicate tests are given.

$$E_{\text{MSE}} = \sqrt{\frac{1}{N}\sum_{i=1}^{N}\left(\left(\theta - \widehat{\theta}\right)^2 + \left(\varphi - \widehat{\varphi}\right)^2\right)} \tag{5.15}$$

Under the conditions that the azimuthal angle was changed from 0° to 360°in steps of 1° at the SNR of −30 dB, 100 duplicate tests were conducted at intervals of 1°. The relation of direction-finding precision changing along with azimuth is illustrated in the diagram (Fig. 8).

As shown in the diagram, the accuracy of angular measurement remained at the level of 0.3° around in every direction when the SNR was −35 dB. It suggested that the proposed algorithm provided favorable direction-finding precision for the signals coming in every direction. By means of comparison to know if the arrival direction of every incoming signal matches the known direction based on ephemeris calculation, it is possible to identify whether the incoming signal is a normal satellite signal or a spoofing signal.

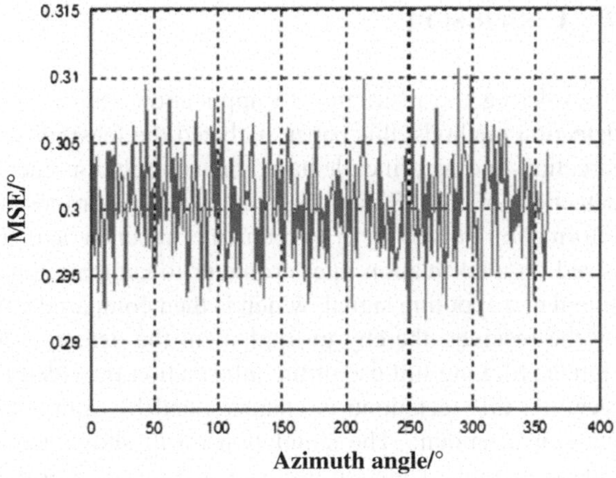

Fig. 8 Direction-finding precisions at different azimuths

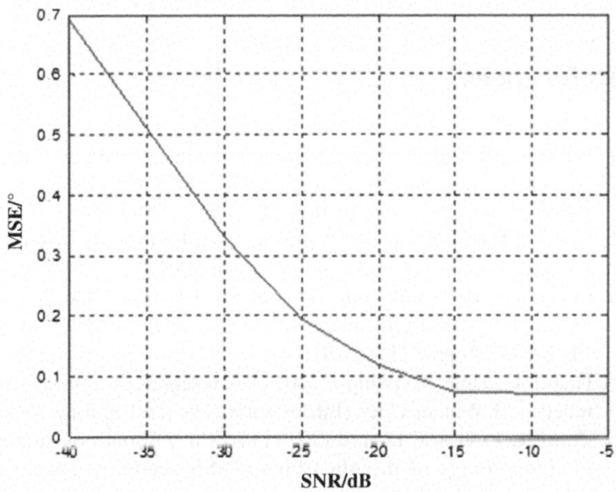

Fig. 9 Direction-finding precisions at different SNRs

To simulate the direction-finding performance of the proposed algorithm at different SNRs, 100 duplicate tests were done with the azimuthal angle of the incoming signal set at 60° and at the SNR that varied from −45 dBm to −5 dBm. The relation of direction-finding precision changing along with SNR is shown in the diagram (Fig. 9).

As shown in the diagram, the proposed algorithm performed well in angle measurement even at a low SNR. Generally, the power of a spoofing signal is 5–10 dB higher than that of a navigation signal, i.e., the SNR of a spoofing signal is often above −40 dB. Since the direction-finding precision of the proposed algorithm is finer than 0.7°, it is capable of identify the spoofing signals precisely. Conclusively, this proposed technique is of favorable detection and identification performance for spoofing signals.

7 Conclusion

Considering the restriction to application scenarios of the conventional spoofing detection methods, this paper analyzed the feasibility of two-channel multi-antenna direction-finding principle based on a correlation interferometer being applied to the detection of GNSS spoofing attack. The proposed method utilizes the phase information solved by a two-channel receiver and the direction-finding principle based on a correlation interferometer to estimate the arrival direction of a satellite signal or a spoofing signal, which is then compared with the satellite position solved by the receiver, thereby to implement the detection and identification of spoofing signals. Making full use of the information provided by the off-the-shelf navigation receiver, this technique is characterized by simplicity in structure and low computational amount. The simulation result shows that the proposed algorithm provides favorable detection performance for all spoofing signals.

References

1. Basker S (2010) Jamming: a clear and present danger. GPS World 21(4):8–9
2. Nielsen J, Broumandan A, Lachapelle G (2012) GNSS spoofing detection for single antenna handheld receivers. Navigation 58(4):335–344
3. Ledvina BM, Bencze WJ, Galusha B, Miller I (2010) An in-line anti-spoofing device for legacy civil GPS receivers. In: Proceedings of ION ITM 2010, San Diego, CA, 26 Jan 2010
4. O'Hanlon BW, Psiaki ML, Humphreys TE, Bhatti JA (2010) Real-time spoofing detection in a narrow-band civil GPS receiver. In: Proceedings of ION GNSS 2010, Portland, OR, PP, Sept 21–24, 2010, pp 2211–2220
5. Huang L, Tang X, Wang F (2011) A research of anti-spoofing methods in satellite navigation receiver. J Wuhan Univ (Information Science Edition) 36(11):1344–1347
6. Zhou X, Li G, Cai D et al (2013) Review and outlook of anti-spoofing technologies in GNSS. In: Proceedings of the 4th China GNSS academic annual conference—S7 BDS/GNSS User Terminal Technologies

Effects of Spectrum Leakage on Frequency Domain Antijamming Performance

Jian Li, Baiyu Li, Junwei Nie, Xiaohui Liu and Feixue Wang

Abstract Frequency domain antijamming (FDAJ) has been widely adopted by most of the military satellite navigation receivers. It is simple in theory and easy for realization. It can effectively suppress single-tone interference, sweep interference, and narrow band interference with Gaussian distribution. However, FDAJ will result in distortion of navigation signal, while suppressing interference. Previous works about effects of FDAJ on navigation signal have not taken interference spectrum leakage induced by time window shaping. It results in effects of FDAJ on navigation signal which has no relationship with interference power, and violates the practical. Therefore, take narrow band interference with Gaussian distribution as an example. First, its spectrum leakage induced by time window shaping is analyzed. Then, extra FDAJ loss induced by interference leakage is analyzed. Finally, FDAJ theoretical model of effects on effective carrier-to-noise ratio (CN0) and pseudocode tracking variance is derived, taking interference spectrum leakage. Numerical computation show that, effective CN0 and code tracking variance increases with jam-to-noise ratio (JNR), when FDAJ suppresses narrow band interference shaped by time window, different from ideal one (without spectrum leakage). When adopting different time windows, rate of receiver performance deterioration with JNR increases caused by FDAJ is different. When adopting Hamming or Hanning window, the rate is far lower than that of rectangular window.

Keywords Navigation satellite system · Frequency domain antijamming · Time domain window · Effective carrier-to-noise ratio · Code tracking variance

J. Li · B. Li · J. Nie · X. Liu · F. Wang (✉)
National University of Defence Technology, Changsha, China
e-mail: wangfeixue_nnc@163.com

J. Sun et al. (eds.), *China Satellite Navigation Conference (CSNC) 2016 Proceedings: Volume I*, Lecture Notes in Electrical Engineering 388,
DOI 10.1007/978-981-10-0934-1_54

1 Introduction

Effects on ranging and speed measurement of navigation equipments are main issues of interference effect evaluation. Corresponding evaluation indicators are effective carrier-to-noise ratio (*CN0*) and code tracking variance [1, 2]. Interference effect evaluation plays an important role in interference threat ranking, interference mitigation research, satellite ground site selection, etc. Thus, it is necessary to establish accurate evaluation model of interference effect on receiver. Betz has acquired abundant findings on receiver's effective *CN0* estimation theory, in presence of partial-band interference with Gaussian distribution [3–5]. With extensive application of mitigation measurements in receivers, their effect on receiver draws increasing attention. Typically, Omar theoretically predicts code tracking error of receivers with time domain blanking (TDB) and frequency domain adaptive filtering (FDAF) [6]. Zhang T. has analyzed frequency domain antijamming (FDAJ) effect on receiver tracking performance [7]. But, FDAJ effect on tracking performance has no business with interference power in [6, 7]. Namely, no matter how strong interference power is, FDAJ can succeed in suppressing the narrow band interference, which is contrary to the practical. Therefore, spectrum leakage and its effect in FDAJ are first analyzed. Then, FDAJ effect on effective *CN0* and code tracking variance are derived, taking interference spectrum leakage into consideration.

2 Frequency Domain Antijamming Flow

FDAJ can be divided into delay choosing and delay adding. Reference [8] points out that 1/2 delay adding FDAJ hardly induce signal-to-noise ratio (SNR) loss, when Hamming time window is adopted. Thus, 1/2 delay adding FADJ is employed here. Its flow is as shown in Fig. 1.

Intermediate frequency (IF) signal $x_{IF}(t)$ is shaped by time window,

$$x_W(t) = x(t)h_W(t) \tag{1}$$

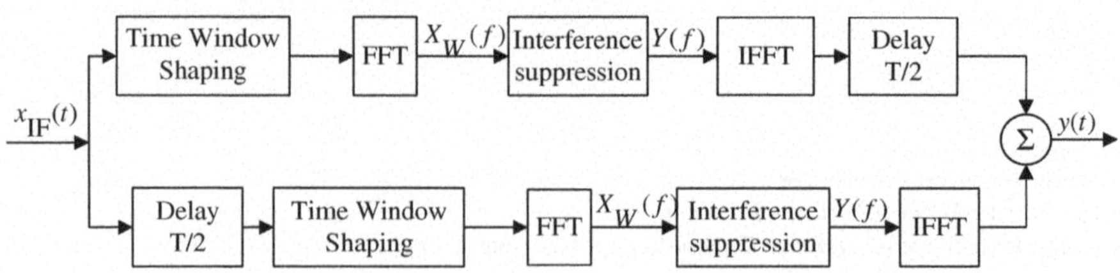

Fig. 1 Flow chart of FDAJ

To Fourier transform property, multiplication in time domain is mapped to convolution in frequency domain. So, spectrum of $x_W(t)$ is as follows:

$$X_W(f) = X(f) * H_W(f) \tag{2}$$

where * is convolution, $X(f)$ and $H_W(f)$ is, respectively, the spectrum of $x_{IF}(t)$ and time window $h_W(t)$. Then, set $X_W(f)$ zero according to threshold λ

$$Y(f) = X_W(f)H(f) \tag{3}$$

$$H(f) = \begin{cases} 0, |X_W(f)| > \lambda \\ 1, |X_W(f)| < \lambda \end{cases} \tag{4}$$

where λ is threshold of interference suppression. Signal in time domain after interference suppression is obtained through carrying inverse Fourier transform of $H(f)$. Sum of time domain signal in the upper and lower branch in Fig. 1 is the output of FDAJ.

3 Time Window Shaping and Spectrum Leakage

Signal in practical is always limited in time domain, namely truncated by time window. This results in spectrum leakage of signal. Direct truncation equals rectangular window shaping. Rectangular time window has narrow main lobe and wide side lobe with small main lobe to side lobe ratio, which results in serious side lobe leakage. Therefore, Hamming or Hanning window with narrow side lobe is employed in engineering instead of the rectangular one. Expression of normal rectangular time window is as follows [9]:

$$w(t) = \begin{cases} 1, & -T_w/2 \le t \le T_w/2 \\ 0, & \text{Others} \end{cases} \tag{5}$$

where T_w is the window width. Spectrum of rectangular time window is as follows:

$$H_{\text{rec}}(f) = T_w \sin c(\pi f T_w) \tag{6}$$

Expression of Hamming time window in time domain and frequency domain is as follows:

$$w(t) = \begin{cases} 0.54 + 0.46\cos(2\pi t/T_w), & -T_w/2 \le t \le T_w/2 \\ 0, & \text{Others} \end{cases} \tag{7}$$

$$H_{\text{ham}}(f) = 0.54T \sin c(\pi f T_w) + 0.23T_w[\sin c(\pi T_w(f + 1/T_w)) + \sin c(\pi T_w(f - 1/T_w))] \quad (8)$$

Expression of Hanning time window in time domain and frequency domain is as follows:

$$w(t) = \begin{cases} 0.5(1 + \cos(2\pi t/T_w)), & -T_w/2 \leq t \leq T_w/2 \\ 0, & \text{Others} \end{cases} \quad (9)$$

$$H_{\text{han}}(f) = 0.5T_w \sin c(\pi f T_w) + 0.25T_w[\sin c(\pi T_w(f + 1/T_w)) + \sin c(\pi T_w(f - 1/T_w))] \quad (10)$$

In engineering, to fulfill request of real-time fast Fourier transform (FFT) in Fig. 1, time window points is usually 2^{10} (namely 1024). Regular sample frequency of GPS or BeiDou receiver is 20–40 MHz. Thus, typical time window width is T_w = 0.05 ms. Power spectrum density of different time window is as shown in Fig. 2, when T_w = 0.05 ms.

Reference [9] points out that the rectangular time window has the narrowest main lobe among all time windows but, its side lobe is wide. As shown in Fig. 2, main lobe of rectangular window is 20 kHz, while that of Hamming or Hanning window with same width reaches 40 kHz. But, main lobe to side lobe ratio of rectangular is 13 dB, while that of Hanning window is 31 dB and that of Hamming window even reaches 41 dB. Equation (2) shows that spectrum of signal shaped by time window is convolution of signal spectrum and time window spectrum. Take ideal narrow band interference with Gaussian distribution as an example and analyze effect of time window on signal. Define spectrum of narrow band interference as follows.

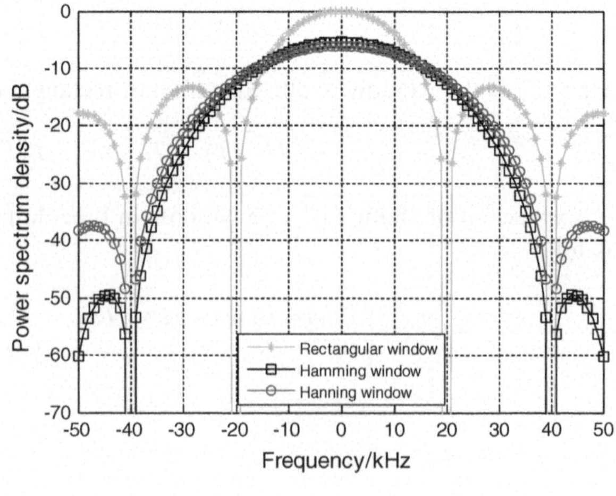

Fig. 2 Spectrum of time domain window (Color figure online)

$$l(f) = \begin{cases} A_j/\sqrt{B_j}, & |f - f_j| < B_j/2 \\ 0, & \text{Others} \end{cases} \tag{11}$$

where A_j is amplitude of interference, B_j is its band. Substituting (11) and (6) into (2) yields spectrum of narrow band interference shaped by time window, as follows.

$$\begin{aligned} X_{\text{rec}}(f) &= l(f) * T_w \sin c(\pi f T_w) \\ &= A_j T_w B_j^{-1/2} \int_{f+f_j-B_j/2}^{f+f_j+B_j/2} \sin c(\pi T_w \tau) \mathrm{d}\tau \\ &= A_j \pi^{-1} B_j^{-1/2} \left[\sin \text{int}\left(\pi T_w (f+f_j+B_j/2)\right) - \sin \text{int}\left(\pi T_w (f+f_j-B_j/2)\right)\right] \end{aligned} \tag{12}$$

where $\sin \text{int} x = \int_0^x \sin t/t \, \mathrm{d}t$. Substituting (11) and (8) into (2) yields spectrum of narrow band interference shaped by Hamming window as follows.

$$\begin{aligned} X_{\text{ham}}(f) &= l(f) * H_{\text{ham}}(f) \\ &= 0.54 A_j X_{\text{rec}}(f+f_j) + 0.23 A_j \left[X_{\text{rec}}(f+f_j+1/T_w) + X_{\text{rec}}(f+f_j-1/T_w)\right] \end{aligned} \tag{13}$$

Substituting (11) and (10) into (2) yields spectrum of narrow band interference shaped by Hanning window as follows:

$$\begin{aligned} X_{\text{han}}(f) &= l(f) * H_{\text{han}}(f) \\ &= 0.5 A_j X_{\text{rec}}(f) + 0.25 A_j \left[X_{\text{rec}}(f+f_j+1/T_w) + X_{\text{rec}}(f+f_j-1/T_w)\right] \end{aligned} \tag{14}$$

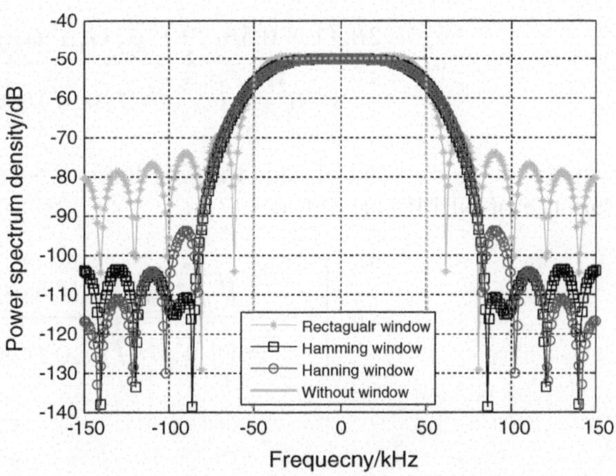

Fig. 3 Spectrum of narrow band signal shaped by time domain window (Color figure online)

Assuming amplitude of narrow band interference is unit, band $B_j = 100$ kHz, frequency $f_j = 100$ Hz and time window width $T_W = 0.05$ ms. Power spectrum density of narrow band interference shaped by rectangular window, Hamming window, and Hanning window is given in Fig. 3. It can be concluded from the results that there exists different degree of spectrum leakage, no matter what time window is adopted. Interference shaped by rectangular has highest level amplitude of side lobe. The second is interference shaped by Hanning window and the last one is interference shaped by Hamming window. However, main lobe band of interference shaped by Hamming window or Hanning window is double of that shaped by rectangular window.

4 Spectrum Leakage Effect on FDAJ Performance

For receiver of pseudocode ranging, main indicators of evaluating receiver performance are effective CN0 and code tracking variance. According to structure of early-late code loop, code tracking loop can be divide into coherent early-late processing (CELP) and noncoherent early-late processing (NELP). Effective CN0 expression shared by CELP and NELP code tracking loop is as follows [7]:

$$\mathrm{CN0_{eff}} = \frac{2T\frac{C_S}{N_0}\left(\int_{-\beta/2}^{\beta/2} H(f)G_s(f)\mathrm{d}f\right)^2}{\int_{-\beta/2}^{\beta/2} G_h(f)G_s(f)\mathrm{d}f + \int_{-\beta/2}^{\beta/2} G_h(f)G_j(f)G_s(f)\mathrm{d}f} \tag{15}$$

where $G_s(f) = T_c \sin c^2(\pi f T_c)$, T_c is code chip width, β is receiver band, T is receiver correlation time, C_S is signal power, N_0 is noise power spectrum density, $H(f)$ is interference suppression filter, $G_h(f) = |H(f)|^2$, $G_s(f)$ is normal navigation signal power spectrum density, $G_j(f)$ is normal interference power spectrum density. Code tracking variance of CELP is as follows [7].

$$\sigma_c^2 \cong \frac{2B_L(1-0.5B_LT)\int_{-\beta/2}^{\beta/2} G_h(f)G_s(f)\sin^2(\pi f\Delta)\mathrm{d}f}{(2\pi)^2\frac{C_s}{N_0}\left(\int_{-\beta/2}^{\beta/2} fH(f)G_s(f)\sin(\pi f\Delta)\mathrm{d}f\right)^2} \tag{16}$$

where B_L is noise band of code tracking loop, Δ is early-late space. Code tracking variance of NELP is as follows [7]:

$$\sigma_n^2 \cong \sigma_c^2\left[1 + \frac{\int_{-\beta/2}^{\beta/2} G_h(f)G_s(f)\cos^2(\pi f\Delta)\mathrm{d}f}{\frac{C_s}{N_0}\left(\int_{-\beta/2}^{\beta/2} H(f)G_s(f)\cos(\pi f\Delta)\mathrm{d}f\right)^2}\right] \tag{17}$$

Usually, $\cos(\pi f \Delta) \approx 1$. Thus, $\sigma_n^2 \approx \sigma_c^2$. It can be concluded from Fig. 1 that filter *H*(*f*) should be set to null FFT of interference shaped by time window according to (4). Last section reveals that time window shaping will induce spectrum leakage. This may result in null band of *H*(*f*) wider than interference band. The stronger interference power is, the wider null band of *H*(*f*) is. However, interference suppression filter in [6, 7] is set as follows:

$$H_O(f) = \begin{cases} 0, & |f - f_j| < B_j/2 \\ 1, & \text{Others} \end{cases} \tag{18}$$

Effect of FDAJ adopting interference filter $H_O(f)$ in [6, 7] on effective CN0 and code tracking variance is uncorrelated with interference power. But, engineering reveals that FDAJ has limited suppression interference power. That is FDAJ that is correlated with interference power. Therefore, modify evaluation models above, taking spectrum leakage arising from time window shaping into consideration. It's necessary to get threshold λ to set filter *H*(*f*), according to (4). Threshold λ setting based on channel noise power is employed here. Effect of threshold λ on FDAJ performance is not focus of this paper. Please read related reference, if you are interested. Channel noise power after time window shaping is defined as follows:

$$\sigma_W^2 = \mathrm{var}\{n(t)w(t)\} = E\left\{|n(t)|^2\right\}E\left\{|w(t)|^2\right\} = \frac{N_O\beta}{T_w}\int_{-T_w/2}^{T_w/2} w^2(t)\mathrm{d}t \tag{19}$$

Threshold is usually set as $\lambda = 3\sigma_W/\sqrt{\beta}$. Thus,

$$\lambda = 3\sqrt{\frac{N_O}{T_w}\int_{-T_w/2}^{T_w/2} w^2(t)\,\mathrm{d}t} \tag{20}$$

Respectively substituting (5), (7), and (9) into (2) yields threshold in different time window.

$$\lambda_{\text{rec}} = 3\sqrt{N_0}, \quad \lambda_{\text{ham}} = 3\sqrt{0.3974N_0}, \quad \lambda_{\text{han}} = 3\sqrt{0.365N_0} \tag{21}$$

Substituting (21) into (4) yields interference suppression filter. Because *H*(*f*) has taken spectrum leakage into consideration, it is assumed that interference can be completely suppressed by *H*(*f*). Reference [8] points out that Hamming or Hanning time window shaping hardly induced SNR loss, if FDAJ adopts 1/2 delay addition structure. Thus, it is assumed SNR has no loss after time window shaping. Therefore, (15)–(17) are still available. Only difference from model in [6, 7] is that *H*(*f*) is set according to (4) not interference band.

5 Numerical Computations

For better interpretation of interference spectrum leakage effect on receiver performance, carry out following numerical computations. Navigation signal and receiver parameters are set as follows: time window $T_w = 0.05$ ms, receiver band $\beta = 2.046$ MHz, code symbol rate $R_c = 1.023$ Mcps, early-late space adopts 1/2 code chip. In numerical computation, effective CN0 is transferred to effective CN0 loss for better receiver performance comparison with different time window.

$$\text{loss}_{\text{CN0}} = \frac{\left(\int_{-\beta/2}^{\beta/2} H(f)G_s(f)\,\mathrm{d}f\right)^2}{\int_{-\beta/2}^{\beta/2} G_h(f)G_s(f)\,\mathrm{d}f} \tag{22}$$

Since code tracking variance of NELP approximates to that of CELP, code tracking variance of CELP is computed in numerical computation. Besides, square root of code tracking variance is employed here.

5.1 JNR Versus Receiver Performance

Narrow band interference parameters are set as follows: band $B_j = 100$ kHz, frequency is zero (that is navigation carrier frequency coincides with interference center frequency), JNR range 20–35 dB with step of 2 dB.

JNR versus effective CN0 loss and code tracking standard derivation induced by FDAJ curves are, respectively, given in Figs. 4 and 5. When FDAJ suppress narrow band interference shaped by time window, effective CN0 loss and code tracking standard derivation increase with JNR. Especially, when rectangular window is

Fig. 4 JNR versus CN0 loss (Color figure online)

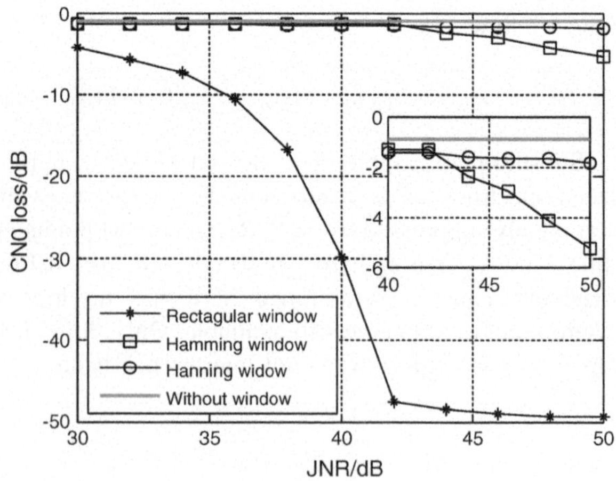

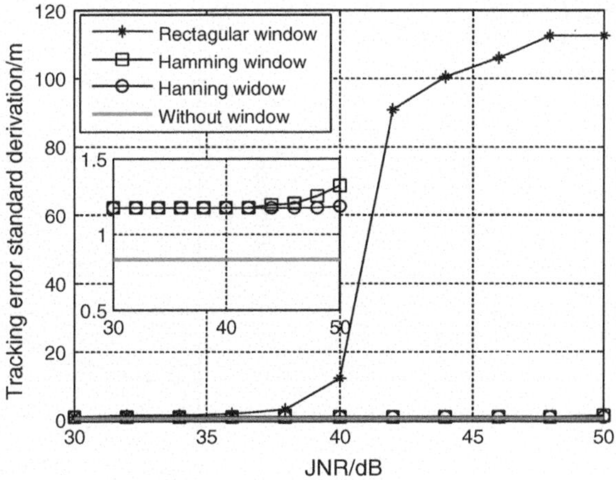

Fig. 5 JNR versus tracking error standard derivation (Color figure online)

adopted, effective CN0 loss and code tracking standard derivation are bigger than other windows. When JNR reaches 40 dB, effective CN0 loss sharply reaches about 50 dB and code tracking standard derivation reaches about 100 m (navigation signal is unavailable). If Hamming or Hanning window is adopted, effective CN0 loss and code tracking standard derivation approximates to that without time window shaping (without interference spectrum leakage), when JNR is below 40 dB. But, when JNR is above 40 dB, maximum extra effective CN0 loss reaches 5 dB and code tracking standard derivation hardly deteriorates, compared to that without time window shaping.

5.2 *Interference Frequency Versus Receiver Performance*

Interference band B_j = 100 kHz, JNR 35 dB, interference frequency range 0–R_c with step of $R_c/40$.

Interference frequency versus effective CN0 loss and code tracking standard derivation induced by FDAJ curves are, respectively, given in Figs. 6 and 7. With interference frequency leaving away from carrier frequency, effective CN0 loss gradually decreases. The reason is that null band of filer $H(f)$ leaves away from carrier frequency, with interference center frequency leaving away from carrier frequency, as (15) indicates. Maximum power spectrum density of navigation signal is at carrier frequency. The further away from carrier frequency is, the smaller power spectrum density is. Thus, power loss arising from FDAJ increases with interference frequency leaving away from carrier frequency. Code tracking standard derivation increases at the beginning and then decreases with interference frequency leaving away from carrier frequency. Code tracking standard derivation reaches maximum when interference frequency is away from carrier frequency $R_c/2$. The reason is that code tracking standard derivation is related not only with

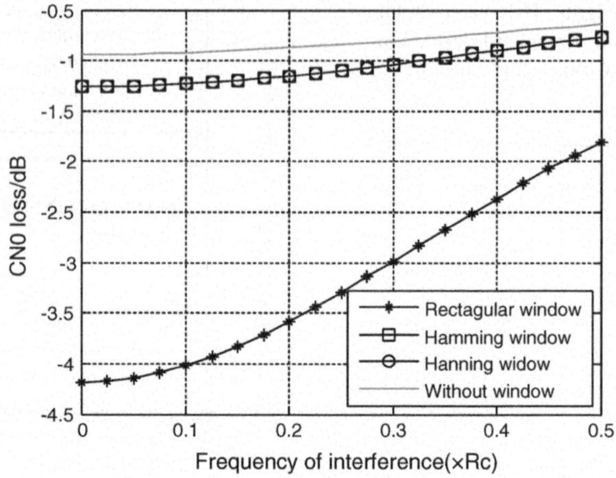

Fig. 6 Interference frequency versus CN0 loss (Color figure online)

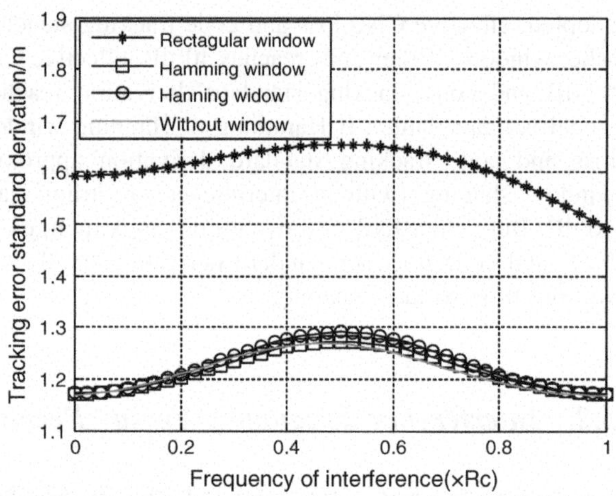

Fig. 7 Interference frequency versus tracking error standard derivation (Color figure online)

navigation signal power spectrum but also with navigation frequency, as (16) and (17) indicate. When adopting rectangular window, effective CN0 loss and code tracking standard derivation induced by FDAJ are bigger than that adopting Hamming or Hanning window. When adopting Hamming or Hanning window, effective CN0 loss or code tracking derivation induced by FDAJ approximates to that without time window shaping (without spectrum leakage).

6 Conclusion

Aiming at problem that FDAJ effect on receiver performance in previous works have not taken spectrum leakage of narrow band interference, interference spectrum leakage arising from time window shaping and its effect on FDAJ are, respectively, theoretically analyzed. Finally, effect of FDAJ adopting time window shaping on receiver performance theoretical model is derived.

Different from conclusion in previous works that interference power has non business with receiver performance, numerical computation results here show that effective CN0 loss or code tracking derivation increase with JNR, when interference spectrum leakage arising from time window shaping is considered. Degree of spectrum leakage arising from different time window is different and extra anti-jamming loss induced by it is different. When rectangular window is adopted, extra loss induced by spectrum leakage is far bigger than that when Hamming or Hanning window is adopted. When JNR is big enough (such as 50 dB), receive may fail to receive navigation signal. However, if Hamming or Hanning window is adopted, effective CN0 and code tracking standard derivation deterioration rate is far slower than rectangular window. Even when JNR is 50 dB, navigation signal is available after FDAJ.

In summary, compared with case without time window shaping (without interference spectrum leakage), interference spectrum leakage arising from time window shaping will result in extra antijamming loss (effective CN0 loss or code tracking standard derivation) increasing with JNR. But, if time window is suitable, deterioration rate of extra antijamming loss arising from spectrum leakage can be slowed down.

References

1. Betz JW (2000) Effect of partial-band interference on receiver estimation of CN0 theory. ION NTM 2001
2. Betz JW (2000) Effect of narrowband interference on GPS code tracking accuracy. ION NTM 2000
3. Betz JW (2009) Generalized theory of code tracking with an early-late discriminator part 1 lower bound and coherent processing. IEEE Trans Aerosp Electron Syst 45(4):1538–1550
4. Betz JW (2009) Generalized theory of code tracking with an early-late discriminator part 2 non-coherent processing and numerical results. IEEE Trans Aerosp Electron Syst 45(4): 1551–1564
5. Zhang J, Lohan ES (2011) Effect of narrowband interference on Galileo E1 signal receiver performance. Int J Navig 2011:1–10
6. Omar AYO, Jesus G, Gustavo LR (2013) Analytical performance of GNSS receivers using interference mitigation techniques. IEEE Trans Aerosp Electron Syst 49(2):885–906
7. Zhang T, Zhang X, Lu M (2013) Effect of frequency domain anti-jamming filter on satellite navigation signal tracking performance. In: Proceedings of China satellite navigation conference (CSNC) 2013, Wuhan, pp 507–516

8. Zeng X, Li Z, Wang F (2004) Study on windowing degradation of frequency domain narrowband interference suppression algorithms in spread spectrum system. J Electron Inf Technol 26(8):1276–1281 (in Chinese)
9. Gao X, Ding Y (2008) Digital signal processing. Xi'an, Xi'an Electronical and Science University Press (in Chinese)

The GPS Spoofing Detection Based on the Joint WSSE of DOA and Pseudorange

Guangteng Fan, Yangbo Huang, Guozhu Zhang, Junwei Nie and Guangfu Sun

Abstract With the development of GPS spoofer techniques, spoofing signals can be consistent with the authentic signals on direction of arrival (DOA) or pseudorange. The receiver will not be effective for detecting such spoofing signals, when using traditional spoofing detection method based on either DOA or pseudorange. In order to suppress this problem, a method based on the joint weighted sum of squares of errors (WSSE) of DOA and pseudorange for GPS spoofing detection is introduced. This method uses the carrier-phase difference and pseudorange observations to form two equations simultaneously, and adds the residuals of two equations after normalization. According to the joint WSSE, it can test the DOA and pseudorange of arrival signals at the same time. Because of this, it can greatly improve the adaptability to different spoofing scenarios. This paper also gives the process to identify and eliminate spoofing signals in the received signals. In conclusion, the performance between the method based on the joint WSSE of DOA and pseudorange and the traditional method based alone on DOA or pseudorange is compared. The simulation results show that the proposed method is superior to the traditional way both in the adaptability to different spoofing scenarios and the detection performance.

Keywords Spoofing detection · DOA · Pseudorange · WSSE

1 Introduction

Traditional direction of arrival (DOA) detection method is based on the inconsistency between the spoofing signals and authentic signals [1–3]. However there is a limitation. Although we can coordinate the antenna array baseline for installation in the motion vector (car, airborne, shipborne application scenarios) in advance, the

G. Fan (✉) · Y. Huang · G. Zhang · J. Nie · G. Sun
Satellite Navigation R&D Center, National University of Defense Technology, Changsha, Hunan, People's Republic of China
e-mail: fanguangteng@163.com

J. Sun et al. (eds.), *China Satellite Navigation Conference (CSNC) 2016 Proceedings: Volume I*, Lecture Notes in Electrical Engineering 388,
DOI 10.1007/978-981-10-0934-1_55

antenna array attitude is real-time changed according to the vehicle motion. Therefore, even if the authentic satellites orbital coordinates are obtained through ephemeris parameters [4], the receiver cannot get DOA of satellite signals relative to the baseline of antenna array. Because of this, the traditional DOA detection method cannot be used in the motion vectors [5]. There are literatures that introduce a kind of carrier-phase double difference detection technology. This method exploits the spatial signature which spoofing transmitters usually transmit several spoofing signals from the same antenna, while the authentic signals are transmitted from different satellites with various directions. This method can be applied to motion vectors [6], but it will fail to detect single spoofing signal or the spoofing transmitters which transmit several spoofing signals from multiple antennas. Literature [7, 8] presents a spoofing detection method based on the SSE during the antenna array attitude determination. While this method solves the problem, the baseline and attitude of the antenna array must be calibrated in advance when using traditional DOA detection method. This method needs the estimation of DOA of every satellite signal in the local coordinate frame of the antenna array, making the receiver use at least three elements in array antenna.

On the other hand, the spoofing signals are different from authentic signal not only on the DOA but also on the pseudorange [9, 10]. For the existing spoofers, it is hard to keep consistent in DOA and pseudorange with the authentic signals at the same time. Therefore, this paper proposes a method based on the joint weighted sum of squares of errors (WSSE) of DOA and pseudorange for spoofing detection.

The paper is organized as follows. The next section of the paper is dedicated to the principle of this method and the decision criterion. Further in the fourth section, the performance between the method based on the joint WSSE of DOA and pseudorange and the traditional method based alone on DOA or pseudorange method is compared by Monte-Carlo simulations. Finally, the paper will be concluded by summarizing the results of this study.

2 Detection Principle

2.1 The Joint WSSE of DOA and Pseudorange

The DOA of signals has a corresponding relationship with the carrier-phase difference between different array antenna elements. Assume that the receiver adopts binary array. The DOA of signal and the baseline and attitude of antenna array in Earth-centered Earth-fixed (ECEF) coordinate system is shown in the Fig. 1.

The baseline and attitude of antenna array vector are defined as $\overrightarrow{AC} = (a, b, c)$ in the ECEF coordinate system. The DOA of *i*th signal is defined as: $\overrightarrow{DOA_i} = [\cos\alpha_i \sin\beta_i, \cos\alpha_i \cos\beta_i, \sin\alpha_i]^{\mathrm{T}}$, where α_i is the *i*th signal elevation angle and β_i is the *i*th signal azimuth angle in the ECEF coordinate system. The relationship between carrier-phase difference and DOA can be written as follows:

Fig. 1 The DOA of signal and the baseline and attitude of antenna array in ECEF coordinate system

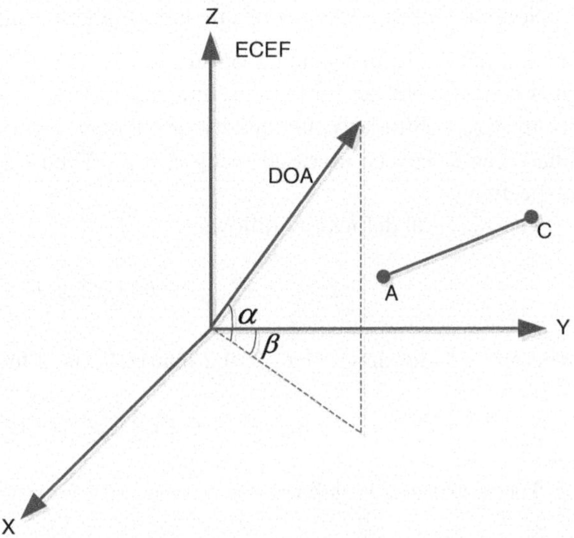

$$d\varphi_i = a\cos\alpha_i\sin\beta_i + b\cos\alpha_i\cos\beta_i + c\sin\alpha_i + \tau + \varepsilon_i^{\varphi}, \tag{1}$$

where τ is time delay between element A and C. ε_i^{φ} is the ith signal measurement error of carrier-phase difference, which follows $N(0, \sigma_i^{\varphi 2})$ distribution.

If we ignore the measurement error of carrier-phase difference, we can get the following four elements linear equations.

$$\begin{cases} a\cos\alpha_1\sin\beta_1 + b\cos\alpha_1\cos\beta_1 + c\sin\alpha_1 + \tau = d\varphi_1 \\ a\cos\alpha_2\sin\beta_2 + b\cos\alpha_2\cos\beta_2 + c\sin\alpha_2 + \tau = d\varphi_2 \\ \ldots \\ a\cos\alpha_N\sin\beta_N + b\cos\alpha_N\cos\beta_N + c\sin\alpha_N + \tau = d\varphi_N \end{cases} \tag{2}$$

Also using pseudorange which is measured by the receiver, we can get the following four elements nonlinear equations [11].

$$\begin{cases} \sqrt{(x_1 - x)^2 + (y_1 - y)^2 + (z_1 - z)^2} + \delta t_u = \rho_1 \\ \sqrt{(x_2 - x)^2 + (y_2 - y)^2 + (z_2 - z)^2} + \delta t_u = \rho_2 \\ \ldots \\ \sqrt{(x_N - x)^2 + (y_N - y)^2 + (z_N - z)^2} + \delta t_u = \rho_N \end{cases}, \tag{3}$$

where x_i , y_i and z_i are the ith satellite position getting from ephemeris. x, y, z, and δt_u represent the position of antenna phase center and receiver clock error. ρ_i is the pseudorange which is measured by receiver from ith satellite signal.

Solving the (2), (3) equations by the weighted residuals of least squares solution, we can get the solutions to equations: $\widehat{a}$, $\widehat{b}$, $\widehat{c}$, $\hat{\tau}$ and $\widehat{x}$, $\widehat{y}$, $\widehat{z}$, $\widehat{\delta t_u}$. The quality of the obtained solution for the baseline and attitude of antenna array can be assessed using $\varepsilon_{\mathrm{WSSE}}^{\varphi}$. Similarly, the quality of obtained solution for the position of antenna phase center can be assessed using $\varepsilon_{\mathrm{WSSE}}^{\rho}$. Then $\varepsilon_{\mathrm{WSSE}}^{\varphi}$ and $\varepsilon_{\mathrm{WSSE}}^{\rho}$ are calculated, respectively.

The $\varepsilon_{\mathrm{WSSE}}^{\varphi}$ is defined as follows:

$$\varepsilon_{\mathrm{WSSE}}^{\varphi} = b_{\varphi}^{\mathrm{T}} S_{\varphi}^{\mathrm{T}} C_{\varphi} S_{\varphi} b_{\varphi}, \tag{4}$$

where $b_{\varphi} = [\mathrm{d}\varphi_1 \mathrm{d}\varphi_2 \ldots \mathrm{d}\varphi_N]^{\mathrm{T}}$, the matrix S_{φ} is defined as:

$$S_{\varphi} = I - G_{\varphi}(G_{\varphi}^{\mathrm{T}} C_{\varphi} G_{\varphi})^{-1} G_{\varphi}^{\mathrm{T}} C_{\varphi}. \tag{5}$$

The matrix G_{φ} is defined as:

$$G_{\varphi} = \begin{bmatrix} \cos\alpha_1 \sin\beta_1 & \cos\alpha_1 \cos\beta_1 & \sin\alpha_1 & 1 \\ \cos\alpha_2 \sin\beta_2 & \cos\alpha_2 \cos\beta_2 & \sin\alpha_2 & 1 \\ \cdots & \cdots & \cdots & \cdots \\ \cos\alpha_N \sin\beta_N & \cos\alpha_N \cos\beta_N & \sin\alpha_N & 1 \end{bmatrix}. \tag{6}$$

The normalized matrix of DOA residuals is defined as:

$$C_{\varphi} = \begin{bmatrix} 1/\sigma_1^{\varphi 2} & & & \\ & 1/\sigma_2^{\varphi 2} & & \\ & & \ddots & \\ & & & 1/\sigma_N^{\varphi 2} \end{bmatrix}, \tag{7}$$

where $\sigma_i^{\varphi 2}$ is the variance of carrier-phase difference of ith signal.

The $\varepsilon_{\mathrm{WSSE}}^{\rho}$ is defined as follows:

$$\varepsilon_{\mathrm{WSSE}}^{\rho} = b_{\rho}^{\mathrm{T}} S_{\rho}^{\mathrm{T}} C_{\rho} S_{\rho} b_{\rho}, \tag{8}$$

where

$$b_{\rho} = \begin{bmatrix} \rho_1 - \sqrt{(x_1 - \widehat{x})^2 + (y_1 - \widehat{y})^2 + (z_1 - \widehat{z})^2} - \widehat{\delta t_u} \\ \rho_2 - \sqrt{(x_2 - \widehat{x})^2 + (y_2 - \widehat{y})^2 + (z_2 - \widehat{z})^2} - \widehat{\delta t_u} \\ \cdots \\ \rho_N - \sqrt{(x_N - \widehat{x})^2 + (y_N - \widehat{y})^2 + (z_N - \widehat{z})^2} - \widehat{\delta t_u} \end{bmatrix}. \tag{9}$$

The matrix S_ρ is defined as:

$$S_\rho = I - G_\rho (G_\rho^{\mathrm{T}} C_\rho G_\rho)^{-1} G_\rho^{\mathrm{T}} C_\rho. \tag{10}$$

The matrix G_ρ is defined as:

$$G_\varphi = \begin{bmatrix} \frac{\partial r_1}{\partial x}\Big|_{x=\hat{x}} & \frac{\partial r_1}{\partial y}\Big|_{y=\hat{y}} & \frac{\partial r_1}{\partial z}\Big|_{z=\hat{z}} & 1 \\ \frac{\partial r_2}{\partial x}\Big|_{x=\hat{x}} & \frac{\partial r_2}{\partial y}\Big|_{y=\hat{y}} & \frac{\partial r_2}{\partial z}\Big|_{z=\hat{z}} & 1 \\ \cdots & \cdots & \cdots & \cdots \\ \frac{\partial r_N}{\partial x}\Big|_{x=\hat{x}} & \frac{\partial r_N}{\partial y}\Big|_{y=\hat{y}} & \frac{\partial r_N}{\partial z}\Big|_{z=\hat{z}} & 1 \end{bmatrix}, \tag{11}$$

r_i is the observation vector length of receiver for ith satellite.
The normalized matrix of pseudorange residuals is defined as:

$$C_\rho = \begin{bmatrix} 1/\sigma_1^{\rho 2} & & & \\ & 1/\sigma_2^{\rho 2} & & \\ & & \ddots & \\ & & & 1/\sigma_N^{\rho 2} \end{bmatrix}, \tag{12}$$

where $\sigma_i^{\rho 2}$ is the variance of pseudorange of ith signal.
The joint WSSE of DOA and pseudorange is defined as:

$$\varepsilon_{\mathrm{WSSE}} = \varepsilon_{\mathrm{WSSE}}^{\varphi} + \varepsilon_{\mathrm{WSSE}}^{\rho}. \tag{13}$$

It shows that the joint WSSE of DOA and pseudorange represents the confidence value for the solutions both of the baseline and attitude of antenna array and the phase center of antenna. The smaller the WSSE the higher the confidence value is. If the spoofing has differences with authentic signals, whether in the DOA or pseudorange, the value of $\varepsilon_{\mathrm{WSSE}}$ will become unusually large.

2.2 *Decision Criterion Based on the Joint WSSE of DOA and Pseudorange*

If the signals received by the receiver are all authentic signals, the residuals of weighted least square solution are unbiased and have zero mean. In this case, the $\varepsilon_{\mathrm{WSSE}}$ as defined by (13), follows a central chi-squared distribution with k equable to $2N - 8$ degrees of freedom. In another case, when some spoofing are received which the DOA or the pseudorange are not consistent with the expected ones, the residuals of weighted least square solution become biased and the $\varepsilon_{\mathrm{WSSE}}$ follows a noncentral chi-squared distribution with the same number of degrees of freedom as in the error-free case, but with some nonzero noncentrality parameter γ. Based on the above analysis, we get the binary hypothesis judgment criteria:

$$\begin{gathered}\begin{cases} H_0: \varepsilon_{\mathrm{WSSE}} \sim \chi^2(\varepsilon_{\mathrm{WSSE}}, k) \\ H_1: \varepsilon_{\mathrm{WSSE}} \sim \chi^2(\varepsilon_{\mathrm{WSSE}}, k, \gamma) \end{cases}, \\ k = 2N - 8 \\ \gamma = \sum_{i=1}^{N} \left[\left(\frac{\Delta_i^{\varphi}}{\sigma_i^{\varphi}} \right)^2 + \left(\frac{\Delta_i^{\rho}}{\sigma_i^{\rho}} \right)^2 \right] \end{gathered} \tag{14}$$

where Δ_i^{φ} is a bias in the ith carrier-phase difference measurement performed by the antenna array, Δ_i^{ρ} is a bias in the ith pseudorange measurement. Thus, the probability of false alarm P_{fa} and the probability of detection P_D are given by:

$$\begin{aligned} P_{fa} &= P\{\varepsilon_{\mathrm{WSSE}} > \rho_{Th} | H_0\} = \int_{\rho_{Th}}^{\infty} \chi^2(x, k) \mathrm{d}x \\ P_D &= P\{\varepsilon_{\mathrm{WSSE}} > \rho_{Th} | H_1\} = \int_{\rho_{Th}}^{\infty} \chi^2(x, k, \gamma) \mathrm{d}x \end{aligned}. \tag{15}$$

Examples of probability density function (PDF) of the $\varepsilon_{\mathrm{WSSE}}$ test statistics for H_0 and H_1 cases are shown in Fig. 2.

As it can be observed in Fig. 2, the PDFs for H_0 and H_1 hypotheses become clearly separated at values of the noncentrality parameter $\gamma = 49$ or larger. Assumed standard deviation σ_{φ} of the carrier-phase difference is 0.002 m, standard deviation σ_{ρ} of the pseudorange is 7 m. If a single inconsistent signal is observed under adopted assumptions, this signal with an inconsistent direction of arrive can be effectively detected if it causes the carrier-phase difference separation between the spoofing signal and the authentic signal larger than 0.014 m. Similarly, this signal with an inconsistent pseudorange can also be effectively detected if it causes the pseudorange separation between the spoofing signal and the authentic signal larger than 49 m.

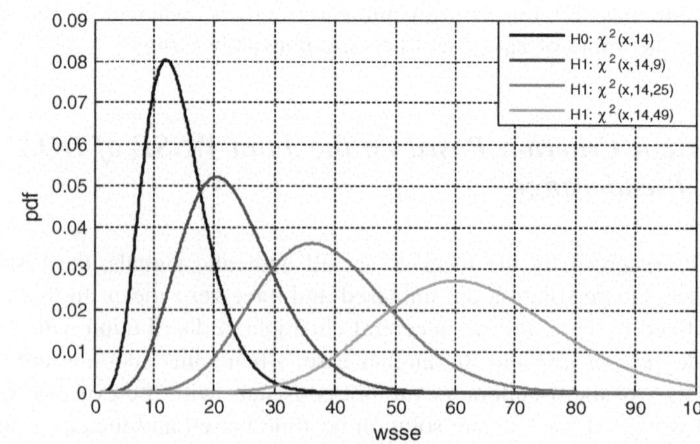

Fig. 2 Probability distribution functions of $\varepsilon_{\mathrm{WSSE}}$ metric without outliers (H_0) and with outliers (H_1) for 11 satellites (Color figure online)

3 Performance Assessment

The DOA vector and the pseudorange observation by the receiver are shown in Table 1, according to the ephemeris data published by IGS.

In order to simplify the analysis, assume all the standard deviations of receiver signals measurement error are the same. The standard deviation σ_φ of the carrier-phase difference is 0.002 m. The standard deviation σ_ρ of the pseudorange is 7 m. The following simulation, respectively, to three different scenarios of spoofing and all the spoofing scenarios just have a single spoofing signal. The scenarios of multiple spoofing signals can be analyzed similarly. The Monte-Carlo software simulations have been carried out to compare the performance between the method based on the joint WSSE of DOA and pseudorange and the traditional method based alone on DOA or pseudorange method. The simulated measurements are for 10,000 Monte-Carlo. The scenarios were designed as following

The Monte-Carlo simulations have been set up of scenario 1 which introduces a bias in DOA and pseudorange for satellite SV12 in order to model the occurrence of spoofing signal. The DOA measurements were biased by 7° for elevation α and −7° for azimuth β. The pseudorange measurements were biased by 35 m. The Monte-Carlo simulations have been set up of scenario 2 which introduce a bias in DOA for satellite SV12 in order to model the occurrence of spoofing signal. The measurements were biased by 7° elevation for α and −7° for azimuth β. The Monte-Carlo simulations have been set up of scenario 3 which introduces a bias in pseudorange for satellite SV12 in order to model the occurrence of spoofing signal. The pseudorange measurements were biased by 35 m.

Table 1 The DOA vector and the satellite position

Satellite number	DOA vector	Satellite position (m)
2	(0.329, 0.749, 0.575)	(8,865,131, 20,172,916, 15,492,487)
3	(−0.508, −0.362, 0.782)	(−13,490,995, −9,614,542, 20,755,685)
5	(0.124, 0.938, −0.323)	(3,289,106, 24,852,150, −8,555,863)
6	(−0.203, 0.553, 0.808)	(−5,385,028, 14,680,539, 21,459,185)
9	(−0.947, 0.253, 0.200)	(−25,149,816, 6,707,523, 5,309,307)
10	(−0.352, 0.880, 0.319)	(−9,235,699, 23,072,063, 8,367,665)
12	(0.419, 0.504, 0.755)	(11,128,360, 13,386,417, 20,050,468)
17	(−0.729, 0.560, 0.394)	(−19,361,175, 14,860,539, 10,475,460)
20	(−0.599, −0.053, 0.799)	(−15,833,492, −1,389,687, 21,124,769)
23	(−0.842, −0.064, 0.535)	(−22,577,185, −1,721,550, 14,350,244)
28	(−0.478, 0.719, −0.505)	(−12,509,431, 18,836,642, −13,218,792)

3.1 Simulation Results

See Figs. 3, 4 and 5.

3.2 The Comparison of Three Detection Methods

As the results of analysis, the method based on the joint WSSE of DOA and pseudorange can achieve a good performance in all cases, unless the spoofing signals are consistent in DOA and pseudorange with the authentic signals at the same time. This is

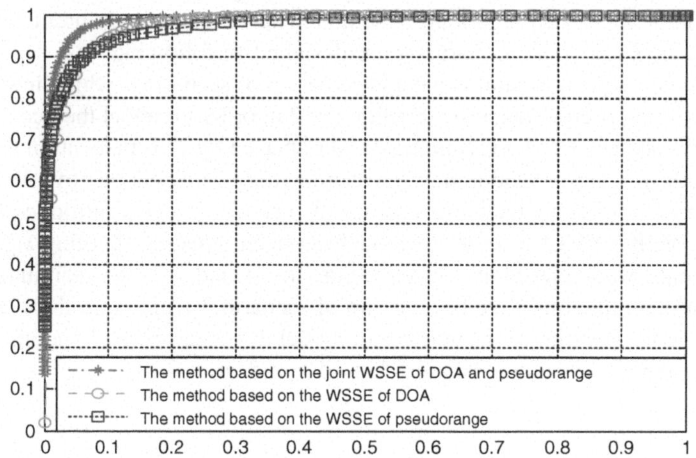

Fig. 3 The ROC curves of three methods in scenario 1 (Color figure online)

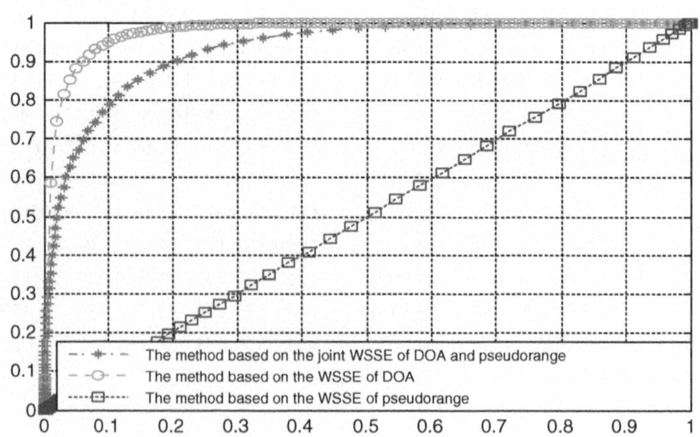

Fig. 4 The ROC curves of three methods in scenario 2 (Color figure online)

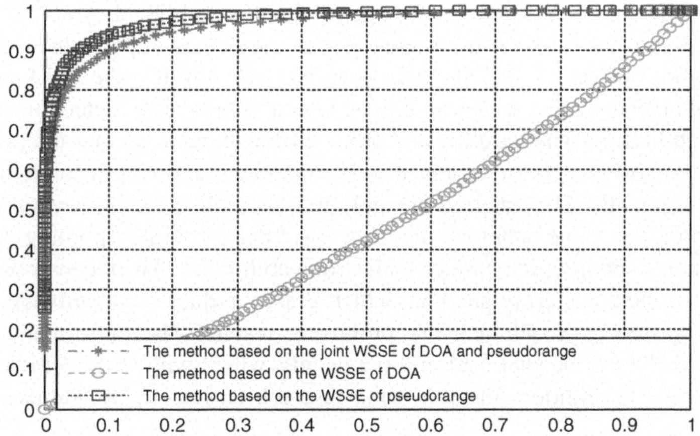

Fig. 5 The ROC curves of three methods in scenario 3 (Color figure online)

Table 2 The performance of three different methods in different scenarios

The scenario of spoofing	Based on the joint residuals of DOA and pseudorange	Based on the residuals of DOA	Based on the residuals of pseudorange
The deviation of spoofing signals and authentic signals is both in DOA and pseudorange	High	Middle	Middle
The deviation of spoofing signals and authentic signals is only in DOA	Middle	High	Low
The deviation of spoofing signals and authentic signals is only in pseudorange	Middle	Low	High

difficult to implement in engineering. Comparing spoofing signals is consistent in both DOA and pseudorange with the authentic signals. The spoofing signals can be consistent in one of the DOA or pseudorange with the authentic signals much easier, which will invalidate the method based alone on DOA or pseudorange. Table 2 compares the performance of three different methods in different scenarios.

4 Summary and Conclusions

In this paper, an approach to joint WSSE of DOA and pseudorange has been proposed. It has been shown that the GPS signals with inconsistent DOA or pseudorange can be detected and excluded from this method. The simulation results show that the proposed method is superior to the traditional way, both in the adaptability to different spoofing scenarios and the detection performance.

The algorithm has the following advantages: (i) it can test the DOA and pseudorange of arrival signals at the same time, which improves the detection capability and adaptability. (ii) The baseline and attitude of antenna array and the position of the antenna phase center are assumed to be unknown, allowing it to be applied to motion vectors. (iii) This method not only has been effective for several spoofing signals from the same antenna, but can also detect single spoofing signal or multi-elements spoofing transmitter with strong ability. (iv) The receiver only needs to use a two-elements array antenna, which greatly reduces the hardware cost.

The proposed approach delivers promising results. The performance in real environments has to be analyzed in more details, which will be the topic of further investigations. In addition, the following research topics can be addressed in the next studies:

1. The influence of measurement error of phase error and pseudorange on the detection performance.
2. The receiver is located in the environment which has both spoofing and multipath signals.

References

1. Montgomery PY, Humphreys TE, Ledvina BM (2009) Receiver-autonomous spoofing detection: experimental results of a multi-antenna receiver defense against a portable civil GPS spoofer. In: Proceedings of the Institute of Navigation—international technical meeting (ITM '09), January Anaheim, Calif, USA
2. Ledvina B, Montgomery P, Humphreys T (2009) A multi-antenna defense: receiver autonomous GPS spoofing detection. Inside GNSS, March/April
3. Cuntz M, Konovaltsev A, Heckler M, Hornbostel A, Kurz L, Kappen G, Noll T (2010) Lessons learnt: the development of a robust multi-antenna GNSS Receiver. In: ION GNSS2010
4. Montgomery P, Humphreys TE (2009) A multi-antenna defense: receiver-autonomous GPS spoofing detection. Inside GNSS, March/April
5. Wesson K, Shepard D et al (2012) Straight talk on anti-spoofing. GPS World
6. Daneshmand S, Jafarnia-Jahromi A, Broumandon A, Lachapelle G (2012) A low-complexity GPS anti-spoofing method using a multi-antenna array. ION GNSS, Nashville, TN
7. Meurer M, Konovaltsev A, Cuntz M, Hättich C (2012) Robust joint multi-antenna spoofing detection and attitude estimation using direction assisted multiple hypotheses RAIM. ION GNSS, Nashville, TN
8. Ledvina BM, Bencze WJ, Galusha B, Miller I (2010) An in-line anti-spoofing device for legacy civil GPS receivers. In: Proceedings of the Institute of Navigation—international technical meeting (ITM '10), San Diego, Calif, USA
9. Juang JC (2011) GNSS spoofing analysis by VIAS. Coordinates Magazine
10. Lee YC (2007) Two new RAIM methods based on the optimally weighted average solution (OWAS) concept. Navigation 54(4)
11. Kaplan ED, Hegarty CJ (2006) Understanding GPS: principles and applications. Artech House, Norwood

Anti-jamming Algorithm Analysis Based on UCA for Attitude High-Dynamic GNSS

Sujiao Li, Dan Lu and Renbiao Wu

Abstract In global navigation satellite systems (GNSS), when the receiver carrier is in the situation of attitude high dynamic, the directions of arrival (DOA) of signal and interference will change rapidly over time. At this point, conventional adaptive anti-jamming algorithms become invalid. In this paper, according to the structure shape of practical attitude high dynamic carrier, the DOAs change of signal and interference in the case of uniform circular array (UCA) are analyzed, and the Capon-Laplace algorithm based on UCA was provided to suppress the interference. The effectiveness of the algorithm is verified by simulation experiments. Through the analysis of attitude high-dynamic parameter, the paper compared the feasibility of the algorithm in different degrees' attitude high-dynamic circumstances giving a further verification on the effectiveness of the algorithm.

Keywords GNSS · Attitude high dynamic · Capon-Laplace · Jamming suppression

1 Introduction

The application of global navigation satellite system (GNSS) penetrated into various fields, such as military, civil aviation, surveying, marine navigation and in-car navigation, and positioning. Because it refers to the national defense security, the GNSS application in the military field is very significant. Since S. Hinedi proposed the definition of high dynamic [1] in 1998, more scholars began to study technologies related to the satellite navigation in the situation of high dynamic to solve the high-dynamic problem. With the emergence of new types of weapons in the military field, GNSS receiver carriers may be in the state of attitude high dynamic, such as missile-borne GNSS receivers for high-speed spin forward movement.

S. Li (✉) · D. Lu · R. Wu
Tianjin Key Lab for Advanced Signal Processing, Civil Aviation University of China, Tianjin 300300, China
e-mail: lisujiao@yeah.net

J. Sun et al. (eds.), *China Satellite Navigation Conference (CSNC) 2016 Proceedings: Volume I*, Lecture Notes in Electrical Engineering 388,
DOI 10.1007/978-981-10-0934-1_56

Since there is a long distance between satellites and the earth, the power of satellite signal is so small when it arrives the GNSS receiver such that GNSS becomes vulnerable to interferences [2]. Therefore, interference mitigation is very significant in attitude high-dynamic GNSS.

It is an effective method to suppress the interference in GNSS to use the adaptive array. Nowadays, there are so many methods using spatial adaptive processing algorithms to suppress the interference in GNSS [3–6]. However, when the receiver carrier is in attitude high dynamic, the receiver platform is instable, so that the direction of arrival (DOA) of signal and interference will change fast over time, and the snapshots which can be used to find the adaptive weights are few. Since nulls formed by the conventional adaptive anti-jamming algorithms are pointed, interferences may shift out from the array pattern nulls fast and could not be suppressed. At the same time, GNSS signals also may shift out from main beams fast, leading to the severe performance deterioration. Therefore, the above methods are failure to reject interferences for attitude high-dynamic GNSS.

Classical nulling widening methods cover derivative constraint (DC) [7] and the covariance matrix taper (CMT) method [8–10]. Mailloux and Zatman methods are the thought of original covariance multiplied by the extended matrix, known as CMT method. The two are not fit for the UCA. While UCA is widely applied for attitude high-dynamic carriers such as guide missile, the two are not suitable for attitude high dynamic. But the Laplace widening nulling method [11] proposed by Tianjin Key Laboratory for Advanced Signal Processing is fit for UCA, the paper proposed the Capon-Laplace method according to the DOA vary of signal and interference in attitude high dynamic. The proposed method can form the widening null, produces gain in the direction of the satellite signal and further improve the anti-jamming performance so that the satellite signal can be acquired correctly. Through the analysis of attitude high-dynamic parameter, the paper compared the feasibility of the algorithm in different degrees' attitude high-dynamic circumstances, giving a further verification on the effectiveness of the algorithm.

The rest of this paper is organized as follows. DOAs change of signal and interference based on UCA in attitude high-dynamic GNSS is analyzed in Sect. 2. Section 3 describes the anti-jamming algorithm for attitude high-dynamic CNSS. Section 4 simulates the algorithm, and compares the feasibility of the algorithm in different degrees' attitude high dynamic. Finally, Sect. 5 concludes the whole paper.

2 Analysis of Attitude High Dynamic Model

2.1 *Attitude High Dynamic Movement and UCA Structure Model*

It assumed that the receiver carrier keeps moving forward at high speed $v(t)$ and rotates with velocity $n = 50-300\,\text{r/s}$ (i.e., spin forward), which can be regarded as

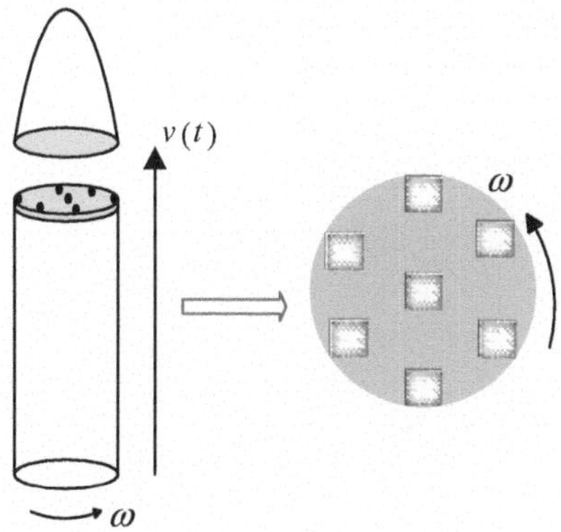

Fig. 1 Carrier and circular array structure diagram

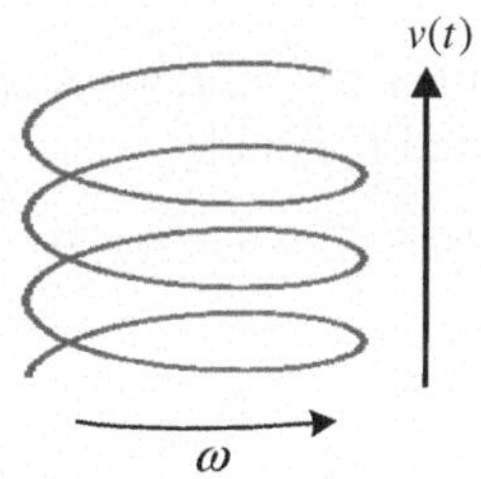

Fig. 2 Carrier's motion trajectory simulated diagram

the most simply motion model of attitude high dynamic. The rotation direction is counterclockwise. Assume the angular velocity as ω.

The paper uses the UCA with an element in the center of the circle to suppress interference, and structure model simulation is shown in Fig. 1.

The simulated motion trajectory of carrier is shown in Fig. 2.

2.2 DOAs Changes of Signal and Interference

According to Fig. 3, it can analyze and calculate the effect on DOAs changes of signal and interference due to the carrier rotation. Figure 3 shows that carrier rotation will make the azimuth change quickly, and the elevation is constant for the satellite signal, while for the interference, because the jammer is far away from the distance of the vehicle relative to the radius of UCA, the effect of rotation can be regarded as consistent with that of the satellite signal. So it simply needs to analyze the effect on the azimuth of the satellite signal.

Assume that the position of the UCA is $OABCDEF$ at the moment t, while the azimuth and the elevation of the satellite signal is θ and φ, respectively. Assume that the position of the UCA is $OA'B'C'D'E'F'$ at the moment t', while the azimuth and the elevation of the satellite signal is θ' and φ', respectively. Then in the period

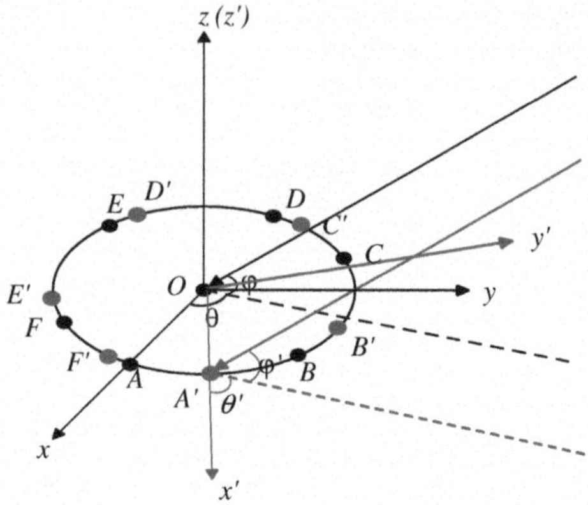

Fig. 3 The diagram of effect on azimuth via forward motion

of $\Delta t = t' - t$, the azimuth change is $\Delta\theta = \theta' - \theta$. The unit of ω is r/s; The unit of θ and θ' is rad.

Case 1: when $\pi \leq \theta' < 2\pi$, the amuzith θ' is 2π to π, it can obtain the formula

$$2\pi + \theta - \theta' = \omega \cdot \Delta t \tag{2.1}$$

Case 2: when $\theta \leq \theta' < \pi$, the amuzith θ' is π to θ, it can obtain the formula

$$2\pi + \theta - \theta' = \omega \cdot \Delta t \tag{2.2}$$

Case 3: when $0 \leq \theta' < \theta$, the amuzith θ' is θ to 0 , it can obtain the formula

$$\theta - \theta' = \omega \cdot \Delta t \tag{2.3}$$

In conclusion, the azimuth of satellite signal and interference change due to the carrier rotation is

$$\begin{cases} 2\pi + \theta - \theta' = \omega \cdot \Delta t, & \theta \leq \theta' < 2\pi \\ \theta - \theta' = \omega \cdot \Delta t, & 0 \leq \theta' < \theta \end{cases} \tag{2.4}$$

According to Fig. 4, it can analyze and calculate the effect on DOAs change of signal and interference due to the carrier forward. Since the satellite is far away from the carrier, the satellite signal elevation changes because the forward motion is very small that it can be negligible. Thus, only the elevation change of interference should be considered.

As shown in Fig. 4, assume the distance of forward trajectory and the jammer is known as *d*. The forward initial speed is set as v_0. Assume that the position of the carrier is at the point *O* at the moment *t*, and the UCA is *OABCDEF*, while the

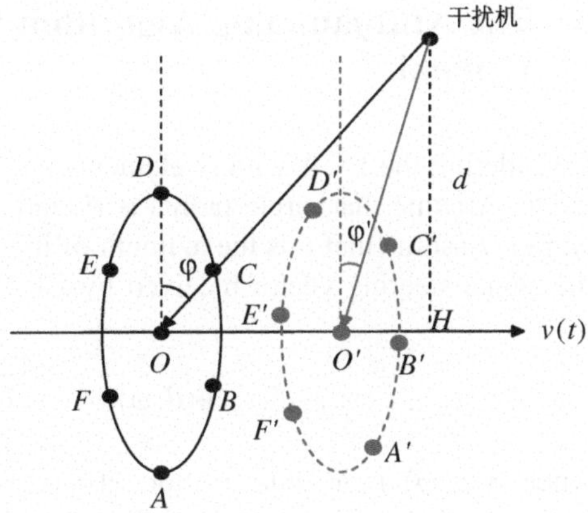

Fig. 4 The diagram of effect on elevation via forward motion

elevation of the interference is φ, and the displacement of the carrier is $-s$. Assume that the position of the carrier is at the point O' at the moment t', and the UCA is $OA'B'C'D'E'F'$, while the elevation of the interference is φ', and the displacement of the carrier is $-s'$. Then in the period $\Delta t = t' - t$, the elevation change of interference is $\Delta\varphi = \varphi' - \varphi$. We can obtain that

$$\tan \varphi = \frac{OH}{d} = \frac{s}{d} \tag{2.5}$$

$$\tan \varphi' = \frac{O'H}{d} = \frac{s - s'}{d} \tag{2.6}$$

Then the elevation of interference at t and t' are, respectively,

$$\varphi = \arctan\left(\frac{s}{d}\right) \tag{2.7}$$

$$\varphi' = \arctan\left(\frac{s - s'}{d}\right) \tag{2.8}$$

So the change of interference due to the forward motion is

$$\Delta\varphi = \varphi' - \varphi = \arctan\left(\frac{s - s'}{d}\right) - \arctan\left(\frac{s}{d}\right) \tag{2.9}$$

In conclusion, the azimuth changes very fast, while the elevation has nearly no change for the satellite signal in the attitude high dynamic. But for the interference, the two changes very fast.

3 The Anti-jamming Algorithm for Attitude High Dynamic

Considering the UCA with N elements and there is an element in the center of the circle. Assume that circle radius is d, and $d = \lambda N/(4\pi)$, where λ is the length of carrier. Assume that θ is the azimuth of incident wave, and φ is the elevation. Then the signal steering vector received by UCA with N elements is

$$\mathbf{a}(\theta, \varphi) = \mathrm{e}^{-j\frac{2\pi}{\lambda}\mathbf{P}^T\mathbf{r}(\theta, \varphi)} \tag{3.10}$$

where $\mathbf{r}(\theta, \varphi)$ is a DOA vector, $\mathbf{r}(\theta, \varphi) = -[\cos\theta\cos\varphi \quad \cos\theta\sin\varphi \quad \sin\varphi]^{\mathrm{T}}$. $\mathbf{P}$ denotes the position matrix of elements, which is a $3 \times M$ dimensions matrix, and the column vector of $\mathbf{P}$ is

$$\begin{cases} \mathbf{p}_0 = [0, 0, 0]^{\mathrm{T}} \\ \mathbf{p}_m = d[\cos r_m, \sin r_m, 0]^{\mathrm{T}}, \quad m = 1, 2, \ldots, M-1 \end{cases} \tag{3.11}$$

where $r_m = \frac{2\pi(m-1)}{M-1}$.

Suppose there are L satellite signals and P interference signals. From above section's analysis about the attitude high-dynamic GNSS, the azimuth of satellite signal received by GNSS receiver change rapidly, while the elevation is constant. The azimuth and elevation of interference change fast. So the signals received by the UCA can be expressed as

$$\mathbf{x}(t) = \sum_{l=1}^{L} s_l(t)\mathbf{a}[\theta_l(t), \varphi_l] + \sum_{p=1}^{P} j_p(t)\mathbf{a}[\theta_p(t), \varphi_p(t)] + \mathbf{n}(t) \tag{3.12}$$

where $s_l(t)$, $\theta_l(t)$ and φ_l respectively denote the l-th satellite signal, its azimuth ($l = 1, 2, \ldots, L$) which is changing with time, and its elevation. $j_p(t)$ represents the p-th interference, $\theta_p(t)$ and $\varphi_p(t)$ respectively denotes the azimuth and the elevation of the p-th interference changing with time ($p = 1, 2, \ldots, P$). $\mathbf{n}(t)$ denotes the Gaussian white noise with zero mean. Assume the satellite signal, the interference and noise are independent from each other.

From Sect. 2, we know that the DOAs of signal and interference change fast. Aiming at this question, the paper proposed the Capon-Laplace algorithm based on UCA. The idea of widening null is to add virtual interference sources, using the covariance matrix to multiply the extended matrix to construct a new covariance matrix. In this paper, Laplace method [11] is used to widen null, and the extended matrix is

$$\mathbf{T}(m, n) = \frac{1}{1 + \left(\Delta_{\max}\frac{\pi}{180}\right)^2} \tag{3.13}$$

where $\mathbf{T}(m, n)$ denotes the element of the extended matrix at the m-th row and the n-th column, $\Delta_{\max} = \frac{2\pi}{\lambda}\xi_{\max} d\sqrt{(\cos r_m - \cos r_n)^2 + (\sin r_m - \sin r_n)^2}$, and $\xi^2_{\max}$ is the variance of Laplace distribution.

The paper uses the Capon beamforming and assumes that the GNSS receiver is hot start, then the initial direction of satellite signal can be regarded as known. Thus, the cost function is written as

$$\begin{aligned} &\min \mathbf{w}^H(\mathbf{R} \odot \mathbf{T})\mathbf{w}^H \\ &\text{s.t.} \quad \mathbf{w}^H \mathbf{a}(\theta_l, \varphi_l) = 1 \end{aligned} \tag{3.14}$$

where $\mathbf{w} = [w_1, w_2, \ldots, w_N]^{\mathrm{T}}$ is the weight vector of the array, θ_l and φ_l, respectively, denotes the initial azimuth and the initial elevation of the satellite signal. $\xi = -\pi \sin \theta$, $m = 1, 2, \ldots, M$. The symbol "$\odot$" denotes Hadamard product. The symbol H denotes conjugate transpose.

The solution to the formula (3.14) can be expressed as

$$\mathbf{w}_{\text{opt}} = \frac{\tilde{\mathbf{R}}_{xx}^{-1} \mathbf{a}(\theta_l, \varphi_l)}{\mathbf{a}^H(\theta_l, \varphi_l)\tilde{\mathbf{R}}_{xx}^{-1} \mathbf{a}(\theta_l, \varphi_l)} \tag{3.15}$$

where $\tilde{\mathbf{R}}_{xx} = \mathbf{R}_{xx} \odot \mathbf{T}$.

4 Numerical Results

Suppose the carrier does spin forward motion at a constant speed, then the DOAs change of interference and signal mainly come from the attitude change of the carrier. This paper compares the beam pattern of Capon beamforming and Capon-Laplace method and the output *signal and interference to noise ratio* (SINR). Then, setting the Laplace parameter, compare the interference suppression feasibility of Capon-Laplace method to the different degree attitude high dynamic.

The data for simulation are generated by GNSS simulator in attitude high dynamic. The signal sampling frequency is 5.714 MHz, and intermediate frequency is 4.309 MHz. The initial DOAs of the satellite signal from PRN2 and the interference are respectively set as (90°, 60°) and (180°, 45°). *Signal to noise ratio* (SNR) is −20 dB and *interference to noise ratio* (INR) are 50 dB.

4.1 Analysis of Beam Pattern

This experiment sets the rotation speed as $n = r/s$, the element number of UCA as $N = 11$, and the Laplace parameter as $\xi_{\max} = 0.8$. Figure 5 is the beam pattern

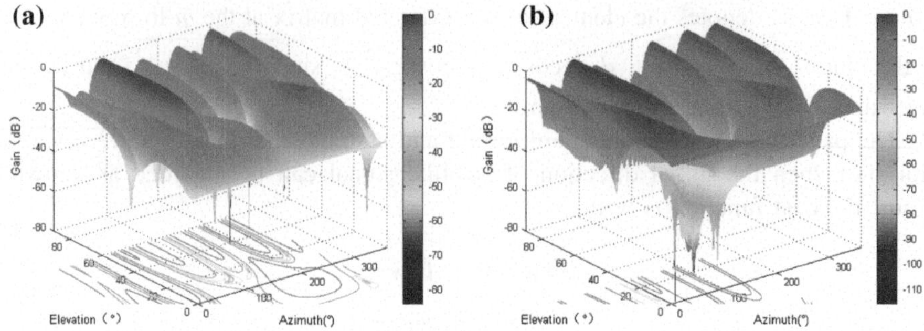

Fig. 5 The beam pattern via Capon and Capon-Laplace. **a** Capon. **b** Capon-Laplace

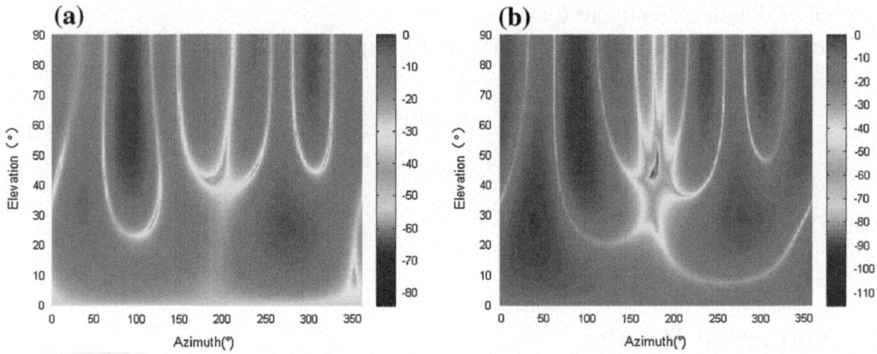

Fig. 6 The elevation view of beam pattern via Capon and Capon-Laplace. **a** Capon. **b** Capon-Laplace

comparison of Capon beamforming and Capon-Laplace method. Figure 5a, b is the 3d beam patterns of Capon beamforming and Capon-Laplace method, respectively. From Fig. 5b, the null can be shown effectively.

For convenient to observation, Fig. 6 gives the top views of the two beam patterns. From the top view, the darker area in Fig. 6a is lesser,namely, that the null of Capon is deep and sharp. While the darker area in Fig. 6b is larger, so the null is widened obviously and the null is shallow.

4.2 Analysis of Anti-jamming Performance

This experiment sets the element number of UCA as $N = 11$. Through setting the rotation speed n and Laplace parameter $\xi_{\max}$, we can estimate the anti-jamming degree of algorithms for the attitude high-dynamic GNSS. The range of the rotation

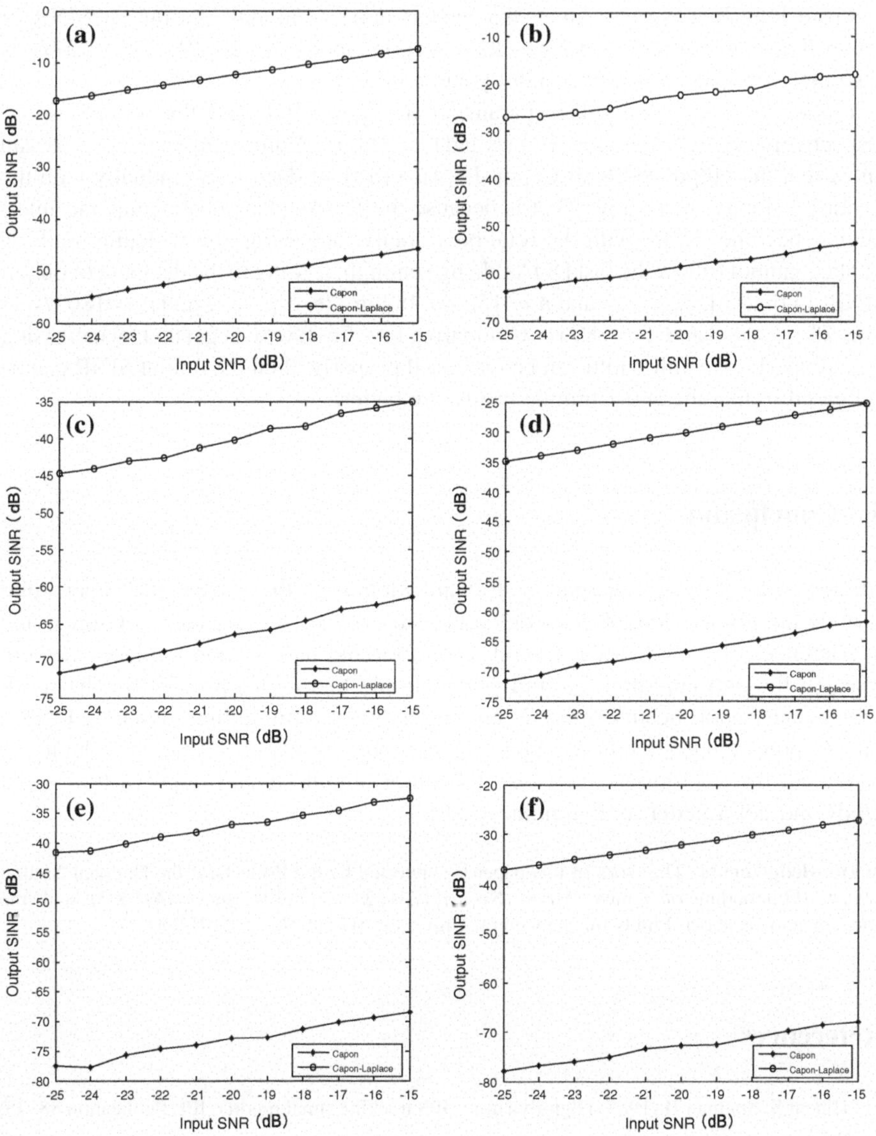

Fig. 7 The output SINR of Capon and Capon-Laplace. **a** $\xi_{max} = 0.8$, $n = 50$ r/s. **b** $\xi_{max} = 0.8$, $n = 100$ r/s. **c** $\xi_{max} = 0.8$, $n = 150$ r/s. **d** $\xi_{max} = 2$, $n = 150$ r/s. **e** $\xi_{max} = 2$, $n = 180$ r/s. **f** $\xi_{max} = 3$, $n = 180$ r/s

speed can be $n = 50$–300 r/s. Figure 7 is the output SINR of Capon and Capon-Laplace method. The result is obtained by 100 times Monte Carlo. We can know the anti-jamming degree by analyzing the output SINR.

From Fig. 7a–f we can know that in the same confident, the output SINR of Capon-Laplace method is bigger than Capon about 20–40 dB. The Capon is absolutely invalid for attitude high-dynamic GNSS.

Figure 7a–c set the Laplace parameter as $\xi_{max} = 0.8$, and the rotation speed respectively as $n = 50$ r/s, $n = 100$ r/s and $n = 150$ r/s. Comparing Fig. 7a–e, we can know that the output SINR of Capon-Laplace method decreases gradually with the rotation velocity increasing. That is because the DOAs change of signal and interference becomes faster with the rotation velocity increasing, the widening null via Laplace cannot follow the fast DOAs change enough, leading to the decrease of output SINR. In Fig. 7d, $\xi_{max} = 2$ and $n = 150$ r/s. In Fig. 7e, $\xi_{max} = 2$ and $n = 180$ r/s. In Fig. 7f, $\xi_{max} = 3$ and $n = 180$ r/s. Comparing Fig. 7c–f, we can get that by increasing ξ_{max} properly, the null width can be increased properly, then the output SINR can be increased so that the anti-jamming results are better.

5 Conclusion

Aiming at the high-speed rotating forward motion of the carrier, the paper build attitude high-dynamic GNSS model, and analysis the DOAs change of signal and interference. By comparing Capon beamforming and Capon-Laplace method, research the anti-jamming feasibility for attitude high-dynamic GNSS. The result shows that Capon beamforming is no longer fit for attitude high-dynamic GNSS. The Capon-Laplace methods' effect is obvious. When the rotating velocity is increased, the appropriate increase of Laplace parameter can improve the output SINR and get a better anti-jamming result.

Acknowledgements The work of this paper is supported by the Projects of the National Natural Science Foundation of China (Grant No. 61172112, 61271404 and 61471363) and the Fundamental Research Funds for the Central University (Grant No. 3122014D003).

References

1. Hinedi S, Statman JI (1988) High-dynamic GPS tracking (final report). JPL Publication 88–35, p 12
2. Nesreen IZ, James LG (1989) Extended Kalman filter-based tracking of weak GPS signals under high dynamic conditions. In: Proceedings of the 17th international technical meeting of the satellite division of the institute of navigation
3. Compton RT (1979) The power-inversion adaptive array: concept and performance. IEEE Trans Aerosp Electron Syst 15(6):803–814
4. Capon J (1969) High resolution frequency-wavenumber spectrum analysis. Proc IEEE 57 (8):1408–1418
5. Amin MG, Sun W (2005) A novel interference suppression scheme for global navigation satellite systems using antenna array. IEEE J Sel Areas Commun 23(5):999–1012

6. Sun W, Amin MG (2005) A self-coherence anti-jamming GPS receiver. IEEE Trans Signal Process 53(10):3910–3915
7. Gershman AB, Serebryakov GV, Boehme JF (1996) Constrained hung-turner adaptive beam-forming algorithm with additional robustness to wideband and moving jammers. IEEE Trans Antennas Propagation 44(5):361–367 (MENG)
8. Guerci JR (1999) Theory and application of convariance matrix tapers for robust adaptive beamforming. IEEE Trans Signal Process 47(4):977–985
9. Zatman M (1995) Production of adaptive array troughs by dispersion synthesis. Electron Lett 31(25):2141–2142
10. Mailloux RJ (1995) Covariance matrix augmentation to produce adaptive array pattern troughs. Electroncs Lett 31(10):771–772
11. Ma Y, Lu D, Wang W et al (2014) A high-dynamic null-widen GPS anti-jamming algorithm based on statistical model of the changing interference doa. In: China satellite navigation conference (CSNC) 2014 proceedings, vol I. Springer, Berlin, pp 695–70

The Anti-jamming Performance Analysis for Vector Tracking Loop

Fusheng Li, Renbiao Wu and Wenyi Wang

Abstract Because Global Positioning System (GPS) is easily interferenced by other signal, this paper designs an anti-jamming GPS vector tracking (or vector lock loop-VLL) receiver, which improve its anti-jamming performance on the tracking loop. The traditional scalar tracking loop (or scalar lock loop-SLL) tracks the satellite independently. There is no information exchange between each channel, which ignores the relationship between the receiver's navigation (position and velocity) and the tracking loop information (code phase and Doppler frequency). The vector tracking loop take the advantage of this relationship, and combine the navigation filter with tracking loop in a Kalman filter, which improve the receiver's anti-jamming ability and robustness. The simulation analyzes the influence on the tracking loop whether there has an interference, and this paper gives the comparison of position and velocity error. Finally, the weighted least squares method is used to compare the variance of predicted pseudorange error of the two tracking loops when the power of the interference changed. This paper concluded that the vector tracking loop has better anti-jamming ability.

Keywords Vector tracking loop · Weighted least squares · Anti-jamming

1 Introduction

The interference generally has different characteristics with the GPS satellite signals. The anti-jamming algorithm is mainly from a different domain to suppress the interference. Some popular algorithms suppress the interference in the spatial domain, such as power inverse algorithm, interference subspace algorithm, and so on [1–3]. They may suppress the interference with perfect results. However, these

F. Li (✉) · R. Wu · W. Wang
Tianjin Key Lab for Advanced Signal Processing,
Civil Aviation University of China, Tianjin, China
e-mail: thxlfs2013@163.com

J. Sun et al. (eds.), *China Satellite Navigation Conference (CSNC) 2016 Proceedings: Volume I*, Lecture Notes in Electrical Engineering 388,
DOI 10.1007/978-981-10-0934-1_57

algorithms usually require additional hardware, which may change the parameters hardly.

Now, with the rapid development of software-defined radio (SDR) technology, GPS software receiver has been widely used, which makes it is possible to improve the anti-jamming capability from the tracking loop. The vector tracking loop was first proposed by Spilker [4], Benson derived an equivalent VDLL model, and concluded that the anti-jamming capability of the VDLL has an improvement with 3-6 dB in 2002 [5]. The Geodesic Institute of Navigation of FAF Munich has published a number of articles on vector tracking loop from 2005 [6]. The GAVLAB lab from Auburn University has a detailed study on vector tracking loop, they have analyzed the performance of the vector tracking loop under different environment [7].

This paper established a vector tracking loop which combines the tracking loop with the navigation solution in a Kalman filter. This paper achieved the code and frequency vector tracking loop and analyzed the anti-jamming performance. The article is organized as follows: The principle of vector tracking loop is introduced in the Sect. 2. In Sect. 3 the simulation results are presented and analyzed. Section 4 summarizes the whole paper.

2 The Principle of Vector Tracking Loop

The traditional scalar tracking loop tracks the satellite independently (Fig. 1).

There is no information exchange in each channel, which ignores the relationship between the receiver's information and the tracking loop parameters. The vector tracking combines the tracking loop with navigation solution in a Kalman filter. The filter tracks the GPS signal and estimates the position and velocity of the receiver. The key to build a vector tracking loop is to choose a suitable filter. Generally, we choose the receiver's position error, clock error, velocity error, and drift error as the states of the central Kalman filter, rather than themselves [8]. Figure 2 is the architecture of vector tracking receiver and the basic step can be divided into: (1) prediction, (2) local signal generated, (3) the correction of predicted states.

2.1 Prediction

The state vector is $\boldsymbol{X} = [x\ y\ z\ ct_b\ v_x\ v_y\ v_z\ ct_d]^T$, the vector $\boldsymbol{p} = [x\ y\ z]^T$ and $\boldsymbol{v} = [v_x\ v_y\ v_z]^T$ is the receiver's three-dimensional position and velocity in ECEF, with the unit m and m/s. t_b, t_d is the receiver's clock bias and drift, which is multiplied by c to give them the unit with m, m/s.

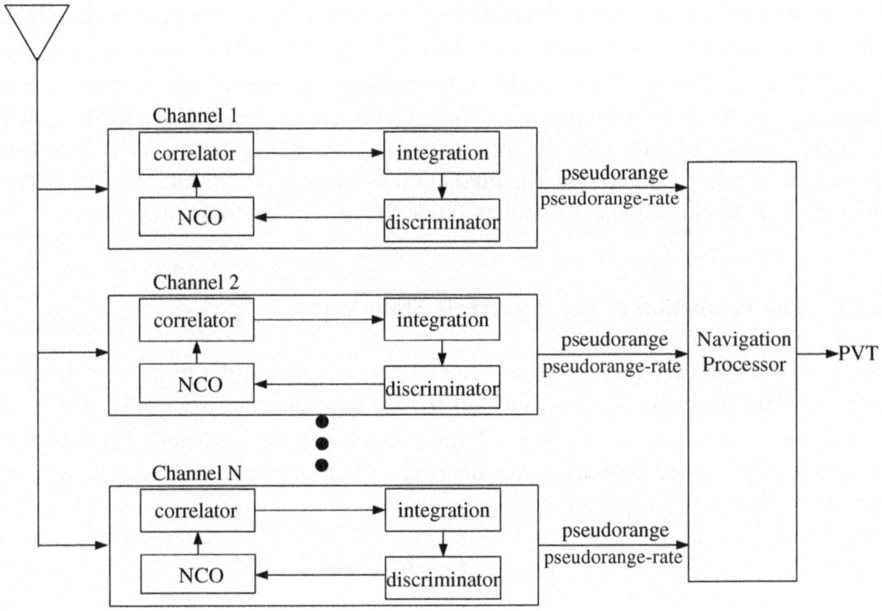

Fig. 1 Traditional receiver architecture

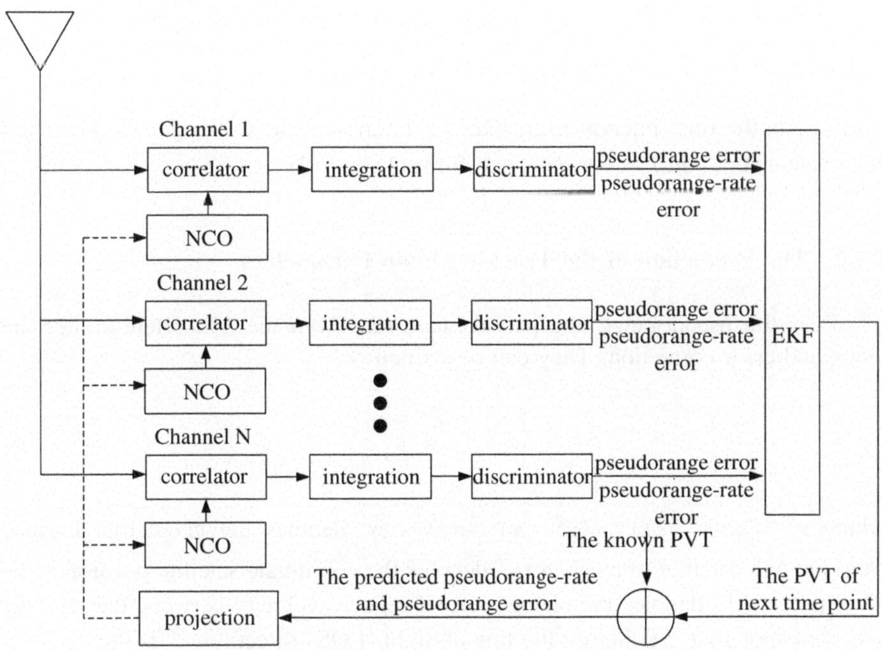

Fig. 2 Vector tracking receiver architecture

The states of the central Kalman filter are the error of position, velocity, clock bias, and drift. $\left(\delta \boldsymbol{X} = \left[\delta x\ \delta y\ \delta z\ c\delta t_b\ \delta v_x\ \delta v_y\ \delta v_z\ c\delta t_d\right]^T\right)$ We need to know the initial value of the state vector and the tracking loop parameters (code phase, carrier frequency, etc.) before the prediction. These values are generally given by the scalar tracking loop or by the other sensors. (such as inertial navigation, etc.) The prediction can be divided into: (1) The prediction of the receiver's state vector (2) The prediction of tracking loop parameters (code phase, carrier frequency, etc.)

2.1.1 The Prediction of the Receiver's State Vector

When the time is very short (1 ms), it can be considered that the receiver is doing a linear motion in a very short period. So if we know the state vector $\boldsymbol{X}_k$ in k time epoch, the state vector $\hat{\boldsymbol{X}}_{k+1}$ in $k+1$ time epoch can be predicted. The position components $\hat{\boldsymbol{p}}_{k+1}$ and velocity components $\hat{\boldsymbol{v}}_{k+1}$ in $k+1$ time epoch can be written as:

$$\begin{aligned} \hat{\boldsymbol{p}}_{k+1} &= \boldsymbol{p}_k + t_{k,k+1}\boldsymbol{v}_k \\ \hat{\boldsymbol{v}}_{k+1} &= \boldsymbol{v}_k \end{aligned} \tag{1}$$

And the clock bias and drift $k+1$ time epoch can be written as

$$\begin{aligned} \hat{t}_{b,k+1} &= t_{b,k} + t_{k,k+1}t_{d,k} \\ \hat{t}_{d,k+1} &= t_{d,k} \end{aligned} \tag{2}$$

$t_{k,k+1}$ is the time interval from k to $k+1$ time epoch, so we got the predicted state vector $\hat{\boldsymbol{X}}_{k+1}$ in $k+1$ time epoch from (1) and (2).

2.1.2 The Prediction of the Tracking Loop Parameters

$\rho^N, \dot{\rho}^N$ is the pseudorange and pseudorange rate from the Nth satellite after the clock and error correction. They can be written as

$$\begin{cases} \rho^N = r^N + ct_b \\ \dot{\rho}^N = \dot{r}^N + ct_d \end{cases} \tag{3}$$

where $r^N = \sqrt{(x^N - x)^2 + (y^N - y)^2 + (z^N - z)^2}$ denotes the geometric distance from the Nth satellite to receiver (x^N denotes the coordinate satellite position in x). $\dot{r}^N = (\boldsymbol{v}^N - \boldsymbol{v})\boldsymbol{l}^N$ denotes the change-rate of relative velocity between the receiver and corresponding satellite on the line-of-sight (LOS) direction, $\boldsymbol{l}^N$ is the

$$\boldsymbol{l}^N = \begin{bmatrix} l_x \\ l_y \\ l_z \end{bmatrix} = \begin{bmatrix} \frac{\partial r^N}{\partial x} \\ \frac{\partial r^N}{\partial y} \\ \frac{\partial r^N}{\partial z} \end{bmatrix} = \frac{1}{r^N} \begin{bmatrix} x^N - x \\ y^N - y \\ z^N - z \end{bmatrix} \tag{4}$$

unit vector from the receiver to Nth satellite, can be written as (4). When we have got the state vector $\hat{\boldsymbol{X}}_{k+1}$ and the position of the satellite (from the known ephemeris) in $k+1$ time epoch, and the pseudorange and pseudorange rate in $k+1$ time epoch can be got from (3), then we can predict the code and carrier frequency with the following equations:

$$\begin{aligned} \hat{f}^N_{\text{code},k+1} &= f_{\text{code}} + \hat{f}^N_{d,k+1} = f_{\text{code}} - \hat{\dot{\rho}}^N_{k+1} \frac{f_{\text{code}}}{c} \\ &= f_{\text{code}} - \left((\mathbf{v}^N_{k+1} - \hat{\mathbf{v}}_{k+1}) \hat{\boldsymbol{l}}^N_{k+1} + c\hat{t}_{d,k+1} \right) \frac{f_{\text{code}}}{c} \end{aligned} \tag{5}$$

$$\begin{aligned} \hat{f}^N_{\text{carr},k+1} &= f_{\text{IF}} + \hat{f}^N_{d,k+1} = f_{\text{IF}} - \hat{\dot{\rho}}^N_{k+1} \frac{f_{L1}}{c} \\ &= f_{\text{IF}} - \left((\mathbf{v}^N_{k+1} - \hat{\mathbf{v}}_{k+1}) \hat{\boldsymbol{l}}^N_{k+1} + c\hat{t}_{d,k+1} \right) \frac{f_{L1}}{c} \end{aligned} \tag{6}$$

f_{code} is the frequency of the C/A code ($f_{\text{code}} = 1.023$ MHz), the intermediate frequency is f_{IF},$\hat{\boldsymbol{l}}^N_{k+1}$ can be predicted as Eq. (4), so the code phase in $k+1$ time epoch can be written as

$$\hat{\varphi}^N_{k+1} = \varphi^N_k + (\Delta \boldsymbol{p}^N_{k,k+1} - t_{k,k+1} \hat{\boldsymbol{v}}_{k+1}) \hat{\boldsymbol{l}}^N_{k+1} \tag{7}$$

2.2 The Generate of the Local Signal

The predicted local signal of $k+1$ time epoch correlated with the input intermediate signal, then the discriminator gives the code phase error $\Delta\tau^N_{k+1}$ and carrier frequency error Δf^N_{k+1}. They are the differences between the predicted state $(\hat{\varphi}^N_{k+1}, \hat{f}^N_{\text{carr},k+1})$ and the real state $(\varphi^N_{k+1}, f^N_{\text{carr},k+1})$ in $k+1$ time epoch:

$$\begin{cases} \Delta\tau^N_{k+1} = \hat{\varphi}^N_{k+1} - \varphi^N_{k+1} + \eta^N_{k+1} \\ \Delta f^N_{k+1} = \hat{f}^N_{\text{carr},k+1} - f^N_{\text{carr},k+1} + w^N_{k+1} \end{cases} \tag{8}$$

η^N_{k+1}, w^N_{k+1} represents the measurement noise of discriminator in Nth channel, then the pseudorange error $(\Delta\rho^N_{k+1})$ and pseudorange rate error $(\Delta\dot{\rho}^N_{k+1})$ can be converted by the code phase error $\Delta\tau^N_{k+1}$ and carrier frequency error Δf^N_{k+1}

multiplied by c (the speed of light). On the other hand, the Eq. (3) do the first-order Taylor expansion get the following equations

$$\begin{cases} \Delta\rho_{k+1}^N = \hat{\rho}_{k+1}^N - \rho_{k+1}^N = l_{x,k+1}^N \delta x + l_{y,k+1}^N \delta y + l_{z,k+1}^N \delta z - c\delta t_b \\ \Delta\dot{\rho}_{k+1}^N = \hat{\dot{\rho}}_{k+1}^N - \dot{\rho}_{k+1}^N = l_{x,k+1}^N \delta v_x + l_{y,k+1}^N \delta v_y + l_{z,k+1}^N \delta v_z - c\delta t_d \end{cases} \tag{9}$$

This will combine the pseudorange error and pseudorange rate error with receiver three-dimensional position and velocity error. So we choose (9) as the measurement equation of the Kalman filter. With the known state Eqs. (1) and (2), the equation of the Kalman filter can be written as

$$\begin{cases} \delta \boldsymbol{X}_{k+1} = F_{k,k+1} \delta \boldsymbol{X}_k + W_k \\ \boldsymbol{Z}_{k+1} = H_{k+1} \delta \boldsymbol{X}_{k+1} + V_k \end{cases} \tag{10}$$

$F_{k,k+1}$ represents a step state transition matrix can be written as

$$F_{k,k+1} = \begin{bmatrix} 1 & 0 & 0 & 0 & t_{k,k+1} & 0 & 0 & 0 \\ 0 & 1 & 0 & 0 & 0 & t_{k,k+1} & 0 & 0 \\ 0 & 0 & 1 & 0 & 0 & 0 & t_{k,k+1} & 0 \\ 0 & 0 & 0 & 1 & 0 & 0 & 0 & t_{k,k+1} \\ 0 & 0 & 0 & 0 & 1 & 0 & 0 & 0 \\ 0 & 0 & 0 & 0 & 0 & 1 & 0 & 0 \\ 0 & 0 & 0 & 0 & 0 & 0 & 1 & 0 \\ 0 & 0 & 0 & 0 & 0 & 0 & 0 & 1 \end{bmatrix} \tag{11}$$

W_k denotes the process noise matrix, V_k represents the measurement noise matrix, H_{k+1} is the project matrix between the measurement vector and state vector, which can be got from the Eq. (9).

2.3 The Correction of the Predicted Value

We get the error of state vector in $k+1$ time epoch, and the state vector of $k+1$ time epoch can be updated as $\boldsymbol{X}_{k+1} = \hat{\boldsymbol{X}}_{k+1} + \delta\hat{\boldsymbol{X}}_{k+1}$. The code phase, the frequency of code and carrier can be updated as following equations:

$$\begin{aligned} \varphi_{k+1}^N &= \hat{\varphi}_{k+1}^N + \delta \boldsymbol{p}_{k+1} \boldsymbol{l}_{k+1}^N + c\delta t_{b,k+1} \\ f_{\text{code},k+1}^N &= \hat{f}_{\text{code},k+1}^N + (c\delta t_{d,k+1} + \delta \boldsymbol{v}_{k+1} \boldsymbol{l}_{k+1}^N) \frac{f_{\text{code}}}{c} \\ f_{\text{carr},k+1}^N &= \hat{f}_{\text{carr},k+1}^N + (c\delta t_{d,k+1} + \delta \boldsymbol{v}_{k+1} \boldsymbol{l}_{k+1}^N) \frac{f_{L1}}{c} \end{aligned} \tag{12}$$

The state vector of receiver and the parameters of tracking loop in $k+1$ time epoch can be got from the correction, then the state vector of the filter is set to zero to prepare for the next time, the closed circuit is formed.

3 Simulation Results and Analysis

The experimental data used in the simulation is produced by a high-fidelity GPS simulation software, the sampling frequency is 5.714 MHz. The validity of the data has been verified by the software and hardware receiver. We suppose the position of static receiver is known, and there are five satellites with known ephemeris, the carrier to noise ratio of signal is 45 dB Hz.

3.1 *The Influences to the Tracking Loop of Interference*

The data is 70 s long, it is normal in the front 21 s. There is an interference with JNR = 20 dB (interference to noise ratio) in 21 s to 26 s. Then the interference disappears in 26 s to 31 s. In 31 s the interference comes again with the same power. We suppress the interference from 45 s with the power inversion algorithm. We process these data with the scalar and vector tracking loop, to compare the navigate performance, the results shown in Figs. 3 and 4.

The red arrow indicates the presence period of interference, the receiver calculated velocity and position every 100 ms. As it can be seen from Fig. 3 the position error quickly becomes large when the interference is added in scalar tracking loop, but the vector tracking loop is relatively stable. When the tracking loop is affected by the interference, the scalar tracking loop loses lock. The position of receiver is in

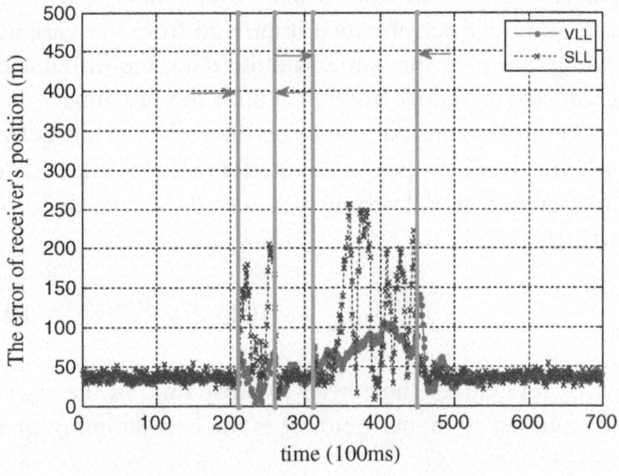

Fig. 3 The position error of the receiver (Color figure online)

Fig. 4 The velocity error of the receiver (Color figure online)

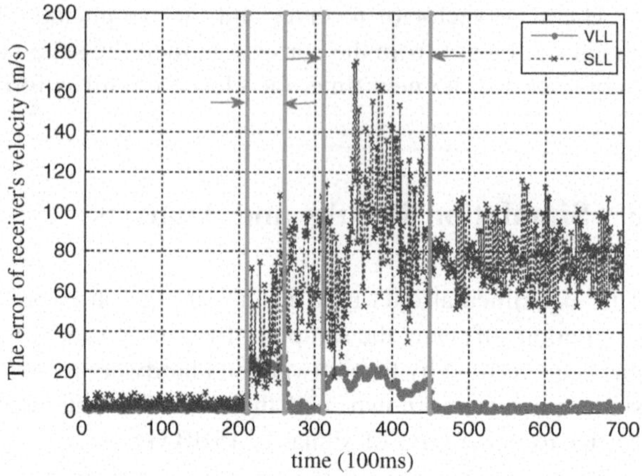

a random walk, but in the vector tracking loop, the position error seems very small. When the interference is suppressed, the position error of the vector tracking loop can be restored to normal levels, which indicated that the vector tracking loop has a better an-jamming performance than scalar tracking loop.

Figure 4 is much clearer, when there is an interference, the velocity error of vector tracking loop becomes large, and so does the scalar tracking loop. But when the interference is suppressed, the vector tracking loop quickly returned to normal. The scalar tracking loop has been losing lock, which also shows the frequency tracking loop is more susceptible to interference than the code tracking loop.

3.2 Analysis of the Variance of Pseudo-range Error

The above simulation explains the vector tracking loop has a better anti-jamming performance than the scalar loop, next part we will analysis how much anti-jamming performance improved from the variance of pseudo-range error. The simulation uses the same satellite data, the difference is the power of interference gradually increases from 21 s until the last time.

The following part analyzes the relationship between the variance of pseudorange (σ_v^2) and the power of interference. In a satellite navigation system if there has interference, it will seriously reduce the carrier to noise ratio, the relationship can be written as [9]:

$$(C/N_0)_{\text{eq}} = -10\lg\left[10^{-(C/N_0)/10} + 10^{(J/S)/10}/QR_c\right] \tag{13}$$

C/N_0 denotes the carrier to noise ratio of the signal without interference. J/S is the jammer to signal ratio, Q is the anti-jamming quality factor (1 for narrowband interference, wideband interference is 2), R_c is the code rate. The relationship between the variance of pseudorange and carrier-to-noise ratio is

$$\sigma_v^2 = \frac{\beta^2}{2\left(T \cdot 10^{(C/N_0)_{eq}/10}\right)^2} + \frac{\beta^2}{4T \cdot 10^{(C/N_0)_{eq}/10}} \tag{14}$$

T is the integration time, we set it to 20 ms in this paper, $\beta = 293.3$ m. Lashley's doctor dissertation analyzes the variance of pseudorange error from the two tracking loop with the weighted least squares method. For the scalar tracking loop,

$$\begin{aligned} E\{(\Delta\hat{\boldsymbol{\rho}} - E\{\Delta\hat{\boldsymbol{\rho}}\})(\Delta\hat{\boldsymbol{\rho}} - E\{\Delta\hat{\boldsymbol{\rho}}\})^T\} &= \left(\boldsymbol{I}_{N\times N}^T \cdot \boldsymbol{R}_v^{-1} \cdot \boldsymbol{I}_{N\times N}\right)^{-1} \\ &= \sigma_v^2 \cdot \boldsymbol{I}_{N\times N} \end{aligned} \tag{15}$$

And the variance of pseudorange error for the vector tracking loop is

$$\begin{aligned} &E\{(\Delta\hat{\rho} - E\{\Delta\hat{\rho}\})(\Delta\hat{\rho} - E\{\Delta\hat{\rho}\})^T\} = \sigma_v^2 \cdot \boldsymbol{W}_{N\times N} \\ &\boldsymbol{W}_{N\times N} = \boldsymbol{H}_{N\times 4} \cdot \left(\boldsymbol{H}_{N\times 4}^T \cdot \boldsymbol{H}_{N\times 4}\right)^{-1} \cdot \boldsymbol{H}_{N\times 4}^T \end{aligned} \tag{16}$$

$\Delta\hat{\rho}$ is predicted pseudorange error matrix, $\boldsymbol{H}$ is the projection matrix. From the above two equation, we can see that the variance of pseudorange error from each channel in the scalar tracking loop are all the same, but in the vector tracking loop the variance of pseudorange error of each channel is related to the position of satellite. The following part shows the relationship between the power of interference and the variance of pseudorange error as Fig. 5. When the JSR = 0, the variance of pseudorange error is about 141 m^2. We choose it as a standard to verify the improvement of the anti-jamming performance in the vector tracking loop.

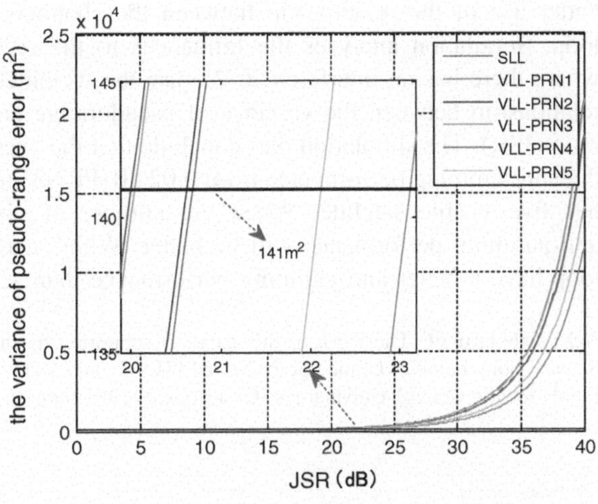

Fig. 5 The relationship between JSR and the variance of pseudorange error (Color figure online)

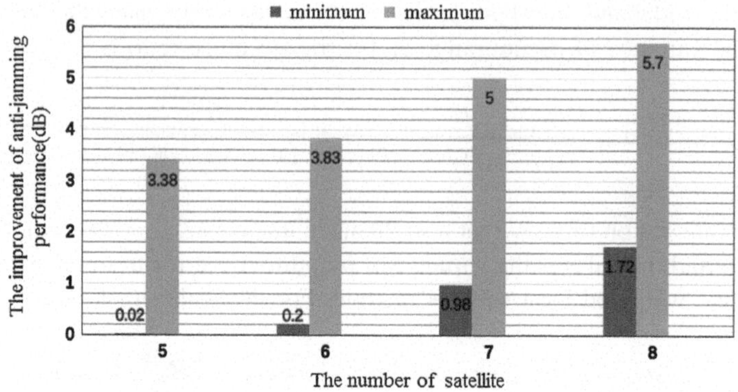

Fig. 6 The anti-jamming improvement of vector tracking loop

The channel is independent in scalar tracking loop, and the variance of pseudorange error for each channel are the same (the red line in Fig. 5), but the variance is different in the vector tracing loop. With the same pseudorange error variance, the vector tracking anti-jamming capability raise from 0.02 dB(PRN3) to 3.38 dB(PRN1) compared to the scalar tracking loop.

It is shown in Fig. 6, the anti-jamming performance of vector tracking is related to the number of satellite. In the vector tracking loop, when the number of satellites is raised, we can get better anti-jamming performance.

4 Conclusion

In this paper, we use the easily implement of GPS software receiver, established a vector tracking loop based on extended Kalman filter. The tracking loop makes better use of the relationship between the channels compared to scalar tracking loop. Simulation analyzes the influences to the receiver's position and velocity where there has an interference. We use the weighted least squares to analyze the relationship between the variance of pseudorange error and the jammer to signal ratio (JSR). The simulation part conclude that the vector tracking loop can improve the anti-jamming performance from 0.02 dB(PRN3) to 3.38 dB(PRN1), when there has five visible satellite. When the number of visible satellites is raised, the anti-jamming performance will be better. Which certified that the vector tracking loop have a better anti-jamming performance than scalar tracking loop.

Acknowledgment The work of this paper is supported by the Projects of the National Natural Science Foundation of China (Grant No. 61471363 and 61271404) and the Fundamental Research Funds for the Central Universities (Grant No. 3122014D008).

References

1. Gecan A, Zoltowski M (1995) Power minimization technique for GPS null steering antennas institute of navigation conference. Palm Springs, CA, pp 13–15
2. Sun W, Amin M (2005) Maximum signal-to-noise ratio GPS anti-jamming receiver with subspace tracking. IEEE International Conference on Acoustics, Speech, and Signal Processing (ICASSP). Philadelphia, PA, pp 1085–1088
3. Sun W, Amin M (2005) A self-coherence anti-jamming GPS receiver. IEEE Trans Signal Process 53(10):3910–3915
4. Spilker JJ Fundamentals of signal tracking theory. In: Parkinson BW, Spilker JJ Global positioning system: theory and applications
5. Pany T, Eissfeller B (2006) Use of a Vector delay lock loop receiver for GNSS signal power analysis in bad signal conditions. ION Annual Meeting/IEEE PLANS, San Diego CA, pp 893–903
6. Lashley M Modelling and performance analysis of GPS vector tracking algorithms. Ph.D. dissertation, Auburn
7. Benson D (2007) Interference benefits of a Vector Delay Lock Loop (VDLL) GPS receiver. In: Proceedings of 63rd Annual Meeting Institute of Navigation, Cambridge, CA, pp 749–756
8. Zhao S, Akos DM (2011) An open source GPS/ GNSS vector tracking loop-implementation, filter tuning, and results. ION ITM, pp 1293–1305
9. Kaplan DE, Hegarty JC (2006) Understanding GPS, principles and application, 2nd edn. Artech House, Norwood, pp 459–552

The manufacturer's authorised representative in the EU is Springer Nature Customer Service Centre GmbH, Europaplatz 3, 69115 Heidelberg, Germany. If you have any concerns regarding our products, please contact ProductSafety@springernature.com

Printed and bound by CPI Group (UK) Ltd, Croydon, CR0 4YY
25/09/2026
02222839-0016